Wine
Science

Food Science and Technology

International Series

A complete list of the books in this series appears at the end of the volume.

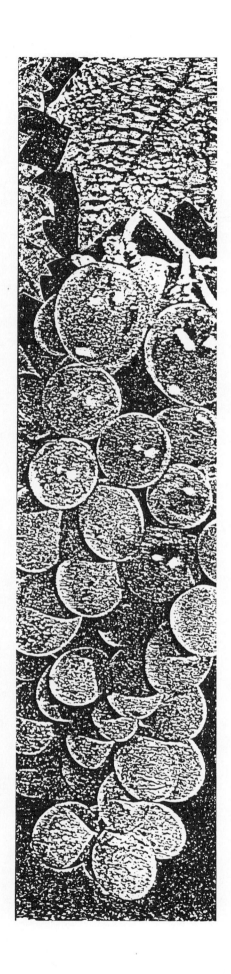

Wine Science

Principles

and

Applications

Ron S. Jackson

Botany Department
Brandon University
Brandon, Manitoba
Canada

Academic Press

A Division of Harcourt Brace & Company

San Diego New York Boston
London Sydney Tokyo Toronto

Front cover photo: From Richard H. Gross, Biological Photography.

This book is printed on acid-free paper. ∞

Copyright © 1994 by ACADEMIC PRESS

Academic Press
525 B Street, Suite 1900, San Diego, California 92101-4495

United Kingdom Edition published by
Academic Press Limited
24–28 Oval Road, London NW1 7DX

Library of Congress Cataloging-in Publication Data

Jackson, Ron S.
    Wine science : principles and applications / Ron S. Jackson.
        p.   cm. - - (Food science and technology series)
    Includes index.
    ISBN 0-12-379060-3
    1. Wine and wine making.   2. Viticulture.   I. Title.   II. Series.
TP548.J15   1993
663' .2--dc20   93-30186
                                                              CIP

PRINTED IN THE UNITED STATES OF AMERICA
            98   99   QW   9 8 7 6 5 4

The book is dedicated
to the miraculous microbes
that can turn a marvelous fruit
into a seraphic beverage,
and to God, who has given us
the ability to savor its
finest qualities and pleasures.

Contents

Preface xi
Acknowledgments xiii

1
Introduction

Grapevine and Wine Origins 1
Commercial Importance of Grapes and Wine 3
Wine Classification 5
   Still Table Wines 6
   Sparkling Wines 7
   Fortified Wines (Dessert and Appetizer
      Wines) 7
Wine Quality 8
Health Related Aspects of Wine Consumption 9
Suggested Readings 9
References 9

2
Grape Species and Varieties

Introduction 11
The Genus *Vitis* 12
Geographic Origin and Distribution of *Vitis* and
   Vitis vinifera 14

Cultivar Origins 18
Grapevine Breeding 20
   Nonstandard Breeding Techniques and Other
      Means of Cultivar Improvement 23
Grapevine Varieties 25
   Vitis vinifera Cultivars 26
   Interspecies Hybrids 28
Suggested Readings 29
References 30

3
Grapevine Structure and Function

Vegetative Structure and Function 32
   The Root System 33
   The Shoot System 37
   Tendrils 40
   Leaves 40
Reproductive Structure and Development 45
   Inflorescence (Flower Cluster) 45
   Berry Growth and Development 50
Suggested Readings 66
References 67

4

Vineyard Practice

Vine Cycle and Vineyard Activity 72
Management of Vine Growth 75
Physiological Effects of Pruning 75
Pruning Options 77
Pruning Procedures and Timing 78
Training Options and Systems 83
Rootstock 95
Vine Propagation and Grafting 98
Multiplication Procedures 99
Grafting 99
Soil Preparation 102
Vineyard Planting and Establishment 102
Irrigation 103
Timing and Level of Irrigation 105
Water Quality 106
Types of Irrigation 106
Fertilization 108
Factors Affecting Nutrient Supply and
Acquisition 109
Assessment of Nutrient Need 111
Nutrient Requirements 111
Organic Fertilizers 117
Disease, Pest, and Weed Control 119
Control of Pathogens 120
Examples of Grapevine Diseases and Pests 124
Harvesting 140
Criteria for Timing of Harvest 141
Sampling 142
Containers for Collecting and Harvesting
Grapes 142
Harvest Mechanisms 142
Suggested Readings 145
References 147

5

Site Selection and Climate

Introduction 154
Soil Influences 155
Geologic Origin 155
Texture 155
Structure 156
Drainage and Water Availability 157
Soil Depth 158
Nutrient Content and pH 158
Color 158
Organic Content 159

Topographic Influences 159
Solar Exposure 159
Wind Direction 160
Frost and Winter Protection 161
Drainage 161
Mesoclimatic and Macroclimatic Influences 162
Temperature 163
Solar Radiation 169
Wind 172
Water 174
Suggested Readings 174
References 175

6

Chemical Constituents of Grapes and Wine

Introduction 178
Overview of Chemical Functional Groups 179
Chemical Constituents 182
Water 182
Sugars 182
Pectins, Gums, and Related
Polysaccharides 183
Alcohols 184
Acids 186
Phenols and Related Phenol (Phenyl)
Derivatives 188
Aldehydes and Ketones 198
Acetals 199
Esters 200
Lactones and Other Oxygen Heterocycles 201
Terpenes and Oxygenated Derivatives 202
Nitrogen-Containing Compounds 203
Hydrogen Sulfide and Organosulfur
Compounds 204
Hydrocarbons and Derivatives 205
Macromolecules and Growth Factors 206
Dissolved Gases 207
Minerals 211
Chemical Nature of Varietal Aromas 211
**Appendix 6.1 Conversion Table for Various
Hydrometer Scales Used to Measure Sugar
Content of Must 213**
**Appendix 6.2 Conversion Table for Various
Measures of Ethanol Content at 20°C 214**
**Appendix 6.3 Interconversion of Acidity
Units 215**
Suggested Readings 214
References 216

7

Fermentation

Introduction 220
Basic Procedures of Wine Production 221
Prefermentation Practices 222
  Stemming 222
  Crushing 223
  Supraextraction 223
  Maceration 223
  Dejuicing 226
  Pressing 226
  Must Clarification 228
  Adjustments to Juice and Must 229
Alcoholic Fermentation 232
  Fermentors 232
  Fermentation 234
  Biochemistry of Alcoholic Fermentation 236
Yeasts 241
  Classification and Life Cycle 241
  Ecology 241
  Succession during Fermentation 243
  Desirable Yeast Characteristics and
    Breeding 244
  Genetic Modification 244
  Environmental Factors Affecting
    Fermentation 246
  Stuck Fermentation 258
Malolactic Fermentation 259
  Lactic Acid Bacteria 260
  Effects of Malolactic Fermentation 261
  Origin and Growth of Lactic Acid Bacteria 263
  Factors Affecting Malolactic Fementation 264
  Control 267
**Appendix 7.1 Partial Synonymy of Several
  Important Wine Yeasts 269**
**Appendix 7.2 Physiological Races of
  Saccharomyces cerevisiae Previously Given
  Species Status 270**
Suggested Readings 271
References 272

8

Postfermentation Treatments and Related Topics

Wine Adjustments 277
  Acidity and pH Adjustment 278
  Sweetening 280
  Dealcoholization 281

  Color Adjustment 281
  Blending 282
Stabilization and Clarification 282
  Stabilization 283
  Fining 288
  Clarification 290
Aging 294
  Effects of Aging 295
  Factors Affecting Aging 298
  Rejuvenation of Old Wines 299
Oak and Cooperage 299
  Oak Species and Wood Properties 299
  Barrel Production 301
  Chemical Composition of Oak 306
  Aeration 309
  In-Barrel Fermentation 309
  Disadvantages of Oak Cooperage 310
  Other Cooperage Materials 310
Cork and Other Bottle Closures 311
  Cork 311
  Cork Faults 316
  Cork Alternatives 318
  Cork Insertion 319
  Leakage Caused by Insertion Problems 320
Bottles and Other Containers 320
  Glass Bottles 321
  Bag-in-Box Containers 323
Wine Spoilage 323
  Cork-Related Problems 323
  Yeast-Induced Spoilage 325
  Bacteria-Induced Spoilage 325
  Sulfur Off-Odors 328
  Additional Spoilage Problems 329
  Accidental Contaminants 330
Suggested Readings 331
References 332

9

Specific and Distinctive Wine Styles

Introduction 338
Sweet Table Wines 338
  Botrytized Wines 339
  Nonbotrytized Sweet Wines 344
Red Wine Styles 345
  Recioto Style Wines 345
  Carbonic Maceraction Wines 347
Sparkling Wines 354
  Traditional Method 355
  Transfer Method 362
  Bulk Method 362

Other Methods   363
   Carbonation   363
   Production of Rosé and Red Sparkling
      Wine   363
   Effervescence and Foam Characteristics   364
Fortified Wines   365
   Sherry and Sherrylike Wines   366
   Port and Portlike Wines   372
   Madeira   375
Suggested Readings   376
References   376

10

Wine Laws, Authentication, and Geography

Appellation Control Laws   380
   Basic Concepts and Significance   380
   Geographic Expression   382
**Detection of Wine Misrepresentation and
   Adulteration   386**
   Validation of Geograhic Origin   387
   Validation of Conformity to Wine Production
      Regulations   388
World Wine Regions   390
   Western Europe   390
   Southern Europe   403
   Eastern Europe   411
   North Africa and the Near East   415
   Far East   415
   Australia and New Zealand   416
   South Africa   419
   South America   420
   North America   423
Suggested Readings   428
References   430

11

Sensory Perception and Wine Assessment

Visual Sensations   432
   Color   432
   Clarity   433
   Viscosity   433
   Sparkle (Effervescence)   434
   Tears   434

Taste and Mouth-Feel   434
   Taste   434
   Factors Influencing Taste Perception   436
   Mouth-Feel   438
   Taste and Mouth-Feel Sensation in Wine
      Tasting   440
Odor   440
   Olfactory System   440
   Odorants and Olfactory Stimulation   442
   Sensations from the Trigeminal Nerve   442
   Odor Perception   443
   Sources of Variation in Olfactory
      Perception   444
   Significance of Odor Assessment in Wine
      Tasting   446
   Off-Odors   446
Wine Assessment and Sensory Analysis   448
   Conditions for Sensory Analysis   448
   Wine Score Cards   450
   Number of Tasters   451
Tasters   452
   Training   452
   Measuring Tasting Acuity and Consistency   453
Wine Tasting Technique   453
   Appearance   453
   Odor In-Glass   454
   In-Mouth Sensations   454
   Finish   455
   Assessment of Overall Quality   455
**Statistical and Descriptive Analysis of Tasting
   Results   455**
   Simple Tests   455
   Analysis of Variance   456
   Multivariate Analysis and Descriptive Analysis of
      Wine   456
**Appendix 11.1 Aroma and Bouquet
   Samples   457**
Appendix 11.2 Basic Off-Odor Samples   458
**Appendix 11.3 Taste and Mouth-Feel
   Samples   458**
**Appendix 11.4 Training and Testing of Wine
   Tasters   458**
**Appendix 11.5 Multipliers for Estimating
   Significance of Difference by Range: One-Way
   Classification, 5% Level   460**
**Appendix 11.6 Minimum Numbers of Correct
   Judgments to Establish Significance at Various
   Probability Levels for the Triangle Test   461**
Suggested Readings   462
References   462

Index   467

Preface

The science of wine involves three major interrelated topics: grapevine growth, wine production, and wine sensory analysis. Although in most situations these topics can be covered separately, joint discussion of certain aspects of viticulture, enology, and wine assessment is valuable and reinforces the natural interrelationships of these subjects.

Consistent with modern biological thought, much of wine science is interpretable in terms of physics and chemistry. Because of the botanical nature of the raw materials and their microbial transformation into wine, the physiology and genetics of vine, yeasts, and bacteria are crucial to a current understanding the origins of wine quality. In addition, physical microclimatology and soil physicochemistry are revealing the vineyard origins of grape quality. Finally, knowledge of human sensory psychophysiology is essential to the interpretation of wine quality data. For those more interested in applications, much of the scientific discussion has been placed so that the practical aspects can be accessed without necessarily reading and understanding the scientific explanations.

I hope that this book will help place the present knowledge on these topics in perspective and illustrate where further study is needed. It was not possible in a text of this size to provide a detailed treatment of all the diverging views on the many topics covered. I have therefore presented those views that in my opinion have the greatest support or significance. With much of our information based on a comparatively few cultivars or yeast strains, making valid generalizations is difficult. In addition, several topics are quite contentious among grape growers and wine makers. For some issues, further study will clarify the topic; for others, personal preference will always be the deciding factor. I extend my apologies to those who may feel that their views have been inadequately represented.

Where no common chemical name is available or currently preferred, IUPAC terminology has been used. For grape cultivar names, single quotes have been used around the name (i.e., 'Pinot noir'), in lieu of the other accepted practice of placing *cv.* before the name to conform to the International Code of Botanical Nomenclature. Except in tables, the current practice of naming rootstock cultivars with a number and the originator's name is used, in lieu of the number and a contraction of the originator's name (i.e., 3309 Couderc versus 3309 C).

A list of suggested readings is given at the end of each chapter to guide further study. Although several are in foreign languages, they are excellent sources of precise information. To have omitted them would have done disservice to those wishing to pursue the topics involved. In addition, references are given in the text where the information is very specific or not readily available in the suggested readings. Further details can be obtained from sources given for the figures and tables.

Wherefore it maye please your . . . gentlines to take these labours in good worthe, not according unto their unworthiness, but accordinge unto my good mind and will, offering and gevinge them unto you.

from William Turner's *Herbal,* 1568

Acknowledgments

Without the astute observations of generations of wine makers and grape growers, and the dedicated research of countless enologists and viticulturalists, this work would have been impossible. Thus, acknowledgment is given to those whose work has not been specifically mentioned. Appreciation also is given to constructive criticism given by W. N. Kliewer, C. Belke, W. Koblet, R. M. Pool, T. Henick-Kling, F. Radler, T. C. Somers, C. Shelton, D. Stanislawski, R. E. Smart, A. Walker, W. H. N. Paton, E. F. Boller, P. G. Goussard, A. Tekauz, A. Noble, A. Rogosin, and D. Oleson. Gratitude is also expressed to the many researchers, companies, institutes, and publishers who freely donated photographs, data, diagrams, or figures reproduced in the book. Appreciation also goes to Mrs. Alicja Brancewicz for great assistance in locating sources of many of the obscure references and the Brandon University Science Research Committee for financial assistance in purchasing the rights to some graphical material. In addition, I am indebted to the invaluable assistance provided by Connie Parks and Michelle Walker, Charlotte Brabants, and the staff of Academic Press in the preparation of the book. Finally, but not least, recognition must also go to Mrs. Mae Jackson for typing much of the text, and to my wife, Suzanne Ouellet, who helped in the translation of French articles and has long suffered as an author's 'widow.'

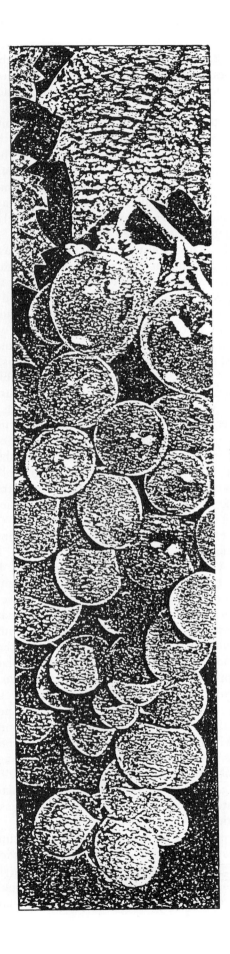

1

Introduction

Grapevine and Wine Origins

Wine has a recorded history stretching back some 6000 years, with the earliest known wine residues dating from the late fourth millennium B.C. (Badler *et al.*, 1990). Hieroglyphic representations of wine presses some 5000 years old have been dated to the reign of Udimu in Egypt (Petrie, 1923). Most researchers think that the discovery of winemaking, or at least its development, began in southern Caucasia. This area presently includes parts of northwestern Turkey, northern Iraq, Azerbaijan, and Georgia. It also is generally thought that the domestication of the wine grape (*Vitis vinifera*) initially occurred within this area. It is here that the natural distribution of *V. vinifera* most closely approaches the probable origin of Western agriculture (Zohary and Hopf, 1988). Domestication also may have occurred independently in Spain (Núñez and Walker, 1989).

Although grapes readily ferment into wine, the wine yeast (*Saccharomyces cerevisiae*) apparently is not an indigenous member of the grape-skin flora. The natural habitat of the ancestral strains of *Sacch. cerevisiae* may be the bark and sap exudate of oak trees (Phaff, 1986). If so, the growth of grapevines on oak trees, or the harvesting of both grapes and acorns as food, may have encouraged the inoculation of grapes and grape juice with *Sacch. cerevisiae*. The fortuitous overlap in the distribution of ancestral forms of *Sacch. cerevisiae* and *Vitis vinifera* with the origin of Western agriculture may have fostered the discovery of winemaking, as well as its subsequent development and spread. It may not be pure coincidence that most of the major yeast-fermented beverages and foods (wine, beer, mead, and bread) have their origins in the same Near Eastern regions.

The hypothesis of the Near Eastern origin and spread of winemaking is supported by the remarkable similarity between the words meaning *wine* in most Indo-European languages (see Table 2.1). It is commonly believed that the spread of agriculture into Europe resulted from the dispersion of peoples speaking Proto-Indo-European languages (Renfrew, 1989). In addition, most eastern Mediterranean myths locate the origin of winemaking in northeastern Asia Minor (Stanislawski, 1975).

Unlike the major cereal crops of the Near East (wheat and barley), cultivated grapes develop a high yeast-inoculum by maturity. Piled unattended for several days, grapes begin to self-ferment. On rupturing, the juice is rapidly colonized by the yeast flora, which convert the sugars in the juice into alcohol (ethanol). This is favored by the ability of *Saccharomyces cerevisiae* to grow in acidic grape juice and selectively ferment sugars to ethanol.

The fermentation of grapes into wine is expedited if the grapes are first crushed. Crushing releases and mixes the juice with yeasts on the grape skins. While yeast fermentation is more rapid in contact with air, continued air contact results in the wine turning vinegary. Although unacceptable as a beverage, vinegar probably was a valuable commodity in its own right. As a source of acetic acid, vinegar facilitated pottery production and the preservation (pickling) of perishable foods.

Of the many fruits gathered by ancient humans, only grapes store carbohydrates predominantly as soluble sugars. Thus, the major sources of nutrients in grapes is in a form readily metabolized by wine yeasts. Most other fleshy fruits store carbohydrates as starch and pectins, nutrients not fermentable by wine yeasts. The rapid production of ethanol by *Sacch. cerevisiae* quickly limits the growth of most bacteria and other yeasts in grape juice. Consequently, wine yeasts generate conditions that essentially give them exclusive access to grape nutrients.

Another unique property of grapes concerns the acids they contain. The major acid found in mature grapes is tartaric acid. This acid seldom occurs in other fruits but is commonly found in the leaves and other parts of plants. Because tartaric acid is seldom metabolized by microbes, fermented grape juice (wine) remains sufficiently acidic to limit the growth of most bacteria and fungi. In addition, the acidity of wine gives it much of its fresh taste. The combined action of grape acidity and the accumulation of ethanol suppresses the growth and metabolism of most wine-spoilage organisms. This is especially true in the absence of air. For ancient humans, the result of grape fermentation was the transformation of a perishable, periodically available fruit into a beverage that could be preserved for several months and had interesting intoxicating properties.

Unlike many crop plants, the grapevine has required little modification to adapt it to cultivation. Its mineral and water requirements are low, permitting it to flourish on soils and hill sites unsuitable for other food crops. Its ability to climb trees and other supports meant it could be grown with little tending in association with other crops. In addition, its immense regenerative potential has allowed it to adapt to intense pruning. Intense pruning turned a trailing climber into a short, shrublike plant suitable for monoculture. The short stature of the shrubby vine minimized the need for supports and may have decreased water stress in semiarid environments. The regenerative powers and woody structure of the vine also have permitted it to withstand considerable winter-kill and still produce reasonable crops in cool climates. This favored the spread of viticulture into central Europe and the domestication of varieties from indigenous grapevines.

The major change that converted wild vines into a domesticated crop was the shift to a bisexual flowering habit. Wild vines are functionally unisexual but often possess both male and female parts. In some instances, conversion to functional bisexuality may have required only the inactivation of a single dominant gene. However, the complexity of sexual differentiation in some cultivars (Carbonneau, 1983) suggests that several independent mutations may have been involved in the development and spread of bisexuality in *Vitis vinifera*. Self-fertile flowers would have significantly improved crop production of vines isolated from insect pollinators and male vines. Although other modifications distinguish wild from domesticated forms of *V. vinifera*, the changes in seed and leaf shape that have been noted are not of obvious selective value (see Chapter 2). The lower acidity and higher sugar content that characterize domesticated forms of *V. vinifera* are not exclusively the properties of domesticated vines.

The evolution of winemaking from an infrequent, haphazard event to a common cultural practice presupposes the development of a settled life-style. A nomadic habit is incompatible with the accumulation of grapes to consistently produce significant quantities of wine. As wine increasingly became associated with religious rites, the development of a steady supply of wine required the planting of grapevines in or around human settlements. Because grapevines begin to bear a significant crop only after 3 to 5 years, and require several additional years to reach full productivity, such an investment in time and effort would be logical only if the planter lived nearby. Under such conditions, grape collection for winemaking could have initiated the beginning of viticulture. If, as seems reasonable, wine production is dependent on a settled agricultural existence, then significant wine production cannot predate the agricultural revolution. Because grapevines are not indigenous to the Fertile Crescent, where Western agriculture had its origin, the beginnings of winemaking probably occurred after agricultural skills moved north into Caucasia. Present evidence suggests the southern Caucasus region, about 4000 B.C.

From its origins in Caucasia, grape growing and winemaking probably spread southward to Palestine, Syria, Egypt, and Mesopotamia. From this base, wine consumption, and its socioreligious connections, spread winemaking around the Mediterranean. However, new evidence suggests that an extensive system of grape culture already existed in southern Spain several centuries before Phoenician colonization (Stevenson, 1985). Nevertheless, colonization from the eastern Mediterranean is still viewed as the predominant source of grape growing and winemaking knowledge in southern and central Europe. In more recent time, European exploration and colonization have spread grapevine cultivation into most regions of the globe (see Fig. 2.7).

Throughout much of this period, current wine styles either did not exist or occurred in forms considerably different from their present expression. Most ancient wines probably resembled dry to semidry table wines, turning vinegary by spring. Protection from oxidation was generally poor and the use of sulfur unknown or infrequent. Thus, prolonged wine storage probably would have been avoided. However, various techniques were used to extend the drinkable life of some early wines. The addition of pitch (resin) was used by the ancient Greeks and Romans. Its use today is limited to the flavoring of a few wines such as *retsina*. Concentration by heating was occasionally used, as well as the addition of boiled-down grape juice and honey. The famous wines of ancient Greece or Rome, such as those from Chios, Lesbos, and Falernum, have often been thought of as being syrupy due to prolonged evaporation. However, the stoppering of wine amphoras with cork, and the prestige ascribed to aged wine by Roman authors, may require a reassessment of the characteristics of wines appreciated by the ancient Greeks and Romans (Tchernia, 1986). Wine amphoras also may have

been stored upside down, thus keeping the cork wet (Grace, 1979). Amphoras, still cork-sealed and containing wine remnants, have been excavated from the Mediterranean (Cousteau, 1954).

Wines began to take on their modern expressions about the seventeenth century. The widespread use of sulfur in barrel treatment seems to have occurred about this time. This would have greatly increased the likelihood of producing better quality wines and extending their life. Stable sweet wines also started to be produced in the mid 1600s, commencing with the famous Tokaji wines of Hungary.

Before the production of sparkling wine was possible, there had to be the invention of a process for producing strong glass. This development occurred in England in the late 1600s. With the production of bottles able to withstand the high internal pressures generated by trapped carbon dioxide, and the reintroduction of cork as bottle stopper, the stage was set for the rapid development of sparkling wines.

Vintage port production also depended on the production of inexpensive glass bottles. The evolution in bottle shape from bulbous to cylindrical permitted bottles to be laid on their sides. As the cork stayed wet in this position, the wine remained isolated from oxygen and potentially was able to develop a smooth character and complex fragrance. The development of port also depended on the perfection of wine distillation. Distilled spirits needed to be added to the fermenting juice to stop its fermentation prematurely. As a consequence, grape sugars were retained along with sufficient pigment to produce a dark red wine. Modern sherries also require the use of grape spirits. Although alcohol distillation was first developed by the Arabs about the eleventh century A.D., adoption of the technique in Medieval Europe was slow. Thus, most fortified wines are of relatively recent origin.

With mechanization, the widespread adoption of glass bottles for both wine transport and maturation was possible. The reintroduction of cork for stoppers in the seventeenth century provided conditions favorable for the production of modern wine maturation. The discovery by Pasteur in the 1860s of the importance of yeasts and bacteria in fermentation set in motion a chain of events that have produced the incredible range of wines which typify modern commerce.

Commercial Importance of Grapes and Wine

From its simple origins, grape production has developed into the most important fresh fruit crop in the world. Worldwide grape production in 1990 was about 60 million metric tons. That compares with about 52,

46, and 40 million metric tons for oranges, bananas, and apples, respectively (FAO, 1991). The area planted to grapevines in 1990 was estimated at about 8.7 million hectares (21.5 million acres). About 71% of the production was fermented into wine, 27% used as a fresh fruit crop, and the remaining 2% dried for raisins. Use varies widely from country to country, often depending on both the physical and politicoreligious (wine prohibition) climate of the region.

Grape production is largely restricted to climate zones similar to that of the indigenous range of *Vitis vinifera* in Eurasia. This zone approximates the region between the 10° to 20° annual isotherms (Fig. 1.1). Extension into cooler and warmer regions is possible when local conditions modify the climate or viticultural practice compensate for less than ideal circumstances. Commercial production even occurs in subtropical regions, where severe pruning stimulates year-round growth of the vine.

In Europe, where 70% of the world's vineyard hectarage is located, about 77% of the crop is fermented into wine. The latter percentage is slightly less for the world production (71%), owing to the predominant use of grapes as a table or raisin crop in Islamic countries. Since the 1970s, wine production has ranged from about 270 to 370 million hl (85 to 98 million gallons), with recent production levels being about 290 million hl. While Spain has the largest area under grape cultivation, France and Italy produce the largest volumes of wine. Together, France and Italy produce about 45% of the world's wine but generate about 60% of world wine exports. Statistics on wine production and export for several countries are given in Fig. 1.2. Interestingly, several major wine-producing countries, such as Argentina and the former Soviet Union are little involved in wine export. In contrast, others with comparatively small productions, such as Germany and Australia, are very important in international wine export.

Although many European countries are major wine exporters, they are also the primary wine-consuming countries. For centuries, wine has been a significant caloric food source in the daily diets of many workers in France, Italy, Spain, and other Christian Mediterranean nations. As wine was an integral part of daily food consumption, drinking to excess has not had the tacit acceptance found in some northern portions of Europe. Alcohol abuse, especially in the United States, spawned both the prohibitionist and, now, neoprohibitionist movements. The views of neoprohibitionists that consuming any beverage containing alcohol is detrimental to one's health is in marked conflict with evidence supporting the benefits of moderate wine consumption (Katz, 1981; Rimm *et al.*, 1991; Friedman, 1992). In some countries such as the United States, statements suggesting possible harmful effects from wine consumption may be required on wine labels. The reticence of some governments to accept the beneficial consequences of moderate wine consumption conflicts with the wide and valid use of wine in medicine.

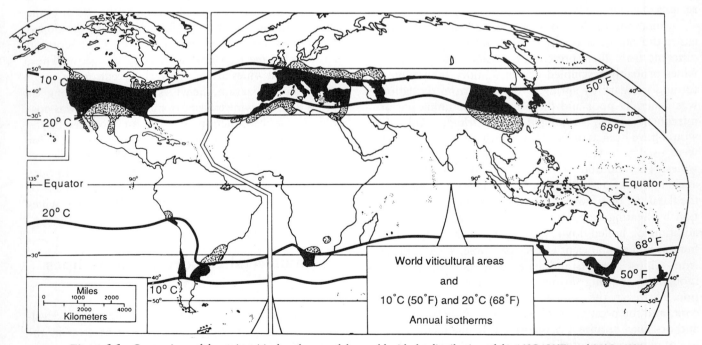

Figure 1.1 Comparison of the major viticultural areas of the world with the distribution of the 10°C (50°F) and 20°C (68°F) annual isotherms. (From de Blij, 1983, reproduced with permission.)

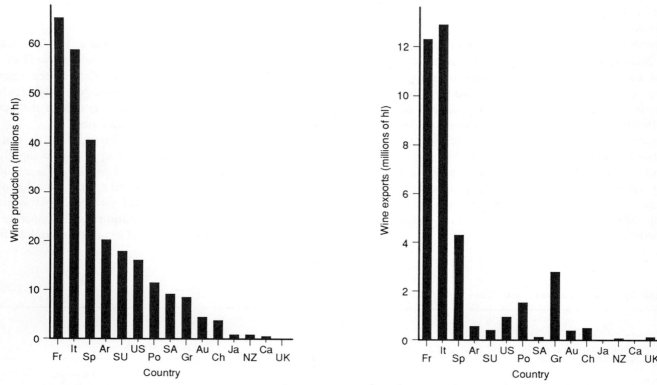

Figure 1.2 Production and export statistics (1990) for several countries: France (Fr), Italy (It), Spain (Sp), Argentina (Ar), Soviet Union (SU), United States (US), Portugal (Po), South Africa (SA), Germany (Gr), Australia (Au), Chile (Ch), Japan (Ja), New Zealand (NZ), Canada (Ca), United Kingdom (UK). (Data from OIV, 1991.)

Except for some New World and Asian countries, per capita consumption of wine is declining (Fig. 1.3). The reasons for these trends are complex and often region-specific. Nevertheless, decline in total consumption has been partially associated with a shift toward the consumption of less but better quality wine. Because of the finer quality of wines, there is less likelihood of their being denigrated as simply an "alcoholic beverage."

Wine Classification

Except in the broadest sense, there is no generally accepted way of classifying wines. They may be grouped by sweetness, alcohol content, carbon dioxide level, color, grape variety used, fermentation or maturation processes involved, or geographic origin. For taxation purposes, wines often are divided into three basic categories, namely, still wines, sparkling wines, and fortified wines; the latter two typically are taxed at a higher rate. This division recognizes significant differences not only in production but also in use.

Wines commonly are subdivided by geographic origin. In many European countries this is associated with the

traditional use of particular grape cultivars, as well as grapegrowing and winemaking techniques. New World wines may be classified similarly, but few regions have become consistently associated with particular styles. Although use of European regional names, such as Cha-

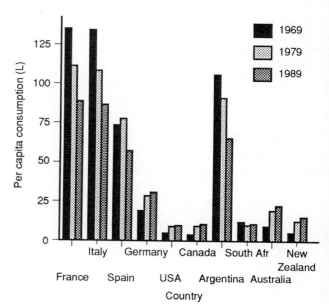

Figure 1.3 Changes in per capita wine consumption in selected countries. (Data from OIV, 1991.)

blis and Burgundy, was extensive in the past in much of the New World this practice is being replaced by more appropriate brand name or grape varietal designations.

Because geographic wine classifications rarely provide the consumer with useful information on the sensory characteristics that might be expected from the wine, the following arrangement is based primarily on stylistic differences. Wines are initially grouped based on alcohol concentration. This commonly is indicated by the terms **table** (alcohol contents commonly ranging between 9 and 14% by volume) and **fortified** (alcohol contents often ranging between 17 and 22% by volume). Table wines are subdivided into **still** and **sparkling** categories, depending on the carbon dioxide level retained in the bottled wine.

Still Table Wines

As most wines fall into the category of still table wines, it requires the most complex classification system (Table 1.1). The oldest division is based on color and separates wines into white, red, and rosé groups. Not only does it have the benefit of long acceptance, it reflects distinct differences in flavor, use, and production.

Because most white wines are intended to be consumed with meals, they are designed to have a slightly acidic finish. Combined with food proteins, the acidic aspect of the wine becomes balanced and can both accentuate and harmonize with food flavors. Most white wines are given little if any maturation in oak cooperage. Only wines with distinct varietal aromas tend to benefit

Table 1.1 Classification of Still Wines Based on Stylistic Differences[a]

White			
Long-aging (often matured and occasionally fermented in oak cooperage)		Short-aging (seldom exposed to oak)	
Typically little varietal aroma (examples)[b]	Varietal aroma commonly detectable (examples)[b]	Typically little varietal aroma (examples)[b]	Varietal aroma commonly detectable (examples)[b]
botrytized wines	'Riesling'	'Trebbiano'	'Müller-Thurgau'
Vernaccia di	'Chardonnay'	'Muscadet'	'Kerner'
San Gimignano	'Sauvignon blanc'	'Folle blanche'	'Pinot blanc'
Vin Santo	'Parellada'	'Chasselas'	'Chenin blanc'
	'Sémillon'	'Aligoté'	'Seyval blanc'

Red			
Long-aging		Short-aging	
Tank oak-aging (many European wines, except those from France are in this category) (examples)[b]	Barrel oak-aging (most French, "new" European, and New World wines in this category) (examples)[b]	Little varietal aroma detectable (examples)[b]	Varietal aroma often detectable (examples)[b]
'Tempranillo'	'Cabernet Sauvignon'	'Gamay'	'Dolcetto'
'Sangiovese'	'Pinot noir'	'Grenache'	'Grignolino'
'Nebbiolo'	'Syrah'	'Carignan'	'Baco noir'
Garrafeira	'Zinfandel'	'Barbera'	'Lambrusco'

Rosé	
Sweet	Dry
Mateus	Tavel
Pink Chablis	Cabernet Rosé
Rosato	White 'Zinfandel'
some Blush wines	some Blush wines

[a] Although predominantly dry, many have a sweet finish. These include both light 'sipping' wines and the classic botrytized wines.

[b] Representative examples in single quotes refer to the names of grape cultivars used in the wines production.

from the association with oak flavors. Those with a sweet finish are intended to be consumed alone as a "sipping" wine, to accompany dessert, or to replace dessert. Most botrytized (late harvest) wines and eisweins fall into the latter category.

Modern red wines are almost exclusively dry. The absence of a detectable sweet taste is consistent with their intended use as a food beverage. The bitter and astringent compounds characteristic of most red wines bind with food proteins, producing a balance that otherwise would not develop. Occasionally, well-aged red wines are saved for enjoyment after the meal. Their diminished tannin content does not require food to develop smoothness. Also, the complex subtle bouquet of aged wines often is appreciated more fully without food flavors.

Most red wines that age well are given the benefit of some maturation in oak cooperage. Storage in small oak cooperage (~225 liter barrels) usually speeds maturation and adds subtle flavors. Following in-barrel maturation, the wines typically receive additional in-bottle aging. Where less oak character is desired, cooperage of over 1000 liter capacity may be used. Alternately, the wine may be matured in inert tanks to avoid oxidation and the uptake of additional flavors.

One of the more common differences between red wines depends on the consumer market for which they are intended. Wines processed for early consumption have light flavors, whereas those processed to enhance aging potential often do so at the expense of early enjoyment and are initially excessively tannic. Beaujolais nouveau is a prime example of a wine designed for early consumption. In contrast, premium 'Cabernet Sauvignon' and 'Nebbiolo' wines illustrate the other extreme, where long aging may be required for development of the finest qualities.

Rosé wines are undoubtedly the most maligned group of wines. This probably results from their mode of production. To achieve the desired rosé color, the grape skins are removed from the juice shortly after fermentation has begun. Thus, the uptake of compounds that give red wines their flavor is also limited. Few if any rosé wines age well, and therefore they have not developed the respect and following of connoisseurs. In addition, most are made to have a sweet finish. Seemingly to counter the stigma attached to the name rosé, they may be termed "blush" wines.

Sparkling Wines

Sparkling wines often are classified by the method used to achieve the high carbon dioxide content. The three primary techniques are the traditional (champagne), transfer, and bulk (Charmat) methods. In these methods, yeasts generate the carbon dioxide that pro-

Table 1.2 Classification of Sparkling Wines and Representative Examples

With added flavors	Natural (without flavors added)	
Coolers (low alcohol)	Highly aromatic (sweet)	Subtly aromatic (dry or sweet)
Fruit-flavored, carbonated wines	Asti-style Muscat-based wines	Traditional-style Champagne Vin Mousseux Cava Sekt Spumante Crackling/carbonated Perlwein Lambrusco Vinho Verde

duces the effervescence. Although precise, these divisions often do not reflect the more significant sensory differences found in each division. For example, the traditional and transfer methods both aim to produce dry to semidry wines, accentuating subtlety, limiting varietal aroma, and ideally possessing a "toasty" fragrance. Sparkling wines often differ more because of variations in the duration of yeast contact and grape varieties used than in the method of production. Although most bulk-method wines tend to be sweet and aromatic (i.e., Asti Spumante), some are dry with subtle fragrances.

Wines deriving their sparkle from carbon dioxide incorporation under pressure show an even wider range of styles. These incude dry white wines, such as *vinho verde* (historically obtaining its sparkle from malolactic fermentation), sweet sparkling red wines, such as Lambrusco, most sparkling rosés, and fruit-flavored "coolers."

Fortified Wines (Dessert and Appetizer Wines)

All terms applied to the category of fortified wines are somewhat misleading. For example, some achieve their elevated alcohol levels without the addition of distilled spirits (e.g., the sherrylike wines from Montilla, Spain). The alternative designation, aperitif and dessert wines, also has problems. While most are used as aperitif or dessert wines, many table wines are used similarly. Some people consider sparkling wines the ultimate aperitif and botrytized wines the preeminent dessert wines.

Regardless of designation, fortified wines typically are consumed in small amounts, and are seldom completely

— older barrells.

Table 1.3 Classification of Fortified Wines and
Representative Examples

With added flavors	Without added flavors
Vermouth	Sherrylike
Byrrh	Jerez-Xerès-Sherry
Marsala (some)	Malaga (some)
Dubonnet	Montilla
	Marsala
	Château-Chalon
	New World solera and submerged
	sherries
	Portlike
	Porto
	New World Ports
	Madeiralike
	Madeira
	baked New World Sherries and ports
	Muscatel
	Muscat-based Wines
	Setúbal
	Samos (some of)
	Muscat de Beaunes de Venise
	Communion wine

consumed on opening. The high alcohol content limits microbial spoilage, and the marked flavor and insensitivity to oxidization often allow them to retain their aromatic features for weeks after opening. These are desirable properties for wines usually consumed in small amounts. Exceptions are fino sherries and vintage ports, which lose their distinctive properties shortly after bottling or on opening, respectively.

Fortified wines are produced in a wide range of styles. Dry and/or bitter-tasting forms are normally consumed as an aperitif before meals. They stimulate the appetite and activate the release of digestive juices. Examples are fino-style sherries and dry vermouths. The latter are flavored with a variety of herbs and spices. More commonly, fortified wines possess a sweet character. Major examples are oloroso sherries, ports, madeiras, and marsalas. These wines are consumed after meals or as a dessert.

Wine Quality

Quality is a perception easier to experience than describe. Appreciation of wine often is modified by experience, and affected by the genetic makeup of the individual. Nevertheless, quality does have components that can be more or less quantified. Negative quality factors (faults) are generally the easiest to identify and control. Positive quality factors tend to be more elusive. Some of the more obvious are the distinctive aromas derived from some grape varieties. Recognizable modifications produced by viticultural practices, climate, winemaking style, processing, and aging also may be highly regarded. When too accentuated, or unfamiliar, these same features may be considered faults. Most faults blur or destroy the subtle characteristics that distinguish wines from one another. More serious faults make the wine undrinkable. Although subtle differences in quality are difficult to describe precisely, most people believe they exist. However, whether some individuals possess the claimed ability to consistently recognize the subtle variations that exist between similar wines from different vintages or vineyard sites has yet to be established beyond reasonable doubt.

For occasional wine drinkers, knowledge of the geographic or varietal origin of the wines is likely to be of little concern. Pleasure is probably measured by strictly subjective, often highly individualistic criteria. In contrast, geographic origin and reputation can strongly influence wine sales to connoisseurs, and presumably the pleasure derived from the wine. For the connoisseur, whether and how well a wine reflects established expectations can be critical to its perceived quality. Historically developed expectations are central to the concepts of quality embodied in most Appellation Control laws.

In addition to the purely subjective and historical views of quality, aesthetic quality is the most highly prized property possessed by premium wines. Aesthetic quality is interpreted similarly, and uses the same language, as other aesthetic endeavors such as sculpture, architecture and literature. These include balance, harmony, symmetry, development, duration, complexity, subtlety, interest, and memorableness. Defining these terms precisely is difficult owing to variability in individual perception. Nevertheless, balance and harmony in wine commonly refer to a smooth taste and mouth-feel, without any aspect interfering with the overall pleasurable sensation. Symmetry refers to the perception of compatibility between the sapid (taste and mouth-feel) and olfactory (fragrant) sensations of a wine. Development typically refers to the changes in intensity and aromatic character of the wine after pouring. When pleasurable, development is important in maintaining interest. Fragrance duration is also important to the aesthetic perception of wine quality. Complexity and subtlety are additional highly valued attributes of the fragrance and flavor of wine. The impact of these factors on the memory of the sensory experience is probably the most significant determinant in the overall perception of wine quality.

Health Related Aspects of Wine Consumption

Until the 1900s, wine was used in the effective treatment of several human ailments and as an important solvent in many medications (Lucia, 1963). In the twentieth century, well-meaning but misguided prohibitionists persuaded several national governments and the medical establishment in many countries that all forms of beverages containing alcohol were undesirable in any use, with the possible exception of religious services.

Currently there is return in interest among medical scientists in assessing the legitimate and health benefits of moderate wine consumption. These effects include reduction in the undesirable effects of stress, enhanced sociability, lowered rates of clinical depression, and improved self-esteem and appetite in the elderly (Baum-Baicker, 1985; Delin and Lee, 1992). In such studies, it is critical for the researchers to be fully cognizant of the social, environmental, and individual factors that may influence the results and to take appropriate steps to compensate for their potential effects. Neoprohibitionists are quick to point out both real and imaginary faults in any study that presents findings contrary to their established beliefs.

In contrast, the benefits of moderate wine consumption on increasing the plasma levels of high-density lipoprotein (HDL), which is helpful in lowering blood cholesterol concentration and the incidence of arteriosclerosis, are well established (Katz, 1981; Kinsella *et al.*, 1993; and Rimm *et al.*, 1991). Even the National Institute on Alcohol Abuse and Alcoholism (NIAAA) was moved to record in the April 1992 issue of *Alcohol Alert* that "there is a considerable body of evidence that lower levels of drinking decrease the risk of death from coronary heart disease."

In an interesting study by Lindman and Lang (1986), only wine among beverages containing alcohol was associated with positive social expectations. Thus, wine appears unique in its being identified with happiness, contentment, and romance. Additional studies also have found that wine is associated with more socially desirable stereotypes than other alcohol-containing beverages (see Delin and Lee, 1992). In addition to revealing the potential benefits of wine consumption, researchers also are beginning to investigate some of the occasionally undesirable consequences of even moderate wine use. The association of red-wine-induced headaches with insufficient production of a platelet phenolsulphotransferase (Littlewood *et al.*, 1992) and headache prevention by prior use of ASA (salicylic acid acetate; Kaufman, 1992) are good examples.

Although wine consumption is counterindicated in a few medical instances such as gastrointestinal ulcerations and cancers, in most situations the daily consumption of wine in moderate amounts as part of the meal is beneficial to human health.

Suggested Readings

Allen, H. W. (1961). "A History of Wine." Faber and Faber, London.

Amerine, M. A., and Singleton, V. L. (1977). "Wine—An Introduction." Univ. of California Press, Berkeley.

de Blij, H. J. (1983). "Wine: A Geographic Appreciation." Rowman and Allanheld, Totowa, New Jersey.

Fregoni, M. (1991). "Origines de la Vigne et de la Viticulture." Musumeci Editeur, Quart (Vale d'Aosta), Italy.

Henderson, A. (1824). "The History of Ancient and Modern Wines." Baldwin, Craddock & Joy, London.

Redding, C. (1851). "A History and Description of Modern Wines," 3rd Ed. Henry G. Bohn, London.

Unwin, T. (1991). "Wine and the Vine: An Historical Geography of Viticulture and the Wine Trade." Routledge, London.

Younger, W. (1966). "Gods, Men and Wine." George Rainbird, London.

References

Badler, V. R., McGovern, P. E., and Michel, R. H. (1990). Drink and be merry! Infrared spectroscopy and ancient Near Eastern wine. *MASCA Res. Pap. Sci. Archaeol.* 7, 25–36.

Baum-Baicker, C. (1985). The psychological benefits of moderate alcohol consumption: A review of the literature. *Drug Alcohol Depend.* 15, 305–322.

Carbonneau, A. (1983). Stérilités mâle et femelle dans le genre *Vitis*. II. Conséquences en génétique et sélection. *Agronomie* 3, 645–649.

Cousteau, J.-Y. (1954). Fish men discover a 2,200-year-old Greek ship. *Nat. Geogr.* 105(1), 1–34.

de Blij, H. J. (1983). "Wine: A Geographic Appreciation." Rowman and Allanheld, Totowa, New Jersey.

Delin, C. R., and Lee, T. L. (1992). Psychological concomitants of the moderate consumption of alcohol. *J. Wine Res.* 3, 5–23.

FAO (1991). "FAO Production Year Book," Vol. 44. Food and Agriculture Organization of the United Nations, Rome.

Friedman, S. (1992). Wine's history as therapy for mankind. *Wines Vines* 72(6), 51–56.

Grace, V. R. (1979). "Amphoras and the Ancient Wine Trade" (Photo 63). Amer. School Classical Studies Athens, Institute of Advanced Study. Princeton, New Jersey.

Katz, H. J. (ed.) (1981). "Proceedings of Wine, Health and Society—A Symposium." GRT Books, Oakland, California.

Kaufman, H. S. (1992). The red wine headache and prostaglandin synthetase inhibitors: A blind controlled study. *J. Wines Res.* 3, 43–46.

Kinsella, J. E., Frankel, E., German, J. B., and Kanner, J. (1993). Possible mechanisms for the protective role of antioxidants in wine and plant foods. *Food Technol.* (April), 85–89.

Lindman, R., and Lang, A. R. (1986). Anticipated effects of alcohol consumption as a function of beverage type: A cross-cultural replication. *Int. J. Psychol.* 21, 671–678.

Littlewood, J. T., Glover, W., Davies, P. T. G., Gibb, C., Sandler, M., and Rose, F. C. (1988). Red wine as a cause for migraine. *Lancet* (Mar. 12), 558–559.

Lucia, S. P. (1963). "A History of Wine as Therapy." Lippincott, Philadelphia, Pennsylvania.

Núñez, D. R., and Walker, M. J. (1989). A review of palaeobotanical findings of early *Vitis* in the Mediterranean and of the origins of cultivated grape-vines, with special reference to prehistoric exploitation in the western Mediterranean. *Rev. Paleobot. Palynol.* **61**, 205–237.

OIV (1991). The state of vitiviniculture in the world and the statistical information in 1990. *Bull. O.I.V.* **64**, 894–954.

Petrie, W. M. F. (1923). "Social Life in Ancient Egypt." Methuen, London.

Phaff, H. J. (1986). Ecology of yeasts with actual and potential value in biotechnology. *Microb. Ecol.* **12**, 31–42.

Renfrew, C. (1989). The origins of Indo-European languages. *Sci. Am.* **261**(4), 106–116.

Rimm, E. B., Giovannucci, E. I., Willett, W. C., Colditz, G. A., Ascerio, A., Rosner, B., and Stampfer, M. J. (1991). Prospective study of alcohol consumption and risk of coronary disease in men. *Lancet* **338**, 464–468.

Stanislawski, D. (1975). Dionysus westward: Early religion and the economic geography of wine. *Geogr. Rev.* **65**, 427–444.

Stevenson, A. C. (1985). Studies in the vegetational history of S. W. Spain. II. Palynological in investigations at Laguna de los Madres, Spain. *J. Biogeogr.* **12**, 293–314.

Tchernia, A. (1986). "Le Vin de l'Italie Romaine: Essal d'Histoire Economique d'après les Amphores." École Fracnçise de Rome, Rome.

Zohary, D., and Hopf, M. (1988). "Domestication of Plants in the Old World." Oxford Univ. Press (Clarendon), Oxford.

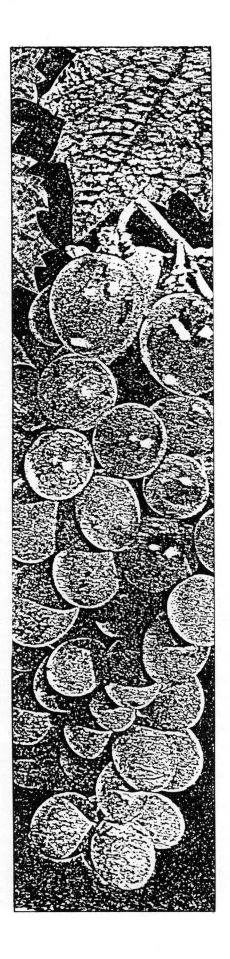

2

Grape Species and Varieties

Introduction

Grapevines are placed in a single genus, *Vitis,* within the family Vitaceae. Other well-known temperate zone members of the family are Boston ivy (*Parthenocissus tricuspidata*) and Virginia creeper (*P. quinquefolia*). Members of the Vitaceae typically show a climbing habit, have leaves that develop alternately on shoots (Fig. 2.1), and possess swollen or jointed nodes that generate tendrils or flower clusters opposite the leaves. The flowers are minute, bi- or unisexual, and occur in clusters. Most flower parts occur in groups of fours or fives, with the stamens developing opposite the petals. The ovary consists of two carpels, partially enclosed by a receptacle, and develops into a two-compartmented berry containing up to four seeds (Fig. 3.16).

The family Vitaceae is predominantly tropical and subtropical and contains some 12 to 14 genera, possessing about a thousand species. In contrast, the genus *Vitis* has a primarily temperate zone distribution, occurring extensively only in the Northern Hemisphere.

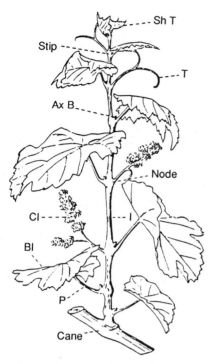

Figure 2.1 *Vitis vinifera* shoot, showing the arrangement of leaves, clusters (Cl), and tendrils (T). Ax B, Axillary buds; Bl, blade; I, internode; P, petiole; Sh T, shoot tip; Stip, stipule. (From von Babo and Mach, 1923, reproduced by permission.)

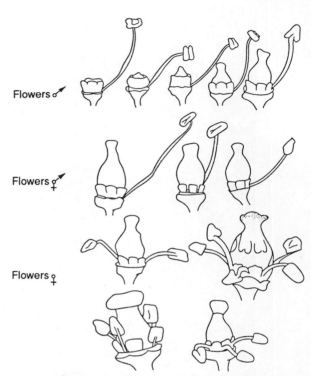

Figure 2.2 Diagrammatic representation of the variety of male, female, and bisexual flowers produced by *Vitis vinifera*. (From Levadoux, 1946, reproduced by permission.)

The Genus *Vitis*

Grapevines are distinguished from related genera primarily by floral characteristics. The flowers are typically unisexual, being either male (possessing erect functional anthers and lacking a fully developed pistil) or female (containing a functional pistil and either producing recurved stamens and sterile pollen or lacking anthers) (Fig. 2.2). The fused petals, called the **calyptra** or cap, remain connected at the apex, while splitting along the base from the receptacle (Fig. 3.13). The petals are shed at maturity. At the base of the ovary are swollen nectaries that generate a mild fragrance which attracts several pollinating insects to the flowers. The sepals of the calyx exist only as vestiges. The fruit is juicy and acidic.

The genus has typically been divided into two sections or subgenera, *Euvitis* and *Muscadinia*. *Euvitis* is the larger, containing all species except three placed in the section *Muscadinia*. The two sections are sufficiently distinct that some taxonomists have separated the muscadine grapes into their own genus, *Muscadinia*.

Members of the section *Euvitis* are characterized by having shredding bark, minute lenticels, a pith interrupted at nodes by woody tissue (the **diaphragm**), tangentially positioned phloem fibers, branched tendrils, elongated flower clusters, fruit that adheres to the fruit stalk at maturity, and predominantly pear-shaped seeds possessing a prominent beak and smooth chalaza. The chalaza is the pronounced, circular, depressed region on the dorsal (back) side of the seed. In contrast, species in the section *Muscadinia* possess a tight, nonshredding bark, prominent lenticels, no diaphragm interrupting the pith at nodes, radially arranged phloem fibers, unbranched tendrils, small floral clusters, berries which individually separate from the cluster as they reach maturity, and boat-shaped seeds with a wrinkled chalaza. Some of these features are illustrated in Fig. 2.3.

The two sections also differ in chromosomal composition. *Euvitis* species contain 38 chromosomes ($2n = 6x = 38$), whereas *Muscadinia* spp. possess 40 chromosomes ($2n = 6x = 40$). The symbol n refers to the number of chromosome pairs formed during meiosis, and x refers to the number of chromosome complements (**genomes**) thought to be involved in the origin of the current *Vitis* sections (Patel and Olmo, 1955). Crosses can be made experimentally between species of the two sections, but their progeny show poor fertility. This probably results from the imprecise pairing and separation of chromosomes during meiosis.

The origin of the genus *Vitis* is thought to have involved the crossing of diploid species, ensued by the crossing of its tetraploid offspring with several diploids

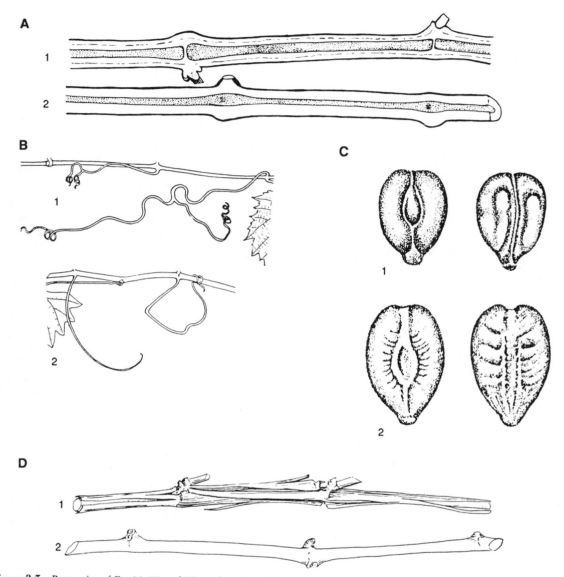

Figure 2.3   Properties of *Euvitis* (1) and *Muscadinia* (2). (A) Internal cane morphology; (B) tendrils; (C) front and back seed morphology; (D) bark shedding. (A, B, and D from Bailey, 1933; C from Rives, 1975; reproduced with permission.)

(Fig. 2.4). In each case, chromosome doubling is considered to have imparted fertility to the infertile progeny. The offspring of most interspecies crosses are infertile owing to improper and/or unbalanced chromosome pairing during meiosis. The initial cross probably involved diploids possessing 6 and 7 chromosome pairs; respectively. Thus, both *Vitis* sections have a common, or similar, set of 13 chromosome pairs. Subsequently, these tetraploid ($4x = 26$) vines crossed with diploid vines possessing either 6 or 7 chromosome pairs. *Euvitis* ($6x = 38$) arose following chromosome doubling of offspring from the tetraploid progenitor ($4x = 26$) crossed with a diploid possessing 6 chromosome pairs. In contrast, hexaploid *Muscadinia* ($6x = 40$) arose from a crossing between the tetraploid progenitor ($4x = 26$)

and a diploid vine possessing 7 chromosome pairs. Although the hypothesized tetraploid species apparently no longer exist, the diploid parents may be related to other members of the Vitaceae, which possess either 6 or 7 chromosome pairs.

While current *Vitis* species have descended from hexaploid progenitors, they may have undergone a process termed diploidization, similar to that in cultivated wheats and other crops (Briggs and Walters, 1986). Diploidization may result from the inactivation of all but one diploid set of parental chromosomes, making a polyploid functionally diploid. Another and more immediately important process following chromosome doubling is the prevention, or precise regulation, of multivalent crossovers between multiple sets of similar

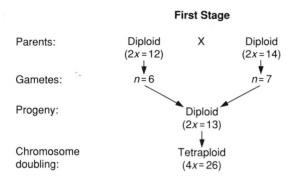

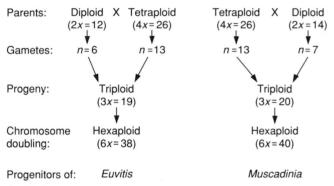

Figure 2.4 Hypothesized evolution of the *Euvitis* and *Muscadinia* sections of *Vitis* involving sequential hybridization and chromosome doubling of the progeny. (Based on work of Patel and Olmo, 1955.)

chromosomes. It is important that chromosome separation occur evenly during meiosis to avoid unequal chromosome complements occurring in the pollen and egg cells. The latter usually results in partial or complete sterility.

In contrast to the relative genetic isolation imposed by the differing chromosome complements of *Euvitis* and *Muscadinia,* hybrids between species within each section are often fertile and vigorous. The ease with which interspecies crossing occurs complicates the task of delimiting species. Many of the criteria commonly used to differentiate species are not applicable to grape species. Most *Vitis* species have similar chromosome numbers, are cross-fertile, often overlap in geographic distribution, and show few distinctive morphological differences. The quantitative differences that do exist between species, such as shoot and leaf hairiness, are often strongly influenced by the environment. Evolution into distinct species appears incomplete, and local populations may be more appropriately considered ecospecies or ecotypes rather than biological species. Establishment of a definitive taxonomic classification of *Vitis* spp. may require a genetic study of the morphological features on which species designation currently is based. Nevertheless, the proper-

ties of some species are sufficiently distinct to be useful in providing sources of genetic variation in grapevine breeding. A recent classification of the eastern North American species of *Vitis* is given by Moore (1991).

Geographic Origin and Distribution of *Vitis* and *Vitis vinifera*

Where and when the genus *Vitis* evolved is unclear. The current distribution of *Euvitis* species includes northern South American (the Andean chain in Colombia and Venezuela), Central and North America, Asia and Europe. In contrast, species in the section *Muscadinia* are restricted to the southeastern United States and northeastern Mexico.

In the nineteenth century, many species were proposed based on fossil leaf impressions (Jongmans, 1939). These are no longer accepted as valid designations, as such evidence is of dubious value. Several plants, only very distantly related to grapes, possess leaves similar in outline. Greater confidence can be placed in the more unique morphology of seeds. Nevertheless, even here, considerable interspecies variation exists (Bailey, 1933). On the basis of seed morphology, two groups of fossilized grapes have been identified, namely, those of the *Vitis ludwigii* and *V. teutonica* types. Seeds of the *Vitis ludwigii* type, resembling those of *Muscadinia* grapes, have been found in Europe from the Pliocene (2 to 10 million years B.P.). Those of the *Vitis teutonica* type, resembling those of *Euvitis* grapes, have been discovered as far back as the Eocene (40 to 55 million years B.P.). However, these identifications, based on comparatively few specimens, remain tenuous because of the morphological variation that occurs in seed samples. In addition, seeds produced by related genera, such as *Ampelocissus* and *Tetrastigma,* are similar to those produced by *Vitis.* Presently, most grape fossil remains come from Europe. However, this may reflect more the availability of appropriate sedimentary deposits, or the distribution of paleobotanical interest, than the historical distribution of grape species.

A proposal of Baranov outlined in Zukovskij (1950) suggests that the ancestral forms of *Vitis* were bushy and inhabited sunny locations. As forests expanded during the more humid Eocene, development of a climbing growth habit allowed *Vitis* to retain its preference for sunny conditions. This may have involved mutations modifying the differentiation of some floral clusters into tendrils, thus improving climbing ability. This hypothesis is plausible since differentiation of bud tissue into flower clusters or tendrils apparently is based on the balance of two plant growth regulators, namely, gibberellins and cytokinins (Srinivasan and Mullins, 1981).

Regardless of the manner and geographic origin of *Vitis,* the genus appears to have reestablished itself throughout its present range by the end of the last major Quaternary glacial period (8000 B.C.). It is believed that the repeated glacial advances and retreats during the Quaternary markedly affected the evolution of *Vitis,* notably *V. vinifera.* The alignment of the major mountain ranges in the Americas and Eurasia also appears to have had an important bearing on the evolution of the genus in these areas. In the Americas the mountain ranges run predominantly north/south, whereas in Europe and western Asia they predominantly run east/west. This would have permitted the displacement of North American and eastern Chinese species southward in advance of expanding ice sheets. Southward movement of grapevines in Europe and western Asia would have been largely restricted by the Pyrenees, Alps, Caucasus, and Himalayas. This may explain the occurrence of but one *Vitis* spp. (*V. vinifera*) in Europe and western Asia, while China possesses 30 plus species (Fengqin *et al.,* 1990) and North and Central America about 34 species (Rogers and Rogers, 1978).

Although glaciation and cold destroyed most favorable habitats in the Northern Hemisphere, major southward displacement was not the only option open for survival. In certain areas, favorable sites, **refuges,** permitted the continued existence of grapes throughout the duration of the glacial advances. In Europe, for example, refuges occurred around the Mediterranean basin and south of the Black and Caspian seas (Fig. 2.5). These refuges may have played a role in the evolution of *Vitis vinifera.*

Of the species that may have existed in Europe before the most recent (Quaternary) ice age, only *Vitis vinifera* survived. It not only survived but reestablished itself throughout central Europe during the various interglacial periods (Planchais, 1972–1973). Although displaced during successive glacial advances, *Vitis vinifera* was again inhabiting southern regions of France by 8000 B.C. For the next several thousand years, the climate slowly ameliorated to about 2° to 3°C warmer than present (Dorf, 1960). Partial domestication appears to have occurred in Europe by about 5000 to 6000 B.C. This is indicated by the discovery of seeds associated with human activity showing properties intermediate between the wild (*silvestris*) and domesticated (*sativa*) forms of *V. vinifera.* This is several thousand years before the agricultural revolution is thought to have spread into western Europe. Semidomesticated grape seed remains have also been discovered in England (2700 B.C.) and in several Neolithic and Bronze Age lake dwellings in Italy, Switzerland, and the former Yugoslavia. In addition, grape seeds and pollen have also been identified from Neolithic sites in southern Sweden and Denmark. Evidence consistent with grape culture has been obtained in southwestern Spain, dating from about 2500 B.C., when the climate was cooler and more humid than currently (Stevenson, 1985).

Seeds from wild vines are elongated, show an average seed index (width/length) of about 0.64, and possess a

Figure 2.5 Distribution of wild *Vitis vinifera* vines about 1850 (dots) superimposed on the forest refuge in the Mediterranean and Transcaucasian regions during the last ice age (line). (From Levadoux, 1956, reproduced by permission.)

prominent beak. In contrast, seeds from domesticated vines are more rounded, possess a nonprominent beak, and have a seed index averaging about 0.55 (Stummer, 1911; Renfrew, 1973). Although differences in floral sexual expression and leaf shape may distinguish wild and domesticated grapevines (Levadoux, 1956), these features are unlikely to be preserved or recognized in vine remains. However, pollen is often sufficiently distinctive and well preserved to establish the presence of a species or subspecies at prehistoric sites.

In *Vitis vinifera*, differences exist between fertile pollen (produced by male and bisexual flowers) and sterile pollen (produced by female flowers) (Fig. 2.6). Fertile pollen is tricolporate (containing three distinct ridges) and produces germ pores, whereas sterile pollen generally is acolporate and does not produce germ pores. Thus far, differences in the fertile/infertile pollen ratio have not been used in indicating the relative frequency of dioecious (wild) and bisexual (cultivated) vines at prehistoric sites. However, size differences between the pollen produced by *silvestris* and *sativa* vines have been used in distinguishing the presence, or absence, of domesticated vines near ancient lake and river sediments (Planchais, 1972, 1973).

It has often been thought that the domestication of *Vitis vinifera* occurred in Transcaucasia or neighboring Anatolia around 4000 B.C. As the climate at that time was similar to that occurring presently, the distribution of *Vitis vinifera* was probably similar to that existent in the mid 1850s before the decimation caused by the phylloxera invasion (Fig. 2.5). In the region between and below the Black and Caspian seas, the indigenous distribution of *Vitis vinifera* most closely approaches the Near

Eastern origin(s) of Western agriculture. Therefore, it seems reasonable to assume that grape domestication may have first occurred in the Transcaucasian Near East.

It is still commonly believed that domesticated vines were carried westward, associated with human migration and the movement of agriculture into the Mediterranean Basin. That the vine was carried into areas such as Palestine and Egypt is indisputable—the vine is indigenous to neither of those regions. Also, the transport of vines into Italy by ancient Greek colonists, and into France, Spain, and Germany by Roman settlers, is suspected from the historical record. However, the significance of these movements to cultivar origin is uncertain. Changes in seed morphology and pollen shape indicate that local grape domestication was in progress before the agricultural revolution reached Europe. Traits such as self-fertilization and larger fruit size often have been attributed to varieties brought in by human activity. However, fruit size often increases under cultivation. In addition, bisexuality occurs, albeit rarely, in several *Vitis* species not exposed to domestication or cultivation (Fengqin *et al.*, 1990). Thus, the occurrence of the bisexual habit in *silvestris* populations of *Vitis vinifera* need not require crossings between wild and domesticated vines. In addition, locally derived cultivars would have had the advantage of being adapted to the prevailing soil and climatic conditions.

Until devastated by foreign pests and disease in the middle of the nineteenth century, wild grapes remained abundant throughout the indigenous range from Spain to Turkmenistan. Although markedly diminished currently, wild vines still occur in significant numbers throughout certain regions of the original range, notably

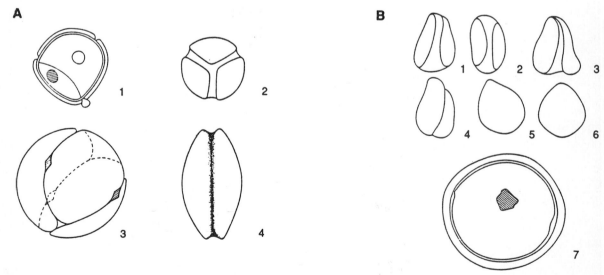

Figure 2.6 Diagrammatic representations of pollen of *Vitis vinifera*. (A) Fertile pollen: I, swollen grain; 2, dry grain, top view; 3, swollen grain in water; 4, dry grain, side view. (B) Infertile pollen: 1 to 6, dry grains; 7, swollen grain in water. (From Levadoux, 1946, reproduced by permission.)

southwestern Russia. Locally occurring and adapted vines may be the progenitors of most European cultivars. These may have been either selections from indigenous vines or the progeny of introgression with cultivars brought in with colonists. Some cultivars, such as the Pinot group, possess numerous traits resembling those of *V. vinifera* f. *silvestris* grapevines.

One of the more distinctive features characterizing domesticated grapes, namely, the lower seed index, has no known agronomic or enological importance. Other properties, such as self-fertility and larger fruit, have indisputable viticultural value. The ancestral unisexual habit would have been an impediment, requiring the cultivation of both fruit-bearing (female) and non-fruit-bearing (male) vines. Planting only fruit-bearing cuttings would have resulted in poor yield by isolating female vines from male, pollen-bearing vines. Mutations leading to the production of self-fertile bisexual flowers would have had strong selective advantage.

The accumulation of agronomically useful mutations probably occurred slowly until the need for a more consistent supply of fruit encouraged their selection. The early association of wine with religious rites could have provided the incentive. Cultivation would have separated fruit-bearing vines from open woodlands, the primary source of pollen from wild vines, and accentuated the productivity of self-fertile vines. Selective propagation of those vines would have led to the progressive elimination of unisexual vines. Viticulture also would have favored the selection of vines showing increased fruit size, as well as improved visual, taste, and aroma characteristics.

Support for the interrelation between winemaking and grapevine domestication comes from the remarkable similarity between the words for "wine" and "vine" in Indo-European languages (Table 2.1). In contrast, little resemblance exists between the words for "grape." The persistence of local terms for grape in European languages may suggest that grape use occurred long before the introduction of viticulture and winemaking and the dispersion of Indo-European languages into Europe (Renfrew, 1989). In contrast, Semitic languages that evolved in regions where grapevines were not indigenous often possess terms for grape and vine similar to those for wine (Forbes, 1965) and are thought to be related to an ancestral term for wine, *woi-no* (Gamkrelidze and Ivanov, 1990). This may indicate that the knowledge of grapes, grapegrowing, and winemaking were gained simultaneously. Although the hypothesis is interesting, the controversial aspect of constructing ancient languages and the movements of prehistoric peoples and ideas makes confidence in these lines of argument tenuous.

One view of the diffusion routes of viticulture into Europe is given in Fig. 2.7. The beginnings of agriculture are thought to have spread to, or developed in, the southern Caucasus (Anatolian) region of Turkey between 5000 and 6000 B.C. (Renfrew, 1989).

Another line of evidence suggesting the antiquity of

Table 2.1 Comparison of Wine-Related Terms in Different Indo-European Languages (Based on information in Buck, 1949)

Language	Wine	Vine	Grape
Greek	OὶνOζ	ἄμπελOζ	βὸτρυζ
Latin	vinum	vitis	uva
Italian	vino	vite, viti	uva
French	vin	vigne	raisin
Spanish	vino	vid	uva
Rumanian	vin	viśa	strugure
Irish	fīn	fīnemain	fín
Breton	gwin	gwinienn	rezinenn
Gothic	wein	weinatriu	weinabasi
Danish	vin	vinranke (-stok)	drue
Swedish	vin	vinranka (-stock)	druva
English	wine	vine	grape
Dutch	wijn	wijnstok	durif
German	wein	weinstock, rebe	traube, weinbeere
Lithuanian	vynas	vynmedis	keke, vynuoge
Lettic	víns	vína kuoks	k'ekars, vínuoga
Serbo-Croatian	vino	losa	grozd
Czech	víno	réva, vinný keř	hrozen
Polish	wino	winorośl	winogrono
Russian	vino	vinograd	vinograd
Sanskrit	drākṣarasa	drākṣā-	drākṣā-

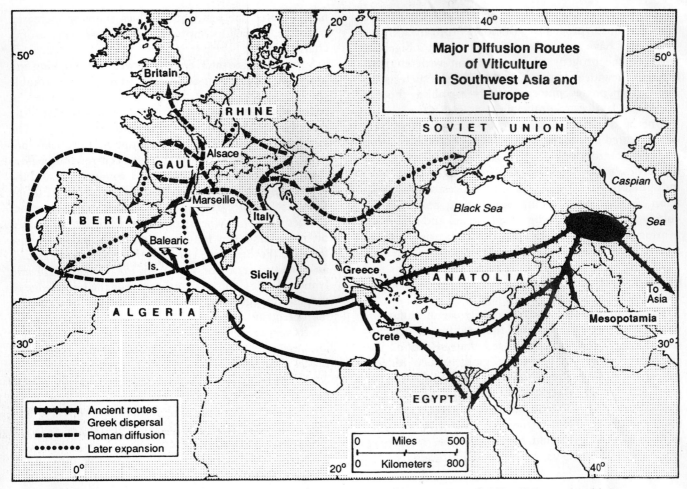

Figure 2.7   Major diffusion routes of viticulture in southwest Asia and Europe. (From de Blij, 1983, reproduced by permission.)

grape culture in the Caucasus comes from the advanced state of cultivar evolution in the region. Local varieties possess many recessive mutants, such as smooth leaves, large branched grape clusters, and medium-sized juicy fruit. The large number of recessive traits is considered an indication of cultivar age. Varieties showing these traits were classified by Negrul (1938) in the cultivar grouping called *proles orientalis*. Varieties showing fewer recessive traits found in Georgia and the Balkans were classified as *proles pontica*. Cultivars found along the nothern Mediterranean and in central Europe are considered of relatively recent origin (possess few recessive traits) and closely resemble wild vines. They are placed in the *proles occidentalis* group. Characteristics of the three groups are given in Table 2.2. Extension of this classification system may be found in Tsertsvadze (1986); Gramotenko and Troshin (1988).

Cultivar Origins

The origin of most cultivars produced within the latter part of the nineteenth and twentieth centuries is a part of the written scientific record. However, for the vast majority of cultivars, information on origin is scarce or nonexistent. Even evidence for the cultivation of most European cultivars seldom goes back more than two centuries. While most European cultivars are generally considered ancient, the supportive evidence is usually absent or of dubious merit. Name derivation has occasionally been used to suggest local origin, for example, 'Sémillon' from *semis* (Fr.), meaning "seed," and 'Sauvignon' from *sauvage* (Fr.), meaning "wild" (Levadoux, 1956). Although indigenous origin may be correct, the existence of multiple unrelated names for many well-known cultivars does not lend credence to this line of argument.

Table 2.2 Classification, Properties, and Distribution of the Proles Groups of Varieties of *Vitis vinifera* According to Negrul (1938)[a]

Proles orientalis	*Proles pontica*	*Proles occidentalis*
Regions		
Central Asia, Afghanistan, Iran, Armenia, Azerbaijan	Georgia, Asia Minor, Greece, Bulgaria, Hungary, Romania	France, Germany, Spain, Portugal
Vine Properties		
Buds glabrous, shiny	Buds velvety, ash-gray to white	Buds weakly velvety
Lower leaf surfaces glabrous to setaceous pubescent	Lower leaf surface with mixed pubescence (webbed and setaceous)	Lower leaf surfaces with webbed pubescence
Leaf edges recurved toward the tip	Leaf edges variously recurved	Leaf edges recurved toward the base
Grape clusters large, loose, often branching	Grape clusters medium-sized, compact, rarely loose	Grape clusters generally very large, compact
Fruit generally oval, ovoid, or elongated, medium- to large-sized, pulpy	Fruit typically round, medium to small, juicy	Fruit often round, more rarely oval, small- to medium-sized, juicy
Varieties mostly white with about 30% rosés	About equal numbers of white, rosé, and red varieties	Varieties commonly white or red
Seeds medium to large with an elongated beak	Seeds small, medium, or large (table grapes)	Seeds small with a marked beak
Fruiting Properties		
Many varieties partially seedless, some seedless	Many varieties partially seedless, some completely so	Seedless varieties rare
Varieties produce few, low-yielding fruiting shoots	Varieties often produce several, highly productive fruiting shoots	Varieties typically produce several, highly productive fruiting shoots
Varieties short-day plants with long growing periods, not cold hardy	Varieties relatively cold hardy	Varieties long-day plants with short growing periods, cold hardy
Most varieties are table grapes, few possess good winemaking properties	Many varieties are good winemaking cultivars, a few are table grapes	Most varieties possessing good winemaking properties
Grapes low in acidity (0.3–0.6%), sugar content commonly 18 to 20%	Grapes acidic (0.6–1.0%), sugar content commonly 18 to 20%	Grapes acidic (0.6–1%), sugar content commonly 18 to 20%
Self-crossed seedlings of certain varieties possessing simple leaves	Self-crossed seedings of certain varieties with dwarfed shoots and rounded form	Self-crossed seedlings of certain varieties having mottled colored leaves

[a] After Levadoux, 1956, reproduced by permission.

More substantial evidence comes from morphological (ampelographic) comparison of cultivars. However, this is useful only when cultivars have diverged from a common ancestor, for example, the color mutants of 'Pinot noir'—'Pinot meunier,' 'Pinot gris,' and 'Pinot blanc.' Vegetative propagation typically maintains the morphological traits of the progenitor. In contrast, sexual (seed) propagation seldom maintains the morphological traits of the parents unchanged (Bronner and Oliveira, 1990). This eliminates morphological traits as reliable indicators of origin when crossing has occurred.

More recently, chemical indicators of genetic origin have been used to investigate varietal origin and classification. These have included analyses of isomeric forms of enzymes as well as the distribution of various anthocyanin, flavonoid, and nonflavonoid phenolic compounds. Even the techniques of DNA cloning are being used to establish degrees of similarities between cultivars (Bourquin *et al.*, 1991, Mauro *et al.*, 1992). Because the latter measures the relatedness of similar portions of the genetic material of each cultivar, investigation of the significance of eastern European cultivars to the evolution of western European cultivars may now be possible.

Until little more than a century ago, deliberate breeding of grapevines was almost nonexistent. Some of the earliest hybridizations occurred in North America between native species of *Vitis* or between them and imported *V. vinifera* cultivars. Regrettably, the parentage of the most well-known American cultivars is speculative. Whereas 'Catawba' and 'Isabella' may be selections from wild *V. labrusca* vines, 'Concord,' and 'Ives,' are thought to be crossings between *V. labrusca* and *V. vinifera*. The latter belief led to the delimitation of *V. labruscana* for *V. labrusca* × *V. vinifera* hybrids. As the involvement of hybridization in the origin of cultivars such as 'Concord' is still in doubt, the *V. labruscana* designation has not been used in the text. Another product of a possible chance crossing is 'Delaware,' between *V. vinifera, V. labrusca,* and *V. aestivalis.* In contrast, 'Dutchess' is a hybrid derived from the crossing of *V. labrusca* and *V. vinifera.* Aside from *V. labrusca,* the involvement of native North American species in the development of indigenous cultivars is limited. Exceptions include 'Noah' and 'Clinton' (*V. labrusca* × *V. riparia*), 'Herbemont' and 'Lenoir' (*V. aestivalis* × *V. cinerea* × *V. vinifera*) and 'Cynthiana' and 'Norton' (possibly *V. vinifera* × *V. aestivalis*). These cultivars are termed **American hybrids** to distinguish them from cultivars derived from crosses originally made in France between *V. vinifera* and *V. rupestris, V. riparia,* or *V. lincecumii.* The latter are variously termed **French–American hybrids, French hybrids,** or **direct producers.** The latter designation comes from their ability to grow ungrafted in phylloxera-infested soils. French–American hybrids generally are of complex parentage, based on subsequent backcrossing to one or more *V. vinifera* cultivars. Correspondingly, they tend to possess vinifera-like winemaking qualities.

The primary aim of French breeders was to develop cultivars containing the wine-producing characteristics of *V. vinifera* and the phylloxera resistance of American species. It was hoped that this would avoid the problems and expense associated with grafting European cultivars onto American rootstocks. Grafting was the only effective control found against the devastation caused by phylloxera (*Daktulosphaira vitifoliae*). It also was hoped that breeding would solve other disease problems, such as oidium and downy mildew. Incidentally, several of these new varieties were more productive than existing cultivars. They became so popular with local grape growers that by 1955 one-third of French vineyards were planted to French–American hybrids. This success, and their different aromatic character, came to be viewed as a threat to the established reputation of famed French viticultural regions. Consequently, laws were passed restricting the use of French–American hybrids and prohibiting new plantings in most French viticultural regions. Similar legislation subsequently was passed in most European countries.

Although largely rejected in the land of their development, French–American hybrids have found broad acceptance in much of North America. In addition to growing well in many areas in the United States and Canada, French-American hybrids have helped in the production of regionally and varietally distinctive wines. They are also extensively grown in several other regions of the world, for example Brazil and Japan.

In the coastal plains of the southeastern United States, commercial viticulture is based primarily on selections of *Vitis rotundifolia.* 'Scuppernong' is the most widely known cultivar, but it has the disadvantage of being unisexual. Newer cultivars such as 'Noble,' 'Magnolia,' and 'Carlos' are bisexual and avoid the need to interplant with male vines to achieve adequate fruit set. The resistance of muscadine grapes to most indigenous diseases and pests in the southern states has permitted a local wine industry to develop where commercial cultivation of *V. vinifera* cultivars is difficult to impossible. Although Pierce's disease has limited the cultivation of most non-muscadine cultivars in this region, exceptions are the cultivars 'Herbemont,' 'Lenoir,' and 'Conquistador.' The first two are thought to be natural *V. aestivalis* × *V. cinerea* × *V. vinifera* hybrids, while 'Conquistador' is a complex cross between several local *Vitis* species and *V. vinifera.*

Although grafting prevented the demise of grape growing and winemaking in Europe, early rootstock cultivars created their own problems. Most of the initial rootstock varieties were selections from *V. riparia* or *V. rupestris* vines. As most were poorly adapted to the high calcium soils frequently found in Europe, they tended to favor the development of lime-induced chlorosis in the grafted shoot system (scion). Nevertheless, both species rooted easily, grafted well, and were relatively phylloxera tolerant. *Vitis riparia* rootstocks also provided some cold hardiness and resistance to *coulure* (unusually poor fruit set), restricted vigor on deep rich soils, and favored early fruit maturity. Some *V. rupestris* selections showed good resistance to lime-induced chlorosis and tended to root deeply. Although the latter provided some drought tolerance on deep soils, it was unsuitable under dry conditions on shallow soils.

Problems associated with early selections, such as 'Gloire de Montpellier' and 'St. George,' led to the development of new rootstocks incorporating desirable properties from other *Vitis* species (Fig. 2.8). For example, *V. berlandieri* can donate resistance to lime-induced chlorosis; *V. candicans* (*V. champinii*) can supply tolerance to root-knot nematodes; and *V. cordifolia* can provide drought tolerance under shallow soil conditions. Alone, these species have major drawbacks as rootstocks: both *V. berlandieri* and *V. cordifolia* are difficult to root; *V. candicans* has only moderate resistance to phylloxera; and both *V. candicans* and *V. cordifolia* are susceptible to lime-induced chlorosis. Important cultural properties of some of the more widely planted rootstocks are given in Tables 4.6 and 4.7.

Grapevine Breeding

The goals of current grape breeding programs have changed little from those of the early breeders such as F. Baco, A. Seibel, and B. Seyve. The focus still involves

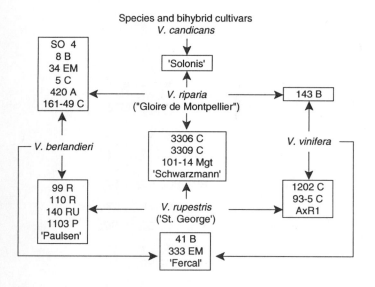

Species and bihybrid cultivars
V. candicans

Multiple-species cultivars

196-17 Ca	— *V. riparia x (V. rupestris x V. vinifera)*
44-53, 106-8	— *V. riparia x (V. rupestris x V. cordifolia)*
1616 C	— *V. rupestris x* 'Solonis'
216-3 Ca	— 'Solonis' x 'Othello' (*V. labrusca x V. riparia x V. vinifera*)
'Harmony,' 'Freedom'	— 'Dog Ridge' x 1613 C

Figure 2.8 Origin of some important rootstock varieties: ( ), selection from single species; □, interspecies crossing. The Champin varieties, 'Dog Ridge' and 'Salt Creek' ('Ramsey') are considered to be either natural *Vitis candicans* × *Vitis rupestris* hybrids, or strains of *Vitis champinii*. 'Solonis' is considered by some authorities to be a selection of *Vitis longii*, than a *V. riparia* × *V. candicans* hybrid. On this basis, some multiple-species noted above are bihybrids. (After Pongrácz, 1983, reproduced by permission.)

improving the agronomic properties of rootstock varieties and enhancing the agronomic and winemaking properties of fruit-bearing (scion) cultivars. The major difference is the availability of analytical techniques and an understanding of genetics unknown to early breeders. Such knowledge can dramatically increase the effectiveness of a breeding program.

Of breeding programs, those developing new **rootstocks** are potentially the simplest. Improvement often involves the incorporation of properties, such as lime or drought tolerance, into existing cultivars already possessing desirable traits, such as resistance to phylloxera. When one or a few linked genes are involved, incorporation may require only a simple breeding program. A **simple cross** program (Fig. 2.9A) involves an initial mating followed by selection of offspring with the desired traits. **Backcross** programs (Fig. 2.9B) are useful when integrating single dominant traits into an existing culti-

var with many desirable properties. However, the repeated backcrossing (required to reestablish expression of the desirable characteristics of the recipient) may result in a loss of seedling viability and vine vigor. The latter phenomenon is termed **inbreeding depression.** This occasionally can be countered if more than one cultivar can serve as a recurrent parent or if the technique of **embryo rescue** can be used. Embryo rescue involves growing embryos isolated early from seed on special culture media (Spiegel-Roy *et al.,* 1985).

Developing new fruiting varieties tends to be more complex than breeding rootstock because of the multigenic nature of features such as aroma composition, color stability, and cold hardiness. This greatly complicates selection by blurring the distinction between environmentally and genetically induced variation. These problems may be partially diminished if the genetics of individual aspects of complex traits is known or if **combining ability** can be assessed. Combining ability refers to the property possessed by some cultivars in being more useful in crosses then others.

Another problem occasionally associated with the breeding of fruit-bearing varieties comes from the requirement to use a new cultivar name. Thus, the progeny of famous cultivars forfeit the marketing advantage of the names of the parents. Most connoisseurs are suspect of new cultivars, generally considering them inferior to the parents. Thus, new cultivars usually are grown in regions where premium varieties cannot be grown profitably, or where increased yield or reduced production costs offset the lost marketing advantage of cultivating established varieties.

Adding to the complexities of grape breeding is the long life cycle of the vine. For many properties, the offspring must be 3 to 6 years old before testing and selection can begin. Detailed assessment can take more than 25 years.

Occasionally, cultivar assessment can be speeded if components of complex genetic traits can be detected early in a breeding program. Chemical and other readily detected features are especially useful in the early elimination of unsuitable offspring. Thus, only potentially valuable progeny are retained for propagation and further study. Early elimination of undesirable progeny also releases vineyard space for use in studying traits, such as winemaking quality, that require prolonged vineyard trials. For example, the presence of methyl anthranilate and other volatile esters has been used in the early elimination of progeny possessing a labrusca fragrance (Fuleki, 1982). Improved knowledge of color stability in red wines should be helpful in the early selection of offspring possessing better color retention. Even selection for disease resistance may be aided by chemical analysis. The production of phytoalexins, or resistance

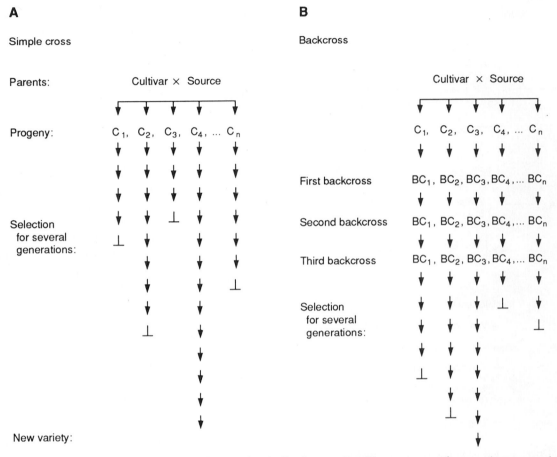

Figure 2.9 Genetic strategies for improving rootstocks. A. Simple cross breeding program with vegetative propagation following the initial cross. B. Backcross breeding program involving a cross followed by several backcrosses (BC) to a common (parental variety) to reestablish the desirable properties of the parental variety.

of fungal toxins, may be useful in the early screening of seedlings for disease resistance. Regrettably, laboratory trials are not always good indicators of field characteristics.

The induction of flowering in cuttings and seedlings can speed the selection for properties detected in small fruit samples (Mullins and Rajasekaran, 1981). Precocious flowering and fruit set also can condense dramatically the interval between crosses. Generation cycling can be shortened further by exposing seed to peroxide and gibberellin prior to chilling (Ellis *et al.*, 1983). The treatment cuts the normally prolonged cold treatment required for seed germination. In theory, a combination of these techniques could reduce the generation time from 3 to 4 years to about 8 months. However, practical problems have limited their use. The induction of flowering in dormant cuttings is another means of accelerating the breeding process by permitting crossing to occur year-round under greenhouse conditions.

Although chemical and other clear indicators of gene expression are highly desirable, few genetic traits of

agronomic value are amenable to early detection. Properties such as winemaking quality can take decades to assess adequately. Thus, realization of the value of new varieties often extends beyond the career of individual breeders. Part of the long "gestation" period results from conservatism in the regulation and marketing of wines. In Europe, the Appellation Control designation is restricted to wines made from pure *Vitis vinifera* cultivars. This automatically places wines produced from interspecies hybrids at a marketing disadvantage. This situation remains despite the demonstration that interspecies crosses can possess vinifera-like aromas (Becker, 1985). Correspondingly, the possibility of incorporating disease and pest resistance into existing *V. vinifera* cultivars by traditional procedures is very limited. Although species other than *V. vinifera* are the primary sources of new disease and pest resistance, untapped resistance may still exist in *V. vinifera* cultivars. For example, the crossing of two 'Riesling' clones gave rise to the cultivar 'Arnsburger' possessing a high degree of resistance to *Botrytis* (Becker and Konrad, 1990).

While most interspecies hybrids possess aromas different from those shown by *V. vinifera* cultivars, the aroma of hybrids is not necessarily less enjoyable. Distinctive interspecific hybrid and non-vinifera aromas can form the basis of regional wine styles. Regional distinctiveness gives European wines much of their marketing image.

New varieties have the best chance of acceptance in new wine-producing regions, especially where existing *V. vinifera* cultivars do not grow well. The increasing popularity of "organically grown" wines may increase the acceptance and cultivation of new varieties because they often require less pesticide and fertilizer use.

Currently, North American *Vitis* species are still the primary source of new properties used in grapevine breeding. Nevertheless, Asian species such as *V. amurensis* and *V. armata* possess several desirable traits. Species such as *V. riparia* and *V. amurensis* are sources of mildew resistance, and both species as well as *V. armata* exhibit *Botrytis* resistance. In addition, *V. amurensis* and northern strains of *V. riparia* are valuable sources of cold hardiness, early maturity, and resistance to *coulure*. Other species possessing effective resistance or tolerance to diseases, nematodes, salinity, and alkalinity are *V. rupestris*, *V. berlandieri*, *V. cinerea*, *V. champinii*, and *V. longii*. For subtropical and tropical regions, sources of disease and environmental stress resistance include *Vitis aestivalis* and *V. shuttleworthii* from the southeastern United States and Mexico, *V. caribaea* from Central America (Jimenez and Ingalls, 1990), and *V. davidii* and *V. pseudoreticulata* from southern China (Fengqin *et al.,* 1990).

Although most interspecies hybridization has involved crossings within the *Euvitis* section, success in producing partially fertile offspring from *V. vinifera* × *V. rotundifolia* crosses is encouraging. Although the progeny of these crosses usually are infertile, some possess partial fertility. Backcrossing to one of the parental species (usually the pollen source) has been used in several field crops to restore fertility to interspecies hybrids. Although the breeding procedure is not simple, genes from *V. rotundifolia* can provide resistance to Pierce's disease, most grapevine viruses, downy and powdery mildews, anthracnose, black rot, and phylloxera as well as tolerance of root-knot and dagger nematodes, and heat endurance. These breeding programs also can be used to introduce vinifera-like winemaking properties into muscadine cultivars.

Nonstandard Breeding Techniques and Other Means of Cultivar Improvement

Standard breeding techniques have been successful but involve considerable time, effort, and expense. Nevertheless, they will probably remain the techniques of choice into the foreseeable future. Although genetic engineering has the potential of introducing specific genes without disrupting the varietal characteristics and name of a cultivar, it is very expensive. Even if the problems of gene transfer and the development of whole plants from transformed cells were solved, the recognition and isolation of desirable genes from a donor would present major obstacles. In addition, many desirable traits are likely to be under complex regulation and may not be easily expressed in another cultivar unmodified.

While genetic engineering possesses great potential, **clonal selection** is the primary means by which cultivar characteristics can be improved without changing the varietal name. However, improvement is limited to the genetic variation already existent or generated within the cultivar and its clones. This variation consists primarily of mutations that have accumulated since the origin of the cultivar. Once a mutation has occurred, it will be reproduced essentially unmodified within cuttings of the clone. Because mutations slowly accumulate, the older the cultivar, the greater the probability that it will consist of several clones differing by one or more mutations. Thus, old cultivars tend to possess more variation than new cultivars. For example, 'Pinot noir' is estimated to consist of more than 300 distinguishable clones.

In addition to isolating strains with particularly useful properties, clonal selection has been used in the elimination of viruses. Because viruses generally invade systemically, that is, infect most or all cells of the vine, simple selection alone is ineffectual in eliminating viruses and other systemically transmitted disease agents. Thermotherapy has been particularly valuable in the elimination of some viruses. The technique involves exposing dormant cuttings to high temperature (38°C) for several weeks. Alternately, excision and propagation of plants from small portions of bud tissue have been effective in eliminating viruses that do not invade meristematic tissues. Elimination of systemic disease-causing agents is particularly significant as grafting has probably been important in the worldwide dispersal of most systemic grapevine pathogens.

Where possible, clonal selection starts with material free of all known systemic pathogens. This alone usually provides cuttings of superior vigor and yield. Other means of improving yield can involve features such as improved pollen fertility and the production of thicker canes. The latter results in better bud break and fruit set (Kliewer and Bowen, 1990). Improvements also may involve the selection of clones with enhanced aroma production (Versini *et al.,* 1990). In most cases, however, the origin of improved yield and quality is unknown.

Clonal selection does not entail a specific set of procedures. It can vary from the simple selection of cuttings from different sources to the analysis of successive pedigreed lines from individual clones. One of the most well-developed systems of clonal selection has been devised in Germany (Becker, 1985; Schöfflinger and Deroo, 1991).

Its success is partially accredited for the dramatic increase in grape yield in German vineyards (Schöfflinger *et al.*, 1985).

One of the central features of the German system involves sequential sets of selections derived from individual vines. This approach attempts to isolate individuals free from the phenotypic instability of chimeric mutations. Chimeras possess meristematic (embryonic) layers that differ genetically from one another (Fig. 2.10). For example, periclinal chimeras possess buds in which the outer tunica layer differs genetically from the inner tunica and corpus layers. Because the outer tunica

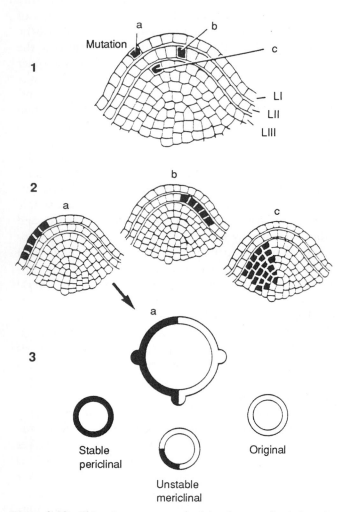

Figure 2.10 Chimeric structures and origins. Structured apical meristems have two or more distinct layers (L-I, L-II, L-III, etc.). A mutation can arise (1) in a single cell of any of the layers (a, b, c), which can develop as single layers or as a core of mutated tissue (2) to produce a stable periclinal chimera, an unstable mericlinal chimera, and a stable reversion to the original type (3). Only the development of a mutation in the L-I layer is shown. (From Hartmann/Kester/Davies, PLANT PROPAGATION: Principles and Practices, 5e, © 1990, p. 179. Reprinted by permission of Prentice Hall, Englewood Cliffs, New Jersey).

produces only epidermal tissue, the epidermis may express properties distinct from other tissues of the vine. Such properties usually cannot be transmitted sexually as only the inner tunica generates cells that develop into gametes. Thus, unique genetic properties possessed by the epidermis are propagated only vegetatively. Although chimeras may be a source of clonal instability, some are quite stable, for example, 'Pinot Meunier,' a periclinal chimera of 'Pinot noir.'

Other features of the German system involve extensive field trials carried out in different regions over many years and the sensory analysis of wines made from the subclones. Once superior clones have been identified, they are propagated for general distribution. Because mutations continue to occur, and clones may become reinfected by systemic pathogens, new subclonal selections are continually in the process of production.

Although clonal selection has been worthwhile in optimizing the inherent yield and quality features of a variety, its value is limited when character expression is sensitive to microclimatic variation. The value of clonal selection may be limited when cultivars are used for the production of different wine styles. The needs of each style can place unique requirements for the ideal clone. For example, upright clones of 'Pinot noir' are preferred for the production of champagne. Here, the ability to produce grapes of limited aroma and sugar content, but elevated acidity, is valuable. In contrast, the lower yielding, prostrate clones of 'Pinot noir' produce the more flavorful red grapes needed to produce the red wines preferred in Burgundy (Bernard and Leguay, 1985). While low-yielding clones generally are considered to produce superior grapes and wines, this is not consistently true. In several instances, higher yielding clones produce grapes of similar quality (Whiting and Hardie, 1981) or equivalent wine quality (McCarthy and Ewart, 1988) as lower yielding clones. Larger fruit size also is not necessarily associated with lower aroma, flavor, and color in the wine (Watson *et al.*, 1988). Choice between clones will depend on the style desired and the sensitivity of these differences to climatic influences.

In most clonal selections, microvinification of grapes from each line is an essential aspect of selection. However, caution is required in the interpretation of judging results based on small samples. It also is important that judges assess the wine based on criteria important to the targeted consumer group. Most consumers rate wines on different grounds than experienced wine judges (Ley, 1980; Williams, 1982). There is also the problem of assessing aging potential in young wines.

One of the current concerns about clonal selection is to avoid the loss of genetic variation. In this regard, some countries are making collections to maintain existing cultivar and species variation.

Although elimination of viral infection and selection have made significant improvements in the planting material available to grape growers, further advances will likely depend on artificial generation of variation or on developing means by which the expression of variation can be enhanced in existing clones. Variation can be enhanced by exposing meristematic tissue (Becker, 1986) or cells in tissue culture to mutagenic chemicals or radiation. In addition, somaclonal selection can enhance the expression and selection of clonal variation. Somaclonal selection involves the selective growth enhancement of particular lines of cells. For example, lines possessing tolerance to salinity or fungal toxins may be isolated by exposing cells to these agents during growth in tissue culture. However, the cellular tolerance selected for in tissue culture may not be expressed as tissue or whole plant tolerance (Lebrun *et al.*, 1985). Culture conditions also may disrupt the normal distinction between the tunica and corpus layers of apical meristems. As a result, cells of the outer tunica may act as inner tunica cells in regenerated plantlets. Consequently, traits found in the epidermis of periclinal chimera may be expressed in other tissues of the vine or transmitted by sexual reproduction.

The major limitation to the extensive use of somaclonal selection is the extremely poor yield of plantlets derived from single cells. Another problem often associated with somaclonal selection, and other means of grapevine propagation involving tissue culture, is variability between individual progeny. As this variability may decrease with vine age, or repeated propagation, it may only be an expression of juvenility in the progeny (Mullins, 1990).

Grapevine Varieties

With the number of named grapevine cultivars approaching 15,000, a comprehensive system of cultivar classification would be useful. Regrettably, no such system exists. In some countries, notably France, there have been attempts to rationalize local cultivars into related groups (Fig. 2.11). These ecogeographic associations are based on ampelographic (structural) and physiological properties. Because of the localized distribution of each group, they may have evolved from one or a few related individuals. Another method of assessing relatedness involves comparing aroma profiles. An example is given in Fig. 2.12. To reduce some of subjective nature of classification, numerical taxonomic procedures such as multivariate analysis have been used (Fanizza, 1980). Nevertheless, the interpretation of the data obtained still requires considerable subjective evaluation. The modern techniques of DNA restriction analysis may offer the

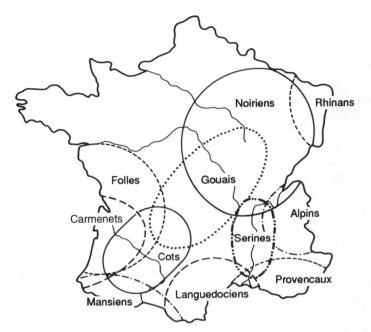

Figure 2.11 Zones of origin or extension of principal groups of French grape varieties. (From Bisson, 1989, reproduced by permission.)

possibility of obtaining a more objective assessment of the evolutionary relationships of grape cultivars (Bowers *et al.*, 1993). For the present, however, ampelographic associations are the most feasible and common means of grouping presumably related cultivars.

On a broader scale, cultivars are grouped relative to their specific or interspecific origin. Most commercial

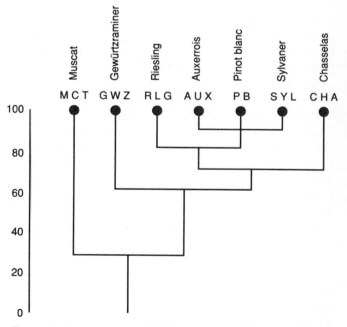

Figure 2.12 Similarity description of varieties based on aromatic profile. (From Lefort, 1980, reproduced by permission).

varieties are pure *Vitis vinifera* cultivars. French–American hybrids constitute the next largest group of cultivars. They are derived from crosses between *V. vinifera* and one or more of the following: *V. riparia, V. rupestris,* and *V. aestivalis.* Early cultivars selected or bred in North America from indigenous grapevines are termed "American hybrids." They may contain some *V. vinifera* parentage. More recent interspecific cultivars are often crosses between *V. vinifera* and species such as *V. amurensis, V. riparia, V. armata,* and *V. rotundifolia.* They generally are designated by the species involved, for example, *V. amurensis* × *V. vinifera* or *V. armata* × *V. vinifera* hybrids.

Vitis vinifera Cultivars

Because of the large number of named cultivars, only a few of the internationally or regionally renowned varieties are discussed below. Although often world famous, these cultivars seldom constitute the primary varieties grown, even in the country of origin. Their productivity is usually lower, and their cultivation more demanding, compared to other cultivars. Their reputation usually comes from their production of fine, varietally distinctive, balanced wines with long-aging potential. In favorable locations, their excellent winemaking properties compensate for the reduced yield and increased production costs.

Many of the varieties listed below were evolved or became famous in France. This is probably a geographic/historical accident, with many other worthy European cultivars being little recognized outside their homeland. The cool climate of the best vineyards in France has been favorable for the retention of subtle fragrances, complete fermentation, and potential for long aging. The former position of France as a major political and cultural power, and proximity to rich, connoisseur-conscious countries, undoubtedly favored the development of the finest of French grape cultivars and wines. The close historical, albeit often strained, relations between England and France, and the wide dispersal of the British colonial influence, fostered the planting of French cultivars throughout much of the English-speaking world. Regrettably, fine cultivars from southern and eastern Europe have not received the same opportunities afforded French, and some German, cultivars.

Examples of several regionally important, but less well-known, aromatically distinctive varieties are 'Arinto' (white) and 'Ramisco' (red) from Portugal; 'Corvina' (r), 'Dolcetto' (r), 'Fiano' (w), 'Garganega' (w), and 'Torbato' (w) from Italy; 'Furmint' (w) from Hungary; 'Malvasia' (w), 'Parellada' (w), and 'Graciano' (r) from Spain; and 'Tămîiosa romînească' (w) from Ro-

mania. These may have widespread value in wine production.

RED CULTIVARS

'**Cabernet Sauvignon**' is undoubtedly the most well-known red cultivar. This is due to both its association with one of Europe's finest red wines (Bordeaux) and its production of equally fine wines in the New World. Under optimal conditions, it produces a fragrant wine possessing a black currant aroma (described as violet in France). Under less favorable conditions, it generates a bell-pepper aroma. The berries are small, acidic, seedy, and possesses a darkly pigmented tough skin. The cultivar is highly susceptible to the fungal diseases, powdery mildew and phomopsis.

In Bordeaux, and increasingly in other regions, wines made from 'Cabernet Sauvignon' is mixed with wines produced from related cultivars, the most common being 'Cabernet Franc' and 'Merlot.' The latter moderates the tannin content of the blended wine and speeds its maturation. Whereas 'Merlot' has the advantage of growing in cooler, more moist soils than 'Cabernet Sauvignon,' it is susceptible to *coulure*.

'Ruby Cabernet' is a cross between 'Carignan' and 'Cabernet Sauvignon' developed at Davis, California. It possesses a Cabernet aroma but grows better in hot climates than 'Cabernet Sauvignon.'

'**Gamay noir à jus blanc**' is a white-juiced Gamay cultivar. Its reputation has risen in accordance with the increase in popularity of Beaujolais wines. Crushed and fermented by traditional procedures, it produces a nondistinctive wine. When processed by carbonic maceration, however, it yields a fruity red wine. 'Gamay noir' donates some varietal flavor, but most of the wine's fragrance comes from the carbonic maceration treatment. It produces medium-sized fruit with tough skins. It is sensitive to most grapevine fungal diseases. The variety is little grown outside of France. Neither the 'Gamay Beaujolais' nor 'Napa Gamay' cultivars grown in California are directly related to 'Gamay noir.' They are, respectively, clones of 'Pinot noir' and 'Valdiguié.'

'**Nebbiolo**' is generally acknowledged as producing the most highly regarded red wines of northwestern Italy. With traditional vinification, it produces very dark wines, high in tannin, that require many years to mellow. Common descriptions of its aroma include tar, violets, and truffles. The variety has not been planted extensively outside northern Italy. It is particularly susceptible to powdery mildew.

'**Pinot noir**' is the famous, but variable, red grape of Burgundy. It appears to be one of the most environmentally sensitive of varieties and consists of a large number of distinctive clones. Fruit-color mutants have given rise

uneven coverage of details

to 'Pinot gris' and 'Pinot blanc.' Usually, the more prostrate, lower yielding clones produce more flavorful wines. The upright, higher yielding clones are more suited to the production of rosé and sparkling wines. Although it produces an aromatically distinctive wine under optimal conditions, its designation in words seems to have eluded all attempts. Various 'Pinot noir' wines are said to show aspects resembling beets, mints, or cherries. The cultivar produces small clusters of small- to medium-sized fruit with large seeds. If the clusters are compact, it is particularly sensitive to bunch rot. Crossed with 'Cinsaut,' it has produced one of the most distinctive of South African cultivars, 'Pinotage.'

'Sangiovese' consists of a wide variety of distinctive clones scattered throughout much of Italy. It is most well known for the light- to full-bodied wines from Chianti. It is involved in the production of many of the finest red wines in Italy. Under optimal conditions, it yields a wine possessing an aroma reminiscent of cherries, violets, and licorice. Like most of the extensive range of Italian grape cultivars, 'Sangiovese' has achieved little of the international recognition it deserves. 'Sangiovese' often suffers from bunch rot under moist conditions.

'Syrah' is the most renowned French cultivar from the Rhône Valley. Lower yielding strains yield a deep red, tannic wine with long aging potential. 'Syrah' is little grown outside France, except in Australia, where, as 'Shiraz,' it constitutes about 40% of the red cultivars grown. Under favorable conditions, 'Syrah' yields fragrant, flavorful wines with aspects reminiscent of violets, raspberries, and currants. 'Syrah' is particularly prone to drought, bunch rot, and the grape berry moth. The identity of 'Petit Sirah' in California is in doubt as recent DNA studies by Bowers *et al.* (1993) suggest that it is not the Rhône cultivar 'Durif' as thought by Galet.

'Tempranillo' is probably the finest red grape variety from Spain. Under favorable conditions, it yields a fine, subtle, well-aging wine. It is the most important red cultivar in Rioja. Outside Spain, it is primarily grown in Argentina. In California, it usually goes under the name 'Valdepeñas.' Its aroma is distinguished by a complex, berry-jam fragrance, with nuances of citrus and incense. 'Tempranillo' produces mid-sized, thick-skinned fruits that are subject to both powdery and downy mildews.

'Zinfandel' is a cultivar extensively grown in California whose European origin and name are uncertain. Some authorities consider it similar to, if not identical with, the southern Italian variety 'Primitivo.' It is used to produce a wide range of wines, from ports to light blush wines. In rosé versions, it expresses a characteristic raspberry fragrance, while full-bodied red wines possess a rich berry flavor. Some of the difficulties with 'Zinfandel'

are the uneven manner in which the fruit ripens and its tendency to form a "second crop" later in the season. Both properties make it difficult to harvest fruit of uniform maturity.

WHITE CULTIVARS

'Chardonnay' is undoubtedly the most prestigious white French cultivar. This stems from both its appealing fruit fragrance and its tendency to grow well in most cool, wine-producing regions of the world. In addition to producing fine table wines, it also yields one of the finest sparkling wines. Under optimal conditions, the wine develops aspects reminiscent of various fruits including apple, peach, and melon. The vine and fruit are predisposed to powdery mildew and bunch rot.

'Chenin blanc' comes from the central Loire Valley of France. There it yields fine sweet and dry table wines, as well as sparkling wines. 'Chenin blanc' also is grown extensively in Australia, California, and South Africa (as 'Steen'). Fine examples of its wine often exhibit a fragrance similar to guava fruit or camellia blossoms. The fruit is tough skinned and of medium size. It is especially susceptible to both downy and powdery mildews, bunch rot, and the grape berry moth.

'Traminer' is a distinctively aromatic cultivar grown throughout the cooler regions of Europe and much of the world. Although possessing a rose blush in the skin, it produces a white wine. 'Traminer' is fermented to produce dry to sweet styles, depending on regional preferences. Intensely aromatic ('Gewürztraminer') clones generally have an aroma strongly resembling the scent of litchi fruit. Mildly aromatic versions of the variety are called 'Savagin' in Southern France. All forms produce modest clusters of small fruit with tough skins. The variety is prone to powdery mildew and bunch rot and often expresses *coulure*.

'Müller-Thurgau' is possibly the most well-known modern *V. vinifera* cultivar. It was developed by H. Müller-Thurgau from a crossing between 'Riesling' and 'Silvaner' in 1882. Its first commercial plantings began in 1903. It is now extensively grown in most cool regions of Europe and in New Zealand. Its mild acidity and fruity–floral fragrance are ideal for light wines. 'Müller-Thurgau' is a high-yielding cultivar that produces lateral (side) clusters of mid-sized fruit. The fruit are subject to both powdery and downy mildew and bunch rot.

'Muscat blanc' is one of many related Muscat varieties grown widely throughout the world. Their aroma, called muscaty, is so marked and distinctive that it is described in terms of the cultivar name. Because of the intense flavor, slight bitterness, and tendency to oxidize, Muscat grapes have most commonly been used in the production of dessert wines. The new muscat cultivar 'Symphony' is less bitter, and its lower susceptibility to oxidation gives it better aging ability. 'Muscat blanc' ('Moscato bianco' in Italy) is the primary variety used in the flourishing sparkling wine industry in Asti. Other named Muscat varieties include the 'Orange Muscat' ('Muscato Fiori d'Arinico' in Italy), 'Muscat of Alexandria,' 'Muscat Ottonel,' and the darkly pigmented 'Black Muscat' ('Muscat Hamburg').

'**Pinot gris**' and '**Pinot blanc**' (respectively, 'Ruländer' and 'Weissburgunder' in Germany) are color mutants of 'Pinot noir.' Both are used extensively in Europe for the production of dry, botrytized, and sparkling wines. Neither have gained much popularity outside Europe. 'Pinot gris' often yields subtly fragrant wines with aspects of passion fruit, whereas 'Pinot blanc' is more fruity, with suggestions of hard cheese.

'**Riesling**' ('White Riesling,' 'Johannisberg Riesling') is without doubt Germany's most highly respected grape variety. It can produce aromatic, well-aged wines that vary from dry to sweet. Its floral aroma, commonly reminiscent of roses, has made it popular throughout central Europe and much of the world. This popularity is reflected in the number of cultivars whose names have incorporated the term Riesling. Examples are 'Grey Riesling,' 'Goldriesling,' 'Frankenriesling,' and 'Wälschriesling.' 'Riesling' produces clusters of small- to medium-sized berries. They are sensitive to powdery mildew and bunch rot.

'**Sauvignon blanc**' is one of the primary white varieties in Bordeaux and the main white cultivar in the upper Loire Valley. It has become popular in California and New Zealand in recent years. It also is grown in northern Italy and eastern Europe. Often, its aroma has elements of green peppers as well as a herbaceous aspect. Better clones possess a subtly floral character. The modest clusters produce small berries. It is particularly sensitive to powdery mildew and black rot, while possessing partial resistance to bunch rot and downy mildew.

'**Sémillon**' is widely grown in Bordeaux and is most well known for its use in producing Sauternes wines. In Bordeaux, it is commonly blended with 'Sauvignon blanc.' 'Sémillon' is grown extensively in Chile, with significant plantings also occurring in Australia. When fully mature, and without the intervention of "noble rot," it yields a dry wine said to contain nuances of fig or melon. 'Sémillon' produces small clusters of medium-sized fruit. It is notably vulnerable to bunch rot and fanleaf degeneration.

Interspecies Hybrids

AMERICAN HYBRIDS

Although decreasing in significance in much of North America, early selections from native American grape-

vines or accidental interspecies hybrids are still of considerable importance. American hybrids constitute the major plantings in eastern North America, and they are grown commercially in South America, Asia, and eastern Europe.

Of American hybrid cultivars, the most important are based on *V. labrusca*. They possess a wide range of flavors. Some, such as 'Niagara' are characterized by the presence of the so-called foxy aspect. Others, however, are more distinguished by strawberry fragrances ('Ives'), the grapey aspect of methyl anthranilate ('Concord'), and a strong floral aroma ('Catawba'). The higher acidity and low sugar content of American hybrid grapes make chaptalization (the addition of sugar to the juice) necessary in wine production.

Various methods have been used to diminish the overabundant flavor. Long aging results in dissipation of labrusca flavors, but this is generally not feasible as most labrusca wines are consumed young. The presence of high levels of carbon dioxide, as in sparkling wines, tends to mask most labrusca fragrances. Processing the grapes via carbonic maceration is another means of reducing the intensity of labrusca flavors. However, the most generally accepted mechanism is early picking and cold fermentation. These, respectively, limit the development and extraction of labrusca flavors, while still producing a wine with fruitiness.

The other major group of American cultivars are those derived from *V. rotundifolia*. Although 'Scuppernong' is the most well-known, new self-fertile varieties possessing different aromatic properties are replacing it in commercial vineyards. The excellent resistance of these cultivars to indigenous diseases, especially Pierce's disease, have allowed them to flourish in the southern coastal states of North America. Like *V. labrusca* cultivars, the low sugar content of the fruit usually requires juice chaptalization before vinification. The pulpy texture, tough skin, differential fruit maturation, and separation of fruit from the pedicel on maturation complicate their use in winemaking.

Most muscadine cultivars have a distinctive and marked fragrance, containing aspects of orange blossoms and roses. Some fertile crossings with *V. vinifera* show viniferalike flavors combined with the fruiting characteristics and disease resistance of the muscadine parent.

FRENCH–AMERICAN HYBRIDS (DIRECT PRODUCERS)

French–American hybrids were developed to avoid the need, complexity, and expense of grafting *V. vinifera* cultivars onto rootstock resistant to phylloxera. The easier cultivation, reduced sensitivity to several leaf pathogens, and higher yields of the hybrids made them popular with many grape growers in France. While the tendency of base buds to grow and bear fruit increased yield, it also tended to result in overcropping and exacerbated the increasingly serious problem of grape overproduction in France. This factor, combined with the nontraditional fragrance of French–American hybrids, led to a general ban on new plantings and a prohibition on their use in Appellation Control (AC) wines. Restrictions against new planting of French–American hybrids subsequently spread throughout the European Economic Community (EEC). The primary exception is the cultivation of 'Baco blanc' for Armagnac production. Nevertheless, existing plantings of French–American hybrids still constitute a significant proportion of French vineyards ($\sim$22.5 $\times$ 10^3 ha), although down from the peak of about 30% coverage in the late 1950s (Boursiquot, 1990).

In North America, with the exception of most of the southern, Gulf, and western coastal states, French–American hybrids have formed the basis of much of the expanding wine industry since the early 1960s. They also are grown extensively in some South American and Asian countries. In Europe, as well as other areas, they are used in breeding programs as valuable sources of resistance to several foliar, stem, and fruit pathogens.

Unlike the American hybrids noted above, few French–American hybrids resulted from crosses with *Vitis labrusca*. *Vitis rupestris*, *V. riparia*, and *V. aestivalis* (*V. lincecumii*) were the primary species used. Although some French–American hybrids are simple crosses between American grapevines and a *V. vinifera* parent, most are derived from complex crosses between American species and several *V. vinifera* cultivars.

Brief descriptions of some of the better American and French–American hybrids follow. '**Baco noir**' is a 'Folle blanche' $\times$ *V. riparia* hybrid. Its acidity, flavor, and pigmentation yield a wine with considerable aging potential. It develops a fruity aroma associated with aspects of herbs. It is particularly sensitive to bunch rot and soil-borne viruses. '**de Chaunac**' is a Seibel crossing of unknown parentage. It yields red wines with a berrylike aroma, associated with hints of tar. It is prone to several soil-borne viruses. '**Chambourcin**' is a Joannes Seyve hybrid of unknown parentage. Its popularity in the Loire Valley increased during the 1960s and 1970s after its release, and it has done well in Australian trials. It produces both full-flavored red and light rosé wines with a somewhat herbaceous aroma. It possesses good resistance to downy and powdery mildews.

'**Delaware**' is one of the finest original white American hybrids. It generally is thought to be a *V. labrusca* $\times$ *V. aestivalis* $\times$ *V. vinifera* hybrid. Nevertheless, susceptibility to phylloxera and various pathogens limits more expanded cultivation of 'Delaware.' '**Dutchess**' is another

highly rated, older, white American hybrid. It has a mild, fruity aroma with little labrusca flavor. As with 'Delaware,' difficulty in growing the cultivar negates much of its enological qualities.

'**Magnolia**' is one of the more popular new muscadine cultivars. It produces bisexual flowers and is self-fertile. It yields sweet, bronze-colored fruit. '**Maréchal Foch**' is a Kuhlmann hybrid derived from crossing a *V. riparia* $\times$ *V. rupestris* selection with 'Goldriesling' ('Riesling' $\times$ 'Courtiller musqué'). It yields deeply colored, berry-scented, early-maturing wines. '**Noble**' is a dark red muscadine cultivar. Its deep red color and bisexual habit have made it popular in the southeastern United States.

'**Seyval blanc**' is a Seyve-Villard hybrid of complex *V. vinifera*, *V. rupestris*, and *V. lincecumii* parentage. The variety yields a mildly fruity white wine with a pomade fragrance. It is susceptible to bunch rot. '**Veeblanc**' is one of the newer white cultivars of complex parentage developed in Ontario. It generates a mildly fruity wine of good quality. '**Vidal blanc**' is possibly the best of the white French–American hybrids. This Vidal hybrid of complex ancestry has both excellent winemaking and viticultural properties. Under optimal conditions it yields a wine of Riesling-like character. As an eiswein, 'Vidal blanc' produces a wine of superior quality.

Suggested Readings

Evolution of *Vitis*

Levadoux, L. (1956). Les populations sauvages et cultivées de *Vitis vinifera* L. *Ann. Amelior. Plant. Sér. B* **1**, 59–118.

Olmo, H. P. (1976). Grapes: *Vitis, Muscadinia* (Vitaceae). *In* "Evolution of Crop Plants" (M. W. Simmonds, ed.), pp. 294–298. Longman, London.

Rives, M. (1980). Ampélographie. *In* "Sciences et Techniques de la Vigne" (J. Ribéreau-Gayon and E. Peynaud, eds.), Vol. 1, pp. 131–170. Dunod, Paris.

Zukovskij, P. M. (1950). "Cultivated Plants and Their Wild Relatives." State Publ. House Soviet Science, Moscow (abridged translation by P. S. Hudson, 1962, Commonwealth Agriculture Bureau, Wallingford, Oxon, England).

Rootstocks

Anonymous (1991). "Alternative Rootstock Update." ASEV Tech. Projects Committee and UC Cooperative Extension, Univ. of California Agriculture Publ., Oakland, California.

Howell, G. S. (1987). *Vitis* rootstocks. *In* "Rootstocks for Fruit Crops" (R. C. Rom and R. F. Carlson, eds.), pp. 451–472. Wiley(Interscience), New York.

Morton, L. T. (1985). The myth of the universal rootstock. *Wines Vines* **66** (February), 35–39.

Morton, L. T., and Jackson, L. E. (1988). Myth of the universal rootstock: The fads and facts of rootstock selection. *Proc. 2nd Int. Symp. Cool Climate Vitic. Oenol., Jan. 11–15, 1988, Auckland,*

N.Z. (R. E. Smart, S. B. Thornton, S. B. Rodriguez, and J. E. Young, eds.), pp. 25–29. New Zealand Soc. Vitic. Oenol., Auckland, New Zealand.

Pongrácz, D. P. (1983). "Rootstocks for Grape-Vines." David Philip, Cape Town, South Africa.

Grape Breeding

Alleweldt, G., and Possingham, J. V. (1988). Progress in grapevine breeding. *Theor. Appl. Genet.* **75**, 669–673.

Becker, H. (1985). White grape varieties for cool climate. *Symp. Cool Climate Vitic. Enol.* (B. A. Heatherbell, P. B. Lombard, F. W. Bodyfelt, and S. F. Price, eds.), OSU Agric. Exp. Stn. Tech. Publ. No. 7628, Oregon State Univ., Corvallis, Oregon.

Becker, H. (1986). Induction of somatic mutations in clones of grape cultivars. *Acta Hortic.* **180**, 121–128.

Becker, H. (1988). Breeding resistant varieties and vine improvement for cool climate viticulture. *Proc. 2nd Int. Symp. Cool Climate Vitic. Oenol., Jan. 11–15, 1988, Auckland, N.Z.,* (R. E. Smart, S. B. Thornton, S. B. Rodriguez, and J. E. Young, eds.), pp. 63–64, New Zealand Soc. Vitic. Oenol, Auckland, New Zealand.

Meredith, C. P., Stamp, J. A., Dandekar, A. M., and Martin, L. A. (1987). New genetic technology for grape improvement. *Proc. 6th Aust. Wine Ind. Tech. Conf.* (T. Lee, ed.), pp. 62–65. Australian Industrial Publ., Adelaide, Australia.

Mullins, M. G. (1987). Strategies for the genetic improvement of wine grapes. *Proc. 6th Aust. Wine Ind. Tech. Conf.* (T. Lee, ed.), pp. 66–70. Australian Industrial Publ., Adelaide, Australia.

Mullins, M. G. (1990). Applications of tissue culture to the genetic improvement of grapevines. *Proc. 5th Int. Symp. Grape Breeding, Sept. 12–16, 1989,* p. 399. St. Martin, Pfalz, Germany (Special Issue of *Vitis*, 1990).

Olmo, H. P. (1986). The potential role of (*vinifera* × *rotundifolia*) hybrids in grape variety improvement. *Experientia* **42**, 921–926.

Proc. 3rd Int. Symp. Grape Breeding, June 15–18, 1980. Dept. of Vitic. Enol., Univ. of California, Davis.

Proc. 4th Int. Symp. Grape Breeding, April 13–18, 1985, Verona, Italy, 13(supplement of *Vignevini* 12, 1986).

Proc. 5th Int. Symp. Grape Breeding, Sept. 12–16, 1989 St. Martin, Pfalz, Germany (Special Issue of *Vitis*, 1990).

Schöffling, H., and Deroo, J. G. (1991). Méthodologie de la sélection clonale en Allemagne. *J. Int. Sci. Vigne Vin* **25**, 203–227.

Grape Cultivars

Allenwelt, G. (1989). "The Genetic Resources of *Vitis:* Genetic and Geographic Origin of Grape Cultivars, their Prime Names and Synonyms." Fed. Res. Center Grape Breed., Geilweilenhof, Germany.

Galet, P. (1979). "A Practical Ampelography—Grapevine Identification," translated and adapted by L. T. Morton. Cornell Univ. Press, Ithaca, New York.

Galet, P. (1990). "Cépages et Vignobles de France." Tome II. L'Ampélographie Française." Published by P. Galet, Montpellier, France.

Morton, L. T. (1985). "Winegrowing in Eastern America." Cornell Univ. Press, Ithaca, New York.

Olien, W. C. (1990). The muscadine grape: Botany, viticulture, history, and current industry. *Hortscience* **25**, 732–739.

Olmo, H. P. (1971). *Vinifera rotundifolia* hybrids as wine grapes. *Am. J. Enol. Vitic.* **61**, 87–91.

Pool, R. M., Kimball, K., Watson, J., and Einset, J. (1979). Grape varieties for New York State. *N.Y. Food Life Sci. Bull.* **80**, 1–6.

References

Bailey, L. H. (1933). The species of grapes peculiar to North America. *Gentes Herbarum* **3**, 150–244.

Becker, H. (1985). White grape varieties for cool climate. *Int. Symp. Cool Climate Vitic. Enol.* (B. A. Heatherbell, P. B. Lombard, F. W. Bodyfelt, and S. F. Price, eds.), OSU Agric. Exp. Stn. Tech. Publ. No. 7628, pp. 46–62. Oregon State University, Corvallis, Oregon.

Becker, H. (1986). Induction of somatic mutations in clones of grape cultivars. *Acta Hortic.* **180**, 121–128.

Becker, H., and Konrad, H. (1990). Breeding of botrytis tolerant *V. vinifera* and interspecific wine varieties. *Proc. 5th Int. Symp. Grape Breeding, Sept. 12–16, 1989,* p. 302. St. Martin, Pfalz, Germany, (Special Issue of *Vitis*).

Bernard, R., and Leguay, M. (1985). Clonal variability of Pinot noir in Burgundy and its potential adaptation under other cooler climates. *Int. Symp. Cool Climate Vitic. Enol.* (B. A. Heatherbell, P. B. Lombard, F. W. Bodyfelt, and S. F. Price, eds.), OSU Agric. Exp. Stn. Tech. Publ. No. 7628, pp. 63–74. Oregon State Univ., Corvallis, Oregon.

Bisson, J. (1989). Les Messiles, groupe ampélographique du bassin de la Loire. *Connaiss. Vigne Vin.* **23**, 175–191.

Bourquin, J. C., Otter, L., and Walter, B. (1991). Identification of grapevine rootstocks by RFLP. *C. R. Acad. Sci. Ser. 3* **312**, 593–598.

Boursiquot, J. M. (1990). Évolution de l'encépagement du vignoble français au cours des trente dernières années. *Prog. Agric. Vitic.* **107**, 15–20.

Bowers, J. E., Bandman, E. B. and Meredith, C. P. (1993). DNA fingerprint characterization of some grape cultivars. *Am. J. Enol. Vitic.* **44**, 266–274.

Briggs, D., and Walters, S. M. (1986). "Plant Variation and Evolution." Cambridge Univ. Press, Cambridge.

Bronner, A., and Oliveira, J. (1990). Creation and study of the Pinot noir variety lineage. *Proc. 5th Int. Symp. Grape Breeding, Sept. 12–16, 1989,* p. 69. St. Martin, Pfalz, Germany, (Special Issue of *Vitis*).

Buck, C. D. (1949). "A Dictionary of Selected Synonyms in the Principal Indo-European Languages." Univ. of Chicago Press, Chicago.

de Blij, H. J. (1983). "Wine: A Geographic Appreciation." Rowman and Allanheld, Totowa, New Jersey.

Dorf, E. (1960). Climatic change of the past and present. *Am. Sci.* **48**, 341–364.

Ellis, R. H., Hong, T. D., and Roberts, E. H. (1983). A note on the development of a practical procedure for promoting the germination of dormant seed of grape (*Vitis* spp.). *Vitis* **22**, 211–219.

Fanizza, G. (1980). Multivariate analysis to estimate the genetic diversity of wine grapes (*Vitis vinifera*) for cross breeding in southern Italy. *Proc. 3rd Int. Symp. Grape Breeding,* pp. 105–110. Univ. of California, Davis, California.

Fengqin, Z., Fangmei, L., and Dabin, G. (1990). Studies on germplasm resources of wild grape species (*Vitis* spp.) in China. *Proc. 5th Int. Symp. Grape Breeding, Sept. 12–16, 1989,* pp. 50–57. St. Martin, Pfalz, Germany, (Special Issue of *Vitis*).

Forbes, R. J. (1965). " Studies in Ancient Technology," 2nd Ed., Vol. 3, E. J. Brill, Leiden, The Netherlands.

Fuleki, T. (1982). The vineland grape flavor index—A new objective method for the accelerated screening of grape seedlings on the basis of flavor character. *Vitis* **21**, 111–120.

Gamkrelidze, T. V., and Ivanov, V. V. (1990). The early history of Indo-European languages. *Sci. Am.* **262**(3), 110–116.

Gramotenko, P. M., and Troshin, L. P. (1988). Improvement in the classification of *Vitis vinifera* L. *In* "Prospects of Grape Genetics and Breeding for Immunity" pp. 45–52. Naukova Dumka, Kiev (in Russian).

Hartmann, H. T., Kester, D. E., and Davies, F. T. (1990). "Plant Propagation: Principles and Practices." Prentice Hall, Englewood Cliffs, New Jersey.

Jimenez, A. L. G., and Ingalls, A. (1990). *Vitis caribaea* as a source of resistance to Pièrce's disease in breeding grapes for the tropics. *Proc. 5th Int. Symp. Grape Breeding, Sept. 12–16, 1989*, pp. 262–270. St. Martin, Pfalz, Germany. (Special Issue of *Vitis*).

Jongmans, W. (ed). (1939). "Fossilium Catalogus. II. Plantae. Pars 24: Rhamnales I: Vitaceae." Dr. W. Junk, Verlag für Naturwissenschaften, 's-Gravenhage (The Hague), The Netherlands.

Kliewer, M., and Bowen, P. (1990). The influence of clonal variation, pruning severity and cane structure on yield components of three Cabernet Sauvignon clones. *Proc. 5th Int. Symp. Grape Breeding, Sept. 12–16, 1989*, p. 512. St. Martin, Pfalz, Germany (Special Issue of *Vitis*).

Lebrun, L., Rajasekaran, K., and Mullins, M. G. (1985). Selection *in vitro* for NaCl-tolerance in *Vitis rupestris* Scheele. *Ann. Bot.* 56, 733–739.

Lefort, P.-L. (1980). Biometrical analysis of must aromagrams: Application to grape breeding. *Proc. 3rd Int. Symp. Grape Breeding*, pp. 120–129, Univ. of California, Davis.

Levadoux, L. (1946). Étude de la fleur et de la sexualité chez la vigne. *Ann. Ec. Natl. Agric. Montpellier* 27, 1–90.

Levadoux, L. (1956). Les populations sauvages et cultivées de *Vitis vinifera* L. *Ann. Amelior. Plant. Sér. B* 1, 59–118.

Ley, R. J. (1980). Statistical analysis of a wine evaluation test with new varieties in the Upper Moselle wine-growing district. *Proc. 3rd Int. Symp. Grape Breeding*. pp. 158–168. Univ. of California, Davis.

McCarthy, M. G., and Ewart, A. J. W. (1988). Clonal evaluation for quality winegrape production. *Proc. 2nd Int. Symp. Cool Climate Vitic. Oenol., Auckland, N.Z.* (R. E. Smart, S. B. Thornton, S. B. Rodriguez, and J. E. Young, eds.), pp. 34–36. New Zealand Soc. Vitic. Oenol., Auckland, New Zealand.

Mauro, M.-C., Strefeler, M., Weeden, N. F., and Reisch, B. I. (1992). Genetic analysis of restriction fragment length polymorphisms in *Vitis. J. Hered.* 83, 18–21.

Moore, M. O. (1991). Classification and systematics of eastern North American *Vitis* L., Vitaceae, north of Mexico. *SIDA* 14, 339–367.

Mullins, M. G. (1990). Applications of tissue culture to the genetic improvement of grapevines. *Proc. 5th Int. Symp. Grape Breeding, Sept. 12–16, 1989*, pp. 399–407. St. Martin, Pfalz, Germany (Special Issue of *Vitis*).

Mullins, M. G., and Rajasekaran, K. (1981). Fruiting cuttings: Revised method for producing test plants of grapevine cultivars. *Am. J. Enol. Vitic.* 32, 35–40.

Negrul, A. M. (1938). Evolution of cultivated forms of grapes. *C. R. (Dokl.) Acad. Sci. URSS* 18, 585–588.

Patel, G. I., and Olmo, H. P. (1955). Cytogenetics of *Vitis.* I. The hybrid *V. vinifera × V. rotundifolia. Am. J. Bot.* 42, 141–159.

Planchais, N. (1972–1973). Apports de l'analyse pollinique à la con- naissance de l'extension de la vigne au quaternaire. *Naturalia Monspeliensia, Sér. Bot.* 23–24, 211–223.

Pongráez, D. P. (1983). "Rootstocks for Grape-vines". David Philip, Cape Town, South Africa.

Pratt, C. (1988). Grapevine structure and growth. Compendium of Grape Diseases. (R. C. Pearson and A. C. Goheen, eds.). APS Press, St. Paul, Minnesota.

Renfrew, J. M. (1973). "Palaeoethnobotany: The Prehistoric Food Plants of the Near East and Europe." Columbia Univ. Press, New York.

Renfrew, C. (1989). The origins of Indo-European languages. *Sci. Am.* 261(4), 106–116.

Rives, M. (1975). Les origines de la vigne. *Recherche* 53, 120–129.

Rogers, D. J., and Rogers, C. F. (1978). Systematics of North American grape species. *Am. J. Enol. Vitic.* 29, 73–78.

Schöfflinger, H., and Deroo, J. G. (1991). Methologie de la sélection clonal en Allemagne. *Bull. O.I.V.* 625, 203–227.

Schöfflinger, H., Fass, K. H., Faber, W., and Duplessis, H. (1985). Étude de nouveaux cépages (*V. vinifera* L.) en allemagne federale. *Connaiss. Vigne Vin* 19, 1–16.

Splegel-Roy, P., Sahar, N., Baron, I., and Lavi, U. (1985). *In vitro* culture and plant formation from grape cultivars with abortive ovules and seeds. *J. Am. Soc. Hortic. Sci.* 110, 109–112.

Srinivasan, C., and Mullins, M. G. (1981). Physiology of flowering in the grapevine—A review. *Am. J. Enol. Vitic.* 32, 47–59.

Stevenson, A. C. (1985). Studies in the vegetational history of S. W. Spain. II. Palynological investigations at Laguna de los Madres, Spain. *J. Biogeogr.* 12, 293–314.

Stummer, A. (1911). Zur Urgeschichte der Rebe und des Weinbaues. *Mitt. der Anthropol. Ges. Wein* 41, 283–296.

Tsertsvadze, N. V. (1986). Classification of cultured grape *Vitis vinifera* L. in Georgia. *Tech. Prog. Vitic. Georgia, Tbilisi*, pp. 229–240. (in Russian).

Versini, G., Rapp, A., Volkmann, C., and Scienza, A. (1990). Flavour compounds of clones from different grape varieties. *Proc. 5th Int. Symp. Grape Breeding, Sept. 12–16, 1989*, pp. 513–524. St. Martin, Pfalz, Germany (Special Issue of *Vitis*).

von Babo, A. F., and Mach, E. (1923). "Handbuch Weinbaues und der Kellerwirtschaft," 4th Ed., Vol. 1, Part 1. Parey, Berlin.

Watson, B., Lombard, P., Price, S., McDaniel, M., and Heatherbell, D. (1988). Evaluation of Pinot noir clones in Oregon. *Proc. 2nd Int. Symp. Cool Climate Vitic. Oenol., Jan. 11–15, 1988, Auckland, N.Z.* (R. E. Smart, S. B. Thornton, S. B. Rodriguez, and J. E. Young, eds.), pp. 276–278. New Zealand Soc. Vitic. Oenol., Auckland, New Zealand.

Whiting, J. R., and Hardie, W. J. (1981). Yield and compositional differences between selections of grapevine cv. Cabernet Sauvignon. *Am. J. Enol. Vitic.* 32, 212–218.

Williams, A. A. (1982). Recent developments in the field of wine flavour research. *J. Inst. Brew.* 88, 43–53.

Zukovskij, P. M. (1950). "Cultivated Plants and Their Wild Relatives." State Publ. House Soviet Science, Moscow (abridged translation by P. S. Hudson, 1962), Commonwealth Agriculture Bureau, Wallingford, Oxon, England).

3

Grapevine Structure and Function

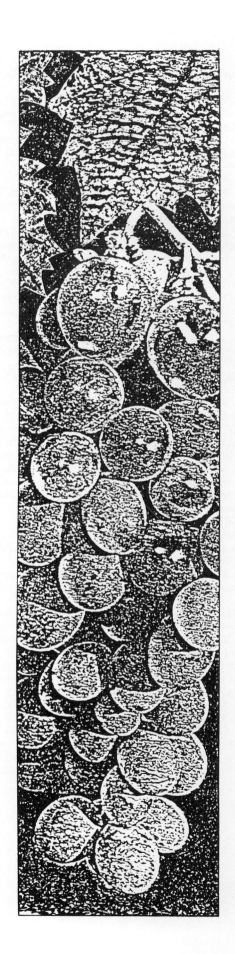

Vegetative Structure and Function

The uniqueness of some aspects of plant structure is obvious even to the casual observer. However, many of the important features of plant structures become apparent only when studied microscopically. Unlike animal cells, plant cells are enclosed in rigid cell walls. Nevertheless, each cell initially possesses direct cytoplasmic connections with adjacent cells through thin channels called **plasmodesmata.** Thus, embryonic tissues resemble one huge "cell," divided into thousands of interconnected compartments, each possessing cytoplasm and a single nucleus. As the cells differentiate, many die and the plant begins to resemble longitudinal, semiindependent cones of tissue connected primarily by specialized conductive (vascular) tissue.

Because most vascular cells elongate longitudinally, movement of water and nutrients is predominantly between sections of root and shoot tissue directly above and below each other. Conduction of water and nutrients laterally between adjacent tissue is limited, and direct translocation between tissues on opposite sides of shoots and roots is nonexistent.

The vascular system consists of two structurally and functionally different components. The main water and mineral conducting elements, the **xylem tracheae**, become functional on the disintegration of their cytoplasmic contents. The empty cell walls act as passive conduits. The primary cells translocating organic nutrients are the **sieve tube elements** of the **phloem.** They become functional only after their nuclei disintegrate.

Plants also show a distinctive growth habit. Growth in length typically occurs behind special reproductive (**meristematic**) cells located in shoot and root tips. Growth in breadth, beyond that initiated by the shoot or root tip, occurs when a circular band of cells—the **vascular cambium**—becomes meristematic and generates cells both laterally and radially. In addition, plants show distinctive growth patterns which generate leaves and their evolutionary derivatives, the flower parts. In these organs, sites of growth (**plate meristems**) occur dispersed throughout the young leaf or flower part. Nonuniform rates and patterns of growth of the plate meristems generate the characteristic shape of the plant parts.

The Root System

The root system possesses several cell types, tissues, and regions, each with a particular function. The root tip performs many of the most significant root functions, such as water and inorganic nutrient uptake, synthesis of growth regulations, and expansion growth into new regions of the soil. In addition, secretions from the root tip significantly influence microbial growth on and around the root. Older, mature portions of the root system transport water and inorganic nutrients upward to the shoot system and organic nutrients downward to growing portions of the root. The outer secondary tissues of mature roots restrict water and nutrient loss into the soil and help to protect the root from parasitic and mechanical injury. Permanent parts of the root system also anchor the vine and act as an important nutrient storage organ in winter (Yang *et al.*, 1980). Because of the significant differences in the structure and function of young and old roots, they are discussed separately.

THE YOUNG ROOT

Structurally and functionally the young root can be divided into several zones along its length. The **root tip** contains the **apical meristem** and **root cap.** The meristem contains cells that remain embryonic as well as those that differentiate into the **primary tissues** of the root. Primary tissues are considered to be those that develop from apically located meristematic tissues, whereas **secondary tissues** develop from laterally located meristematic tissues (**cambia**). The embryonic cells are concentrated in the center of the meristematic zone, called the **quiescent region.** This region is additionally important as a major site for cytokinin and gibberellin synthesis. These growth regulators are translocated upward to the shoot system. They help to maintain a favorable shoot/root ratio and may influence inflorescence (flower cluster) initiation and fruit development.

Surrounding the quiescent zone are rapidly dividing meristematic cells and cells in the initial stages of cell differentiation. Cells at the tip apex develop into root cap cells. These are short-lived cells which produce mucilaginous polysaccharides that ease root penetration into the soil. The cap also appears to cushion the embryonic cells from physical damage during soil penetration. To the sides, cells differentiate into the **epidermis,** the **hypodermis,** and the **cortex,** the latter consisting of several layers of undifferentiated parenchyma cells. Behind the apical meristem, vascular tissues begin to differentiate and elongate. Cell enlargement occurs primarily along the axis of the root and produces root elongation in a short (~2mm) region, called the **elongation zone.** Except for some enlargement in cell diameter, most lateral root expansion results from the production of new (**secondary**) tissue behind the root apex.

Behind the elongation zone is a region in which most cells differentiate into their mature form (the **differentiation zone**). The initial portion is commonly called the **root hair zone.** In this region, localized extensions of the epidermal cells produce **root hairs.** The formation, length, and period of activity of root hairs depend on many factors. For example, alkaline conditions suppress root hair development, and mycorrhizal associations often inhibit root hair formation. Root hairs usually have been considered important in increasing root/soil contact and, thereby, water and mineral uptake. Although important, the release of organic nutrients into the immediate soil surroundings may be equally significant. Nutrient release greatly enhances microbial growth on and around the root. The flora helps to protect the root from soil-borne pathogens and stimulates solubilization of inorganic soil nutrients. Root hairs are generally short-lived, and their collapse provides additional nutrients for microbial activity.

The first tissue in the vascular cylinder to differentiate is the phloem. It contains the cells that conduct organic nutrients throughout the vine. Early phloem development aids the translocation of organic nutrients to the dividing and differentiating cells of the root tip. The xylem, involved in long-distance transport of water and inorganic nutrients to other parts of the plant, develops further back in the differentiation zone. This sequence probably evolved as the water and inorganic nutrients absorbed in the apical region of the root are required locally for cell growth.

Adjacent to where the primary xylem differentiates, the innermost cortical layer differentiates into the **endodermis**. The endodermis deposits a band of wax and lignin around its radial walls called the **Casparian strip**. This prevents the diffusion of water and nutrients between the cortex and the vascular cylinder. Movement of water and nutrients between these two regions must pass through the cytoplasm of the endodermal cells. This gives the root a degree of metabolic control over the transport of material into and out of the vascular tissues.

Further back along the differentiation zone, the **pericycle** develops just inside the endodermis. It generates a protective cork layer if the root continues to grow and differentiate. The pericycle also initiates lateral root growth.

Although still considered the major site for water and mineral uptake, the root hair zone is no longer considered the only site involved. Behind the root hair zone, the epidermis eventually dies and the underlying hypodermis becomes encased in waxy suberin. Suberization occurs rapidly during the summer and may advance to include the root tip under dry conditions (Pratt, 1974). The endodermis also becomes largely, but nonuniformly, suberized back from the root tip. These changes reduce water uptake to a low but relatively constant level. The influence of suberization on mineral uptake varies with the inorganic ion involved. Water and potassium uptake by heavily suberized roots can approach 20 and 4%, respectively, of that absorbed by young unsuberized roots (Queen, 1968). Older parts of the root lose their cortical tissues and become covered with cork. However, as only the inner wall surfaces of cork cells are suberized, diffusion of water and solutes through pores in the region of the wall between adjacent cells is possible (Atkinson, 1980).

Regardless of the site, water uptake is predominantly induced by passive forces. In the spring, absorption and movement may be driven by the conversion of stored carbohydrates into soluble, osmotically active sugars. The negative osmotic potential generated produces what is termed **root pressure**. Root pressure produces the "bleeding" that occurs at the cut ends of spurs and canes in the spring. Once the leaves have expanded sufficiently, transpiration from their surfaces creates the negative pressures that maintain the upward flow of water and inorganic nutrients throughout most of the season. In contrast, mineral uptake and unloading into the xylem involve the expenditure of metabolic energy. Subsequent movement appears to be passive, along with the water being drawn up the xylem. Only rarely are organic molecules absorbed from the soil and translocated in the conducting elements of the root system. Exceptions include systemic fungicides and herbicides, some fungal toxins, and growth regulators produced by soil microorganisms.

Another important factor influencing water and mineral uptake is the formation of mycorrhizal associations between the root and a fungus. Grapevines are not unique in this regard, as more than 80% of plant species develop mycorrhizal associations. Of the three major groups of mycorrhizal fungi, only the vesicular–arbuscular group invades grapevine roots (Schubert *et al.*, 1990).

Glomus is the primary genus associated with grapevine mycorrhizae, although other genera occasionally involved are *Acaulospora, Gigaspora,* and *Sclerocystis*. The fungi produce spores from which infective hyphae invade the epidermis of host roots. From the epidermis, the fungus penetrates and colonizes the cortex. Here, the fungus produces large swollen **vesicles** and highly branched **arbuscules** (Fig. 3.1), which give rise to the **vesicular–arbuscular** (VA) designation of the group. The fungi do not markedly alter root morphology as other mycorrhizal associations tend to do. However, they do restrict root hair development, increase lateral root production, and result in more dichotomous root branching. Infection results in the normally white root tip becoming yellowish. While reducing root elongation and soil exploration, mycorrhizal infection produces a more economical root system for nutrient acquisition (Schellenbaum *et al.*, 1991).

The mycorrhizal association normally advances apically as the root grows and produces new cortical tissue. Mycorrhizal activity declines and ceases rearward as the root matures and the cortex is sloughed off.

As the fungus establishes itself in the cortex, wefts of hyphae grow outward and ramify extensively into the surrounding soil. The hyphal extensions absorb and translocate water and minerals back to the root. Mycorrhizal fungi generally are more effective at taking up minerals, notably phosphorus, than the root itself. The fungus produces hydroxyamates (peptides) that combine with and facilitate the uptake of nutrients. This is especially marked with poorly soluble inorganic nutrients such as phosphorus, zinc, and copper. In the cortex, sugars from the root are exchanged for minerals, notably phosphorus, from the fungus. The augmented phosphorus supply facilitates the phospholipid synthesis required for rapid root growth. Improved root production, in turn, supplies additional organic nutrients to the fungus and surrounding soil.

Both root and mycorrhizal exudates appear to influence the soil flora and texture. One of the effects is the selective favoring of nitrogen-fixing and ethylene-synthesizing bacteria around the root (Meyer and Linderman, 1986). These changes usually increase root resistance, or at least tolerance, to fungal and nematode infections. Mycorrhizal associations also tend to reduce vine sensitivity to water stress, salinity, and mineral deficiencies and toxicities. However, in some instances,

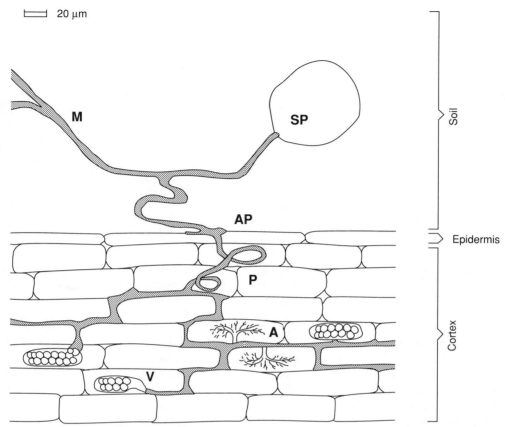

Figure 3.1 Schematic representation of a vesicular–arbuscular mycorrhiza in a grapevine root. A, Arbuscule; AP, appressorium; M, external mycelium; P, cortex parenchymal cell; SP, mycorrhizal spore; V, vesicle. (From Schubert, 1985, reproduced by permission.)

mycorrhizae may increase the availability of and toxicity to trace elements, such as aluminum. By affecting the level of growth regulators, such as cytokinins, gibberellins, and ethylene, mycorrhizal fungi may further influence vine growth. An additional effect may be the production of chelators, such as catechols, by soil bacteria. These help to keep minerals such as iron available to the plant.

In most vineyard soils, mycorrhizal associations develop spontaneously from the naturally occurring soil inoculum. Artificial inoculation of cuttings is complicated and often fails, as most rootlets die on replanting (Conner and Thomas, 1981). Artificial inoculation may be beneficial only when apical tissues are micropropagated or when vines are replanted into fumigated soils. Although spontaneous infection is poorly understood, it is known to be influenced by soil type, fungicidal sprays, and the mycorrhizal species present. Low nutrient soils generally favor mycorrhizal symbiosis, while rich soils limit development.

SECONDARY TISSUE DEVELOPMENT

As noted above, the epidermis and root cortex often become infected by mycorrhizal fungi. These tissues also

may succumb to microbial and nematode attack. The latter probably generates the brown and collapsed regions commonly observed along otherwise healthy roots. Regardless of health, the cortex and outer tissues soon die if the central region of the root commences secondary (lateral) growth (Fig. 3.2). Secondary growth includes the production of a cork layer next to the endodermis that cuts off nutrient flow to the cortex and epidermis. A meristem (**vascular cambium**) develops between the initial (**primary**) xylem and phloem and produces new (**secondary**) xylem and phloem to the inside and outside of the cambium, respectively. Periodically, **ray cells** are generated from the vascular cambium that facilitate movement of material between adjacent regions of the phloem and xylem. As the root enlarges in diameter, new cork layers may develop within the secondary phloem, cutting off the older cork layers and nonfunctional phloem.

ROOT SYSTEM DEVELOPMENT

In exploiting the soil, the root system employs both **extension growth** and **branching.** Extension growth involves the rapid growth of thick, leader roots into unoccupied soil. If the site is favorable, many thin, highly branched lateral roots develop from the leader root.

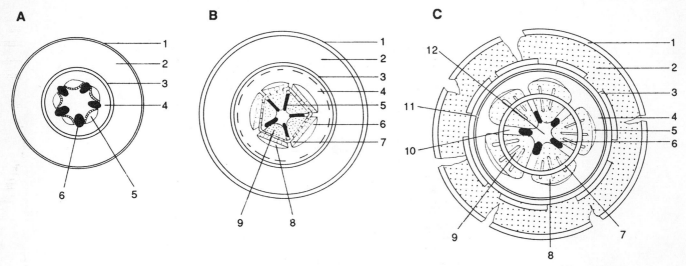

Figure 3.2 Diagrammatic representation of the secondary growth in grapevine roots. 1, epidermis; 2, cortex; 3, endodermis; 4, pericycle; 5, primary phloem; 6, primary xylem; 7, vascular cambium; 8, secondary phloem; 9, secondary xylem; 10, medullary rays; 11, periderm; 12, pith. (From Swanepoel and de Villiers, 1988, reproduced by permission.)

Most of the lateral (side) roots are short-lived and are replaced by new laterals. Few of the laterals survive and add to the permanent structural framework of the root system. Large roots are retained much longer than most young roots but are themselves subject to being sloughed off periodically (McKenry, 1984).

The largest roots generally develop about 0.3 to 0.35 m below ground level. Their number and distribution tend to stabilize a few years after planting (Branas and Vergnes, 1957). From this basic framework, smaller roots spread out horizontally and vertically into the surrounding soil.

Root system expansion is dependent on both environmental and genetic factors. Spread may be limited by layers or regions of compacted soil, high water tables, or saline, mineral, or acidic zones. Tillage, mulching, and irrigation all influence the positioning of major root development. Increasing planting density tends to decrease root mass but increases root density (roots per soil volume). Differences in the angle and depth of penetration vary with the genetic properties of the rootstock and rootstock/shoot interaction. For example, *Vitis rupestris* rootstocks sink their roots at a steep angle and tend to penetrate more deeply than *V. riparia* rootstocks. The latter produce a shallower, more spreading root system (Fig. 3.3).

Most root growth develops within a region about 4 to 8 m around the trunk. Depth of penetration varies widely but is largely confined to the top 1 m of soil. Nevertheless, roots up to 1 cm in diameter can penetrate more than 6 m deep. The finest roots, which do most of the absorption, generally occur within the first 0.1 to 0.6 m of the soil. Despite extensive root production, roots com-

monly constitute only about 0.05% of the available soil volume (McKenry, 1984).

Unlike the situation in many other woody perennial plants, resumption of root growth in the spring lags significantly behind shoot growth. New root production usually begins after bud break, with peak growth occuring during or following fruit set (Fig. 3.4). A second, autumnal, root growth period may occur in warm climatic regions (van Zyl and van Huyssteen, 1987). In young vines, the total root mass appears to increase at a fairly constant rate throughout the growing season (Araujo and Williams, 1988).

Reactivation of new root production may be affected by varietal traits and the soil temperature (Conradie, 1990). Root growth also is markedly influenced by soil moisture. Much of the effect of moisture may be indirect, by affecting the aeration and mechanical resistance of the soil. Some cultivars appear to be particularly sensitive to

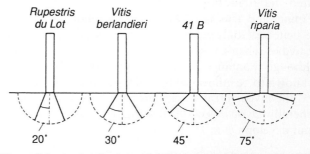

Figure 3.3 Angle of root production of various types of rootstocks. (From Bouard, 1980, reproduced by permission.)

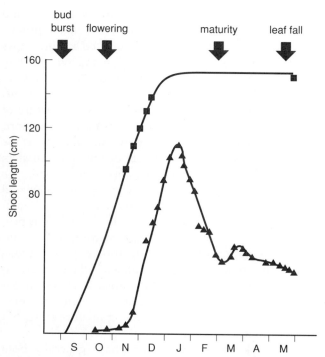

Figure 3.4 Relationship between the state of shoot growth (■) and root growth (▲) for the variety 'Shiraz' grown in the Southern Hemisphere. (After Richards, 1983, from Freeman and Smart, 1976, reproduced by permission.)

anaerobiosis, with oxygen tensions as low as 2% being toxic (Iwasaki, 1972). Waterlogging also enhances salt toxicity (West and Taylor, 1984). Furthermore, soil moisture content affects the rate at which soil warms in the spring, as well as the availability and movement of nutrients in soil.

The Shoot System

The shoot system of the grapevine has an unusually complex developmental pattern. The complexity provides the shoot system with a remarkable ability to grow throughout much of the growing season. Three or more successive sets of shoots may develop in a single year, depending on the climate. In addition, dormant buds from previous seasons may become active. In recognition of this complexity, shoots may be designated according to their origin, age, position, and length. The buds from which the shoots arise are identified relative to their position, germination sequence, and fertility.

BUDS

All grapevine buds are **axillary buds,** that is, they form in the **axils** of foliar leaves or their modifications, **bracts.**

Axils are defined as the positions along a shoot where leaves develop. As such, the axil is part of the circular region of the stem called the **node** (see Fig. 2.1). Most structures that develop from shoots—leaves, buds, tendrils, and inflorescences—develop at nodes. Most stem elongation occurs in the region separating adjacent nodes, the **internodes.** Buds occurring in the axils of the bracts (leaves modified for bud protection) occasionally may be termed **accessory buds.** When located at the base of mature canes, however, they are more commonly called **base buds.** Buds of any kind that remain dormant for several seasons are termed **latent buds.** Depending on their location, latent buds may give rise to **water sprouts** or **suckers** (see Fig. 4.6). Differentiation is based on whether the latent bud originated above or below ground, respectively.

Each shoot node potentially can develop an axillary bud complex consisting of four buds (immature shoot systems). These include the **lateral bud,** positioned to the dorsal side of the shoot, and the **compound bud,** which is positioned more ventrally (Fig. 3.5). The compound bud possesses three buds in differing states of development, namely, the **primary, secondary,** and **tertiary buds.**

Depending on genetic and environmental factors, lateral buds may differentiate into shoots during the year they are produced. Alternately, they may remain dormant and become a latent bud. Shoots developing from lateral buds are variously termed **lateral** or **summer shoots.** Because they seldom produce flower clusters, lateral shoots normally contribute only to the photosynthetic activity of the vine. However, fruit-bearing lateral shoots may produce what is called a **second crop.** Lateral buds formed in the leaf axils of lateral shoots also possess the potential to develop during the current year, and they may generate a second series of lateral shoots.

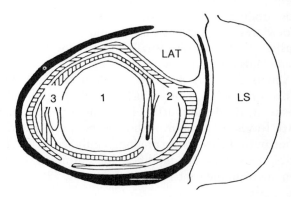

Figure 3.5 Transverse section through an axillary bud complex. LS, Leaf scar; LAT, lateral bud; primary (1), secondary (2), and tertiary (3) buds of the compound bud. Bracts of the lateral (solid) and primary (hatched) structures are also shown. (From Pratt, 1959, reproduced by permission.)

The primary bud of a compound bud develops in the axil of the bract (**prophyll**) produced by the lateral bud. Secondary and tertiary buds develop in the axils of the two bracts produced by the primary bud. All three buds typically remain dormant during the growing season in which they form. Consequently, the compound bud often is termed the **dormant bud.** Loss of dormancy usually requires exposure to cold temperatures during the late fall and winter months. In tropical and subtropical regions, severe pruning usually substitutes for the cold treatment. Hydrogen cyanamide also can effectively advance bud maturity, as well as promote uniform bud break (McColl, 1986).

Assuming that the primary buds are not destroyed by freezing temperatures, insect damage, or other causes, they generate the **primary shoots** (major shoot systems) of each year's growth. The secondary and tertiary buds, respectively, become active only if the primary and secondary buds die.

The primary bud (primordium) is the most developed in the compound bud and typically possesses several primordial leaves, inflorescences, and lateral buds before becoming dormant in the year of formation (Morrison, 1991). The secondary bud occasionally is fertile, but the tertiary bud is typically infertile (does not bear inflorescences) and is the least differentiated. Inflorescence differentiation is a function of the genetic characteristics of the cultivar, vigor of the rootstock, bud location, and the surrounding environment during bud formation.

Buds that bear nascent inflorescences (primordia) are termed **fruit (fertile) buds.** whereas those that do not are called **leaf (sterile) buds.** In most *V. vinifera* cultivars, fruit buds form with increasing frequency up from the basal leaf, often reaching a peak between the fourth and tenth nodes. However, this property varies with the cultivar (Huglin, 1986). Some cultivars, such as 'Nebbiolo,' do not form fruit buds at the base of the shoots. Such cultivars are not spur-pruned as it would drastically limit fruit production. Bud fruitfulness also influences pruning procedures. For example, the production of small fruit clusters by 'Pinot noir' usually requires the retention of many buds on long canes.

Shoots generally produce two (one to four) flower clusters per shoot. Flower clusters typically develop opposite the third and fourth, fourth and fifth, or fifth and sixth leaves on a shoot. Although the number and position of fruit clusters are predominantly scion characteristics, they may be modified by hormonal changes coming from the rootstock (see Richards, 1983).

SHOOTS AND SHOOT GROWTH

Components of the shoot system obtain their names from the type and position of the bud from which they are derived and their age, position, and relative length.

Once the outer photosynthetic tissues degenerate and turn brown, shoots are termed **canes.** If a cane is pruned to possess but a few buds, it is designated a **spur.** Canes retained for 2 or more years and supporting fruiting wood (spurs and canes) are called **arms.** When arms are positioned horizontally, they are referred to as **cordons.** The **trunk** is the major permanent upright structure of the vine. When old and thick, trunks need no support. Nevertheless, most vines require trellising to support the arms, canes, and growing shoots of the vine. All stem tissue 2 or more years old is classified as **old wood.**

Shoot growth is controlled by many environmental and genetic factors. Growth is favored by warm conditions, especially warm nights. Low light conditions promote shoot elongation but are detrimental to inflorescence induction. Genetic factors, probably acting through hormone production, influence the characteristic growth patterns of different cultivars. These tendencies can be modified by pruning and other cultural practices. For example, minimal pruning induces more, but shorter thinner shoots.

Bud break and shoot growth have generally been thought to begin when the mean daily temperature reaches 10°C or above, following loss of endogenous dormancy. However, bud break in several varieties may commence as low as 0.4°C (average 3.5°C) and leaf production begin at 5°C (average 7°C) (Moncur *et al.*, 1989). The rate of bud break and shoot growth increases rapidly above the minimum temperatures (Fig. 3.6). Once initiated, growth rapidly reaches a maximum, after which it progressively slows. Further shoot growth dur-

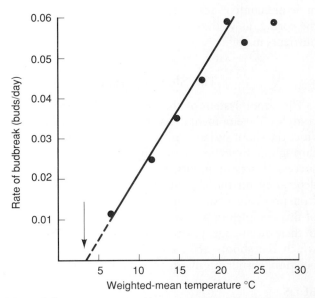

Figure 3.6 Determination of base temperature (↓) (arrow) for bud break of 'Pinot noir' from the effect of temperature on the rate of bud break. (From Moncur *et al.*, 1989, reproduced by permission.)

ing the summer generally originates from the activation of lateral buds.

The slow initiation of growth in the spring, and the potential for lateral shoot growth throughout the season, probably reflect the ancestral growth habit of the vine. Slow bud activation in the spring would have delayed leaf production until the support tree for the vine had completed most of its foliage production. Thus, the vine could position its foliage, relative to the foliage of the host, in locations optimally suited for its own photosynthesis. In addition, the potential to generate several series of lateral shoots allows the vine to continuously position new leaves in favorable sites for photosynthesis.

Older portions of the vine provide the support and translocation needs of the growing shoots, leaves, and fruit. These functions are provided by the woody parts of the grapevine that constitute the vast majority of its structure. Mature wood also acts as a significant storage organ, thereby helping to cushion the effects of unfavorable growth conditions on fruit production. These reserves also tend to increase average vine productivity. Although sugar content and fruit acidity are little affected by vine age, or the proportion of old wood, fruit aroma and flavor may be altered. For example, berry aroma and fruit flavor may increase with vine age (Heymann and Noble, 1987).

Shoot growth early in the season depends primarily on previously stored nutrients (May, 1987) (see Fig. 4.4). Nitrogen mobilization initiates in the canes and progresses downward into older parts of the vine, finally reaching the roots (Conradie, 1991b). Significant mobilization and translocation of nutrient reserves to the developing fruit also may occur following *véraison* (the onset of color change in the fruit). The vine stores organic nutrients, primarily starch and arginine, as well as inorganic nutrients, such as potassium, phosphorus, zinc, and iron. Movement of nutrients such as nitrogen into woody parts of the grapevine probably occurs throughout the growing season, but it is most marked following fruit ripening (Conradie, 1990, 1991a). It appears that some of the nutrients first mobilized in the spring are the last stored in the fall.

TISSUE DEVELOPMENT

Shoot growth originates from an apical meristem consisting of the **tunica** cells that cover the inner **corpus** cells (Fig. 3.7). The outer tunica layer produces the epidermal tissues of the stem, leaves, tendrils, flowers, and fruit, as well as initiate development of leaf and flower primordia. The corpus generates the inner tissues of the shoot and associated organs. The shoot apex, unlike its root equivalent, has a highly complex morphology. Buds, leaves, tendrils, and flower clusters all begin their development within a few millimeters of the apex. Subsequent cell division and elongation produce the mature structures of each organ.

The extension of vascular tissue, called **traces**, into the structures borne by the shoot is equally complex. As in

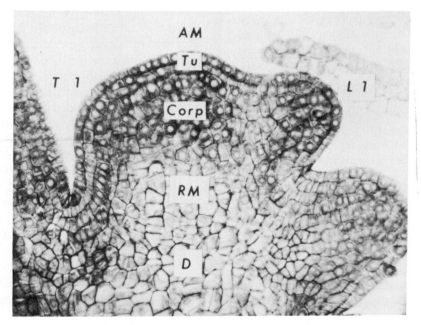

Figure 3.7 Median longitudinal section through a dormant shoot apex of *Vitis labrusca*. AM, Apical meristem; Tu, two-layered tunica; Corp, corpus; RM, rib meristem; D, diaphragm; T, tendril initiation; L, leaf primordium. ×360. (From Pratt, 1959, reproduced by permission.)

the root, phloem is the first vascular tissue to differentiate. The need for organic nutrients by the rapidly dividing apical tissues undoubtedly explains the early differentiation of the phloem. Cell elongation, which primarily entails water uptake, occurs later. Correspondingly, xylem differentiation can occur later. As tissue elongation ruptures connections between existing phloem cells, additional sieve tubes differentiate to replace those inactivated.

The outer tissues of the young stem consist of a layer of epidermis and several layers of cortical cells. The epidermis is initially photosynthetic and bears stomata, hair cells, and pearl glands. Most cortical cells also contain chloroplasts and are photosynthetic. Collections of cells with especially thickened side walls (**collenchyma**), differentiate opposite the vascular bundles. These regions produce the ridges (ribs) of the young shoot.

When cell elongation ceases, a layer of cells between the phloem and xylem differentiates into a lateral meristem, the **vascular cambium.** It generates the secondary vascular and ray tissues of the maturing stem. The secondary phloem, like the primary, contains translocating cells (**sieve tubes**), companion cells, fibers, and storage parenchyma. The secondary xylem consists primarily of large diameter xylem vessels and structural fibers. Ray cells elongate horizontally along the stem radius and transport nutrients into and between the xylem and phloem. In the phloem, ray tissue expands outward to form V-shaped segments. These cells are extensively involved in the storage of starch, along with xylem parenchyma. The innermost stem tissue, the **pith**, consists predominantly of thin-walled parenchyma cells. In species belonging to the section *Euvitis*, the pith soon disintegrates in all but the nodal region, where it develops into the woody **diaphragm** (Fig. 2.3a).

During shoot maturation, a cork cambium develops from parenchyma cells in the phloem. The cork generated cuts off nutrient and water supplies to the outer tissues (mostly cortex and epidermis). These subsequently die and produce the brown appearance of maturing shoots.

In most regions, with the probable exception of the tropics, the end walls (**sieve plates**) of the sieve tubes become plugged with **callose** in the autumn, stopping translocation. This occurs as the leaves die and are shed.

When buds become active in the spring, dormant phloem cells progressively regain the ability to translocate nutrients. Activation progresses longitudinally up and down the cane from each bud, and outward from the vascular cambium (Aloni and Peterson, 1991). Release of auxins from buds appears to be involved in stimulating callose breakdown (Aloni *et al.*, 1991). While cambial reactivation also is associated with bud activation, it shows marked apical dominance; that is, activation commences with the uppermost buds and progress downward, and laterally around the canes and trunk until the enlarging discontinuous patches meet (Esau, 1948). Unlike either phloem or cambium, mature xylem vessels contain no living material and are potentially functional whenever the temperature premits water to exist in a liquid state.

Phloem cells in the trunk may remain viable for up to 4 years, but most sieve tubes are functionally active only during the year in which they form (Aloni and Peterson, 1991). In the xylem, vessel inactivation begins about 2 to 3 years after formation but is complete only after 6 to 7 years. Inactivation involves tyloses, which grow into the vessel cavity from surrounding parenchyma cells.

New cork cambia develop at infrequent intervals in nonfunctional regions of the secondary phloem. In *Euvitis*, the outer tissues split and are eventually shed, whereas in the *Muscadinia*, the cork cambium forms under the epidermis and persists for several to many years. Thus, *Muscadinia* species do not form shedding bark like *Euvitis* species (see Fig. 2.3D).

Tendrils

Tendrils are considered to be modified flower clusters, which in turn are viewed as modified shoots. Not surprisingly, tendrils bear obvious morphological similarities to shoots. Unlike vegetative shoots, tendril growth is determinant, that is, its growth is strictly limited. Tendrils also pass through three developmental and functional phases. Initially, tendrils develop water-secreting openings called **hydathodes** at their tips. Subsequently, the hydathodes degenerate, and pressure-sensitive cells develop along the tendril. On contact with solid objects, the pressure-sensitive cells activate the elongation and growth of cells opposite the site of contact. This induces the twining of the tendril around the object contacted. At maturity, the tendril becomes woody and rigid, with development of collenchyma cells in the cortex and xylem and lignification of the ray cells.

With the exception of *Vitis labrusca* and its cultivars, tendril production develops in a discontinuous manner in *Euvitis*. In other words, tendrils are produced opposite the first two of every three leaves, beyond the first two to three basal leaves. In *V. labrusca*, tendrils are produced opposite most leaves. On bearing shoots, flower clusters replace the tendrils in the lower two or more locations in most *V. vinifera* cultivars, and in the basal three to four tendril locations in *V. labrusca* cultivars.

Leaves

Leaves develop initially as localized outgrowths in the inner tunica layer of growing shoots. Corpus cells differentiate to connect the developing vascular system of the stem with that of the expanding leaf. The first two leaf-like structures of a shoot develop into bracts. The internodes between the bracts are very short. Subsequent internodes elongate normally, and the leaf primordia subtended expand into mature leaves typical of the cultivar.

Maturing leaves consist of a broad photosynthetically active **blade,** the supportive and conductive **petiole,** and two basal semicircular **stipules.** The latter soon die and fall off, leaving only the petiole and blade. Unlike the growth of most plant structures, leaf growth is not induced by apical or lateral meristems. Leaf growth involves the action of many, variously positioned **plate meristems.** These generate the flat, lobed appearance of the leaf.

The leaf blade consists of an upper and lower epidermis, a single palisade layer, about three layers of spongy mesophyll, and a few large (and several small) veins (Fig. 3.8). The **upper epidermis** is covered by a waxy **cuticle** consisting of an inner layer of cutin and several waxy plates at the surface. The cuticle retards water and solute loss, helps limit the adherence of pathogens, minimizes mechanical abrasion, and slows diffusion of chemicals into the leaf. The **lower epidermis** shows a less well-developed cuticle, but it contains more stomata and commonly possesses leaf hairs and pearl glands. The latter derive their name from the small, beadlike secretions they produce. Water-secreting hydathodes commonly develop at the pointed tips (teeth) of the blade.

The **palisade layer** consists of cells directly below, and elongated at right angles to, the upper epidermis. When young, the cells are tightly packed, but intercellular spaces develop as the leaf matures. Cells of the palisade layer are the primary photosynthetic cells of the plant.

Directly below the palisade layer are several layers of **spongy mesophyll.** The cells of the mesophyll are extensively lobed and generate large intracellular spaces in the lower half of the leaf. The large surface area so generated, along with the stomata, enables the rapid exchange of water and gases between the inner leaf cells and the surrounding air. Without this efficient exchange, the leaf would rapidly overheat in full sun, suppressing photosynthesis. The exchange is equally important for the rapid interchange of CO_2 and O_2. Carbon dioxide is an essential ingredient in photosynthesis, and oxygen, one of its by-products, is inhibitory to the crucial action of ribulose bisphosphate carboxylase (RuBPCase) in photosynthesis.

The branching vascular network of the leaf consists of a few large veins, containing several **vascular bundles,** and many smaller veins consisting of only a single vascular bundle. Each vascular bundle consists of xylem on its upper side and phloem on the lower portion. The vascular tissues are surrounded by a set of thickened cells called the **bundle sheath.** Xylem provides most of the mineral and water needs of the leaf, while the phloem transports organic compounds, such as sugars, amino acids, and hormones, as well as some inorganic ions to and from the leaf.

In the autumn, an abscission layer forms at the base of the petiole. Connections between the shoot and the leaf cease when a periderm forms between the stem and the abscission layer.

PHOTOSYNTHESIS AND OTHER LIGHT-ACTIVATED PROCESSES

The production of organic compounds from carbon dioxide and water, using the energy of sunlight, is the quintessential property of plant life. Correspondingly,

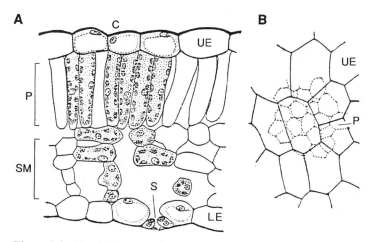

Figure 3.8 Transverse (A) and epidermal view (B) of a grape leaf. C, Cuticle; UE, upper epidermis; LE, lower epidermis; P, palisade layer; SM, spongy mesophyll; S, stomatal apparatus with guard cells. (After Mounts, 1932, reproduced by permission.)

the molecular and biophysical details of photosynthesis have been extensively studied. Although sugar is the major organic by-product, photosynthesis also generates intermediates that can be directly diverted into fatty acid, amino acid, and nitrogen-base biosynthesis. From these, all the organic constituents of the vine can be derived.

By absorbing sunlight, chlorophyll activates the splitting of water into oxygen and hydrogen. Oxygen escapes into the surrounding air, while hydrogen dissolves as free ions in the cytoplasm. Energized electrons, removed from hydrogen atoms, are passed along an electron-transport chain to nicotinamide adenine dinucleotide phosphate ($NADP^+$), generating reducing power (NADPH). In the process, ADP is phosphorylated to adenosine triphosphate (ATP). Both NADPH and ATP are required in the Calvin cycle for the regeneration of ribulose bisphosphate (RuBP), the substrate to which carbon dioxide is fixed in the "dark" reactions of photosynthesis. The immediate by-product of fixation is very unstable, however, and splits into two molecules of 3-phosphoglyceric acid. As sufficient carbon dioxide is incorporated via fixation, intermediates of the Calvin cycle can be extracted for the synthesis of sucrose and other compounds. Quantitatively, sucrose is the most important by-product of photosynthesis, as well as being the major organic compound translocated in the phloem. Sucrose may be stored temporarily as starch in the leaf but is usually translocated out of the leaf to other parts of the plant. It may be subsequently stored as starch, polymerized into structural components of the cell wall, metabolized into any of the other organic components of the vine, or consumed as an energy source for cellular metabolism.

Because photosynthesis is fundamental to plant function, the viticulturally important aspects of photosynthesis relate to providing an optimal environment for its occurrence. This is one of the primary functions of canopy management. A favorable light environment also influences other light-activated processes. These include such vital aspects as inflorescence initiation, fruit ripening, and cane maturation.

For photosynthesis, the light intensity in the blue and red regions of the visible spectrum is particularly important. For most other light-activated processes, though, it is the balance between the red and far-red portions of the light spectrum that is critical. These differences relate to both the energy requirements and pigments involved in the processes activated.

In photosynthesis, the important pigments are **chlorophylls** and **carotenoids,** whereas in most other processes **phytochrome** is involved. Both chlorophylls *a* and *b* absorb optimally in the red and blue regions of the light spectrum; carotenoids absorb significantly only in the blue. The splitting of water, the major light-activated

process in photosynthesis, is energy intensive but becomes maximal at about one-third the intensity of full sunlight ($\sim$800 μmol/m^2/sec). This is apparently due to rate limitations imposed by the speed of the dark reactions in photosynthesis.

In contrast, phytochrome-activated processes require very little energy. For these, conversion between the two states of phytochrome depends on the relative energy available in the red and far-red portions of the spectrum. These regions correspond to the absorption peaks of the two states of phytochrome (P_r and P_{fr}). Red light converts P_r to the physiologically active P_{fr} form, while far-red radiation transforms it back to the physiologically inactive P_r state. In sunlight, the natural red/far-red balance of 1.1 to 1.2 generates a 60 : 40 balance between P_{fr} and P_r. However, when light passes through a leaf canopy, the strong red absorbency of chlorophyll shifts the red/far-red balance toward the far-red. This probably means that processes active in sunlit tissue are inactive in shaded tissue. The precise levels and actions of phytochrome in grapevine tissues are unknown.

Because of the negative influence of shading on photosynthesis and phytochrome-induced phenomena, most modern pruning and training systems are designed to maximize light exposure. The effects of shading on spectral intensity and the red/far-red balance are illustrated in Fig. 3.9A. In contrast, cloud cover produces little spectral modification, although markedly reducing light intensity (Fig. 3.9B). Cloud cover also significantly reduces the intensity of solar infrared (heat) radiation received directly from the sun. Because photosynthesis is usually maximal at light intensities equivalent to that provided by a lightly overcast sky, cloud cover most effects the photosynthesis of shaded leaves. Canopy shading rapidly attenuates the intensity of sunlight down to or below the compensation point (the light intensity at which the rates of CO_2 fixation and respiratory release are equal). The lower respiratory rate of cooler, shaded leaves does not compensate for the poorer spectral quality and reduced intensity of shade light. As a result, shade leaves generally do not photosynthesize sufficiently to export sucrose or contribute significantly to vine growth. This situation might change if the shaded leaves were exposed to sufficient **sunflecking,** periodic exposure to sunlight through the canopy. Although sunflecking is highly variable, average rates of 0.6 sec per 2-sec intervals could significantly improve net photosynthesis (Kriedemann *et al.*, 1973). Sunflecks also appear to delay premature leaf senescence and dehiscence.

When young and rapidly unfolding, leaves act as a sink for carbohydrates and show no photosynthetic export. Leaves begin to export photosynthate when they reach about 30% of their full size, but they continue to import carbohydrates until they have reached 50 to 75% of their

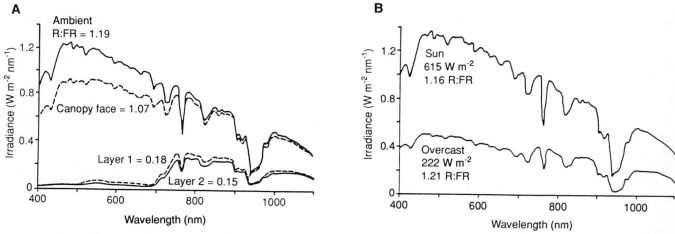

Figure 3.9 Spectral distribution of ambient sunlight radiation, at the canopy surface, and under one- and two-leaf canopies (A) and between sunny and overcast days (B). (From Smart, 1987, reproduced by permission.)

mature size (Koblet, 1969). Maximal photosynthesis and sugar export occur about 40 days after unfolding, when the leaves are fully expanded (Fig. 3.10). Old leaves continue to photosynthesize, but their contribution to the vine varies significantly, depending on shading and the activity of new leaves. Leaves cease to be net producers of photosynthate when they lose their typical dark green coloration. Leaves formed early in the season may photosynthesize at up to twice the rate of those produced later on.

Although the angle of the sun relative to the leaves can influence the rate of photosynthesis, even leaf blades aligned parallel to the sun's rays may photosynthesize at rates up to 50% that of perpendicularly positioned leaves (Kriedemann *et al.*, 1973). This may reflect the importance of diffuse skylight to photosynthesis. The effect of temperature on photosynthesis may vary slightly throughout the growing season. In the summer, optimal synthesis occurs at about 25° to 30°C, while in the autumn the range may fall to between 20° and 25°C (Stoev and Slavtcheva, 1982).

Fruit load as well as canopy size affect photosynthesis, presumably through a feedback system which, within limits, adjusts the rate of photosynthesis to demand. Thus, removing leaves to improve fruit light exposure, and to diminish disease incidence, can be partially compensated by increased photosynthetic efficiency of the remaining canopy. Improved photosynthetic efficiency appears to be associated with wider stomatal openings and increased gas exchange. In addition, greater sugar demand could activate export, reducing the concentration of Calvin cycle intermediates in the leaf that may act as feedback inhibitors in photosynthesis.

The direction of carbohydrate translocation from the leaves varies with leaf position and season of the year.

Export from young maturing leaves generally is upward to the new growth. Subsequently, photosynthate from the upper maturing leaves is directed downward to the fruit. By the end of berry ripening, the basal leaves are no longer involved in export. This function is taken over by the upper leaves. Translocation typically is restricted to the side of the cane and trunk on which the leaves are produced.

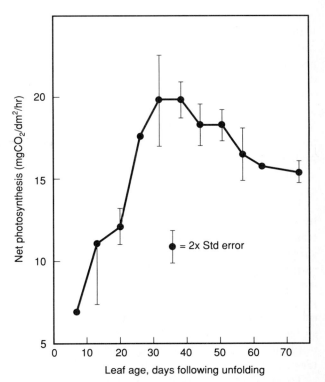

Figure 3.10 Relationship between leaf age and net photosynthesis of 'Thompson seedless' vines. (From Kriedemann *et al.*, 1970, reproduced by permission.)

Phytochrome is associated with a wide range of photoperiodic and photomorphogenic responses in plants. Typically, grapevines are relatively insensitive to photoperiod, but they may show several phytochrome-induced responses. Several important enzymes in plants are known to be partially activated by P_{fr}. These include enzymes involved in the synthesis and metabolism of malic acid, the production of phenols and anthocyanins, and possibly sucrose hydrolysis, nitrate reduction, and phosphate accumulation. The actual importance of the shift in the P_{fr}/P_r ratio in shaded grapevine leaves is currently unclear. Also, whether intermittent penetration of sunlight into the leaf canopy (sunflecking) significantly affects the phytochrome balance in grape leaves is unknown.

A third light-induced phenomenon is activated by ultraviolet radiation, namely, the toughening of the cuticle. The strong absorption of ultraviolet radiation by the epidermis results in almost complete removal of ultraviolet wavelengths from shade light. Consequently, the cuticle of shaded leaves and fruit is softer than that of sun-exposed counterparts. This factor, combined with the slower rate of drying and higher humidity in the undercanopy, may explain the greater sensitivity of shaded tissue to fungal infection.

TRANSPIRATION AND STOMATAL FUNCTION

To facilitate photosynthesis, the leaf must act as an efficient light trap and receive an adequate supply of water and carbon dioxide. Because of the importance of diffuse radiation coming from the sky, leaves usually provide a broad surface for light impact. Even on clear days, up to 30% of the incident radiation impacting a leaf may come from diffuse skylight. The large surface area so provided also makes the leaf blade an effective heat trap. Heating can be reduced by evaporative cooling, but if unrestricted this would place an unacceptably high water demand on the root system. The cuticular coating of the leaf limits water loss, but it also retards gas exchange with the surrounding air.

Transpiration and gas exchange are regulated largely by stomatal function. When water supply is adequate, transpirational cooling minimizes leaf heating on exposure to sunlight (Millar, 1972). Even under mild transpiration conditions, however, water may be lost from leaves at rates of 10 mg/cm²/hr. Except for the small proportion of water (<1%) used in photosynthesis and other processes, most of the water absorbed by the root system is lost by transpiration. Water stress usually does not limit transpiration, or photosynthesis, until leaf water potentials fall below -13 to -15 bars.

To regulate the processes of transpiration and gas exchange, leaves depend primarily on stomatal function.

Stomatal closure results in both a rise in leaf temperature and O_2 content and a reduction in CO_2 content. These influences suppress net carbon fixation by increasing photorespiration and decreasing photosynthesis. The effects of water stress on stomatal function and photosynthesis often linger after the return of turgor. The slow return to normal leaf function may result from the accumulation of abscisic acid produced during water stress, or from disruption of the photosynthetic apparatus in the chloroplasts. *Vitis labrusca* cultivars appear to be less sensitive in this regard than *V. vinifera* cultivars. When water stress develops slowly, increased root growth may offset the effects of reduced water supply in the soil (Hofäcker, 1976). Some varieties also appear to make osmotic adjustments to drought conditions (Düring, 1984).

The **stoma,** or more correctly the **stomatal apparatus,** consists of two guard cells and associated accessory cells. The lower epidermis of grapevine leaves possesses about 10 to 15×10^3 stomata/cm² (Kriedemann, 1977). Although stomata constitute only about 1% of the lower leaf surface, they permit transpiration at a rate equivalent to about 25% of the total leaf surface. This paradoxical finding is explained by the spongy mesophyll functioning as the actual surface over which transpiration occurs. The stomata act as openings through which the evaporated water can escape into the atmosphere. The same mesophyll surfaces provide efficient carbon dioxide and oxygen exchange with the atmosphere.

Control of stomatal opening and closing is complex, as befits its central role in leaf function. Carbon dioxide and water stress probably have the most direct effect on stomatal function. Low CO_2 tensions induce opening, while high CO_2 levels and water stress activate closing. Split-root experiments have shown that the water-stressed component of a root system can affect stomatal conductance throughout the vine (Düring, 1990). High temperatures indirectly induce closing, possibly by activating abscisic acid synthesis (via increased water stress) and increasing the CO_2 concentration (by suppressing photosynthesis and spurring respiration). Light tends to activate stomatal opening by inducing malic acid synthesis from starch as well as by reducing the CO_2 level via photosynthesis.

Stomatal opening and closing, respectively, involve the active uptake and release of potassium ions by the guard cells. Malic acid synthesis may provide the hydrogen ions needed for K^+ ion exchange in the guard cells. As a result of the K^+ influx or efflux, water, respectively, moves into or out of the guard cells. Water uptake provides the turgor pressure that forces the guard cells to elongate and curve in shape, causing stomatal opening. Water loss from the guard cells conversely induces closing.

Reproductive Structure and Development

Inflorescence (Flower Cluster)

INDUCTION

As noted above, only some buds differentiate inflorescences. At inception, fertile and sterile buds appear identical. The precise factors that induce inflorescence inception are unclear, but some of the prerequisite conditions are known.

A young vine typically does not begin to bear flowers until the second or third season. Juvenility is not based simply on size, as cytokinin application can induce flowering in young seedlings. Thus, flowering appears to be controlled by physiological rather than chronological age.

The induction of inflorescence differentiation in developing compound buds generally coincides with blooming and the slowing of vegetative growth. Induction usually occurs about 2 weeks before morphological signs of differentiation become evident. A period of several weeks may separate the initiation of the first and second inflorescence primordia (*anlagen*) (Fig. 3.11).

Environmental conditions around the bud and closely associated leaves can markedly affect induction. This probably results from changes in the hormonal and/or nutrient status of the bud. Cool conditions favor gibberellin synthesis, which promotes vegetative growth and limits nutrient accumulation. In contrast, warm conditions (25° to 30°C) promote cytokinin synthesis, which favors inflorescence differentiation. Optimum nitrogen, potassium, and phosphorus supplies also favor synthesis of cytokinins in roots. The action of the growth retardant chlormequat in favoring fruit bud initiation may be due to inhibition of gibberellin synthesis and stimulation of cytokinin production. Gibberellin application also can shift inflorescence development into tendril formation. In contrast, auxin application improves inflorescence induction as well as the number of flowers formed per cluster.

The timing and deposition of carbohydrates in the shoot correlate well with the period and node position of fruit bud development. This period corresponds to the interval between rapid shoot growth and the acceleration of fruit development. Correspondingly, factors such as high vigor or untimely drought that could disrupt carbohydrate accumulation can seriously diminish inflorescence initiation.

Full exposure to the sun is generally more important than day length in inducing fruit bud development in *V. vinifera* cultivars. In *V. labrusca* varieties, long days appear to improve inflorescence induction. Depending on the cultivar, the frequency and size of the fruit buds

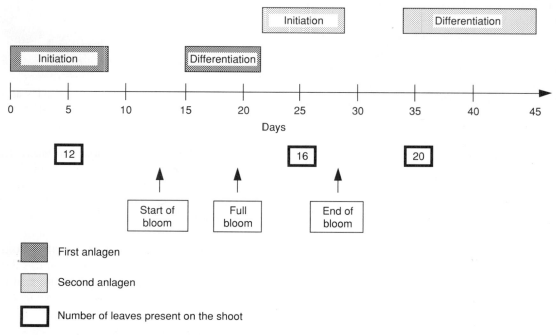

Figure 3.11 Diagrammatic representation of events associated with the initiation and primary development of the two inflorescence primordia of 'Chenin blanc' fruit buds. (From Swanepoel and Archer, 1988, reproduced by permission.)

progressively increase from the shoot base, often reaching a maximum about the tenth node. Fruitfulness tends to decline beyond this pont (Huglin, 1986). While lateral shoots seldom produce inflorescences, woody laterals may favor the fruitfulness of the compound buds associated with the lateral (Christensen and Smith, 1989).

As indicated above, fruitfulness is primarily associated with the primary bud. Fruitfulness of secondary buds is generally important only if the primary bud dies. Formation of inflorescence primordia in base buds is seldom of importance as they remain inactive. However, base bud activation in French–American hybrid cultivars frequently leads to overcropping, poor wood maturation, and weakening of the vine.

In contrast to the information known about conditions favoring inflorescence induction in the pimary bud, little is known about the conditions affecting flowering of lateral shoots produced during the growing season. Production of a second crop by lateral shoots can seriously complicate attempts to regulate crop yield and quality.

CLUSTER MORPHOLOGY

The grapevine inflorescence is a complex, highly modified branch system containing reduced shoots and flowers (Fig. 3.12). The complete branch system is called the **rachis.** It is composed of a basal stem, the **peduncle,** two main branches, the **inner** and **outer (lateral) arms,** and various subbranches terminating in **pedicels.** Pedicels give rise to the individual flowers. Flowers commonly occur in groups of threes, called a **dichasium,** but may occur singly or in clusters of modified dichasia. Individual flowers are considered to be reduced shoots, with leaves modified for reproductive functions.

The branching of the inflorescence into arms occurs very early in differentiation, similar to that of tendrils. The outer arm develops into the lower and smaller branch of the inflorescence. Occasionally it develops into a tendril, or fails to develop at all. The inner arm develops into the major branch of the flower cluster. In the buds that develop the most rapidly, usually those between the fourth and twelfth nodes, branching is complete by the end of the growing season. Such buds also may contain up to 6 to 10 leaf primordia by fall. The rudiments of flower primordia usually are present at this time.

Fully developed buds form nascent flower clusters possessing the typical shape of the variety. Less mature buds generally form atypically shaped smaller clusters. Flower development may continue during warm winter spells.

Unlike most perennial plants, in grapevines the inflorescence develops opposite a leaf, usually replacing a tendril in that location. The vine may produce up to four flower clusters per shoot, though two is common. They

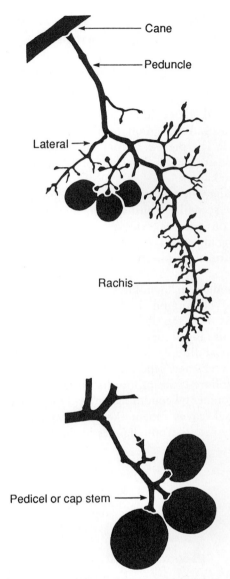

Figure 3.12 Structure of the grape cluster and its attachment to the cane. (From Flaherty *et al.,* 1981, reproduced by permission.)

are typically positioned opposite adjacent leaves at the third and fourth, fourth and fifth, and fifth and sixth, sixth and seventh, or, nodes, depending on the variety. Although buds formed from the fourth to twelfth nodes are generally the most fruitful, flower buds may develop beginning with the first leaf (count) node along a cane. Many European cultivars are fruitful from the base, whereas varieties of western and central Asiatic origin are often barren at the base. Varieties that do not produce basal fruit buds are not amenable to spur pruning.

FLOWER DEVELOPMENT

At the end of the growing season, flowers may have developed to the point of producing **receptacles**—the swollen base of the pedicel from which the flower parts originate. As buds swell in the spring, cells in the flower primordia begin to divide. Differentiation progresses inward from the receptacle rim as the shoot and rachis elongate.

Shortly after leaves begin to unfold, **sepal** differentiation becomes evident. These leafy structures do not develop significantly and are barely visible in mature flowers. About 1 week later, **petals** begin to differentiate. They rapidly arch upward, past the sepals, and fuse into a unified enclosing structure, the **calyptra**. Petal fusion occurs through the action of special interlocking cells that form on the edges of the petals (Fig. 3.13A). Within about 3 weeks, **stamens** have formed, their elongating **filaments** pushing the pollen-bearing **anthers** upward. Each of the five anthers contains two elongated **pollen sacs**, attached at a central point to the tip of the filament. Each anther produces thousands of **pollen grains**, containing a **generative** and a **pollen-tube nucleus**. About 1 week after stamen genesis, two **carpels** form, each producing two **ovules**. The two carpels fuse to form the developing **pistil**, composed of a basal, swollen **ovary**, a short **style**, and a slightly flared **stigma**. **Egg** development in the ovary closely parallels that of the pollen in the anthers.

As the pollen matures, the base of the petals separate from the receptacle (Fig. 3.13B) and curve outward and upward. The freed calyptra dries and falls, or is blown off. Shedding of the calyptra occurs most frequently in the early morning. Shedding often triggers rupture of the pollen sacs (**anthesis**) and pollen discharge, but anthesis can occur prior to shedding of the calyptra. Rupture occurs along a line of weakness adjacent to where the pollen sacs are attached to the filament. The region contains a layer of thickened cells, which can rip open the pollen sac epidermis on drying. The violent release of pollen sheds pollen onto the stigma. The short style of several varieties further favors the likelihood of self-fertilization.

TIMING AND DURATION OF FLOWERING

Flowering normally occurs within 8 weeks of bud break. Precise timing varies with weather conditions and cultivar characteristics. In warm temperate zones, flowering often begins when the mean daily temperature reaches 20°C. In cooler northern and southern climates, increasing day length may be an important stimulus. Where flowering is staggered over several weeks, cyanamide treatment may be used to improve synchronization. This can favor more uniform maturation and quality in the fruit of cultivars that tend to have asynchronous flowering, such as 'Merlot.'

Pollen release, as measured in the surrounding air, has been recently recommended as a means of assessing probable fruit yield (Besselat and Cour, 1990). Presumably the warm, sunny, dry conditions that favor aerial pollen dispersal correlate well with the conditions that favor self-pollination.

Flowering usually begins on the uppermost shoots, similar to the sequence of bud break. For individual clusters, however, blooming commences from the base of the inflorescence. Under favorably warm sunny conditions, individual flower clusters may complete blooming within a few days. Because of timing differences throughout the vine, blooming usually lasts about 7 to 10 days. Under cold, rainy conditions, flowering may last over 1 month. Under such conditions, the calyptra may not be lost. Although not preventing pollination, these conditions usually diminish the level of fertilization. In some varieties, such as 'Malbec,' poor calyptra release commonly occurs at the end of the flowering period. In other varieties, such as 'Zinfandel,' irregular cluster blooming may lead to asynchronous fruit set. Whether induced by varietal or environmental conditions, poor synchrony can lead to an undesirable range of fruit maturity within grape clusters at harvest.

POLLINATION AND FERTILIZATION

Self-pollination appears to be the rule for most grapevine cultivars. Under vineyard conditions, wind and insect pollination appear to be of little significance. Even in areas where grapes are the dominant agricultural crop, pollen levels in the air may be low during flowering. Yields of about 1.4×10^4 pollen grains/m^2/day during July have been recorded in Montpellier, France (Cour *et al.*, 1972–1973). The low level of insect pollination may be due to the nearly simultaneous blooming of innumerable vines over wide areas. Nevertheless, syrphid flies, long-horned and tumbling flower beetles, and occasionally bees visit grape flowers. The primary attractant appears to be the scent produced by the **nectaries**. These are located between the stamens and pistil (Fig. 3.13C) and are particularly prominent in male flowers on wild *V. vinifera* vines. Nectaries are modified for scent, rather than nectar production as the name might imply. Visiting insects feed on the pollen. Pollen fertility has no influence on flower attractiveness, but the presence of anthers does affect the duration of visits to female flowers (Branties, 1978).

Shortly after landing on the stigma, the pollen begin to swell. The sugary solution produced by the stigma is required both for pollen growth and to prevent osmotic lysis of the **germ tube**. The stigmatic fluid also occurs in the intercellular spaces of the style. This may explain why rain does not significantly inhibit or delay pollen germination or penetration of the germ tube into the style. However, pollen germination and germ-tube growth are markedly affected by temperature (Fig. 3.14),

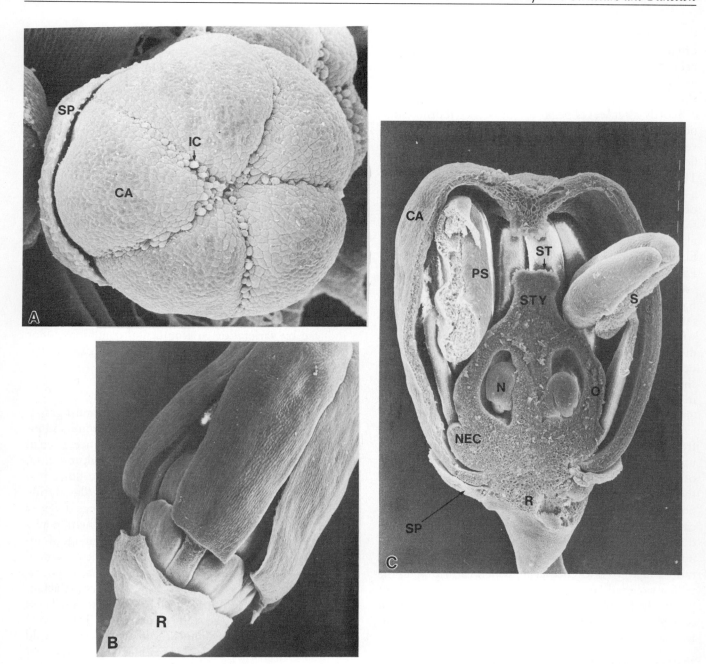

Figure 3.13 Grape flower. (A) Young calyptra (CA) with interlocking cells (IC); (B) loosening of calyptra; and (C) longitudinal flower cross section with stamens (S), anthers (PS), stigma (ST), style (STY), ovary (O), nucellus (N), nectary (NEC), sepals (SP), and receptacle (R). (From Swanepoel and Archer, 1988, reproduced by permission.)

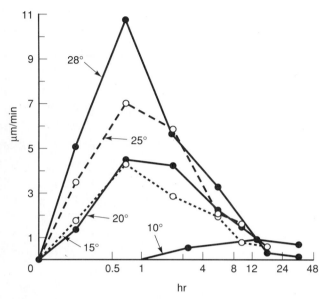

Figure 3.14 Velocity of pollen germ tube growth in relation to temperature. (From Staudt, 1982, reproduced by permission.)

even though viability is less affected. In contrast, ovules show obvious signs of degeneration within 1 week at temperatures below 10°C.

As the pollen tube grows down the style, the **germinative nucleus** divides into two **sperm nuclei,** if this has not occurred previously. Upon reaching the opening of the ovule, the **micropyle,** one sperm nucleus fuses with the **egg nucleus,** while the other fuses with two **polar nuclei.** Fertilization of the egg nucleus initiates **embryo** development, while fusion with polar nuclei initiates **endosperm** differentiation. Fertilization also sets in motion the series of events that turns the ovules into seeds and the ovary wall into the skin and flesh of the berry. Fertilization is usually complete within 2 to 3 days of pollination.

In certain varieties, abnormalities result in the absence of viable seed following pollination. In **parthenocarpic** cultivars, ovules fail to develop in the flower. Although pollination stimulates sufficient auxin production to prevent abscission of the fruit (**shatter**), it is inadequate to permit normal berry enlargement. 'Black Corinth' is the most important parthenocarpic variety and is the primary commercial source of dried currants. Although parthenocarpic varieties produce no seed, other so-called seedless varieties, such as 'Thompson seedless,' contain seeds. As these usually abort within a few weeks of fertilization, the seeds are empty, small, and typically have soft seed coats. Because of the partial development of the seeds, greater auxin production induces medium-sized fruit development. This situation is called **stenospermocarpy.** If abortion occurs even later, as in the cultivar 'Chaouch,' normal-sized fruit develop but contain hard, empty seeds. In contrast to the well-known

examples of parthenocarpy and stenospermocarpy in grapevines, the development of fruit and viable seeds in the absence of fertilization (**apomixis**) is unconfirmed.

FLOWER TYPE AND ITS GENETIC CONTROL

Most cultivated grapevines produce predominantly perfect flowers, that is, they contain both functional male and female parts. The flowers are also self-fertilizing. Nevertheless, many varieties produce a range of flowers, from strictly male to solely female (Fig. 2.2). Male (**staminate**) flowers have erect stamens and a reduced pistil, or ovaries that abort before forming a mature embryo. Female (**pistillate**) flowers have a well-developed functional pistil but produce reflexed stamens possessing sterile pollen. Sterile pollen is often characterized by one or more of the following properties: disruption of chromosome separation during nuclear division, inner-wall abnormalities, absence of surface furrows, and failure to produce germ pores under suitable conditions. Such wide variation suggests that sexual expression in grapevines is under complex genetic control. Recent studies on the genetics of sexuality in *Vitis vinifera* support this view.

Because both male and female parts develop in unisexual flowers on wild vines, the ancestral state of the genus was presumably bisexual (hermaphroditic). Early in the evolution of the genus, there must have been strong selection for functional unisexuality (individual plants being either male or female). It would have had the advantage of imposing cross-fertilization and maximizing genetic variability. Conversely, self-fertilization is preferable in cultivated vines as it is less susceptible to environmental disruption. Because of the many agricultural advantages of self-fertilization, different mutations reinstating full functionality to both male and female flower parts may have frequently been selected during domestication. This accounts for the different origins of the bisexual, self-fertilizing habit of most grapevine cultivars.

Sexual determination in grapevines, unlike many organisms, is not associated with morphologically distinct chromosomes. In *V. vinifera*, sexuality may be controlled by a pair of loci positioned on a single chromosome (Table 3.1). Each locus (gene) is polymorphic, that is, can exist in a number of physiologically distinct forms (**alleles**). Normally, no more than two alleles of any gene function together in diploid cells.

The primary locus exists in at least five distinguishable allelic states. These show a dominance series, with the male allele (M) dominant over all others. Second in dominance is the hermaphroditic allele (H). The female alleles (F_m, F_h, and F'_m) are recessive and are distinguished by how the second locus affects their expression. The second locus is an epistatic gene that modifies the expres-

Table 3.1 Control of Sexual Expression in *Vitis*[a,b]

	ME	Me	HE	He	$F_m E$	$F_m e$	$F_h E$	$F_h e$
ME	$\dfrac{ME}{ME}$ ♂	$\dfrac{ME}{Me}$ ♂	$\dfrac{ME}{HE}$ ♂	$\dfrac{ME}{He}$ ♂	$\dfrac{ME}{F_m E}$ ♂	$\dfrac{ME}{F_m e}$ ♂	$\dfrac{ME}{F_h E}$ ♂	$\dfrac{ME}{F_h e}$ ♂
Me		$\dfrac{Me}{Me}$ ♂	$\dfrac{Me}{HE}$ ♂	$\dfrac{Me}{He}$ ♂	$\dfrac{Me}{F_m E}$ ♂	$\dfrac{Me}{F_m e}$ ♂	$\dfrac{Me}{F_h E}$ ♂	$\dfrac{Me}{F_h e}$ ♂
HE			$\dfrac{HE}{HE}$ ♀	$\dfrac{HE}{He}$ ⚥	$\dfrac{HE}{F_m E}$ ♂	$\dfrac{HE}{F_m e}$ ♂	$\dfrac{HE}{F_h E}$ ⚥	$\dfrac{HE}{F_h e}$ ⚥
He				$\dfrac{He}{He}$ ⚥	$\dfrac{He}{F_m E}$ ♂	$\dfrac{He}{F_m e}$ ⚥	$\dfrac{He}{F_h E}$ ⚥	$\dfrac{He}{F_h e}$ ⚥
$F_m E$					$\dfrac{F_m E}{F_m E}$ ♂	$\dfrac{F_m E}{F_m e}$ ♂	$\dfrac{F_m E}{F_h E}$ ♂	$\dfrac{F_m E}{F_h e}$ ♂
$F_m e$						$\dfrac{F_m e}{F_m e}$ ♀	$\dfrac{F_m e}{F_h E}$ ♂	$\dfrac{F_m e}{F_h e}$ ♀
$F_h E$							$\dfrac{F_h E}{F_h E}$ ⚥	$\dfrac{F_h E}{F_h e}$ ⚥
$F_h e$								$\dfrac{F_h e}{F_h e}$ ♀

[a] Sexual expression in grape is based on a primary locus consisting of a dominance series of alleles ($M > H > F_m, F_h$) and a secondary, epistatic locus possessing two alleles (E and e). ♂, Functionally male; ⚥, functionally hermaphroditic; ♀, functionally female.

[b] After Carbonneau (1983), reproduced by permission. Copyright 1983 INRA, Paris.

sion of the alleles of the primary locus. The epistatic locus exists in dominant (E) and recessive (e) alleles. With the $F_m' F_m'$ genotype, for example, the epistatic gene modifies expression to male, hermaphrodite and female with EE, Ee, and ee, respectively.

In other varieties, genetic data are compatible with control based on a single gene pair, with, respectively, a dominant allele suppressing pistil development (maleness) and a recessive allele suppressing anther development (femaleness) (Negi and Olmo, 1971). The molecular mechanism by which these genes regulate sexuality is unclear. However, because application of cytokinin can induce pistil development in male vines, growth regulators are likely involved.

Berry Growth and Development

Following fertilization, the ovary begins its growth and transformation into the berry. Typically, a slightly double-sigmoid growth curve is observed (Fig. 3.15). In phase I, the berry shows rapid cell division, followed by cell enlargement and seed endosperm growth. Phase II is a transitional period in which growth may slow, the embryo develops, and the seed coats harden. At the end of phase II, the berry begins to lose its green color. This stage, called *véraison,* signifies the beginning of the fundamental physiological change that culminates in berry maturation. Phase III is associated with this change and the final enlargement of the berry. Ripening is associated with tissue softening, a decrease in acidity, the accumulation of sugars, the synthesis of anthocyanins in red-skinned varieties, and the development of aroma compounds. Although the differences between phases I and III are clear, the so-called lag component of phase II may be minimal or even undetectable (Staudt *et al.,* 1986). Overripening of fruit on the vine is occasionally called phase IV.

BERRY STRUCTURE

As with other aspects of vine morphology, the terms applied to berry structure can vary from author to author. According to standard botanical usage, the ovary wall develops into the fruit wall or **pericarp.** The pericarp is in turn subdivided into an outer **exocarp,** a middle **mesocarp,** and an inner **endocarp.** Various authors have applied these terms differently to the skin and various fleshy portions of the berry (pulp), depending on their respective views on berry development. One representation is given in Fig. 3.16. Because of the uncertainty in the correct use of technical anatomical terms, the common terms "skin" and "flesh" will be used in the text.

The grape skin consists of two anatomically distinguishable regions: the outer **epidermis** and an inner **hypodermis** consisting of several cell layers (Fig. 3.17). The internal fleshy tissues of the berry also may be subdivided into two regions—the outer portion (**outer wall**), comprising the tissues between the peripheral vascular strands and the hypodermis, and the inner section (**inner**

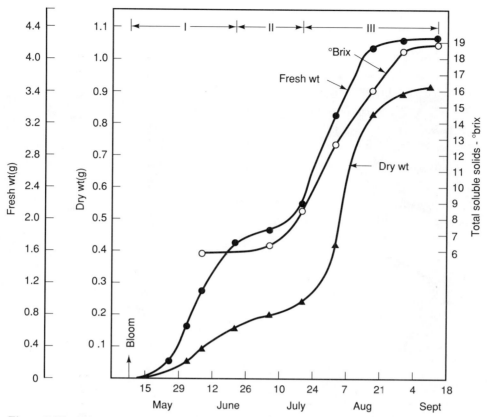

Figure 3.15 Relationship of growth phases I, II, and III and the accumulation of total soluble solids in 'Tokay' berries. (From Winkler *et al.,* 1974, reproduced by permission.)

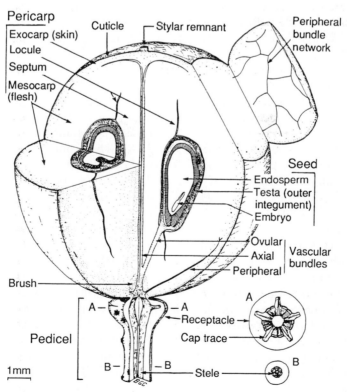

Figure 3.16 Diagrammatic representation of a grape berry. (After Coombe, 1987a, reproduced by permission.)

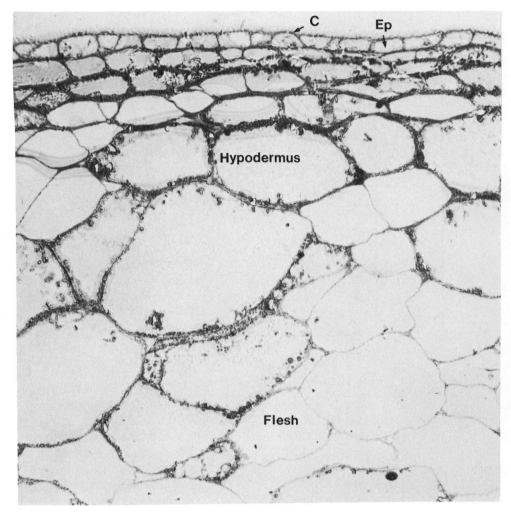

Figure 3.17 Cross section through the epidermis, hypodermis, and outermost flesh cells of 'Chardonnay' fruit (×400). C, Cuticle; Ep, epidermis. (Photograph courtesy of J. C. Audran.)

wall), being delimited by the peripheral and axial vascular strands. The layer of cells that corresponds to the boundary between the flesh and the seed locules is occasionally termed the **inner epidermis.** The base of the berry, where it attaches to the pedicel, is designated the **brush.** The **septum** comprises the central region where the two carpels of the pistil join.

Cell division in the flesh occurs most rapidly about 1 week after fertilization. It subsequently slows and may stop within 3 weeks. Cell division ceases earliest close to the seeds and last next to the skin. Cell enlargement occurs throughout the growth phases of berry development but is most marked after cell division ceases.

The epidermis consists of a layer of flattened, disk-shaped cells, with irregularly undulating edges. Depending on the variety, they may develop thickened or lignified walls. The epidermis does not differentiate hair cells and produces few stomata. The latter are sunken within

raised (peristomatal) regions on the fruit surface (Fig. 3.18). The stomata soon become nonfunctional. The surface of the peristomatal region accumulates high quantities of silicon and calcium, and possesses more phosphorus, chlorine, and sulfur than surrounding epidermal regions (Blanke, 1991). Microfissures may develop next to the stomata during periods of rapid fruit enlargement (Fig. 4.26A).

As with most other aerial plant surfaces, a relatively thick layer of cuticle and epicuticular wax develops over the epidermis (Fig. 3.19A). The cuticle is present almost in its entirety before anthesis. It forms a series of compressed ridges a few weeks prior to anthesis (Fig. 3.19B). During subsequent growth, the ridges spread apart and eventually form a relatively flat layer over the epidermis (Fig. 3.19C). At anthesis, plates of epicuticular wax begin to form over the cuticle. Initially, the wax platelets are small and upright, occurring both between and on

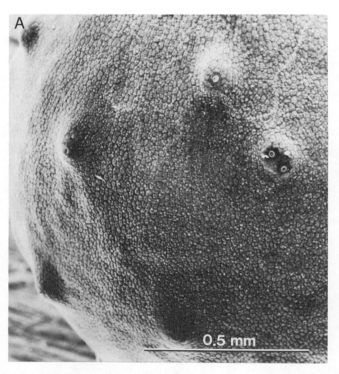

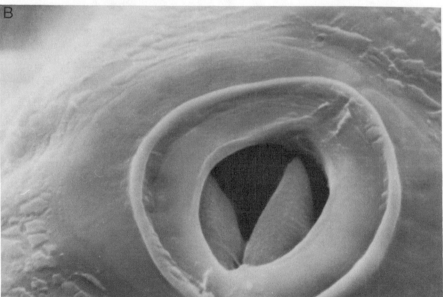

Figure 3.18 Scanning electron micrograph of (A) the epidermal surface of berries at *véraison* showing raised regions with one or two stomata (from Pucheu-Plante & Mércier, 1983, reproduced by permission) and (B) open stoma (×2300). (From Blanke and Leyhe, 1987, reproduced by permission.)

cuticular ridges. The number, size, and structural complexity of the platelets increase during berry growth (Fig. 3.19D). Changes in platelet structure during berry maturation may correspond to changes in the chemical nature of the wax deposits. The cuticular covering is not necessarily entire, as evidenced by the occurrence of **microfissures** in the epidermis (Fig. 4.26A) and **micropores** in the cuticle (Fig. 4.26B). The occurrences of such features, as well as the cuticular thickness and wax plate structure, are affected by both hereditary and environmental factors. Thus, their influence on disease resistance is complex and may vary from season to season.

The hypodermis consists of a variable number of tightly packed layers of cells (Fig. 3.17). The cells are

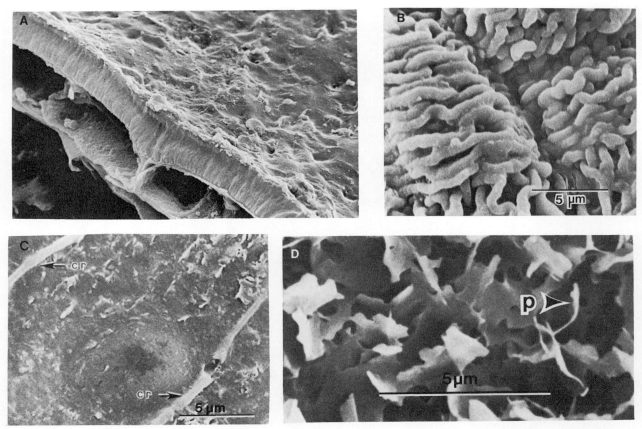

Figure 3.19   Berry cuticular structure. (A) Cuticle and epidermis cross section. (From Blaich *et al.*, 1984, reproduced by permission.) Cuticular ridges (cr) with wax removed, 2 weeks before anthesis (B) and at maturity (C), and (D) epicuticular wax plates (p). (From Rosenquist and Morrison, 1988, reproduced by permission.)

flattened, with especially thickened corner walls. The hypodermis commonly contains 10 cell layers but can vary from 1 to 17, depending on the cultivar. When young, the cells are photosynthetic. After *véraison,* they lose their chlorophyll and may accumulate anthocyanins and tannins, notably in the outermost layers. In "slip-skin" varieties, the cells of the flesh adjacent to the hypodermis become thinner and lose much of their pectinaceous cell wall material. This produces a zone of weakness, permitting the skin to separate readily from the flesh.

Most cells of the flesh are round to ovoid, containing large vacuoles. These pericarp cells make up about 65% of the berry volume. Cell division begins several days before anthesis and continues for about another 3 weeks. Subsequent growth results from cell enlargement. Fruit size appears to be predominantly regulated by cell enlargement, as the number of pericarp cells is little influenced by the external environment. The most rapid period of pericarp enlargement occurs shortly after anthesis. Although associated with rapid cell division, this period is also when water uptake is most rapid. This may explain the high sensitivity of fruit set to water stress during this period (Nagarajah, 1989).

The vascular tissue of the fruit develops directly from that of the ovary. It consists primarily of a series of **peripheral bundles** that ramify throughout the outer circumference of the berry and **axial bundles** that extend directly up through the septum. The **locules** of the fruit correspond to the ovule-containing cavities of the ovary and are almost undetectable in mature fruit. The locular space is filled either by the seeds or by growth of the septum.

Fruit abscission may occur either at the base of the pedicel or where the fruit joins the receptacle. Separation at the pedicel base (**shatter**) results from the localized formation of thin-walled parenchymatous cells. This commonly occurs if the ovules are unfertilized or the seeds abort early. Some varieties, such as 'Muscat Ottonel,' 'Grenache,' and 'Gewürztraminer,' are particularly susceptible to fruit abscission shortly after fertilization, a physiological disorder called inflorescence necrosis (*coulure*). In contrast, dehiscence that follows fruit ripening in muscadine cultivars results from an abscission layer that forms in the vascular tissue at the apex of the pedicel, next to receptacle.

SEED MORPHOLOGY

As noted earlier, seed development is important in synthesizing growth regulators vital to fruit growth. Therefore, fruit size is partially a function of the number of seeds formed. The seeds themselves make up only a small fraction of the fruit fresh weight. Equally, the embryo comprises only a small proportion of the seed volume (Levadoux, 1951). The embryo consists of two seedling leaves (**cotyledons**), a nascent shoot (**epicotyl**), and an embryonic root (**radicle**) at the tip of the **hypocotyl**. The embryo is surrounded by a nutritive **endosperm** which makes up the bulk of the seed. The endosperm is enclosed in a pair of seed coats (**integuments**), of which only the outer integument (**testa**) develops significantly. The testa consists of an hard inner section, a middle parenchymatous layer, and an outer papery component or epidermis.

CHEMICAL CHANGES DURING BERRY MATURATION

Owing to the major importance of both the quantitative and qualitative aspects of berry growth and maturation, the subject has received extensive investigation. However, as most studies have been directed primarily at improving grape yield and quality, no unified theory of berry development has emerged. More recently, several investigators have been pursuing the latter goal. A general theory of berry development might permit grape growers to predict the likely consequences of various viticultural options. In this regard, the following discussion will incorporate, where possible, physiological explanations of berry development. Because of the chemical nature of certain topics, some readers may wish to refer to Chapter 6 for clarification.

As noted above, a major shift in metabolism occurs simultaneously with berry color change (*véraison*). That these changes are controlled by plant growth regulators is beyond doubt. Regrettably, the specific actions of the growth regulators are unclear. Unlike animal hormones, plant growth regulators have many, and differing, actions. Their effects can depend on the tissue involved, its physiological state, and the relative and absolute concentration of the growth regulators present. This plasticity has made precise prediction of growth regulator effects exceedingly difficult.

During phase I of berry growth, auxin, cytokinin, and gibberellin content tends to increase. They undoubtedly stimulate cell division and enlargement up until *véraison*. Their role in subsequent cell enlargement is unclear, as their concentration declines thereafter. The decline in auxin and gibberellin concentration coincides with the increasing concentration of the growth inhibitor abscisic acid. The decline in auxin content also correlates with the drop in fruit acidity, accumulation of sugars, and rapid cessation of shoot growth (Alleweldt and Hifny, 1972). Subsequently, the level of gibberellins may rise. Ethylene does not appear to play a significant natural role in grape ripening, unlike the case in many fruits. Ethylene concentration rises slightly until *véraison*, but it declines back to a low level by maturity.

Because of their influence on berry development, growth regulators have been used in regulating fruit development and berry spacing within clusters. Artificial growth regulators have generally been more effective than their natural counterparts. Artificial hormones are less likely to be affected by feedback mechanisms that control the levels of natural growth regulators. Artificial auxins, such as benzothiazole-2-oxyacetic acid, can delay ripening up to several weeks when applied to immature fruit. Conversely, the artificial growth retardant methyl-2-(ureidooxy)propionate markedly hastens ripening when applied before *véraison* (Hawker *et al.*, 1981). Ripening also can be shortened marginally by applying growth regulators, such as ethylene or abscisic acid, after *véraison*.

During development, structural changes in the skin and vascular tissues influence the types and amount of substances transported into the berry. Degeneration of stomatal function and development of a thicker waxy coating combine to dramatically reduce transpirational water loss from the fruit (Blanke and Leyhe, 1987). This also reduces gas exchange between the berry and the atmosphere and increases the tendency of fruit to develop high temperatures on sun exposure and experience marked day/night temperature fluctuations.

During *véraison*, xylem vessels in the peripheral vascular tissue commonly rupture. This restricts water and mineral supply to the peripheral portions of the fruit. The most significantly affected mineral elements are those carried predominantly in the xylem, such as calcium. In contrast, potassium, which is transported equally in the phloem and xylem, continues to accumulate throughout ripening (Hrazdina *et al.*, 1984). During maturation, the xylem acts more as a conduit for the movement of water out of the berry than for water uptake (Lang and Thorpe, 1989). After *véraison*, the phloem becomes the primary source of water uptake. This probably results from sucrose transport to, and accumulation in, the fruit. The increasing importance of phloem transport during ripening is found in other fruits, including apples and tomatoes.

Sugars Sucrose is quantitatively the most significant organic compound translocated into the fruit. Not surprisingly, it is also the primary organic compound transported in the phloem. Very early in berry development, much of the carbohydrate used in growth is photosynthesized by the berry itself. As the berry enlarges and approaches *véraison*, however, the surrounding leaves take over as the major suppliers of carbohydrate. The trunk and arms of the vine also may act as significant carbohydrate sources. It has been estimated that up to 40% of the carbohydrate accumulated in fruit can, on occasion, be supplied from permanent woody parts of the grapevine (Kliewer and Antcliff, 1970).

Sugar uptake by the fruit becomes particularly marked following *véraison*. As this coincides with a pronounced decline in glycolysis, there is a dramatic increase in sugar accumulation. As primary shoot growth usually ceases about the same time, a major redirection of sugars to the fruit is possible. Root growth also declines following *véraison*, reducing its drain on photosynthate (Fig. 3.4). Nevertheless, a fully adequate explanation of the dramatic redirection of sugar translocation toward the fruit, the decline in berry glycolysis, and fruit sugar accumulation has yet to be formulated (see Coombe, 1992). However, the drop in sugar metabolism may involve a switch to malic acid respiration. Malic acid accumulates in considerable amounts early in fruit development, and it is a major intermediary in the TCA (tricarboxylic acid) cycle. Thus, malic acid may act as a substitute for glucose in respiration during the later stages of ripening.

On reaching the berry, sucrose is split into fructose and glucose. Although the enzyme invertase generates equal amounts of glucose and fructose, the levels of the two sugars are seldom equal in the fruit. In young berries, the proportion of glucose is generally higher. During ripening, the glucose/fructose ratio often falls, with fructose accumulation/retention being slightly higher than that of glucose. By maturation, if not before, fructose is usually the predominant sugar in grapes (Kliewer, 1967). The reasons for this disequilibrium are unknown, but it may originate from differential metabolism of glucose and fructose or selective fructose synthesis from malic acid.

As with other cellular constituents, the degree of sugar accumulation varies with the cultivar and prevailing environmental conditions. Depending on these factors, sugar content may vary from 12 to 28% by maturity. Further increases during overripening appear to reflect concentration, owing to water loss from the fruit, rather than additional sugar uptake. For winemaking, the desired sugar content generally ranges from about 21 to 25%, depending on the type of wine to be produced.

In some *Vitis* species and interspecies hybrids, a significant proportion of the translocated sucrose is stored as such. This is most marked in muscadine cultivars, where the sucrose concentration can reach 10 to 20% of the sugar content. In some French–American hybrids, su-

crose constitutes about 2% of the sugar content. Little sucrose is stored in the grapes of *Vitis vinifera*. Small amounts of other sugars are found in mature berries, such as raffinose, stachyose, melibiose, maltose, galactose, arabinose, and xylose. Their presence is considered insignificant, as they are neither metabolized by yeast cells nor influence the sensory characteristics of wine.

Sugar storage occurs primarily in cell vacuoles. Glucose and fructose accumulation occurs predominantly in the flesh, with lesser amounts being deposited in the skin. Often, the sugar content is highest in the flesh next to the skin, but this can vary with the cultivar and stage of ripening. In contrast, the small amount of sucrose that occurs in *V. vinifera* fruit is restricted to the axial vascular bundles and the skin adjacent to the peripheral vascular strands.

During ripening, much of the sugar comes from leaves located on the same side of the shoot as the fruit cluster, and directly above the cluster. Additional supplies come from the shoot tip as it develops (Fig. 3.20), from associated lateral shoots and from old wood. Estimates of the foliage cover required to fully ripen grape clusters vary considerably, from 6.2 cm^2/g (Smart, 1982) to 10 cm^2/g (Jackson, 1986); the latter appears to be the more common. The variation probably reflects varietal and environmental influences on fruit load and leaf photosynthetic efficiency.

Other substances transported into the berry include nitrogenous compounds (notably amino acids), organic phosphates, vitamins, growth regulators, and inorganic ions. Details on the types, amounts, and significance of many of these are unavailable. Tartaric and malic acids, once thought to be translocated into the developing berry, are now known to be largely synthesized *in situ*.

Acids Next to sugar accumulation, the decline in juice acidity is the most significant quantitative change in berry organic content during maturation. Tartaric and malic acids account for about 70 to 90% of the berry acid content. The remainder consists of variable amounts of organic acids, such as citric and succinic acids, phenolic acids, such as quinic and shikimic acids, and amino and fatty acids.

Although structurally similar, tartaric and malic acids are synthesized and metabolized differently. The slow production of tartaric acid from radioactively labeled carbon dioxide suggests that it is a secondary metabolite. In leaves, it appears to be derived from pretaric acid, via the pentose phosphate pathway. In fruit, it appears to be synthesized from galacturonic acid (Ruffner, 1982). In contrast, malic acid is an important intermediate in the TCA cycle. As such, it can be variously synthesized from sugars (via glycolysis and the TCA cycle) or via carbon dioxide fixation from phosphoenolpyruvate (PEP). Ma-

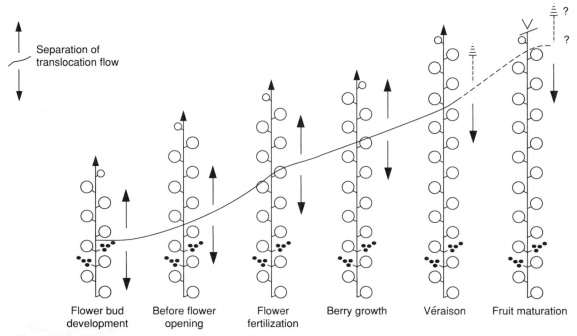

Separation of
translocation flow

Flower bud
development

Before flower
opening

Flower
fertilization

Berry growth

Véraison

Fruit maturation

Figure 3.20 Development in the translocation of photosynthate from leaves during shoot growth. (After Koblet, 1969, reproduced by permission.)

lic acid also can be readily metabolized in respiration or decarboxylated via oxaloacetate to PEP in the gluconeogensis of sugars. Not surprisingly, the malic acid content of berries changes more rapidly and strikingly than that of tartaric acid.

The degree and nature of conversions of acids can vary widely, depending on the cultivar and environmental factors during maturation. Nevertheless, several major trends are generally observed. After an initial, intense synthesis of tartaric acid in the ovary, both tartaric and malic acids slowly increase in concentration up until *véraison*. Subsequently, the amount of tartaric acid tends to stabilize, while that of malic acid declines. It is hypothesized that the initial accumulation of malic acid acts as a nutrient reserve to replace glucose as the major respired substrate following *vérasion*. This would explain the rapid drop in malic acid concentration during the later stages of ripening.

It has long been known that grapes grown in hot climates often metabolize all their malic acid prior to harvest. Conversely, grapes grown in cool climates may retain most of their malic acid into maturity. Although exposure to high temperature activates enzymes that catabolize malic acid, this alone is insufficient to explain the effect of temperature on berry malic acid content. Reduction in the level of synthesis, and possibly heightened gluconeogensis, also may play a role in the decline in malic acid content.

The drop in acid concentration during ripening is particularly marked when based on berry fresh weight. This results from the combined effects of the rate of water uptake during the second major berry enlargement period (phase III) exceeding the slow accumulation of tartaric acid, and the metabolism of malic acid.

Besides environmental influences, hereditary factors greatly affect berry acid content. Some varieties, such as 'Zinfandel,' 'Cabernet Franc,' 'Chenin blanc,' 'Syrah,' and 'Pinot noir,' are proportionally high in malic acid, whereas others, such as 'Riesling,' 'Sémillon,' 'Merlot,' 'Grenache,' and 'Palomino,' are high in tartaric acid content (Kliewer *et al.*, 1967).

Tartaric acid tends to accumulate primarily in the skin and outer flesh, with smaller amounts distributed throughout the flesh (Fig. 3.21). This difference may slowly diminish during ripening. In contrast, malic acid deposition is low in the skin and increases to a maximum near the berry center. As ripening advances, the differences across the flesh become less marked. By maturity, the malic acid concentration in the skin, though low, may surpass that in the flesh. This may result from malate metabolism initiating around the axial vascular bundles and progressing outward.

The free versus salt state of tartaric acid generally changes throughout maturation. Initially, most of the tartaric acid exists as the free acid. During ripening, progressively more tartaric acid combines with cations,

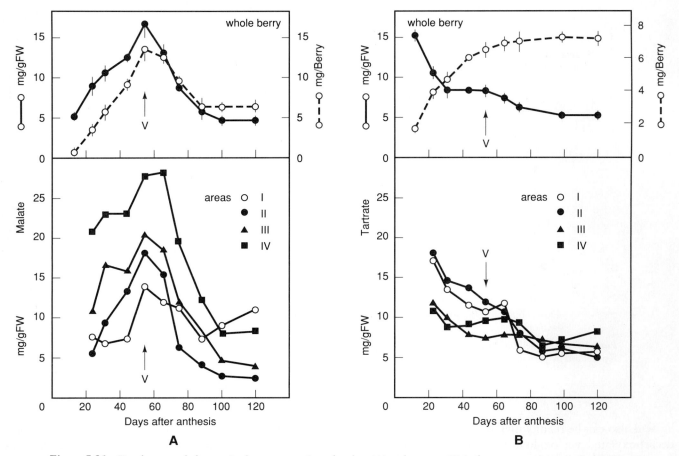

Figure 3.21 Developmental changes in the concentration of malate (A) and tartrate (B) in four concentric areas: I, skin; II, outer flesh; III, inner flesh; IV, area around the seeds; V, *véraison*. (From Possner and Kliewer, 1985, reproduced by permission.)

predominantly K^+. Salt formation with Ca^{2+} ions usually results in the deposition of calcium tartrate crystals in the vacuoles of skin cells (Ruffner, 1982). In contrast, malic acid generally remains as the free acid. The combination of the changing acid concentrations and their salt states results in the lowest titratable acidity occurring next to the skin. The highest titratable acidity normally exists next to the seeds in the center of the berry.

The factors regulating the synthesis of tartaric and malic acids are poorly understood. The effect of temperature on malic acid degradation has already been mentioned. Nutritional factors considerably influence berry acidity and acid content.

Increased potassium fertilization results in slight increases in the level of both acids. The increase is probably driven by the need of cells to produce additional acids to permit potassium uptake. Transport of K^+ into the cell requires the simultaneous export of H^+. Although potassium uptake may enhance acid synthesis, it also results in a rise in pH, presumably due to salt formation. This effect may not be simple, though, as the proportion of free tartaric acid in skin cells may rise during maturation, even though up to 40% of potassium accumulation

occurs in the skin. This apparent anomaly may result from the differential compartmentalization of potassium in the skin cells (Iland and Coombe, 1988).

Nitrogen fertilization enhances malic acid synthesis but may induce a decline in tartaric acid content. The increase in malic acid synthesis is probably involved with cytoplasmic pH stability. Nitrate reduction to ammonia tends to raise cytoplasmic pH by releasing OH^- ions. The synthesis of malic acid could neutralize this effect. However, because of the low level of malic acid ionization in grape cells, and its metabolization via respiration during ripening, the early synthesis of malic acid does not permanently prevent a rise in pH during ripening.

Although both malic and tartaric acids appear to be involved in maintaining a favorable ionic and osmotic balance in berry cells, they tend to be stored in cell vacuoles to avoid excessive cytoplasmic acidity. Here, the distinctive transport characteristics of the vacuolar membrane isolate the acids from the cytoplasm.

Potassium   As noted above, potassium uptake during maturation can affect both acid synthesis and ionization. The accumulation of potassium in berry tissues is poorly understood. Its uptake consistently increases after *véraison,* especially in the skin. The skin, often constituting about 10% of the berry weight, may contain up to 30 to 40% of the potassium content. Potassium is thought to provide osmotic balance between the skin cells, which are relatively low in sugar content, and the sugar-rich cells of the flesh. The high correlation between potassium uptake and sugar accumulation fits this interpretation. Because potassium accumulation in the vacuole increases its permeability, potassium uptake eventually may participate in the release of malic acid and its subsequent metabolism in the cytoplasm. Potassium is well known as an enzyme activator. However, the reason for the nonuniform distribution of potassium in skin cells remains unclear. Whether it relates to poorly understood aspects of skin anatomy, cytoplasmic compartmetalization, or differential functioning of potassium in various cells of the skin is unknown.

Phenols In red grapes, phenols constitute the third most significant group of organic compounds. Phenols not only generate the color, but give the characteristic taste and aging properties to red wines. White grapes have lower total phenol contents and do not synthesize anthocyanins. Aside from seed phenolics, both red and white grapes contain most of their phenolics in the skin. Because seed phenolics are slowly extracted, the primary source of grape phenolics in most wines is the skin. Hydroxycinnamic acid esters of tartaric acid form the predominant phenolic components of the flesh.

Pigmentation in most white and red cultivars is restricted to the outer hypodermal layers of the skin. In white grapes, the color comes from the presence of carotenes, xanthophylls, and flavonols such as quercetin. In red grapes, similar pigments occur, but the predominant pigments are anthocyanins. Carotenoids are located predominantly in plastids, whereas flavonoids are deposited in cell vacuoles.

Only the first hypodermal layer of the skin of red varieties is normally darkly pigmented. The next two hypodermal layers may contain smaller amounts of anthocyanins, and subsequent layers tend to be sporadically and weakly pigmented. Pigmentation seldom occurs deeper than the sixth hypodermal layer, except in *teinturier* varieties which possess weakly pigmented flesh (Hrazdina and Moskowitz, 1982). Otherwise, anthocyanins occur in the flesh only in overripe fruit: as skin cells senesce after ripening, their pigmentation can diffuse into the flesh. Because the anthocyanin content in the outer hypodermal layer(s) approaches saturation, the anthocyanins exist primarily in self-associated complexes or in complexes with tannins.

In addition to anthocyanins and tannins, red-skinned cultivars contain flavonols and hydroxycinnamic acid esters of tartaric acid. The predominant flavonol in *V.*

vinifera cultivars is kaempferol, whereas in *V. labrusca* cultivars quercitin appears to dominate. The flavonols are commonly glycosidically linked to glucose, rhamnose, and glucuronic acid. The predominant hydroxycinnamic acid ester is caffeoyl tartrate, with smaller amounts of coumaroyl tartrate and feruloyl tartrate. As with most other chemical constituents, the specific concentrations can vary widely from season to season and cultivar to cultivar.

Phenol synthesis begins with the condensation of erythrose, derived from the pentose phosphate pathway (PPP), and phosphoenolpyruvate, derived from glycolysis. The phenylalanine so generated is converted to cinnamic acid through the shikimic acid pathway. Subsequent conversion of cinnamic acid generates various nonflavonoid phenols, or it may be incoporated as the B ring in the synthesis of flavonoid phenols (Hrazdina *et al.,* 1984). The A ring of flavonoid phenols comes from acetate via malonyl-CoA. The initial anthocyanin synthesized is cyanidin-3-glucose. All other anthocyanins are derived from cyanidin (Roggero *et al.,* 1986).

Phenol synthesis begins early during berry development. Some anthocyanins apparently are synthesized early, but most production involves other flavonoid or nonflavonoid phenols. Anthocyanin synthesis becomes pronounced only after *véraison.* The specific timing of anthocyanin synthesis depends on conditions such as temperature and sugar accumulation and on hereditary factors. After reaching a maximum, anthocyanin concentration tends to decline slightly.

Although the anthocyanin content of red grapes increases during ripening, the proportion of the various anthocyanins, and their acylated derivatives, can change markedly (González-SanJosé *et al.,* 1990). As the final distribution can significantly influence the hue, intensity, and stability of the color of a wine, these changes have great practical importance. Methylation of the oxidation-sensitive *o*-diphenol sites improves anthocyanin stability. Correspondingly, the monophenolic anthocyanins, peonin and malvin, are often the most stable. Susceptibility to oxidation also is decreased by bonding with sugars and acyl groups (Robinson *et al.,* 1966).

The precise conditions that initiate anthocyanin synthesis during *véraison* are currently unknown. One hypothesis suggests that sugar accumulation provides the substrate needed for synthesis. The marked correlation between sugar accumulation in the skin and anthocyanin synthesis is consistent with this hypothesis. However, sugar accumulation may act indirectly through its effect on the osmotic potential of the skin cells. In culture, high osmotic potentials in the medium can induce anthocyanin synthesis and methylation in grape cells (Do and Cormier, 1991). Alternately, anthocyanin synthesis may be triggered by changes in the potassium and calcium contents of the skin. Potassium accumulates in the skin following *véraison*, and its cellular distribution is as nonuniform as initial anthocyanin synthesis. Conversely, calcium content decline is inversely correlated with anthocyanin synthesis.

Phenol synthesis tends to decline and may cease following *véraison*. As a result, the concentrations of hydroxycinnamic acid esters and catechins usually decline, probably due to dilution during berry enlargement. Nevertheless, some low molecular weight phenolic compounds (i.e., benzoic and cinnamic acid derivatives) may increase during ripening (Fernández de Simón *et al.*, 1992). With flavonoid tannins, the decline in solubility following procyanidin polymerization contributes to a drop in fruit astringency.

In white grapes, the phenolic changes during ripening primarily involve nonflavonoid hydroxycinnamic tartrates and catechins. Although the dynamics of hydroxycinnamic tartrate metabolism are unclear, the levels of the compounds tend to fall strikingly during ripening (Lee and Jaworski, 1989). As the degree and nature of the decline are influenced by genetic and environmental factors, their role in oxidative browning can change markedly from variety to variety and year to year. Catechin levels may rise following *véraison* but decline again to low levels by maturity. Catechin monomers tend not to polymerize into large tannins in white cultivars.

The predominant phenols in seeds are flavonoids. Although they constitute the major source of phenols in grapes, they seldom are extracted in significant quantities in most wines. After a pronounced increase early in berry development, the catechin content declines as the concentration of their polymers, the procyanidins, increase.

Pectins One of the more obvious changes during berry ripening is the softening of berry texture. Softening is important as it facilitates the release of juice from the flesh and the extraction of phenolic and flavor components from the skin. Softening is associated with an increase in the content of water-soluble pectins and the concomitant loosening of the bonds between fruit cells. The changes may result from enzymatic degradation or a decline in calcium content of cell wall pectins. Calcium is important in maintaining the solid pectin matrix that characterizes most plant cell walls, and calcium uptake essentially ceases following *véraison*. The role of cellulases and hemicellulases in addition to that of pectinases, in softening is unknown.

Lipids The lipid content of grapes consists of cuticular and epicuticular waxes, cutin fatty acids, membrane phospho- and glycolipids, and seed oils. Seed oils are important as an energy source during seed germi-

nation, but they are seldom found in juice or wine. Their presence in significant amounts could generate a rancid odor. Because the lipid concentration of the berry changes little during growth and maturation, synthesis appears to be equivalent to berry growth.

The most abundant fraction of grape lipids are the phospholipids, followed by neutral lipids. Glycolipids constitute the least common group. The predominant fatty components of grape lipids are the long-chain fatty acids linoleic, linolenic, and palmitic acids (Roufet *et al.*, 1987).

Most of the cuticular layer is deposited in folds by anthesis and expands during subsequent berry growth. In contrast, the epicuticular wax plates are deposited throughout berry growth, commencing about anthesis (Rosenquist and Morrison, 1988). Most of the epicuticular coating consists of hard wax, composed primarily of oleanolic acid. The softer, underlying cuticle contains the fatty acid polymer called **cutin**. It consists primarily of C_{16} and C_{18} fatty acids, such as palmitic, linoleic, oleic, and stearic acids and their derivatives. Cutin is embedded in a complex mixture of relatively nonpolar waxy compounds.

About 30 to 40% of the long-chain fatty acids of the berry are located in the skin. They could act as an important source of unsaturated fatty acids for synthesis of yeast cell membranes during wine fermentation. Although the relative concentrations of the fatty acids change little during maturation, the concentration of individual components may alter considerably. The most significant change involves the decline in linolenic acid content during ripening (Roufet *et al.*, 1987). This could reduce the involvement of its oxidation products in the generation of herbaceous odors in wine.

Another important class of lipids are the carotenoids. In red varieties, the concentration of both major (β-carotene and lutein) and minor (5,6-epoxylutein and neoxanthin) carotenoids falls markedly after *véraison* (Razungles *et al.*, 1988). The decline appears to be the result of dilution during berry growth in phase III, following cessation of synthesis. Because of their insolubility, most carotenoids remain with the skins and pulp following crushing. Nevertheless, enzymatic and acidic hydrolysis during maceration may release water-soluble carotenoid derivatives. These can include important aromatic compounds such as damascenone and β-ionone. For example, the increasing concentration of norisoprenoids is correlated with a decrease in carotenoid content during the maturation of grapes such as 'Muscat of Alexandra' (Razungles *et al.*, 1993).

Nitrogen-Containing Compounds Although soluble proteins may induce haze formation in wine, comparatively little is known about their precise chemical nature, or the conditions affecting their production during berry ripening. The concentration of soluble proteins in grapes can increase up to five times during maturation, reaching levels from 200 to 800 mg/liter. Besides yearly variation, cultivars differ considerably in protein contents (Tyson *et al.*, 1982).

The free amino acid content of grapes also rises at the end of ripening, the primary amino acids being proline and arginine. Their concentration can vary up to 20-fold, depending on the variety and fruit maturity (Sponholz, 1991). Their ratio may shift during maturation from a preponderance of arginine to proline (Polo *et al.*, 1983), or the reverse (Bath *et al.*, 1991). In many plants, proline acts as a cytoplasmic osmoticum (Aspinall and Paleg, 1981), similar to potassium. Whether proline accumulation has the same role in grapes is unknown.

Amino acids usually constitute the principle group of soluble organic nitrogen compounds released during crushing and pressing. Nevertheless, up to 80% of the berry nitrogen content may remain with the pomace (structural components of the fruit).

Changes during ripening also occur in the levels of inorganic nitrogen, primarily ammonia. Ammonia may show either a decline (Solari *et al.*, 1988) or an increase during ripening (Gu *et al.*, 1991). Although a decline might appear to negatively affect fermentation, it probably is more than compensated by the simultaneous increase in the amino acid content. Amino acids are incorporated by yeasts at up to 5 to 10 times the rate of ammonia. Their incorporation also alleviates the requirement of diverting metabolic intermediates to amino acid synthesis. However, as individual amino acids are accumulated by yeast cells at different rates, the specific amino acid content of the juice could significantly affect its fermentability.

Aromatic Compounds Until recently, most research on ripening focused on the major constituents of grapes, such as sugars and acids. Investigation of aroma compounds was beyond the resolving power of the equipment available. Thus, long-held views about the synthesis and location of aromatic compounds in grapes remained unsubstantiated. The development and refinements in gas chromatography are permitting these beliefs to be tested. Of greater practical importance, though, is the possibility that harvesting may be timed to coincide with the accumulation of important aroma compounds in the fruit.

Most of the current research on grape aroma has involved Muscat cultivars, whose aroma is based primarily on monoterpenes. In general, monoterpenes accumulate during ripening as expected. Along with synthesis, however, they may subsequently be converted to nonvol-

atile forms. The nonvolatile forms may exist as glycosides, oxidized derivatives, or polymers of the parent compound (Wilson *et al.*, 1986). Thus, mature grapes often generate less aroma than the level of their aromatic compounds would suggest. Similar trends have been noted with norisoprenoids and nonflavonoid phenols (Strauss *et al.*, 1987a).

Details of the accumulation and "masking" of aroma compounds are most well known in Muscat varieties. In 'Muscat of Alexandria' (Fig. 3.22), the very young berry possesses a high concentration of geraniol. This rapidly declines with a subsequent rise in glycosidically bound geraniol. After *véraison*, the concentration of the major terpene, linalool, rises. During ripening, the proportion of glycosidically bound terpenes increases. After reaching a maximum at maturity, both the free and bound fractions tend to decline.

Although most terpenes accumulate in the skin, notably geraniol and nerol, not all do. For example, free linalool and diendiol I occur more frequently in the flesh than the skin. Often, the proportion of free and glycosidically bound terpene differs between the skin and the flesh.

'Gewürztraminer' and 'Riesling' also show nonuniform terpene distributions in the fruit. In 'Riesling,' particular attention has been paid to the norisoprenoid content. Compounds such as TDN (1,1,6-trimethyl-1, 2-dihydronaphthalene), vitispirane, and damascenone are localized predominantly in the flesh. They primarily occur in glycosidically bound forms, with only trace amounts of damascenone occurring as free, volatile molecules. As with monoterpenes, their concentrations increase during ripening (Strauss *et al.*, 1987b).

In addition to monoterpenes and norisoprenoids, other aromatic compounds are present in nonvolatile, conjugated, or glycosidically bound forms. One group consists of low molecular weight aromatic phenols. Examples that have been found in the variety 'Riesling' include vanillin, propiovanillone, methyl vanillate, zingerone, and coniferyl alcohol. Even phenolic alcohols, such as 2-phenylethanol and benzyl alcohol, may become glycosylated during ripening. Conjugated aromatic phenols have also been isolated from 'Sauvignon blanc,' 'Chardonnay,' and 'Muscat of Alexandria' (Strauss *et al.*, 1987a).

Unlike the compounds just mentioned, some aroma compounds remain in the free volatile form right up to maturity. For example, methyl anthranilate increases in concentration thoughout the later stages of maturation (Robinson *et al.*, 1949). In other instances, the concentrations of impact compounds decline during maturation, for example, methyloxypyrazines (Allen *et al.*, 1989).

In several varieties, maturity affects the production of

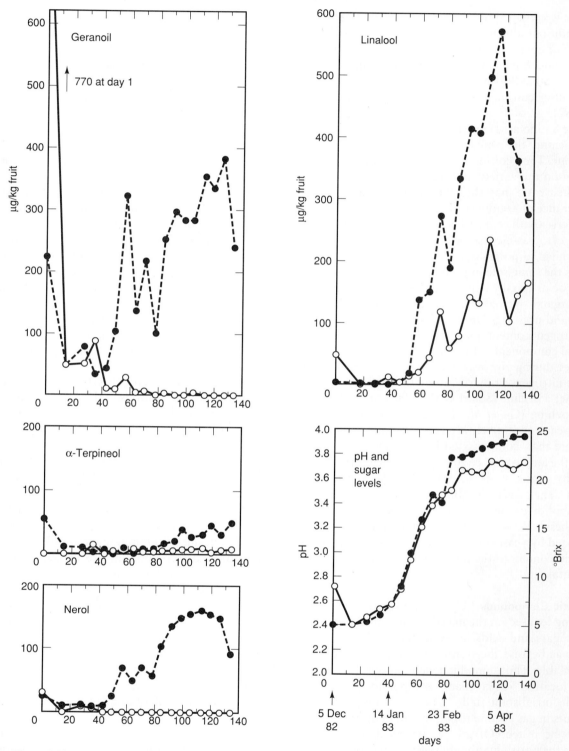

Figure 3.22 Changes in juice concentration of free (solid lines) and glycosidically bound (dashed lines) monoterpenes in developing 'Muscat of Alexandria' grapes. (Reprinted by permission from Wilson *et al.*, 1984. Copyright 1984 American Chemical Society.)

fusel alcohols and esters during fermentation. This has normally been interpreted as a result of their differing sugar contents. However, as Fig. 7.21 shows, other changes during ripening also must affect the formation of esters during fermentation.

Comparable studies to those on white grapes have not been conducted on red grapes. In some varieties, however, aroma development has been correlated with ripeness and sun exposure. In 'Cabernet Sauvignon,' the total volatile component of the fruit was found to be about equally distributed between the skin and flesh (Bayonove *et al.*, 1974). Because the skin constitutes only about 5 to 12% of the fruit mass, its aromatic concentration must be considerably higher than that of the flesh.

CULTURAL AND CLIMATIC INFLUENCES ON BERRY MATURATION

Any factor affecting grapevine growth and health directly or indirectly influences the ability of the vine to nourish and ripen its fruit. Consequently, most viticultural practices are directed at regulating those factors to achieve the maximum yield, consistent with quality and long-term grapevine productivity. Although macroclimatic factors directly affect berry maturation, they are often beyond the control of the grape grower.

Yield   Because of the obvious importance of grape yield to vineyard commercial success, much research has focused on increasing fruit production. However, yield increases can reduce the ability of the vine to mature the fruit, or its potential to produce subsequent crops. This has led to a debate over the appropriate balance between grape yield, fruit quality, and long-term vine health.

In France, yield is considered so directly connected with wine quality that maximum crop yields have been set for Appellation Control regions. Although yield is undoubtedly important, concern about yield can deflect attention from other factors of equal or greater importance to wine quality. The quadrupling of yield in German vineyards during the twentieth century (Anonymous, 1979) is a dramatic, but not isolated, example of yield increases without a noticeable loss in wine quality. Elimination of viral infection often improves the ability of vines to produce and ripen fruit. Clonal selection has also identified lines better able to produce more and better quality fruit. Improved fertilization, irrigation, and weed and pest control, combined with appropriate canopy management, can often further improve the capacity of a vine to produce fully ripened fruit.

Definite overcropping has long been associated with reduced wine quality. Although specific cause/effect relationships are poorly understood, overcropping delays fruit maturity, retains acidity, retards anthocyanin synthesis and sugar accumulation, and limits flavor development. Achieving the "ideal" yield is one of the most perplexing demands of grape growing. It depends not only on the grape variety and soil characteristics, but also on the type of wine desired and the particularities of the prevailing climate.

Although overcropping reduces grape and wine quality, may suppress subsequent yield, and may shorten vine life, low yield does not necessarily improve quality. Undercropping can prolong shoot growth and leaf production, increase shading, depress fruit acidity, and undesirably influence berry nitrogen and inorganic nutrient contents. Reduced yield also tends to induce the vine to produce larger berries. The diminished skin/flesh ratio can negatively effect properties derived from the skin. Even small changes in berry size can have detectable effects on wine quality. Figure 3.23 shows that, although reduced yield increases the level of good wine aroma, taste intensity also rises. In this instance, the increased taste intensity of the red wine partially offset the benefits of the enhanced aroma. Equally, an overly intense aroma can destroy the desirable subtlety of a wine, resulting in diminished appreciation.

It is generally believed that wines produced from vines bearing light to intermediate crops are preferred (Cordner and Ough, 1978; Gallander, 1983; Ough and Nagaoka, 1984). Although the yield/quality ratio may appear relatively constant within the mid-yield range, this has not consistently been found (Fig. 3.23).

An improved understanding of the yield/quality relationship will require better insight into the factors that lead to grape and wine quality. In addition, it is often difficult to separate yield effects from the impact of canopy shading often associated with high yield and vine vigor. Canopy shading can influence fruit quality even when the grapes themselves are not shaded (Schneider *et al.*, 1990). It is also essential that studies on yield/quality relationships investigate more than just effects on young wines. Yield influences on wine aging potential also require investigation.

Another commonly held belief is that "stressing" the vines increases their ability to produce grapes yielding fine quality wine. This view may have arisen from the reduced vigor/improved grape quality associated with the low nutrient status of some renowned European vineyards. The consequential balancing of the leaf/fruit ratio could improve the microclimate within and around the vine, increase photosynthetic efficiency, and promote fruit maturation and flavor development. Appropriate balancing of the vegetative and reproductive growth of the vine also appears to affect favorably fruit acidity and pH.

Sunlight   As the energy source for photosynthesis, sunlight is without doubt the most important climatic factor affecting berry development. Sunlight contains both visible and infrared (heat) radiation, and much of the absorbed radiation is released as heat. Thus it is often difficult to separate light from temperature effects. Because of the early cessation of berry photosynthesis, most of the effects of light on berry maturation are either thermal or phytochrome induced.

Because of the selective absorption of light by chlorophyll, there is a marked increase in the proportion of far-red light in vine canopies. This shifts the proportion of physiologically active phytochrome (P_{fr}) from 60% in sunlight to below 20% in shade (Smith and Holmes,

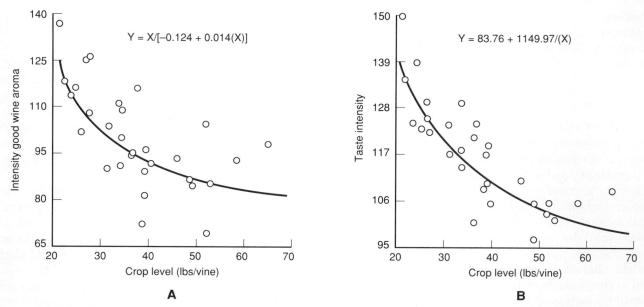

Figure 3.23 Relationship between crop level of 'Zinfandel' vines and (A) good wine aroma and (B) taste intensity. (From Sinton *et al.*, 1978, reproduced by permission.)

1977). Evidence suggests that this may delay the initiation of anthocyanin synthesis, decrease sugar accumulation, and increase ammonia and nitrate content in the fruit (Smart *et al.*, 1988). The red/far-red balance of sunlight and occasionally ultraviolet radiation are involved in the control of flavonoid biosynthesis in several plants (see Hahlbrock, 1981). Whether sunflecks can offset the effect of shading is unknown. It is theoretically possible, as P_r is more efficiently converted to P_{fr} by exposure to red light than is the reverse reaction by far-red radiation. As some plants also show high intensity blue and far-red light responses, the effects of leaf shading may be more complex than the red/far-red balance alone might suggest.

Although fruit exposure is essential for anthocyanin synthesis in most red grape varieties, full sunlight is not always required (Kliewer, 1977b). Occasionally, coloration is almost as intense in berries covered to prevent direct light exposure as in sun-exposed fruit (Weaver and McCune, 1960). Deeply pigmented varieties seem less dependent on light exposure than moderately pigmented cultivars such as 'Pinot noir.' Berry phenol content also tends to follow the trends set by anthocyanin synthesis in response to light exposure. The influence of light exposure on the specific anthocyanin composition of grapes is unknown. Because anthocyanins differ in susceptibility to oxidation, changes in composition during ripening may be more significant to wine color stability than total anthocyanin accumulation.

Berry pigmentation, and its change following *véraison*, can influence the P_r/P_{fr} ratio in maturing fruit

(Blanke, 1990). Whether this significantly influences fruit ripening is unknown, but it could affect the activity of light-activated enzymes such as phosphoenolpyruvate carboxylase (PEPC), a critical enzyme in the metabolism of malic acid (Lakso and Kliewer, 1975).

Sun-exposed fruit may show higher titratable acidity and concentrations of tartaric and malic acid than shaded fruit (Smith *et al.*, 1988). The increase in acidity may or may not be reflected in a decline in pH and potassium accumulation. Shading also tends to increase magnesium and calcium accumulation (Smart *et al.*, 1988). The lower malic acid content occasionally observed in sun-exposed fruit may result from sun-induced heating. The lower sugar concentration in shaded fruit is thought to result from dilution associated with increased berry volume (Rojas-Lara and Morrison, 1989). Sun exposure also may hasten ripening.

Berry size generally increases with fruit and/or leaf shading, but it is little affected when only the fruit is shaded (Morrison, 1988). These differences may result from reduced transpiration from the fruit under shade conditions.

There are several reports of sun exposure affecting grape aroma. Usually the level of grassy or herbaceous odors in 'Sauvignon blanc' (Arnold and Bledsoe, 1990) and 'Sémillon' (Pszczolkowski *et al.*, 1985) are reduced by sun exposure, while other fruit flavors may be unaffected or augmented. Sun exposure also tends to increase the level of norisoprenoids (Marais *et al.*, 1992a) and monoterpenes (Reynolds and Wardle, 1989, Smith *et al.*, 1988) in several cultivars. The higher levels of monoterpenes generated in sun-exposed 'Sémillon' and 'Sauvignon blanc' grapes appear to mask the vegetative odors produced by alkylmethyl pyrazines (Reynolds and Wardle, 1991). In 'Cabernet Sauvignon,' aromatic differences between shaded and sun-exposed fruit could be detected but the differences were not consistently associated with verbal descriptors (Morrison and Noble, 1990). These effects also may depend on the degree and timing of leaf removal. For 'Sauvignon blanc,' the effect was most marked when leaves around the cluster were removed several weeks before *véraison* (Arnold and Bledsoe, 1990).

In addition to direct light-activated influences, sun exposure indirectly influences fruit maturation through associated heating effects. As berries mature, their stomata cease to open, and the cooling induced by transpirational water loss is reduced. Although the grape skin is more water permeable than those of many other fleshy fruits (Nobel, 1975), and thereby could show considerable transpirative cooling, sun exposure occasionally can generate temperatures up to 15°C above ambient (Smart and Sinclair, 1976). The level of heating depends primarily on the light intensity, angle of incidence, and wind velocity. Densely packed clusters of dark fruit show greater heating than loose clusters of light-colored fruit. Shaded fruit is often cooler than the surrounding air owing to the reflection of heat by the canopy (Fig. 5.20).

Temperature   Temperature is known to markedly influence enzyme function and cell membrane permeability. However, how these apply to the many effects of temperature on fruit development is unclear. Some of the influences are summarized in Fig. 3.24.

Although the effect of temperature on malate respiration is well known, its precise mode of action is unknown. While temperature enhances the action of malic enzymes, this influence appears inadequate to explain the marked affect of temperature on malate respiration. Whether warm temperatures induce the leakage of malic acid from cell vacuoles appears not to have been studied. In contrast, tartaric acid content is little affected by temperatures below 30°C.

Temperature has a pronounced effect on anthocyanin synthesis and stability in grapes. In several varieties, synthesis is favored by warm daytime temperatures and cool nights (20° to 25°C and 10° to 15°C, respectively). However, varieties such as 'Kyoho' may show optimal anthocyanin synthesis at temperatures as low as 15°C. Cool conditions can result in some white varieties remaining green up to harvest, whereas high temperatures (≥35°C) may suppress or inhibit anthocyanin synthesis in some red cultivars (Kliewer, 1977a).

The high temperatures that can develop in sun-exposed fruit also can lead to color loss in some varieties. This may be more important in varieties that terminate anthocyanin synthesis early during berry development.

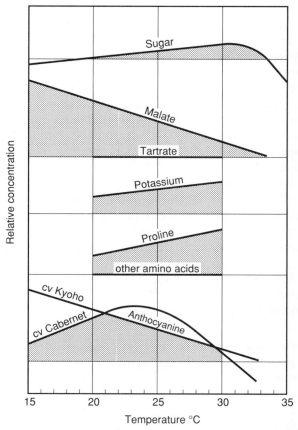

Figure 3.24   Summary of some of the effects of temperature on the concentration of chemcial compounds in grapes. (Data from Buttrose *et al.*, 1971; Hale, 1981; Hale and Buttrose, 1973, 1974; Kliewer, 1971, 1977a; Kliewer and Lider, 1970; Kobayashi *et al.*, 1965a,b; Lavée, 1977; Radler, 1965; Tomana *et al.*, 1979a,b; graphed by Coombe, 1987b, reproduced by permission.)

Whether temperature influences the compositional change in grape anthocyanins during maturation (González-SanJosé *et al.*, 1990) has been little investigated.

Increasing temperature generally has a beneficial influence on sugar accumulation, except above 32° to 35°C. The upper limit may result from the disruption of photosynthesis at high temperatures (Kriedemann, 1968). In overmature fruit, high temperatures can increase the sugar concentration by enhancing water loss, and thereby give the impression that the sugar content has risen.

Warm conditions can increase the amino acid content of developing fruit. Most of the change is expressed in the accumulation of proline, arginine, or other amino acid amides (Sponholz, 1991). Potassium accumulation usually increases at warmer temperatures (Hale, 1981). Whether this is driven by the increased uptake of water, sugar accumulation, or some other factor is uncertain.

The effect of temperature on the accumulation of aromatic compounds has thus far drawn little attention. The common belief is that cool conditions enhance flavor development, or its retention. For example, production of aromatic compounds in 'Pinot noir' is reported to be higher in grapes grown in long, cool seasons than in short, hot seasons (Watson *et al.*, 1991). However, cool conditions suppressed the expression of berry aromas and fruit flavors, and enhanced the development of less desirable vegetable odors, in 'Cabernet Sauvignon' (Heymann and Noble, 1987). Cool climate conditions also appear to limit the production/accumulation of 1,1,6-trimethyl-1,2-dihydronaphthaline (TDN) in 'Riesling' grapes (Marais *et al.*, 1992b).

High temperatures, during or shortly after pollination, negatively affect fertility and fruit development. Reduced fertility may arise from direct effects on inflorescence suppression, or indirectly, from increasing vine water stress (Matthews and Anderson, 1988). Heat stress early in fruit development appears to irreversibly restrict berry enlargement (Hale and Buttrose, 1974), leading to smaller fruit production as well as delayed maturation.

Inorganic Nutrients   Low soil nitrogen content appears to favor anthocyanin synthesis, whereas high nitrogen levels suppress production. These effects are probably consequences of the influence of nitrogen on vegetative growth and canopy shading and the impact on fruit maturation and sugar accumulation. Increasing nitrogen availability also increases the accumulation of free amino acids, notably arginine.

Phosphorus and sulfur deficiencies are known to increase anthocyanin content in several plants. In grapevines, phosphorus deficiency induces interveinal reddening in leaves, but how it affects berry anthocyanin synthesis is unclear. Nevertheless, high potassium levels can reduce berry pH and thereby lower fruit color and color stability in red wines.

Water   It is often believed that mild water stress following *véraison* is beneficial to grape quality. This may result from a reduction in vegetative growth, leading to a more balanced leaf/fruit ratio. A direct effect on the acceleration of ripening also appears possible (Freeman and Kliewer, 1983). However, water stress early in the season can lead to depressed berry set and reduced berry size. Thus, it may be advisable to limit water stress by irrigation up to *véraison* but subsequently irrigation should be restricted up until harvest. In "excess," irrigation can more than double fruit yield, increase fruit size, and delay maturity, all potentially detrimental to fruit quality.

The effect of irrigation on sugar content is usually minor (<10%). Although typically resulting in a decline, irrigation can increase grape sugar content under certain conditions. Commonly, irrigation results in an initial increase in grape acidity, presumably by permitting sufficient transpirational cooling to reduce malic acid respiration. However, there is seldom an accompanying drop in pH. This anomaly may partially be explained by the dilution associated with fruit enlargement. When harvest is sufficiently delayed, the high acidity of the fruit may dissipate, presumably from the slower rate of malate respiration being compensated by the longer maturation period.

The poor wine pigmentation often associated with irrigation may result from reduced anthocyanin synthesis. However, poor wine color also can arise from dilution associated with increased berry volume. Poor coloration in red cultivars is one of the prime reasons why irrigation is generally ill-advised after *véraison*.

Little is known about the effects of irrigation on aroma development. It has been noted, though, that irrigation diminishes the rate of monoterpene accumulation and the development of juice aroma in 'Riesling' berries (McCarthy and Coombe, 1985).

Suggested Readings

General References

Champagnol, F. (1984). "Élements de Physiologie de la Vigne et de Viticulture Générale." Published by F. Champagnol, Montpellier, France.

Huglin, P. (1986). "Biologie et Écologie de la Vigne." Payot Lausanne, Paris.

Mullins, M. G., Bouquet, A., and Williams, L. E. (1992). "Biology of the Grapevine." Cambridge Univ. Press, Cambridge.

Pongrácz, D. P. (1978). "Practical Viticulture." David Philip, Cape Town, South Africa.

Weaver, R. J. (1976). "Grape Growing." Wiley, New York.

Winkler, A. J., Cook, J. A., Kliewer, W. M., and Lider, L. A. (1974). "General Viticulture." Univ. of California Press, Berkeley.

Root System

Linderman, R. G. (1988). Mycorrhizal interactions with the rhizosphere microflora: The mycorrhizosphere effect. *Phytopathology* **78**, 366–371.

Menge, J. A., Raski, D. J., Lider, L. A., Johnson, E. L. V., Jones, N. O., Kissler, J. J., and Hemstreet, C. L. (1983). Interaction between mycorrhizal fungi, soil fumigation, and growth of grapes in California. *Am. J. Enol. Vitic.* **34**, 117–121.

Pratt, C. (1974). Vegetative anatomy of cultivated grapes. A review. *Am. J. Enol. Vitic.* **25**, 131–148.

Richards, D. (1983). The grape root system. *Hortic. Rev.* **5**, 127–168.

Van Zyl, J. L. (compiler) (1988). "The Grapevine Root and Its Environment." Dept. Agric. Water Supply Tech. Commun. No. 215. Pretoria, South Africa.

Shoot System

Gerrath, J. M., and Posluszny, U. (1988). Morphological and anatomical development in the Vitaceae. I. Vegetative development in *Vitis riparia. Can. J. Bot.* **66**, 209–224.

Koblet, W., and Perret, P. (1982). The role of old vine wood on yield and quality of grapes. *Grape Wine Centennial Symp. Proc. 1980*, pp. 164–169. Univ. of California, Davis.

May, P. (1987). The grapevine as a perennial plastic and productive plant. *Proc. 6th Aust. Wine Ind. Tech. Conf.* (T. Lee, ed.), pp. 40–49. Australian Industrial Publ., Adelaide, Australia.

Morrison, J. C. (1991). Bud development in *Vitis vinifera*. *Bot. Gaz.* 152, 304–315.

Pratt, C. (1974). Vegetative anatomy of cultivated grapes. A review. *Am. J. Enol. Vitic.* 25, 131–148.

Photosynthesis and Transpiration

Düring, H. (1987). Stomatal responses to alterations of soil and air humidity in grapevines. *Vitis* 26, 9–18.

Düring, H. (1988). CO_2 assimilation and photorespiration of grapevine leaves: Responses to light and drought. *Vitis* 27, 199–208.

Koblet, W. (1985). Influence of light and temperature on vine performance in cool climates and applications to vineyard management. *Int. Symp. Cool Climate Vitic. Enol.* (B. A. Heatherbell, P. B. Lombard, F. W. Bodyfelt, and S. F. Price, eds.), OSU Agric. Exp. Stn. Tech. Publ. No. 7628, pp. 139–157. Oregon State Univ., Corvallis, Oregon.

Kriedemann, P. E. (1977). Vineleaf photosynthesis. *Int. Symp. Quality Vintage*, pp. 67–87. Oenol. Viticult. Res. Inst., Stellenbosch, South Africa.

Smart, R. E. (1983). Water relations of grapevines. *In* "Water Deficits and Plant Growth, Volume VII. Additional Woody Crop Plants." (T. T. Kozlowski, ed.), pp. 137–196. Academic Press, New York.

Reproductive System

Branties, N. B. M. (1978). Pollinator attraction of *Vitis vinifera* subsp. *silvestris*. *Vitis* 17, 229–233.

Carbonneau, A. (1983a). Stérilités mâle et femelle dans le genre *Vitis*. I. Modélisation de leur hérédité. *Agronomie* 3, 635–644.

Carbonneau, A. (1983b). Stérilités mâle et femelle dans le genre *Vitis*. II. Conséquences en génétique et sélection. *Agronomie* 3, 645–649.

Gerrath, J. M., and Posluszny, U. (1988). Morphological and anatomical development in the Vitaceae. II. Floral development in *Vitis riparia*. *Can. J. Bot.* 66, 1334–1351.

Pratt, C. (1971). Reproductive anatomy of cultivated grapes. A review. *Am. J. Enol. Vitic.* 21, 92–109.

Srinivasan, C., and Mullins, M. G. (1981). Physiology of flowering in the grapevine—A review. *Am. J. Enol. Vitic.* 32, 47–63.

Swanepoel, J. J., and Archer, E. (1988). The ontogeny and development of *Vitis vinifera* L. cv. Chenin blanc inflorescence in relation to phenological stages. *Vitis* 27, 133–141.

Berry Maturation

Champagnol, F. (1986). L'acidité des moûts et des vins. 2ᵉ Partie. Facteurs physiologiques et agronomiques de variation. *Prog. Agric. Vitic.* 103, 361–374.

Coombe, B. G. (1992). Research on development and ripening of the grape berry. *Am. J. Enol. Vitic.* 43, 101–110.

Coombe, B. G., and Iland, P. F. (1987). Grape berry development. *Proc. 6th Aust. Wine Ind. Tech. Conf.* (T. Lee, ed.), pp. 50–54. Australian Industrial Publ., Adelaide, Australia.

Darriet, Ph., Boidron, J.-N., and Dubourdieu, D. (1988). L'hydrolysis des hétérosides terpéniques du Muscat à petits grains par les enzymes périplasmiques de *Saccharomyces cerevisiae*. *Connaiss. Vigne Vin.* 22, 189–195.

Hrazdina, G., Parsons, G. F., and Mattick, L. R. (1984). Physiological and biochemical events during development and maturation of grape berries. *Am. J. Enol. Vitic.* 35, 220–227.

Romeyer, F. M., Macheix, J. J., and Sapis, J. C. (1986). Changes and importance of oligomeric procyanidins during maturation of grape seeds. *Phytochemistry* 25, 219–221.

Ruffner, H. P. (1982a). Metabolism of tartaric and malic acids in *Vitis*: A review. Part A. *Vitis* 21, 247–259.

Ruffner, H. P. (1982b). Metabolism of tartaric and malic acids in *Vitis*: A review. Part B. *Vitis* 21, 346–358.

Silacci, M. W., and Morrison, J. C. (1990). Changes in pectin content of Cabernet Sauvignon grape berries during maturation. *Am. J. Enol. Vitic.* 41, 111–115.

Williams, P. J., Strauss, C. R., Aryan, A. P., and Wilson, B. (1987). Grape flavour—A review of some pre- and postharvest influences. *Proc. 6th Aust. Wine Ind. Tech. Conf.* (T. Lee, ed.), pp. 111–116. Australian Industrial Publ., Adelaide, Australia.

Williams, P. J., Strauss, C. R., and Wilson, B. (1988). Developments in flavour research on premium varieties. *Proc. 2nd Int. Symp. Cool Climate Vitic. Oenol., Jan. 11–15, 1988, Auckland, N.Z.* (R. E. Smart, S. B. Thornton, S. B. Rodriguez, and J. E. Young, eds.), pp. 331–334. New Zealand Soc. Vitic. Oenol., Auckland, New Zealand.

Factors Affecting Berry Maturation

Clingeleffer, P. R. (1985). Use of plant growth regulating chemicals in viticulture. "Chemicals in the Vineyard," pp. 95–100. Aust. Soc. Vitic. Oenol., Mildura, Australia.

Coombe, B. G. (1987). Influence of temperature on composition and quality of grapes. *Acta Hortic.* 206, 23–35.

Freeman, B. M. (1983). Effects of irrigation and pruning of Shiraz grapevines on subsequent red wine pigments. *Am. J. Enol. Vitic.* 34, 23–26.

Hepner, Y., and Bravdo, B. (1985). Effect of crop level and drip irrigation scheduling on the potassium status of Cabernet Sauvignon and Carignane vines and its influence on must and wine composition and quality. *Am. J. Enol. Vitic.* 36, 140–147.

Jackson, D. I. (1988). Factors affecting soluble solids, acid, pH, and color in grapes. *Am. J. Enol. Vitic.* 37, 179–183.

Morrison, J. C., and Noble, A. C. (1990). The effects of leaf and cluster shading on the composition of Cabernet Sauvignon grapes and on fruit and wine sensory properties. *Am. J. Enol. Vitic.* 41, 193–200.

Sinton, T. H., Ough, C. S., Kissler, J. J., and Kasimatis, A. N. (1978). Grape juice indicators for prediction of potential wine quality. I. Relationship between crop level, juice and wine composition, and wine sensory ratings and scores. *Am. J. Enol. Vitic.* 29, 267–271.

Smart, R. E. (1987). Influence of light on composition and quality of grapes. *Acta Hortic.* 206, 37–47.

Smart, R. E., and Robinson, M. (1991). "Sunlight into Wine. A Handbook for Winegrape Canopy Management." Winetitles, Adelaide, Australia.

References

Allen, M. S., Lacey, M. J., Brown, W. V., and Harris, R. L. N. (1989). Occurrence of methoxypyrazines in grapes of *Vitis vinifera* cv. Cabernet Sauvignon and Sauvignon blanc. *Actualités Oenologiques 89: C. R. 4th Symp. Int. Oenol., June 15–17, 1989, Bordeaux, France* (P. Ribéreau-Gayon and A. Lonvaud, eds., pp. 25–30, Bordas, Paris.

Alleweldt, G., and Hifny, H. A. A. (1972). Zur Stiellähme der Reben. II. Kausalanalytische Untersuchungen. *Vitis* 11, 10–28.

Aloni, R., and Peterson, C. A. (1991). Seasonal changes in callose levels and fluorescein translocation in the phloem of *Vitis vinifera*. *IAWA Bull.* **12**, 223–234.

Aloni, R., Raviv, A., and Peterson, C. A. (1991). The role of auxin in the removal of dormancy callose and resumption of phloem activity in *Vitis vinifera*. *Can. J. Bot.* **69**, 1825–1832.

Anonymous (1979). "The Wine Industry in the Federal Republic of Germany." Evaluation and Information Service for Food, Agriculture and Forestry, Bonn, Germany.

Araujo, F. J., and Williams, L. E. (1988). Dry matter and nitrogen partitioning and root growth of young 'Thompson Seedless' grapevines grown in the field. *Vitis* **27**, 21–32.

Arnold, R. A., and Bledsoe, A. M. (1990). The effect of various leaf removal treatments on the aroma and flavor of Sauvignon blanc wine. *Am. J. Enol. Vitic.* **41**, 74–76.

Aspinall, D., and Paleg, L. G. (1981). Proline accumulation: Physiological aspects. *In* "The Physiology and Biochemistry of Drought Stress in Plants" (L. G. Paleg and D. Aspinall, eds.), pp. 205–241. Academic Press, New York.

Atkinson, D. (1980). The distribution and effectiveness of the roots of tree crops. *Hortic. Rev.* **2**, 424–490.

Bath, G. E., Bell, C. J., and Lloyd, H. L. (1991). Arginine as an indicator of the nitrogen status of wine grapes. *Proc. Int. Symp. Nitrogen Grapes Wine, Seattle, WA, June, 1991* (J. M. Rantz, ed.), pp. 202–205. Am. Soc. Enol. Vitic., Davis, California.

Bayonove, C., Cordonnier, R., and Ratier, R. (1974). Localisation de l'arome dans la baie de raisin: Variétés Muscat d'Alexandrie et Cabernet-Sauvignon. *C. R. Seances Acad. Agric. Fr.* **6**, 1321–1328.

Besselat, B., and Cour, P. (1990). La prévison de la production viticole à l'aide de la technique de dosage pollinique de l'atmosphère. *Bull. O.I.V.* **63**, 721–740.

Blaich, R., Stein, U., and Wind, R. (1984). Perforation in der Cuticula von Weinbeeren als morphologischer Faktor der Botrytisresistenz. *Vitis* **23**, 242–256.

Blanke, M. M. (1990). Carbon economy of the grape inflorescence. 4. Light transmission into grape berries. *Wein-Wiss.* **45**, 21–23.

Blanke, M. M. (1991). Kohlenstoff-Haushalt der Infloreszenz der Rebe. 7. Oberflächenanalyse der Spaltöffnungen der Weinbeere. *Wein-Wiss.* **46**, 8–10.

Blanke, M. M., and Leyhe, A. (1987). Stomatal activity of the grape berry cv. Riesling, Müller Thurgau and Ehrenfelser. *J. Plant Physiol.* **127**, 451–460.

Bouard, J. (1980). Tissues et organes de la vigne. *In* "Sciences et Techniques de La Vigne. Tome 1. Biologie de la Vigne, Sols de Vignobles." (J. Ribéreau-Gayon and E. Peynaud, eds.), pp. 3–130. Dunod, Paris.

Branas, J., and Vergnes, A. (1957). Morphologie du système radiculaire de la vigne. *Prog. Agric. Vitic.* **74**, 1–47.

Brantes, N. B. M. (1978). Pollinator activation of *Vitis vinifera* subsp. *silvestris*. *Vitis* **17**, 229–233.

Buttrose, M. S., Hale, C. R., and Kliewer, W. M. (1971). Effect of temperature on the composition of Cabernet Sauvignon berries. *Am. J. Enol. Vitic.* **22**, 71–75.

Carbonneau, A. (1983). Stérilités mâle et femelle dans le genre *Vitis*. I. Modélisation de leur hérédité. *Agronomie* **3**, 635–644.

Christensen, L. P., and Smith, R. J. (1989). Effects of persistent woody laterals on bud performance of Thompson Seedless fruiting canes. *Am. J. Enol. Vitic.* **40**, 27–30.

Conner, A. J., and Thomas, M. B. (1981). Re-establishing plantlets from tissue culture: A review. *Proc. Int. Plant Prop. Soc.* **31**, 342–357.

Coombe, B. G. (1987a). Distribution of solutes within the developing grape berry in relation to its morphology. *Am. J. Enol. Vitic.* **38**, 120–128.

Coombe, B. G. (1987b). Influence of temperature on composition and quality of grapes. *Acta Hortic.* **206**, 23–35.

Coombe, B. G. (1992). Research on development and ripening of the grape berry. *Am. J. Enol. Vitic.* **43**, 101–110.

Conradie, W. J. (1990). Distribution and translocation of nitrogen absorbed during late spring by two-year-old grapevines grown in sand culture. *Am. J. Enol. Vitic.* **41**, 241–250.

Conradie, W. J. (1991a). Distribution and translocation of nitrogen absorbed during early summer by two-year-old grapevines grown in sand culture. *Am. J. Enol. Vitic.* **42**, 180–190.

Conradie, W. J. (1991b). Translocation and storage by grapevines as affected by time of application. *Proc. Int. Symp. Nitrogen Grapes Wine, Seattle, WA, June, 1991* (J. M. Rantz, ed.), pp. 32–42. Am. Soc. Enol. Vitic., Davis, California.

Cordner, C. W., and Ough, C. S. (1978). Prediction of panel preference for Zinfandel wine from analytical data: Using difference in crop level to affect must, wine, and headspace composition. *Am. J. Enol. Vitic.* **29**, 254–257.

Cour, P., Duzer, D., and Planchais, N. (1972–1973). Analyses polliniques de l'atmosphère de Montpellier: Document correspondant à la phénologie de la floraison de la vigne, en 1972. *Naturalis Nonspeliensia, Ser. Bot.* **23–24**, 225–229.

Do, C. B., and Cormier, F. (1991). Accumulation of peonidin-3-glucoside enhanced by osmotic stress in grape (*Vitis vinifera* L.) cell suspension. *Plant Cell Tissue Organ Cult.* **24**, 49–54.

Düring, H. (1984). Evidence for osmotic adjustment to drought in grapevines (*Vitis vinifera* L.). *Vitis* **23**, 1–10.

Düring, H. (1990). Stomatal adaptation of grapevine leaves to water stress. *Proc. 5th Int. Symp. Grape Breeding, Sept. 12–16, 1989*, pp. 366–370. St. Martin, Pfalz, Germany (Special Issue of *Vitis*).

Esau, K. (1948). Phloem structure in the grapevine, and its seasonal changes. *Hilgardia* **18**, 217–296.

Fernández de Simón, B., Hernández, T., Estrella, I., and Gómez-Cordovés, C. (1992). Variation in phenol content in grapes during ripening: Low-molecular-weight phenols. *Z. Lebensm. Unters. Forsch.* **194**, 351–354.

Flaherty, D. L., Jensen, F. L., Kasimatis, A. N., Kido, H., and Moller, W. J. (1981). "Grape Pest Management," Publ. 4105. Cooperative Extension, Univ. of California, Oakland.

Freeman, B. M., and Kliewer, W. M. (1983). Effect of irrigation, crop level and potassium fertilization on Carignane vines. II. Grape and wine quality. *Am. J. Enol. Vitic.* **34**, 197–207.

Freeman, B. M., and Smart, R. E. (1976). A root observation laboratory for studies with grapevines. *Am. J. Enol. Vitic.* **27**, 36–39.

Gallander, J. F. (1983). Effect of grape maturity on the composition and quality of Ohio Vidal Blanc wines. *Am. J. Enol. Vitic.* **34**, 139–141.

González-SanJosé, M. L., Barron, L. J. R., and Diez, C. (1990). Evolution of anthocyanins during maturation of Tempranillo grape variety (*Vitis vinifera*) using polynomial regression models. *J. Sci. Food Agric.* **51**, 337–343.

Gu, S., Lombard, P. B., and Price, S. F. (1991). Inflorescence necrosis induced by ammonia incubation in clusters of Pinot Noir grapes. *Proc. Int. Symp. Nitrogen Grapes Wine. Seattle, Washington, June, 1991.* (J. M. Bantz, ed.). pp. 67–77. Am. Soc. Enol. Vitic., Davis, California.

Hahlbrock, K. (1981). Flavonoids. *In* "The Biochemistry of Plants" (P. K., Strumpf and E. E. Conn, eds.), Vol. 7, pp. 425–456. Academic Press, New York.

Hale, C. R. (1981). Interaction between temperature and potassium in grape acids. *CSIRO Div. Hort. Res. Rep. 1979–1981*, pp. 87–88. Glen Osmond, South Australia.

Hale, C. R., and Buttrose, M. S. (1973). Effect of temperature on anthocyanin content of Cabernet Sauvignon berries. *CSIRO Div.*

Hort. Rep. 1971–1973, pp. 98–99. Glen Osmond, South Australia.

Hale, C. R., and Buttrose, M. S. (1974). Effect of temperature on ontogeny of berries of *Vitis vinifera* L. cv. Cabernet Sauvignon. *J. Am. Soc. Hortic. Sci.* **99**, 390–394.

Hawker, J. S., Hale, C. R., and Kerridge, G. H. (1981). Advancing the time of ripeness of grapes by the application of methyl 2-(ureidooxy)propionate (a growth retardant). *Vitis* **20**, 302–310.

Heymann, H., and Noble, A. C. (1987). Descriptive analysis of Pinot noir wines from Carneros, Napa and Sonoma. *Am. J. Enol. Vitic.* **38**, 41–44.

Hofäcker, W. (1976). Untersuchungen über den Einfluss wechselnder Bodenwasserversorgung auf die Photosyntheseintensität und den Diffusionswiderstand bei Rebblättern. *Vitis* **15**, 171–182.

Hrazdina, G., and Moskowitz, A. H. (1982). Subcellar status of anthocyanins in grape skins. *Grape Wine Centennial Symp. Proc. 1980*, pp. 245–253. Univ. of California, Davis.

Hrazdina, G., Parsons, G. F., and Mattick, L. R. (1984). Physiological and biochemical events during development and maturation of grape berries. *Am. J. Enol. Vitic.* **35**, 220–227.

Huglin, P. (1986). "Biologie et Écologie de la Vigne." Payot Lausanne, Paris.

Iland, P. G., and Coombe, B. G. (1988). Malate, tartrate, potassium, and sodium in flesh and skin of Shiraz grapes during ripening: Concentration and compartmentation. *Am. J. Enol. Vitic.* **39**, 71–76.

Iwasaki, K. (1972). Effects of soil aeration on vine growth and fruit develpoment of grapes. *Mem. Coll. Agric., Ehime Univ.* (*Ehime Daigaku Nogakubu Kiyo*) **16**, 4–26 (in Japanese).

Jackson, D. I. (1986). Factors affecting soluble solids, acid pH, and color in grapes. *Am. J. Enol. Vitic.* **37**, 179–183.

Kliewer, W. M. (1967). The glucose–fructose ratio of *Vitis vinifera* grapes. *Am. J. Enol. Vitic.* **18**, 33–41.

Kliewer, W. M. (1971). Effect of day temperature and light intensity on concentration of malic and tartaric acids in *Vitis vinifera* L. grapes. *J. Am. Sci. Hortic. Sci.* **96**, 372–377.

Kliewer, W. M. (1977a). Influence of temperature, solar radiation and nitrogen on coloration and composition of Emperor grapes. *Am. J. Enol. Vitic.* **28**, 96–103.

Kliewer, W. M. (1977b). Grape coloration as influenced by temperature. *Int. Symp. Quality Vintage*, pp. 89–106. Oenol. Viticult. Res. Inst., Stellenbosch, South Africa.

Kliewer, W. M., and Antcliff, A. J. (1970). Influence on defoliation, leaf darkening, and cluster shading on the growth and composition of Sultana grapes. *Am. J. Enol. Vitic.* **21**, 26–36.

Kliewer, W. M., Howorth, L., and Omori, M. (1967). Concentrations of tartaric acid and malic acids and their salts in *Vitis vinifera* grapes. *Am. J. Enol. Vitic.* **18**, 42–54.

Kliewer, W. M., and Lider, L. A. (1970). Effect of day temperature and light intensity on growth and composition of *Vitis vinifera* L. fruits. *J. Am. Soc. Hortic. Sci.* **95**, 766–769.

Kobayashi, A., Yukinaga, H., and Itano, T. (1965a). Studies on the thermal conditions of grapes. III. Effects of night temperature at the ripening stage on the fruit maturity and quality of Delaware grapes. *J. Jpn. Soc. Hortic. Sci.* **34**, 26–32.

Kobayashi, A., Yukinaga, H., and Matsunaga, E. (1965b). Studies on the thermal conditions of grapes. V. Berry growth, yield, quality of Muscat of Alexandria as affected by night temperature. *J. Jpn. Soc. Hortic. Sci.* **34**, 152–158.

Koblet, W. (1969). Wanderung von Assimilaten in Rebtrieben und Einfluss der Blattfläche auf Ertrag und Qualität der Trauben. *Wein-Wiss.* **24**, 277–319.

Kriedemann, P. E. (1968). Photosynthesis in vine leaves as a function of light intensity, temperature and leaf age. *Vitis* **7**, 213–220.

Kriedemann, P. E. (1977). Vineleaf photosynthesis. *Int. Symp. Quality Vintage*, pp. 67–87. Oenol. Viticult. Res. Inst., Stellenbosch, South Africa.

Kriedemann, P. E., Kliewer, W. M., and Harris, J. M. (1970). Leaf age and photosynthesis in *Vitis vinifera* L. *Vitis* **9**, 97–104.

Kriedemann, P. E., Törökfalvy, E., and Smart, R. E. (1973). Natural occurrence and photosynthetic utilization of sunflecks by grapevine leaves. *Photosynthetica* **7**, 18–27.

Lakso, A. N., and Kliewer, W. M. (1975). Physical properties of phosphoenolpyruvate carboxylase and malic enzyme in grape berries. *Am. J. Enol. Vitic.* **26**, 75–78.

Lang, A., and Thorpe, M. R. (1989). Xylem, phloem and transpirational flow in a grape: Application of a technique for measuring the volume of attached fruits to high resolution using Archimedes' principle. *J. Exp. Bot.* **40**, 1069–1078.

Lavée, S. (1977). The response of vine growth and bunch development to elevated winter, spring and summer day temperature. *Int. Symp. Quality Vintage*, pp. 209–226. Oenol. Viticult. Res. Inst., Stellenbosch, South Africa.

Lee, C. Y., and Jaworski, A. (1989). Major phenolic compounds in ripening white grapes. *Am. J. Enol. Vitic.* **40**, 43–46.

Levadoux, L. (1951). La sélection et l'hybridation chez la vigne. *Ann. Ec. Natl. Agric. Montpellier* **28**, 9–195.

McCarthy, M. G., and Coombe, B. G. (1985). Water status and winegrape quality. *Acta Hortic.* **171**, 447–456.

McColl, C. R. (1986). Cyanamide advances the maturity of table grapes in cental Australia. *Aust. J. Exp. Agric.* **26**, 505–509.

McKenry, M. V. (1984). Grape root phenology. Relative to control of parasitic nematodes. *Am. J. Enol. Vitic.* **35**, 206–211.

Marais, J., van Wyk, C. J., and Rapp, A. (1992a). Effect of sunlight and shade on norisoprenoid levels in maturing Weisser Riesling and Chenin blanc grapes and Weisser Riesling wines. *S. Afr. J. Enol. Vitic.* **13**, 23–32.

Marais, J., Versini, G., van Wyk, C. J., and Rapp, A. (1992b). Effect of region on free and bound monoterpene and C_{13}-norisoprenoid concentrations in Weisser Riesling wines. *S. Afr. J. Enol. Vitic.* **13**, 71–77.

Matthews, M. A., and Anderson, M. M. (1988). Fruit ripening in *Vitis vinifera* L.: Responses to seasonal water deficits. *Am. J. Enol. Vitic.* **39**, 313–320.

May, P. (1987). The grapevine as a perennial plastic and productive plant. *Proc. 6th Aust. Wine Ind. Tech. Conf.* (T. Lee, ed.), pp. 40–49. Australian Industrial Publ., Adelaide, Australia.

Meyer, J. R., and Linderman, R. G. (1986). Selective influences on populations of rhizosphere or rhizoplane bacteria and actinomycetes by mycorrhizas formed by *Glomus fasciculatum*. *Soil Biol. Biochem.* **18**, 191–196.

Millar, A. A. (1972). Thermal regime of grapevines. *Am. J. Enol. Vitic.* **23**, 173–176.

Moncur, M. W., Rattigan, K., Mackenzie, D. H., and McIntyre, G. N. (1989). Base temperatures for budbreak and leaf appearance of grapevines. *Am. J. Enol. Vitic.* **40**, 21–26.

Morrison, J. C. (1988). The effects of shading on the composition of Cabernet Sauvignon grape berries. *Proc. 2nd Int. Symp. Cool Climate Vitic. Oenol. Jan. 11–15, 1988, Auckland, N.Z.* (R. E. Smart, S. B. Thornton, S. B. Rodriguez, and J. E. Young, eds.), pp. 144–146. New Zealand Soc. Vitic. Oenol., Auckland, New Zealand.

Morrison, J. C. (1991). Bud development in *Vitis vinifera*. *Bot. Gaz.* **152**, 304–315.

Morrison, J. C., and Noble, A. C. (1990). The effects of leaf and cluster shading on the composition of Cabernet Sauvignon grapes and on fruit and wine sensory properties. *Am. J. Enol. Vitic.* **41**, 193–200.

Mounts, B. T. (1932). Development of foliage leaves. *Univ. Iowa Stud. Nat. Hist.* **14**, 1–19.

Nagarajah, S. (1989). Physiological responses of grapevines to water stress. *Acta Hortic.* **240**, 249–256.

Nobel, P. S. (1975). Effective thickness and resistance of the air boundary layer adjacent to spherical plant parts. *J. Exp. Bot.* **26**, 120–130.

Ough, C. S., and Nagaoka, R. (1984). Effect of cluster thinning and vineyard yield on grape and wine composition and wine quality of Cabernet Sauvignon. *Am. J. Enol. Vitic.* **35**, 30–34.

Polo, M. C., Herraiz, M., and Cabezudo, M. D. (1983). A study of nitrogen fertilization and fruit maturity as an approach for obtaining the analytical profiles of wines and wine grapes. *Instrum. Anal. Foods: Recent Prog. Proc. Symp. Int.* Flavor Cont. *3rd* **2**, 357–374.

Possner, D. R. E., and Kliewer, W. M. (1985). The localisation of acids, sugars, potassium and calcium in developing grape berries. *Vitis* **24**, 229–240.

Pratt, C. (1959). Radiation damage in shoot apices of Concord grape. *Am. J. Bot.* **46**, 102–109.

Pratt, C. (1974). Vegetative anatomy of cultivated grapes. A review. *Am. J. Enol. Vitic.* **25**, 131–148.

Pszczolkowski, Ph., Morales, A., and Cava, S. (1985). Composicion quimica y calidad de mostos y vinos obtenidos de racimos diferentemente asoleados. *Cienc. Invest. Agrar.* **12**, 181–188.

Pucheu-Plante, B., and Mércier, M. (1983). Étude ultrastructurale de l'interrelation hôte–parasite entre le raisin et le champignon *Botrytis cinerea:* Exemple de la pourriture noble en Sauternais. *Can. J. Bot.* **61**, 1785–1797.

Queen, W. H. (1968). Radial movement of water and ^{32}P through suberized and unsuberized roots of grape. *Diss. Abstr. Sect. B* **29**, 72–73.

Radler, F. (1965). The effect of temperature on the ripening of 'Sultana' grapes. *Am. J. Enol. Vitic.* **16**, 38–41.

Razungles, A., Bayonove, C. L., Cordonnier, R. E., and Sapis, J. C. (1988). Grape carotenoids: Changes during the maturation period and localization in mature berries. *Am. J. Enol. Vitic.* **39**, 44–48.

Razungles, A., Gunata, Z., Pinatel, S., Baumes, R., and Bayonove, C. (1993). Étude quantitative de composés terpéniques, norisoprénoïds et de leurs précurseurs dans diverses variétés de raisins. *Sci. Aliments* **13**, 59–72.

Reynolds, A. G., and Wardle, D. A. (1989). Influence of fruit microclimate on monoterpene levels of Gewürztraminer. *Am. J. Enol. Vitic.* **40**, 149–154.

Reynolds, A. G., and Wardle, D. (1991). Effects of fruit exposure on fruit composition. *Am. J. Enol. Vitic.* **42**, 89 (Abstract).

Richards, D. (1983). The grape root system. *Hortic. Rev.* **5**, 127–168.

Robinson, W. B., Shaulis, N., and Pederson, C. S. (1949). Ripening studies of grapes grown in 1948 for juice manufacture. *Fruit Prod. J. Am. Food Manuf.* **29**, 36–37, 54, and 62.

Robinson, W. B., Weirs, L. D., Bertino, J. J., and Mattick, L. R. (1966). The relation of anthocyanin composition to color stability of New York State wines. *Am. J. Enol. Vitic.* **17**, 178–184.

Roggero, J. P., Coen, S., and Ragonnet, B. (1986). High performance liquid chromatography survey on changes in pigment content in ripening grapes of Syrah. An approach to anthocyanin metabolism. *Am. J. Enol. Vitic.* **37**, 77–83.

Rojas-Lara, B. A., and Morrison, J. C. (1989). Differential effects of shading fruit or foliage on the development and composition of grape berries. *Vitis* **28**, 199–208.

Rosenquist, J. K., and Morrison, J. C. (1988). The development of the cuticle and epicuticular wax of the grape berry. *Vitis* **27**, 63–70.

Roufet, M., Bayonove, C. L., and Cordonnier, R. E. (1987). Étude de la composition lipidique du raisin, *Vitis vinifera:* Evolution au cours de la maturation et localisation dans la baie. *Vitis* **26**, 85–97.

Ruffner, H. P. (1982). Metabolism of tartaric and malic acids in *Vitis:* A review. Part A. *Vitis* **21**, 247–259.

Schellenbaum, L., Berta, G., Ravolanirina, F., Tisserant, B., Gianinazzi, S., and Fitter, A. H. (1991). Influence of endomycorrhizal infection on root morphology in a micropropagated woody plant species (*Vitis vinifera* L.) *Ann. Bot.* **68**, 135–141.

Schneider, A., Mannini, F., Gerbi, V., and Zeppa, G. (1990). Effect of vine vigour of *Vitis vinifera* cv. Nebbiolo clones on wine acidity and quality. *Proc. 5th Int. Symp. Grape Breeding, Sept. 12–16, 1989,* pp. 525–531. St. Martin, Pfalz, Germany, Special Issue of *Vitis.*

Schubert, A. (1985). Les mycorhizes à vesicules et arbuscules chez la vigne. *Connaiss. Vigne Vin* **19**, 207–214.

Schubert, A., Mazzitelli, M., Ariusso, O., and Eynard, I. (1990). Effects of vesicular–arbuscular mycorrhizal fungi on micropropagated grapevines: Influence of endophyte strain, P fertilization and growth medium. *Vitis* **29**, 5–13.

Sinton, T. H., Ough, C. S., Kissler, J. J., and Kasimatis, A. N. (1978). Grape juice indicators for prediction of potential wine quality. I. Relationship between crop level, juice and wine composition, and wine sensory ratings and scores. *Am. J. Enol. Vitic.* **29**, 267–271.

Smart, R. E. (1982). Vine manipulation to improve wine grape quality. *Grape Wine Centennial Symp. Proc. 1980,* pp. 362–375. Univ. of California, Davis.

Smart, R. E. (1987). Influence of light on composition and quality of grapes. *Acta Hortic.* **206**, 37–47.

Smart, R. E., and Sinclair, T. R. (1976). Solar heating of grape berries and other spherical fruits. *Agric. Meteorol.* **17**, 241–259.

Smart, R. E., Smith, S. M., and Winchester, R. V. (1988). Light quality and quantity effects on fruit ripening for Cabernet Sauvignon. *Am. J. Enol. Vitic.* **39**, 250–258.

Smith, H., and Holmes, M. G. (1977). The function of phytochrome in the natural environment. III. Measurement and calculation of phytochrome photoequilibria. *Photochem. Photobiol.* **25**, 547–550.

Smith, S., Codrington, I. C., Robertson, M., and Smart, R. (1988). Viticultural and oenological implications of leaf removal for New Zealand vineyards. *Proc. 2nd Int. Symp. Cool Climate Vitic. Oenol. Jan. 11–15, 1988, Auckland, N.Z.* (R. E. Smart, S. B. Thornton, S. B. Rodriguez, and J. E. Young, eds.), pp. 127–133. New Zealand Soc. Vitic. Oenol., Auckland, New Zealand.

Solari, C., Silvestroni, O., Guidici, P., and Intrieri, C. (1988). Influence of topping on juice composition of Sangiovese grapevines *V. vinifera. Proc. 2nd Int. Symp. Cool Climate Vitic. Oenol. Jan. 11–15, 1988, Auckland, N.Z.* (R. E. Smart, S. B. Thornton, S. B. Rodriguez, and J. E. Young, eds.). pp. 147–151. New Zealand Soc. Vitic. Oenol., Auckland, New Zealand.

Sponholz, W. R. (1991). Nitrogen compounds in grapes, must, and wine. *Proc. Int. Symp. Nitrogen Grapes Wine, Seattle, WA, June, 1991* (J. M. Rantz, ed.), pp. 67–77. Am. Soc. Enol. Vitic., Davis, California.

Staudt, G. (1982). Pollenkeimung und Pollenschlauchwachstum *in vivo* bei *Vitis* und die Abhähgigkeit von der Temperatur. *Vitis* **21**, 205–216.

Staudt, G., Schneider, W., and Leidel, J. (1986). Phases of berry growth in *Vitis vinifera. Ann. Bot.* **58**, 789–800.

Stoev, K., and Slavtcheva, T. (1982). La photosynthése nette chez la vigne (*V. vinifera*) et les facteurs écologiques. *Connais. Vigne Vin* **16**, 171–185.

Strauss, C. R., Gooley, P. R., Wilson, B., and Williams, P. J. (1987a). Application of droplet countercurrent chromatography to the analysis of conjugated forms of terpenoids, phenols, and other constituents of grape juice. *J. Agric. Food Chem.* **35**, 519–524.

Strauss, C. R., Wilson, B., Anderson, R., and Williams, P. J. (1987b). Development of precursors of C_{13} nor-isoprenoid flavorants in Riesling grapes. *Am. J. Enol. Vitic.* **38**, 23–27.

Swanepoel, J. J., and Archer, E. (1988). The ontogeny and development of *Vitis vinifera* L. cv. Chenin blanc inflorescence in relation to phenological stages. *Vitis* **27**, 133–141.

Swanepoel, J. J., and de Villiers, C. E. (1988). The anatomy of *Vitis* roots and certain abnormalities. *In* "The Grapevine Root and Its Environment" (J. L. Van Zyl, ed.), Dept. Agric. Water Supply Tech. Commun. No. 215, pp. 138–146. Pretoria, South Africa.

Tomana, T., Utsunomiya, N., and Kataoka, I. (1979a). The effect of environmental temperatures on fruit ripening on the tree. II. The effect of temperatures around whole vines and clusters on the colouration of 'Kyoho' grapes. *J. Jpn. Soc. Hortic. Sci.* **48**, 261–266.

Tomana, T., Utsunomiya, N., and Kataoka, I. (1979b). The effect of environmental temperatures on fruit ripening on the tree—the effect of temperatures around whole vines and clusters on the ripening of 'Delaware' grapes. *Stud. Inst. Hortic. Kyoto Univ. (Engeigaku Kenkyu Shuroku)* **9**, 1–5.

Tyson, P. J., Luis, E. S., and Lee, T. H. (1982). Soluble protein levels in grapes and wine. *Grape Wine Centennial Symp. Proc. 1980*, pp. 287–290. Univ. of California, Davis.

van Zyl, J. L., and van Huyssteen, L. (1987). Root pruning. *Decid. Fruit Grow.* **37**(1), 20–25.

Watson, B. T., McDaniel, M., Miranda-Lopez, R., Michaels, N., Price, S., and Yorgey, B. (1991). Fruit maturation and flavor development in Pinot noir. *Am. J. Enol. Vitic.* **42**, 88–89 (Abstract).

Weaver, R. J., and McCune, S. B. (1960). Influence of light on color development in *Vitis vinifera* grapes. *Am. J. Enol. Vitic.* **11**, 179–184.

West, D. W., and Taylor, J. A. (1984). Response of six grape cultivars to the combined effects of high salinity and rootzone waterlogging. *J. Am. Soc. Hortic. Sci.* **109**, 844–851.

Wilson, B., Strauss, C. R., and Williams, P. J. (1984). Changes in free and glycosidically bound monoterpenes in developing Muscat grapes. *J. Agric. Food Chem.* **32**, 919–924.

Wilson, B., Strauss, C. R., and Williams, P. J. (1986). The distribution of free and glycosidically bound monoterpenes among skin, juice, and pulp fractions of some white grape varieties. *Am. J. Enol. Vitic.* **37**, 107–114.

Winkler, A. J., Cook, J. A., Kliewer, W. M., and Lider, L. A. (1974). "General Viticulture." Univ. of California Press, Berkeley.

Yang, Y., Hori, I., and Ogata, R. (1980). Studies on retranslocation of accumulated assimilates in Delaware grape vines. II and III. *Tokyo J. Agric. Res.* **31**, 109–129.

4

Vineyard Practice

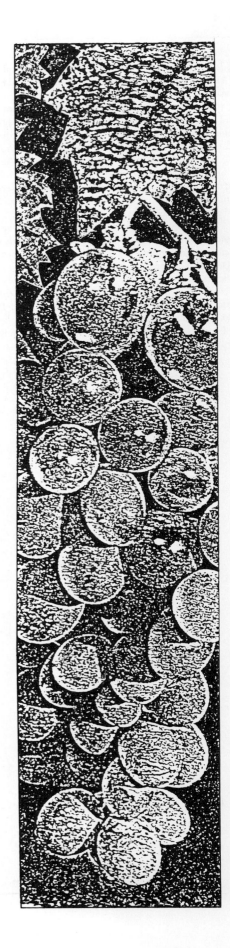

In the previous chapter, details were given concerning many of the major physiological processes in grapevines. In this chapter, vineyard practice, and how it impacts on grape yield and quality, is discussed. Because vineyard practice is so closely allied to the yearly cycle of the vine, a brief description of the growth cycle, and its association with vineyard activities, are given below.

Vine Cycle and Vineyard Activity

The ending of one growth cycle, and the beginning of another, coincides in temperate regions with the onset of winter dormancy. Winter dormancy provides grape growers with the opportunity to do many of the less urgent vineyard activities for which there was insufficient time during the growing season. In addition, pruning is conducted more conveniently when the vine is dormant, and the absence of foliage permits easier wood selection and tying of canes.

On the return of warmer weather, both the metabolic activity of the vine and the pace of vineyard endeavors quicken. Usually, the first sign of renewed activity is the "bleeding" of sap from the cut ends of canes and spurs. The sap contains organic nutrients, such as sugars and amino acids, as well as growth regulators. When the temperature rises above a critical value, dependent on the variety, buds begin to burst (Fig. 4.1). Activation progresses downward from the tips of canes and spurs as the phloem again becomes functional. The cambium subsequently resumes its meristematic activity and produces new xylem and phloem to the interior and exterior, respectively.

Typically, only the primary bud in the dormant bud becomes active in the spring. The secondary and tertiary buds remain inactive, unless the primary bud is killed or severely damaged. Once initiated, shoot growth rapidly reaches its maximum as the climate continues to improve. The primordial leaves, tendrils, and inflorescence clusters reinstitute their development and begin to enlarge (Fig. 4.2). As growth continues, the shoot differentiates new leaves and tendrils. Simultaneously, new buds arise in the leaf axils. Those that form early may give rise to lateral shoot systems, as well as compound buds. Inflorescence initiation occurs early during bud development.

Root growth lags behind shoot growth, often coinciding with flowering (five to eight-leaf stage). Peak root development commonly occurs between the end of flowering and the initiation of fruit coloration (*véraison*). Subsequently, root development declines, with possibly a second growth period in the autumn.

Flower development in the spring progresses from the outermost ring of flower parts, the sepals, inward to the pistil. By the time the anthers mature and split open (anthesis), the cap of fused petals (calyptra) separates from the ovary base and is shed. In the process, self-pollination typically occurs as liberated pollen falls into the stigma. If followed by fertilization, embryo and berry development commences.

Several weeks after bloom, many small berries dehisce (**shatter**). This is a normal process in all varieties. It sheds nascent fruit from the cluster in which fertilization did not occur or the embryos aborted. At least one seed is required for berry development to continue.

Once the shoots have developed sufficiently, they typically are tied to a support system. Shoots derived from latent buds on old wood (water sprouts and suckers) are usually removed early in the season. Spraying for disease and pest control also commences early. Spraying or cultivation for weed control often occurs before bud burst. Irrigation, if desired and permissible, is usually terminated by *véraison*, as its subsequent use affects fruit quality negatively. When required, irrigation may be reinitiated after harvest to permit optimal cane maturation.

When berry development has reached *véraison*, primary shoot growth has usually ceased. Because lateral shoot growth may continue, causing both a nutrient drain and excessive shade production, shoot topping may be practiced. Conversely, fruit cluster thinning may be performed if fruit yield appears excessive. Flower cluster thinning earlier in the season also may achieve a desired yield reduction. Basal leaf removal

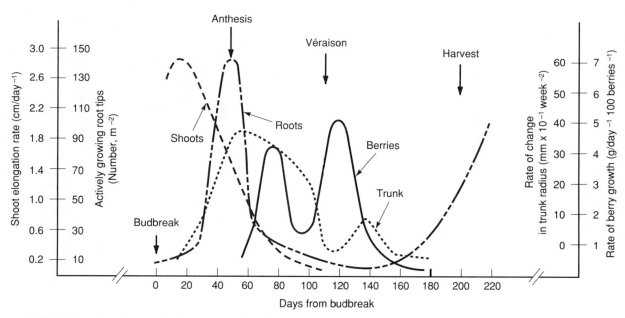

Figure 4.1 Growth rate of various organs of 'Colombar' grapevines grown in South Africa throughout the season. (After van Zyl, 1984, from Williams and Matthews, 1990, reproduced by permission.)

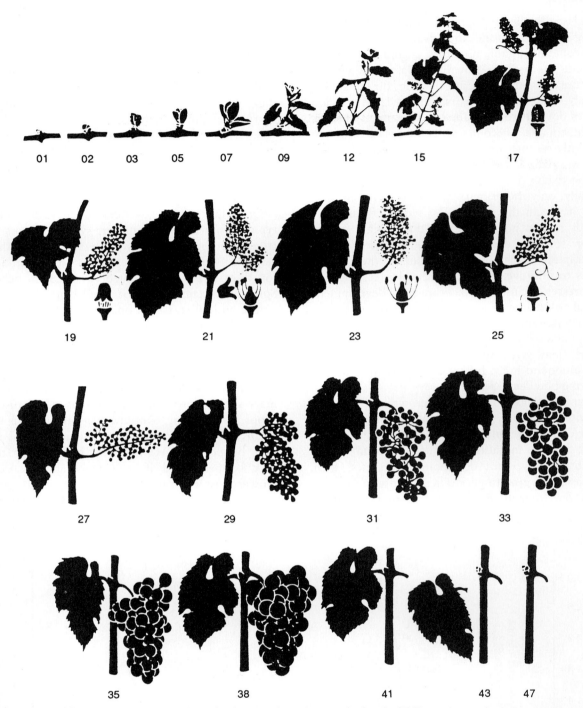

Figure 4.2 Stages in grapevine shoot development from dormant bud to leaf fall: 01, winter dormancy; 02, bud swelling; 03, wool stage; 05, bud burst; 07, first leaf stage; 09, two to three leaf stage; 12, five to six leaf stage, inflorescence visible; 15, inflorescence elongating; 17, flowers separating; 19, start of flowering; 21, early flowering (25% caps fallen); 23, full flowering (50% caps fallen); 25, late flowering (80% caps fallen); 27, fruit set; 29, cluster begins to hang; 31, pea-size berry stage; 33, berries begin to touch; 35, *véraison;* 38, fruit ripening; 41, fruit maturity; 43, canes mature; 43, beginning of leaf fall; 47, canes bare. (After Eichhorn and Lorenz, 1977, reproduced by permission.)

may be employed to improve cane and fruit exposure to light and air, as well as access to disease and pest control chemicals.

As the fruit approaches maturity, vineyard activity shifts in preparation for harvest. Fruit samples are taken throughout the vineyard for chemical analysis. Based on the results, environmental conditions, and the desires of the winemaker, a tentative harvest date is set. In cool climates, measures are put in place for frost protection.

Once harvest is complete, vineyard activity is directed toward preparing the vines for winter. In cold climates, this may vary from mounding soil up around the shoot/rootstock union to removing the whole shoot system from its support system for burial.

Management of Vine Growth

Although vineyard practice is directed toward obtaining the maximum fruit yield relative to desired quality, the primary means of achieving this goal involves training and pruning. As such, they have been extensively analyzed for the last century, and empirically studied for millennia. The result is a bewildering array of systems, discussion of which is beyond the scope of this text. However, the following sections look at aspects that help to explain why much of the diversity still exists. For detailed analyses of various training and pruning systems, the reader is directed to the suggested readings at the end of the chapter.

In discussing vine management, it is necessary to define several commonly used terms. **Training** refers to the development of a permanent vine structure and the location of the renewal wood, so that shoot growth occurs in desired locations. **Renewal wood** consists of spurs used to generate canes (**bearing wood**) from which bearing shoots will originate the subsequent year. Training usually is associated with a support (**trellis**) system. It takes into consideration factors such as prevailing climate, harvesting practices, and the fruiting characteristics of the cultivar. **Canopy management** is generally viewed as positioning and maintaining **bearing shoots** and their fruit in a microclimate optimal for grape quality, inflorescence initiation, and cane maturation. **Pruning** may involve the partial removal of canes, shoots, wood, or leaves, or the severing of roots, to obtain the goals of training and canopy management. However, the most common use of the term "pruning" refers to the removal of excess shoot growth before the beginning of each new year's growth. **Thinning** involves the removal of whole or parts of flower and fruit clusters to improve the berry microclimate and leaf/fruit balance. Finally, **vigor** refers to vegetative growth rate,

while **capacity** refers to the amount of growth and its ability to mature fruit.

Physiological Effects of Pruning

Pruning is usually employed for several separate but related functions. It permits the grape grower to establish a particular training system and regulate individual vine yield. Pruning can permit precise selection of bearing wood (spurs and canes) and, thereby, influence the location and development of canopy growth. This in turn can affect grape yield, health, and maturation, as well as pruning and harvesting costs.

To understand the effects of pruning, it is essential to remember the basic properties of grapevine growth. The vine, in common with most other perennial fruit crops, shows inflorescence initiation a season in advance of development. Unlike many other fruit crops, though, fruit production occurs opposite leaves on vegetative shoots, rather than on specialized, short, determinant shoots called flower spurs, as, for example, in apples, pears, and peaches. The property of inflorescence initiation a year in advance of flower development allows the grape grower to limit fruiting capacity by bud removal, long before flowering. Winter pruning has, and still remains, the primary means of restricting grapevine yield. However, as a perennial crop, the vine stores considerable energy reserves in its woody parts. Thus, pruning removes nutrients that limit the ability of the vine to initiate rapid growth in the spring. Vines with little old wood are less able to fully mature crops in poor years than vines possessing large cordons and/or trunks.

Since the 1940s, it has become obvious that *both* excessive pruning *and* excessive overcropping suppress vine growth. The reduced growth caused by overcropping has long been known and has been part of the rationale for removing 85 to 90% of the vine's annual growth. However, as illustrated by Fig. 4.3, severe pruning also can cause excessive growth reduction. This apparent anomaly results from several factors. Severe pruning not only reduces shoot growth, but it also decreases and delays leaf production. The result is a reduction in vine photosynthesis. Because base and lateral buds become active, leaf production continues into the fruiting period, causing a drain on the photosynthate available for fruit ripening. The typical relationship of shoot to berry growth and the concentration of carbohydrate in the shoot and fruit are illustrated in Fig. 4.4. In contrast, the absence of pruning permits rapid development of the leaf canopy and initiation of fruit development. Subsequent growth reduction results from the excessive carbohydrate demands placed on the vine as the fruit matures. Al-

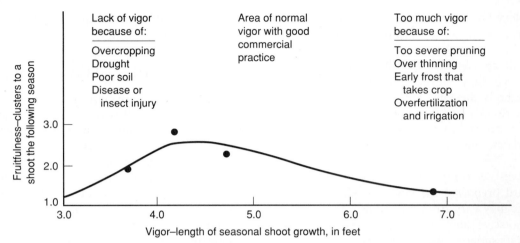

Figure 4.3   The relation of vigor (length) of shoot growth to fruitfulness of buds of 'Muscat of Alexandria' and 'Alicante Bouschet' at Davis, California. (From Winkler *et al.*, 1974, reproduced by permission.)

though berry maturation is delayed, sugar content reduced, and fruit quality diminished in nonpruned vines, yield can be increased by up to 2.5 times that of normally pruned vines. However, fruit yield in succeeding years may be suppressed in nonpruned vines by diminished inflorescence initiation and reduction in vine nutrient reserves (Fig. 4.3).

Because both inadequate and excessive pruning can result in reduced growth, decreased fruit quality, and diminished fruitfulness, it is important to match pruning to vine capacity. Winter pruning is the most practical but is the least discriminatory. Capacity can be only imprecisely predicted. In contrast, fruit thinning is the most precise, but also the most labor intensive. This results from the difficulty of selectively removing fruit from clusters partially obscured by foliage. The prefera-

ble compromise, where feasible, is a combination of winter pruning followed by remedial flower cluster thinning in the spring. This allows removal of most of the vine's excessive fruiting potential, while delaying final adjustment until a better measure of probable fruit load can be determined. It also has the advantage that leaf production occurs early, maximizing carbohydrate production over the year and minimizing competition between foliage and fruit during berry ripening.

Additional increases in fruit yield and quality may be obtained with better canopy management by improving the microclimate in and around the grape cluster. Positioning the basal region of shoots in a favorable light environment often improves inflorescence induction and helps maintain fruitfulness from year to year.

Clonal selection, the elimination of debilitating vi-

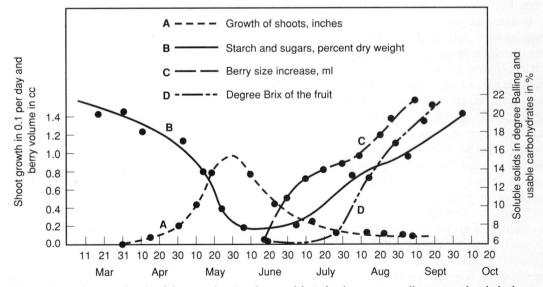

Figure 4.4   The annual cycle of the vine, showing shoot and fruit development as well as seasonal carbohydrate changes in berries and at the bases of 1-year-old wood. (From Winkler *et al.*, 1974, reproduced by permission.)

ruses, improved disease and pest control, irrigation, and fertilization have significantly improved vine vigor. One of the modern challenges of viticulture is to convert these improvements into enhanced fruit yield and quality. More recently, interest has been shown in minimal pruning. This is based on the potential of the vine to self-regulate its fruit production after several years. However, as the applicability of this technique to most cultivars and climates is currently unknown, most of the following discussion deals with the fundamentals of pruning.

In general, the principles of pruning enunciated by Winkler *et al.* (1974) remain valid:

1. Pruning reduces vine capacity by removing both buds and stored nutrient in canes. Thus, pruning should be kept to the minimum necessary to permit the vine to fully ripen the fruit it bears.

2. Excessive or inadequate pruning depresses vine capacity for several years. To avoid this, pruning should attempt to match bud removal to vine capacity.

3. Capacity partially depends on the quality of the vine to generate a leaf canopy rapidly. This usually requires light pruning, followed by cluster thinning to balance crop production to the existing canopy.

4. Increased crop load and shoot number depress shoot elongation and leaf production during fruit development and, up to a point, favor full ripening. Moderately vigorous shoot growth is most consistent with vine fruitfulness.

5. In establishing a training system, retention of one main vigorous shoot enhances growth and suppresses early fruit production. Both speed development of the permanent vine structure. With established vines, balanced pruning augments yield potential.

6. Cane thickness is a good indicator of bearing capacity. Thus, to balance growth throughout the vine, the level of bud retention should reflect the diameter of the cane. Alternately, where bud number should remain constant, canes or spurs retained should be relatively uniform in thickness.

7. Optimal capacity refers to the maximal fruit load the vine can ripen fully within the normal growing season. Reduced fruit load has no effect on the rate of ripening. In contrast, overcropping delays maturation, increases fruit shatter, and decreases berry quality. Capacity is a function of the current environmental conditions, those of the past few years, and the genetic potential of the cultivar.

Pruning Options

When considering pruning options, it is necessary to evaluate other features than just the principles noted above. The prevailing climate, genetic characteristics of the rootstock, soil fertility, and training system can sig-

nificantly influence the consequences of pruning. Several of these factors are discussed below.

The prevailing climate imposes constraints on where and how grapevines can be grown. In cool continental climates, severe winter temperatures may kill most buds. Equally, late frosts can destroy newly emerged shoots and inflorescences. Where such losses occur frequently, more buds should be retained than theoretically ideal to compensate for the potential bud kill. Subsequent disbudding, or flower cluster thinning, can adjust potential yield downward if necessary. Alternately, tender varieties may be pruned in late winter or very early spring to permit pruning level to reflect actual bud viability.

As deep fertile soil enhances vegetative vigor, light pruning is often employed to restrain vegetative growth. Vines on poorer soils usually benefit from more extensive pruning. This channels nutrient reserves into the buds retained, assuring sufficient vigor for canopy development, fruit ripening, and inflorescence initiation.

Varietal fruiting characteristics, such as the number of bunches per shoot, number of flowers per cluster, berry weight, and bud position along the cane, influence both the extent and type of pruning. Some cultivars, such as 'Sultana' and 'Nebbiolo,' commonly produce sterile buds near the base of the cane. As spur pruning would leave few fruit buds, these varieties are cane-pruned. Varieties showing strong apical dominance also may fail to bud out at the base or middle portion of the vine (Fig. 4.5). Arching and inclining the cane downward often minimizes apical dominance and balances bud break along the cane. Alternately, spur pruning limits the expression of apical dominance. Varieties tending to produce small clusters are left with more buds than those bearing large clusters to balance yield to varietal fruiting characteristics. Many of the more renowned cultivars produce small clusters, for example, 'Riesling,' 'Pinot noir,' 'Chardonnay,' 'Cabernet Sauvignon,' and 'Sauvignon blanc.' Large clustered varieties include 'Chenin blanc,' 'Grenache,' and 'Carignan.' The susceptibility of varieties such as 'Gewürztraminer' to *coulure* can influence how many buds should be retained. In addition, the ideal number of buds retained depends on the tendency of the variety to produce fertile buds. Examples of cultivars having a high percentage of fertile buds are 'Muscat of Alexandria' and 'Rubired.'

As noted below, the type of harvesting (manual versus mechanical) can influence the pruning and training system used. Spur pruning is generally more adapted to mechanical harvesting than cane pruning.

Not least among grower concerns is the cost of pruning. Although the most economical is winter pruning, it is the most difficult to employ skillfully, since it is im-

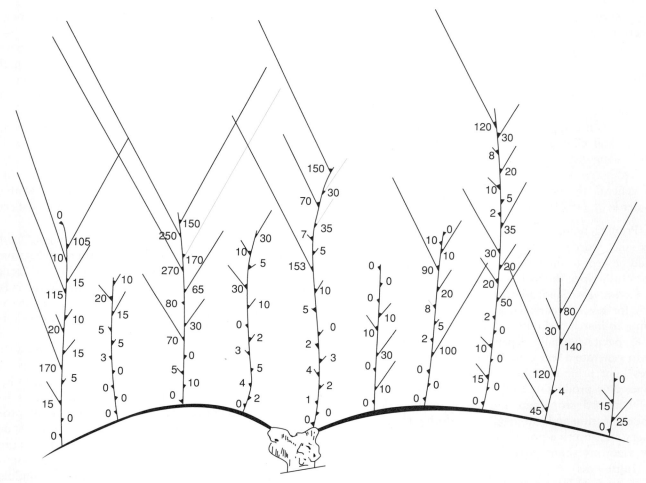

Figure 4.5 Apical dominance of shoot growth on unpruned vertically positioned canes. (From Huglin, 1958, reproduced by permission.)

possible to predict bud viability precisely. Cluster and fruit thinning permit the most precise regulation of yield but are the most expensive. Economic pragmatism generally favors winter pruning as the method of choice.

Pruning Procedures and Timing

Widely spaced vines of average vigor and pruning may generate up to 25 canes, possessing about 30 buds each. Thus, before pruning, the vine may possess upward of 750 buds. Even without pruning, only a fraction (about 100 to 150) of the buds would burst the following spring. As this is considerably above the ability of the vine to mature fully the fruit so generated, pruning is required. However, as noted above, excessive removal undesirably impacts on fruit maturity and subsequent fruitfulness. Thus, determining the appropriate number of buds to retain is one of the more complex yearly tasks of the grape grower. The task is made more difficult since actual fruitfulness is known only

months after completing winter pruning. This has led to several attempts to assess yield potential more accurately.

One method of measuring yield potential involves the direct observation of sectioned buds. Although reliable, it is not easily used by growers. Another procedure pioneered by Shaulis and co-workers is **balanced pruning**, which involves an initial estimate of the degree of pruning required. Based on this estimate, the vine is pruned to retain slightly more than the optimal number of buds. Only buds separated by obvious internodes are counted. Base (*noncount*) buds usually remain dormant and are not counted. The prunings are weighed, and a final pruning brings the bud number down to the recommended level. The recommended number of buds to retain is indicated in increments of growth (**weight-of-prunings**) above a minimum established empirically for each variety (see Table 4.1). Pruning recommendations are usually given in parentheses, for example, (30 + 10). The first value indicates the number of buds (*count nodes*) retained for the first pound of prunings, and the

Table 4.1 Suggested Pruning Severity for Balanced Pruning of Mature Vines of Several American (*Vitis labrusca*) and French–American Varieties in New York State[a]

Grape variety	Number of nodes to retain per pound (0.45 kg) of cane prunings		
	First pound (0.45 kg)		Each extra pound (0.45 kg)
American cultivars			
'Concord'	30	+	10
'Fredonia'	40	+	10
'Niagara,' 'Delaware,' 'Catawba'	25	+	10
'Ives,' 'Elvira,' Dutchess'	20	+	10
French–American hybrid Cultivars[b]			
Small-clustered varieties ('Maréchal Foch,' 'Leon Millot,' etc.)	20	+	10
Medium-clustered varieties ('Aurore,' 'Cascade,' 'Chelois,' etc.)	10	+	10
Large-clustered varieties ('Seyval,' 'de Chaunac,' 'Chancellor,' etc.)	20	+	10

[a] After Jordan *et al.* (1981), reproduced by permission.

[b] All require suckering of the trunk, head, and cordon during the spring and early summer.

second value refers to the number of buds retained for each additional pound of prunings. Vines producing less than a minimum weight-of-prunings are pruned to less than the specified number of buds. Skilled pruners need weigh the prunings from only a few vines per hectare to check their estimate of pruning weight.

In balanced pruning, the current year's growth is used as a measure of the fruitfulness and capacity of the vine. The technique has been widely adapted for *V. labrusca* varieties in eastern North America. Application of the technique to other varieties has been somewhat less successful. When used, pruning recommendations must account for cultivar fertility and prevailing climatic conditions. For example, varieties of low fertility need more buds to improve yield potential, while bud retention must be adjusted downward in regions with short growing seasons to permit full ripening of the crop. In addition, the tendency of noncount (base) buds to augment fruit production has limited the value of balanced pruning with most French–American hybrids (Morris *et al.*, 1984).

Pruning is often done in midwinter, but it may be conducted from leaf fall until or after bud break. Pruning should begin only after the phloem has sealed itself with callose for the winter. This avoids nutrient loss and activation of partially dormant buds. Pruning during the winter has many advantages, the most impor-

tant of which may be that most other vineyard activities are at a minimum. Also, bud counting and cane selection are more easily performed on bare vines. Finally, prunings can be quickly disposed of either by chopping for soil incorporation or collected for burning.

For tender varieties, pruning is normally performed late, when the danger of killing frost has passed. The proportion of dead buds can then be assessed directly by cutting open a sampling of buds. Dead buds show blackened primary and secondary buds. An alternative procedure involves **double-pruning.** Most of the pruning and cleanup occur during the winter, with the final removal delayed until bud break. The delay in bud burst induced by this technique often has been used for frost protection. It can retard bud break by over 1 week in precocious varieties. Late pruning also has been used to delay bud break in regions experiencing serious early season storms, such as the Margaret River region of Western Australia. Another factor that can favor late pruning is a reduction in the incidence of Eutypa dieback. Because the delay induced often persists through berry maturity, harvesting may be equally retarded. Thus, the advantages of delayed bud break must be weighed against the disadvantages of deferred ripening.

Pruning and thinning in late spring and early summer involve adjustments that can improve vine microclimate, minimize wind damage, and limit fruit production. These activities may involve disbudding, pinching, suckering, topping, and thinning of flower or fruit clusters.

Disbudding is used commonly in the initial training of a young vine into its desired mature form. It also may be used early in the year to remove unwanted base or latent buds that may have become active. Their early removal diminishes unproductive expenditure of nutrient reserves and favors early vigor in the remaining shoots.

Once growth has commenced, "summer" pruning may involve the partial or complete removal of shoots. Complete removal, called **suckering,** usually involves the removal of suckers and water sprouts (Fig. 4.6). Repeated activation of suckers and water sprouts throughout the growing season is often a sign of overpruning and insufficient energy being directed to fruit production.

Although suckers are undesirable because they direct photosynthate away from the fruit and disrupt training, some water sprouts can be useful. If favorably positioned, water sprouts can act as **replacement spurs** in repositioning growing points of the vine (Fig. 4.6). As the vine grows, the location of **renewal wood** tends to move outward. Replacement spurs reestablish, and thus maintain, the shape and training system of the vine.

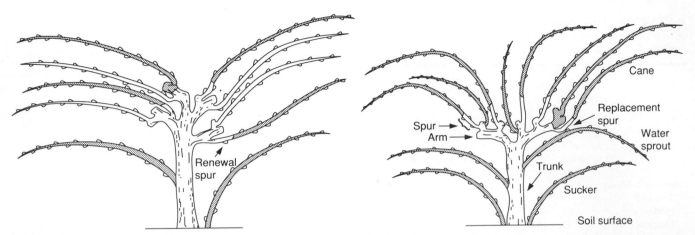

Figure 4.6   Illustration showing a head-trained vine cane-pruned (left) and spur-pruned (right). Shaded areas refer to water sprouts, suckers, and portions of canes removed at pruning. (From *Grape Growing*, R. J. Weaver, Copyright © 1976 John Wiley & Sons, Inc. Reprinted by permission of John Wiley & Sons, Inc.)

Depending on their position, water sprouts also may supply nutrients for fruit development.

Partial removal, called **trimming,** can vary widely in degree and timing. **Pinching** refers to the removal of the uppermost few centimeters of shoot growth. More extensive trimming is variously called **tipping, topping,** or **hedging,** depending on the length of shoot removed.

Pinching is usually conducted early in the season. When pinching is conducted during flowering, fruit set may be enhanced. The procedure is useful in reducing *coulure* in varieties affected by that physiological disorder. This effect may be explained as diminishing the competition between the developing leaves and embryonic fruit. The lateral bud activation that follows partially replaces the leaf canopy that would have been produced at the primary shoot apex. Pinching also can help to keep shoots positioned upright. The subsequent growth of lateral shoots may produce fruit shading that can minimize fruit sunburn in hot sunny climates.

Tipping is usually performed later and may be repeated periodically throughout the growing season. In windy environments, it produces shorter, more sturdy, upright shoots that limit damage caused by the whipping of long shoots. Topping also can improve canopy microclimate by removing draping leaf cover. Hedging is commonly used to facilitate movement of machinery through the vineyard by removing entangling vegetation. This is necessary where vines are densely planted in narrow rows (≤2 m), but it also can be useful with vigorous vines grown in wide rows (3.6–4 m).

Both early (during or before fruit set) and late (after *véraison*) trimming tends to be undesirable. Early trimming stimulates bud activation and creates new carbohydrate demands that compete with fruit set and in-florescence initiation. Late trimming, by promoting further vegetative growth, can delay both fruit and cane maturation, as well as decrease cold hardiness. These negative influences become more pronounced as the level of trimming is increased.

Although trimming removes photosynthetically active vegetation, the vine often compensates for the foliage loss. This commonly involves the induction of additional shoot and leaf growth. The early termination of such growth associated with the fruit zone can aid berry maturation. More subtle physiological compensation occurs through delayed leaf senescence and higher photosynthetic rates in the remaining leaves.

Trimming can variously affect fruit composition, depending on the timing and degree of shoot removal. Many of the effects resemble those of basal leaf removal discussed below. The increase in fruit potassium content occurs more slowly and peaks at a lower value (Solari *et al.*, 1988). The rise in berry pH, and the decline in malic acid content associated with ripening, may be less marked. °Brix (a measure of total soluble solids) may be little influenced, or reduced, in trimmed vines. Anthocyanin synthesis may be adversely affected in varieties, such as 'de Chaunac' (Reynolds and Wardle, 1988). This tendency is presumably not found with 'Cabernet Sauvignon' in Bordeaux, where hedging is commonly practiced. Improvements in amino acid and ammonia nitrogen levels associated with light trimming shortly after fruit set, if common, may be one of the more desirable features induced by trimming (Solari *et al.*, 1988).

Considerable interest has been shown recently in single or repeated leaf removal around fruit clusters (**basal leaf removal**). Leaf removal selectively improves air movement in, and light exposure around, the clusters.

It also eases the effective application of protective chemicals to the fruit. Reduced incidence of bunch rot was one of the early benefits detected after basal leaf removal. This has been associated with the increased evaporation potential, wind speed, higher temperature, and improved light exposure in and around the fruit (Thomas *et al.*, 1988).

Additional benefits of basal leaf removal generally include a reduction in titratable acidity (Fig. 4.7), which is associated with a reduction in the uptake of potassium and enhanced degradation of malic acid. Tartaric and citric acid levels are seldom affected. Anthocyanin levels generally are significantly enhanced, whereas total phenol content may rise or remain unchanged. Levels of grassy, herbaceous, or vegetable odors decline, while fruity aromas may rise or remain unaffected. Levels of terpenes (Table 4.2) and norisoprenoids (Marais *et al.*, 1992a) rise with increased fruit exposure in some cultivars. While usually beneficial, excessive production of some norisoprenoids [e.g., 1,1,6-trimethyl-1,2-dihydronaphthalene (TDN)] and their precursors can generate an undesirable kerosenelike fragrance in 'Riesling' wines (Marais *et al.*, 1992b). Berry sugar content is seldom affected significantly.

Most studies have noted no change in cold hardiness or fruitfulness as a result of basal leaf removal. This may result because leaf removal usually is conducted after inflorescence initiation and the more mature nature of the buds near the cane base.

The benefits of basal leaf removal generally show most clearly when detachment occurs between blooming and *véraison.* Earlier removal increases *coulure,* reduces inflorescence initiation, decreases berry quality, and disrupts subsequent bud break (Candolfi-Vasconcelos and Koblet, 1990). These effects probably

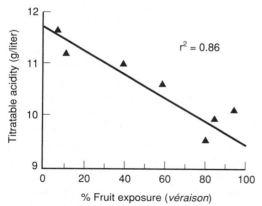

Figure 4.7 Relationship between fruit exposure at *véraison* and titratable acidity at harvest of the cultivar 'Sauvignon blanc.' (From Smith *et al.*, 1988, reproduced by permission.)

Table 4.2 Aroma Profile Analysis and Protein Content of 'Sauvignon Blanc' Fruit with and without Basal Leaf Removal[a]

	Control	Basal leaf removal
Free aroma constituents (μg/liter)		
Geraniol	1.6	9.5
trans-2-Octen-1-al	0.3	3.1
1-Octen-3-ol	2.4	6.6
trans-2-Octen-1-ol	6.5	11.7
trans-2-Penten-1-al	4.6	12.7
α-Terpineol	1.9	4.1
Potential volatiles (μg/liter)		
Linalool	23	49
trans-2-Hexen-1-ol	29	61
cis-3-Hexen-1-ol	5.1	6.3
trans-2-Octen-1-ol	321	830
2-Phenylethanol	17.9	50
β-Ionone	26	66
Protein (mg/liter)[b]		
Molecular weight > 66,000	32	33
Molecular weight < 20,000	62	81

[a] From Smith *et al.* (1988), reproduced by permission.
[b] Based on bovine serum albumin as standard.

develop because the basal leaves are the first to export carbohydrates to the growing shoot and embryonic fruit. Late removal shows few benefits but does stimulate root growth (Hunter and Le Roux, 1992).

The impact of basal leaf removal is also influenced by the number of leaves removed. Removal usually involves the leaves positioned immediately above and below, and opposite, the fruit cluster. The vine can compensate for the leaves lost by increasing the number of leaves produced on lateral shoots, delaying leaf senescence, and/or increasing the photosynthetic efficiency of the remaining leaves (Candolfi-Vasconcelos and Koblet, 1991).

The practice of basal leaf removal, currently a labor-intensive activity, may become more common with the development of efficient mechanical leaf-removers. One model involves suction that draws leaves toward the cutting blades.

Another form of spring/summer pruning involves flower or fruit cluster thinning. In this practice, the purpose is to prevent overcropping. Flower cluster thinning has a similar effect to delayed winter pruning, in permitting a more precise adjustment of fruit load to vine capacity. Fruit cluster thinning can be even more precise in this regard, but it is more difficult owing to the more advanced state of canopy development. Fruit cluster thinning is especially useful with French–American hybrid cultivars, as base bud activation can add considerably to fruit load. As fruit set tends to suppress growth of lateral shoots, fruit cluster thinning achieves the desired fruit production without promoting undesirable bud activation.

Occasionally, flower cluster thinning favors the development of more, but smaller, berries in the remaining clusters. This can be valuable in reducing bunch rot in varieties forming compact clusters, such as 'Seyval blanc' and 'Vignoles' (Reynolds *et al.*, 1986). However, within the usual range of vine capacity, fruit cluster thinning often shows only marginal improvement in fruit quality. Thus, the yield losses incurred by cluster thinning must be balanced against the benefits of its use. It may be of value only with cultivars or clones that tend to overproduce regularly (Bavaresco *et al.*, 1991).

Although balancing yield to capacity is important, it is equally important to choose the best canes for fruit bearing. The proper choice of bearing wood is especially significant in cool climates. Fully matured, healthy canes produce buds that are the least susceptible to winter injury. Canes that develop in well-lit regions of the vine also tend to be the most fruitful. Browning of the bark is the most visible sign of cane maturity. Typical internode distances are a varietally useful indicator of fruitfulness. The presence of mature lateral shoots on a cane may signify good bearing wood. Canes of moderate thickness (1 to 1.5 cm) are generally preferred as their buds tend to be the most fruitful. Canes of similar diameter tend to be of equal vigor and, thus, maintain balanced growth throughout the vine.

Where canes of different diameter are retained, the number of buds (nodes) left per cane should reflect the capacity of the individual canes. Correspondingly, weaker canes are pruned shorter than thicker canes.

Also important to balanced growth is an appropriate spacing of the bearing wood. Canes retained should permit the optimal positioning of the coming season's bearing shoots for photosynthesis and fruit production. Thus, canes are normally selected that originate from similar positions on the vine. If the vine is cordon trained, canes originating on the upper side of the cordon are generally preferred.

Proper manual pruning requires a knowledge of how varietal traits and the prevailing climate influence vine capacity. An assessment of the health of each vine and the appropriate location, size, and maturity of individual canes is also critical. This requires considerable skill in a labor force that is becoming harder to locate and retain. Because of the savings obtained with the implementation of mechanical harvesting, pruning has become one of the major costs in maintaining a vineyard. These factors have led to increased interest in **mechanical pruning**. In addition, **minimal pruning** is becoming

popular in some regions. It involves only **skirting** around the base of the vine to prevent shoot and fruit contact with the ground. By 1987, minimal pruning was being practiced on over 1500 hectares of commercial vineyards in Australia (Clingeleffer and Possingham, 1987).

Mechanical pruning is most successfully used on cordon-trained spur-pruned vines. The cutting planes can be adjusted easily to remove all growth except that in a designated zone around the cordon. The cutting planes also can be adjusted for skirting or hedging foliage just before harvest to ease mechanical harvesting. If subsequent manual pruning is required to obtain the desired results, the mechanical component is termed **prepruning**. Prepruning can reduce pruning time by up to 80%.

With several grape varieties, mechanical pruning has the additional advantage of generating vines of more uniform size that produce smaller, but more numerous, bunches. Yield is generally unaffected or increased. However, with varieties such as 'Concord,' mechanical pruning can give rise to uneven ripening, low soluble solids, and poor coloration.

Problems with the tendency to overprune, associated with excessive shoot vigor and reduced grape quality, led to a study of the use of **minimal pruning.** Minimal pruning is based on the realization that some cultivars can regulate their own growth and yield good quality fruit without pruning. Initially developed by Clingeleffer (1984) for 'Sultana' grown in hot, irrigated vineyards, minimal pruning has been used with considerable success with several premium grape varieties. It has been successful in both cool and warm viticultural regions of Australia (Clingeleffer and Possingham, 1987). Initially, vines may overcrop, but this diminishes as the vines reduce the number of buds that mature and burst in succeeding years. Spontaneous abscission of most immature shoots in the autumn eliminates the need for winter pruning. Long, trailing vegetative growth can be removed mechanically by skirting.

Minimally pruned vines produce more but smaller shoots, which possess fewer but more closely positioned nodes. Most cultivars maintain their shape and vigor when cordon trained, then minimally pruned. Crop yield is either sustained or considerably enhanced, depending on the variety, clone, and rootstock involved. The fruit is borne on an increased number of bunches, each containing fewer and smaller berries. Commonly, the fruit is borne more uniformly over the outer portion of the vine, and in well-exposed locations. Fruit maturity is generally delayed about 1 week, as indicated by the slightly reduced degrees Brix, lower pH, and higher acidity (McCarthy and Cirami, 1990).

Fruit color may be reduced, but this may be compensated for in winemaking by the smaller berry size.

The improved yield of minimally pruned vines may result from nutrients not lost by pruning. This permits rapid canopy development and limits leaf/fruit competition during berry development. An additional advantage of minimal pruning comes from the easier removal of fruit during mechanical harvesting. In addition, fewer leaves are removed along with the fruit because of the more flexible canopy.

A potential benefit of minimal pruning may come from a reduced incidence of some diseases, like Eutypa dieback. The elimination of most pruning wounds should decrease the incidence of the diseases. However, retention of canes infected with *Phomopsis viticola* could lead to the increased incidence of Phomopsis cane and leaf spot, a rare disease in conventionally pruned vineyards (Pool *et al.*, 1988).

Currently, minimal pruning appears to be best suited to situations where the vines are vigorous, owing to properties of the clone or rootstock, fertilizer application, or irrigation. It appears to be less applicable to vines grown in poor soils, under humid conditions, or in cool climates. In cool regions, ripening may be critically delayed and vine self-regulation be less pronounced. Wines produced from the fruit are well balanced but generally lighter in color than those derived from vines trained and pruned more traditionally.

In viticulture, pruning normally refers to the removal of aerial parts of the vine. In other perennial crops, such as fruit trees, pruning frequently involves root trimming to restrict excessive shoot vigor. In most instances, **root pruning** is employed to restrict excessive shoot vigor. Clear cultivation in the vineyard, by repeatedly severing surface feeder roots, can have the same effect. However, cutting large diameter roots of grapevines to a depth of 60 cm can promote shoot growth, under conditions where growth is retarded by soil compaction. Under such conditions, root pruning is recommended in adjacent rows in alternate years (van Zyl and van Huyssteen, 1987). Susceptibility of the roots to pathogenic attack does not seem enhanced by pruning. New lateral root development may occur up to 5 cm back from the pruning wound.

Training Options and Systems

Training involves the development and maintenance of the woody structure of the vine in a particular form. The form attempts to achieve optimal fruit quality and yield, consistent with prolonged vine health. Because grapevines have remarkable regenerative powers, established vines can buffer the effects of training for several years. Thus, studies on training systems need to be conducted over several years to assess the long-term effects. In addition, factors such as varietal fruiting habits, prevailing climate, disease prevalence, desired fruit quality, and grape pricing all can influence the choice of training system. Many of these factors are similar to those that affect pruning choices discussed above. In addition, tradition and Appellation Control legislation can affect the acceptance of newer, more efficient systems.

Because many training systems are only of local interest, and because a better understanding of vine physiology is suggesting new and better ways of training, the following discussion focuses on an understanding of the various components that distinguish training systems. Excellent illustrative diagrams outlining the establishment of some popular training systems are given in Pongrácz (1978). Details on locally used training systems are generally available from regional viticultural research stations.

There is no universally accepted system of classifying training systems. Nevertheless, most systems can be characterized by the origin of the bearing wood (**head** versus **cordon**) and length of the bearing wood (**canes** versus **spurs**). These groupings can be further subdivided by the height and position of the bearing wood, placement of the bearing shoots, and the number of trunks retained. Ancient systems of training on trees do not conveniently fit into the above classification and resemble slightly modified natural grapevine growth.

BEARING WOOD ORIGIN

One of the most distinguishing features of a training system relates to the origin of the bearing wood. On this basis, most training systems can be classed as either head trained or cordon trained (Fig. 4.8). Head training positions the canes or spurs that generate the fruit-bearing shoots radially on a swollen apex, or several radially positioned short arms at the trunk apex. In contrast, cordon training positions the bearing wood, from which the fruit bearing shoots arise, equidistantly along the upper portion(s) of the trunk, the cordon(s). In most cordon-trained systems, the cordon is developed along a horizontal plane, usually parallel to the row. Occasionally, however, the cordon is inclined at an intermediate angle or is developed vertically.

Head training was previously much more common than at present. Its simplicity of form makes it easy and inexpensive to develop. Also, as the bearing shoots of one season often can be used to provide the bearing wood for the next year's crop, the system is easy to maintain. As the trunk thickens, the vine becomes self-supporting, avoiding the expense and complexity of a trellis and the shoot tying required by more complicated training systems. The lack of wiring along rows

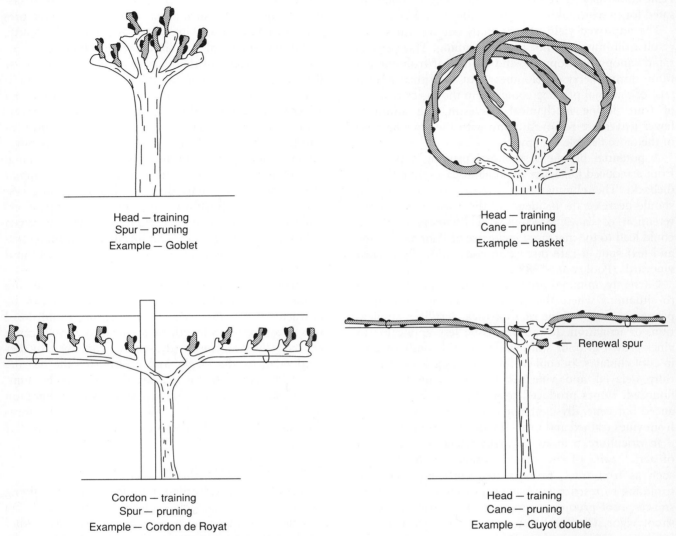

Head — training
Spur — pruning
Example — Goblet

Head — training
Cane — pruning
Example — basket

Cordon — training
Spur — pruning
Example — Cordon de Royat

Head — training
Cane — pruning
Example — Guyot double

◄— Renewal spur

Figure 4.8 Diagrammatic sketches illustrating head and cordon training systems. Unshaded areas represent old wood, shaded areas bearing wood, and black areas buds. (After *Grape Growing*, R. J. Weaver, Copyright © 1976 John Wiley & Sons, Inc. Reprinted by permission of John Wiley & Sons, Inc.)

also permits cross-cultivation, thus facilitating tillage for weed control.

The primary drawback to head training is the potential for shoot crowding. Shoot crowding leads not only to undesirable fruit, leaf, and cane shading, but also to higher canopy humidity. The latter favors disease development, whereas the former leads to poor fruit maturation, as well as reduced photosynthesis, bud fruitfulness, and cane maturation. However, avoiding these problems with severe pruning creates additional problems, notably curtailed yield and delayed development of shoot growth. Severe pruning also can accentuate vigorous vegetative growth, to the detriment of fruit development and maturation. Furthermore, limited bud

retention places the vine at greater risk to the effects of killing frosts.

The disadvantages of head training are often augmented when associated with spur pruning. Spur pruning increases the tendency to produce compact canopies around the head. While providing useful fruit shading in hot, dry climates, it favors disease development and poor maturation in cool, moist climates. These effects may be partially offset by using fewer, but longer, spurs. However, this makes maintaining the vine shape more difficult. Moreover, head-trained, spur-pruned systems are unsuited to mechanical harvesting. They most commonly have been used in dry Mediterranean-type climates, even though they may lead to higher evapotranspiration rates than other training systems (van Zyl and van Huyssteen, 1980).

The *Goblet* training system used extensively in southern Europe and elsewhere (Fig. 4.8) is an example of a head-trained, spur-pruned system. It is adapted to cultivars with medium-sized fruit clusters that produce fruitful buds to the base of the cane.

In contrast, cane pruning associated with head training has been more popular in cooler, more moist climates. In addition to providing a more favorable canopy microclimate by dispersing the fruit-bearing shoots away from the head, cane pruning permits more buds to be retained. This is important to improving yield in cultivars bearing small fruit clusters, for example, 'Riesling,' 'Chardonnay,' and 'Pinot noir.' Larger-clustered varieties, such as 'Chenin blanc,' 'Carigan,' and 'Grenache,' or those with very fruitful buds, such as 'Muscat of Alexandra' and 'Rubired,' tend to overproduce with cane pruning. However, the shoot dispersion provided by cane pruning usually requires the added expense of wiring and trellis supports. The *Mosel Arch* (Fig. 4.10A) is a novel and effective approach by which cane arching supplies both shoot support and diminished apical dominance along the cane.

In cane pruning, it is necessary to select renewal spurs (Fig. 4.6). They are required because the canes of one season usually cannot be used as bearing wood the next season. Their use would quickly result in relocating the bearing wood away from the head. The same tendency occurs with spur pruning, but at a much slower rate. The conscious selection of renewal shoots may be reduced if water sprouts develop from the head. Common examples of head-trained, cane-pruned systems are the *Guyot* in France and the *Kniffin* in eastern North America.

In contrast, to the central location of the bearing wood in head training, **cordon training** positions the bearing wood along the upper portion of an elongated trunk. Most systems possess either one (**unilateral**) or two (**bilateral**) horizontally positioned cordons. Occasionally, the trunk is divided into two horizontal trunks, directed at the right angles (laterally) to the vine row. These subsequently branch into two or more cordons running in opposite directions parallel with the row. This is **quadrilateral** cordon training. Vertical (upright) cordon systems are uncommon as apical dominance and shading combine to promote growth at the top. These features complicate maintaining a balanced vertical cordon system. Vertical cordons are also poorly adapted to mechanized pruning and harvesting. Several obliquely angled cordon-trained systems are popular in northern Italy.

Horizontal cordons experience little apical dominance because of the uniform distance of the buds from the ground. However, the horizontal positioning of the cordon places considerable stress on its junction with the trunk. This requires the use of one or more support wires to carry the weight of the shoot system and crop. Because the bearing wood is selected to be uniformly spaced along the cordon, the shoots generally develop a canopy microclimate favorable to optimal fruit maturation. Higher yields can result from improved net photosynthesis, production of more fruitful buds, and increased nutrient reserves located in the enlarged woody structures of the vine.

Location of the fruit in a common zone along the row makes cordon training well suited to mechanical harvesting. It also provides relatively homogeneous growing conditions, which favor the development of uniform fruit color, maturity, size, and chemical composition. Furthermore, positioning the bearing wood in a narrow region above or below the cordon makes the vine more amenable to mechanized pruning.

The disadvantages of cordon training include the higher costs involved in using a stronger trellis and support wires, the additional time and expense required in its establishment, and the greater skill demanded in selecting and positioning the arms that bear the spurs or canes. In cool climates, the more synchronous bud break can increase the damage caused by late spring frosts. Because cordon-trained vines tend to be more vigorous, they are commonly spur-pruned to minimize overcropping. This feature limits the use of cordon training to varieties that bear fruit buds down to the base of the cane.

Cane pruning is combined with cordon training only for strong vines capable of maturing heavy fruit crops. Because of the large permanent woody structure of the trunk and cordon(s), cordon training is often inappropriate for French–American hybrids. The woody component increases the number of base (noncount) buds that may become active and accentuates the tendency to overcrop.

Because of the many advantages of cordon training, most modern training systems, such as the *Geneva Double Curtain*, the *Ruakura Twin Two Tier*, *Lyre*, and *Tatura* systems, employ it. Several older systems also are examples of cordon training, for example the *Cordon de Royat* in France, the *Hudson River Umbrella* in eastern North America, the *Dragon* in China, and pergolas in Italy.

BEARING WOOD LENGTH

As noted above, the choice of spur versus cane pruning often depends on the training system used. Conversely, the training system may be chosen based on the advantages provided by spur or cane pruning, respectively.

Cane pruning is especially suited to cultivars producing small clusters that need to retain extra buds. However, to permit the development of a desirable canopy microclimate, the variety also must possess relatively long internodes to limit shoot overcrowding. In addition, cane pruning enhances vine capacity by retaining more apically positioned buds. These are more fruitful than basally positioned buds. This is especially important for cultivars that produce sterile (leaf) buds at the base of the canes. Cane pruning not only allows the precise positioning of shoots along the vine row, but also facilitates the development of wide-topped trellises that can extend both laterally and parallel to the row. Cane-pruned vines tend to develop their canopy sooner in the season. This is particularly valuable in varieties susceptible to *coulure,* such as 'Gewürztraminer' and 'Muscat of Alexandra,' or in vigorous vines.

Disadvantages of cane pruning include the expense involved in the trellising and tying shoots. Because the crop develops from but a few canes, particular care must be taken in choosing those retained. Correspondingly, pruning must be done manually by skilled workers. In addition, damage or death of even one cane can seriously reduce individual vine yield. A further complicating factor is the removal of the current year's bearing wood. Bearing wood for the next season's crop typically comes from shoots that develop from renewal spurs. The long length of the bearing wood can result in uneven shoot development owing to apical dominance. This can lead to nonuniform canopy development and asynchronous fruit ripening. Arching or positioning the canes obliquely downward can often minimize apical dominance but itself places the bearing shoots and fruit in diverse environments. Finally, converting a vine from spur to cane pruning often leads, temporarily, to overcropping.

Spur pruning tends to show the inverse set of advantages and disadvantages to cane pruning. Because of its greater simplicity, it is potentially more amenable to mechanical pruning. Correspondingly, it requires less skill when conducted by manual labor. Because spur pruning tends to restrict fruit production to predetermined locations, it can be particularly useful when designing training systems appropriate for mechanical harvesting. Assuming that all spurs are located equidistantly from the ground, there is little likelihood of apical dominance. The tendency for spur pruning to restrict fruit production can be either beneficial or detrimental, depending on the vigor and capacity of the vine.

Restricting yield may be desirable in cool climates, or under poor nutrient conditions, but less important in warm climates on deep, rich soils. Spur pruning is inappropriate for varieties that bear sterile buds at the base of the bearing wood or those susceptible to *coulure.* The delay in leaf production associated with spur pruning can result in increased competition between expanding leaves and the young developing fruit. Finally, without removal of malpositioned spurs, shoot crowding is likely.

SHOOT POSITIONING

As with training of the woody structure of the vine, shoot placement can be designed to promote vertical, horizontal, or inclined growth (Fig. 4.9). In addition, the shoots may be prevented from, permitted to, or encouraged to arch over and grow downward (Fig. 4.10). Currently, vertical positioning is generally favored for its suitability to mechanical harvesting and pruning. Mechanical harvesting is facilitated by the central location of the fruit on the vine. Mechanical pruning and trimming are equally aided by the largely unobstructed access to the bearing shoots.

Trailing growth usually requires little shoot tying and has the added advantage of limiting vigorous growth. The tendency of most *V. vinifera* cultivars to grow upright complicates early formation of trailing vertical canopies. Trailing growth has the further advantage of

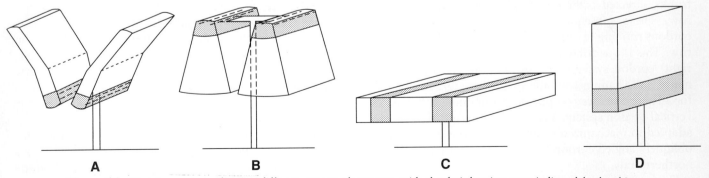

   A   **B**   **C**   **D**

Figure 4.9 Training systems showing different canopy placements, with the fruit-bearing zone indicated by hatching. (A) Inclined upright (Lyre), (B) trailing (Geneva Double Curtain), (C) horizontal (Lincoln), and (D) vertical upright (Guyot).

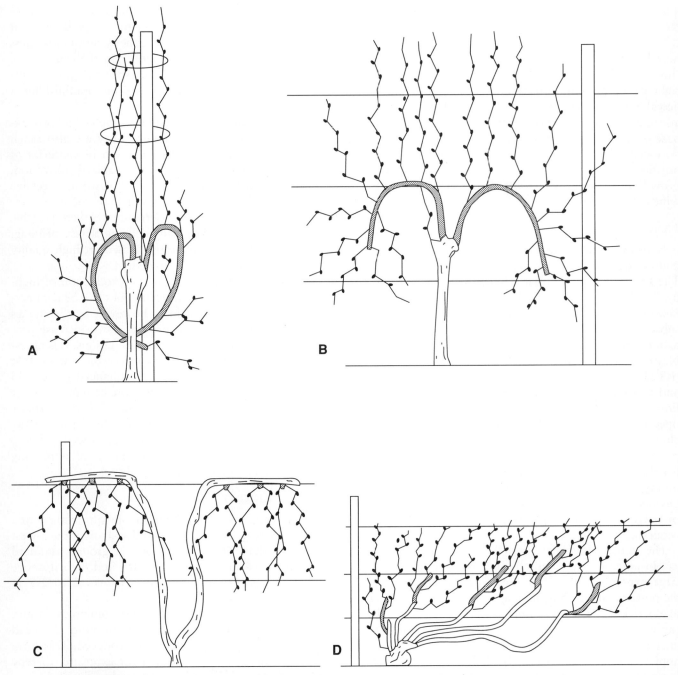

Figure 4.10 Diagrammatic illustrations of vines showing different shoot positioning. (A) Upright and horizontal (Mosel Arch), (B) upright and procumbent (Umbrella Kniffin), (C) procumbent (Hudson River Umbrella), and (D) upright (Chablis). Old wood is unshaded, and bearing wood is shaded.

requiring limited trellising and wire supports. Arching or direct trailing exposes the basal portion of the shoot to sun and wind. This favors cane maturation, bud fruitfulness, fruit maturation, and vine health. Only in hot, sunny climates is some fruit shading desirable. Vertical canopies are readily accessible to most vineyard practices and, therefore, are preferred over horizontally

(*Lincoln* and Italian *Tendone*) or inclined (*Dragon* and *Tatura Trellis*) canopies. The latter systems can be useful where increased fruit shading is considered beneficial.

Upright vertical canopies can be more expensive to maintain and often produce increased fruit shading. Positioning long shoots in an upright position usually requires tying the shoots to several wires. This is costly in terms of materials and complicates pruning because of cane entanglement in the support wires. These features may be avoided by developing short, stout shoots by shoot trimming. Another potential disadvantage of upright canopies is the positioning of the fruit and renewal zones at the base of the canopy. However, with adequate canopy division, sufficient light reaches the base of the shoot to permit adequate fruit ripening, inflorescence initiation, and cane maturation. The basal positioning of the fruit zone can prove advantageous in providing protection against hail, sunburn, strong winds, and bird damage.

CANOPY DIVISION

Division of the canopy to improve canopy microclimate is one of the newer concepts in training design. The first and most popular of these systems, pioneered by N. Shaulis in New York, is known as the *Geneva Double Curtain* (GDC) (Shaulis *et al.*, 1966). It has subsequently been modified to make it more applicable to mechanical harvesting and pruning (Baldini, 1982). Newer examples are the *Ruakura Twin Two Tier* (RT2T) (Smart *et al.*, 1990c), the *Lyre* (Carbonneau and Casteran, 1987), and the *Tatura Trellis* (van den Ende *et al.*, 1986). Such training systems have developed largely out of fundamental studies on vine microclimate. By dividing the canopy into separate components, exposure of the fruit and vegetation to the sun is enhanced, fluctuation in temperature augmented, humidity decreased, and transpiration rate increased. Divided-canopy systems possess many of the properties of narrow-row dense plantings, but tend to be more economical.

The changes involved favor fruit maturity and quality. In areas of high rainfall and low evaporation, the large exposed leaf canopy increases transpiration and the potential development of water stress. This can help to restrict shoot growth and favor fruit ripening. Because more buds (nodes) can be retained without producing excessive within-canopy shading, divided canopies can permit increased yield without quality loss. In addition, the retention of more buds can restrict excessive vine vigor, a problem in many New World vineyards. The high shoot number limits shoot growth and decreases internode length, leaf number, leaf size, and lateral shoot activation, features likely to create a favorable canopy microclimate.

Although divided canopies are often efficient and valuable means of restricting vine vigor, this may be an undesirable means of achieving this goal in low nutrient soils. Also, large exposed canopies could cause excessive water stress in areas where rainfall, or the possibility of irrigation, is limited. Although the complex trellis required for divided-canopy training systems are expensive, this may be offset by the increased yield and fruit quality obtained. Furthermore, canopy division is a less expensive means of vine devigoration than increasing planting density. Planting vines, especially grafted cuttings, is often the major expense in establishing a vineyard.

Undivided canopies are usually simpler and less expensive to develop and maintain. They are also much more common and have the sanction of centuries of use. Prior to the modern use of mechanized cultivation, irrigation, fertilization, and disease control, excessive vigor probably was less common than today. Correspondingly, favorable vine microclimates were possible with undivided canopies using hedging, severe pruning, and high-density vine planting. However, high quality of fruit came at the price of low yield.

Where the market will bear the production of high-cost, high-quality wine, old techniques can be commercially viable. In most regions, however, market forces require the use of modern procedures to increase production and reduce costs. In these situations, it is necessary to employ measures that direct the energy that could go into increased vigor into enhanced grape yield and quality. Where new vineyards are being planted, or old vineyards being replanted, use of divided-canopy training systems may be advisable. In existing vineyards, leaf and lateral shoot removal, as well as hedging, are less costly but valuable means of improving canopy microclimate and fruit quality.

CANOPY HEIGHT

Trunk height is one of the more obvious features of a training system. Training is considered low if the lowest arm is less than 0.6 m (2 ft) above the ground, standard if between 0.6 and 1.2 m (2–4 ft), and high if above 1.2 m (4 ft). Arbors often have trunks 2 to 2.5 m (6–7 ft) long.

Low trunks have been used most commonly in hot, dry climates. Combined with spur pruning and head training, short globular vines are produced. This form often minimizes leaf and fruit light exposure, thereby reducing drought stress and fruit sunburn. Soil shading also may reduce water evaporation from the soil and limit soil heating. However, once the soil is heated, the vine is in immediate proximity to the heat radiated from the soil. This may be valuable in minimizing day/night fluctuations in temperature, and it can be an important heat source for grape maturation and frost protection in the fall. Low trunks occasionally are used in cool climatic regions. Small vine size could minimize the portion of the vine damaged by freezing, as well as limit vine vigor on poor soils. The location of the renewal region close to the ground could provide the additional protection of snow cover. By limiting the need for vine trellising, short, stout trunks can lower vineyard operating expenses.

Although low training is occasionally used in cool regions, for example, Champagne, high training systems are more common. By positioning the buds and shoots away from the ground, they are partially protected from cold air that accumulates at soil level. Another advantage of high training is the greater access direct sunlight has to the ground in the spring. This can speed soil warming and encourage early vine growth. Medium to high trunks are required for training systems with trailing shoots. High trunks reduce, or eliminate, the need for skirting long vegetative growth. High trunks also facilitate herbicide application by positioning buds and shoots away from the ground. Furthermore, high trunks make most manual and mechanized vineyard practices easier by locating the canopy about breast height. Finally, high trunks possess a large woody structure that permits improved nutrient storage. This can result in earlier bud burst and increase fruit yield. However, by raising the canopy, greater stress is placed on the trunk. Thus, stronger trellising and wiring become necessary to support the vine.

TRUNK NUMBER

In most situations, the vine is trained to have a single trunk. Subsequently, it may be divided at the apex into two or more cordons. In some cold climates, such as the northeastern United States, two or more trunks are commonly developed.

Dual and multiple trunks have several advantages in cold regions. They provide protection against the consequences arising from cold damage or from fungal and bacterial infections that may kill one trunk. Also, by dividing the vine's energy between several trunks, each trunk grows more slowly and remains more flexible. This is important if the vine needs to be removed from its trellis and buried as protection against frigid cold weather every winter.

PLANTING DENSITY AND ROW SPACING

Although not directly an aspect of training, the effect of planting density on row spacing often significantly influences the choice of training system. Both planting density and row spacing have marked effects on the economic operation of a vineyard, as well as influencing vine growth.

In practice, it is often difficult to separate the direct effects of planting density, such as root competition, from indirect influences on canopy microclimate or from the impact of unrelated factors such as training system or soil type. Further complicating the situation is the fact that vineyards having the same average planting density may possess different between- and within-row vine spacing. Thus, vines inhabiting similar average soil volumes may experience markedly different degrees of root and shoot crowding. These factors undoubtedly help explain the diversity of opinion on the relative merits of various planting densities.

The planting densities commonly used in Europe have changed considerably over the past century and a half. Before the phylloxera epidemic, planting densities are reported to have reached up to 30,000 to 40,000 vines per hectare in Burgundy (Freese, 1986). Current values for narrow rows in Europe often vary from 4000 and 5000 vines per hectare, and occasionally rise above 10,000 vines/ha. In California and Australia, common figures for wide-row plantings range from about 1100 to 1600 vines/ha (2700 to 4000 vines/acre).

Vineyards planted at higher vine densities often, but not consistently, show desirable features such as improved fruit yield and quality (Table 4.3). The lower productivity of individual vines is usually more than compensated by the yield from the greater number of vines. Improved grape quality is usually explained in terms of limited vegetative vigor (lower level of bud activation and restricted shoot elongation) and the improved canopy microclimate that results. If true, the beneficial effects on yield and quality produced by higher planting density are similar in nature to that of canopy division, namely, by vine devigoration.

One of the advantages of high vine density planting is the greater rapidity with which vineyards come into full production. This suggests that some element of intervine competition is involved in inducing earlier physiological vine maturity. Higher vine (and bud) numbers per hectare also may provide some protection against yield loss due to winter kill.

The suppression of vegetative vigor associated with dense planting is probably explained in part by restricted root growth. Shoot and root growth are

Table 4.3 Effect of Plant spacing on Yield of Three-Year-Old 'Pinot noir' Vines[a]

Plant spacing (m)	Yield per vine	Yield per hectare
1.0 × 0.5	0.58	11.64
1.0 × 1.0	1.03	10.33
2.0 × 1.0	1.43	7.15
2.0 × 2.0	2.60	6.54
3.0 × 1.5	2.50	5.51
3.0 × 3.0	4.12	4.57
D-value ($p \leq 0.01$)	0.80	4.064

[a] From Archer *et al.* (1988), reproduced by permission.

strongly interrelated. Dense planting promote deeper root penetration, but the roots are confined laterally (Table 4.4). The proportions of fine, medium, and large roots are generally unaffected by planting density. Although root production per vine is reduced, high density planting results in an overall increase in root density per soil volume. The confined, but dense, root development associated with high density planting can result in high water stress under drought conditions. Large vines under low density conditions possess a more extensive root system, often with several deeply penetrating major roots. Thus, they are more likely to be able to extract water from deep soil horizons under drought conditions than are densely planted vines.

The major disadvantage of dense vine planting is the markedly increased cost of establishing a vineyard. The cost of planting grafted vines, even at low density, can exceed the cost of all other aspects of vineyard establishment. Thus, the expense of planting at high density may offset the potential economic benefits of increased yield and quality. In addition, improved grape quality is not always observed (Eisenbarth, 1992). Because increased planting density usually requires the use of narrower rows, additional costs may be incurred by the purchase of narrow wheelbase equipment when shifting from low to high density planting. Close planting complicates soil cultivation and may increase the need for herbicide use. The requirement for more severe and precise pruning can add further to the costs of high density planting. Increased expenditures also can result from treating more vines per hectare with protective chemicals. Finally, deep, rich soils may counteract the devigoration produced by vine competition.

An important feature favoring the retention of the wide-row plantings that typify most New World vineyards is their adaptation to existing vineyard machinery. With new training systems, widely spaced vines can achieve or surpass the yield and quality of the traditional narrow-row, dense plantings of Europe.

These features are achieved at lower planting costs, both initially as well as during replanting. Large vines also appear to live longer than the smaller vines characteristic of densely planted vineyards. Finally, the deeper and more extensive root system of large vines may limit the development of vine water stress.

The major disadvantage of wide-row spacing is its potential for crowding and poor canopy microclimate development. As noted below under vigor control, these can be limited by various procedures and the greater vine capacity used to economic advantage.

ROW ORIENTATION

Many factors can influence row orientation. Often the slope of the land dictates row orientation. To minimize soil erosion, it is usually advisable to position vine rows at right angles to the slope. Slope orientation also can influence the practicality of the normally preferred north/south alignment of vine rows.

Where possible, row orientation is positioned to achieve the maximal canopy exposure to direct sunlight. For most training systems possessing vertical canopies, this is obtained by a north/south row orientation. The north/south alignment provides the maximum direct exposure of the canopy's largest surface (the sides) during the midmorning and midafternoon hours (Smart, 1973). However, angling of the rows toward the southwest can improve photosynthesis by increasing exposure to the early morning sun. Conversely, a southeast angling could improve heat accumulation by the fruit in the autumn. Conditions such as the prevailing direction of the wind also can influence optimal vine row orientation.

CANOPY MANAGEMENT AND TRAINING SYSTEM DEVELOPMENT

In the discussion of training system components above, several divided-canopy systems were noted. These systems have developed out of fundamental stud-

Table 4.4 Effect of Plant Spacing on Root Pattern of 3-Year-Old Vines of 'Pinot noir' on '99 Richter' Rootstock[a]

Parameter	Plant spacing (m)					
	3 ×3	3 ×1.5	2 × 2	2 × 1	1 × 1	1 × 0.5
Primary roots (m)	2.21 (37%)	1.76 (38%)	1.67 (35%)	1.63 (39%)	1.09 (37%)	0.89 (36%)
Secondary roots (m)	2.99 (50%)	2.31 (49%)	2.58 (53%)	1.95 (47%)	1.38 (47%)	1.12 (46%)
Tertiary roots (m)	0.77 (13%)	0.61 (13%)	0.59 (12%)	0.56 (14%)	0.46 (16%)	0.46 (18%)
Total root length (m)	5.96×10^3	4.68×10^3	4.84×10^3	4.13×10^3	2.93×10^3	2.45×10^3
Root density (m/m^3)	1.10×10^3	1.73×10^3	2.02×10^3	3.44×10^3	4.89×10^3	8.21×10^3
Angle of penetration	15.3°	22.6°	30.9°	41.1°	58.6°	77.5°

[a] From Archer and Strauss (1985), reproduced by permission.

ies on vine microclimate. Smart *et al.* (1990a) have condensed many of these findings into a series of canopy management principles:

1. The rapid development of a large canopy surface to volume ratio increases photosynthetic efficiency, as well as fruit set and ripening. Tall, thin, vertical canopies aligned along a north/south axis permits maximal sun exposure.

2. To avoid both excessive interrow shading and energy loss by insufficient canopy development, the ratio of canopy height to interrow width between canopies should approximate unity.

3. Shading in the renewal/fruit-bearing zone of the canopy should be minimized. Shading has several undesirable influences on fruit maturation and health. These include augmented potassium levels, increased pH and herbaceous character, retention of malic acid, enhanced susceptibility to powdery mildew and bunch rot, and reduced sugar, tartaric acid, monoterpene, anthocyanin, and tannin levels. Shading also reduces inflorescence initiation, favors primary bud necrosis, suppresses fruit set, and slows berry growth and ripening. Current information does not permit a precise indication of the level of shading at which undesirable influences begin.

4. Excessive and prolonged shoot growth, causing a drain on the carbohydrate available for fruit maturation and vine storage, should be restrained by trimming or devigoration procedures (see below).

5. Location of different parts of the vine in distinct regions not only favors uniform growing conditions and even fruit maturation, but also facilitates mechanized pruning and harvesting practices.

These principles have been used to develop an ideotype of a good training system (Table 4.5).

NEW TRAINING SYSTEMS

The first training system based primarily on microclimate investigation was the **Geneva Double Curtain** (GDC) (Shaulis *et al.*, 1966). It is a tall (1.5 to 1.8 m) bilateral cordon system pruned to short spurs (four to six buds). The cordons diverge laterally and then bend, to be held about 1.2 m apart by parallel wires running along the row. Alternately, two short lateral cordons are pruned to four long canes. The latter are supported on cordon wires. The bearing shoots are positioned downward about flowering time. The rows are widely spaced, about 3 to 3.6 m apart, with interrow canopy spacings of about 2.4 m. Some further shoot positioning during the season may be necessary to keep the two canopies separate and minimize shading. Initially devel-

oped for *V. labrusca* varieties such as 'Concord,' the GDC has been used with French–American hybrids and *V. vinifera* cultivars in several parts of the world. Consistent with its divided canopy, and increased bud retention, fruit quality and yield tend to be increased. Although the GDC system demands more in terms of skill and materials, the higher yield and adaptation to mechanical pruning and harvesting usually more than offset the higher establishment costs.

Because the GDC positions the fruit/renewal zone at the apex of the canopy, it is especially valuable where maximal direct-sun exposure is desired. In some regions, this can result in increased fruit sunburn, rain damage, and bird damage. Locating the arms on the upper portion of the cordon or less shoot positioning may increase leaf protection of the fruit. Because trailing shoots need little support, the GDC has the lowest wiring costs of any divided-canopy system.

The **Lyre** system is considered the best of a series of divided-canopy systems developed in France (Carbonneau, 1985). The system consists of a short trunk branching into bilateral cordons that diverge laterally, and then bend along parallel cordon wires positioned about 0.7 m apart (Fig. 4.11). The bearing wood consists of equidistantly positioned spurs. The shoots are trained up onto two inclined trellises, supported by fixed and movable catch wires. Rows are placed about 3 to 3.6 m apart, with about 2.4 m separating vines within the row. The Lyre training system has been described as being an inverted GDC. Trimming excessive growth may be needed to keep the canopies separate and minimize basal shading.

The inclined canopies are ideally suited to maximizing direct sun exposure in the early morning and late evening, when photosynthetic efficiency is at a maximum. However, this advantage comes at the cost of extensive shading at the exterior base of the canopy by the overhanging inclined vegetation.

The Lyre system is considered to produce more fruit of equal to or better quality than that produced using traditional systems, such as the double Guyot (Carbonneau and Casteran, 1987). In Bordeaux, the value of the Lyre system is particularly evident on less favorable sites and during poor vintage years. These advantages probably arise from the increased canopy size and the favorable microclimate generated. Also, it has been noted that Lyre-trained vines are less susceptible to winter injury and inflorescence necrosis (*coulure*). The system is reported to be amenable to mechanical harvesting and pruning with slight adjustment of locally existing equipment.

The **Ruakura Twin Two Tier** (RT2T) system is based on principles noted above (see also Table 4.5). It differs from the previous systems by dividing the ca-

Table 4.5   Canopy Characteristics Promoting Improved Grape Yield and Quality[a]

Character assessed	Optimal value	Justification of optimal value
Canopy characters		
Row orientation	North/south	Promotes radiation interception (Smart, 1973), although Champagnol (1984) argues that hourly interception should be integrated with other environmental conditions (i.e., temperature) which affect photosynthesis to evaluate optimal row orientation for a site; wind effects can also be important (Weiss and Allen, 1976a,b)
Ratio of canopy height to alley width	~1 : 1	High values lead to shading at canopy bases, and low values lead to inefficiency of radiation interception (Smart *et al.*, 1990a)
Foliage wall inclination	Vertical or nearly so	Underside of inclined canopies is shaded (Smart and Smith, 1988)
Renewal/fruiting area location	Near canopy top	A well-exposed renewal/fruiting area promotes yield and, generally, wine quality, although phenols may be increased above desirable levels
Canopy surface area	~21,000 m²/ha	Lower values generally indicate incomplete sunlight interception; higher values are associated with excessive cross-row shading
Ratio of leaf area to surface area	<1.5	An indication of low canopy density is especially useful for vertical canopy walls (Smart, 1982; Smart *et al.*, 1985a)
Shoot spacing	~15 shoots/m	Lower values are associated with incomplete sunlight interception, higher values with shade; optimal value is for vertical shoot orientation and varies with vigor (Smart, 1988)
Canopy width	300–400 mm	Canopies should be as thin as possible; values quoted are minimum likely width, but actual value will depend on petiole and lamina lengths and orientation
Shoot and fruit characters		
Shoot length	10–15 nodes, ~600–900 mm length	These values are normally attained by shoot trimming; short shoots have insufficient leaf area to ripen fruit, and long shoots contribute to canopy shade and cause elevated must and wine pH
Lateral development	Limited, say, less than 5–10 lateral nodes total per shoot	Excessive lateral growth is associated with high vigor (Smart *et al.*, 1985a, 1989, 1990b; Smart and Smith, 1988; Smart, 1988)
Ratio of leaf area to fruit mass	~10 cm²/g (range 6–15 cm²/g)	Smaller values cause inadequate ripening, and higher values lead to increased pH (Shaulis and Smart, 1974; Peterson and Smart, 1975; Smart, 1982; Koblet, 1987); a value around 10 is optimal
Ratio of yield to canopy surface area	1–1.5 kg fruit/m² canopy surface	This is the amount of exposed canopy surface area required to ripen grapes (Shaulis and Smart, 1974); values of 2.0 kg/m² have been found to be associated with ripening delays in New Zealand, but higher values may be possible in warmer and more sunny climates
Ratio of yield to total cane mass	6–10	Low values are associated with low yields and excessive shoot vigor; higher values are associated with ripening delays and quality reduction
Growing tip presence after *véraison*	Nil	Absence of growing tip encourages fruit ripening since actively growing shoot tips are an important alternate sink to the cluster (Koblet, 1987)
Cane mass (in winter)	20–40 g	Values indicate desirable vigor level; leaf area is related to cane mass, with 50–100 cm² leaf area/g cane mass, but values will vary with variety and shoot length (Smart and Smith, 1988; Smart *et al.*, 1990b)
Internode length	60–80 mm	Values indicate desirable vigor level (Smart *et al.*, 1990b) but will vary with variety
Ratio of total cane mass to canopy length	0.3–0.6 kg/m	Lower values indicate canopy is too sparse, and higher values indicate shading; values will vary with variety and shoot length (Shaulis and Smart, 1974; Shaulis, 1982; Smart, 1988)
Microclimate characters		
Proportion of canopy gaps	20–40%	Higher values lead to sunlight loss, and lower values can be associated with shading (Smart and Smith, 1988; Smart, 1988)
Leaf layer number	1–1.5	Higher values are associated with shading and lower values with incomplete sunlight interception (Smart, 1988)
Proportion of exterior fruit	50–100%	Interior fruit has composition defects
Proportion of exterior leaves	80–100%	Shaded leaves cause yield and fruit composition defects

[a] From Smart *et al.* (1990a), reproduced by permission.

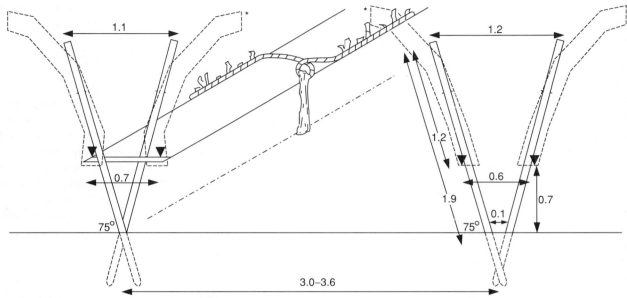

Figure 4.11 Diagrammatic representation of the Lyre training system, showing the inclined upright positioning of the divided canopies, with dimensions in meters. (From Carbonneau, 1985, reproduced by permission.)

nopy both vertically and laterally (Fig. 4.12). Each cordon bends along and is supported by wires running parallel to the row. Spur pruning facilitates both equal and uniform distribution of the canopies along the rows. The RT2T system also is compatible with mechanical pruning. The vertical canopy division is achieved by training alternate vines high and low, to higher and lower cordon wires, respectively. This is necessary to avoid gravitrophic effects on growth with individual vines. Buds positioned higher on a vine tend to grow more vigorously than those nearer the ground. Rows are placed 3.6 m apart. As the between-row canopies are positioned 1.8 m apart, the same as the within-row canopies, all canopies are equally separated. Also, as the combined height of the two vertical canopies (tiers) is equivalent to the width between the canopies, the ratio of canopy height to interrow canopy separation is unity. Individual vines are planted about 2 m apart within the rows.

To limit shading of the lower tier by the upper tier in the RT2T system, a gap of about 15 cm (6 inches) is maintained between the canopies by trimming. An alternative technique places the two cordons of each tier about 15 cm apart, with the upper canopy trained upward and the lower trailing downward. By positioning the fruit-bearing regions of both tiers under approxi-

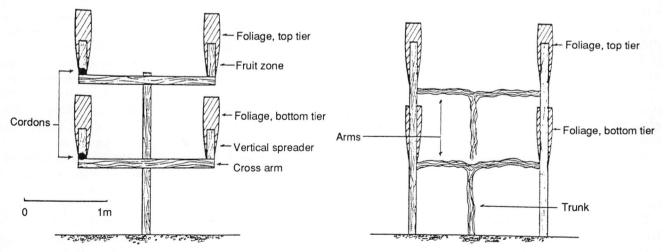

Figure 4.12 Two versions of support systems for the Ruakura Twin Two Tier training system. (From Smart *et al.*, 1990c, reproduced by permission.)

mately the same environmental conditions, chemical differences between the fruit of both tiers should be minimized.

Advantages of RT2T training involve a high ratio of leaf surface area to canopy volume and extensive cordon development. The former favors the creation of limited water stress that helps to restrict shoot growth following blooming. As a result, most of the carbohydrate is available for fruit growth or storage in the trunk. The formation of 4 m of cordon per meter of row provides many well-spaced shoots per vine. These further act to limit vine vigor by restricting internode elongation and leaf enlargement, thereby lessening canopy shading. Increased shoot numbers also enhance vine productivity, while improved canopy microclimate maintains or improves grape quality. Because of the strong vigor control provided by RT2T training, it is particularly useful for vigorous vines grown on deep, rich soils with ample water supply.

Although RT2T systems are more complex and expensive to establish than more typical systems, the wide row spacing limits planting costs. Also, the narrow vertical canopies ease mechanical pruning and harvesting, as well as increase the effectiveness of protective chemical application.

Several training systems have been developed from canopy management principles designed for tree fruit crops. An example is the *Tatura Trellis*, initially developed for peaches and pears (van den Ende, 1984). It possesses a 2.8 m high, V-shaped trellis, arranged with support wires to hold six tiered, horizontally arranged cordons on both inclined planes. Each vine is divided near the base into two inclined trunks. Each trunk gives rise to six short cordons, three on each side that run parallel to the row, or the vines are trained alternately high and low to limit gravitrophic effects while still providing six cordon tiers. The vines are spur-pruned. A third placement system consists of using the bilateral trunks directly as inclined cordons (van den Ende *et al.*, 1986). The vines are then pruned with alternate regions of the trellis used for fruit and replacement shoot development.

In the Tatura Trellis system, the vines are densely planted at one vine per meter along the row, with the rows spaced 4.5 to 6 m apart. Because of root competition between the closely spaced vines, and the large number of shoots developed, excessive vigor is restricted, shading limited, and fruit productivity increased. The Tatura Trellis favors early development of the fruiting potential of the vine.

The most serious limitation to the Tatura Trellis is its tendency to concentrate fruit production in the upper part of the trellis. In addition, the vine must be trained, pruned, and harvested manually, as the trellising is currently unsuited to mechanical harvesting and pruning equipment.

Although different in several aspects, these training systems are designed to direct the benefits of improved plant health and nutrition toward enhanced fruit yield and quality. This also is true for minimal pruning. While appearing to lose their advantages under situations of marked water and disease stress, poor drainage, or salt buildup, divided-canopy systems provide long-term economic benefits in several situations. Whether the yield and quality improvements justify their additional costs will depend on the level of increased profit derived, compared with more traditional procedures.

CONTROL OF VINE VIGOR

As noted previously, rich soils of many New World vineyards stimulate vegetative vigor to the detriment of fruit ripening. This feature has been accentuated by the use of vigorous rootstock, irrigation, fertilization, weed control, and the elimination of viral infections. The problem is not that the vines grow too well, but that the vines produce excessive growth which must be pruned away at the end of each season. The intention of vigor control is to either restrict vegetative growth or redirect it into increased fruit yield and improved ripening.

An old technique of restricting vine vigor is repeated hedging. However, the effect expresses itself slowly and risks including excessive loss in capacity. Another procedure used extensively in Europe has been enhanced intervine competition by increasing grapevine density. Ground covers restrict root growth and may induce vine devigoration. However, restricted vigor is most easily achieved when scions are grafted to devigorating rootstocks, such as '3309 Couderc,' '420 A,' '101-14 Mgt,' and 'Gloire de Montpellier.' In contrast, rootstock cultivars, such as '99 Richter' and '140 Ruggeri,' accentuate vigorous growth (see Table 4.6).

Additional measures employed to restrict vine vigor are limiting nitrogen fertilization and irrigation. Limiting fertilization, notably nitrogen, minimizes vegetative growth, as does restricted water availability, particularly after *véraison*. For example, shoot growth may stop more than 1 month earlier under conditions of water stress (Matthews *et al.*, 1987). Water stress also tends to have a more pronounced effect on reducing leaf growth on lateral than on primary shoots (Williams and Matthews, 1990). Promoting a trailing growth habit also can retard shoot elongation and restrain lateral shoot initiation. Although soil type can indirectly affect vine vigor, choosing a soil type is only an option when selecting a vineyard site. Stony to sandy soils provide sufficient, but restricted, water and nutrient supply and thereby can limit vegetative vigor.

Another alternative devigoration technique involves the application of growth regulators such as ethephon and paclobutrazol. Although effective, they may have undesirable secondary effects. For example, ethephon reduces the photosynthesis rate of sprayed leaves (Shoseyou, 1983).

Ideally, devigoration should be effected by directing the potential for excessive vegetative vigor into additional fruit production and improved grape quality. For example, the increased fruit yield associated with higher bud retention in divided-canopy and minimal pruning systems limits vegetative growth. As long as a favorable canopy microclimate is developed, the increased fruit load has a good chance of maturing fully, without adversely affecting subsequent vine fruitfulness and longterm viability.

Rootstock

The original reason for grafting grapevines was to stop the destruction being caused by phylloxera in Europe. Although the threat of phylloxera remains one of the most important reasons for grafting, rootstocks permit cultivars to be grown in soils damaging to the vine's own root system for other reasons. In addition, some cultivar characteristics are modified by interaction with the rootstock. Thus, rootstock selection provides an opportunity for changing the expression of varietal traits without genetically modifying the scion (shoot system of the vine).

One of the problems in choosing the "right" rootstock is predicting how the two components will interact. Interaction results from the mutual translocation of nutrients and growth regulators between the scion and rootstock. An obvious example of interaction is the influence of particular rootstocks on scion vigor. More subtle examples are the induced susceptibility of *V. vinifera* scions to phylloxera leaf galling when grafted on rootstocks susceptible to leaf galling (Wapshere and Helm, 1987) and the reduction in sensitivity of some rootstock varieties to lime-induced chlorosis when grafted to *V. vinifera* scions (Pouget, 1987). In the latter instance, less citric acid transported from the scion to the roots where less iron from the soil is chelated to ferric citrate and translocated up to the leaves. Because climatic and soil conditions can modify the expression of both rootstock and scion traits, their interaction may vary from year to year and location to location. Thus, although general trends can be noted in Tables 4.6 and 4.7 (see also Howell, 1987), the applicability of particular rootstocks must be assessed based on local experience.

In selecting a rootstock, ranking desired properties is necessary as each rootstock has its benefits and deficits. Selection cannot be taken lightly since, once a rootstock has been chosen, it remains a permanent component of the vineyard until replanting.

The most essential criterion is the affinity between the rootstock and scion. Affinity refers to the formation and stability of the rootstock/scion union. Early and complete fusion of the adjoining cambial tissues is critical to the effective union of the vascular and cortical tissues across the graft site. Areas that do not join shortly after grafting never fuse. Such gaps leave weak points that provide sites for invasion by various disease-causing agents.

For many rootstock varieties, empirical data are available on basic properties (Table 4.6). In some regions, desirable rootstock combinations have already been identified for local cultivars. However, for new scion/rootstock combinations, or in new viticultural regions, fresh trials are necessary. Because of the large number of rootstock/scion combinations possible, it is important to be able to predict unsuitable combinations. Although the parentage of a rootstock gives clues to its potential suitability, prediction of affinity has been more difficult. This may be eased by the use of a new technique based on the electrophoretic similarity of scion and rootstock phosphatases (Masa, 1989). Generally, compatible unions show similar phosphatase electrophoretic patterns. By rapidly screening out potentially incompatible unions, time and space can more effectively be used in assessing the performance of promising combinations.

In the initial rootstock trials conducted over a century ago, the most successful were selections from *Vitis riparia*, *V. rupestris*, or crosses between them or *V. berlandieri*. Their progeny still constitute the bulk of rootstock cultivars (Howell, 1987). Subsequent breeding has incorporated traits from species such as *V. vinifera*, *V. candicans*, and *V. rotundifolia*.

In regions where phylloxera (*Daktulosphaira vitifoliae*) occurs, grafting *V. vinifera* cultivars to resistant rootstock is generally essential. Even where phylloxera is not currently present, serious consideration should be given to using phylloxera-resistant rootstock. The past history of accidental phylloxera introduction suggests that eventual infestation of compatible sites is only a matter of time.

The occurrence of several *Daktulosphaira vitifoliae* biotypes, along with the presence of differential tissue sensitivity in different grapevine species, indicates that phylloxera resistance is complex (Wapshere and Helm,

Table 4.6 Important Cultural Characteristics, Other than Resistance[a] to Phylloxera, of Commercially Cultivated Rootstocks[b]

Rootstock	Vigor of grafted vine	Vegetative cycle	Propagation by			Affinity with *V. vinifera*
			Cutting (rooting)	Bench grafting	Field grafting	
'Rupestris du Lot'	xxxx	Long	xxx	xxx	xxx	xxxx
'Riparia Gloire'	xx	Short	xxx	xxx	xxx	xx
'99 Richter'	xxxx	Medium	xxxx	xxxx	xxxx	xxxx
'110 Richter'	xxx	Very long	xxx	xxx	xxx	xxxx
'140 Ruggeri'	xxxx	Very long	xxx	xxx	xxx	xxxx
'1103 Paulsen'	xxx	Long	xxx	xxx	xxx	xxxx
'SO 4'	xx	Medium	xx	xx	xx	xxx
'5 BB Teleki'	xx	Medium	xx	xx	xx	x
'420 A Mgt'	xx	Long	xx	xx	xxx	xx
'44–53 Malègue'	xxx	Medium	xxxx	xxxx	xxxx	xxxx
'3309 Couderc'	xx	Medium	xxx	xx	xxx	xx
'101–14 Mgt'	xx	Short	xxx	xx	xx	xx
'196–17 Castel'	xxx	Medium	xxx	xxx	xxx	xxx
'41 B Mgt'	xx	Short	x	xx	xxx	xxx
'333 EM'	x	Medium	x	xx	xxx	xxx
'Salt Creek'	xxxx	Very long	x	xx	xx	x

[a] Rootstocks that have proved insufficiently resistant to phylloxera, and for this reason abandoned nearly everywhere (e.g., '1202 C,' 'ARG,' and '1613 C'), are not included.

[b] From Pongrácz (1983), reproduced by permission. Summarized from data of Branas (1974), Boubals (1954, 1980), Cosmo *et al.* (1958), Galet (1971, 1979), Mottard *et al.* (1963), Pàstena (1972), Pongrácz (1978), and Ribéreau-Gayon and Peynaud (1971).

1987). Phylloxera biotypes often are distinguished on the basis of differential rates of multiplication on particular rootstocks. For example, the recently discovered biotype B phylloxera multiplies twice as rapidly as biotype A on 'A×R#1' ('Ganzin 1') (Granett *et al.*, 1987). Because 'A×R#1' is the predominant rootstock used in much of northern California, the appearance of a biotype capable of causing severe damage on 'A×R#1' is causing considerable concern. Although most commercial rootstock cultivars possess some phylloxera resistance derived from *Vitis riparia, V. rupestris,* and/or *V. berlandieri,* additional sources of genetic resistance are *V. rotundifolia, V. candicans, V. cinerea,* and *V. cordifolia.*

While phylloxera resistance is the prime reason for rootstock grafting in most vineyards, nematode resistance can be paramount in other regions. Grapevine roots may be attacked by several pathogenic nematodes, but the most important are the root-knot (*Meloidogyne* spp.) and dagger (*Xiphinema* spp.) nematodes. Dagger nematodes are additionally damaging as transmitters of the causal virus of fanleaf degeneration. Because *Vitis rotundifolia* is particularly resistant to fanleaf degeneration and nematode damage, the species

has been used in breeding several new rootstock cultivars resistant to these maladies, notably 'VR 039-16' and 'VR 043-43' (Walker *et al.*, 1989). Another valuable source of nematode resistance is *V. champinii.*

The importance of local soil factors limiting to grapevine growth can significantly influence rootstock choice. For example, tolerance to high levels of active lime ($CaCO_3$) is essential in many European soils, while low sensitivity to aluminum is crucial in some acidic Australian and South African soils. Because of the importance of soil factors, most commercial rootstocks have been studied to determine their tolerance of such factors (see Table 4.7). Where conditions vary considerably within a single vineyard, use of several rootstocks may be required for optimal results.

Drought tolerance can be another factor critical in choosing a rootstock, especially in arid sites where irrigation is not possible or permissible. As with most traits, drought tolerance is based on complex physiological, developmental, and anatomical properties. Differences in root depth, distribution, and density appear to be partially involved. Several drought-tolerant varieties produce fewer and smaller stomata than drought-sensitive varieties (Scienza and Boselli, 1981). With

Table 4.7 Important Cultural Characteristics of Commercially Cultivated Rootstocks[a]

Rootstock	Humidity ("wet feet")	Dry shallow clay	Deep silt or dense loam	Deep, dry, sandy soil	Nematode resistance	Drought	Active lime (%)	Salt[b]
'Rupestris du Lot'	x	xx	xxx	x	xx	xx	14	0.7 g/kg
'Riparia Gloire'	xxx	x	xx	xx	xx	x	6	—[c]
'99 Richter'	x	xx	xxxx	xx	xxx	xx	17	—
'110 Richter'	xxx	xxxx	xxx	xxx	xx	xxxx	17	—
'140 Ruggeri'	xx	xxx	xxx	xxxx	xxx	xxxx	20	—
'1103 Paulsen'	xxx	xxx	xxx	xxx	xx	xxx	17	0.6 g/kg
'SO 4'	xxx	x	xx	x	xxxx	x	17	0.4 g/kg
'5 BB Teleki'	xxx	xx	xx	x	xxx	x	20	—
'420 Mgt'	xx	xxx	xx	xx	xx	xx	20	—
'44–53 Malègue'	xxx	xx	xxx	xx	xxx	xx	10	—
'3309 Couderc'	xxx	xx	xx	xx	x	x	11	0.4 g/kg
'101–14 Mgt'	xxx	xx	xx	x	xx	x	9	—
'196–17 Castel'	xx	x	xx	xxx	x	xxx	6	—
'41 B Mgt'	x	x	x	x	x	xxx	40	Nil
'333 EM'	x	x	x	x	x	xx	40	Nil
'Salt Creek'	xx	x	xxx	xxxx	xxx	xx	?	—

[a] From Pongrácz (1983), reproduced by permission. Summarized from data of Branas (1974), Cosmo *et al.* (1958), Galet (1971), Mottard *et al.* (1963), Pàstena (1972), Pongrácz (1978), and Ribéreau-Gayon and Peynaud (1971).

[b] Approximate levels of tolerance are as follows: American species, 1.5 g/kg absolute maximum; *V. vinifera*, 3 g/kg absolute maximum.

[c] —, not available.

some rootstocks such as '110 Richter,' this tendency also is expressed in the grafted scion (Düzenlí and Ergenoğlu, 1991). Because high density plantings restrict root growth, the merits of using drought-tolerant rootstock may be diminished. Although *V. berlandieri* is one of the most drought-tolerant grapevine species, expression of this trait varies considerably in *V. berlandieri*-based rootstocks. *Vitis cordifolia*-based rootstocks are often particularly useful on shallow soils in drought situations.

In regions having short growing seasons, early fruit ripening and wood maturation are essential. Most rootstocks favoring early maturity have *V. riparia* in their parentage. Where yearly variation in cold severity is marked, random grafting of vines to more than one rootstock may provide some protection from climatic vicissitudes (Hubáčková and Hubáček, 1984).

Another vital factor influencing rootstock selection is its effect on grapevine yield. Although rapid establishment of a vineyard is aided by vigorous vegetative growth, this property may be undesirable in the long run. Thus, rootstocks may be chosen specifically to in-

duce devigoration. This usually limits yield and may improve fruit quality and repress physiological disorders such as inflorescence and bunch-stem necrosis. Fruit yield generally shows a weak negative correlation with quality, as measured by sugar content (Huglin, 1986). The specific yield/quality influence of any rootstock/scion combination varies considerably (Fig. 4.13), and is influenced by conditions such as vineyard layout, canopy management, and irrigation (Whiting, 1988; Foott *et al.*, 1989).

Some of the specific effects of rootstocks on vine vigor and fruit quality may arise from differential uptake of nutrients from the soil. For example, the preferential accumulation of potassium can antagonize the uptake of other cations. For rootstocks such as 'SO 4' and '44-53 M,' this can lead to magnesium deficiency in scion cultivars such as 'Cabernet Sauvignon' and 'Grenache' (Boulay, 1982). Limited zinc uptake by 'Rupestris St. George' may be the source of poor fruit set occasionally associated with use of that rootstock (Skinner *et al.*, 1988). Because the root system supplies most of nitrogen required in the early part of the grow-

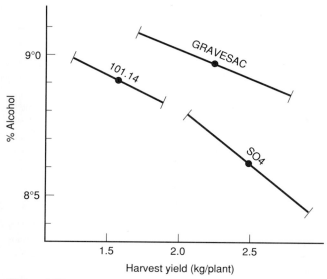

Figure 4.13 Linear relationship between probable alcohol content (ordinate) and yield (abscissa) of 'Cabernet Sauvignon' on three different rootstocks. (From Pouget, 1987, reproduced by permission.)

ing season (Conradie, 1988), variation in rootstock nitrogen uptake and storage can influence scion fruitfulness. The importance of rootstock selection in limiting lime-induced chlorosis has already been mentioned. Although most of these effects are undoubtedly under the direct genetic control of the rootstock, some variation may result from differential colonization of the roots by mycorrhizae.

Rootstock selection also can significantly alter the composition of the fruit produced by a scion. By affecting berry size, rootstocks can influence the skin/flesh ratio and thereby wine characteristics. Additional indirect effects on fruit composition may result from increased vegetative growth, augmenting leaf/fruit competition and shading. Nevertheless, some rootstock effects are probably direct, via differential mineral uptake from the soil (Fig. 4.21). The impact of the rootstock on potassium uptake (Ruhl, 1989) and differential accumulation in the vine (Failla *et al.*, 1990) can be especially influential. Potassium distribution affects not only growth, but also juice pH and potential wine quality. Rootstock modifications of fruit amino acid content have been correlated with the rate of juice fermentation (Huang and Ough, 1989). Although few studies have investigated the significance of rootstocks on grape aroma, one study has noted a decrease in monoterpene content associated with rootstocks promoting high yield (McCarthy and Nicholas, 1989).

Although rootstock grafting is beneficial if not essential in most regions, it is expensive. In addition, the cost of special rootstocks may be higher, owing to limited demand or difficulty in propagation. These features tend to make them difficult to obtain as well. Neverthe-

less, the long-term benefits of using the most suitable rootstock will usually outweigh the additional expenditure associated with obtaining and purchasing it.

An unintentional consequence of the widespread adoption of grafting probably has been the unsuspected propagation and distribution of grapevine viruses and viroids (Szychowski *et al.*, 1988). Graft unions also may act as sites for invasion by several pathogens and pests. Nevertheless, the grafting procedure itself can induce, at least temporarily, resistance to infection and transmission of the tomato ringspot virus (Stobbs *et al.*, 1988).

In the popular press, there is continual musing over the possible (undesirable) influence of grafting on wine quality. In most situations, the question is purely academic as grape culture would be impossible without grafting. As in other aspects of grape production, choice of a rootstock can either enhance or diminish the quality of the grapes and the wine produced therefrom.

Vine Propagation and Grafting

Grapevines typically are propagated by vegetative means to retain their unique genetic makeup. Sexual reproduction, by rearranging grapevine traits, disrupts desirable gene combinations. Thus, seed propagation is limited to breeding cultivars where new genetic combinations are desired.

Grapevines may be vegetatively propagated by several techniques. The method of choice depends primarily on pragmatic matters, such as the number of plants required, the rapidity of multiplication, when propagation is conducted, whether grafting is involved, and, if so, the thickness of the trunk. Regardless of the method, some degree of callus formation occurs. Callus tissue consists of undifferentiated cells that develop in response to physical damage. Callus develops most prominently in and around meristematic tissues. In grafting, the callus establishes the union between adjacent vascular and cortical tissues of the rootstock and scion. For the union to persist, it is essential that the thin cambial layers of both rootstock and scion be aligned opposite one another. Callus formation also is associated with, but not directly involved in, the formation of roots from cuttings. New roots typically develop from or near the cambial cells between the vascular bundles of the cane. Most roots develop adjacent to the basal node of the cutting. Finally, callus cells developed on tissue culture media may differentiate into shoot and root meristems, from which whole plants can develop.

Callus tissue is particularly metabolically active, and its formation is favored by warm conditions and ample oxygen. Because of the predominantly undifferentiated state of the callus, its thin-walled cells are very sensitive to drying and sun exposure. Therefore, moist conditions and minimal light exposure are required until differentiation has developed protective layers over the graft union.

Multiplication Procedures

The simplest means of vegetative propagation is **layering.** The technique involves bending a cane down to the ground and burying a section in the soil. Once rooting has developed sufficiently, connection to the parent plant can be severed. Layering has the advantage that the buried section is continuously supplied with water and nutrients during root development. This has been particularly useful with difficult-to-root vines, such as muscadine cultivars. Layering is seldom used currently, as other means are often as effective and do not disrupt viticultural practices such as cultivating and chemical weed control.

The most common means of grapevine propagation involves cane **cuttings.** The cuttings are usually taken from the prunings removed in the winter. The best cane sections are usually those 8 to 13 mm in diameter, uniformly brown, and possessing internode lengths typical of the variety. These features indicate that the cane development occurred under favorable conditions and is well matured. Appropriate cane wood is cut into pieces about 35 to 45 cm in length. The length often depends on the water retention properties of the soil into which the vine is to be planted and the availability of irrigation water subsequent to planting. It is essential that the buried portion remain continuously moist.

A lower perpendicular cut is made just below a node, and an upper 45° diagonal cut is made about 20 to 25 mm above the uppermost bud. The diagonal cut permits rapid identification of the top of a cutting. It is important that the original apical/basal orientation of the cutting be retained when rooted. Cane polarity restricts root initiation to the basal region of cuttings. The internode section retained at the apex also helps to protect the apical bud from physical damage. Protecting the apical bud is especially important in rootstock varieties, where all other buds are removed before rooting to limit subsequent rootstock suckering. In this case, the apical bud is the sole natural source of auxin to activate root development.

If the cutting is to be rooted directly in the vineyard, it is important to leave about 10 cm of cane above ground. This length permits grafting at a height sufficient to minimize the likelihood of scion rooting. If scion rooting were to occur, the scion roots might initially outgrow the rootstock but subsequently succumb to the conditions for which the rootstock was selected.

With the exception of muscadine cultivars, the canes of most commercial wine grape varieties root easily.

The same is true for most rootstock cultivars that are selections of *V. rupestris* or *V. riparia*, hybrids between them, or hybrids with *V. vinifera*. Most rootstocks containing *V. berlandieri*, *V. cinerea*, *V. candicans*, *V. cordifolia*, or *V. rotundifolia* parentage are to varying degrees difficult to root (see Table 4.6). Because the latter varieties contain many useful properties, considerable effort has been spent to increase rooting success. Generally, the most effective activators of rooting include soaking in water for 24 hr, dipping in a solution of about 2000 ppm indolebutyric acid (IBA), applying bottom heat (25° to 30°C) to the rooting bed, and periodic misting to maintain high humidity. Additional factors of potential value are spraying source plants with chlormequat (CCC) in the spring (Fabbri *et al.*, 1986) and water culturing after callus formation (Williams and Antcliff, 1984).

Rooting success is improved when the cuttings are produced and planted directly after cane harvesting. If rooting cannot be initiated immediately, the canes are best refrigerated (1° to 5°C) and kept from drying or molding. Upright storage of the canes in moist sand or sawdust is common.

An alternative method of cane propagation involves **green cuttings.** These are single-node pieces cut from growing shoots, rooted under mist propagation in the greenhouse. This method is useful when the supply of desirable canes is limited and the need for rapid multiplication is high. The technique is more complex and demanding, both in equipment and in care after rooting. Because of the tender nature of green cuttings, considerable cuation is required in preparing the plants to withstand vineyard conditions.

Although cuttings are the most common means of grapevine propagation, it may be inadequate for the rapid multiplication of special stock. **Micropropagation** from axillary buds is the simplest of tissue culture methods. If the financial return is sufficient, vines may be multiplied even more quickly by employing **shoot-apex fragmentation** (Barlass and Skene, 1978) or **somatic embryogenesis** (Monnier *et al.*, 1990). More complex and demanding than other reproduction techniques, tissue culture is the only means of mass propagating a cultivar rapidly. Also, because strict hygiene is required for successful plantlet development, infection of disease-free stock is avoided.

Grafting

In locations where conditions obviate the need for grafting, rooted cuttings can be directly planted in the vineyard. However, in most viticultural areas profitable grape culture is dependent on grafting fruiting cultivars onto suitable rootstocks. This typically involves grafting one-bud sections of the desirable scion onto a rootstock cutting. When grafting is done indoors, as in a nursery or greenhouse, it is called **bench grafting**. When grafting occurs at, or shortly following, planting of the rootstock in the vineyard, it is termed **field grafting**. The other major use of grafting is to convert existing vines over to another fruiting variety. When the scion piece consists of a cane segment, the process is called **grafting** to distinguish it from the use of only a small side piece from a cane, termed **budding**.

Bench grafting has the advantage of being more adaptable to mechanization. It also can be performed over a longer period, as it commonly uses dormant cuttings. To facilitate proper cambial alignment when using grafting machines, it is necessary to presort the rootstock and scion pieces by size. After making the cuts and joining the two sections, the grafted cutting is placed under moist, warm conditions. This favors rapid callus development and graft union. Grafting machines permit junctions of sufficient strength that grafting tape is not needed while the union forms. If the grafted rootstock has already been rooted, the vine is ready for planting shortly after the union has taken hold and the exposed callus hardened off. If difficult-to-root dormant rootstock is grafted, the base may be treated with IBA and placed in a heated rooting bed while the upper grafted region is kept cool. This favors root development before the scion bud bursts and places water demands on the young root system. With easily rooted rootstock, canes usually root sufficiently rapidly to supply the needs of the developing scion without special treatment.

Occasionally, actively growing shoots are grafted directly onto growing rootstocks in a process called **green grafting**. Graft take is usually rapid and successful, but the higher labor costs and more demanding environmental controls usually do not warrant its use. Nevertheless, modern developments may reduce the expense of green grafting (Alleweldt *et al.*, 1991; Collard, 1991). This may make the speed of propagating planting stock by green grafting more commercially attractive.

Where labor and timing are appropriate, field grafting is the least expensive means of grafting new vines. Field grafting is also the only means of converting existing vines over to another variety.

Field grafting preferably occurs shortly after growth has commenced in the spring. By this time, the cambium in the rootstock has become active, and graft union develops quickly. This permits prompt growth of the scion. The rootstocks are planted leaving about 8 to 13 cm projecting above the ground. This both limits scion rooting and places the root system sufficiently deep to minimize damage during manual weeding. Grafting unrooted rootstock in the field is not recommended as the success rate is poor.

Commonly used manual techniques for grafting are **whip** grafting and **chip** budding. Whip grafting involves joining scion and rootstock canes of equivalent diameter. Two cuts are made about 5 mm above and below a scion bud. The upper cut is shallowly angled and directed away from the bud to identify the polarity of the scion piece. The lower cut is long, straight, and steep (15–25°). The lower cut is usually 2.5 times longer than the diameter of the cane. A "tongue" is produced in the lower cut by making a upward slice, and gently pressing outward away from the bud with the pruning knife. A set of cuts matching those in the lower end at the scion piece is made in the rootstock to receive the scion. After connecting and alignment, the union is secured with grafting tape, raffia, or other appropriate material. Plastic grafting tape is popular as it is both rapid and easy to apply, will not cause girdling, and helps limit drying of the graft union. This obviates the need of mounding, and subsequent removal, of moist soil over the union site.

Whip grafting provides an extensive area over which the union can establish itself. Its prime disadvantage is the skill and time required in carrying out the procedure.

Chip budding provides less union surface than whip grafting, but the smaller size of the chip requires less contact area. It is often preferred as it requires less skill in preparing matching cuts. Also, because the scion source does not need to be of identical diameter to the rootstock, time is saved in avoiding the requirement to select matching scion and rootstock pieces.

In chip budding, two oblique downward cuts are made above and below the scion bud (Fig. 4.14). The upper cut is more acute and meets the lower incision, making a wedge-shaped chip about 12 mm long and 3 mm deep at the base. A matching section is cut out about 8 to 13 cm above ground level on the rootstock. The chip is held in position with grafting tape or equivalent material.

For either grafting procedure, it is imperative that each set of cuts and the insertion of the scion piece be performed rapidly to avoid drying. Drying of the cut surfaces dramatically reduces the chances of a successful union.

Various techniques are used for converting existing vines over to another fruit-bearing cultivar. For trunks less than 2 cm in diameter, whip grafting is commonly used; for trunks between 2 and 4 cm in diameter, side-whip grafting is often employed. Trunks over 4 cm may be **notch, wedge, cleft,** or **bark grafted** (Alley, 1975). In all size classes, chip budding can be used, whereas T-budding is largely limited to trunks over 4 cm in diameter. Budding techniques are often preferred to the use of larger scion pieces because they require less skill and can be just as successful (Steinhauer *et al.*, 1980). Con-

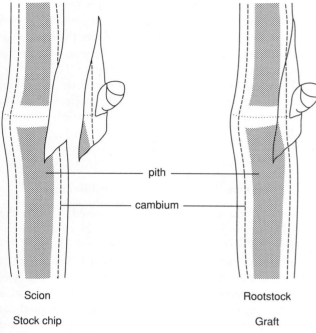

Figure 4.14 Chip-bud grafting.

version high on the trunk allows most of the existing trunk to be retained, thus speeding the return of the vine to full productivity.

T-budding derives its name from the shape of the two cuts produced in the trunk of the vine being converted. After the cuts are made, the bark is pulled back to form two flaps, and open a gap into which the scion piece (**bud shield**) is slid. The bud shield is produced by making a shallow downward cut behind the bud on the scion cane. The slice begins and ends about 2 cm above and below the bud. A second oblique incision below the bud liberates the bud shield. To facilitate better union, T-cuts in the trunk are recommended to be inverted (Alley and Koyama, 1981). With the inverted-T (Fig. 4.15), the rounded upper end of the bud shield is pushed upward, and the bark flaps cover the bud shield. In the original version of the procedure, failure of the top of the shield to join well with the trunk created a critical zone of weakness. Because winds could easily split the growing shoot from the trunk, shoots had to be tied to a support shortly after emergence.

Depending on the trunk diameter, several bud shields may be grafted around the trunk. They are necessary to nourish the associated trunk and root sections of the grapevine. Each bud shield is grafted at the same height to conserve grafting tape and speed grafting.

T-budding has the advantage of requiring the least skill of any vine conversion technique. In addition, demands on cold storage space are minimal because bud

shields rather than cuttings can be stored (Gargiulo, 1983). It has the disadvantage, though, that the period in the spring during which it can be performed most effectively is short. The grape grower must wait until the trunk cambium has become active after bud break. Only then can the bark be easily separated from the wood to permit insertion of a bud shield. Postponing T-budding much beyond this point delays bud break and may result in poor shoot maturation by the fall.

In contrast, chip budding can be performed earlier, thereby allowing the grape grower greater flexibility in timing budding-over. With chip budding, it is necessary to use a section of the trunk with a curvature similar to that of the scion chip. Otherwise, cambial alignment may be inadequate and the union will fail. Depending on trunk thickness, two or more buds are grafted per vine. Because of the hardness of the wood, cutting out slots for chip budding is more difficult than for T-budding. Nevertheless, the longer period over which chip budding can be performed may make it preferable. The success rate of chip budding is equivalent to that of T-budding (Alley and Koyama, 1980).

Although vine conversion is usually performed in the spring, this has the disadvantage that the crop is lost for the year. It is necessary to remove the existing top to permit the newly grafted scion pieces to develop into the new top. To offset the resulting financial loss, chip budding may be performed in early fall, when the buds have matured but weather conditions are still favorable (≥15°C) for graft union (Nicholson, 1990). Full union

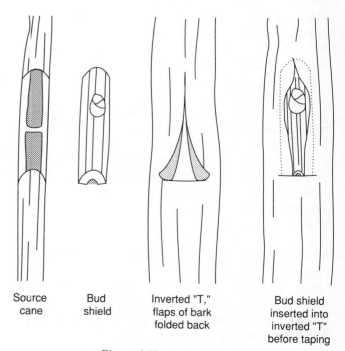

| Source cane | Bud shield | Inverted "T," flaps of bark folded back | Bud shield inserted into inverted "T" before taping |

Figure 4.15 Inverted-T graft.

is usually complete by leaf fall, but the bud remains dormant until spring. Angling the upper and lower cuts away from the bud produces a bud chip that slides into a matching slot made in the host vine. Although the technique is more complex, the interlocking of the chip in the trunk assures a firm connection with the vine. Grafting tape protects the graft site from drying. In the spring, the tape is cut to permit the bud to sprout. An encircling incision above the grafted buds stimulates early bud break while restraining growth of the existing top. This technique allows the vine to bear a crop while the grafted scion establishes itself. At the end of the season, the existing top can be removed and the grafted cultivar trained as desired.

The older techniques of cleft, notch, and bark grafting are still used but less frequently than in the past. Not only do the older techniques require more skill in cutting and aligning the scion and trunk cambia, but they also take longer and require grafting compound to protect to graft while the union develops. Readers desiring detailed instructions on these and other grafting techniques are directed to standard references (e.g., Winkler *et al.*, 1974; Weaver, 1976; Alley, 1975; Alley and Koyama, 1980, 1981).

Soil Preparation

Before rooted cuttings are planted in the vineyard, the soil must be prepared to receive the new vines. The degree of preparation depends on the soil texture, degree of compaction, drainage conditions, previous use, irrigation needs, and prevailing diseases and pests. If the land is virgin, the soil needs to be prepared for use, that is, noxious weeds and rodents eliminated and obstacles to efficient cultivation removed. Where the soil has already been under cultivation, providing sufficient drainage and soil loosening for excellent root development are the primary concerns.

Where inadequate, drainage is most effectively improved by laying drainage tile. Winkler *et al.* (1974) recommend draining soil to a depth of about 1.5 m in cool climates and to 2 m in hot climates. Narrow ditches may be effective but can complicate vineyard mechanization. In addition, the ditches may remove valuable vineyard land from production. Drainage efficiency may be further improved by breaking hardpans or other impediments to water percolation. Deep plowing, used to break hardpans, can equally be used to loosen deep soil layers. This is especially useful in heavy nonirrigated soil where greater soil access can minimize water stress under drought conditions (van Huyssteen, 1988b). Homogeneity of soil loosening also is important in favoring effective soil use by vines (Saayman, 1982; Saayman and van Huyssteen, 1980).

Where nematodes are a problem, it is often beneficial to fumigate the soil, even when nematode-resistant rootstocks are used. Fumigation reduces the level of infestation and enhances the effectiveness of resistant rootstocks in maintaining healthy vines.

Where surface (furrow) irrigation is desired, the land must be flat or possess only a slight slope. Thus, land leveling may be required if such an irrigation technique is planned.

Vineyard Planting and Establishment

Various planting procedures are used, but mechanized planting is favored because of its time and cost savings. Where bare-rooted cutting are planted, it is critical to protect the plants from drying. Because roots are trimmed relative to the size of the planting hole, holes of sufficient size should be used to retain most of the existing root system. Direct planting of rooted cuttings in tubes or pots maximizes root retention. If sufficiently acclimated to field conditions before use, potted vines suffer minimal transplantation shock. This approach also gives the grape grower more flexibility in scheduling planting time.

Where permanent stakes are not already in position, it is advisable to angle the planting hole away from the stake location and parallel to the rows. This minimizes subsequent root damage from the use of posthole diggers and cultivators. Often soil is hilled around the exposed portion of vine until shoot development is well developed. Alternately, planting may take place on mounds of earth covered by meter-wide sheets of black plastic. The latter minimize the manual weeding normally required during the first year and promote root development as well.

Transplanted vines are watered at least once after planting, and additionally if drought conditions develop. Irrigation is avoided after midsummer to restrict continued vegetative growth and favor cane maturation. Cane maturation is required to permit the vine to withstand early frosts and winter cold.

During the first growing season, the vine is permitted to grow largely at will to facilitate the establishment of an effective root system. Most vineyard activities are limited to weed, disease, and pest control to protect the young succulent tissues. Topping is conducted only if watering is insufficient to prevent severe water stress. Pruning occurs after growth has ceased and the leaves have fallen. For the majority of training systems, only one strong, well-positioned cane is retained, and it is often pruned to four buds.

In the spring of the second season, the strongest shoot is retained and tied to the trellising stake to form the future vine trunk. Subsequent pruning varies depending on the training systems desired.

Irrigation

Grape growing possess one of the longest historical records of irrigation of any crop. Reports of vineyard irrigation occur as far back as 2900 B.C. in Mesopotamia (Younger, 1966). Mesopotamian agriculture also provides an early example of improper irrigation, leading to salinization and loss of soil productivity.

In Europe, irrigation is prohibited in most Appellation Control regions. This has given rise to the impression in the popular wine press that irrigation is *ipso facto* undesirable. If used excessively, irrigation can have undesirable effects on fruit quality and cane maturation. Used wisely, it not only permits grape culture in arid and semiarid regions, but also can regulate vine growth and favor quality grape production.

The use of irrigation to regulate grape production is another of the complex aspects of viticulture. The difficulty partially arises from dealing with the subterranean portion of viticulture, namely, the root system and its microclimate. As these components cannot be easily measured or visualized, knowledge about them is more limited than any other aspect of viticulture.

Soil acts not only as an anchorage and source of inorganic nutrients, but also as a reservoir of water. Of the water present in soil, only a variable fraction is available for plant use. This portion depends on the textual properties of the soil (stones, sand, silt, and clay) and its organic components (humus and plant and animal remains).

When soil becomes saturated after a rain or irrigation, all its cavities are filled with water (Fig. 4.16). Within several hours, water contained in the larger soil voids drain out of the root zone. Because **gravitational water** is lost so rapidly, it seldom plays a significant role in grapevine growth. The amount of water that remains is termed the **field capacity**. This component is held by forces sufficient to counteract the action of gravity. These forces include those involved in the adsorption and hygroscopic bonding of water to soil particles as well as the capillary action of small soil pores and fissures. Some of the water is held weakly and can be readily absorbed by roots. As the readily available portion is absorbed, the roots increasingly must draw on water held more strongly in the soil (Fig. 4.16). If dry conditions prevail, the roots extract all the "available" water and the **permanent wilting percentage** is reached. At this point, plants can no longer extract water from the soil. Consequently, the **available water** component

of soil refers to the difference between the field capacity and the permanent wilting percentage.

Different instruments measure various aspects of soil/water relations, but none directly assesses water availability. **Tensiometers** measure soil water potential, an indicator of the force required to extract water. Accurate only to about -0.1 MPa, however, tensiometer readings are of value only when the grower wishes to maintain the soil at near field capacity. Various forms of **resistance blocks** measure the electrical resistance of the soil and are accurate to water potentials between -0.7 to -1.5 MPa. **Neutron probes** do not directly measure water content, but rather estimate its presence by the slowing of fast neutrons by the soil's hydrogen content. The neutron probe is valuable as it functions over the full range of water contents in soil and can estimate water content at any desired soil depth. This instrument is particularly useful in soil low in organic content where essentially all the hydrogen is associated with water molecules.

The primary function of water measurement is to assess the effectiveness of irrigation in raising the moisture level in the soil horizon occupied by grapevine roots. When water flows into soil, it raises the immediate wetted area to its field capacity, before it moves vertically or laterally into the surrounding soil. Thus, once the field capacity of the effective rooting region has been determined, a neutron probe can chart water loss through this zone, and estimate the amount of water required to reestablish the soil to field capacity.

The proportion and amount of available water vary widely with soil type (Fig. 4.17). Sandy soils have the lowest water retentive properties but have the highest proportion (upward of 70%) in an available form. Clay soils often possess the highest field capacities, but only about 35% may be held weakly enough to permit easy removal by plants (Milne, 1988). Silt soils often retain slightly less water than clay soils, but more of the aqueous component is in an available form (up to 200 mm/m).

Another important aspect of soil structure and composition relating to water availability is their influence on the ease with which available water can be removed. Up to 90% of the available water in sandy soils can be readily removed at tensions equal or less than -0.2 MPa (2 bar) (Hagan, 1955). This corresponds to the typical osmotic tension (water potential) of root cytoplasm. In contrast, about 30 to 50% of the available water can be removed at such low tensions in clay soils. Below a soil moisture tension of -0.2 MPa, leaf transpiration usually provides the force required to draw water up the vine. Under periods of water stress, vine root and leaf water potentials can decline by an

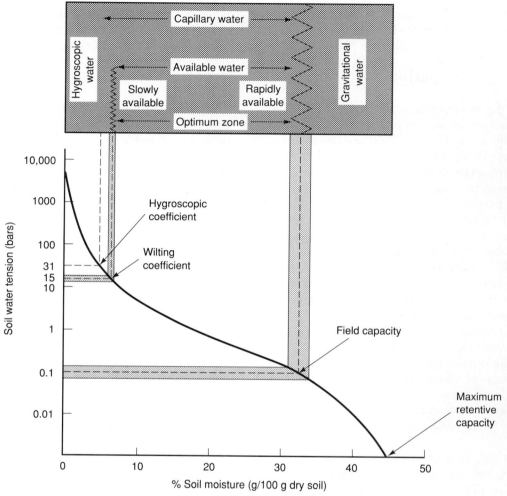

Figure 4.16 Water tension curve of a loam soil as related to different terms describing the various states of water in soil. Wavy lines indicate that measurements such as field capacity are imprecise. (Reprinted with the permission of Macmillan Publishing Company from THE NATURE AND PROPERTIES OF SOILS, Eighth Edition by Nyle C. Brady. Copyright © 1974 by Macmillan Publishing Company.)

additional −0.4 MPa (Düring, 1984). This facilitates water uptake while retarding transpirational water loss.

Another significant aspect of soil/plant water relations concerns the rate at which the soil water potential declines. The more rapidly the soil water potential falls, the sooner vines are likely to experience water stress. This means that during drought conditions, water stress tends to develop both earlier and more suddenly on sandy than on silt or clay soils. Grapevines, similar to most plants, cannot extract significant amounts of water from soil below −1.5 MPa. Leaf stomata generally close when the water potential, developed by transpiration, falls below −1.3 MPa (Kriedemann and Smart, 1971).

Leaf wilting, the standard sign of high water stress in most flowering plants, typically is not shown by grapevines. When it does occur, it is restricted to young leaves and the shoot apex. Nevertheless, an early sign of water stress is a decrease in the angle subtended by the petiole and the plane of the leaf blade (Smart, 1974). If leaf wilt occurs, it develops when the whole plant suddenly experiences high water stress. This typically occurs only on shallow or sandy soils. On deep or silty and clayey soils, moisture stress tends to build up slowly, and osmotic adjustments and stomatal closure limit leaf wilting. Although wilting is seldom expressed in grapevines, other signs of water stress can develop early and under conditions of mild water stress.

The sensitivity of different tissues and physiological processes to water stress varies widely in grapevines. Because the shoot tip is particularly sensitive, suppression of shoot elongation is one of the earliest signs of water stress. As shoot growth slows, the yellow-green color of young leaves and shoot tips changes to the

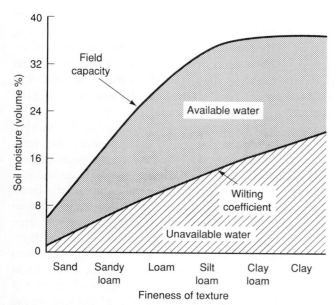

Figure 4.17 General relationship between soil moisture characteristics and soil texture. Note the continuous increase in wilting coefficient as the texture becomes heavier, while field capacity generally levels off beyond silt soils. (Reprinted with the permission of Macmillan Publishing Company from THE NATURE AND PROPERTIES OF SOILS, Eighth Edition by Nyle C. Brady. Copyright © 1974 by Macmillan Publishing Company.)

grey-green of mature leaves. The root system reacts markedly differently to water stress. Initially, growth and fine root production are stimulated by mild water stress (van Zyl, 1988), whereas severe drought restricts root growth, especially at the soil surface. Both roots and leaves adjust osmotically by increasing their solute concentration (Düring, 1984).

Stomata also open later and close earlier in the day as soil water potential falls. This results in a rise in leaf temperature and suppression of photosynthesis. Cultivars showing drought tolerance may make better use of their water supply. For example, 'Riesling' shows earlier stomatal closure than do drought-sensitive varieties. The responsiveness of the leaves to water stress allows them to minimize water loss while optimizing carbon dioxide uptake for photosynthesis (Düring, 1990).

Water stress can reduce fruit set, berry size, inflorescence initiation and subsequent development. The degree of the effects depends on both the timing and duration of the water stress (Smart and Coombe, 1983). The most sensitive period is during the first few weeks following flowering. As much of the increase in berry size occurs shortly after anthesis (Harris *et al.*, 1968), water stress can permanently limit berry enlargement. The effect of water stress on berry size occasionally has been used to restrict cluster compactness and diminish the incidence of bunch rot (Winkler *et al.*,

1974). In contrast, excessive irrigation may diminish berry quality by reducing the skin to flesh ratio.

Subsequent to *véraison,* fruit expansion is little influenced by water stress (Matthews and Anderson, 1989). Mild stress following *véraison* can be beneficial in hastening fruit ripening, enhancing sugar and anthocyanin contents, and diminishing excessive acidity. Effects on pH tend to be more variable, probably because pH depends not only on the concentration of the various organic acids, but also on the accumulation and location of potassium in the berry. However, marked stress appears to have the reverse effects on fruit composition and maturation, and can lead to shoot tip death, basal leaf drop, and the curling and spotting of young leaves.

By suppressing shoot growth, water stress can limit competition between vegetative growth and fruit development and thus favor the formation of a more open and desirable canopy microclimate. Moderate water stress also favors periderm formation in shoots. Although photosynthesis is not markedly affected by moderate water stress, the transport and accumulation of sugars in berries and old wood are favored (Schneider, 1989).

After harvest, both drought and overwatering need to be avoided to favor leaf function into the fall, restrict late vegetative growth, and favor cane maturation.

Although problems associated with water stress are more common, water saturation also can have undesirable consequences. If occurring late in the season, it can induce skin cracking and favor bunch rot. Protracted periods of soil saturation suppress root growth. Even maintaining the soil surface at or near field capacity by irrigation limits the growth of surface roots (van Zyl, 1988).

Timing and Level of Irrigation

Irrigation has the potential in arid and semiarid conditions of permitting the grape grower another means of influencing fruit yield and quality. The difficulty in applying this potential is the lack of simple and inexpensive means of accurately assessing vine water stress, water availability in soil, and the significance of these measures to vine physiology.

Possibly the best indicator of grapevine water stress is the differential between the ambient and canopy temperature, and the rapidity of its development. The difference between ambient and leaf temperature is a measure of stomatal opening and, thereby, the development of low water potentials in the vine. As water potential falls, stomata close, the cooling produced by transpiration diminishes, and leaf temperature rises. The rapidity with which temperature differentials develop during the day indicates the degree to which the roots are experiencing difficulty in extracting water from the soil. Thus, the dynamics of the temperature differential is a biological indicator to how soil, atmosphere, and canopy conditions impact vine water demand. Although less sensitive to water stress than shoot elongation, the ambient/canopy temperature differential can be more easily and frequently measured. Hand-held infrared thermometers have made leaf temperature measurement relatively simple, regrettably proper interpretation of the data is less so (Stockle and Dugas, 1992). More complex and instrumentally demanding indicators, based on estimates of evapotranspiration potential and canopy temperature, are also being assessed. Evapotranspiration refers to the water lost both by leaf transpiration and by evaporation from the soil.

Because most water stress indicators are influenced by current and past vineyard conditions, proper interpretation of the results needs to include an assessment of these conditions. The influence of intermittent sunny periods on these indicators is clear. Less obvious are factors such as the lingering effect of wind exposure. Windy conditions can affect stomatal opening, and thereby canopy temperature, several days after the winds have ceased (Kobriger *et al.*, 1984). Also, soil texture and depth can markedly affect the speed of water stress development. Moreover, canopy size greatly affects water demand. Finally, more needs to be known about the physiological significance of water stress on vine growth and fruit development throughout the growing season.

Water Quality

Most water supplies contain a variety of dissolved salts, ions, and suspended particles. Unless absorbed by the plant, precipitated in insoluble forms, or leached from the soil, the salts and ions can lead to soil salinization. The importance of dissolved salts in irrigation water depends not only on the concentration and chemical nature, but also on soil texture, depth, and drainage, as well as annual precipitation and irrigation method.

The most common toxic salts found in water are borates and chlorides. Sodium occasionally can reach toxic levels but is generally more significant through the displacement and subsequent loss of calcium and magnesium from the soil surface. The exchange of divalent cations with monovalent sodium ions weakens the association between clay particles that help generate soil aggregate formation. Over time, this can lead to the establishment of a hardpan. When the aggregate structure is lost, clay particles flow downward with water and plug soil capillaries. On drying, the soil surface cakes into a rigid layer difficult to cultivate and becomes impermeable to water infiltration. Sodium accumulation also can result in a rise in pH, releasing caustic carbonate and bicarbonate ions into the soil solution. Such a condition can result in the creation of sodic soils. Salt accumulation decreases water availability by decreasing soil water potential and increasing the force required by roots to extract water.

Because of the many variables affecting the buildup of salts in soil, it is difficult to make general statements on water quality. Nevertheless, waters possessing electrical conductivity (EC_e) values below 0.75 mmho/cm,[1] a ratio of sodium to calcium and magnesium levels (SAR) below 8, and slightly acid to alkaline pH (6.5 to 8.5) generally do not create problems. Low chloride (<100 ppm) and boron (≤1 ppm) levels are also desirable. Water with higher levels of these indicator values can be safely used in some circumstances, where natural conditions, or increased irrigation, leach them out of the root zone. Addition of powered gypsum to the water also can counteract high SAR values.

Types of Irrigation

Where availability of irrigation water is limited because of scarcity, price, or quality, use of drought-tolerant rootstock and scion varieties is crucial to vineyard success. However, where irrigation is permissible, required, and of adequate quality, factors such as soil texture, depth, and slope, as well as heat, wind, cost, and tillage practice can influence the choice of an appropriate system. Additional factors affecting choice may be its use for frost and heat protection and the benefits of simultaneous fertilizer and pesticide application.

Of the factors influencing irrigation decisions, water pricing and availability are typically beyond the control of the grape grower. Where irrigation water is in ample supply and of low cost, systems requiring low initial costs, such as furrow irrigation, may be viable options. However, where water is costly or availability low, use of systems such as drip irrigation are more cost-effective. Water quality also greatly affects system feasibility. For example, use of saline water is most satisfactory with either broad bottom furrows or drip irrigation. Broad bottom furrows disperse the water over a large surface area and thereby delay the accumulation of high levels of salt deposits. Natural rainfall, or additional irrigation, may be sufficient to prevent a serious salt buildup in the soil. With drip irrigation, the slow but frequent addition tends to move the salts to the

---

[1] A EC_e value of 1 mmho/cm is produced by about 640 ppm of salt.

edge of the wetted area, and away from the region of root concentration.

Furrow irrigation is an ancient but effective means of irrigation. Where the water supply and cost are not major limiting factors, its inefficient use of water may not be critically important. Plant use of the water applied may be as low as 30% but commonly is around 60 to 70%. Where the water is marginally saline, additional irrigation or rainfall often flushes out the salt additions. In such situations, it is essential to have adequate drainage to avoid raising the water table to the root zone.

Typically, furrow irrigation involves several evenly spaced, shallow, V-shaped trenches or a few wide, flat-bottomed furrows between each row. Where the soil is sandy and penetration rapid, the furrows are relatively short and filled quickly to avoid uneven water penetration. Broad flat-bottomed furrows often are used to offset the minimal lateral movement of water in sandy soils and favor uniform irrigation (Fig. 5.2). Silt and clay soils, with their considerable lateral water flow, can effectively use narrow furrows. Furrow irrigation in blocks often have been used in California (Fig. 4.18).

In furrow irrigation, sufficient water is added to moisten the effective rooting depth of the soil to field capacity. Typically this is about 1 to 1.5 m. Additional water may be added periodically to flush out salt accumulations.

Furrow irrigation is most easily used where the soil is flat, has been leveled, or possesses no more than a minimal slope. Otherwise, a series of checks along the furrows are required to divide each channel into self-contained segments, each showing an acceptable drop in soil level. Furrow irrigation is not feasible on hilly terrain because of water runoff and erosion. In fine textured soils, there is a tendency to cause dispersion of clay particles and hardening of the soil surface. This causes increasingly long retention times for water infiltration and enhanced evaporative water loss. Periodic cultivation is required to improve water permeability.

Because of the large area of soil wetted, root growth is promoted through much of the upper soil volume. Where the soil alone acts as the primary source of inorganic nutrients, this is desirable. However, it can be wasteful when chemical fertilizer is applied. The large soil volume requires more fertilizer application than is required with drip irrigation. Thus, furrow irrigation often requires the addition of more fertilizer and results in greater nutrient loss by leaching, volatilization, or uptake by weeds and microbes compared to drip irrigation. Furrow irrigation, by moistening most of the soil surface, may enhance surface salt accumulation and often accentuates weed problems.

Sprinkler irrigation has distinct advantages in sloped terrain where runoff and erosion are potential prob-

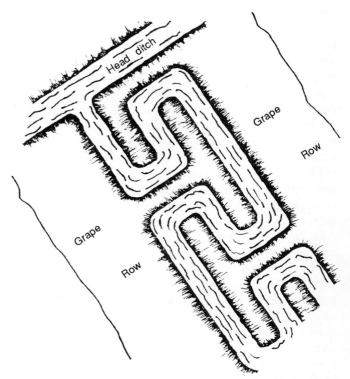

Figure 4.18 Diagrammatic illustration of a block furrow irrigation system. (From Bishop *et al.*, 1967, reproduced by permission.)

lems. Sprinklers have advantages under both highly and poorly porous soil conditions. In porous soils, there is little lateral movement of water and nonuniform water distribution can be a problem. The widespread dispersal of water possible with sprinklers can assure uniform soil moistening. With heavy soils, possessing low infiltration rates, the choice of nozzles for sprinkler irrigation permits the application of water slowly and as fine droplets. Fine droplets tend not to disrupt soil aggregate structure and, thereby, do not aggravate existing permeability problems. Because of uniform water application, sprinkler irrigation is especially valuable in leaching saline soils and minimizing wind erosion. In addition, sprinkler systems can be used for the foliar application of micronutrients and pesticides. Although initial installation costs of a fixed sprinkler system are high, subsequent labor costs are low.

Despite the advantages, sprinklers owe much of their popularity to their use in frost control. Sprinkler irrigation has also been investigated as a means of heat control. However, its use in cooling requires that the water quality be high; otherwise, the vegetation may become coated with toxic levels of salts, notably borates and chlorides.

The major drawbacks to sprinkler irrigation are the high costs of installation and operation. Its property of uniform soil moistening also can lead to increased evaporative water loss from soil and increased weed growth. Where fairly saline water must be used, irrigation should occur at night, or on overcast days, to avoid toxic salt buildup on the foliage and fruit. By prolonging foliage wetting and enhancing pesticide removal, sprinkler irrigation potentially favors disease development. While the strong drying associated with the arid climates typically counteracts the development of most disease problems, washing off insecticides may increase pesticide use.

Movable sprinkler systems, such as the **wheel line** and **center pivot** systems frequently used with annual crops, rarely are used in vineyards. Their movement requires gaps in trellised vineyards. In addition, the periodic, but heavy, application of water can lead to soil compaction, disease, and erosion problems.

For most new installations, especially where water costs are high or availability low, **drip (trickle) irrigation** is preferred. Water is supplied under low pressure and released through special emitters that generate a slow trickle. Emitters are spaced to produce a uniform zone of irrigation along the length of each row. Consequently, the emitters themselves do not need to be close to vine trunks. Water is supplied to the emitters through an array of surface or buried plastic pipes throughout the vineyard. Because the root system quickly becomes focused within the moist zone created by drip irrigation, the system can be used even in established vineyards (Safran *et al.*, 1975). Root zone concentration permits the efficient application of fertilizers and nematicides to grapevines.

Root concentration limits the amount of water that needs to be applied and permits the application to be more specifically related to the needs of the vine. Not only can this improve efficient water use, but it also permits the potential use of irrigation in regulating vegetative growth and modifying fruit quality. The slow rate of water release is especially useful in soils possessing slow infiltration properties, and it minimizes runoff. A slow trickle also means that little water leaches out of the root zone. The efficient use of water can be further enhanced by pulse application and burial of the system. For example, pulse application every 2 hr minimizes percolative loss, while burial limits surface moistening. Both reduce surface evaporative loss and minimize salt accumulation at the soil surface.

Drip irrigation is as effective on steep slopes as on rolling or flat surfaces. It is uninfluenced by wind conditions, which can restrict the timing of sprinkler irrigation. Localization of water application facilitates the control of all but drought-tolerant weeds. Furthermore, drip irrigation permits the better use of shallow soils or those with saline water tables close to the soil surface.

Other advantages of drip irrigation include limited energy consumption (due to low wattage pumps), avoidance of salt accumulations on leaves, and improved efficiency of nutrient uptake (and thereby reduced fertilizer costs and nitrate contamination of groundwater). Moreover, it does not offset the benefits of arid environments in limiting most disease development.

Although possessing many advantages, drip irrigation is not without its problems or limitations. Primary among these is the tendency of emitters to plug. Plugging caused by particulate matter in the water is usually limited by the use of filtration. Growth of slime-producing microbes in the system can usually be controlled by the periodic addition of chlorine to the water line. Plugging from the deposition of calcium carbonate (lime) on the emitters may be minimized by the inclusion of a homopolymer of maleic anhydride (Meyer *et al.*, 1991). Finally, obstruction by root growth around the emitters in buried systems can be deterred by the incorporation of minute amounts of herbicides and the use of acid fertilizers. Because of limited soil leaching, salts can accumulate around the wetted zone if rainfall or irrigation is insufficient to flush them out.

Water application by drip irrigation must be frequent because the root system is concentrated in a comparatively small region of the soil. This is especially important in sandy soils, where the wetted zone is narrow owing to limited lateral movement of water (Fig. 5.2).

Fertilization

In the previous discussion, the potential use of irrigation water in applying fertilizers was mentioned. Although possible with any system, fertilization via drip irrigation has a unique potential. Bravdo and Hepner (1987) have stressed the ability of combined fertilization and irrigation (fertigation) to regulate vine growth and grape quality. It provides the opportunity, under field conditions, to achieve some of the control possible with hydroponics. Appropriate use of fertigation to regulate vine growth requires knowledge of both the factors that affect water and nutrient availability as well as their effect on different stages of vine growth. In the previous section, the influences of water availability were discussed. In this section, nutrient availability and its effects on grapevine growth are discussed.

Based on relative need, inorganic nutrients may be grouped into macro- and micronutrient classes. Macronutrients include the three elements typically found in most commercial fertilizers—nitrogen (N), phosphorus (P), and potassium (K)—as well as calcium (Ca), magnesium (Mg), and sulfur (S). The other major elements required by living cells—carbon (C), hydrogen (H), and oxygen (O)—are not discussed as they come from the atmosphere and water. They are also required in much higher levels than any of the mineral elements. Micronutrients are required only in trace amounts and include boron (B), chlorine (Cl), copper (Cu), iron (Fe), manganese (Mn), molybdenum (Mo), and zinc (Zn). Most are involved as catalysts in pigments, enzymes, and vitamins, or in their activation.

Factors Affecting Nutrient Supply and Acquisition

Although nutrient availability is dependent on the microbial activity and mineral and organic makeup of the soil, nutrient uptake is dependent on the varying physiological characteristics of the scion and rootstock. In comparison with most crops, the nutritional requirements (net annual accumulation) of the grapevine are relatively small (Olson and Kurtz, 1982). This, combined with the storage of nutrient from year to year by the vine, makes determining nutrient requirements and response to fertilizer application more difficult than in most other crops.

Although the soil acts as a reservoir of plant nutrients, only a small fraction is in a readily absorbed form. Most assimilable nutrients occur in dilute form in the soil solution (10^{-6} to 10^{-3} M) and amount to less than 0.2% of that present in the soil. Nearly all nutrients (about 98%) are bound in unavailable forms in the humus or mineral fractions of the soil. These become available only slowly as the humus decomposes and the mineral fraction weathers. The remaining 2% exists adsorbed onto the surfaces of the soil colloidal fraction (humus and clay particles) or as chelates with organic compounds. These nutrients become available as a result of shifts in the equilibria between adsorbed and dissolved forms, as well as through ion exchange. Because of the immense surface area of soil colloids (600 to 800 m^2/g for montmorillonite clays and upward to 700 m^2/g for humus; Brady, 1974), they significantly influence plant nutrient availability and acquisition.

Because of the predominantly negative charge on both inorganic and organic colloidal particles, they act as a reservoir of extractable positively charged ions (cations). The degree to which cations dissolve into the soil solution depends on the strength of their charge, their tendency to become hydrated, the soil pH, and the presence of other ions. Most negatively charged (anions) exist organically bound in the humus. As free ions, they do not adsorb significantly to soil particles. Thus, anions such as nitrates and sulfates are comparatively mobile and readily leached out of soil (Fig. 4.19). Nevertheless, this property allows them to be applied

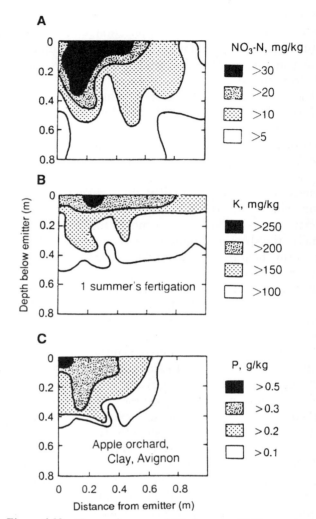

Figure 4.19 Measured pattern of (A) nitrogen—N (B) potassium—K^+ and (C) phosphorus—PO_4^{3-} in soil after one summer's application of soluble fertilizer from a drip emitter. (Redrawn from Guennelon *et al.*, 1979, by Elrick and Clothier, 1990, reproduced by permission.)

effectively on the soil surface, where rainfall or irrigation water can move them down into the root zone. In contrast, the low solubility of phosphate salts and their rapid combination with aluminum, iron, and calcium ions, restrict phosphate movement in soil. Thus, although phosphate is usually in adequate supply, application requires direct deposition in furrows within the root zone if needed. Most nutrient cations show limited movement in soils.

The tendency of mineral cations to adsorb to soil colloids decreases in order of Ca^{2+}, Mg^{2+}, NH_3^+, and K^+, whereas anions decrease in the order of PO_4^{3-}, SO_4^{3-}, NO_3^-, Cl^-. Heavy metal ions such as Zn^{2+} are adsorbed, but only in trace amounts. Because of these relationships, one ion can influence the adsorption/desorption equilibria of other ions, affecting their retention and plant availability. For example, liming soil provides active calcium which replaces hydrogen and other cations, while liberal application of potassium fertilizer liberates calcium and other ions. In addition, solutes in irrigation water both add and remove nutrients from the soil.

The adsorptive binding of most mineral elements to soil colloids has many advantages. It limits the leaching of most nutrient cations out of the root zone and retains them in a form available to the vine. Adsorption also helps to keep the concentration of most nutrients in the soil solution low (10^{-6} to 10^{-3} M). Consequently, the water potential of the soil remains high, easing vine access to water. It also minimizes the development of toxic levels of elements in the soil solution. Finally, the dynamic equilibria between adsorbed and free forms help to provide a relatively stable supply of nutrient cations.

Roots may gain access to nutrients by several means. Usually nutrients are directly assimilated from dissolved ions in the soil solution. This shifts the balance of the equilibria between free and adsorbed forms, permitting the release of more ions and replenishing the readily available nutrient supply. In addition, the release of H^+ ions, carbon dioxide, and organic acids from roots make H^+ ions available for cation exchange with adsorbed cations. This releases extractable nutrient cations into a free state accessible for root uptake. Hydrogen ions are also involved in reducing the less available ferric ions (Fe^{3+}) to the more accessible ferrous (Fe^{2+}) state. A somewhat similar release of negatively charged nutrients occurs in alkaline soils by anion exchange but is largely restricted to phosphate salts. Organic phosphates are released through the action of extracellular phosphatases released by microbes and roots. Finally, the release of chelating and reducing compounds by roots and microbes may help keep metallic ions, such as iron and zinc, in readily available forms. Most metallic cations tend to be in limited supply in neutral and alkaline soils, owing to the formation of insoluble oxides, sulfides, silicates, and carbonates.

In addition to the uptake of nutrients for growth, cations such as potassium may be incorporated to maintain the electrical and osmotic balance of the cytoplasm. This may be required to counter the negative charges associated with the uptake of the major nutrient anions, NO^{3-} and PO_4^{3-}.

In spite of the processes releasing soil nutrients, plant demand may outstrip the ability of the soil to replenish the nutrient supply in the immediate vicinity of feeder roots. Consequently, root extension into new regions of the soil is vital for maintaining an adequate nutrient supply. This is especially important for nutrients such as phosphates, zinc, and copper that do not migrate significantly in soil.

While uptake increases the liberation of adsorbed ions into the soil solution, addition of fertilizer can reverse the reaction, resulting in the precipitation of soluble nutrients. This, combined with deep rooting and nutrient storage in the vine (Conradie, 1988), helps to explain why grapevines often respond slowly and marginally to fertilizer application.

Of soil factors, pH probably has the greatest influence on soil nutrient availability. Depending on the chemical nature of the parental rock, the degree of weathering, and the organic content of the soil, most soils are buffered within a narrow pH range. In calcareous (lime) soils, the primary buffering salts are $CaCO_3$ and $Ca(HCO_3)_2$. By buffering the soil within an alkaline range, the salts can restrict Fe^{2+}, Mn^{2+}, Zn^{2+}, Cu^{2+}, Cu^+, and PO_4^{3-} availability. Through prolonged leaching, most, but not all, soils in high rainfall areas are acidic. Such soils are often deficient in available Ca^{2+}, Mg^{2+}, K^+, PO_4^{3-}, and MoO_4^{2-}. In contrast, the low rainfall of arid regions tends to generate saline or sodic soils, especially in the subsoil where leachates often accumulate.

Soil pH also influences the activity of microbes, especially bacteria, most of which are particularly sensitive to acidic conditions. The consequential slowing of microbial action retards the oxidation of NH_4^+ to NO_3^-. Soil temperature also influences microbial activity, and thus the oxidation of ammonia to nitrate. Cool temperatures also slow the liberation of organically bound phosphates into the soil solution. Consequently, if fertilization is performed in the spring, it is usually most effective with nitrate and inorganic phosphate, as soil microbial activity is still minimal to oxidize ammonia and liberate organically bound phosphates.

Where vineyard soils are acidic, they are often treated with crushed limestone (limed) to raise their pH. The amount required depends on the soil pH, organic content, and texture (Fig. 4.20), as well as the neutralizing power of the lime source. Acid-tolerant rootstocks, such as '140 Ruggeri,' '110 Richter,' '99 Richter,' or 'Gravesac,' can minimize the detrimental effects of acidic soils.

Alkaline and sodic soils may be modified by the addition of sulfur and gypsum (calcium sulfate), respectively. Elemental sulfur is microbially oxidized to sulfuric acid, which neutralizes hydroxides in the soil. Calcium sulfate both neutralizes hydroxides and permits sodium leaching by displacing it from soil colloids and carbonates. Lime-tolerant rootstocks are extensively used on calcareous soils; most American rootstocks are sensitive to lime-induced chlorosis.

Rootstock choice also can differentially affect nutrient uptake. Although rootstocks differ little in the accumulation of nitrogen, phosphorus, and zinc, they vary

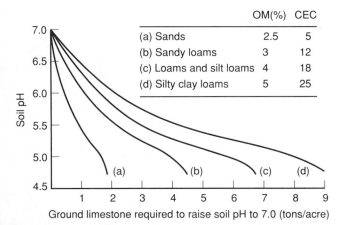

	OM(%)	CEC
(a) Sands	2.5	5
(b) Sandy loams	3	12
(c) Loams and silt loams	4	18
(d) Silty clay loams	5	25

Figure 4.20 Relationship between soil texture and amount of limestone required to raise the soil pH to 7.0. OM, Organic matter; CEC, cation exchange capacity. (From Peech, 1961, reproduced by permission.)

markedly in the adsorption of potassium, calcium, magnesium, and chlorine (see Fig. 4.21A). Such variation probably arises from the selective activity of transport systems in root cell membranes. Differences in nutrient concentration in grapevines also may develop because of differential accumulation in tissues of the scion (Fig. 4.21B).

Although nutrient uptake may be initially passive, all nutrients must pass into cytoplasm on or before reaching the endodermis. Direct access to the vascular system is prevented by the Casparian strip. The latter makes the endodermal cell wall impermeable to water-soluble substances. Thus, transport into the vascular tissues is under the metabolic control of one or more transport systems in the endodermis. Unloading into the xylem and phloem also is likely to be under metabolic, and thus genetic, control. Accumulation within different tissues in the vines is undoubtedly under genetic control as well.

Some rootstock-derived differences in mineral uptake may actually arise indirectly from the association of roots with particular mycorrhizal fungi. This is especially likely for phosphate, where mycorrhizal association favors enhanced uptake from low phosphate soils. Mycorrhizae also may contribute to zinc and copper uptake.

Assessment of Nutrient Need

Deficiency and toxicity symptoms are often sufficiently distinctive to permit the grape grower to diagnose the problem. However, detrimental effects can occur before diagnostic symptoms appear. Thus, the need for more sensitive indicators of nutrient stress has long been known. The primary method of nutrient analysis is based on sampling plant tissue. Most standard soil analyses for nutrient status are unreliable in assessing nutrient stress because of marked differences between nutrient availability and measurable presence. Grapevine requirements also vary throughout the season. Nevertheless, soil nutrient and pH analyses can be useful in predicting major deficiency and toxicity problems.

The new procedure of **electroultrafiltration** (EUF) may reestablish soil nutrient analysis as an important technique in determining nutrient availability (Schepers and Saint-Fort, 1988). The technique can measure desorption of nitrogen, phosphorus, potassium, calcium, magnesium, manganese, and zinc and the presence of phytotoxic levels of aluminum in soil. Although EUF is not widely used in viticulture, EUF values have shown high correlation with grapevine nutrient uptake in several soil types (Eifert **et al.**, 1982, 1985).

Currently, nutrient availability normally is assessed by means of tissue analysis. Cook (1966) found that leaf petioles were ideal owing to their storage of nutrients and the ease with which large, statistically valid samples could be handled. In California, one sampling at full bloom from leaves opposite clusters is considered sufficient. In France, both leaf and petiole analyses are taken at the end of flowering and at *véraison*.

Except for quick nitrogen analyses, samples are dried for 48 hr at 70°C and ground to 20 mesh fineness. Fresh leaves can be assessed for nitrate content by cutting the petiole and leaf base lengthwise. A drop of indicator solution (1 g diphenylamine/100 ml H_2SO_4) is added to the basal 2 cm of the cut surface. Deficiency is indicated when less than 25% of a 20-leaf sample shows bluing. When more than 75% of a 20-leaf sample shows a positive reaction, nitrogen availability may be in excess. Other nitrogen and nutrient assessments are usually conducted in the laboratory on dried samples by colorimetry and atomic absorption spectrometry.

Another component influencing the level of fertilization, or any other component in vineyard practice, is the relative economic return on investment. For example, improvement in fruit yield and quality may be optimal when fertilizer addition is low (root uptake high and runoff minimal). Optimal fertilizer addition is roughly equivalent to grapevine need, minus that naturally available in soil (Löhnertz, 1991). Supply in excess of need seldom improves vine growth or fruit quality, and may be detrimental.

Nutrient Requirements

Although analysis of the soil nutrient status can be fairly accurately assessed, interpretation of the data in

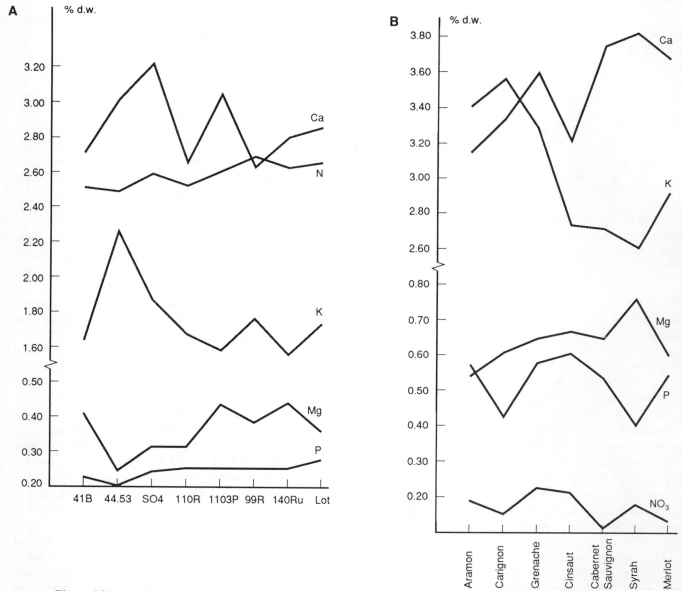

Figure 4.21 Level of mineral nutrients (percentage dry weight) in the leaves of 'Grenache' as a function of eight rootstocks (A) and of the petioles of eight scion varieties on 'SO 4' rootstock (B). Values of P and Mg are doubled in both A and B and for NO₃ in B. Data for A and B are average values for different sets of years, from different locations and growth conditions. (From Loué and Boulay, 1984, reproduced by permission.)

terms of nutrient availability is more difficult. Nutrient availability, uptake, and accumulation often vary considerably throughout the growing season (Fregoni, 1985) and can differ significantly between cultivars (Christensen, 1984). Thus, critical values need to be established for each cultivar and nutrient.

At below a certain value, deficiency becomes progressively more severe. Within mid-range concentrations, nutrient supply is adequate, and plant response is typically unmodified by increasing nutrient availability (Fig. 4.22). However, plants may show "luxury" storage beyond need. Different tissues may respond differently in this regard. For example, fruit nitrogen levels com-

monly rise with increasing nitrogen fertilization, while leaf and petiole values may remain only slightly modified. Eventually, though, excessive nutrient accumulation becomes toxic.

Adding to the complexity of determining nutrient need is the extensive storage and mobilization of nutrients in the woody parts of grapevines. This may explain the occasional delay in vine response to fertilization. Under deficiency conditions, the response to fertilization is usually rapid, but may decline in subsequent years (Skinner and Matthews, 1990). Nutrient balance in the soil and vine tissues also can be important. For example, phosphorus can limit the uptake of potassium

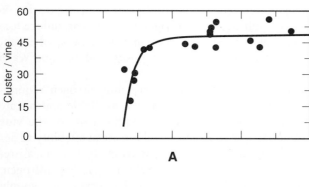

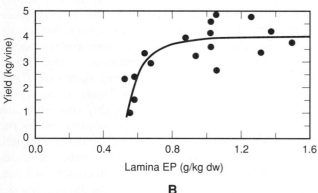

Figure 4.22 Relationship between extractable phosphorus of leaf blades (EP) and cluster number (A) and yield (B) in the second season after phosphorus application. (From Skinner *et al.*, 1988, reproduced by permission.)

Table 4.8 Approximate Range of Nutrients Needed by Grapevines[a,b]

Nutrient	Deficiency	Adequate	Excess
Nitrogen (NO_3^-)	<50 ppm	600–1200 ppm	>2000 ppm
Phosphorus	<0.15%	0.15–0.2%	>0.3–0.6%
Potassium	<1%	1.2–2.5%	>3%
Magnesium	<0.3%	0.5–0.8%	>1%
Zinc	<15 ppm	25–150 ppm	>450 ppm
Boron	<25 ppm	25–60 ppm	>300 ppm (leaves)
Chlorine	<0.05%	0.05–0.15	>0.5%
Iron	<50 ppm	100–200 ppm	>300 ppm
Magnesium	<20 ppm	30–200 ppm	>500 ppm
Copper	<4 ppm	5–30 ppm	>40 ppm

[a] Based on petiole analysis of leaves opposite clusters in full bloom except where noted.
[b] Data from Christensen *et al.*, 1978; Cook and Winkler, 1976; and Fregoni, 1985.

Although constituting about 78% of the earth's atmosphere, nitrogen is unavailable to plants in the gaseous dinitrogen (N_2) form. It must first be reduced to ammonia (NH_4^+) by one of several nitrogen-fixing bacteria. Ammonia can be absorbed directly by grapevines, but the nitrogen is usually assimilated as nitrate (NO_3^-). This occurs after ammonia has been oxidized

(Conradie and Saayman, 1989a), whereas potassium antagonizes the adsorption of calcium and magnesium (Scienza *et al.*, 1986). In addition, phosphorus deficiency can induce magnesium deficiency by limiting its translocation from the roots (Skinner and Matthews, 1990).

Because of these factors, only general ranges for most nutrient requirements are presently possible for grapevines (see Table 4.8). Deficient, adequate, and toxic levels for several micronutrients have not been determined for grapevines. If vine reaction is similar to that of most plants, then the response for boron and zinc should be similar to that given in Fig. 4.23.

Nitrogen Nitrogen is an essential constituent in amino acids and nucleotides and, thereby, proteins and nucleic acids, respectively. Nucleotides also function as essential electron carriers in cellular metabolism. In addition, nitrogen is a constituent element of several growth regulators and chlorophyll. Although nitrogen is required in larger quantities than any other inorganic soil nutrient, the requirements of grapevines are considerably less than those of most other agricultural crops. Yearly use is estimated to vary from about 40 to 70 kg N/ha (Champagnol, 1978; Löhnertz, 1991).

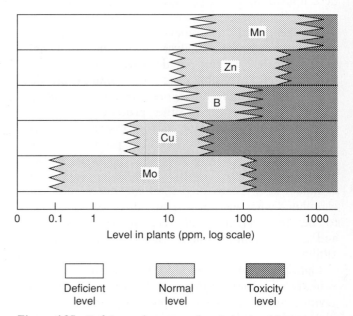

Figure 4.23 Deficient, adequate, and toxic levels of five micronutrients in plants. (From data in Allaway, 1968, graphed by Brady, 1974. Reprinted with the permission of Macmillan Publishing Company from THE NATURE AND PROPERTIES OF SOILS, Eighth Edition by Nyle C. Brady. Copyright © 1974 by Macmillan Publishing Company.)

to nitrate by nitrifying bacteria in the soil. Because of its negative charge, nitrate is more mobile in soil than positively charged ammonia. Ammonia commonly adheres to soil colloids and remains relatively immobile under most conditions. Thus, nitrogen supplied as ammonia should be applied early enough for its microbial conversion to nitrate to coincide with the primary nutrient uptake period of grapevine growth.

Until the use of inorganic nitrogen fertilizers, vineyard nitrogen supply was dependent on the activity of nitrogen-fixing bacteria and the addition of manure. Unlike other nutrients derived from the soil, nitrogen is not a component of the mineral makeup of soil. Its availability, unlike that of potassium, phosphorus, and magnesium, is particularly dependent on the effect of seasonal factors, such as soil moisture, aeration, and temperature, and the activity of soil microorganisms and cover crops. In addition, as nitrate is poorly adsorbed by soil colloids, nitrate tends to be leached out if not rapidly assimilated by plants and microbes. Nitrogen tends to be the nutrient most frequently found deficient in vineyard soils.

The lower cost of ammonia combined with its adsorption by soil particles tends to make it the preferred form of nitrogen fertilizer. However, the greater tendency of ammonia to induce vegetative growth can predispose the vine to increased disease susceptibility (Rabe, 1990). In contrast, most organic fertilizers have the advantage of delaying nitrogen liberation. In addition, as nitrogen liberation is dependent on microbial decomposition, its release may correspond more closely to grapevine uptake and need.

Nitrogen status in grapevines commonly has been assessed by petiole analysis using the diphenylamine–sulfuric acid test noted above. The test is particularly useful as visible deficiency symptoms are slow to develop and may not express themselves until deficiency is severe. Several other procedures have been used to assess grapevine nitrogen status (Kliewer, 1991). Although arginine has been proposed as such an indicator, it is also a general indicator of stress in many plants (Rabe, 1990). Arginine appears to accumulate as a means of detoxifying the ammonia accumulated when growth is retarded. There also appears to be an interaction between potassium nutrition and grape arginine content (Conradie and Saayman, 1989b).

Light green foliage and reduced shoot growth are the primary symptoms of nitrogen deficiency. Apical shoot regions, as well as cluster stems and petioles, may take on a pink to red discoloration. Nitrogen deficiency in the fruit can lead to premature termination (sticking) of fermentation during winemaking. Deficiency problems are enhanced when vines are lightly pruned, or a ground cover grown, on nitrogen-poor soils. Although

nitrogen deficiency is a problem in some areas, the excessive application of nitrogen fertilizer can induce toxicity. This is most noticeable in older leaves, where the edges accumulate saltlike deposits and become water soaked and finally necrotic.

An additional means of assessing nitrogen requirement involves determining if the available soil nitrate levels are below the known requirements of the vine (Löhnertz, 1991). Because soil nitrate availability varies throughout the year, it is important to take measurements when the vine is actively assimilating nitrogen. Although mobilization of stored reserves may supply much of the nitrogen required for early growth (Löhnertz et al., 1989), most of the nitrogen subsequently comes directly from the soil. Nitrogen uptake occurs slowly until root growth commences in the spring, peaks about véraison, and declines thereafter (Löhnertz, 1991; Araujo and Williams, 1988). Thus, uptake is improved when increased availability coincides with bud break, and especially around véraison when the nitrogen demand is greatest. During véraison, adequate nitrogen supply increases the average number of berries per bunch reaching maturity. In warm climatic regions, the association of renewed root production in the autumn with fall nitrogen application can provide much of the nitrogen requirements for spring growth (Conradie, 1992). Appropriate timing and degree of fertilization are components of maximizing nitrogen uptake efficiency while minimizing groundwater pollution with nitrate.

Nitrogen fertilization enhances vegetative growth but does not induce "luxury" accumulation in the foliage. Nevertheless, fruit may show enhanced accumulation with increased nitrogen availability. Nitrogen accumulation is most noticeable in the increase in free amino acid content, notably proline and arginine, and soluble proteins. Although enhanced nitrogen content facilitates rapid juice fermentation, the accompanying increase in pH may be undesirable. In addition, an increase in arginine level may augment the production of ethyl carbamate (a suspected carcinogen) in wines subsequently heated during processing (Ough, 1991). Fruit maturation also may be delayed by competition with enhanced vegetative growth under high nitrogen conditions, leading to poor fruit quality.

High nitrogen levels are suspected to be involved, at least partially, in the development of several grapevine physiological disorders. These include a "false potassium deficiency" (Christensen et al., 1990) as well as inflorescence and bunch-stem necroses. Excessive nitrogen availability also tends to enhance susceptibility to several fungal infections, especially bunch rot. This may result from suppressed phytoalexin synthesis (Bavaresco and Eibach, 1987) and/or increased canopy density.

Phosphorus Phosphorus is an important component of cell membrane lipids, nucleic acids, energy carriers such as ATP, and some proteins, and it is required for sugar metabolism. Phosphorus accumulates primarily in meristematic regions, seeds, and fruit.

Phosphorus deficiency is rare in grapevines. This probably results from the limited phosphorus requirements of the vine, remobilization of phosphorus within the vine, and the ample presence of inorganic and organic phosphates in most vineyard soils. Nevertheless, phosphorus deficiency has been recognized in grapevines grown on acidic soils and low-phosphorus hillside soils.

Deficiency symptoms include the formation of dark green leaves, down turning of leaf edges, reduced shoot growth, purple coloration of the main veins and older leaves, and premature fruit ripening and reduced yield. More subtle consequences of phosphorus deficiency involve reduced initiation, development, and maintenance of flower primordia (Skinner and Matthews, 1989). Phosphorus fertilization can alleviate these problems, increase grape color and sugar content, and augment free monoterpene accumulation in some cultivars (Bravdo and Hepner, 1987).

Because phosphorus deficiency in grapevines is uncommon, phosphorus fertilization should be minimal to avoid interfering with potassium, manganese, magnesium, and iron acquisition. Disruption of cation uptake from the soil may result directly from metabolic effects on the vine or indirectly from effects on cation exchange in the soil. For example, addition of alkaline rock phosphate ($3Ca_3(PO_4)_2 \cdot CaF_2$) can retard manganese solubilization, whereas the calcium released by the more soluble superphosphate ($CaCH_2PO_4)_2$ and $CaHPO_4$) can liberate manganese, making it more readily available for root uptake (Eifert *et al.*, 1982).

Potassium Potassium is the only macronutrient not a structural component in cellular macromolecules. Its presence is required for osmotic and ionic balance in cells, neutralization of organic acids, regulation of stomatal function, cell division, enzyme activation, protein synthesis, and synthesis and translocation of sugars.

Potassium is absorbed as free K^+ ions dissolved in the soil solution. As mineral weathering only slowly replenishes the potassium supply, soil depth is important to potassium fertility. Removal of fertile topsoil during land leveling, or erosion on slopes, can expose less productive shallow subsoil. This can produce patchy areas of potassium deficiency in vineyards. Sandy soils in high rainfall regions also are often potassium deficient. In addition, potassium deficiency occurs frequently on virgin soil, known in parts of Europe as "pasture burn" (*Wasenbrand*) (P. Perret, 1992, personal communication).

Grapevines may express one or possibly two foliar forms of potassium deficiency. "Leaf scorch" occurs initially in the middle of primary shoots, and apically on lateral shoots. The symptoms begin as a loss of green color (chlorosis), or bronzing, along the edges of the leaf that progresses inward. The margins subsequently dry and roll. In contrast, "black leaf" begins with interveinal necrosis on the upper surface of sun-exposed leaves and may spread to cover the leaf surface.

When adequate, potassium favors grape quality by enhancing fruit coloration and sufficient acidity. However, excess potassium may raise juice pH undesirably and can potentially suppress magnesium uptake by the roots. Potassium deficiency may be induced if phosphate availability is excessive.

Because potassium migrates poorly in soil, potassium fertilizers are added deeply to position the element within the root zone. Potassium sulfate is generally preferred to potassium chloride to avoid increasing the chloride content in the soil solution.

Calcium Calcium is a vital constituent of plant cell walls, where it reacts with pectins, making them relatively water insoluble and rigid. Calcium also plays important roles in regulating cell membrane permeability, ion and hormone transport, and enzyme function. Along with potassium, calcium helps to detoxify organic acids in cell vacuoles by inducing acid precipitation.

Calcium deficiency typically occurs only on strongly acidic quartz gravel. It is expressed as a narrow zone of necrosis along the edge of leaves that may progress toward the petiole. Minute, brown, slightly raised regions may develop in the bark of shoots. Clusters may show dieback from the tip. When in excess, as in calcareous soils, it can cause lime-induced chlorosis with sensitive rootstocks. In the form of gypsum (calcium sulfate), it may be added to acidic soils to raise soil pH and limit aluminum toxicity (Kotzé, 1973), or improve water permeability in sodic soils.

Magnesium Magnesium is a vital cofactor in the splitting of water by chlorophyll in photosynthesis. It stabilizes ribosome, nucleic acid, and cell membrane structure and is involved in the activation of phosphate-transfer enzymes in metabolism.

Deficiencies are frequently associated with sandy soils in high rainfall regions, poorly drained sites, and high pH soils. This results from the relative ease with which magnesium is leached from the soil. Magnesium deficiency appears to be involved in the European expression of the physiological disorder termed bunch-stem necrosis. Symptoms first begin to develop in basal leaves. Interveinal regions develop a straw-yellow chlorotic discoloration, while regions bordering the veins remain green. Early in the season, symptoms may appear as small brownish spots next to leaf margins.

Sulfur Sulfur is an integral component of the amino acids cysteine and methionine. Cross-linking of their sulfur atoms is often crucial in the functional structure of many proteins. Sulfur also is an integral component of the vitamins thiamine and biotin, and it is present in coenzyme A.

Although an essential nutrient, sulfur appears in ample supply in most vineyard soils. It is often applied to vines, not as a nutrient but as a fungicide. Sulfur also may be applied unintentionally as sulfate with potassium or magnesium fertilizers, or with calcium in gypsum. Sulfur is not known to occur naturally in soil at toxic levels.

Zinc Zinc plays important roles as cofactors in several enzymes and in the synthesis of the growth regulator indoleacetic acid (IAA).

Although zinc is required in small amounts, suppression of solubility in alkaline soils can lead to the development of deficiency symptoms. Zinc insufficiency also can develop in sandy soils where low levels of inorganic or humus colloidal material limit zinc availability. High levels of phosphate may precipitate the metal as zinc phosphate and restrict its availability to plants. Land leveling also may lead to scattered areas of zinc deficiency throughout a vineyard. Rootstocks such as 'Salt Creek' and 'Dog Ridge' are particularly susceptible to zinc deficiency.

Zinc deficiency can generate a series of distinctive symptoms. It produces a mottled interveinal chlorosis, with an irregular green border along clear veins. Modified leaf development produces atypically larger angles between the main leaf veins. Leaves at the apex of primary shoots, and along lateral shoots, are much reduced in size and often asymmetrically shaped. The latter symptom is called "little leaf." Zinc deficiency also can lead to poor fruit set and clusters containing small, green, immature, "shot" berries. Symptoms may resemble fanleaf degeneration.

For spur-pruned vines, application of zinc sulfate to freshly cut spur ends can be beneficial. Alternately, leaves may be sprayed with a foliar fertilizer containing zinc.

Although less common, zinc toxicity can affect root growth, occasionally being associated with the use of contaminated compost. Compost also may be a source of toxic levels of other heavy metals, such as lead (Pb), cadmium (Cd), and copper (Cu) (Perret and Weissenbach, 1991).

Manganese Manganese functions in the activation of, or as a cofactor in, several enzymes and is essential to the membrane structure of chloroplasts. Manganese is directly involved in the synthesis of fatty acids, in the neutralization of toxic oxygen radicals, and in the reduction of nitrates to ammonia.

As with most bivalent cations, the availability of manganese to plants is reduced in alkaline soils and poorly drained soils. Deficiency symptoms produce a chlorosis, associated with a green border along the veins, similar to that shown by zinc deficiency. However, manganese deficiency neither modifies leaf vein angles nor induces "little leaf." In addition, chlorosis develops early in the season and commences with the basal leaves. Leaves may take on a geranium-like appearance.

Iron Iron plays a role in the development of chloroplasts, acts as a cofactor in redox reactions involving cytochrome, and is a component in catalase and peroxidase enzymes. As with other nutrients closely associated with photosynthesis, deficiency of iron results in chlorosis. Chlorosis usually begins apically and early in the season. Yellowing often is marked, with only fine veins remaining green. Severely affected leaves may wither and fall.

Lime-induced chlorosis is most commonly observed on calcareous soils. While the high pH of calcareous soils may limit iron solubility in the soil solution, this in itself may not induce iron-deficiency chlorosis. Leaves showing chlorosis may have iron contents as high as healthy green leaves (Mengel *et al.*, 1984). Thus, chlorosis may result from other factors associated with the high lime (calcium carbonate) content of soil. This could include disruption of the cellular incorporation or utilization of iron by the accumulation of bicarbonate in leaf tissue. Chlorosis also may be associated with the tendency of lime soils to compact easily. This could aggravate conditions that limit root growth, such as anaerobiosis following heavy rains, increased ethylene synthesis in the soil (Perret and Koblet, 1984), heavy metal toxicity, or overcropping. Young roots are the primary sites of iron acquisition. In a study by Bavaresco *et al.* (1991), iron uptake and resistance to chlorosis were closely correlated with root diameter and root hair development.

Most North American grapevines are sensitive to lime-induced chlorosis on calcareous soils. Thus, careful selection of rootstock is required when grafted *V. vinifera* cultivars are planted on calcareous soils. Alternatively, the foliage may be sprayed with iron chelate or, if the soil is not too calcareous or alkaline, the pH raised with sulfur addition.

Boron Boron is required in nucleic acid synthesis, in the maintenance of cell membrane integrity, and in calcium utilization. As boron is required in trace amounts, the range separating deficiency and toxicity is narrow (Fig. 4.23). Deficiency symptoms develop as blotchy yellow chlorotic regions on terminal leaves. Chlorosis soon spreads to the leaf margin. Apical regions of the shoot may develop slightly swollen, dark green bulges. Early dieback may activate the development of stunted lateral shoots. Boron deficiency also induces poor fruit set, the development of many "shot" berries, or excessive fruit drop. Deficiency symptoms tend to develop on sandy soils in high rainfall areas, on soils irrigated with water low in boron, or on strongly acidic soils.

Application of borax to counteract deficiency can lead to toxicity if not evenly distributed. Toxicity symptoms begin with the development of brown to black specks on the tips of leaf serrations. The necrotic regions spread and become continuous. Inhibition of growth may result in leaves wrinkling and puckering along the margin.

Copper Copper acts primarily as a cofactor in oxidative reactions during respiration and the synthesis of proteins, carbohydrates, and chlorophyll. Copper deficiency symptoms are only rarely observed on grapevines, possibly because of the use of copper in fungicides such as Bordeaux mixture. When deficiency occurs, the leaves are dwarfed and pale green, shoots develop short internodes, cane bark has a rough texture, and root development is poor.

Toxicity produces a leaf chlorosis similar to that of iron-deficiency chlorosis. Toxicity most commonly arises in soils where copper has accumulated following prolonged use of copper-containing fungicides such as Bordeaux mixture (Scholl and Enkelman, 1984). Copper buildup in the soil may be the cause of some "replant problems" in vineyard soils.

Molybdenum Molybdenum is required in nitrate reduction and the synthesis of proteins and chlorophyll. Only rarely has molybdenum been noted to be deficient in vineyard soils. Molybdenum deficiency produces a necrosis that rapidly spreads from the leaf margins inward. The demarcation zone between healthy and necrotic tissue is pronounced, and the unaffected area appears normal. Affected leaves often remain attached to the shoot and fall with difficulty.

Chlorine Chlorine is involved in both the osmotic and ionic balance in cells as well as the splitting of wa-

ter during photosynthesis. Although required in fairly large amounts, chlorine is not known to be limiting in vineyard soils. More commonly, it is associated with toxicity under saline conditions. It produces a progressive, well-defined necrosis that moves from the edges of the leaf toward the midvein and petiole. Because physiological disruption occurs at levels well below those causing visible toxicity, use of chlorine-tolerant rootstocks such as 'Salt Creek' and 'Dog Ridge' on saline soils is advisable (Table 4.7).

Organic Fertilizers

Until comparatively recently, the addition of manure to vineyards was standard practice. With the modern concentration and localization of many agricultural industries, ready access to animal manures became more difficult, while the production cost of inorganic fertilizers fell. This led to a shift to the convenience of commercial inorganic fertilizers. Currently, several trends are encouraging a return to animal and green manures. Increased energy costs, combined with contraction in grape demand, have made inorganic fertilizers less cost-effective. In addition, environmental and health concerns have led to a growing market for "organically" grown grapes and wine. Several governments also are enacting laws designed to limit nitrate pollution of water from urban, industrial, and agricultural sources. This has induced viticulturalists to investigate means by which vineyard production may be maintained with limited or no dependence on inorganic fertilizers.

Organic fertilizers have several advantages beyond the supply of nutrients for vine growth. Primary among these is the supply of humic materials. Humus consists of a wide variety of complex organic compounds that often decompose slowly. They form colloidal particles that bind with clay colloids, through the action of bivalent cations such as Ca^{2+} and Fe^{3+}, to create soil aggregates. Humus also acts as a food base or surface on which microbes and soil invertebrates can grow. The mucilage they produce aids in the formation and stability of a friable soil structure. The soil aggregate structure is important not only for water infiltration, but also for minimizing soil erosion. Manures also increase the activity of nitrogen-fixing bacteria, such as *Azotobacter*. In contrast, the activity of nitrogen-fixing bacteria is repressed by the application of nitrogen fertilizers.

The organic colloids of humus also play vital roles in the maintenance of soil fertility through effects on water and nutrient availability. As the humus decomposes, adsorbed and bound nutrients are released into the soil solution. This is especially important in soil nitrogen availability. The microbial activity associated with humus formation also binds nutrients otherwise lost by leaching. However, manures high in liquid components must be applied during warm weather to assure that microbial activity is sufficient to bind the nutrients and limit runoff. Additional benefits of manure application include improved soil aeration (Baumberger, 1988) and a shift in soil pH toward neutrality (Fardossi *et al.*, 1990).

Besides animal and green manures, chopped vine prunings and winery wastes commonly have been incorporated into vineyard soils. Although moderate applications are usually beneficial, massive application of grape by-products is undesirable. Use of winery or distillery waste is not recommended on acid soils where manganese can build up to toxic levels (Boubals, 1984). Additionally, pomace can be important in the spread of viral contamination in soil and may increase the incidence of nematode-transmitted viral infections.

Animal Manures Although animal manures have many advantages, they possess several inconveniences beyond local unavailability. Variability in composition is of major concern as it greatly complicates the calculation of appropriate application rates (Table 4.9). Important factors influencing nutritional composition are the source animal, the nature and amount of litter (bedding) incorporated, and the handling and storage before application. Examples of additional potential difficulties are as follows: (1) toxicity due to copper supplementation of pig feed, (2) development of zinc deficiency following the use of poultry manure, (3) induction of nitrogen deficiency associated with the decomposition of large amounts of incorporated straw, and (4) nitrogen loss connected with ammonia volatilization during urine breakdown. Problems of odor production can be minimized by early application.

Because animals utilize only a portion of the organic and nutrient content of their feed, much nutritive value remains in the manure. About half of the nitrogen, most of the phosphates, and nearly 40% of the potassium found in the original feed remain in the feces. The average distribution of the three main plant nutrients in the solid and liquid components of cow, sheep, and pig manure is noted in van Slyke (1932). Not all the nutrients in animal manures are readily available for plant uptake, however. Brady (1974) estimates that roughly 1000 kg (~1 ton) of manure supplies about 2.5 kg of nitrogen, 0.5 kg phosphorus, and 2.5 kg of potassium. Most of the plant remains are either hemicellulose, lignins, or lignin–protein complexes. Additional organic materials consist of the cell remains of bacteria. They can constitute up to 50% of manure dry mass.

Ground Covers, Green Manures, and Compost Use of green manures is another long-established procedure for improving soil structure and nutrient content. The practice involves growing a crop and plowing it under while still green. If plowing occurs while the crop is still succulent, its nitrogen content is typically adequate to permit its rapid decay without limiting the available nitrogen supply. Sewage sludge occasionally has been used as a vineyard compost, but contamination of sludge with heavy metals can make its use undesirable (Perret and Weissenbach, 1991).

Green manures (cover crops) commonly are planted in midsummer to early fall to limit or avoid inducing water and nutrient stress in the vines. This is especially important under nonirrigated, arid conditions, or on poor soils. However, some competition may be desired to limit vine vigor under conditions of high rainfall, or on rich soils. Green manures are also useful in limiting water, fertilizer, and soil loss from slopes. For example, ground covers can dramatically restrict water runoff during downpours—15 versus 80% (Rod, 1977).

Table 4.9 Average Values for Moisture and Nutrient Content of Farm Animal Manures[a]

Animal	Feces/urine ratio	Water (%)	Level in manure Kg/metric ton (lb/ton)		
			Nitrogen	Phosphate	Potassium
Dairy cattle	80:20	85	20 (10.0)	5.4 (2.7)	15.0 (7.5)
Feeder cattle	80:20	85	23.8 (11.9)	9.4 (4.7)	14.2 (7.1)
Poultry	100:0	62	49.8 (29.9)	28.6 (14.3)	14.0 (7.0)
Swine	60:40	85	25.8 (12.9)	14.2 (7.1)	21.8 (10.9)
Sheep	67:33	66	46 (23.0)	14.0 (7.0)	43.4 (21.7)
Horse	80:20	66	29.8 (14.9)	9.0 (4.5)	26.4 (13.2)

[a] Reprinted with the permission of Macmillan Publishing Company from THE NATURE AND PROPERTIES OF SOILS, Eighth Edition by Nyle C. Brady, 1974. Copyright © 1974 by Macmillan Publishing Company.

While loss is less marked when measured over the whole season, ground covers may limit water runoff to 1.5 versus 19% on slopes over 40° (Rod, 1977). In addition, ground covers can facilitate weed control, bind nutrients otherwise lost by leaching, and keep dust down. If green manure crops are left until the spring, they can trap snow and limit frost penetration into the ground.

Several leguminous and cereal crops have been used. Legumes have the advantage of forming rhizobial root nodules that fix nitrogen. Legume cover crops can add significantly to the soil nitrogen supply (Fig. 4.24). However, high seed cost, need for *Rhizobium* inoculation, and higher water demands can limit their feasibility. Cereal crops require less water and can often penetrate and loosen compacted soil. Consequently, legumes and grains are frequently combined together (Winkler *et al.*, 1974).

Disease, Pest, and Weed Control

Changes similar to those affecting vineyard fertilization are influencing the practice of disease, pest, and weed control. At the extreme edge of this change are those dedicated to "organic" or "ecological" viticulture (Jenkins, 1991). Although based on a laudable philosophy, "natural" pesticides are not guaranteed to be inherently safer than synthetic equivalents. Also, various

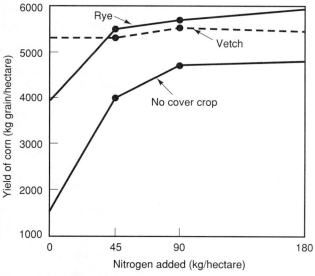

Figure 4.24 Effect of legume (vetch) and nonlegume (rye) cover crops on the 7-year average yield of corn receiving three levels of nitrogen fertilizer on a sandy loam soil. Cover crops were grown each year. (Data from Adams *et al.*, 1970, graphed in Brady, 1974. Reprinted with the permission of Macmillan Publishing Company from THE NATURE AND PROPERTIES OF SOILS, Eighth Edition by Nyle C. Brady. Copyright © 1974 by Macmillan Publishing Company.)

oils used in "organic" pest control can affect the taste and odor of grapes and wine (Redl and Bauer, 1990). Thus, "natural" pesticides should require the same exhaustive assessment as "chemical" agents now require. While health concerns about the risks of pesticide use have induced several governments to contemplate pesticide deregistration, little thought has gone into the health risks of a likely increase in mycotoxin contamination of food. Nevertheless, the inability of current pesticides and herbicides to provide adequate crop protection and the need to reduce production costs have spurred research into alternative control measures.

One of the main concepts in providing better and less expensive pest and disease control is called **integrated pest management** (IPM). The term "management" reflects a change in attitude, where it is recognized that limiting damage to an economically acceptable level is more feasible and prudent than "control."

Integrated pest management often combines the expertise of specialists in fields such as plant pathology, economic entomology, plant nutrition, weed control, and soil science. Coordinated programs often can reduce pesticide use, while improving effectiveness. IPM programs involve factors such as environmental modification, biological control, and better timing of pesticide application. Timing pesticide application to coincide with vulnerable periods in the life cycle of the pathogen reduces the frequency of application and increases its effectiveness (Sall, 1982; Bulit *et al.*, 1985). As a consequence, there is less disruption of natural control agents in the soil or surroundings. Modification of canopy development, as well as adjustment to fertilization and irrigation practices, can markedly reduce the disease and pest susceptibility of vines and, thereby, the need for chemical protection. Research on pesticide combinations is helping to avoid antagonism and minimize the number of treatments required (Marois *et al.*, 1987). Improvements in nozzle design that provide better and more uniform chemical spread can reduce runoff and achieve better contact with the intended pest or pathogen. Assuring that a greater percentage of the target receives a toxic dose should delay the development of resistant strains. Sublethal doses selectively favor the survival (reproduction) of pathogens and pests that inherently possess some resistance to the control agent.

Integrated pest management is placing greater emphasis on multiple means of control, rather than reliance on any one technique (Basler *et al.*, 1991). Experience has indicated that dependence on chemical controls is ultimately unsatisfactory. The same will probably be true if overconfidence in biological or genetic controls results in abandoning other forms of control. Because most pathogenic agents multiply exceedingly rapidly, they are disappointingly adept at developing resistance to single selective pressures, whether man-made or "natural" (i.e., the toxin produced by *Bacillus thuringiensis*).

Control of Pathogens

Chemical Methods Primary among procedures designed to improve pesticide effectiveness are those intended to retard the development of pesticide resistance (Staub, 1991). For fungal pathogens, the use of relatively nonselective fungicides has much value. Their broad-spectrum action provides general protection against the vast majority of pathogenic fungi and bacteria. This reserves selective fungicides for situations where their curative action is necessary to prevent the eminent outbreak of serious epidemics. Disease forecasting, the prediction of disease outbreaks based on local meteorological data, is particularly useful in this regard. It reduces the need for frequent (prophylactic) pesticide application, increases the effectiveness of what is applied, and can retard the rate at which resistance builds up in the pathogen or pest populations.

The buildup of resistance to nonspecific pesticides is particularly difficult. Single mutations are unlikely to provide sufficient protection against the nonselective disruption of membrane structure and enzyme function induced by most contact pesticides. The effectiveness of nonselective fungicides was so great that the disease potential of some pathogens was forgotten for decades. Their real potential became evident only when the use of nonselective fungicides declined in the early 1960s (Mur and Branas, 1991). Similarly, the abandonment of nonselective pesticides against grape leafhoppers is thought to be partially attributable for the increased incidence of omnivorous leafrollers in California (Flaherty *et al.*, 1982).

Selective fungicides and insecticides often have the advantage of producing less damage against harmless insects and fungi. Selective pesticides also tend to require fewer applications to be effective as systemic uptake by the plant allows them to kill pathogens within host tissues. Regrettably, the selectivity often permits neutralization of the toxic action by single mutations in the pest or pathogen. The long-term effectiveness of these chemicals probably requires restricting use to situations where their unique toxicity is required (Delp, 1980; Northover, 1987).

Another approach to extending the effective "life span" of selective pesticides involves rotational use of the compounds. By changing from one pesticide to another, the advantage of resistance genes against any one chemical may be counteracted. Most resistance genes are in some way detrimental to the growth or reproduction of the pest or pathogen. Thus, absence of constant

pesticide "pressure" may result in selection against specific resistance genes. Continual pesticide pressure tends to select for mutations that carry little selective disadvantage. A similar, but alternative, approach involves the combination of nonselective and selective pesticides.

The prime function of the nonselective fungicides is to reduce the size of the pest or pathogen population, and thereby reduce the frequency of resistant gene occurrence. This approach is accredited with delaying the appearance of resistance genes in some pathogen populations (Delp, 1980).

In both approaches mentioned above, it is important that the pesticides not belong to the same family of chemical compounds and, therefore, have similar modes of action (Table 4.10). If the two chemicals have similar modes of toxicity, the selective pressures they produce on the pathogen will be similar. Development of resistance to one member of a family of pesticides usually provides resistance against all members. The phenomenon is called **cross-resistance**.

Biological Control While chemical control remains one of the primary means of controlling pests and disease-causing agents, increasing stress is being placed on biological control. Although some predators of insect pests have developed pesticide resistance (Englert and Maixner, 1988), most remain sensitive to insecticides. Thus, for the use of natural insect or mite control agents, pesticide application must be delayed, minimized, or avoided.

In addition to restricting pesticide use, it is often necessary to maintain a broad diversity of plants in the

Table 4.10 Fungicides Used in Combating *Botrytis cinerea* on Grapes, Arranged According to Chemical Family[a]

Chemical group	Common name	Trade name or synonym
Benzimidazoles	Benomyl	Benlate [DuPont]
	Triophanate methyl	Topsin-M [Pennwalt]
	Carbendazim	Delsene [DuPont]
Dicarboximides	Vinclozolin	Ronilan [BASF]
	Iprodione	Rovral [Rhône-Poulenc]
	Clozolinate	Serinal [Agrimont]
	Procymidone	Sumisclex [Sumitomo]
Dithiocarbamates	Thiram	Various [i.e., DuPont]
	Mancozeb	Various [i.e., Rohm and Haas]
Pthalimides	Captafol	Difolatan [Chevron]
	Captan	Various [i.e., Chevron]
	Folpet	Phaltan [Chevron]
Pthalonitriles	Chlorothalonil	Bravo [Diamond]
Sulfamides	Dichlofluanid	Euparen [Bayer]

[a] From O'Conner (1987), reproduced by permission.

vicinity of the vineyard. This helps to provide the range of alternate hosts necessary for maintaining or building up the population of desired pest-controlling agents.

Occasionally, the release of competitive species can provide adequate control for serious pests. By establishing themselves on the host, the competitors restrict the colonizing potential of the pest species. An example of this phenomenon, called **competitive exclusion,** is the action of the Willamette spider mite (*Eotetranychus willamettei*) against the Pacific spider mite (*Tetranychus pacificus*) (Karban *et al.*, 1991).

Although insects are susceptible to a wide range of bacterial, fungal, and viral pathogens, few have shown much promise in becoming practical control agents. A major exception is *Bacillus thuringiensis*. Forms of the toxin have been used successfully in the control of several lepidopteran pests, such as the omnivorous leafroller (*Platynota stultana*) and the grape leaffolder (*Desmia funeralis*). A potentially useful control agent in combating the western grape leaf skeletonizer (*Harrisina brillians*) is the granulosis virus (Stern and Federici, 1990).

Another technique employed in insect management is the use of sex pheromones. These are species-specific airborne hormones used by insects in locating receptive members of the opposite sex. Examples of pheromone use include incorporation in insecticide-containing traps to attract and kill males of a pest population, as well as pheromone spread throughout vineyards to disrupt pest mating. The release of large numbers of artificially reared, sterile individuals during the mating season also has been used to limit the success of pest mating.

Last, but definitely not least, is the use of resistant rootstock, which was developed initially to control the phylloxera infestation in Europe. Biological control via grafting is used in limiting the damage caused by several nematodes, viruses, and environmentally induced toxicities.

Biological control of fungal pathogens is less developed than that for arthropod pests. This partially reflects the growth of fungal pathogens within plants, away from exposure to parasites. Nevertheless, inoculation of leaf surfaces with epiphytic microbes may limit the germination and penetration of fungal pathogens (Redmond *et al.*, 1987). Occasionally, fungal mycoparasites such as *Trichoderma harzianum* may be used to reduce the need for fungicide application in the control of *Botrytis cinerea* (Elad and Zimand, 1992). For fungal pathogens that may survive in the soil for short periods, plowing under of a green cover crop often can significantly reduce survival.

One of the more intriguing examples of biological control involves **cross-protection.** Cross-protection is a

phenomenon in which virally infected cells are immune from further infection by the same or related viruses. The incorporation of genes coding for the viral protein coat into plants can induce a similar phenomenon, called **coat protein-mediated resistance** (Beachy *et al.*, 1990). Thus, the incorporation of coat protein genes from grapevine viruses may provide cultivars with immunity to infection by the source viruses.

Environmental Modification Modifying the microclimate around plants has long been known to potentially minimize disease and pest incidence. By improving the light and air exposure of the grapevine, canopy management can increase the toughness and thickness of fruit and leaf cuticular coverings. A more open canopy structure also facilitates rapid drying of fruit and foliage surfaces. This reduces the time available for fungal penetration and may reduce or inhibit spore production. Furthermore, an open canopy facilitates the more efficient application of control chemicals, or the access of biological control agents.

Exposure of the fruit to the sun and drying action of air can be further enhanced by applying gibberellic acid. This is particularly valuable with tight-clustered varieties. Gibberellin induces cluster stem elongation and opens up the fruit cluster. The reduction of berry compactness also permits the production of a typical epicuticular wax coating on the fruit (Marois *et al.*, 1986). Fruit exposure can be enhanced still further by the removal of leaves opposite, and just above and below, berry clusters (**basal leaf removal**). This technique has been so successful in reducing the need for fungicidal sprays that several wineries have written its use into their contracts with growers (Stapleton *et al.*, 1990). Basal leaf detachment also removes most of the first-generation nymphs of the grape leafhopper (*Erythroneura elegantula*) (Stapleton *et al.*, 1990). This can improve biological control by the leafhopper parasitic wasp (*Anagrus epos*) by allowing time for its population buildup and spread to grapevines from overwintering sites on wild blackberries.

Balanced plant nutrition generally favors disease and pest resistance by promoting the development of the inherent anatomical and physiological resistance properties of the vine. In contrast, nutrient excess and deficiency may have the reverse effect. For example, high nitrogen levels suppress the synthesis of a major group of antifungal compounds, the phytoalexins, by grapevine leaves (Bavaresco and Eibach, 1987).

Irrigation can decrease the consequences of nematode damage by providing an ample water supply to the damaged root system. Irrigation also may favor healthy vine growth by reducing water stress. However, excessive irrigation can favor disease development by promoting luxurious canopy development and increasing berry size and cluster compactness. This is especially important for cultivars, such as 'Zinfandel' and 'Chenin blanc,' that normally produce compact clusters.

Weed control generally reduces disease incidence. This may result from the removal of alternate hosts on which pests and disease-causing agents may survive and propagate. For example, dandelions and plantain often are carriers of tomato and tobacco ringspot viruses, and Bermuda grass is a reservoir for sharpshooter leafhoppers, the primary vectors of Pierce's disease.

Soil tillage occasionally can be beneficial in disease control. For example, burial of the survival stages of *Botrytis cinerea* can promote their decomposition in soil. In addition, the emergence of adults of the grape root borer (*Vitacea polistiformis*) is restricted by the burial of pupae that results from soil cultivation (All *et al.*, 1985).

Although environmental modification can limit the severity of some diseases and pests, it can itself enhance other problems. For example, soil acidification used in the control of Texas root rot (*Phymatotrichum omnivorum*) has increased the incidence of phosphorus deficiency in Arizona vineyards (Dutt *et al.*, 1986). Elimination of weeds as carriers of pests and disease-causing agents may inadvertently limit the effectiveness of some forms of biological control. Carriers of one pest may be reservoirs of parasites and predators of other grapevine pests.

Genetic Control Improved disease resistance is one of the primary aims of grapevine breeding. It was first seriously investigated as a means of avoiding the expense of grafting in phylloxera control. Subsequently, breeding has largely focused on providing better rootstocks possessing improved drought, salt, lime, virus, and nematode tolerance or resistance. Work has also progressed, but more slowly, on developing new fruit varieties with improved pest and disease resistance. In the premium wine market, consumer resistance to new varieties has limited their acceptance. However, at the lower end of the market, new varieties have a distinct advantage because of their reduced production costs. Resistant varieties may be even more acceptable for "organic" viticulture, where synthetic pesticides are forbidden. Here, freedom from pesticide residues is probably more critical to commercial success than varietal origin.

In most European countries, there are legal restrictions against the use of interspecies crosses in the production of "quality" (Apellation Control) wines. This is to their disadvantage as the best sources of disease and pest resistance come from species other than *Vitis vinifera*. The best sources of resistance genes within *V. vinif-era* probably reside in the few remaining populations of wild vines (*V. vinifera* f. *silvestris*) growing in Europe and southwestern Asia (Avramov *et al.*, 1980). While "hidden" sources of resistance still reside in existing cultivars of *V. vinifera* f. *sativa* (Becker and Konrad, 1990), they must be limited in supply.

In most instances, complete resistance (immunity) to infection is desired. Nevertheless, slowing the infection rate in some diseases can retard the spread of the disease sufficiently to adequately limit disease damage in most years. Tolerance to infection also can be valuable, except where systemic infection can lead to the spread of the pathogen by grafting or other forms of mechanical transmission (e.g., tolerance to *Xiphinema index* damage can still permit transmission of the grapevine fanleaf virus).

Eradication and Sanitation In most situations, eradication of established pathogens is impossible owing to their survival on alternate hosts, such as weeds or native plants. Eradication can be useful, though, in the elimination of systemic pathogens. Seed propagation is often used for this purpose, but it is impractical with grapevines as it would disrupt the combination of characteristics that make each cultivar unique. Thus, except where disease-free individuals can be found, elimination of systemic pathogens involves either **thermotherapy** or **meristem culture**.

Thermotherapy involves placing vines or dormant canes at 35° to 38°C for 2 to 3 months. It is effective against several viruses, such as the grapevine fanleaf, tomato ringspot, and fleck viruses, and the leafroll agent (typically associated with a closterovirus). For most other viruses, culturing small meristematic fragments of the shoot apex may permit the isolation and propagation of uninfected vines. Meristem culture has been used in the elimination of the viruslike agents of stem pitting, corky bark, and leafroll, the viroid of yellow speckle, and the bacterium *Agrobacterium tumefaciens*. The latter may also be eliminated by a short, high-temperature treatment (Bazzi *et al.*, 1991). For vines infected by several systemic pathogens, both techniques may be required.

Use of thermotherapy and meristem culture has caused concern with some growers because of the apparent modification of morphological and physiological traits in treated vines. These effects usually disappear as the vines age or are propagated repeatedly (Mullins, 1990).

Once freed of systemic pathogens, vines usually remain disease-free if grafted to disease-free rootstock and planted on soil free of either, or both, pathogen and vector. Most serious grapevine viruses are not transmitted by leaf-feeding insects. Where the soil is infested with nematodes, use of resistant rootstock is advisable. Alternatives are leaving the land fallow for upward of 10 years or fumigating the soil.

Sanitation does not eliminate disease or pest problems, but it may reduce their severity by destroying resting stages or removing survival sites.

Quarantine Most, if not all, wine-producing countries possess laws regulating the importation of grapevines. Some of the best examples illustrating the need for quarantine laws involve the transmission of grapevine diseases and pests. Two of the major grapevine diseases in Europe (downy and powdery mildew) were imported unknowingly from North America in the nineteenth century. The phylloxera root louse was also accidentally introduced into Europe from North America, probably on rooted cuttings. Several viral and viruslike agents are now widespread in all major winegrowing regions. They are thought to have spread unsuspectedly through grafting. Several grapevine cultivars and rootstocks are symptomless carriers of the agents causing leafroll, corky bark, and stem pitting.

Thankfully, some other potentially devastating diseases have as yet to become widespread. For example, Pierce's disease is still largely confined to southeastern North America and Central America. Even phylloxera has not spread throughout all the countries in which it now occurs. Thus, limits to grapevine movement within regions of a single country still can have a significant impact in limiting the spread of pathogens with low potential for self-dispersal.

Because it is difficult to detect the presence of most pests and disease-causing agents, only dormant canes can be imported into most countries. This should avoid the introduction of root pathogens and foliar pests, but fungal spores, insect eggs, and other minute dispersal agents may go undetected. Correspondingly, imported stock is typically quarantined for several years, until determined to be free, or freed, of known pathogens. Most pests, fungal, and bacterial pathogens express their presence within this period when propagated at the quarantine location. However, the presence of latent viruses and viroids usually requires grafting, or mechanical transmission, to indicator plants. The detection of systemic pathogens through transmission to sensitive (indicator) plants is called **indexing**. Analysis techniques employing ELISAs (enzyme-linked immunosorbent assays) (Clark and Adams, 1977) or complementary DNA (cDNA) probes (Koenig *et al.*, 1988) may speed future detection of systemic pathogens and permit the earlier release of imported stock.

Consequences of Pathogenesis on Fruit Quality

The negative influence of pests and diseases is obvious in symptoms like blighting, distortion, shriveling, decay, and tissue destruction. More subtle are effects on vine vigor, berry size, and fruit ripening. Sequelae such as reduced root growth, poor grafting success, reduced photosynthesis, or increased incidence of bird damage on weak vines (Schroth *et al.*, 1988) can be easily overlooked. In some instances, detection of disease-causing agents is impeded by minimal symptoms and the absence of uninfected individuals. For example, the almost universal prevalence of viroids in grapevine cultivars was not recognized until relatively recently, as uninfected vines for comparison were rare.

Most pest and disease research is concerned with understanding the pathogenic state and how it can be managed. However, in making practical decisions on disease control, it is important to know the effects of disease not only on vine health and yield, but also on grape and winemaking quality.

All pests and disease agents disrupt vine physiology and, thereby, can influence fruit yield and quality to some degree. However, agents that attack berries directly have the greatest impact on fruit quality. These include three of the major fungal grapevine pathogens, namely, *Botrytis cinerea*, *Plasmopara viticola*, and *Uncinula necator*. Grapevine viruses and viroids, being systemic, can both directly and indirectly affect berry characteristics. Insect pests can cause fruit discoloration and malformation as well as create lesions favoring invasion by other organisms.

Of fruit-infecting fungi, *Botrytis cinerea* has been the most extensively studied. Under special environmental conditions, the infection progresses toward a "noble" rot, often yielding superb wines. Normally, however, the fungus produces an ignoble (bunch) rot. The early invasion of infected fruit by acetic acid bacteria probably explains the high levels of fixed and volatile acidity in the fruit. Under moist conditions, additional secondary invaders such as *Penicillium* and *Aspergillus* may contribute other off-flavors. More serious to human health is the potential production of mycotoxins by these organisms, and especially *Trichothecium roseum* (Schwenk *et al.*, 1989). *Mucor* species produce compounds inhibitory to malolactic fermentation (San Románo and Silva Alemáo, 1986), but their toxicity to mammals, if any, is unknown. *Botrytis cinerea* does not produce mycotoxins (Krogh and Carlton, 1982), but it can synthesize several antimicrobial compounds active against yeasts and other fungi (Blakeman, 1980). In addition to increased fixed and volatile acidity, bunch rot produces reduced nitrogen and sugar levels that create difficulties during juice fermentation. Large accumulations of β-glucans create clarification problems. Finally, the action of laccases in browning and the generation of phenol flavors generally makes infected red grapes unacceptable for winemaking.

Little information is available on the direct consequences of fungal pathogens other than noble and bunch rot on fruit composition. However, fruit infected with powdery mildew may produce wines with higher pH values and elevated alcohol and phenol contents. Their possession of bitter and other off-tastes partially explains their low appreciation by tasters (Ough and Berg, 1979). Browning is common but can be partially offset by the addition of sulfur dioxide. In *flavescence dorée*, the production of a dense bitter pulp makes wine production from diseased fruit virtually impossible.

Leafroll has been the most investigated of virus and viruslike infections relative to grape and winemaking quality. With potassium transport being affected, berry titratable acidity increases. This typically generates wines of higher pH and poor color. Sugar accumulation in the berries is usually decreased owing to suppression of transport from the leaves. Ripening is often delayed.

The physiological disorder bunch-stem necrosis (*dessèchement de la rafle*) causes grape shrivelling and fruit fall about and after *véraison*. Wines produced from vines so affected often are imbalanced, high in acidity, and low in ethanol and several higher alcohols and esters (Ureta *et al.*, 1982).

Effects on aroma have seldom been reported. Exceptions are the reduction in varietal character of grapes infected by *Botrytis cinerea* or modified by the ajinashika virus (Yamakawa and Moriya, 1983).

Examples of Grapevine Diseases and Pests

Grapevines are damaged by many fungal, bacterial, and viral pathogens, insect and mite pests, rodents, birds, as well as mineral deficiencies and toxicities, air pollutants, and other afflictions. There is insufficient space in this text to deal adequately with these maladies. Thus, only a few important and/or representative examples of the major categories of grapevine disorders are given below. Detailed discussions of grapevine maladies may be found in specialized works such as Pearson and Goheen (1988) and Flaherty *et al.* (1992) (North America), and Galet (1991) and Larcher *et al.* (1985) (Europe).

FUNGAL PATHOGENS

With few exceptions, fungal pathogens grow vegetatively as long, thin, branched, microscopic filaments, individually called **hyphae** and collectively termed **mycelia**. Most fungi produce cell wall ingrowths regularly along the hyphae termed **septa**. The ingrowths are usually incomplete and leave a central opening through which nutrients, cytoplasm, and cell organelles may pass. Thus, fungi possess the potential to adjust the number and proportion of nuclei within the hyphae

system of the organism. This gives fungi a degree of genetic flexibility unknown in other organisms. The filamentous growth habit also provides them with the ability to physically puncture plant cell walls. This property, combined with their degradative powers and prodigious spore production, helps to explain why fungi are the major disease-causing agents of plants.

Botrytis Bunch Rot   Unlike many grape pathogens, *Botrytis cinerea* is not a specialized grapevine parasite and infects an extensive variety of plants. As a consequence, spores may come from plant sources other than grapevines. Nevertheless, most early infections probably develop from spores produced on overwintered fungal tissue in the vineyard (Fig. 4.25). Spores may develop from mycelia in leaves, "mummified" fruit, and bark. In addition, the black, multicellular resting structures called **sclerotia** may generate spores by both asexual (conidia) and sexual (ascospores) means.

Initial infections usually develop on aborted and senescing flower parts. When the remnants of flowers are trapped within growing fruit clusters, they may initiate fruit infections later in the season. Another source of fruit infection comes from latent infections that occur in the late spring. These form when mycelium invades the vessels of young green berries (Pezet and Pont, 1986). The fungus subsequently becomes inactive until the level of acidity and other antifungal compounds in the fruit decline during ripening. Latent infections likely act as the primary cause of bunch rot under dry autumn conditions. Under moist conditions, most bunch rot probably arises from *de novo* infections in the fall. Under protracted wet conditions, infection can rapidly progress into a bunch rot complex, involving secondary invaders such as species of *Penicillium, Aspergillus, Cladosporium, Rhizopus,* and *Acetobacter*. The European grape berry moth (*Lobesia botrana*) can aggravate the situation by transporting and infecting fruit with *B. cinerea* conidia (Fermaud and Le Menn, 1989). Infections of leaves, shoots, and other vine parts occur but are primarily important as sites for winter survival.

During the growing season both physiological and anatomical changes can increase the susceptibility of the fruit to fungal attack. **Microfissures** develop around stomata (Fig. 4.26A), and **micropores** form in the cuticle (Fig. 4.26B). Both provide sites for fungal penetration and the release of plant nutrients that can aid spore germination. Weathering of the waxy plates on the cuticle also favors infection by facilitating spore adherence. The loss of wax is most noticeable where berries press and rub against one another (Marois *et al.*, 1986). Rapid berry enlargement, especially during heavy rains, can induce splitting and the release of juice and thus favors rapid development of bunch rot.

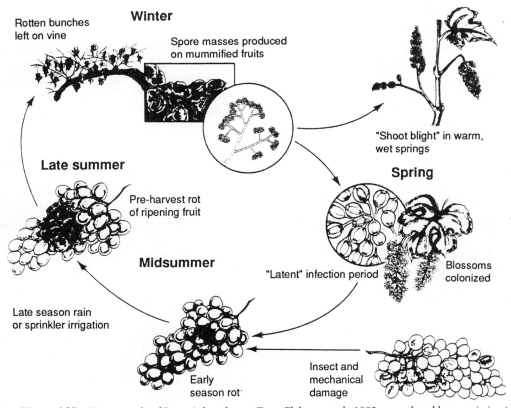

Figure 4.25 Disease cycle of Botrytis bunch rot. (From Flaherty *et al.*, 1982, reproduced by permission.)

Many factors affect bunch rot susceptibility. Skin toughness and open fruit clusters reduce bunch rot incidence, while heavy rains, protracted periods of high humidity, and shallow rooting of the vine increase susceptibility. Shallow rooting exposes the vine to waterlogging, which can favor rapid water uptake and berry splitting. Berry splitting also can result from the osmotic uptake of water through the skin under rainy,

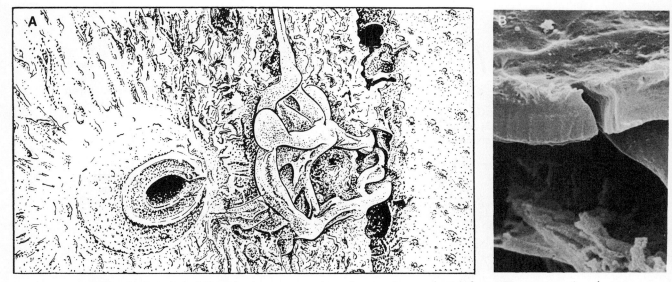

Figure 4.26 (A) Drawing of *Botrytis cinerea* penetrating a berry peristomatal microfissure (From a scanning electron micrograph by Bessis, 1972, in Ribéreau-Gayon *et al.*, 1980) (B) Scanning electron micrograph of a section through a cuticular micropore. (From Blaich *et al.*, 1984, reproduced by permission.)

cloudy conditions (Lang and Thorpe, 1989). In addition, protracted moist periods provide conditions that favor spore germination and production.

On germination, spores produce one or more **germ tubes** that grow out through the spore wall. Fruit penetration occurs shortly thereafter, often through microfissures in the epidermis. Subsequent ramification progresses more or less parallel to the berry surface through the hypodermal tissues.

Depending on the temperature and humidity, new spores are produced within a few days or weeks. Spores are borne on elongated, branched filaments that erupt either through the stomata or epidermis. The white/gray color of the young spores gives rise to the common name for most *Botrytis* diseases, gray mold. Subsequently, the spores turn brown and often are so densely packed as to give the infected tissues a feltlike appearance. Early in infection, white grapes may take on a pale purplish coloration. All infected fruit eventually turns brown.

Effective control often requires both fungicidal sprays and environmental modification. Some fungicides remain localized on the surface and act protectively; others are incorporated into the leaf tissues and possess both protective and curative properties. With

the development of resistance to benzimidazole and dicarboximide fungicides, chemicals such as tebuconazole and dichlofluanid are being investigated as substitutes (Brandes and Kaspers, 1989). Effective fungicide application is enhanced by leaf removal around the clusters.

Tilling under infested plant remains, along with a green manure, helps to reduce survival of the fungus in the vineyard. Use of less vigorous rootstocks, canopy management, and basal leaf removal can help to generate a more open canopy and speed drying of vine surfaces. Application of gibberellic acid also can favor drying by opening tight fruit clusters. Only limited success has been achieved through the action of mycoparasites, such as *Trichoderma viride* (Dubois *et al.*, 1982), or the suppression of spore germination by epiphytic yeasts and bacteria. Several commonly used fungicides can suppress the growth and action of these biological control agents (Ferreira, 1990).

Powdery Mildew (Oidium) Powdery mildew or oidium is induced by a specialized grapevine pathogen, **Uncinula necator.** Specialized hyphal extensions, called **haustoria,** grow only into living epidermal cells (Fig. 4.27). Nevertheless, adjacent cells are physiologically

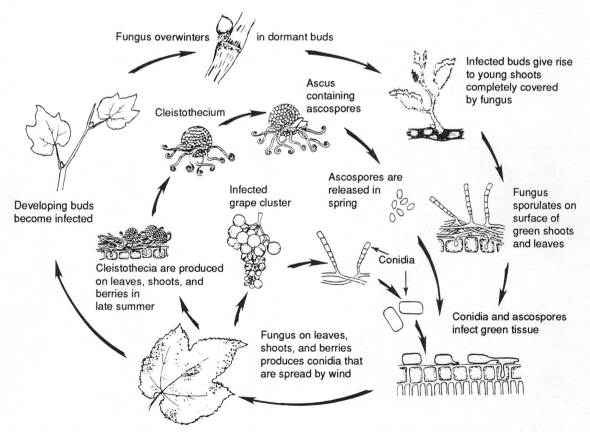

Figure 4.27 Disease cycle of powdery mildew. [Drawing by R. Sticht (Kohlage), from Pearson and Goheen, 1988, reproduced by permission.]

disrupted and soon become necrotic. Most of the fungus remains external to the vine.

Fungal overwintering often depends on dormant hyphae that survive on the inner bud scales of grapevines. In cool climates, survival also may involve microscopic, round, reddish black, resting structures called **cleistothecia.** In the spring, surviving dormant hyphae become active and produce spores. After rains, overwintering cleistothecia may swell, rupture and eject spores. These wash or are blown onto young tissues and initiate early infection following bud burst (Pearson and Gadoury, 1987).

The ability of the fungus to attack all green grapevine tissues, and rapidly to produce prodigious numbers of spores, makes powdery mildew a serious disease of grapevines. Infection can result in leaf and fruit distortion by killing surface tissues before they reach maturity. Severe infection leads to leaf and fruit drop and shoot tip death. Spore production produces a white powdery appearance on infected surfaces. Later in the season, the production of cleistothecia can give infected tissues a distinctive red to black speckled appearance.

Disease management is based primarily on fungicidal sprays. Development of an open canopy tends to reduce disease incidence and improve spray contact with plant surfaces. Biological control is currently inadequate. Where cleistothecia are the primary mode of hibernation, lime-sulfur applied before bud break in the spring can significantly delay disease outbreak (Pearson and Gadoury, 1987). During the growing season, wettable or sulfur dusts have commonly been used. Although not incorporated into plant tissues, sulfur acts both as a preventative and curative agent. This results from most of the fungal mycelium being located exterior to the plant and, thereby, directly exposed to the fungicide. Newer fungicides, such as ergosterol biosynthesis inhibitors (EBI), are more effective and less phytotoxic, but they are more expensive.

Downy Mildew (Peronospora) Downy mildew or peronospora is induced by *Plasmopara viticola,* a fungus unrelated to that causing powdery mildew. The term "mildew" refers to the white cottony growth that develops on infected tissue under moist conditions in both diseases.

The spores, called **sporangia,** germinate to produce several flagellated **zoospores.** Zoospores possess a short motile stage, during which they swim toward stomata. After adhering to the plant surface, spores begin to penetrate the host. Sporangial production usually occurs at night, and spores remain viable for only a few hours after sunrise. Consequently, downy mildew is a serious problem only under conditions where rainfall is prevalent throughout much of the growing season.

Like powdery mildew, downy mildew attacks all green parts of the vine and produces haustoria. However, *Plasmopara viticola* hyphae do not remain exterior to the plant, but ramify extensively throughout the host tissues. Under moist conditions, sporulation rapidly develops as a white cottony growth. On leaves, sporulation occurs preferentially through the stomata on the lower surface. Leaf invasion is the primary source for spores inducing subsequent fruit infections. Infected shoot tips become white with spore production and show a distinct S-shaped distortion. The shoot subsequently turns brown and dies. The fruit is most vulnerable to infection when young, but all parts of the fruit cluster remain susceptible up until maturity. As with leaf infection, severe development of the disease can result in premature abscission.

During the summer, the fungus produces a resting stage called the **oospore.** Oospores remain dormant until the following spring, when they germinate to produce spores. Under mild conditions, both oospores and dormant mycelia in infected leaves may initiate infections in the spring.

No effective biological control measures are known against *Plasmopara viticola.* Production of a more open canopy has only a minor beneficial effect on disease incidence. Consequently, chemicals remain the only effective treatment for this pathogen under conditions favorable to disease development.

Bordeaux mixture and several other nonsystemic fungicides are toxic to *P. viticola,* but they are only preventative in action. Newer systemic fungicides, such as fosetyl aluminum and phenylamides (i.e., metalaxyl), are especially effective owing to their incorporation in plant tissues. Uptake of the fungicide by the plant also limits its removal by rain.

Eutypa Dieback Eutypa dieback is a serious disease showing a slow but insidious attack on the woody components of grapevines. The pathogen, *Eutypa lata* (*E. armeniacae*), preferentially invades wounds in the perennial shoot system. It begins by infecting the xylem exposed by graft conversion or the pruning of old wood required to maintain, or change, training systems. It normally does not invade wounds produced during the pruning of annual shoot growth.

Growth of the fungus in the xylem is slow, and the enlarging, elongated, lens-shaped canker can remain hidden by overlying bark for years. Thus, disease presence is typically first noticed through leaf and shoot symptoms. These are most apparent when shoots are about 25 to 50 cm long. The young leaves of affected shoots are reduced in size and may be upturned, chlorotic, and possess a tattered margin. Shoot internodes are markedly dwarfed. The associated subtending canker is revealed by removing the outer bark.

The cankers contain rows of flask-shaped **perithecia,** structures within which the spores are produced. Cutting tangentially along the wood through the canker exposes the perithecia as round structures appearing to contain a jellylike material. The associated infected wood is brown.

Following sufficient rainfall (>1 mm), spore discharge occurs throughout the summer and into the fall in mild climates. In cold climates, spore production peaks during the later part of winter.

Control of Eutypa dieback is particularly difficult for many reasons. Except in areas where the grapevine is the major woody plant, sanitation has relatively little effect in reducing spore load in the air. This is because *E. lata* has a very wide host range, including some 80 woody species indigenous in grape growing regions. Spores also can be effectively wind dispersed over 100 km. Finally, normal (surface) fungicide spraying is ineffective in preventing spore germination within the xylem.

Control procedures include the early removal and destruction of infected wood. Destruction is recommended as the wood can remain a source of spore production long after removal from the vine. Treating pruning wounds on wood over 2 years old with a sus-

pension of benomyl is desirable. The fungicide must soak into the exposed wood to provide adequate protection. Of additional benefit is pruning when the xylem is active and spore production low. The formation of lignin and suberin in reaction to wounding can markedly reduce disease incidence (Munkvold and Marois, 1990). Thus, very early or very late pruning can be useful in control.

BACTERIAL PATHOGENS

Bacteria are an ancient group of microorganisms, existing predominantly as colonies of independent cells. Most possess a rigid cell wall that gives them spherical to rod shapes. They may or may not be motile, depending on the species and strain. They are restricted to entering plants either via natural openings (stomata, lenticels, etc.) or through wounds. Overwintering occurs as dormant cells in soil, on plant parts, or in vectors. The few bacteria that invade grapevines generally enter the host through grafting wounds or are transmitted via leafhoppers.

Crown Gall   The causal agent of crown gall, *Agrobacterium tumefaciens,* can invade a wide range of plants (Fig. 4.28). Of the strains affecting grapevines,

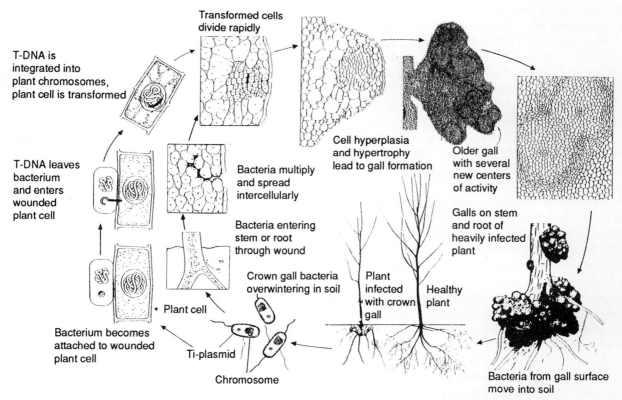

Figure 4.28   General disease cycle of crown gall. (From Agrios, 1988, reproduced by permission.)

biovar 3 is the most common and serious. Biovar 3 appears to infect exclusively grapevines and, other than vines, has been isolated only from vineyard soils. Exterior to the vine, biovar 3 grows only in the rhizosphere around young grape roots. Biovar 3 has recently been considered sufficiently distinct to warrant being recognized as a distinct species—*Agrobacterium vitis* (Ophel and Kerr, 1990).

Agrobacterium tumefaciens biovar 3 can cause lesions on young roots (Burr *et al.*, 1987). These may serve as the means by which the bacterium invades the xylem of healthy roots. From the roots, the bacterium can be translocated up into the shoot system. Invasion of annual shoot growth appears to occur in late summer or early fall (Burr *et al.*, 1988). Thus, although the bacterium grows systemically throughout the vine, its distribution is nonuniform and most commonly confined to the roots. Alternate sources of infection probably come from mechanical wounds produced at the trunk base and during grafting. Because the bacterium may reside saprophytically within the xylem without producing obvious symptoms, rootstock importation may be the primary reason for the widespread occurrence in vineyards.

Despite the more frequent occurrence of the bacterium in the root system, its most serious consequences occur in the trunk. The bacterium can induce uncoordinated cell division in the phloem and xylem, generating gall formation. Individual galls are commonly self-limiting and may subsequently rot and separate from the stem. New galls often originate next to old galls. In young vines, the galls can girdle the trunk, killing the vine. In large vines, the consequence of infection often depends on the number, size, and distribution of the galls. Heavy galling significantly reduces vine vigor and fruit yield (Schroth *et al.*, 1988). Galls are also sites for the invasion of other severe grapevine pathogens, such as *Pseudomonas syringae* f. *syringae* and *Armellariella mellea*.

Wound formation acts as the stimulus inducing gall formation. Frost damage appears to be the most important source of wounding, as gall formation is most severe in cold climatic regions in the spring. Galls develop most commonly near the trunk base (crown), giving rise to the name "crown gall." Graft conversion of existing vines also can be a significant activator of gall formation, resulting in graft failure.

Wounding indirectly stimulates gall induction by activating the division of associated tissue cells. This in turn provokes the transfer of a tumorigenic (T_i) plasmid from the bacterium into dividing host cells. The host cells are subsequently transformed into localized plant tumors (galls). The plasmid induces the transformed cells to produce unique amino compounds,

octapine and cucumopine (Huss *et al.*, 1990), that serve as nutrients for the growth of *A. tumefaciens.*

No fully effective means of crown gall control is currently available. The best option is to plant disease-free stock in soil not previously having been planted with grapevines. However, disinfested clones planted in soil only slightly infested with biovar 3 generally remain disease free. Allowing vineyard land to lie fallow might be of value in reducing the population of *A. tumefaciens* biovar 3, but its effectiveness and the duration required are unknown. Heat treating dormant cuttings at above 54°C appears to hold promise as an effective means of eliminating most systemic infection in existing nursery stock (Bazzi *et al.*, 1991). Apical shoot micropropagation is a more complex, but effective, means of eliminating systemic bacterial infection. Use of pasteurized soil and equipment when propagating disease-free clones should stop the spread of *A. tumefaciens* during bench grafting. Most reports indicate that biological control through inoculation with the agrocin 84 strain of *A. tumefaciens* (*A. radiobacter*) is of little value. Where crown gall is serious, as in northeastern North America, multiple trunking is often a valuable precaution. As a management practice, it diminishes the damage caused by crown gall, as it is unlikely that all trunks of a vine will be killed in the same year. The disease is typically more serious on *Vitis vinifera* cultivars than on *V. labrusca* and French–American hybrid cultivars.

Pierce's Disease Pierce's disease is produced by another xylem-inhabiting bacterium, *Xylella fastidiosa*. Like several other grapevine pathogens, it affects a wide range of plants including crops, such as alfalfa and apricot, and several common weeds. A number of the latter are symptomless carriers of the bacterium. It is transmitted from plant to plant through the action of several insect vectors, predominantly sharpshooter leafhoppers (Cicadellidae) and spittle bugs (Cercopidae).

The geographical distribution of the disease is limited by the presence of suitable vectors and the bacterium. Except for an isolated case in Europe (Boubals, 1989), Pierce's disease is restricted to the coastal gulf plains of the United States, Mexico, and Central America and a few sites in California. In areas where the pathogen is endemic, indigenous species of *Vitis* are resistant or relatively tolerant to infection. However, its occurrences severely limits the commercial cultivation of *V. vinifera* varieties in these areas.

In sensitive vines, invasion of the xylem induces the release of gums and the production of tyloses in the vessels, both of which disrupt water flow and place the vine under potential water stress. Correspondingly, the disease is more severe in hot and/or arid climates, where vines are more likely to experience water stress. The bacterium also produces several phytotoxins that disrupt cellular function.

Leaf symptoms develop as a progressive inward browning and desiccation of the blade. In advance of the necrosing region, concentric areas of discoloration commonly develop, which are yellow in white and red-purple in red cultivars. The blade eventually may drop, leaving the petiole still attached to the shoot. Late in the season, green "islands" may remain on canes surrounded by brown mature bark.

In severely affected vines, bud break may be delayed, and shoot growth slow and stunted. The first four to six leaves are dwarfed, and the main veins are bordered by dark green bands. Subsequent leaves generally are more typical in size and appearance.

Infected *Vitis vinifera* cultivars may survive for 1 to 5 years, depending on the age of the vine when infected, the variety, and local conditions. Muscadine (*V. rotundifolia*) grapevines may be resistant to, or tolerant of, infection, depending on the cultivar.

In warm climates, where the pathogen and vector are common, the only effective control is growing resistant or tolerant cultivars. Where the disease is established, but localized, vineyard planting should avoid areas where reservoirs of the pathogen and vector are common. Transfer is predominantly from symptomless carriers, rather than from vine to vine. Insecticide control of insect vectors has generally been unsuccessful in halting disease spread.

Flavescence Dorée   Flavescence dorée and other yellows diseases are believed to be induced by several mycoplasma-like bacteria (Prince *et al.*, 1993). Although the disease is most damaging in Europe, the current outbreak in southern Europe may have originated in the northeastern United States. A similar, but milder, yellows disease occurs on *V. vinifera* and interspecific hybrids in New York State (Pearson *et al.*, 1985). The primary vector of the pathogen, *Scaphoideus littoralis*, is endemic to the region. The disease may have been accidentally introduced into Europe in the late 1940s along with *S. littoralis*. Several other leafhopper species also may act as vectors of the pathogen(s) (Caudwell, 1990). Some forms of the disease are transmitted at low frequency by grafting. Related forms of the disease may be *Bois noir* and *Vergilbungskrankheit*, Mediterranean grapevine yellows, and subtropical grapevine yellows.

Two distinct expressions of *flavescence dorée* occur in Europe. In the *Nieluccio* type, the disease becomes progressively more severe each year until the vine dies. In the *Baco 22A* type, the vine recovers after symptoms develop the year after infection. If reinfected within a few years of a previous infection, recovered vines show only a localized rather than a systemic reaction.

Newly infected vines show delayed bud burst and shoot growth in the spring. Leaf blades may become partially necrotic and roll downward, more or less overlapping one another; they also become brittle. The foliage may turn yellow in white, and red in red cultivars. Angular spots of the same color may develop on leaf blades. In sensitive cultivars, the most distinctive symptom develops in the summer, as the vine takes on a drooping posture. This results from poor development and lignification of xylem vessels and phloem fibers. Shoot tips may die and develop black pustules. The fruit tends to shrivel and develop a dense, fibrous, bitter pulp. Symptoms typically begin to develop only in the year following infection, the "crisis" year.

No effective control of the disease is known for regions where both vector and pathogen are established, other than growing varieties that show recovery (*Baco 22A* expression). Insecticide spraying can delay, but probably does not stop, vector spread of the disease from vine to vine. Much care must be taken in vine propagation as the pathogen can be spread by grafting. The risk of graft spread is most serious during the year of infection, when the vine is symptomless. Recovered vines are apparently not infectious and can be used safely as stock for graft propagation.

VIRUSES, VIRUSLIKE, AND VIROID PATHOGENS

Viruses and related pathogens are submicroscopic noncellular objects capable of invading all forms of life. Those that attack grapevines possess only RNA as the genetic material. Differentiation between viruses and viroids is based on the presence or absence, respectively, of a protein coat enveloping the nucleic acid. Viruslike diseases are those resembling the transmission characteristics of viruses, but for which no consistent association with a particular viral agent has been established. Infection is usually systemic, affecting all tissues, with the possible exception of the apical meristem and/or pollen and seeds. The absence of apical meristem infection permits the elimination of viruses by apical tissue culture (micropropagation). This is particularly important in grapevines, where vegetative propagation can lead to the spread and perpetuation of systemic infections. Because these agents are translocated systemically in grafted vines, the current widespread occurrence of most viruses and viroids may have developed as a result of the extensive use of grafting to control phylloxera and other root problems. Although all grapevine viruses and viroid infections are graft transmissible, some are also spread by nematode, aphid, mealybug, and fungal vectors (Walter, 1991). Some also can be mechanically transmitted on pruning equipment.

Besides elimination by micropropagation, cultivars may be cured of some viral infections by heat treating dormant stem cuttings or young vines. Thermotherapy is ineffective in the elimination of viroids (Duran-Vila *et al.*, 1988).

Production and propagation of virus- and viroid-free nursery stock are now ongoing projects in most wine-producing countries. This is based on the belief that cured clones grow better, and produce finer quality grapes than infected vines. Although commonly valid, clones free of all known systemic pathogens do not consistently perform better than infected counterparts (Woodham *et al.*, 1984). Nevertheless, elimination of all systemic infections is a desirable goal as transmission to other cultivars may cause development of a debilitating disease or limit grafting success.

Detection of viral and viroid infection has historically been based on indexing, that is, the transmission of the agent to a plant that characteristically produces distinctive symptoms on infection. Identification by serological techniques, especially with enzyme enhancement (ELISA), or with cDNA probes can confirm indexing results or occasionally replace the long indexing procedure. Because disease-free vines micropropagated in culture vessels remain free of pathogens, their use may reduce, if not eliminate, the need for quarantining and indexing vines transported from country to country.

Fanleaf Degeneration Fanleaf degeneration is the oldest known viral infection of grapevines, being identifiable from herbarium specimens over 200 years old. Because of its long history in Europe, and the absence of infection in free-living North American grapevines, grapevine fanleaf virus (GFLV) is assumed to be of European or Near Eastern origin. Its current widespread occurrence is believed due to grafting and the spread of European cultivars throughout much of the world. Natural spread in vineyards is slow because of the limited movement in soil of the major nematode vectors, *Xiphinema index* and *X. italiae* (about 1.5 m/year). Because grapevine roots do not form natural grafts with one another, vine to vine transfer does not occur. Although transferable to other plants, the disease is limited to grapevines under field conditions. This may result from the limited host range of the nematode vectors.

The impact of infection varies widely, depending on the tolerance of the cultivar and environmental conditions. Tolerant cultivars are little affected by infection, whereas susceptible varieties show progressive decline. Yield losses may be up to 80% and the fruit of poor quality. Infected vines have a shortened life span, increased sensitivity to environmental stress, and reduced grafting and rooting potential. Three distinctive syndromes of infection have been recognized, based on particular strain/cultivar combinations, namely, malformation, yellow mosaic, and veinbanding.

Fan-shaped leaf malformation is the most distinctive, but not only, foliage expression of disease. Chlorotic speckling commonly accompanies leaf distortion. Shoots may be misshapen, showing fasciation (stem flattening), double nodes, and other distortions. Fruit set is poor and bunches reduced in size. The yellow mosaic (chromatic) syndome develops as a strikingly bright yellow mottling of the leaves, tendrils, shoots, and inflorescences in the early spring. Discoloration can vary from isolated chlorotic spots to uniform yellowing. The third expression, veinbanding, develops as a speckled yellowing on mature leaves, bordering the main veins in mid- to late summer. In both discoloration syndromes, leaf shape is normal, but fruit set is poor with many shot berries. Fanleaf degeneration also shows a characteristic intracellular development of trabeculae. These appear as strands of cell wall material spanning the lumen of xylem vessels.

Where both vector and virus are established, planting vines grafted on rootstock tolerant or resistant to both agents is usually required. Fumigation of the soil with nematicides can, to varying degrees, reduce but not eradicate nematode populations. Allowing the land to lie fallow can be useful in reducing nematode populations, but this strategy often requires 6 to 10 years. The long requisite fallow period probably results from pathogen survival on undislodged roots that may remain viable for up to 6 years following vine uprooting.

Leafroll Leafroll is a widespread viruslike disease most commonly associated with one or more closteroviruses. Symptomatic expression of infection varies considerbly, but it does not lead to vine degeneration. Many cultivars are symptomless carriers of the infectious agent(s) of the disease. As with fanleaf degeneration, the pathogen(s) probably originated in Europe or the Near East; free-living North American grapevines are not infected.

Spread of the infectious agent(s) appears to depend on graft transmission. Insect vectors are unknown, other than a report of experimental transmission by mealybugs (Rosciglione and Gugerli, 1989). Healthy and infected vines usually coexist side by side without transmission.

The most distinctive symptom of infection occurs late in the season, as the basal leaf blades develop a marked down-rolling. The interveinal areas of the leaves also may turn yellow or red, depending on the cultivar, while the main veins remain green. Whole vines, as well as individual shoots and leaves, are dwarfed in comparison to healthy plants. Fruit production may be depressed by about 20%, and berries may show delayed ripening and reduced sugar levels.

Control is dependent entirely on the destruction of the infected vines and their replacement by disease-free stock. Disease-free nursery stock may be generated by thermotherapy or by micropropagation. As the identity of the causal agent(s) remains uncertain, confirmation of elimination of the infectious agent(s) is performed by grafting to sensitive cultivars (indexing).

Yellow Speckle   Yellow speckle is a widespread, but relatively minor, disease of grapevines. Other viroids occur in grapevines, but their economic significance and causal relationship to recognized grapevine diseases are unclear. Symptoms of infection by grapevine yellow speckle viroid (GYSV) are often short-lived and develop only under special climatic conditions. Foliar symptoms generally develop at the end of the summer and consist of leaf spotting. When sufficiently marked, the scattered chlorotic spots may resemble the veinbanding symptom of fanleaf degeneration.

Control is dependent on planting viroid-free vines. Elimination of the causal agent can be achieved by micropropagation. Thermotherapy is ineffective (Barlass and Skene, 1987).

NEMATODE PATHOGENS

Nematodes are a large group of microscopic unsegmented roundworms that predominantly live saprophytically in soil. Several, however, are parasitic on plants, fungi, and animals. Those attacking grapevines are restricted to feeding on the root system. They derive their nutrition from drawing the cytoplasmic fluids out of root cells. This is accomplished with a spearlike stylet that punctures plant cells.

Feeding may be restricted to the surface of roots or, following burrowing, occur in the root cortex. Besides the damage caused by feeding, and the resultant root disruption, nematodes may transmit viruses and facilitate infection by other pathogens. Active dispersal of nematodes in soil is both slow and very limited, with most long-distance transport being through the action of wind and water, or by the translocation of infected whole plants.

Reproduction occurs via egg production, with or without the interaction of males. The prolonged survival of eggs in a dormant state often markedly reduces the effectiveness of fallowing in nematode control. Soil fumigation, especially in combination with fallow, can dramatically reduce nematode populations in shallow sandy soils, but it is of limited value in deep or clayey soils, which restrict fumigant penetration. Generally, the most effective means of limiting nematode damage involves the use of nematode-resistant or tolerant rootstocks. Regrettably, rootstock cultivars resistant to all grapevine-attacking nematodes are currently unavailable.

Detection and identification of nematode problems usually require microscopic examination of the root system, as above ground symptoms are insufficiently distinctive for diagnostic use. Many nematode species can attack grapevine roots, but few induce marked damage. The most serious are the root-knot (*Meloidogyne*) and dagger (*Xiphinema*) nematodes. Other nematodes occasionally found feeding on grapevine roots include species of genera such as *Pratylenchus*, *Tylenchulus*, and *Criconemella*.

Root-Knot Nematodes   Root-knot nematodes (*Meloidogyne* spp.) are sedentary endoparasites that penetrate young feeder roots. On penetration, adjacent cells are stimulated to divide and increase in size, producing gall-like swellings (Fig. 4.29). Where multiple infections occur, knot formation may give the root a chain-of-beads appearance. Reproduction commonly is timed to the spring and fall root-growth periods of the vine (de Klerk and Loubser, 1988). Each female may produce up to 1500 eggs. The most important species are *M. incognita*, *M. javanica*, *M. arenaria*, and *M. hapla*.

Infection results in a decline in vigor and yield and increased susceptibility to water and nutrient stress. Severity of these expressions can be limited by increasing the water supply and pruning to prevent overcropping. Damage is typically most marked in young developing vines planted in highly infested, light-textured soils. Older vines with deep root systems seem to be less affected by root-knot nematodes.

Dagger Nematodes   In contrast to root-knot nematodes, dagger nematodes (*Xiphinema* spp.) are migratory and feed on epidermal cells near the root tip. Concentrated feeding may initiate root bending followed by lesion darkening. Extensive attack causes root death and induces the production of tufts of lateral roots. In addition to destroying roots, *X. index* can act as a vector of the grapevine fanleaf virus. The combined action of both nematode and viral infections can quickly make a vineyard commercially unproductive. Other species of *Xiphinema* may transmit a range of other, but less serious, viruses to grapevines (see Walter, 1991).

INSECT AND MITE PESTS

Insects and mites are arthropods that can cause extensive damage to grapevines. While most species primarily infest one part of the vine, all parts are attacked by one or more species. Until fairly recently, most control was through the use of pesticides. Because of problems associated with pesticide use, greater emphasis is being placed on cultural and biological controls.

As a group, insects and mites are distinguished by a hard exoskeleton. The latter is shed several times during growth to the adult stage. Mites and many insects

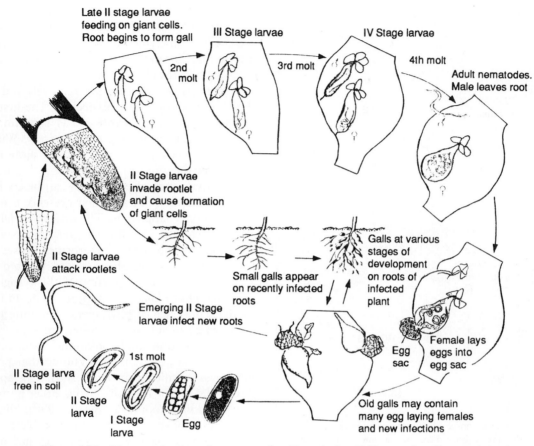

Figure 4.29 Disease cycle of root-knot nematodes. (From Agrios, 1988, reproduced by permission.)

pass through several adultlike, but immature stages, whereas other insects pass through wormlike (larval) stages and a pupal stage before reaching adulthood.

Insects are distinguished by possessing three main body parts (head, thorax, and abdomen), sensory antennae, three pairs of legs attached to the central thorax, and a segmented abdomen. In contrast, mites possess a fused cephalothorax, broadly attached to an unsegmented abdomen, lack antennae, and possess four pairs of legs in all but the initial immature stage. Two pairs of legs are attached to the cephalothorax and the other two pair located on the abdomen. Both groups reproduce primarily by egg production.

While some insects have evolved intimate relations with grapevines, notably phylloxera, others infest a wide range of hosts. As with fungal pathogens, selective infestation on grapevines does not necessarily correlate with severity. Severity depends more on the specific genetic properties of individual cultivars and pest strains, the macro- and microclimate, soil conditions, and the level of resistance to applied pesticides.

Phylloxera Of all grapevine pests, phylloxera (*Daktulosphaira vitifoliae*) has had the greatest impact

on viticulture and wine production. A relatively minor endemic pest of eastern North American *Vitis* species, phylloxera devastated the majority of *V. vinifera* vineyards when introduced into Europe about 1860. This event both demonstrated the need for quarantine laws and illustrated the effectiveness possible with biological control. Grafting sensitive fruiting varieties (scions) to resistant rootstock has been so successful that the danger still posed by the pest is often forgotten. The mutation, or accidental introduction, of a different biotype of phylloxera has recently caused considerable concern in California. The new strain (biotype) of phylloxera is pathogenic to 'AxR$^{\#}$ 1' roots, the predominant rootstock previously used to control phylloxera in California.

The life cycle of the aphidlike phylloxera is illustrated in Fig. 4.30. For the sexual stage to be expressed, the insect must pass through both leaf- and root-galling phases. However, most species of *Vitis* are resistant to either the leaf- or root-galling phase (Table 4.11). Thus, absence of species susceptible to the initial (**fundatrix**) leaf-galling phase probably explains the absence of the sexual cycle in California, and most other viticulture areas where phylloxera is not indigenous. Occasionally,

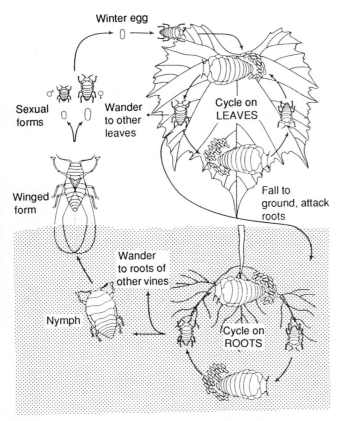

Figure 4.30 Life cycle of grape phylloxera. Shaded area, below-ground portion of life cycle. (From Fergusson-Kolmes and Dennehy, 1991, reproduced by permission.)

the leaves of *V. vinifera* are attacked and galled by phylloxera nymphs that develop on the leaves of American rootstocks in Europe (Borner and Heinze 1957).

Phylloxera biotypes, differentially pathogenic on grapevine cultivars, appear to be common (King and Rilling, 1991), despite the absence of the sexual stage in most infested areas. Differences in pathogenicity may be expressed in the ability of the pests to feed, stimulate gall formation, or reproduce rapidly. The insect usually goes through several asexually reproduced generatio.is per year. Long-distance spread without human involvement is slow.

Susceptible hosts respond to feeding by increasing the supply of nutrients to the damaged region and by forming extensive galls. Feeding on young roots induces distinctive hook-shaped bends and swellings called **nodosities**. These soon die and decay. Older roots produce semispherical swellings, called **tuberosities**. The latter give the root a roughened, warty appearance. The development of tuberosities is the more serious expression of infection. The insect population increases primarily on tuberosities, as nodosities quickly rot and do not support dense populations of feeding insects (Williams

and Granett, 1988). In contrast, resistant vines show limited production of nodosities and healing following attack. Tolerant vines show more galling but are little affected by infestation and support only limited insect reproduction (Boubals, 1966).

Symptoms of infection are relatively indistinct, and the effect is primarily vine decline. The first indication of infection usually is expressed as a stunting of growth and premature yellowing of the leaves. With sensitive cultivars, these effects yearly become more marked until the vine dies.

The occurrence of phylloxera can be confirmed only by root examination, and their presence is most readily observed on young roots. The distinctive lemon-yellow eggs and clusters of yellowish green nymphs and adults aid identification. Many tuberosities possess few phylloxera, these having already been vacated by the insects. However, the presence of the tyroglyphid mite (*Rhizoglyphus elongatus*) appears to be an indicator of the past presence of phylloxera. The mite lives on the decaying cortical tissues of tuberosities.

Many environmental conditions can affect the severity of vine attack. It is well known that phylloxera is much less significant on sandy soils. This appears to be due to the higher level of silicon, either in the soil solution or in vine roots (Ermolaev, 1990). High soil temperatures (above 32°C) are unfavorable to phylloxera and limit its damage (Foott, 1987). Irrigation and fertilization can occasionally diminish the significance of phylloxera damage, while drought increases severity (Flaherty *et al.*, 1982).

Quarantine is still useful in limiting phylloxera dispersal into uninfested areas. Even where the pest is present, preventing the importation of new biotypes may be beneficial.

In nurseries, vines can be disinfected by placing the washed root systems into hot water (52° to 54°C) for about 5 min. Nursery soil can be disinfected by pasteurization or fumigation. Where vineyard soil is already infested, the major control measure remains grafting sensitive scion varieties onto resistant rootstock; although use of chemicals such as aldicarb (Temik) (Loubser *et al.*, 1992) and sodium tetrathiocarbonate (Enzone) show some promise.

Leafhoppers Several genera of leafhoppers are important destructive pests on grapevines as well as vectors of several bacterial pathogens. The species of significance tend to vary from region to region. In much of the grape growing region of North America the western and eastern grape leafhoppers (*Erythroneura elegantula* and *E. comes*, respectively) are the most significant species, in southern California the variegated leafhopper (*E. variabilis*) is the dominant species, while the potato

Table 4.11 Susceptibility of *Vitis* Species to Root and Leaf Expressions of Phylloxera[a]

Host reaction	Resistant (bearing none to few galls on roots)	Tolerant (bearing many galls on roots)
***Vitis* species already exposed to phylloxera** (eastern North America)		
Resistant (bearing none to few galls on leaves)	V. rotundifolia	V. aestivalis
	V. berlandieri	V. girdiana
	V. candicans	V. labrusca
	V. cinerea	V. lincecumii
	V. cordifolia	V. monticola
	V. rubra	
Tolerant (bearing many galls on leaves)	V. riparia (= V. vulpina)	None
	V. rupestris	
***Vitis* species not previously exposed to phylloxera** (western North America, Asia, and Europe/Middle East)		
Resistant (bearing none to few galls on leaves)	V. coignetiae	V. arizonica
		V. californica
		V. davidii
		V. romanetti
		V. ficifolia (= V. flexuosa)
		V. vinifera (incl. V. sylvestris)
Susceptible (bearing many galls on leaves)	V. betulifolia	V. amurensis
	V. reticulata	V. piazeskii

[a] From Wapshere and Helm (1987), reproduced by permission.

leafhopper (*Empoasca fabae*) is particularly important in the southeastern United States. About 20 species of sharpshooter leafhoppers are vectors of *Xylella fastidiosa,* the causal agent of Pierce's disease. In Europe, *Empoasca* leafhoppers tend to be the most important group, especially *E. vitis,* while *Scaphoideus littoralis* (synonym *S. titanus*) is the primary vector of *flavescence dorée.* In South Africa, *Acia lineatifrons* is the most significant species. Most leafhoppers have fairly similar life cycles and are controlled by similar techniques.

Depending on the species and prevailing conditions, leafhoppers may go through one to three generations per year. Those passing through several generations per year are generally the most serious, as their numbers can increase dramatically throughout the season.

Leafhopper eggs typically are laid under the epidermal tissues of the leaves. On hatching, the young (nymphs) often pass through five molts before reaching the adult stage (Fig. 4.31). All stages feed on the cytoplasmic fluid of leaf and fruit tissue. Feeding results in the formation of white spots which, on heavily infested leaves, leads to a marked loss in color. Growth of fungi on escaped sap and insect "honey dew" can produce a sooty appearance on leaf and fruit surfaces. Pronounced damage is usually caused only with heavy infestation (>10 to 15 leafhoppers/leaf). Severe infestation can lead to leaf necrosis, premature defoliation, delayed berry ripening, and reduced fruit quality. Some varieties, notably late maturing cultivars, tend to sustain greater damage than early maturing varieties. Such cultivars not only endure leafhopper infestation for longer periods, but they also may suffer from the migration of leafhoppers from early maturing varieties. Most leafhopper species affecting grapevines do not infest them exclusively. Thus, in the fall and early spring, they often survive and multiply on other plants. Overwintering occurs as adults under leaves, weeds, or debris in and adjacent to vineyards.

Current control measures are increasingly being based on enhancing the populations of insect and mite parasites and predators. In California, the wasp *Anagrus epos* is an effective parasite on leafhopper eggs. Its short life cycle permits up to 10 generations per year. By July, the parasite population may reach levels sufficient to destroy 90–95% of leafhopper eggs. Another parasitic wasp, *Aphelopsis cosemi,* attacks nymphs, resulting in sterilization of the adult. Several predatory insects, such as lacewings and ladybugs, as well as the general predatory mite *Anystis agilis,* attack leafhoppers. Of cultural practices, basal leaf removal is particularly useful in removing most first-generation leafhoppers.

Tortricid Moths Grapevines are attacked by a wide variety of tortricid moths, the species varying from region to region and country to country. In southern regions of Europe, *Lobesia botrana* is the major significant species, whereas in more northern regions *Eupoecilia ambiguella* is especially important. In much

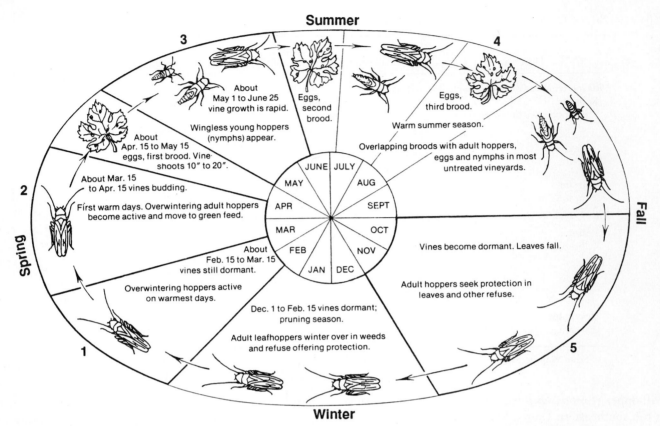

Figure 4.31 Life cycle of grape leafhoppers in California. (From Flaherty *et al.*, 1982, reproduced by permission.)

of eastern and central North America, the important grape berry moth is *Endopiza viteana;* in California, the omnivorous leafroller (*Platynota stultana*) and the orange tortrix (*Argyrotaenia citrana*) are the significant forms in warmer and cooler regions, respectively.

Because they belong to the same taxonomic grouping, tortricid moths possess similar life cycles. The adult females lay bunches of eggs on or close to flowers and grape clusters. Those less specialized to grapevines also lay eggs on the leaves. The eggs hatch into pale-colored larvae that feed predominantly on the tissues of flowers or the developing fruit. Following several molts, the larvae form a weblike cocoon within which they metamorphose into pupae. After a variable period, the adult moths emerge and mate, and the female initiates the next generation. Adults are small, relatively inconspicuous brown moths, possessing a bell-shaped wing profile and projecting snoutlike mouth parts. Depending on the species and prevailing conditions, two to three generations may develop per year.

Although tortricid moths possess many properties in common, significant differences occur. For example, European berry moths form cocoons under rough bark and in crevices and cracks on trellis posts. American berry moths form folds in leaves in which they spin their cocoons; these eventually fall to the ground. Al-

though tortricid moths affecting grapes in California form cocoons in leaf folds, the insects do not hibernate as pupae as do other American and European berry moths. The Californian tortricid moths survive as larvae within web nests formed within mummified grape clusters, or on other vine and vineyard debris. These hibernating characteristics influence the type and success of sanitation in reducing overwintering success.

Second and third generation larvae are the most damaging, as their numbers typically increase throughout the season. Besides the direct damage produced by feeding, the larvae produce wounds subsequently infected by bunch rot fungi, yeasts, and bacteria. Their action further attracts infestation by fruit flies (*Drosophila* spp.). Not only can the tortricid larvae produce wounds facilitating bunch rot, but the European berry moth (*Lobesia botrana*) may itself be a carrier of *Botrytis cinerea* spores (Fermaud and Le Menn, 1989).

In most viticultural regions, increased emphasis is being placed on control by endemic pests and parasites. Diminished use of pesticides, and the establishment of habitats for sustaining populations of indigenous parasites and predators, is essential to successful biological control (Sengonca and Leisse, 1989). In addition, synthetic pesticides may be replaced by commercial preparations of *Bacillus thuringiensis* or by pheromone applications to disrupt mating success. Release of artificially reared *Trichogramma embryophagum*, an egg parasite of European grape moths, has been experimentally successful in controlling tortricid pests (Filip and Isac, 1989). Egg parasitism kills the insect before damage can be done, while larval parasites limit population buildup by preventing reproduction. Where insects overwinter on the ground, row cultivation to bury hibernating pupae or larvae in the spring can be particularly useful. It often takes several years for tortricid populations to build up to critical levels (Flaherty *et al.*, 1982).

Spider Mites Spider mites are a minute but serious pest wherever grapes are grown. They are especially a problem under dusty conditions, where vines are often water stressed. The important species generally differ from region to region. The European red spider mite (*Panonychus ulmi*) and the two-spotted spider mite (*Tetranychus urticae*) tend to be the most important species in much of Europe, while the yellow vine spider mite (*Eotetranychus carpini*) is the primary species in Mediterranean France and Italy. In eastern North America, the European red spider mite is the principal damaging species, and in California the Pacific spider mite (*Tetranychus pacificus*) is the most serious form. The Willamette spider mite (*Eotetranychus willamettei*) is commonly found in California but is less damaging than the Pacific spider mite. It has been released in vineyards as a biological control agent against infestation by the Pacific spider mite. In Chili, *Oligonychus vitis* is the most significant species found.

Spider mites typically overwinter as females under rough bark on trunks and cordons. They begin to emerge early in the spring. If present in high numbers, the mites can kill the margins of growing leaves, permanently stunting leaf growth.

The initial (larval) stage of spider mites resembles the adult, except in size and the possession of but three pairs of legs. After feeding, the larva molts and subsequently passes through the eight-legged protonymph and deutonymph states, to become either a male or female adult. Under favorable conditions, spider mites can pass through the life cycle in about 10 days. This can lead to an explosive increase in populations.

Spider mites typically feed on the undersurfaces of leaves by injecting their mouth parts into epidermal cells. Initial damage results in fine yellow spots on the leaf. With extensive feeding, the foliage turns yellow in white varieties and bronze in red cultivars. Web formation is more or less pronounced, usually occurring in the angles formed by leaf veins. If attack is heavy, leaves usually drop prematurely. Although spider mites seldom attack the fruit, foliage damage may result in delayed ripening or, in severe cases, fruit shriveling and dehiscence.

Control is achieved primarily by favoring conditions that diminish vine susceptibility and enhance natural predation. Grass ground covers, when water and fertilization are ample, diminishes vine susceptibility by limiting dust production. Sprinkler irrigation is unfavorable to spider mite development but not to its predators. Planting vegetation that maintains high levels of spider mite predators, and the avoidance of pesticides toxic to their predators promote effective biological control. In Europe, the most effective predator is the phytoseiid mite (*Typhlodromus pyri*), while in California the primary predator is another mite (*Metaseiulus occidentalis*). For *T. pyri* weed pollen is an important food source in maintaining high predator populations in vineyards, whereas for *M. occidentalis* tydeid mites act as important alternate prey when spider mite populations are low. A wide variety of insect predators feed on spider mites, including *Amblyseius andersoni* and *A. addoensis* in Europe and South Africa, respectively, and the minute pirate bug (*Orius vicinus*) (Duso, 1989). In most cases, the importance of natural predators in control is uncertain.

PHYSIOLOGICAL DISORDERS

Grapes are susceptible to several physiological disorders of ill-defined etiology. This has made differentiation difficult and has led to a wide range of terms whose exact equivalents are unclear. Nevertheless, two groups of phenomena appear fairly distinct. These include the excessive flower and fruit drop associated with flowering, variously referred to as inflorescence necrosis, shelling, early bunch-stem necrosis, or *coulure;* and premature fruit shriveling and drop following *véraison;* variously termed bunch-stem necrosis, shanking, waterberry, *dessèchement de la rafle*, or *Stiellähme.*

Inflorescence necrosis is associated with cold, wet weather during the flowering period and with high vine vigor. It results from the formation of an abscission layer at the base of the pedicel. Ammonia as well as ethylene accumulation have been implicated with induction of this disorder (Gu *et al.*, 1991; Bessis and Fournioux, 1992). *Coulure* also has been associated with nitrogen deficiency (Perret, 1992).

Bunch-stem necrosis is also associated with vine vigor and heavy or frequent rains. It develops as a progressive enlargement of necrotic regions on the rachis and its branches. The European expression usually has been associated with magnesium or calcium deficiency in the rachis. It is treated by spraying the fruit, beginning at *véraison,* with one or more magnesium sulfate applications (Bubl, 1987), occasionally supplemented with calcium chloride. Development of bunch-stem necrosis outside Europe appears to associate ammonia accumulation and high nitrogen fertilization with the malady (Christensen *et al.*, 1991).

AIR POLLUTION

The economic damage caused by air pollution has been little studied in grapevines. Of known air pollutants, ozone and hydrogen fluoride appear to be the most significant in producing visible injury. Sulfur dioxide produces injury, but at atmospheric levels higher than typically found in vineyards. Even in the form of acid rain, sulfur dioxide is not known to severely affect grapevines (see Weinstein, 1984).

The severity of damage caused by air pollution has been difficult to assess and predict owing to marked differential cultivar sensitivity at various growth stages. In addition, the duration, concentration, and environmental conditions of exposure significantly influence sensitivity. Furthermore, it is generally considered that vineyard influences, such as overcropping and other stresses (water, nutrient, weeds, and disease), are more significant in determining the degree of damage than pollutant concentration.

Ozone Ozone is the most injurious of air pollutants to grapevines. The grape also was one of the first plants found to be sensitive to the pollutant.

Ozone is produced naturally in the upper atmosphere by exposure of oxygen to short-wave ultraviolet radiation, and during lightning in the lower atmosphere. However, most of the ozone in the lower atmosphere comes indirectly from automobile emissions. Nitrogen dioxide (NO_2) released in automobile exhaust is photochemically split into nitric oxide (NO) and singlet oxygen (O). The oxygen radical released reacts with molecular oxygen (O_2) in the atmosphere to form ozone (O_3):

$$NO_2 \overset{light}{\rightleftharpoons} NO + O$$
$$O + O_2 \rightarrow O_3$$
$$NO + hydrocarbons \rightarrow PAN$$

Nitrogen dioxide would reform by the reverse reaction were it not for the associated release of hydrocarbons in automobile emission. The hydrocarbons react with nitric oxide forming peroxyacetyl nitrate (PAN). This reaction limits the reformation of nitrogen dioxide and results in the accumulation of ozone. Ozone is the most significant air pollutant affecting grapevines in North America and can severely affect grape yields in some areas.

Ozone primarily diffuses into grapevine leaves through the stomata. Ozone, or a by-product, reacts with membrane constituents, disrupting cell function. The palisade cells appear to be the most sensitive, often collapsing after exposure. Their collapse and death gen-

erate small brown lesions on the upper leaf surface. These coalesce to produce the interveinal spotting called **oxidant stipple**. A severe reaction produces a yellowing or bronzing of the leaf and premature leaf drop. Basal leaves, and most mature portions of new leaves, are the most susceptible to ozone injury. Damage may result in reduced yield both in the current year and, by depressing inflorescence induction, in the subsequent year. Severity of damage is markedly affected by cultivar sensitivity and prevailing climatic conditions.

Sensitivity to ozone damage may be reduced by maintaining relatively high nitrogen levels, avoiding water stress, planting cover crops, and spraying with antioxidant compounds, such as the fungicide benomyl or ethylene diurea (EDU).

Hydrogen Fluoride Hydrogen fluoride is an atmospheric contaminant derived from emissions released in several industrial processes, such as aluminum and steel smelting and in the production of ceramics and phosphorus fertilizer. Although leaves do not accumulate fluoride in large amounts, grapevines are one of the more sensitive plants to this pollutant. Of grapevines, *V. vinifera* cultivars tend to be particularly sensitive to hydrogen fluoride.

Symptoms begin with the development of a gray-green discoloration at the margins of younger leaves. Subsequently, the affected region expands and turns brown, often being separated from healthy tissue by a dark red, brown, or purple band and a thin chlorotic transition zone. Young foliage is more severely affected than older leaves.

Vines may often be protected from airborne fluoride damage by the application of calcium salts. Thus, the presence of slaked lime [$Ca(OH)_2$] in the fungicide Bordeaux mixture unintentionally may provide protection against hydrogen fluoride injury.

Chemical Spray Phytotoxicity Herbicide drift onto grapevine vegetation can cause a wide variety of leaf malformations and injury, depending on the herbicide involved. Several fungicides also produce phytotoxic effects, notably sulfur and Bordeaux mixture. Additionally, some pesticides induce leaf damage under specific environmental conditions or if applied improperly, notably endosulfan, phosalone, and propargite. Details are given in Pearson *et al.* (1988).

WEED CONTROL

Weed control is probably as ancient as viticulture itself. The oldest form of weed control, manual hoeing, may still be practiced in commercial viticulture, notably on slopes where mechanical tillage is impractical. Until comparatively recently, tillage was the principal method of weed control. Increasing energy and labor costs, combined with the development of effective herbicides during the 1950s, resulted in a decline in tillage use. Presently, environmental concerns have again shifted interest toward other methods of weed control, notably plastic mulches, ground covers, and biological controls. No one system of weed control is ideal for all situations, as different systems variously affect water and nutrient supply, disease control, vine growth, and fruit quality. In addition, the selection of one system over another depends on economic concerns, the relative benefits and disadvantages of clean cultivation, and restrictions imposed by governments and "organic" grower associations.

Tillage Tillage to a depth of 15 to 20 cm is still used widely in some countries. In addition to weed control, tillage is used to break up compacted soil, incorporate organic and inorganic fertilizers into the soil, prepare the soil for sowing cover crops, and bury diseased and infested plant remains for pest and disease control. However, awareness of its disadvantages have combined with other factors to curtail the former use of tillage. Among its drawbacks is disruption of the aggregate structure of the soil. This can lead to "puddling" under heavy rain and the progressive formation of a hardpan under the tilled layer. The latter results from the transport, and subsequent accumulation, of clay particles deeper in the soil. Hardpans delay water infiltration, increase soil erosion, reduce water conservation, limit root penetration, and promote root development near the soil surface. Soil cultivation also destroys infiltration and aeration channels produced by cracks in the soil, earthworm action, and root decay. This effect is most marked when the soil is cultivated while wet. Tillage increases oxygen infiltration into the soil and the rate of microbic action. This augments the mineralization of soil organic material and the potential loss of soil nutrients by leaching. Finally, soil compaction by heavy equipment can limit root growth between rows in shallow soils (van Huyssteen, 1988a).

Herbicides No-till cultivation has become increasingly popular in many parts of the world. The use of herbicides has been largely instrumental in permitting the shift away from tillage. No-till avoids most of the disadvantages of cultivation, while achieving the benefits of clean cultivation.

Herbicides fall into one of a number of functional categories. Some, such as simazine, are preemergent chemicals that kill seedlings on germination but do not affect existing weeds. They are typically nonselective and primarily useful in controlling annual weeds. Herbicides, such as diquat and paraquat, kill plant vegetation on contact, while others, such as aminotriazole and glyphosate, are translocated throughout the plant.

To limit vine damage, herbicides are usually applied before bud break and with special rigs designed to direct application to the base of the vines. Mid to high training and limited use during the first few years of vineyard establishment, help to limit herbicide injury.

In cool climatic regions, bindweed (*Convolvulus* spp.) and quack grass (*Agropyrons repens*) tend to be the most widely distributed of noxious weeds. In warmer regions, Bermuda grass (*Cynodon dactylon*) tends to be the most common noxious weed.

Although some herbicides, such as paraquat, slowly decompose in soil, most degrade readily and neither accumulate in the soil, disrupt fermentation, nor contaminate the finished wine. Nevertheless, increasing resistance to herbicide use is being noted in several countries. In addition, increases in the incidence of some pest problems, such as omnivorous leafroller, have been attributed to increasing reliance on herbicide use (Flaherty *et al.*, 1982). In addition, decreased activity of a wide range of microbes and soil invertebrates have been noted in soil associated with herbicide use (Encheva and Rankov, 1990). Consequently, several alternative, nonchemical approaches to weed control are being actively investigated.

Mulches Straw mulches have long been used for weed control and water conservation in several crops. However, mulches have seen limited use in viticulture. Nevertheless, successful use of plastic mulches has led to their renewed consideration in viticulture. Plastic mulches have proved particularly useful in establishing vineyards, where they maintain higher moisture levels near the soil surface and promote faster root development. Although surface rooting is promoted, development of deep roots is not diminished (van der Westhuizen, 1980). Plastic mulches also enhance vine vigor, fruit yield, and eliminate the need for hoeing or herbicide application around individual vines (Stevenson *et al.*, 1986).

Plastic mulches may consist of either impermeable or porous woven sheets of polyethylene. Porous sheeting has the benefit of better air and water permeability, but it increases evaporative water loss. Black plastic has been most commonly used, but colored plastic may provide better heat and photosynthetic light reflection up into the foliage.

Cover Crops Planting cover crops is another but more complex means of weed control. Depending on the relative benefits of a ground cover, they may be restricted to between-row strips or be permitted to form a complete cover. Although cover crops usually contain a particular selection of grasses and legumes, natural vegetation may be used. Where natural vegetation is used, early application of a dilute herbicide solution can restrict excessive and undesirable seasonal growth (Summers, 1985). It also restricts seed production and, thereby, limits self-seeding of the cover crop under the vines. Another option is mowing or mulching the cover crop, sometimes in alternate rows, to restrict nitrogen demand and limit seed production (W. Koblet, 1992, personal communication).

Seeding a cover crop may occur in the fall or winter months, depending on the periodicity of rainfall and the desirability of a winter ground cover. Cover crops usually possess a mixture of one or more grasses (rye, oats, barley) and legumes (vetch, bur clover, subterranean clover). Strain selection can be as important as species selection, because water and nutrient utilization vary considerably between strains. Water and nutrient competition is of particular concern on shallow, nonirrigated, dryland locations, or where cover crops are used to regulate vine vigor (Lombard *et al.,* 1988). Seeded cover crops usually require one mowing per year to prevent self-seeding and limit competition with grapevine growth.

In addition to weed control, ground covers aid against soil erosion. This is particularly valuable on steep slope but also is useful on level locations. The vegetation breaks the force of water droplets that can destroy soil aggregate structure. Cover crop roots help bind soil together and limit sheet erosion. Furthermore, as the roots decay, water infiltration is improved and organic material is added to the soil. Poorly mobile nutrients, such as potassium, are translocated down into the soil by the roots (Saayman, 1981). Besides limiting soil erosion, cover crops can improve water conservation by reducing water runoff. In addition, cover crops facilitate machinery access when the soil is wet, and they reduce the incidence of some pests. Although being reservoirs of the parasites and predators of several vineyard pests, they also may be carriers of several pest and disease-causing agents.

Because cover crops markedly suppress vine root growth near the soil surface (Fig. 4.32), the applicability of cover crop use often depends on soil depth, water availability, and desired vine vigor. In some locations, the possibility of ground covers increasing the likelihood of frost occurrence, by lowering the rate of heat radiation from the ground, may be important. During the growing season, however, ground covers appear to have little significance on the interrow temperature of vineyards (Lombard *et al.,* 1988).

Biological Control   One of the newest techniques in weed control involves the action of pests and diseases of weeds. Although less investigated than in insect control, such biological control has the potential of being an effective adjunct to current control measures. Time will indicate whether this potential can be realized.

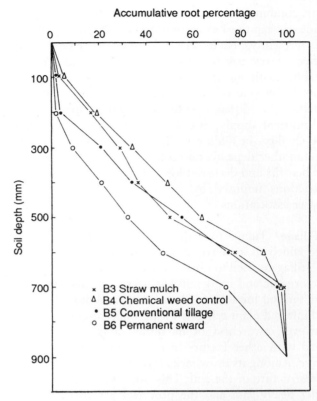

Figure 4.32   Root distribution with depth under different tillage practices in a dryland vineyard. (From van Huyssteen, 1988b, reproduced by permission.)

Harvesting

The timing of harvest is probably the single most important viticultural decision taken each season. The properties of the fruit when harvested set limits on the potential quality of the wine produced. Certain winemaking practices can ameliorate deficiencies in grape quality but cannot fully offset inherent defects. Timing is most critical when, as usual, all the fruit is harvested concurrently. Only rarely is it economically feasible to selectively and repeatedly harvest a vineyard for fruit of a particular quality.

Where the grape grower is also the winemaker, and premium quality a priority, there is little difficulty in justifying the time and effort involved in precisely assessing fruit quality in the vineyard. However, where the grape grower and winemaker are different, adequate compensation for the development and harvesting of fruit at its optimal quality is required. The practice of basing grape payment on variety, weight, and sugar content is inadequate. Greater recognition and remuneration for practices enhancing grape quality should improve wine quality beyond the considerable standards current today.

Criteria for Timing of Harvest

The problem facing the grape grower in choosing the optimal harvest date is knowing how to most appropriately assess grape quality. This results from the chemical basis of wine quality itself being ill-defined and varying with the style desired. For example, grapes of intermediate maturity may produce fruitier but less complex wines than fully mature grapes (Gallander, 1983).

Because of the importance of sugar and acid content to vinification, and their ease of measurement, these constituents have received the major attention as indicators of grape maturity. Often the sugar/acid ratio is the preferred indicator in temperate climates, where desirable changes in both factors occur more or less concurrently during ripening. The sugar/acid ratio is a good indicator of when both constituents are reaching the optimal balance. Because acid content changes more slowly in cool climates, sugar content is often the primary indicator used. In hot climates, adequate degree Brix levels are typical, but avoiding high pH musts is essential. Thus, acidity is of particular concern in hot regions.

In addition to the importance of berry sugar content in vinification, sugar accumulation has been associated by generations of winemakers with berry flavor development. While this belief is largely supported by data on the synthesis of anthocyanins, monoterpenes, and vitispirane precursors, the association is far from simple (Roggero *et al.*, 1986; Dimitriadis and Williams, 1984). Correspondingly, means of more accurately assessing the levels of grape flavor are being developed.

Because of the importance of anthocyanins and tannins to red wine quality, their presence is obviously an indicator of grape quality. Regrettably, association between grape phenol measurements and subsequent wine quality are weak (Somers and Pocock, 1986; Roggero *et al.*, 1986). This arises from difficulties in both rapidly and accurately measuring their presence in small grape samples as well as predicting the influences of vinification practices on the retention and physicochemical state of phenols in wine. However, average berry size appears to be a useful indicator of potential color and flavor in red wines (Singleton, 1972; Somers and Pocock, 1986). This presumably results from an inverse relationship between berry volume and surface area, and the localization of anthocyanin pigments in the skin. Knowledge of the relationship between berry size and grape quality can be put to practical use by restricting irrigation after *véraison* to limit berry enlargement. Alternately, simulation of small berry size can be achieved by procedures, such as cryoextraction or reverse osmosis, as both remove water from the juice. Small berry size also can be an important feature in

white grapes if most of the flavor compounds are located in the skin.

Greatest success has been achieved in correlating grape flavor content with wine quality in cultivars dependent on terpenes for much of the varietal aroma. In some varieties, the volatile monoterpene content of grapes continues to increase for several weeks after appropriate sugar and acid levels have been reached (Fig. 4.33). In others, terpene content may continue to increase during and after ripening, but an increasing proportion is converted to nonvolatile forms. Although free terpenes are important in the fragrance of some young wines, high levels of nonvolatile terpenes have been correlated with subsequent flavor and quality development. Where flavor development is not concurrent with optimal sugar/acid balance, the grape grower is placed in a difficult situation. Solution requires a knowledge of the significance of particular flavor constituents to the wine desired as well as of the acceptability and applicability of sugar and acid amelioration procedures. For example, where alcohol content is an important legal measure of quality, and chaptalization illegal, harvesting at an appropriate degrees Brix may be more important than grape flavor content. However,

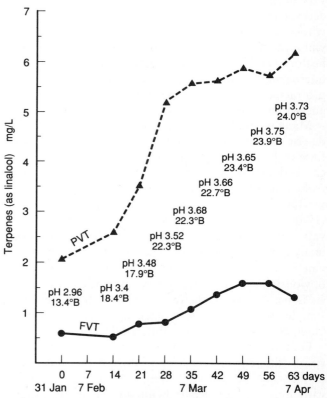

Figure 4.33 Change in free (FVT) and potential (PVT) volatile terpenes after *véraison* for 'Muscat of Alexandria' grapes. Degrees Brix (°B) and pH are shown at each sampling. (From Dimitriadis and Williams, 1984, reproduced by permission.)

where flavor is the primary quality indicator, harvesting when flavor content is optimal, and adjusting the sugar and acid content after crushing, may be preferable.

As the chemical nature of aroma development in wine becomes clearer, grape flavor assessment will probably become an increasingly important factor in setting harvest dates. Such information also would aid the study of vineyard practice on flavor development. For example, irrigation does not significantly affect free terpene levels in 'Gordo Muscat,' but it does lower the rate at which terpenes become bound in nonvolatile complexes (McCarthy and Coombe, 1985). Also, the level of canopy fruit shading affects anthocyanin synthesis, and presumably flavor, in some red cultivars.

In addition to assessing properties correlated with quality, grape growers must also consider factors that may lower fruit quality. These are usually difficult to predict as they depend on local climatic conditions at harvest time. The detrimental effects of early frosts and protracted rainy periods on grape quality are well known, but forecasting their imminent occurrence is still regrettably imprecise.

Sampling

While the criteria used to assess grape maturity are very important, it is also crucial that the measurement accurately represent fruit characteristics within the whole vineyard. Variation in maturity throughout the vineyard can arise from differences in the age and health of the vines, as well as nonuniform soil type and microclimate. Additional variation may arise due to protracted flowering and positioning of the cluster on, and within, the vine canopy.

Many methods of sampling have been investigated to determine the best combination of adequacy and ease. Ease of performance is important as vineyards may be checked almost daily for several weeks prior to harvest. Although different sampling procedures are used worldwide, that proposed by Amerine and Roessler (1958) appears both adequate and simple to use. It involves the collection of about 100 to 200 berries from many grape clusters, selected at random throughout the vineyard. Berries from clusters at row ends, or from obviously aberrant vines, are avoided. Chemical analysis is performed on the juice extracted from the berry sample. Basic methods of must analysis are given in Ough and Amerine (1988) and Zoecklein *et al.* (1990).

Because of variation in the rate of maturation between different cultivars, and between the individuals of the same cultivars in different locations, each vineyard must be sampled individually. Sampling usually begins 2 to 3 weeks before the grapes are likely to reach the optimal state of maturity.

Containers for Collecting and Harvesting Grapes

Many different types of containers have been employed in collecting and transporting grapes to the winery. In Europe, traditional wicker and wooden containers reflect cultural traditions and the terrain of the region. In the New World, plastic or aluminum boxes are commonly used. From these, grapes are transferred into vessels of various sizes for transport to the winery. Mechanically harvested grapes are often directly conveyed into transport containers. Any system is adequate if bruising or breaking the fruit are minimized and rapid movement to the winery permits processing shortly after harvesting.

Harvest Mechanisms

Until the late 1960s, essentially all grapes were harvested manually. Subsequently, market forces as well as labor shortages have combined to make mechanical harvesting progressively more cost-effective. Premium quality grapes may command prices sufficiently high to permit the retention of manual harvesting, but most hectarage in the New World is mechanically harvested.

MANUAL HARVESTING

Manual harvesting wine grapes still has several advantages over mechanical harvesting. This is especially true for thin-skinned cultivars that break open easily. Hand harvesting also permits the rejection of immature, raised, or diseased clusters or berries, as well as the selection of grapes at particular states of maturity. Except for special wines, such as those produced from the outer arm of grape clusters in Amarone production or noble-rotted grapes for botrytized wine production, the latter option is seldom employed.

For the advantages of manual harvesting to be realized, the clusters must be collected and placed in containers with all due care to minimize breakage. The grapes also need be transported quickly to the winery for rapid processing and avoid fruit heating and the growth of undesirable microorganisms on berries or released juice.

Although hand harvesting is beneficial, or required, in special circumstances, the disadvantages of manual harvesting often outweigh its advantages. Besides labor costs and inadequate availability, manual harvesting is slower, does not proceed 24 hours a day, and does not proceed during inclement weather. Correspondingly, the development of better and smaller mechanical harvesting and crushing equipment is continuing to expand their use.

MECHANICAL HARVESTERS

All mechanical harvesters ostensibly use the same means to remove fruit. Force is applied to one or more parts of the vine, inducing the rapid and abrupt fruit swinging that detaches fruit clusters and/or berries. Harvesters are usually classified relative to the mechanism by which they apply to the force. Most harvesters currently in use fall into one of two categories. Those that apply force directly to the bearing shoot are variously called "cane shaker," "slapper," "striker," or "impactor" machines. Those that apply force directly to the vine trunk are termed either "pulsator," "trunk shaker," or simply "shaker" machines. Some harvesters combine both actions and may be referred to as "pivotal pulsators."

Most machines are designed to harvest vines trained as single vertical canopies. Special modification is required to harvest vines trained on T-trellises, such as the Geneva Double Curtain, and major changes in design are required for training systems such as the Tendone or pergolas (Cargnello and Piccoli, 1978).

Lateral Striker Harvesters Lateral striker harvesters possess a double bank of upright flexible rods arranged parallel to, and on each side of, the vine (Fig. 4.34). The banks of rods oscillate back and forth, striking the vine canopy and shaking the fruit loose. For vines possessing less foliage, having both front and back banks oscillate together tends to be preferable; under heavy foliage conditions, having them oscillate alternately tends to be more effective.

Striker machines are generally most suitable for cane-pruned vines, when cordons are young, or with other systems where the fruit is borne away from the permanent vine structure. It is not efficient at dislodging fruit

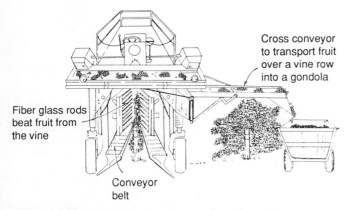

Figure 4.34 Diagrammatic representation of a striker-type mechanical harvester. (From *Grape Growing*, R. J. Weaver, Copyright © 1976 John Wiley & Sons, Inc. Reprinted by permission of John Wiley & Sons, Inc.)

close to the vine head. Increasing striking velocity to dislodge such fruit accentuates vine damage and increases fruit contamination with leaves, leaf fragments, and other vine parts. (MOG, material other than grapes). However, striker machines are easier on the trellis structure and accommodate themselves more to vine rows of imperfect alignment.

Trunk-Shaker Harvesters The trunk-shaker machine typically possesses two parallel oscillating rails that impart vibration to the cordon and/or upper trunk a few centimeters below the fruit zone (Fig. 4.35). This shakes the vine back and forth several centimeters.

Shaker (pulsator) machines are most effective in removing fruit in close proximity to cordons and/or the trunk. Insufficient force tends to spread along flexible canes to efficiently remove fruit on cane-pruned vines. However, the machines do have the advantage of dislodging fewer leaves or other vine material. The major drawback is the strict training required for proper use. Stakes must be flexible and in perfect alignment to withstand the force applied during shaking. The cordons must be firmly attached to support vines and these in turn securely affixed to stakes. Magnets are typically present in the harvester to remove nails, staples, or other metallic parts shaken loose from the trellis.

Striker/Shaker Combination Harvesters Because both striker and shaker equipment types have significant limitations, machines combining both principles have been introduced. They possess a pair of oscillating rails and several short pulsating rods, and operate at lower speeds (50 to 75 rpm). They tend to produce less vine and trellis damage, may be used with a wide range of training systems, and generate lower levels of MOG contamination.

Horizontal Impactor The horizontal impactor is designed for use with training systems employing wide-topped (T) trellises. For use of the harvester, canopy wires must be attached to movable crossarm supports, or the canopy wire must be held loosely in a slot on rigid crossarms. This is necessary as the jarring action to dislodge the fruit comes from a vertically rotating wheel whose spokes strike the canopy wire from below. Because of the extra stress placed on the canopy wire, stronger wire than usual is required.

FACTORS AFFECTING HARVESTER EFFICIENCY

Besides the training system, the growth and fruiting habits of the cultivar are the primary factors affecting harvester applicability. The two major fruiting charac-

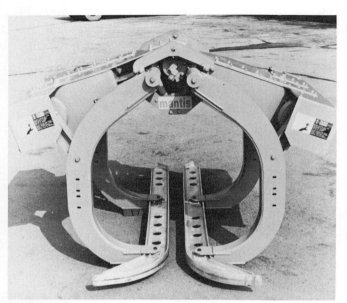

Figure 4.35 Operational component of a trunk-shaker type of mechanical harvester. (Courtesy of American Grape Harvesters, Selma, CA.)

into, the fruit. Brittle wood can produce spur breakage and conveyor plugging. Preharvest thinning often diminishes the severity of undesirable growth characteristics on harvester function.

Other factors influencing harvester efficiency are the time of harvesting, vineyard slope, and condition of the soil. As grapes and other vine parts are typically more turgid at night, berry and cluster separation usually requires less force. Night harvesting also has the advantage of removing fruit at a cool temperature in warm climates. However, poor visibility, even using headlights, makes adequate observation of harvesting performance difficult. Slopes of more than about 7% require harvesters whose wheels can be adjusted individually to level the catching frame of the harvester. Soil grading also may be required to facilitate steering and to remove ridges that can interfere with keeping the catching frame close to the ground.

teristics affecting varietal suitability for machine harvesting are ease of separation from the vine and ease of berry rupture. Most effectively harvested cultivars separate readily as whole bunches, fragment easily at rachis divisions, or dehisce with facility at the pedicle. Ease of detachment releases the stress applied by the harvester and minimizes the likelihood of fruit rupture. Because wine grapes are predominantly juicy, firm attachment and fibrous rachis and cluster structure can lead to extensive berry rupture and juice loss, as, for example, with 'Emerald Riesling' and 'Zinfandel.' Soft-berried varieties, such as 'Sémillon' and 'Muscat Canelli,' also lose much juice if mechanically harvested (see Table 4.12).

The vegetative growth characteristics of the vine most affecting harvester function are cane flexibility and foliage density and size. Dense canopies can interfere with force transmission and fruit dislodgement, as well as plug and disrupt conveyor belt operation. The detrimental effects of excessive MOG contamination can be partially offset by fans blowing over the collector belts. However, small leaf fragments tend to become wetted with juice and stick to, or are propelled

Table 4.12 Adaptability of Several Grape Varieties to Mechanical Harvesting[a]

Variety	Ease of harvest	Amount of juicing
White varieties		
'Chardonnay'	Hard	Medium
'Chenin blanc'	Medium	Medium
'Emerald Riesling'	Very hard	Very heavy
'Flora'	Easy	Light
'French Colombard'	Medium	Medium
'Gewürztraminer'	Easy/medium	Light
'Gray Riesling'	Easy/medium	Light
'Malvasia bianca'	Medium/hard	Medium
'Muscat Canelli'	Hard	Heavy
'Palomino'	Easy/medium	Medium
'Pedro Ximénez'	Easy/medium	Heavy
'Pinot blanc'	Medium	Medium
'Riesling'	Medium	Medium
'Sauvignon blanc'	Medium	Medium
'Sémillon'	Medium/hard	Heavy
'Silvaner'	Medium	Light
'Trebbiano'	Medium	Heavy
Red varieties		
'Aleatico'	Hard	Heavy
'Barbera'	Medium/hard	Medium
'Cabernet Sauvignon'	Easy/medium	Light/medium
'Carignan'	Medium/hard	Medium/heavy
'Gamay'	Medium	Medium
'Grenache'	Hard	Medium/heavy
'Petite Sirah'	Medium	Medium/heavy
'Pinot noir'	Medium	Medium
'Rubired'	Easy	Medium
'Tempranillo'	Medium/hard	Medium/heavy
'Zinfandel'	Hard	Medium/heavy

[a] Data from Christensen *et al.* (1973).

RELATIVE MERITS OF MECHANICAL HARVESTING

With improved harvester design and increasing awareness of the importance of vine training and uniform trellising, fruit of equivalent quality often can be obtained by either manual or mechanical means (Table 4.13). Wines produced from grapes harvested by different means can occasionally be distinguished, but no clear sensory preference has developed either in North America or in Europe (see Clary *et al.*, 1990).

With cultivars suitable for mechanical harvesting, the choice of preferable method often depends on factors other than direct effects of harvesting method on fruit quality. Features such as potential for night and rapid harvesting, cost and availability of manual labor, juice loss, and vineyard size become the deciding factors. For example, where grapes grown in hot climates must be transported long distances to the winery, harvesting during the cool of the night can result in grapes arriving in a healthier condition. Field crushing and storage under cool anaerobic conditions, immediately following harvest, is another solution to preserving juice quality of fruit picked distant from the winery. For high priced, low yielding cultivars, juice losses between 5 and 10% may negate the economic benefits of mechanical harvesting. Also, for French–American hybrids that may produce an extensive second crop, the detrimental effects of harvesting immature grapes on wine quality may outweigh the cost benefits of mechanical harvesting. Whether the lower stem content, marginally higher

MOG contents (Clary *et al.*, 1990), increased potential for juice oxidation, and increased fungal growth on juice-soaked canes of the vine (Bugaret, 1988) associated with mechanical harvesting are of practical significance is uncertain. If so, the significance probably varies considerably, depending on the cultivar, harvesting conditions, and style of wine desired.

Suggested Readings

General References

Coombe, B., and Dry, P. (1988, 1992). "Viticulture. Vol 1 and 2." Winetitles, Adelaide, Australia.

Mullins, M. G., Bouquet, A., and Williams, L. E. (1992). "Biology of the Grapevine." Cambridge Univ. Press, Cambridge.

Pongrácz, D. P. (1978). "Practical Viticulture." David Philip, Cape Town, South Africa.

Smart, R. E., and Robinson, M. (1991). "Sunlight into Wine. A Handbook for Winegrape Canopy Management." Winetitles, Adelaide, Australia.

Weaver, R. J. (1976). "Grape Growing." Wiley, New York.

Winkler, A. J., Cook, J. A., Kliewer, W. M., and Lider, L. A. (1974). "General Viticulture." Univ. California Press, Berkeley.

Pruning

Bailey, L. H. (1919). "The Pruning-Manual." Macmillan, New York.

Bledsoe, A. M., Kliewer, W. M., and Marois, J. J. (1988). Effects of timing and severity of leaf removal on yield and fruit composition of Sauvignon blanc grapevines. **Am. J. Enol. Vitic. 39**, 49–54.

Clingeleffer, P. R., and Possingham, J. V. (1987). The role of minimal pruning of cordon trained vines (MPCT) in canopy management and its adoption in Australian viticulture. *Aust. Grapegrower Winemaker* **280**, 7–11.

Jordan, T. D., Pool, R. M., Zabadal, T. J., and Tomkins, J. P. (1981). "Cultural Practices for Commercial Vineyards." Miscellaneous Bull. 111, Cornell Cooperative Extension Publ., Cornell Univ., Ithaca, New York.

Smith, S., Codrington, I. C., Robertson, M., and Smart, R. (1988). Viticultural and oenological implications of leaf removal for New Zealand vineyards. *Proc. 2nd Int. Symp. Cool Climate Vitic. Oenol., Jan. 11–15, 1988. Auckland, N.Z.* (R. E. Smart, S. B. Thornton, S. B. Rodriguez, and J. E. Young, eds.), pp. 127–133. New Zealand Soc. Vitic. Oenol., Auckland, New Zealand.

Training

Archer, E., Swanepoel, J. J., and Strauss, H. C. (1988). Effect of plant spacing and trellising systems on grapevine root distribution. *In* "The Grapevine Root and Its Environment" (J. L. van Zyl, compiler), pp. 74–87. Tech. Commun. 215, Dept. Agric. Water Supply, Pretoria, South Africa.

Carbonneau, A., and Huglin, P. (1982). Adaptation of training systems to French regions. *Grape Wine Centennial Symp. Proc. 1980*, pp. 376–385. Univ. of California, Davis.

Howell, G. S., Mansfield, T. K., and Wolpert, J. A. (1987). Influence of training system, pruning severity, and thinning on yield, vine size, and fruit quality of Vidal blanc grapevines. **Am. J. Enol. Vitic. 38**, 105–112.

Table 4.13 Effect of Harvesting Method on Crop and Juice Characteristics[a,b]

	Treatment		
Property	Striker harvester	Shaker harvester	Manual harvester
Yield (kg/ha)	12,509	12,475	12,800
Stem content (kg/ha)	155	288	524
Ground loss (kg/ha)	317	249	338
Juice loss (%)	5.7	8.0	0
MOG[c] (%)	1.3	0.7	0.5
°Brix	22.2	22.2	22.6
Total acidity (g/100 ml)	0.75	0.76	0.74
pH	3.26	3.28	3.27
Malic acid (g/liter)	4.552	4.409	4.137

[a] The slightly higher degrees Brix and lower total and malic acidity of the hand-harvested grapes are likely explained by the absence of second crop fruit.

[b] Data from Clary *et al.* (1990).

[c] MOG, Material other than grapes.

Kliewer, W. M. (ed.) (1987). Symposium on Grapevine Canopy and Vigor Management. 22nd Int. Horticultural Congress, Davis California, August 11–20, 1986. *Acta Hortic.* **206**, 1–168.

Morlat, R., Remoué, M., and Pinet, P. (1984). Influence de la densité de plantation et du mode d'entretien du sol sur l'enracinement d'un peuplement de vigne plante en sol favorable. *Agronomie* **4**, 485–491.

Smart, R. E., Dick, J. K., Gravett, I. M., and Fisher, B. M. (1990). Canopy management to improve grape yield and wine quality—Principles and practices. *S. Afr. J. Enol. Vitic.* **11**, 3–18.

Smart, R. E., and Robinson, M. (1991). "Sunlight into Wine. A Handbook for Winegrape Canopy Management." Winetitles, Adelaide, Australia.

Rootstock

Catlin, T. (Chairman) (1991). "Alternative Rootstock Update." Am. Soc. Enol. Vitic., Davis, California.

Wolpert, J. A., Walker, M. A., and Weber, E. (eds.) (1992). "Rootstock Seminar: A Worldwide Perspective." Am. Soc. Enol. Vitic., Davis, California.

Howell, G. S. (1987). *Vitis* rootstocks. *In* "Rootstocks for Fruit Crops" (R. C. Rom and R. F. Carlson, eds.), pp. 451–472. Wiley, New York.

Morton, L. T., and Jackson, L. E. (1988). Myth of the universal rootstock: The fads and facts of rootstock selection. *Proc. 2nd Int. Symp. Cool Climate Vitic. Oenol., Jan. 11–15, 1988, Auckland, N.Z.* (R. E. Smart, S. B. Thornton, S. B. Rodriguez, and J. E. Young, eds.), pp. 25–29. New Zealand Soc. Vitic. Oenol., Auckland, New Zealand.

Pongrácz, D. P. (1983). "Rootstocks for Grape-Vines." David Philip, Cape Town, South Africa.

Pouget, R. (1987). Usefulness of rootstocks for controlling vine vigour and improving wine quality. *Acta Hortic.* **206**, 109–118.

Southey, J. M., and Archer, E. (1988). The effect of rootstock cultivar on grapevine root distribution and density. *In* "The Grapevine Root and Its Environment" (J. L. van Zyl, compiler), pp. 57–73. Tech. Commun. 215, Dept. Agric. Water Supply, Pretoria, South Africa.

Propagation and Grafting

Conner, A. J., and Thomas, M. B. (1981). Re-establishing plantlets from tissue culture: A review. *Proc. Int. Plant Prop. Soc.* **31**, 342–357.

Kurl, W. R., and Mowbray, G. H. (1984). Grapes. *In* "Handbook of Plant Cell Culture" (W. R. Sharp, D. A. Evans, P. V. Ammirato, and Y. Yamada, eds.), Vol. 2, pp. 396–434. Macmillan, New York.

Mullins, M. G., and Rajasekaran, K. (1981). Fruiting cuttings: Revised method for producing test plants of grapevine cultivars. *Am. J. Enol. Vitic.* **32**, 35–40.

Steinhauer, R. E., Lopez, R., and Pickering, W. (1980). Comparison of four methods of variety conversion in established vineyards. *Am. J. Enol. Vitic.* **31**, 261–264.

Irrigation

Hsiao, T. C. (1990). Measurement of plant water status. *In* "Irrigation of Agricultural Crops" (B. A. Stewart and D. R. Nielsen, eds.), Agronomy Series Monograph No. 30, pp. 243–279. American Society of Agronomy, Madison, Wisconsin.

Nagarajah, S. (1989). Physiological responses of grapevines to water stress. *Acta Hortic.* **240**, 249–256.

Nakayama, F. S., and Bucks, D. A. (1986). "Trickle Irrigation for Crop Production: Design, Operation, and Management." Elsevier, Amsterdam.

Smedema, L. K., and Rycroft, D. W. (1983). "Land Drainage." Cornell Univ. Press, Ithaca, New York.

Smart, R. E., and Coombe, B. G. (1983). Water relations of grapevines. *In* "Water deficits and Plant Growth" (T. T. Kozlowski, ed.). Vol. 7, pp. 137–195. Academic Press, New York.

Williams, L. E., and Matthews, M. A. (1990). Grapevines. *In* "Irrigation of Agricultural Crops" (B. A. Stewart and D. R. Nielsen, eds.), Agronomy Series Monograph No. 30. pp. 1019–1055. Madison, Wisconsin.

van Zyl, J. L. (1988). Response of grapevine roots to soil water regimes and irrigation systems. *In* "The Grapevine Root and Its Environment" (J. L. van Zyl, compiler), Tech. Commun. 215, pp. 30–43. Dept. Agric. Water Supply, Pretoria, South Africa.

Nutrition

Christensen, L. P., Kasimatis, A. N., and Jensen, F. L. (1978). "Grapevine Nutrition and Fertilization in the San Joaquin Valley," Publ. No. 4087. Div. Agric. Nat. Res., Univ. of California, Oakland.

Fregoni, M. (1985). Exigences d'éléments nutritifs en viticulture. *Bull. O.I.V.* **58**, 416–434.

Gehlen, P., Neu, J., and Schröder, D. (1988). Bodenchemische und bodenbiologische Vergleichsuntersuchungen konventionell und biologisch bewirtschafteter Weinstandorte an der Mosel. *Wein-Wiss.* **43**, 161–173.

Hill, G. K. (1988). Organic viticulture in Germany—Practical experiences and organization. *Proc. 2nd Int. Symp. Cool Climate Vitic. Oenol., Jan. 11–15, 1988, Auckland, N.Z.* (R. E. Smart, S. B. Thornton, S. B. Rodriguez, and J. E. Young, eds.), pp. 83–84. New Zealand Soc. Vitic. Oenol., Auckland, New Zealand.

MacRae, R. J., and Mehuys, G. R. (1985). The effects of green manuring on the physical properties of temperate-area soils. *Adv. Soil Sci.* **3**, 71–94.

Marschner, H. (1986). "Mineral Nutrition in Higher Plants." Academic Press, London.

Rantz, J. M. (1991). *Proc. Int. Symp. Nitrogen Grapes Wine, Seattle, WA, June 18–19, 1991.* Am. Soc. Enol. Vitic., Davis, California.

Disease, Pest, and Weed Control

Barlass, M., and Skene, K. G. M. (1987). Tissue culture and disease control. *Proc. 6th Austr. Wine Ind. Tech. Conf.* (T. Lee, ed.), pp. 191–193. Australian Industrial Publ., Adelaide, Australia.

Bovey, R., Gärtel, W. B., Hewitt, W. B., Martelli, G. P., and Vuittenez, A. (1980). "Virus and Virus-like Diseases of Grapevines." Colour Atlas of Symptoms, Editions Payot, Lausanne, Switzerland.

Fergusson-Kolmes, L., and Dennehy, T. J. (1991). Anything new under the sun? Not phylloxera biotypes. *Wines Vines* **72**(6), 51–56.

Flaherty, D. L., Christensen, L. P., Lanini, W. T., Marois, J. J., Phillips, P. A., and Wilson, L. T. (1992). "Grape Pest Management," 2nd Ed., Publ. No. 3343. Div. Agric. Nat. Res., Univ. of California, Oakland.

Hogland, R. E. (ed.) (1990). "Microbes and Microbial Products as Herbicides." American Chemical Soc., Washington, D.C.

Milholland, R. D. (1991). Muscadine grapes: Some important diseases and their control. *Plant Dis.* **75**, 113–117.

Pearson, R. C., and Goheen, A. C (eds.) (1988). "Compendium of Grape Diseases." APS Press, St. Paul, Minnesota.

Weinstein, L. H. (1984). Effects of air pollution on grapevines. *Vitis* **23**, 274–303.

Harvesting

Carnacini, A., Amati, A., Capella, P., Casalini, A., Galassi, S., and Riponi, C. (1985). Influence of harvesting techniques, grape crushing and wine treatments on the volatile components of white wines. *Vitis* **24**, 257–267.

Christensen, L. P., Kasimatis, A. N., Kissler, J. J., Jensen, F., and Luvisi, D. (1973). "Mechanical Harvesting of Grapes for the Winery." Agricultural Extension, Univ. of California, Oakland.

Clary, C. D., Steinhauer, R. E., Frisinger, J. E., and Peffer, T. E. (1990). Evaluation of machine- vs. hand-harvested Chardonnay. *Am. J. Enol. Vitic.* **41**, 176–181.

du Plessis, C. S. (1984). Optimum maturity and quality parameters in grapes: A review. *S. Afr. J. Enol. Vitic.* **5**, 35–42.

Morris, J. R. (1985). Approaches to more efficient vineyard management. *HortScience* **20**, 1008–1013.

Olmo, H. P. (1982). Mechanical harvest of grapes. *Grape Wine Centennial Symp. Proc. 1980*, pp. 187–190. Univ. of California, Davis.

Petrucci, V. E., Clary, C. D., and O'Brien, M. (1983). Grape harvesting systems. *In* "Principles and Practices for Harvesting and Handling Fruits and Nuts" (M. O'Brien, B. F. Cargill, and R. B. Fridley, eds.), pp. 525–574. AVI, Westport, Connecticut.

Somers, T. C., and Pocock, K. F. (1986). Phenolic harvest criteria for red vinification. *Aust. Grapegrower Winemaker* **258**, 24, 26–27, and 29–30.

References

Adams, W. E., Morris, H. D., and Dawson, R. N. (1970). Effects of cropping systems and nitrogen levels on corn (*Zea mays*) yields in the southern Piedmont region. *Agron. J.* **62**, 655–659.

Agrios, G. N. (1988). "Plant Pathology," 3rd Ed. Academic Press, New York.

All, J. N., Dutcher, J. D., Saunders, M. C., and Brady, U. E. (1985). Prevention strategies for grape root borer (Lepidoptera: Sesiidae) infestations in Concord grape vineyards. *J. Econ. Entomol.* **78**, 666–670.

Allaway, W. H. (1968). Agronomic controls over the environment cycling of trace elements. *Adv. Agron.* **20**, 235–274.

Alleweldt, G., Reustle, G., and Binzel, A. (1991). Die maschinelle Grünveredlung—Entwicklung eines Verfahrens für die Praxis. *Dtsch. Weinbau* **46**, 976–978.

Alley, C. J. (1975). Grapevine propagation. VII. The wedge graft—a modified notch graft. *Am. J. Enol. Vitic.* **26**, 105–108.

Alley, C. J., and Koyama, A. T. (1980). Grapevine propagation. XVI. Chip-budding and T-budding at high level. *Am. J. Enol. Vitic.* **31**, 60–63.

Alley, C. J., and Koyama, A. T. (1981). Grapevine propagation. XIX. Comparison of inverted with standard T-budding. *Am. J. Enol. Vitic.* **32**, 20–34.

Amerine, M. A., and Roessler, E. B. (1958). Field testing of grape maturity. *Hilgardia*, **28**, 93–114.

Araujo, F. J., and Williams, L. E. (1988). Dry matter and nitrogen partitioning and root growth of young field-grown Thompson seedless grapevines. *Vitis* **27**, 21–32.

Archer, E., and Strauss, H. C. (1985). Effect of plant density on root distribution of three-year-old grafted 99 Richter grapevines. *S. Afr. J. Enol. Vitic.* **6**, 25–30.

Archer, E., Swanepoel, J. J., and Strauss, H. C. (1988). Effect of plant spacing and trellising systems on grapevine root distribution. *In* "The Grapevine Root and Its Environment" (J. L. van Zyl, com-

piler), Tech. Commun. 215, pp. 74–87. Dept. Agric. Water Supply, Pretoria, South Africa.

Avramov, L., Pemovski, D., Lovic, R., Males, P., Ulicevic, M., and Jurcevic, A. (1980). Germ plasm of *Vitis vinifera* in Yugoslavia. *Proc. 3rd Int. Symp. Grape Breeding*, pp. 197–203. Univ. California, Davis.

Barlass, M., and Skene, K. G. M. (1978). *In vitro* propagation of grapevine (*Vitis vinifera* L.) from fragmented shoot apices. *Vitis* **17**, 335–340.

Barlass, M., and Skene, K. G. M. (1987). Tissue culture and disease control. *Proc. 6th Austr. Wine Ind. Tech. Conf.* (T. Lee, ed.), pp. 191–193. Australian Industrial Publ. Adelaide, Australia.

Basler, P., Boller, E. F., and Koblet, W. (1991). Integrated viticulture in eastern Switzerland. *Pract. Winery Vineyard* **May/June**, 22–25.

Baumberger, I. (1988). Regenwürmer—Schütenwerte Nützlinge im Boden. *Weinwirtschaft Anbau* **124**(6), 19–21.

Bavaresco, L., and Eibach, R. (1987). Investigations on the influence of N fertilizer on resistance to powdery mildew (*Oidium tuckeri*), downy mildew (*Plasmopara viticola*) and on phytoalexin synthesis in different grape varieties. *Vitis* **26**, 192–200.

Bavaresco, L., Zamboni, M., and Corazzina, E. (1991). Comportamento produttivo di alcune combinazioni d'innesto di Rondinella e Corvino nel Bardolino. *Rev. Vitic. Enol.* **44**, 3–20.

Bazzi, C., Stefani, E., Gozzi, R., Burr, T. J., Moore, C. L., and Anaclerio, F. (1991). Hot-water treatment of dormant grape cuttings: Its effects on *Agrobacterium tumefaciens* and on grafting and growth of vine. *Vitis* **30**, 177–187.

Beachy, R. N., Loesch-Fries, S., and Tumer, N. E. (1990). Coat protein-mediated resistance against virus infection. *Annu. Rev. Phytopathol.* **28**, 451–474.

Becker, H., and Konrad, H. (1990). Breeding of *Botrytis* tolerant *V. vinifera* and interspecific wine varieties. *Proc. 5th Int. Symp. Grape Breeding, Sept. 12–16, 1989*, p. 302. St. Martin, Pfalz, Germany (Special Issue of *Vitis*).

Bessis, M. R. (1972). Étude en microscopie electronique à balayage des rapports entre l'hôte et le parasite dans le cas de la pourriture grise. *C. R. Acad. Sci. Paris. Ser. D* **274**, 2991–2994.

Bessis, R., and Fournioux, J. C. (1992). Zone d'abscission en coulure de la vigne. *Vitis* **31**, 9–21.

Bishop, A. A., Jensen, M. E., and Hall, W. A. (1967). Surface irrigation systems. *In* "Irrigation of Agricultural Lands" (R. M. Hagan, H. R. Haise, and T. W. Edminster, eds.), Agronomy Monogr. 11, pp. 865–884. Am. Soc. Agronomy, Madison, Wisconsin.

Blaich, R., Stein, U., and Wind, R. (1984). Perforation in der Cuticula von Weinbeeren als morhologischer Faktor der Botrytisresistenz. *Vitis* **23**, 242–256.

Blakeman, J. P. (1980). Behaviour of conidia on aerial plant surfaces. *In* "The Biology of *Botrytis*" (J. R. Coley-Smith, K. Verhoeff, and W. R. Jarvis, eds.), pp. 115–151. Academic Press, New York.

Borner, C. B., and Heinze, K. (1957). Phylloxeridae, Zwergläuse. *In* "Handbuch der Pflanzenkrankheiten, Tierische Schädlinge an Nutzpflanzen 2," Teil. 4 Lief. Homoptera II. Teil, pp. 355–375. Parey, Berlin.

Boubals, D. (1954). Les németodes parasites de la vigne. *Prog. Agric. Vitic.* **71**, 141–173.

Boubals, D. (1966). Étude de la distribution et des causes de la résistance au *Phylloxera radicicole* chez les Vitacées. *Ann. Amelior. Plant.* **16**, 145–184.

Boubals, D. (1980). Conduite pour établer une vigne dans un milieu infecté par l'anguillule ou nématode des racines (*Meloidogynae* sp.). *Prog. Agric. Vitic.* **3**, 99.

Boubals, D. (1984). Note on accidents by massive soil applications of grape by-products. *Prog. Agric. Vitic.* **101**, 152–155.

Boubals, D. (1989). La maladie de Pierce arrive dans les vignobles d'Europe. *Bull O.I.V.* **62**, 309–314.

Boulay, H. (1982). Absorption différenciée des cépages et des porte-greffes en Languedoc. *Prog. Agric. Vitic.* **99**, 431–434.

Brady, N. C. (1974). "The Nature and Properties of Soils," 8th Ed. Macmillan, New York.

Branas, J. (1974). "Viticulture." J. Branas, Montpellier, France.

Brandes, W., and Kaspers, H. (1989). Tebuconazole—ein neues *Botrytis*-Fungizid für den Weinbau. *Pflanzenschutz-Nachr. Bayer (Ger. Ed.)* **42**, 149–161.

Bravdo, B., and Hepner, Y. (1987). Irrigation management and fertigation to optimize grape composition and vine performance. *Acta Hortic.* **206**, 49–67.

Bubl, W. (1987). Control of stem necrosis with magnesium and micronutrient fertilizers during the period 1983 to 1985. *Mitt. Klosterneuburg* **37**, 126–129.

Bugaret, Y. (1988). L'influence des traitements anti-mildiou et de la récolte méchanique sur l'état sanitaire des bois. *Phytoma* **399**, 42–44.

Bulit, J., Molot, B., Riffiod, G., and Strizyk, S. (1985). Raisonnement des traitements de la vigne. *In* "Fungicides for Crop Protection: 100 Years of Progress, Smith, I. M. ed." British Crop Protection Council (BCPC) Monogr. No. 31, Vol. 1, pp. 199–208. Croydon, England.

Burr, T. J., Bishop, A. L., Katz, B. H., Blanchard, L. M., and Bazzi, C. (1987). A root-specific decay of grapevine caused by *Agrobacterium tumefaciens* and *A. radiobacter* Biovar 3. *Phytopathology* **77**, 1424–1427.

Burr, T. J., Katz, B. H., Bishop, A. L., Meyers, C. A., and Mittak, V. L. (1988). Effect of shoot age and tip culture propagation of grapes on systemic infestations by *Agrobacterium tumefaciens* Biovar 3. *Am. J. Enol. Vitic.* **39**, 67–70.

Candolfi-Vasconcelos, M. C., and Koblet, W. (1990). Yield, fruit quality, bud fertility and starch reserves of the wood as a function of leaf removal in *Vitis vinifera*—Evidence of compensation and stress recovering. *Vitis* **29**, 199–221.

Candolfi-Vasconcelos, M. C., and Koblet, W. (1991). Influence of partial defoliation on gas exchange parameters and chlorophyll content of field-grown grapevines—Mechanisms and limitations of the compensation capacity. *Vitis* **30**, 129–141.

Carbonneau, A. (1985). Trellising and canopy management for cool climate viticulture. *Int. Symp. Cool Climate Vitic. Enol.* (B. A. Heatherbell, P. B. Lombard, F. W. Bodyfelt, and S. F. Price, eds.), OSU Agric. Exp. Stn. Tech. Publ. No. 7628, pp. 158–183. Oregon State Univ., Corvallis, Oregon.

Carbonneau, A., and Casteran, P. (1987). Optimization of vine performance by the lyre training systems. *Proc. 6th Austr. Wine Ind. Tech. Conf.* (T. Lee, ed.), pp. 194–204. Australian Industrial Publ., Adelaide, Australia.

Cargnello, G., and Piccoli, P. (1978). Vendemmiatrice per vigneti a pergola e a tendone. *Inf. Agrario* **35**, 2813–2814.

Caudwell, A. (1990). Epidemiology and characterization of *flavescence dorée* (FD) and other grapevine yellows. *Agronomie* **10**, 655–663.

Champagnol, F. (1978). Fertilisation optimale de la vigne. *Prog. Agric. Vitic.* **95**, 423–440.

Champagnol, F. (1984). "Elements de Physiologie de la Vigne et du Viticulture Générale." F. Champagnol, Montpellier, France.

Christensen, L. P. (1984). Nutrient level comparisons of leaf petioles and blades in twenty-six grape cultivars over three years (1979 through 1981). *Am. J. Enol. Vitic.* **35**, 124–133.

Christensen, L. P., Kasimatis, A. N., Kissler, J. J., Jensen, F., and Luvisi, D. (1973). "Mechanical Harvesting of Grapes for the Winery." Agricultural Extension. Publ. No. 2365. Univ. of California, Berkeley.

Christensen, L. P., Boggero, J., and Bianchi, M. (1990). Comparative leaf tissue analysis of potassium deficiency and a disorder resembling potassium deficiency in Thompson Seedless grapevines. *Am. J. Enol. Vitic.* **41**, 77–83.

Christensen, L. P., Boggero, J., and Adams, D. O. (1991). The relationship of nitrogen and other nutritional elements to the bunch stem necrosis disorder "waterberry." *Proc. Int. Symp. Nitrogen Grapes Wine, Seattle, WA, June, 1991* (J. M. Rantz, ed.), pp. 108–109. Am. Soc. Enol. Vitic., Davis, California.

Clark, M. F., and Adams, A. N. (1977). Characteristics of the microplate method of enzyme-linked immunosorbent assay (ELISA). *J. Gen. Virol.* **34**, 475–483.

Clary, C. D., Steinhauer, R. E., Frisinger, J. E., and Peffer, T. E. (1990). Evaluation of machine- vs. hand-harvested Chardonnay. *Am. J. Enol. Vitic.* **41**, 176–181.

Clingeleffer, P. R. (1984). Production and growth of minimal pruned Sultana vines. *Vitis* **23**, 42–54.

Clingeleffer, P. R., and Possingham, J. V. (1987). The role of minimal pruning of cordon trained vines (MPCT) in canopy management and its adoption in Australian viticulture. *Aust. Grapegrower Winemaker* **280**, 7–11.

Collard, B. (1991). Greffe en vert: C'est au point. *Vigne* **14**, 26–27.

Conner, A. J., and Thomas, M. B. (1981). Re-establishing plantlets from tissue culture: A review. *Proc. Int. Plant Prop. Soc.* **31**, 342–357.

Conradie, W. J. (1988). Effect of soil acidity on grapevine root growth and the role of roots as a source of nutrient reserves. *In* "The Grapevine Root and Its Environment" (J. L. van Zyl, compiler). Tech. Commun. No. 215, pp. 16–29. Dept. Agric. Water Supply, Pretoria, South Africa.

Conradie, W. J. (1992). Partitioning of nitrogen in grapevines during autumn and the utilization of nitrogen reserves during the following growing season. *S. Afr. J. Enol. Vitic.* **13**, 45–51.

Conradie, W. J., and Saayman, D. (1989a). Effects of long-term nitrogen, phosphorus, and potassium fertilization on Chenin blanc vines. I. Nutrient demand and vine performance. *Am. J. Enol. Vitic.* **40**, 85–90.

Conradie, W. J., and Saayman, D. (1989b). Effects of long-term nitrogen, phosphorus, and potassium fertilization on Chenin blanc vines. II. Leaf analyses and grape composition. *Am. J. Enol. Vitic.* **40**, 91–97.

Cook, J. A. (1966) Grape nutrition. *In* "Nutrition of Fruit Crops Temperate, Sub-tropical, Tropical" (N. F. Childers, ed.), pp. 777–812. Somerset Press, Somerville, New Jersey.

Cook, J. A., and Wheeler, D. W. (1976). Use of tissue analysis in viticulture. Soil and tissue testing in California. (H. M. Reisenauer, ed.), pp. 14–16. *Calif. Div. Agric. Sci. Bull. No. 1879.*

Cosmo, I., Comuzzi, A., and Polniselli, M. (1958). "Portinesti della Vite." Agricole, Bologna.

de Klerk, C. A., and Loubser, J. T. (1988). Relationship between grapevine roots and soil-borne pests. *In* "The Grapevine Root and Its Environment" (J. L. van Zyl, compiler), pp. 88–105. Tech. Commun. No. 215, Dept. Agric. Water Supply, Pretoria, South Africa.

Delp, C. J. (1980). Coping with resistance to plant disease. *Plant Dis.* **64**, 652–657.

Dimitriadis, E., and Williams, P. J. (1984). The development and use of a rapid analytical technique for estimation of free and potentially volatile monoterpene flavorants of grapes. *Am. J. Enol. Vitic.* **35**, 66–71.

Dubois, B., Jailloux, F., and Bulit, J. (1982). L'antagonisme microbien dans la lutte contre la pourriture grise de la vigne. *Bull. OEPP* **12**, 171–175.

Duran-Vila, N., Juárez, J., and Arregui, J. M. (1988). Production of viroid-free grapevines by shoot tip culture. *Am. J. Enol. Vitic.* **39**, 217–220.

Düring, H. (1984). Evidence for osmotic adjustment to drought in grapevines (*Vitis vinifera* L.). *Vitis* **23**, 1–10.

Düring, H. (1990). Stomatal adaptation of grapevine leaves to water stress. *In Proc. 5th Int. Symp. Grape Breeding, Sept. 12–16, 1989*, pp. 366–370. St. Martin, Pfalz, Germany (Special Issue of *Vitis*).

Duso, C. (1989). Role of the predatory mites *Amblyseius aberrans* (Oud.) *Typhlodromus pyri* Scheuten and *Amblyseius andersoni* (Chant) (Acari, Phytoseiidae) in vineyards. I. The effects of single and mixed phytoseiid population releases on spider mite densities (Acari, Tetranychidae). *J. Appl. Entomol.* **107**, 474–492.

Dutt, E. C., Olsen, M. W., and Stroehlein, J. L. (1986). Fight root rot in the border wine belt. *Wines Vines* **67**(3), 40–41.

Düzenli, S., and Ergenoğlu, F. (1991). Studies on the density of stomata of some *Vitis vinifera* L. varieties grafted on different rootstocks trained up various trellis systems. Doğa-Tr. *J. Aric. For.* **15**, 308–317.

Eichhorn, K. W., and Lorenz, D. H. (1977). Phänologische Entwicklongsstadien der Rebe. *Nachrichtenbl. Dtsch. Pflanzenschutzdienstes* (*Braunschweig*) **29**, 119–120.

Eifert, J., Varnai, M., and Szöke, L. (1982). Application of the EUF procedure in grape production. *Plant Soil* **64**, 105–113.

Eifert, J., Varnai, M., and Szöke, L. (1985). EUF-nutrient contents required for optimal nutrition of grapes. *Plant Soil* **83**, 183–189.

Eisenbarth, H. J. (1992). Der Einfluss unterschiedlicher Belastungen auf die Ertragsleistung der Rebe. *Dtsch. Weinbau* **47**, 18–22.

Elad, Y., and Zimand, G. (1992). Integration of biological and chemical control for grey mould. Recent advances in *Botrytis* research. *Proc. 10th Int. Botrytis Symp., Heraklion, Crete, Greece, Apr. 5–10, 1992* (K. Verhoeff, N. E. Malathrakis, and B. Williamson, eds.), pp. 272–276. Pudoc Sci. Publ., Wageningen, The Netherlands.

Elrick, D. E., and Clothier, B. E. (1990). Solute transport and leaching. *In* "Irrigation of Agronomic Crops" (B. A. Stewart and D. R. Nielsen, eds.), Agronomy Monogr. 30, pp. 93–126. Am. Soc. Agronomy, Crop Sci. Soc. Am., and Soil Sci. Soc. Am. Publ., Madison, Wisconsin.

Encheva, Kh., and Rankov, V. (1990). Effect of prolonged usage of some herbicides in vine plantation on the biological activity of the soil. *Soil Sci. Agrochem.* (*Pochvozn. Agrokhim.*) **25**, 66–73 (in Bulgarian).

Englert, W. D., and Maixner, M. (1988). Biologische Spinnmibenbekämpfung im Weinbau durch Schonung der Raubmilbe *Typhlodromus pyri*. Schonung and Förderung von Nützlingen. *Schriftenr. Bundesminist. Ernährung, Landwirtschaft Forsten, Reihe A: Angew. Wiss.* **365**, 300–306.

Ermolaev, A. A. (1990). Resistance of grape to phylloxera on sandy soils. *Agrokhimiya* **2**, 141–151 (in Russian).

Fabbri, A., Lambardi, M., and Sani, P. (1986). Treatments with CCC and GA3 on stock plants and rootings of cuttings of the grape rootstock 140 Ruggeri. *Am. J. Enol. Vitic.* **37**, 220–223.

Failla, O., Scienza, A., Stringari, G., and Falcetti, M. (1990). Potassium partitioning between leaves and clusters: Role of rootstock. *In Proc. 5th Int. Symp. Grape Breeding, Sept. 12–16, 1989*, pp. 187–196. St. Martin, Pfalz, Germany (Special Issue of *Vitis*).

Fardossi, A., Barna, J., Hepp, E., Mayer, Ch., and Wendelin, S. (1990). Einfluss von organischer Substanz auf die Nährstoffaufnahme durch die Weinrebe im Gefässversuch. *Mitt. Klosterneuberg* **40**, 60–67.

Fergusson-Kolmes, L., and Dennehy, T. J. (1991). Anything new under the sun? Not phylloxera biotypes. *Wines Vines* **72**(6), 51–56.

Fermaud, M., and Le Menn, R. (1989). Association of *Botrytis cinerea* with grape berry moth larvae. *Phytopathology* **79**, 651–656.

Ferreira, J. H. S. (1990). *In vitro* evaluation of epiphytic bacteria from table grapes for the suppression of *Botrytis cinerea*. *S. Afr. J. Enol. Vitic.* **11**, 38–41.

Filip, I., and Isac, G. (1989). Results concerning the efficiency of using the oophag wasp *Trichogramma embryophagum* HTG. In controlling *Lobesia botrana*, Den. & Schiff. *Ann Inst. Cercetări Pentru Viticult Vinificatie valea Călugărească, Bucarest* **12**, 195–202 (in Romanian).

Flaherty, D. L., Jensen, F. L., Kasimatis, A. N., Kido, H., and Moller, W. J. (1982). "Grape Pest Management," Publ. No. 4105. Cooperative Extension Univ. of California, Oakland.

Flaherty, D. L., Christensen, L. P., Lanini, W. T., Marois, J. J., Phillips, P. A., and Wilson, L. T. (1992). "Grape Pest Management," 2nd Ed., Publ. No. 3343. Division of Agriculture and Natural Resources, Univ. of California, Oakland.

Foott, J. H. (1987). A comparison of three methods of pruning Gewürztraminer. *Calif. Agric.* **41**(1), 9–12.

Foott, J. H., Ough, C. S., and Wolpert, J. A. (1989). Rootstock effects on wine grapes. *Calif. Agric.* **43**, (4), 27–29.

Freese, P. (1986). Here's a "close" look at vine spacing. *Wines Vines* **67**(4), 28–30.

Fregoni, M. (1985). Exigences d'élémente nutritifs en viticulture. *Bull. O.I.V.* **58**, 416–434.

Galet, P. (1971). "Précis d'Ampélographie." Dehan, Montpellier.

Galet, P. (1979). "A Practical Ampelography—Grapevine Identification" (translated and adapted by L. T. Morton). Cornell Univ. Press, Ithaca, New York.

Galet, P. (1991). "Précis de Pathologie Viticole." Dehan, Montpellier.

Gallander, J. F. (1983). Effect of grape maturity on the composition and quality of Ohio Vidal Blanc wines. *Am. J. Enol. Vitic.* **34**, 139–141.

Gargiulo, A. A. (1983). Woody T-budding of grapevines—Storage of bud shields instead of cuttings. *Am. J. Enol. Vitic.* **34**, 95–97.

Granett, J., Goheen, A. C., Lider, L. A., and White, J. J. (1987). Evaluation of grape rootstocks for resistance to Type A and Type B grape phylloxera. *Am. J. Enol. Vitic.* **38**, 298–300.

Gu, S., Lombard, P. B., and Price, S. F. (1991). Inflorescence necrosis induced by ammonium incubation in clusters of Pinot noir grapes. *Proc. Int. Symp. Nitrogen Grapes Wine, Seattle, WA, June, 1991* (J. M. Rantz, ed.), pp. 259–261. Am. Soc. Enol. Vitic., Davis, California.

Guennelon, R., Habib, R., and Cockborn, A. M. (1979). Aspects particuliers concernant la disponibilité de N, P et K en irrigation localisée fertilisante sur arbres fruitiers. *In* "Seminaires sur l'Irrigation Localisée I," pp. 21–34. L'Institut d'Agronomie de l'Université de Bologne, Italy.

Hagan, R. M. (1955). Factors affecting soil moisture–plant growth relations. *Rep. 14th Int. Hortic. Congr., The Hague*, pp. 82–102.

Harris, J. M., Kriedemann, P. E., and Possingham, J. V. (1968). Anatomical aspects of grape berry development. *Vitis* **7**, 106–119.

Howell, G. S. (1987). *Vitis* rootstocks. *In* "Rootstocks for Fruit Crops" (R. C. Rom and R. F. Carlson, eds.), pp. 451–472. Wiley, New York.

Huang, Z., and Ough, C. S. (1989). Effect of vineyard locations, varieties, and rootstocks on the juice amino acid composition of several cultivars. *Am. J. Enol. Vitic.* **40**, 135–139.

Hubáčkova, M., and Hubáček, V. (1984). Frost resistance of grapevine buds on different rootstocks. *Vinohrad* **22**, 55–56 (in Russian).

Huglin, P. (1958). Recherches sur les bourgeons de la vigne. Initiation florale et développement végétatif. Ph.D. Thesis, Univ. of Strasbourg, France.

Huglin, P. (1986). "Biologie et Ecologie de la Vigne." Éditions Payot Lausanne, Technique & Documentation, Paris.

Hunter, J. J., and Le Roux, D. J. (1992). The effect of partial defoliation on development and distribution of roots of *Vitis vinifera* L. cv. Cabernet Sauvignon grafted onto rootstock 99 Richter. *Am. J. Enol. Vitic.* **43**, 71–78.

Huss, B., Tinland, B., Paulus, F., Walter, B., and Otten, L. (1990). Functional analysis of complex oncogene arrangement in biotype III *Agrobacterium tumefaciens*. *Plant Mol. Biol.* **14**, 173–186.

Jenkins, A. (1991). Review of production techniques for organic vineyards. *Austr. Grapegrower Winemaker* **328**, 133, 135–138, and 140–141.

Jordan, T. D., Pool, R. M., Zabadal, T. J., and Tomkins, J. P. (1981). "Cultural Practices for Commercial Vineyards." Miscellaneous Bull. 111, New York State College of Agriculture, Cornell Univ., Ithaca, New York.

Karban, R., English-Loeb, G., and Verdegall, P. (1991). Vaccinating grapevines against spider mites. *Calif. Agric.* **45**(1), 19–21.

King, P. D., and Rilling, G. (1991). Further evidence of phylloxera biotypes: Variations in the tolerance of mature grapevine roots related to the geographical origin of the insect. *Vitis* **30**, 233–244.

Kliewer, W. M. (1991). Methods for determining the nitrogen status of vineyards. *Proc. Int. Symp. Nitrogen Grapes Wine, Seattle, WA, June, 1991* (J. M. Rantz, ed.), pp. 133–147. Am. Soc. Enol. Vitic., Davis, California.

Koblet, W. (1987). Effectiveness of shoot topping and leaf removal as a means of improving quality. *Acta Hortic.* **206**, 141–156.

Kobriger, J. M., Kliewer, W. M., and Lagier, S. T. (1984). Effects of wind on water relations of several grapevine cultivars. *Am. J. Enol. Vitic.* **35**, 164–169.

Koenig, R., An, D., and Burgermeister, W. (1988). The use of filter hybridization techniques for the identification, differentiation and classification of plant viruses. *J. Virol. Methods* **19**, 57–68.

Kotzé, W. A. G. (1973). The influence of aluminum on plant growth. *Decid. Fruit Grow.* **23**, 20–22.

Kriedemann, P. E., and Smart, R. E. (1971). Effects of irradiance, temperature, and leaf water potential on photosynthesis of vine leaves. *Photosynthetica* **5**, 6–15.

Krogh, P., and Carlton, W. W. (1982). Nontoxicity of *Botrytis cinerea* strains used in wine production. *Grape Wine Centennial Symp. Proc. 1980*, pp. 182–183. Univ. of California, Davis.

Kurl, W. R., and Mowbray, G. H. (1984). Grapes. *In* "Handbook of Plant Cell Culture" (W. R. Sharp, D. A. Evans, P. V. Ammirato, and Y. Yamada, eds.), Vol. 2, pp. 396–434. Macmillan, New York.

Lang, A., and Thorpe, M. R. (1989). Xylem, phloem and transpirational flow in a grape: Application of a technique for measuring the volume of attached fruits to high resolution using Archimedes' principle. *J. Exp. Bot.* **40**, 1069–1078.

Larcher, W., Häckel, H., and Sakai, A. (1985). "Handbuch der Pflanzenkrankheiten," Vol. 7. Parey, Berlin.

Löhnertz, O. (1991). Soil nitrogen and uptake of nitrogen in grapevines. *Proc. Int. Symp. Nitrogen Grapes Wine, Seattle, WA, June 18–19, 1991* (J. A. Rantz, ed.), pp. 1–11. Am. Soc. Enol. Vitic., Davis, California.

Löhnertz, O., Schaller, K., and Mengel, K. (1989). Nährstaffdynamik in Reben. III. Mitteilung: Stickstoffkonzwntration und Verlauf der Aufnahme in der Vegetation. *Wein-Wiss.* **44**, 192–204.

Lombard, P., Price, S., Wilson, W., and Watson, B. (1988). Grass cover crops in vineyards. *Proc. 2nd Int. Symp. Cool Climate Vitic. Oenol., Jan. 11–15, 1988, Auckland, N.Z.* (R. E. Smart, S. B. Thornton, S. B. Rodriguez, and J. E. Young, eds.), pp. 152–155. New Zealand Soc. Vitic. Oenol., Auckland, New Zealand.

Loubser, J. T., van Aarde, I. M. F., and Höpper, G. F. J. (1992). Assessing the control potential of aldicarb against grapevine phylloxera. *S. Afr. J. Enol. Vitic.* **13**, 84–86.

Loué, A., and Boulay, N. (1984). Effets des cépages et des portegreffes sur les diagnostics de nutrition minérale sur la vigne. *6th Colloq. Int. Nutrition Plantes, P. Martin-Prevel Montpellier*, pp. 357–364. France.

McCarthy, M. G., and Cirami, R. M. (1990). Minimal pruning effects on the performance of selections of four *Vitis vinifera* cultivars. *Vitis* **29**, 85–96.

McCarthy, M. G., and Coombe, B. G. (1985). Water status and winegrape quality. *Acta Hortic.* **171**, 447–456.

McCarthy, M. G., and Nicholas, P. R. (1989). Terpenes—A new measure of grape quality? *Aust. Grapegrower Winemaker* **297**, 10.

Marais, J., van Wyk, C. J., and Rapp, A. (1992a). Effect of sunlight and shade on norisoprenoid levels in maturing Weisser Riesling and Chenin blanc grapes and Weisser Riesling wines. *S. Afr. J. Enol. Vitic.* **13**, 23–32.

Marais, J., van Wyk, C. J., and Rapp, A. (1992b). Effect of storage time, temperature and region on the levels of 1,1,6-trimethyl-1,2-dihydronaphthalene and other volatiles, and on quality of Weisser Riesling wines. *S. Afr. J. Enol. Vitic.* **13**, 33–44.

Marois, J. J., Nelson, J. K., Morrison, J. C., Lile, L. S., and Bledsoe, A. M. (1986). The influence of berry contact within grape clusters on the development of *Botrytis cinerea* and epicuticular wax. *Am. J. Enol. Vitic.* **37**, 293–295.

Marois, J. J., Bledsoe, A. M., Bostock, R. M., and Gubler, W. D. (1987). Effects of spray adjuvants on development of *Botrytis cinerea* on *Vitis vinifera* berries. *Phytopathology* **77**, 1148–1152.

Masa, A. (1989). Affinité biochimique entre le greffon du cultivar Albariño (*Vitis vinifera* L.) et différents porte-greffes. *Connaiss. Vigne Vin* **23**, 207–213.

Matthews, M. A., and Anderson, M. M. (1989). Reproductive development in grape (*Vitis vinifera* L.): Responses to seasonal water deficits. *Am. J. Enol. Vitic.* **40**, 52–59.

Matthews, M. A., Anderson, M. M., and Schultz, H. R. (1987). Phenological and growth responses to early and late season water deficiency in Cabernet Franc. *Vitis* **26**, 147–160.

Mengel, K., Bubl, W., and Scherer, H. W. (1984). Iron distribution in vine leaves with HCO_3^- induced chlorosis. *J. Plant Nutr.* **7**, 715–724.

Meyer, J. L., Snyder, M. J., Valenzuela, L. H., Harris, A., and Strohman, R. (1991). Liquid polymers keep drip irrigation lines from clogging. *Calif. Agric.* **45**(1), 24–25.

Milne, D. (1988). Vine responses to soil water. *Proc. 2nd Int. Symp. Cool Climate Vitic. Oenol., Jan. 11–15, 1988, Auckland, N.Z.* (R. E. Smart, S. B. Thornton, S. B. Rodriguez, and J. E. Young, eds.), pp. 57–58. New Zealand Soc. Vitic. Oenol., Auckland, New Zealand.

Monnier, M., Faure, O., and Sigogneau, A. (1990). Somatic embryogenesis in *Vitis*. *Bull. Soc. Bot. Fr.* **137**, 35–44.

Morris, J. R., Sims, C. A., Bourque, J. E., and Oakes, J. L. (1984). Influence of training system, pruning severity, and spur length on yield and quality of six French–American hybrid grape cultivars. *Am. J. Enol. Vitic.* **35**, 23–27.

Mottard, G., Nespoulus, J., and Marcout, P. (1963). Les port-greffes de la vigne. *Bull. Inf. Agric. Paris*, 182.

Mullins, M. G. (1990). Applications of tissue culture to the genetic improvement of grapevines. *Proc. 5th Int. Symp. Grape Breeding, Sept. 12–16, 1989*, pp. 399–407. St. Martin, Pfalz, Germany (Special Issue of *Vitis*).

Munkvold, G., and Marois, J. J. (1990). Relationship between xylem wound response in grapevines and susceptibility to *Eutypa lata*. *Phytopathology* **80**, 973 (Abstract).

Mur, G., and Branas, J. (1991). La maladie de vieux bois: Apoplexie et eutypiose. *Prog. Agric. Vitic.* **5,** 108–114.

Nicholson, C. (1990). 'Birebent' graft is hailed as a breakthrough in viticulture. *Wines Vines* **71**(9), 16–18.

Northover, J. (1987). Infection sites and fungicidal prevention of *Botrytis cinerea* bunch rot of grapes in Ontario. *Can. J. Plant Pathol.* **9,** 129–136.

O'Conner, B. P. (ed.) (1987). "New Zealand Agrichemical Manual," 2nd Ed. Agpress/Novasearch, Wellington/Manawatu, New Zealand.

Olsen, R. A., and Kurtz, L. T. (1982). Crop nitrogen requirements, utilization and fertilization. *In* "Nitrogen in Agricultural Soils" (F. J. Stevenson, ed.), Agronomy Ser. 22. Am. Soc. Agronomy, Crop. Sci. Soc., and Soil Sci. Soc. Am., Madison, Wisconsin.

Ophel, K., and Kerr, A. (1990). *Agrobacterium vitis* sp. nov. for strains of *Agrobacterium* biovar 3 from grapevines. *Int. J. Syst. Bacteriol.* **40,** 236–241.

Ough, C. S. (1991). Influence of nitrogen compounds in grapes on ethyl carbamate formation in wines. *Proc. Int. Symp. Nitrogen Grapes Wine, Seattle, WA, June, 1991* (J. M Rantz, ed.), pp. 165–171. Am. Soc. Enol. Vitic., Davis, California.

Ough, C. S., and Amerine, M. A. (1988). "Methods for Analysis of Musts and Wines." Wiley, New York.

Ough, C. S., and Berg, H. W. (1979). Powdery mildew sensory effect on wine. *Am. J. Enol. Vitic.* **30,** 321.

Pàstena, B. (1972). "Trattato di Viticultura Italiana." Palermo, Italy.

Pearson, R. C., and Gadoury, D. M. (1987). Cleistothecia, the source of primary inoculum for grape powdery mildew in New York. *Phytopathology* **77,** 1509–1608.

Pearson, R. C., and Goheen, A. C. (eds.) (1988). "Compendium of Grape Diseases." APS Press, St. Paul, Minnesota.

Pearson, R. C., Pool, R. M., Gonsalves, D., and Goffinet, M C. (1985). Occurrence of flavescence dorée-like symptoms on 'White Riesling' grapevines in New York, USA. *Phytopathol. Mediterr.* **24,** 82–87.

Pearson, R. C., Pool, R. M., and Jubb, G. L., Jr. (1988). Pesticide toxicity. *In* "Compendium of Grape Diseases" (R. C. Pearson and A. C. Goheen, eds.), pp. 69–71. APS Press, St. Paul, Minnesota.

Peech, M. (1961). Lime requirements vs soil pH curves for soils of New York State [mimeographed]. Dept. of Agronomy, Cornell Univ., Ithaca, New York.

Perret, P. (1992). "Adaptive Nitrogen-Management" as a tool for the optimisation of N-availability in vineyards. *3rd Int. Cool Climate Symp., June 8–12, 1992, Geisenheim, Germany.*

Perret, P., and Koblet, W. (1984). Soil compaction induced iron-chlorosis in grape vineyards: Presumed involvement of exogenous soil ethylene. *J. Plant Nutr.* **7,** 533–539.

Perret, P., and Weissenbach, P. (1991). Schwermetalleintrag in den Rebberg aus organischen Düngern. *Schweiz. Z. Obst- Weinbau* **127,** 124–130.

Peterson, J., and Smart, R. E. (1975). Foliage removal effects on Shiraz grapevines. *Am. J. Enol. Vitic.* **26,** 119–124.

Pezet, R., and Pont, V. (1986). Infection florale et latence de *Botrytis cinerea* dans les grappes de *Vitis vinifera* (var. Gamay). *Rev. Suisse Vitic. Arboric. Hortic.* **18,** 317–322.

Pongrácz, D. P. (1978). "Practical Viticulture." David Philip, Cape Town, South Africa.

Pongrácz, D. P. (1983). "Rootstocks for Grape-vines." David Philip, Cape Town, South Africa.

Pool, R. M., Kasimatis, A. N., and Christensen, L. P. (1988). Effects of cultural practices on disease. *In* "Compendium of Grape Diseases" (R. C. Pearson and A. C. Goheen, eds.), pp. 72–73. APS Press, St. Paul, Minnesota.

Pouget, R. (1987). Usefulness of rootstocks for controlling vine vigour and improving wine quality. *Acta Hortic.* **206,** 109–118.

Prince, J. P., Davis, R. E., Wolf, T. K., Lee, I. -M., Mogen, B. D., Dally, E. L., Bertaccini, A., Credi, R., and Barba, M. (1993). Molecular detection of diverse mycoplasmalike organisms (MLOs) associated with grapevine yellows and their classification with aster yellows, X-disease, and elm yellows MLOs. *Phytopathology* **83,** 1130–1137.

Rabe, E. (1990). Stress physiology: The functional significance of the accumulation of nitrogen-containing compounds. *J. Hortic. Sci.* **65,** 231–243.

Redl, H., and Bauer, K. (1990). Prüfung alternativer Mittel auf deren Wirkung gegenüber *Plasmopara viticola* und deren Einfluβ auf das Ertagsgeschehen unter österreichischen Weinbaubedingungen. *Mitt. Klosterneuburg, Rebe Wein, Obstbau Früchteverwertung* **40,** 134–138.

Redmond, J. C., Marois, J. J., and MacDonald, J. D. (1987). Biological control of *Botrytis cinerea* on roses with epiphytic microorganisms. *Plant Dis.* **71,** 799–802.

Reynolds, A. G., Pool, R. M., and Mattick, L. R. (1986). Effect of shoot density and crop control on growth, yield, fruit composition, and wine quality of 'Seyval Blanc' grapes. *J. Am. Soc. Hortic. Sci.* **111,** 55–63.

Reynolds, A. G., and Wardle, D. A. (1988). Canopy microclimate of Gewürztraminer and monoterpene levels. *Proc. 2nd Int. Symp. Cool Climate Vitic. Oenol., Jan. 11–15, 1988, Auckland, N.Z.* (R. E. Smart, S. B. Thornton, S. B. Rodriguez, and J. E. Young, eds.), pp. 116–122. New Zealand Soc. Vitic. Oenol., Auckland, New Zealand.

Ribéreau-Gayon, J., and Peynaud, E. (1971). "Traité d'Ampélologie. Sciences et Technques du Vin"—Tome 1 and II. Dunod, Paris.

Ribéreau-Gayon, J., Ribéreau-Gayon, P., and Seguin, G. (1980). *Botrytis cinerea* in enology. *In* "The Biology of *Botrytis*" (J. R. Coley-Smith, K. Verhoeff, and W. R. Jarvis, eds.), pp. 251–274. Academic Press, London.

Rod, P. (1977). Observations sur le ruissellement dans le vignoble vaudois. *Bulletin Assoc. Romande Protect. Eau Air* **82,** 41–45.

Roggero, J. P., Coen, S., and Ragonnet, B. (1986). High performance liquid chromatography survey on changes in pigment content in ripening grapes of Syrah. An approach to anthocyanin metabolism. *Am. J. Enol. Vitic.* **37,** 77–83.

Rosciglione, B., and Gugerli, P. (1989). Transmission of grapevine leafroll disease and an associated closterovirus to healthy grapevine by the mealybug *Planococcus ficus* Signoret. *Proc. 9th Meet. Int. Council Virus Grapevine, Kiryat Anavim, Israel, Sept. 6–11, 1987,* pp. 67–69.

Ruhl, E. H. (1989). Uptake and distribution of potassium by grapevine rootstocks and its implication for grape juice pH of scion varieties. *Aust. J. Exp. Agric.* **29,** 707–712.

Saayman, D. (1981). Wingerdvoeding. *In* "Wingerdbou in Suid-Afrika" (J. Burger and J. Deist, eds.), pp. 343–383. Oenol. Vitic. Res. Inst., Stellenbosch, South Africa.

Saayman, D. (1982). Soil preparation studies: II. The effect of depth and method of soil preparation and organic material on the performance of *Vitis vinifera* (var. Colombar) on Clovelly/Hutton soil. *S. Afr. J. Enol. Vitic.* **3,** 61–74.

Saayman, D., and van Huyssteen, L. (1980). Soil preparation studies: I. The effect of depth and method of soil preparation and of organic material on the performance of *Vitis vinifera* (var. Chenin blanc) on Hutton/Sterkspruit soil *S. Afr. J. Enol. Vitic.* **1,** 107–121.

Safran, R. M., Bravdo, B., and Bernstein, Z. (1975). L'irrigation de la vigne par goutte à goutte. *Bull. O.I.V.* **531,** 405–429.

Sall, M. A. (1982). The application of mathematical models to grape disease management. *Grape Wine Centennial Symp. Proc. 1980,* pp. 36–38. Univ. of California, Davis.

San Romáo, M. V., and Silva Alemáo, M. F. (1986). Premières observations sur les activitiés enzymatiques developpées dans le mout de raisins par certains champignons filamenteux. *Connaiss. Vigne Vin* 20, 39–52.

Schepers, J. S., and Saint-Fort, R. (1988). Comparison of the potentially mineralized nitrogen using electroultrafiltration and four other procedures. *Proc. 3rd Int. EUR Symp., Mannheim, Germany*, pp. 441–450.

Schneider, C. (1989). Introduction à l'écophysiologie viticole. Application aux systèmes de conduite. *Bull. O.I.V.* 62, 498–515.

Scholl, W., and Enkelmann, R. (1984). Zum Kupfergehalt von Weinbergeböden. *Landwirtsch. Focschung* 37, 286–297.

Schroth, M. N., McCain, A. H., Foott, J. H., and Huisman, O. C. (1988). Reduction in yield and vigor of grapevine caused by Crown Gall Disease. *Plant Dis.* 72, 241–245.

Schwenk, S., Altmayer, B., and Eichhorn, K. W. (1989). Untersuchungen zur Bedeutung toxischer Stoffwechselprodukte des Pilzes *Trichoderma roseum* Link ex Fr. für den Weinbau. *z. Lebens. unters Forsch.* 188, 527–530.

Scienza, A., and Boselli, M. (1981). Fréquence et caractéristiques biométriques des stomates de certains porte-greffes de vigne. *Vitis* 20, 281–292.

Scienza, A., Failla, O., and Romano, F. (1986). Untersuchungen zur sortenspezifischen Mineralstoffaufnahme bei Reben. *Vitis* 25, 160–168.

Sengonca, C., and Leisse, N. (1989). Enhancement of the egg parasite *Trichogramma semblidis*, AURIV., Hym., Trichogrammatidae, for control of both grapevine moth species in the Ahr valley. *J. Appl. Entomol.* 107, 41–45 (Abstract).

Shaulis, N. (1982). Responses of grapevines and grapes to spacing of and within canopies. *Grape Wine Centennial Symp. Proc.*, pp. 353–361. University of California, Davis.

Shaulis, N., and Smart, R. (1974). Grapevine canopies: Management, microclimate and yield responses. *Proc. 19th Int. Hortic. Cong., Sept. 1974*. pp. 254–265. Warsaw, Poland.

Shaulis, N., Amberg, H., and Crowe, D. (1966). Response of Concord grapes to light, exposure and Geneva Double Curtain training. *Proc. Am. Soc. Hortic. Sci.* 89, 268–280.

Shoseyov, O. (1983). Out of season grape production of one year old cuttings. M.Sc. Thesis. Faculty of Agriculture, Hebrew Univ. Jerusalem, Rehovot (in Hebrew, with English summary).

Singleton, V. L. (1972). Effects on red wine quality of removing juice before fermentation to stimulate variation in berry size. *Am. J. Enol. Vitic.* 23, 106–113.

Skinner, P. W., and Matthews, M. A. (1989). Reproductive development in grape (*Vitis vinifera*) under phosphorus-limited conditions. *Sci. Hortic.* 38, 49–60.

Skinner, P. W., and Matthews, M. A. (1990). A novel interaction of magnesium translocation with the supply of phosphorus to roots of grapevine. *Vitis vinifera. Plant Cell Environ.* 13, 82–926.

Skinner, P. W., Cook, J. A., and Matthews, M. A. (1988). Responses of grapevine cvs. Chenin blanc and Chardonnay to phosphorus fertilizer applications under phosphorus-limited soil conditions. *Vitis* 27, 95–109.

Smart, R. E. (1973). Sunlight interception by vineyards. *Am. J. Enol. Vitic.* 24, 141–147.

Smart, R. E. (1974). Aspects of water relations of the grapevine (*Vitis vinifera*). *Am. J. Enol. Vitic.* 25, 84–91.

Smart, R. E. (1982). Vine manipulation to improve wine grape quality. *Grape Wine Centennial Symp. Proc.*, pp. 362–375. University of California, Davis.

Smart, R. E. (1988). Shoot spacing and canopy light microclimate. *Am. J. Enol. Vitic.* 39, 325–333.

Smart, R. E., and Coombe, B. G. (1983). Water relations of grape-

vines. *In* "Water Deficits and Plant Growth" (T. T. Kozlowski, ed.), pp. 137–195. Academic Press, New York.

Smart, R. E., Robinson, J. B., Due, G. R., and Brien, C. J. (1985). Canopy microclimate modification for the cultivar Shiraz II. Effects on must and wine composition. *Vitis* 24, 119–128.

Smart, R. E., and Smith, S. M. (1988). Canopy management: Identifying the problems and practical solutions. pp. 109–115. *Proc. 2nd Int. Symp. Cool Climate Vitic. Oenol., Jan. 11–15. Auckland, N.Z.*, (R. E. Smart, S. B., Thornton, S. B. Rodriguez, and J. E. Young, eds.). pp. 109–115. New Zealand Soc. Vitic. Oenol., Auckland, New Zealand.

Smart, R. E., Dick, J. K., Gravett, I. M., and Fisher, B. M. (1990a). Canopy management to improve grape yield and wine quality—Principles and practices. *S. Afr. J. Enol. Vitic.* 11, 3–18.

Smart, R. E., Dick, J., and Gravett, I. (1990b). Shoot devigoration by natural means. *Proc. 7th. Austral. Wine Indust. Tech. Conf., Aug. 13–17, 1989*. Adelaide, SA., (P. J. Williams *et al.*, eds.). pp. 58–65. Winetitles, Adelaide, Australia.

Smart, R. E., Dick, J. K., and Smith, S. M. (1990c). A trellis for vigorous vineyards. *Wines Vines* 71(6), 32–36.

Smith, S., Codrington, I. C., Robertson, M., and Smart, R. (1988). Viticultural and oenological implications of leaf removal for New Zealand vineyards. *Proc. 2nd Int. Symp. Cool Climate Vitic. Oenol., Jan. 11–15, 1988, Auckland, N.Z.* (R. E. Smart, S. B. Thornton, S. B. Rodriguez, and J. E. Young, eds.), pp. 127–133. New Zealand Soc. Vitic. Oenol., Auckland, New Zealand.

Solari, C., Silvestroni, O., Giudici, P., and Intrieri, C. (1988). Influence of topping on juice composition of Sangiovese grapevines (*V. vinifera* L.). *Proc. 2nd Int. Symp. Cool Climate Vitic. Oenol., Jan. 11–15, 1988, Auckland, N.Z.* (R. E. Smart, S. B. Thornton, S. B. Rodriguez, and J. E. Young, eds.), pp. 147–151. New Zealand Soc. Vitic. Oenol., Auckland, New Zealand.

Somers, T. C., and Pocock, K. F. (1986). Phenolic harvest criteria for red vinification. *Aust. Grapegrower Winemaker* 256, 24–30.

Stapleton, J. J., Barnett, W. W., Marois, J. J., and Gubler, W. D. (1990). Leaf removal for pest management in wine grapes. *Calif. Agric.* 44(4), 15–17.

Staub, T. (1991). Fungicide resistance: Practical experience with antiresistance strategies and the role of integrated use. *Annu. Rev. Phytopathol.* 29, 421–442.

Steinhauer, R. E., Lopez, R., and Pickering, W. (1980). Comparison of four methods of variety conversion in established vineyards. *Am. J. Enol. Vitic.* 31, 261–264.

Stern, V. M., and Federici, B. A. (1990). Granulosis virus: Biological control of western grapeleaf skeletonizer. *Calif. Agric.* 44(3), 21–22.

Stevenson, D. S., Neison, G. H., and Cornelsen, A. (1986). The effect of woven plastic mulch, herbicides, grass sod, and nitrogen on 'Foch' grapes under irrigation. *HortScience* 21, 439–441.

Stobbs, L. W., Potter, J. W., Killins, R., and van Schagen, J. G. (1988). Influence of grapevine understock in infection of de Chaunac scion by tomato ringspot virus. *Can. J. Plant Pathol.* 10, 228–231.

Stockle, C. O., and Dugas, W. A. (1992). Evaluating canopy temperature-based indices for irrigation scheduling. *Irrig. Sci.* 13, 31–38.

Summers, P. (1985). Managing cover. It requires care just like any other crop. *Calif. Grape Grower* (June), 26–27.

Szychowski, J. A., Goheen, A. C., and Semancik, J. S. (1988). Mechanical transmission and rootstock reservoirs as factors in the widespread distribution of viroids in grapevines. *Am. J. Enol. Vitic.* 39, 213–216.

Thomas, C. S., Marois, J. J., and English, J. T. (1988). The effects of wind speed, temperature, and relative humidity on development of aerial mycelium and conidia of *Botrytis cinera* on grape. *Phytopathology* **78**, 260–265.

Ureta, F., Boidron, J. N., and Bouard, J. (1982). Influence of "dessèchement de la rafle" on wine quality. *Grape Wine Centennial Symp. Proc. 1980*, pp. 284–286. Univ. of California, Davis.

van den Ende, B. (1984). The Tatura trellis. A system of growing grapevines for early and high production. *Am. J. Enol. Vitic.* **35**, 82–87.

van den Ende, B., Jerie, P. H., and Chalmers, D. J. (1986). Training Sultana vines on the Tatura trellis for early and high production. *Am. J. Enol. Vitic.* **37**, 304–305. .

van der Westhuizen, J. H. (1980). The effect of black plastic mulch on growth, production and root development of Chenin blanc vines under dryland conditions. *S. Afr. J. Enol. Vitic.* **1**, 1–6.

van Huyssteen, L. (1988a). Soil preparation and grapevine root distribution—A qualitative and quantitative assessment. *In* "The Grapevine Root and Its Environment" (J. L. van Zyl, compiler), Tech. Commun. No. 215, pp. 1–15. Dept. Agric. Water Supply, Pretoria, South Africa.

van Huyssteen, L. (1988b). Grapevine root growth in response to soil tillage and root pruning practices. *In* "The Grapevine Root and Its Environment" (J. L. van Zyl, compiler), Tech. Commun. No. 215, pp. 44–56. Dept. Agric. Water Supply, Pretoria, South Africa.

van Slyke, L. L. (1932). "Fertilizers and Crop Production." Orange Judd, New York.

van Zyl, J. L. (1984). Response of Colombar grapevines to irrigation as regards quality aspects and growth. *S. Afr. Enol. Vitic.* **5**, 19–28.

van Zyl, J. L. (1988). Response of grapevine roots to soil water regimes and irrigation systems. *In* "The Grapevine Root and Its Environment" (J. L. van Zyl, compiler), pp. 30–43. Tech. Commun. No. 215, Dept. Agric. Water Supply, Pretoria, South Africa.

van Zyl, J. L., and van Huyssteen, L. (1980). Comparative studies on wine grapes on different trellising systems: I. Consumptive water use. *S. Afr. J. Enol. Vitic.* **1**, 7–14.

van Zyl, J. L., and van Huyssteen, L. (1987). Root pruning, *Decid. Fruit Grow.* **37**(1), 20–25.

Walker, M. A., Wolpert, J. A., Vilas, E. P., Goheen, A. C., and Lider, L. A. (1989). Resistant rootstocks may control fanleaf degeneration of grapevines. *Calif. Agric.* **43**(2), 1314.

Walter, B. (1991). Sélection de la vigne: Le dépistage des maladies de la vigne transmissibles par les bois et plants. *Bull. O.I.V.* **64**, 691–701.

Wapshere, A. J., and Helm, K. F. (1987). Phylloxera and *Vitis*: An experimentally testable coevolutionary hypothesis. *Am. J. Enol. Vitic.* **38**, 216–222.

Weaver, R. J. (1976). "Grape Growing." Wiley, New York.

Weinstein, L. H. (1984). Effects of air pollution on grapevines. *Vitis* **23**, 274–303.

Weiss, A., and Allen, L. H., Jr. (1976a). Air-flow patterns in vineyard rows. *Agric. Meteorol.* **16**, 329–342.

Weiss, A., and Allen, L. H., Jr. (1976b). Vertical and horizontal air flow above rows of a vineyard. *Agric. Meteorol.* **16**, 433–452.

Whiting, J. R. (1988). Influences of rootstocks on yield, juice composition and growth of Chardonnay. *Proc. 2nd Int. Symp. Cool Climate Vitic. Oenol., Jan. 11–15, 1988, Auckland, N.Z.* (R. E. Smart, S. B. Thornton, S. B. Rodriguez, and J. E. Young, eds.), pp. 48–50. New Zealand Soc. Vitic. Oenol., Auckland, New Zealand.

Williams, L. E., and Matthews, M. A. (1990). Grapevines. *In* "Irrigation of Agricultural Crops" (B. A. Stewart and D. R. Nielsen, eds.), Agronomy Monogr. No. 30, pp. 1019–1055. American Society of Agronomy, Madison, Wisconsin.

Williams, P. L., and Antcliff, A. J. (1984). Successful propagation of *Vitis berlandieri* and *Vitis cinerea* from hardwood cuttings. *Am. J. Enol. Vitic.* **35**, 75–76.

Williams, R. N., and Granett, J. (1988). Phylloxera. *In* "Compendium of Grape Diseases." (R. C. Pearson and A. C. Goheen, eds.), p. 63. APS Press, St. Paul, Minnesota.

Winkler, A. J., Cook, J. A., Kliewer, W. M., and Lider, L. A. (1974). "General Viticulture." Univ. of California Press, Berkeley.

Woodham, R. C., Emmett, R. W., and Fletcher, G. G. (1984). Effects of thermotherapy and virus status on yield, annual growth and grape composition of Sultana. *Vitis* **23**, 268–273.

Yamakawa, Y., and Moriya, M. (1983). Ripening changes in some constituents of virus-free Cabernet franc grape berries. *J. Jpn. Soc. Hortic. Sci.* **52**, 16–21.

Younger, W. (1966). "Gods, Men and Wine." George Rainbird, London.

Zoecklein, B., Fugelsang, K. C., Gump, B. H., and Nury, F. S. (1990). "Production Wine Analysis." Van Nostrand-Rheinhold, New York.

5

Site Selection and Climate

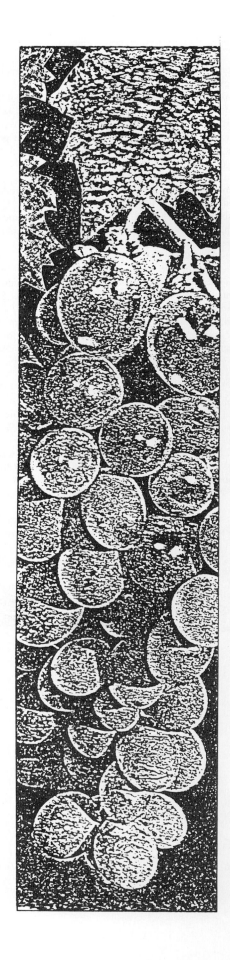

Introduction

The view that the vine needs to suffer to produce fine quality fruit is long established in wine folklore. If interpreted as restrained grapevine vigor, open canopy development, and fruit yield consistent with capacity, the concept of vine suffering has more than just an element of truth. Regions where local conditions have tended to produce these results have come to be noted for their better quality wines. Empirically, grape growers and winemakers came to recognize vineyard practices that enhanced these natural tendencies. It also was realized that some sites possessed undesirable properties that viticultural practice could not offset. These beneficial and detrimental aspects of soil, topography, microclimate, and macroclimate now form the basis for choosing favorable viticultural sites. This knowledge allows grape growers not only to produce better grapes in traditional wine-producing regions, but also to expand production into new viticultural areas.

Probably the first feature recognized as favoring finer grape production was limited soil fertility. Poor soils restrict vegetative growth and permit a higher proportion of photosynthate to be directed toward fruit development and ripening. This is particularly important as flavor development tends to develop near the end of fruit ripening. In addition, many low nutrient soils are highly porous. The latter feature improves drainage and favors rapid warming of, and heat radiation from the soil. This, in turn, can improve the microclimate around the vine and delay or minimize frost severity. Excellent drainage also promotes early spring growth and limits fruit cracking following heavy rains. Finally, vines grown on well-drained soil develop fewer micro- and macrofissures in the berry skin that favor the growth and penetration of parasitic fungi and bacteria.

Another feature recognized as beneficial is medium to low rainfall. Dry conditions enhance the resistance of vines to several pathogens. In Europe, most of the southern regions receive most of their precipitation during the winter. Thus, sufficient moisture is available for early growth. During the summer, however, the vines may be exposed to varying degrees of water stress. It is now known that avoidance of water stress is most important in the spring and early summer, up to the beginning of ripening (*véraison*). Subsequently, restricted water availability tends to improve fruit quality and advance ripening. With limited vegetative growth, more nutrients are available for berry ripening. Because grapevines tend to root deeply, they may avoid serious water stress on deep soil, even during periods of drought. The ability to root deeply probably has helped limit the development of nutrient deficiencies on impoverished soils by locating subsoil nutrient sources. Grapevines are one of the few crops to do well on relatively poor soils.

Possibly during the move of grape culture into central Europe during Roman times, it became apparent that growing cultivars near the northern limit for fruit ripening was beneficial to quality. Cooler conditions are now known to retain fruit acidity, which improves the microbial and color stability of wines. The cool conditions also appear to favor the development and retention of grape aroma compounds. In addition, cool climates have values in relation to winemaking and storage that are independent of the effects on viticulture. However, cultivating grapevines at the northern limit of the growing range heightens the risks of crop failure due to the shorter growing season and increased likelihood of frost damage. This is probably how the benefits of sites on hillside slopes and/or in proximity to large bodies of water were discovered.

Soil Influences

Of climatic influences, soil type generally is the least significant factor affecting grape and wine quality

(Rankine *et al.*, 1971) or is poorly correlated with wine characteristics (Morlat *et al.*, 1983). Soil influences tend to be expressed indirectly through factors such as heat retention, water-holding capacity, and nutritional status. For example, color and textural composition affect heat absorption and, thereby, fruit ripening and frost protection. Consequently, when discussing soil, and its effects on grapevine growth, it is important to distinguish between the various physicochemical properties of soil: the texture composition, aggregate structure, nutrient availability, organic content, effective depth, pH, drainage, and water availability.

Geologic Origin

The geologic origin of the parental material of the soil has little direct influence on grape quality. Fine wines are produced from grapes grown on soils derived from all three basic rock types: igneous (derived from molten magma, i.e., granite), sedimentary (derived from consolidated sediments, i.e., shale, chalk, limestone), or metamorphic (derived from transformed sedimentary rock, i.e., slate, quartzite, schist). Examples of wine regions where the soils are primarily derived from a single rock type are Champagne and Chablis (chalk), Jerez (limestone), and Porto and Mosel (schist). However, equally famous regions have soils derived from a mix of rock types, namely, Rheingau, Bordeaux, and Beaujolais (Wallace, 1972; Seguin, 1986). Some cultivars are reported to do better on soils composed of specific rock types (Fregoni, 1977; Seguin, 1986), but the evidence is predominantly circumstantial. Convincing experimental evidence for these claims appears lacking.

Texture

Soil texture refers to the size and proportion of the mineral component. Internationally, four size categories are recognized: coarse sand, fine sand, silt, and clay. Chapman's (1965) recognition of a larger number of categories, including gravels, pebbles, cobbles, etc. (Table 5.1) is particulary pertinent in several important viticultural regions. Nevertheless, most agricultural soils are classified only on the relative proportions of sand, silt, and clay. "Heavy" soils have a high proportion of clay, whereas "light" soils have a high proportion of sand.

Particles larger than clay and silt consist of unmodified parental rock material. In contrast, clay particles are chemically and structurally transformed minerals bearing little resemblance to the parental material. Clay consists primarily of microscopic silica and alumina plates. Silt particles are partially weathered rock material, possessing properties transitional between sand and clay (Table 5.1).

Table 5.1 Particle Size and Surface Relations of the Different Fractions of Soil[a]

Fraction	Diameter limits	Length of side	Surface exposed by subdivision of a cube 1 cm on a side — Area cm²	Surface exposed by subdivision of a cube 1 cm on a side — Area ft²	Sample (1 cm³) broken down into fractions corresponding to a heavy loam — Fraction, %	Sample (1 cm³) broken down into fractions corresponding to a heavy loam — Surface, cm²	Sample (1 cm³) broken down into fractions corresponding to a heavy loam — % Surface
Gravel	2 mm–8 cm	1.0 cm	6.0	0.0065	—	—	—
Coarse sand	2.0–0.2 mm	1.0 mm	60	0.065	13.6	650	0.01
Fine sand	0.2–0.02 mm	0.1 mm	600	0.65	17.4	8360	0.12
Silt	0.02–0.002 mm	0.01 mm	6,000	6.50	24.7	127,300	1.807
Clay	<0.002 mm	0.001 mm	60,000	65.0	35.1	1,793,000	27.00
Colloidal clay	<0.0005 mm	0.0001 mm (0.5 μm)	600,000	650	9.2	4,757,000	71.00
Upper limit of molecular sizes	0.000001 mm (1.0 nm)	1.0 nm	60,000,000	65,000	—	—	—
Total	—	—	—	—	100	6,686,310	100.0

[a] From Chapman (1965), reproduced by permission.

Clay, with a large surface area to volume (SA/V) ratio, platelike structure, and negative charge, has a major influence on the physical and chemical attributes of soil. Clay particles are so minute that they have colloidal properties. Thus, they are gelatinous and slippery when wet (the plates slide relative to one another) but hard and cohesive when dry. On wetting, clay particles expand like a sponge. Water forces the plates apart and constricts soil pores. This can markedly reduce water infiltration into soils high in clay content. The large SA/V ratio, combined with the negative charge of the clay plates, cause them to attract, retain, and exchange large quantities of positively charged ions (i.e., Ca^{2+}, Mg^{2+}, H^+), as well as water. Both bivalent ions and water help bind clay plates together. This is critical to the formation and maintenance of the aggregate structure of good agricultural soils. The large SA/V ratio also allows a clayey soil to absorb large quantities of water. However, because clay strongly bonds water molecules, much of the water is unavailable to plants (Fig. 4.16). In contrast, soils with a coarse texture allow most of the water to percolate through the soil, but what remains is readily available.

Because texture has a marked bearing on important features such as water and nutrient availability, as well as soil aeration, soil texture significantly affects grapevine growth and fruit maturation. Nevertheless, there are comparatively few reports where one can directly compare the effects of soil texture on vine growth (Nagarajah, 1987).

An important property based on the textural character of the soil is heat retention. In fine-textured soils, much of the heat absorbed is transferred to water as it evapo-rates. This energy is subsequently lost as the water escapes into the air. In contrast, stony soils retain most of the heat absorbed because of their minimal moisture content. This subsequently may be radiated back to the air during the night. The heat so derived can significantly reduce the likelihood of frost damage and accelerate fruit ripening in the autumn (Verbrugghe et al., 1991). Soil compaction also can moderate the temperature within vine rows, and potentially reduce frost damage on cool nights (Bridley et al., 1965).

Structure

Structure often refers to the association of soil particles into complex aggregates. Aggregate formation results from the binding of mineral (clay) and organic (humus) colloids by bivalent ions, water, microbial filamentous growths, and plant, microbial, and invertebrate mucilages. The aggregates are worked by the burrowing action of the soil fauna, root growth, and frost action.

Soils high in aggregate structure are friable, well-aerated, and easily penetrated by roots, have a high water-holding capacity, and are generally considered superior agriculturally. Heavy clay soils are more porous, but the small diameter of the particles makes them poorly aerated when wet. As a consequence, roots remain at or near the soil surface, exposing vines to severe water stress under drought conditions. Lighter soils are well-drained and aerated, but the large pores retain less water. Nevertheless, vines on light soils may experience less severe water stress under drought conditions if the soil is sufficiently deep to permit roots access to ground water. Soil depth also may offset the poor nutrient status of many light soils. The negative effects of the respectively small and large pores of heavy and light soils may be counteracted by humus. Humus modulates pore size, facilitating the upward and lateral movement of water, increases water absorbency, and retains moisture at tensions that permit roots ready access to the water.

Although soil structure affects aeration and mineral and water availability, these features may be modified by vineyard practices such as tillage. Consequently, they are not a constant feature of a site, and their significance to grape and wine quality is difficult to assess accurately. Under zero tillage, the number of pores and pore area are significantly higher than under cultivation. Conventional tillage results in greater total porosity, but this consists primarily of a few large, irregularly shaped cavities (Pagliai *et al.*, 1984). Generally, root development is better under zero tillage (Soyer *et al.*, 1984). Under no-till conditions, most root development occurs in the upper portion of the soil, whereas conventional cultivation limits root growth to deeper portions of the soil. Under grass cover, root distribution is relatively uniform in the top 1 m of soil. Cultivated vineyards show lower levels of organic material (Pagliai *et al.*, 1984). This may result from enhanced aeration and solar heating that stimulate microbial decomposition of the organic content of the soil.

Drainage and Water Availability

As mentioned above both soil texture and structure have effects on water infiltration. Both properties also affect water availability. Once water has moved into the soil, water may be bound to colloidal materials by electrostatic forces, adhere to pore surfaces by cohesive forces, or percolate through the soil under the action of gravity. Only water retained by cohesive forces, or adsorbed to soil colloids, is normally available for root uptake. Capillary water is important as cohesive forces permit both upward (Fig. 5.1) and lateral (Fig. 5.2) movement. Water that fills the large soil pores (free water) rapidly percolates through the soil and is lost to the plant, unless the root system penetrates deeply. Hygroscopically bound water is largely unavailable for root extraction. During drought, however, hygroscopic water may evaporate and move upward in the soil. If it condenses as the soil cools at night, it may become available to the roots. This may be of importance in sandy soils.

In Bordeaux, the ranking of *cru classé* estates has been correlated with the presence of deep, coarse-textured soils located on small rises close to rivulets or drainage channels (Seguin, 1986). These features promote rapid drainage and are thought to promote deep root penetration. Free water occasionally can percolate through soil

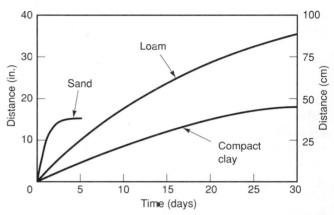

Figure 5.1 Upward movement of moisture from a water table through soils of different textures and structures. Note the rapid but limited rise in sand. The pores of loam are more favorable to movement than those in compact clay. The rate of movement often is more important than the distance raised. (Reprinted with the permission of Macmillan Publishing Company from THE NATURE AND PROPERTIES OF SOILS, Eighth Edition by Nyle C. Brady. Copyright © 1974 by Macmillan Publishing Company.)

to a depth of 20 m within 24 hr. Thus, grapevines are less likely to suffer damage from heavy rains or drought. Although some cultivars are relatively tolerant of waterlogged soils, high soil moisture content accentuates berry cracking and promotes subsequent rotting (Seguin and Compagnon, 1970).

Even when the soil is shallow, there may be factors that diminish the likelihood of water stress under drought conditions, or waterlogging during rainy weather. For example, the compact limestone that underlies the shallow soils in St. Émilion permit the effective upward flow of water from the water table. It is

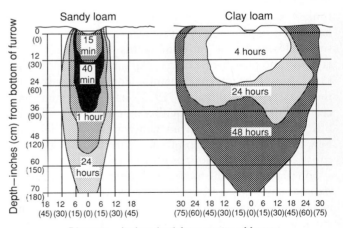

Figure 5.2 Comparative rates and direction of irrigation water movement into and through sandy and clay loams. (From Coony and Pehrson, 1955, reproduced by permission.)

estimated that 70% of the water uptake by vines in 1985 came from this source (Seguin, 1986).

Soil Depth

Soil depth, as well as soil texture and structure, can influence water availability. Shallow hardpans reduce the usable soil depth and enhance the tendency of soil to become waterlogged under heavy rains and fall below the permanent wilting percentage under drought conditions. Limiting root growth to surface layers also can influence nutrient access. For example, potassium and available phosphorus tend to predominate near the surface, especially in clay soils, while magnesium and calcium more commonly characterize the lower horizons. Different soils vary in the accumulation of nutrients through the soil horizons.

Nutrient Content and pH

Nutrient availability in soils is influenced by many, often interrelated, factors. These include the parent material, particle size, humus content, pH, water content, aeration, temperature, root surface area, and mycorrhizal development. Nevertheless, the mineral content of soil is primarily derived from the parental rock substrata. Consequently, it has been thought that the superiority of certain sites for wine production might be explained in terms of the nutrient status of soils. Differences in the accumulation of nitrogen by fruit have been associated with wine quality (Ough and Nagaoka, 1984). In addition, potassium availability and accumulation are well-known to influence wine quality. In Bordeaux, prestigious vineyard sites have been noted to possess higher humus and available nutrient contents than less highly ranked sites (Seguin, 1986). Occasionally, such data have been used to explain the historical ranking of the sites. However, the nutrient status of the sites may equally, and possibly be better, explained in terms of the ranking, and the improved maintenance and fertilization that has resulted. Fregoni (1977) has interpreted perceived differences in wine quality on the basis of mineral differences in the soils from which they were derived. Few researchers will accept such correlations as being causually related without experimental evidence.

Soil pH is one of the most well-known factors affecting mineral solubility and availability. Nevertheless, actual absorption by the vine is primarily regulated by rootstock genotype and the mycorrhizal association of the root. Correspondingly, grapevine mineral content does not directly reflect the mineral content of the soil in which it grows. In addition, excellent wines are produced from grapes grown on acidic, neutral, and alkaline soils. Thus, except where deficiency or toxicity is involved, there seems little justification for assuming that wine quality is dependent on either a specific soil pH or mineral composition.

Color

Soil color is influenced by the parental mineral origin, organic content, and moisture level. For example, soils high in calcium tend to be white, those high in iron are reddish, and those high in humus are dark brown to black.

Color influences the rate of soil warming in the spring, and that of cooling in the fall. Dark soils, irrespective of moisture content, absorb more heat than do the more reflective light-colored soils. Soils of higher moisture content, being darker, absorb more heat (Fig. 5.3) but warm more slowly than drier soils. This apparent anomaly arises from high specific heat of water, which leads to a requirement for a large amount of energy to heat the water content of the soil. Consequently, the surfaces of sandy and coarse soils both warm and cool more rapidly than clay soils of the same color and moisture content. Rapid cooling can significantly warm the air and fruit close to the ground during the night. Reflective ground covers slow the rise in soil temperature during the spring, but also slow its decline in the autumn. Conversely, plastic mulches can enhance early soil warming in vineyards (Ballif and Dutil, 1975).

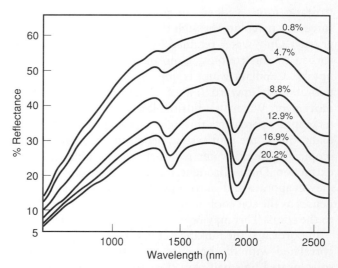

Figure 5.3 Reflectivity of a loam as a function of wavelength and water content. (From S. A. Bowers and R. J. Hanks, Reflection of radiant energy from soils, *Soil Sci.* 100, 130–138, © by Williams and Wilkins, 1965.)

The microclimatic effects of soil color and moisture content on temperature are most significant during the spring and fall. In the summer, temperature differences caused by soil surface characteristics and shading generally have little effect on vine growth and fruit maturity (Wagner and Simon, 1984). However, warm soils may enhance microbial nitrification, enhance potassium uptake, and depress magnesium and iron absorption by the vine.

Occasionally, red varieties have been selectively grown in dark soils and white varieties in light-colored soils. In marginally cool climates, this could provide the greater heat needs required for full color development in red cultivars. Nevertheless, excellent results can be obtained where the growth of red and white cultivars, on dark and light soils, is reversed (Seguin, 1971).

Organic Content

The organic content of soil improves water retention and permeability, as well as aggregate structure and nutrient availability. However, as roots rarely absorb organic material from soil, the organic content only indirectly affects vine growth and potential wine quality. There is no evidence supporting the common contention that soil directly influences the aromatic character of wine. The origin of the "earthy," "barnyardy," and "flinty" qualities of certain regional wines undoubtedly arise during wine production and maturation, not from the soil. Thankfully, the aromatic compounds produced by manure and the earthy odors generated by actinomycetes are not absorbed and translocated to ripening grapes.

Topographic Influences

Like the data on soil attributes, much of the information on slope effects are circumstantial. Nevertheless, the influence of slope on the local microclimate is clear. The effects tend to be most marked at high latitudes or altitudes. The beneficial influences of a sunward-sloped site on microclimate include better exposure to the photosynthetic and heat radiation of the sun, enhanced and earlier soil warming, diminished frost severity, and improved drainage. For the grapevine, photosynthetic potential may be increased, fruit ripening advanced, berry color and sugar to acid balance improved, and the growing season lengthened. Microclimatic disadvantages may include increased soil erosion, nutrient loss, water stress, and early loss of snow cover. Performance of vineyard activities become more difficult, bark split-

ting may occur during the winter, and cold acclimation may be lost prematurely. The net benefit of a sloped site depends on its inclination (vertical deviation), aspect (compass orientation), latitude, soil type, grape cultivar, and viticultural practices.

Solar Exposure

When the beneficial influences of a sloped vineyard are sufficient, they can outweigh the difficulties created in the performance of vineyard activities. Of particular significance is the improvement in solar exposure created by a sun-facing aspect. The benefits of an inclined location often become progressively more important as the latitude or altitude of the vineyard site increases. Not surprisingly, Germany—the most northerly major wine producing region in Europe—is renowned for its steeply sloped, south-facing vineyards.

The primary advantage created by a favorable slope aspect (orientation) and inclination (angle) relate to its reduction of the angle at which the rays of the sun impact the vineyard. This increases both light intensity and solar heating within the vineyard.

At the highest latitudes for commercial viticulture ($\sim$50°), the optimal equatorial inclination for light exposure is about 50°. Although slopes this steep are too difficult to work, solar exposure is only slightly less at 30° (Pope and Lloyd, 1974). This is generally considered the upper limit for manual vineyard work. Machines seldom work well at angles much above 6° (a slope of 10.5%). Sun exposure on east- and west-facing slopes is little affected at inclinations except above 50°. North-facing slopes have a distinctly negative effect on solar input. Figure 5.4 illustrates the importance of seasonal influences on the solar input of sloped sites. The influence is most marked when the altitude of the sun (position above the horizon) is lowest (winter) and least when the solar altitude is highest (summer).

Another factor influencing the importance of sloped sites on light incidence is the frequency of clear skies. Cloud cover, by dispersing solar radiation across the sky, eliminates the benefit of an equatorial-facing slope on solar input. The effect is illustrated in Figure 5.4 by the difference between spring and autumn radiation inputs. The diminished solar exposure observed in the spring occasionally may be beneficial by avoiding excessively early bud break and the associated increase in the likelihood of frost damage. In contrast, sunny weather in the fall is desirable to allow maximal heat accumulation during ripening and harvest.

For solar exposure, the best slopes are those directed toward the equator. In practice, the ideal aspect can be

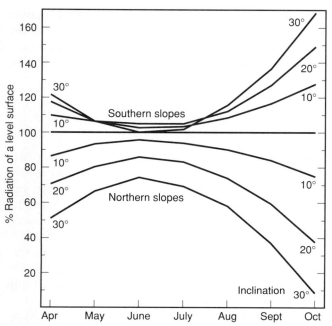

Figure 5.4 Reception of direct sunlight relative to position and inclination of slope (48°15′ N) in the upper Rhine Valley. (From Becker, 1985a, reproduced by permission.)

Reflection of solar radiation from water does not directly augment heating as most of the infrared radiation is absorbed by the water, even at low sun altitudes. However, heating may result indirectly from the absorption of the additional visual radiation received.

Although augmenting solar exposure is generally beneficial at high latitudes and altitudes, the opposite may be true at low latitudes and altitudes. Here, diminished sun exposure may favor a cooler microclimate, leading to retention of more acidity and flavor in the grapes. Correspondingly, east-facing slopes may be preferable. An eastern aspect exposes the vines to the cooler morning sun and provides increased shading from the hot afternoon sun.

Wind Direction

The prevailing wind direction can significantly influence the expression of properties donated by a slope. Heat accumulation achieved on sunward slopes can be lost if winds greater than 7 km/hr blow down the vineyard rows. Crosswinds require twice the velocity to produce the same effect (Brandtner, 1974). Updrafts through vineyards also may diminish the heat accumulation potential of sunward-facing slopes (Geiger, 1966).

At high latitudes, vine rows commonly are planted directly up steep slopes to facilitate cultivation. Offsetting row orientation to minimize the negative influences of the prevailing wind direction is generally impractical. Terracing vineyards can permit alignment of vine rows relative to the prevailing winds, but land leveling eliminates many of the advantages of steeply sloped sites. Terracing also may increase soil erosion problems (Luft *et al.*, 1983).

In humid climates, positioning vine rows at right angles to the prevailing winds can increase surface drying of the foliage and fruit by enhancing wind turbulence. This can enhance the action of fungicides in controlling fungal diseases. If the vineyard faced sunward, the enhanced solar radiation would further speed the drying action of the wind. In dry environments, rows aligned parallel to the prevailing winds may reduce foliage wind drag and potentially reduce evapotranspiration (Hicks, 1973), whereas a perpendicular alignment may lead to increased water stress, owing to stomata remaining open longer during the day (Freeman *et al.*, 1982). Thus, the most appropriate row alignment will depend on the climatic limitations it is designed to alleviate. The presence of natural or artificial shelter belts, by modifying the velocity, turbulence, and flow of the wind, may further influence optimal row alignment and slope orientation.

influenced by local factors. If fog commonly develops during cool autumn mornings, the preferred aspect may be southwest. The scattering of light by fog eliminates the radiation advantage of equatorial-facing slopes. In the late afternoon, when skies are more typically clear, a southwest aspect would provide optimal solar exposure in the Northern Hemisphere. Such situations are not uncommon in river valleys such as the Mosel and Rhine valleys in Germany, or along the Neusiedler See in Austria.

Another important property of sunward-facing slopes is the increased exposure of radiation reflected from water and soil surfaces (albedo). This is particularly significant at low sun angles. At high solar elevations, the albedo off water is very low (2–3% for a smooth surface and 7–8% with a rough surface). However, at low sun elevations (<10°), the reflection can reach over 50% (Büttner and Sutter, 1935). Consequently, reflection of solar energy from water bodies is especially significant during the spring and fall. This has particular importance for sloped vineyards in high latitudes: the steeper the slope, the greater the potential interception of reflected radiation. Radiation reflected off the Main River in Germany (49°49′ N) can constitute 39% of total radiation received on south-facing vineyards in early spring (Volk, 1934). This level of additional exposure could significantly improve early initiation of growth and enhance photosynthesis and fruit ripening in the autumn.

Frost and Winter Protection

Another advantage of sun-facing slopes in cool climates comes from the additional number of frost-free days provided. Part of this microclimatic benefit results from improved heat accumulation, but of greater significance is the flow of cold air away from the vines.

Under cool, clear atmospheric conditions, heat radiation from the soil and grapevines can be considerable. Without wind turbulence, a temperature inversion can form, resulting in temperatures falling near or below the freezing point. Sloped sites often experience some protection from this phenomenon, since the cold air can flow downward and away from the vines. The movement of cold air into low-lying areas can lengthen the frost-free season on slopes by several days or weeks (Fig. 5.5). Depending on the height of the slope, maximal protection may be achieved either at the top or in the mid-region of the slope. The degree of protection often depends on wind barriers, such as tree shelters, or topographical features that facilitate the movement of air between and away from the vines.

The flow of cool air away from grapevines can be important even at temperatures above freezing. Chilling to temperatures below 10°C has been reported to permanently disrupt fruit maturation in some varieties (Becker, 1985b).

Where sloped vineyard sites are associated with lakes or rivers, the water can further modify vineyard microclimate. By acting as a heat sink, water can buffer major temperature fluctuations. Large lakes and oceans generate even more marked modification of the climate.

In regions frequently experiencing severe winter conditions, east- or west-facing slopes may be preferred. For example, in the Finger Lakes region of New York, south-facing slopes promote the early loss of an insulating snow cover (Pool and Howard, 1985). South-facing slopes also may increase the likelihood of bark splitting owing to sudden fluctuations in temperature caused by rapid changes in sun exposure.

Drainage

Because of erosion, soils on slopes tend to be coarsely textured. This provides better drainage and permits soil surfaces to dry more quickly. Correspondingly, less heat is expended in the vaporization of soil moisture from the surface, and sun-facing slopes warm more readily. For

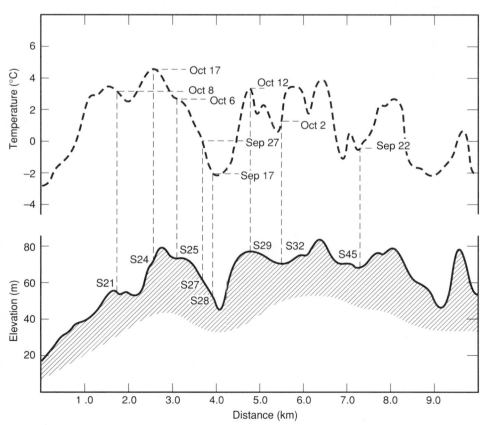

Figure 5.5 Relationship between the estimated average frost date, topography, and temperature on a clear night. S# represents particular site locations. (From Bootsma, 1976, reproduced by permission.)

example, it can take twice as much heat to raise the temperature of a wet soil compared to a dry equivalent. On the negative side, slopes show greater tendencies to be nutrient deficient and require periodic addition of top soil. Enhanced drainage also may increase the likelihood of ground water pollution with nutrients added as fertilizer.

Mesoclimatic and Macroclimatic Influences

Historically, cultivars appropriate for particular regions were chosen empirically. How well they grew and the quality of the fruit produced indicated suitability. Improvements in measuring the physical parameters of a region have led to attempts to classify regions relative to compatible grape varieties. The most well-known system is that devised by Amerine and Winkler (1944) for California. They divided California into five climatic regions based on heat-summation units over the growing season (Fig. 5.6). The units, called **degree-days,** are calculated by summing the product of the number of days in each month with an average monthly temperature above 10°C (50°F) by the average temperature of the month minus 10°C (50°F). The 10°C cutoff point was chosen because few grape varieties initiate significant growth below 10°C. A comparison of the Winkler and Amerine viticul-

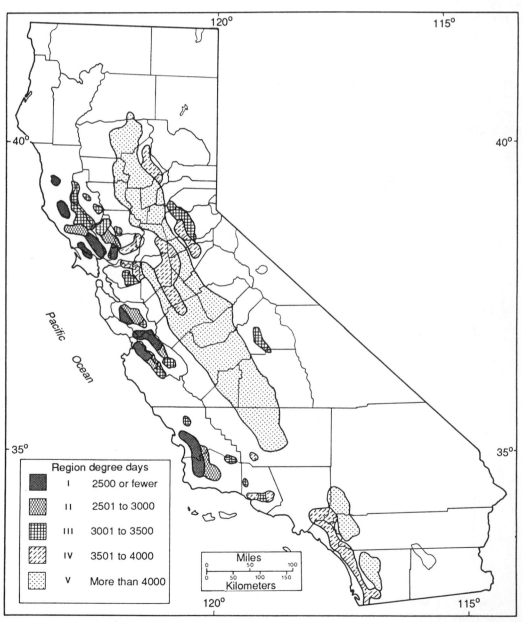

Figure 5.6 Wine districts of California based on Fahrenheit heat-summation units. (From de Blij, 1983, reproduced by permission.)

tural climatic regions, based on Celsius and Fahrenheit degree-day ranges, is given in Table 5.2.

Since its introduction, the degree-day formula has been used widely in many regions. However, it has not met with universal success, even in California (McIntyre *et al.*, 1987), owing to the significance of additional climatic factors. In an attempt to find a more generally applicable measure, modifications to the degree-day formula have been investigated. These have included the incorporation of factors such as indicators of humidity and water stress, or modifications to the temperature formula. One of the latest formulas is the **latitude–temperature index** proposed by Jackson and Cherry (1988). It is calculated as the product of the mean temperature (°C) of the warmest month, multiplied by 60 minus the latitude. Another system proposed by Bentryn (1988) plots sites relative to the mean lowest and highest monthly temperatures, or the mean temperature and daily relative humidity of the warmest month. The latter is useful in predicting the potential severity of several fungal disease problems.

While useful, these formulas give only a rough indication of varietal suitability. For example, regions affected by continental influences may have more that twice the average day/night temperature variation as an equivalent maritime region. Additionally, differences in north/south slope orientation become increasingly important with increasing latitude. In an attempt to obtain more sensitive indicators of local climatic conditions, some researchers have recommended the use of floristic maps to determine cultivar/site compatibility (Becker, 1985a). Because of the sensitivity of some native plants to climatic conditions, they can be accurate indicators of the macro- and microclimatic conditions of a site.

In long-established vineyard regions, climatic indicators are less important as the suitability of most well-known cultivars is already known. However, for new viticultural regions, prediction of cultivar/site suitability is essential to avoid the expense of errors and the need to replant. For some *Vitis vinifera* cultivars, there are empirical data indicating minimum and preferred climatic conditions. For example, cool-adapted cultivars typi-

cally require above 1000 Celsius degree-days annually, a 180-day frost-free period, an average coldest monthly temperature not less than −1°C, temperatures below −20°C occurring less than once every 20 years, and annual precipitation greater than 400–500 mm (Becker, 1985a). Phenologic stages such as bud burst, flowering, and fruit maturation also are well established for several major *V. vinifera* cultivars (Galet, 1979; McIntyre *et al.*, 1982). Regrettably, similar data are not available for the majority of grapevine cultivars.

Members of the genus *Vitis* grow over a wide range of climates, from the continental extremes of northern Canada, Russia, and China to humid subtropical climates in Central and South America. Nevertheless, most individual species are limited to a much narrower latitude range and environmental extremes. For *Vitis vinifera*, adaptation includes the latitude range between 35° to 50° N, including Mediterranean, maritime, and moderate continental climates. These are characterized, respectively, by wet winters and hot dry summers, relatively mild winters and dry cool summers, and cool winters, warm summers with comparatively uniform annual precipitation.

Temperature

Temperature and grapevine growth are markedly affected by site latitude, which are in turn controlled by the periodicity and intensity of light and heat received from the sun. Nevertheless, grapevine growth is primarily regulated by the annual temperature cycle. For example, cold acclimation is primarily influenced by cooling autumn temperatures, not the shortening photoperiod. In addition, bud activation in the spring does not require a specific cold treatment as in many temperate zone plants. Bud activation responds progressively to temperatures above a cultivar-specific minimum temperature Moncur *et al.*, 1989. Above this temperature, bud break and other phenological responses become increasingly likely and rapid up to an optimum temperature (Fig. 5.7). This type of response suggests that temperature control is relatively nonspecific and functions through its effects on the shape and/or flexibility of specific regulator proteins and cell membrane lipids. This interpretation is strengthened by the nonspecific enhancement of cellular respiration during bud break by a diverse range of treatments (Shulman *et al.*, 1983). Up to a maximum value, every 10°C increase in temperature doubles the reaction rate of biochemical processes. As cellular reactions have dissimilar temperature response curves, the overall response will depend on the combined effects of temperature on the individual reactions involved.

The slow activation of bud growth in the spring may reflect the ancestral trailing or climbing habit of the vine. For several cultivars, the average minimal temperatures

Table 5.2 Comparison of Viticultural Climatic Regions Based on Equivalent Ranges of Celsius and Fahrenheit Degree-days

Region	Celsius degree-days	Fahrenheit degree-days
I	≤1390	≤2500
II	1391–1670	2501–3000
III	1671–1940	3001–3500
IV	1941–2220	3501–4000
V	≥2220	≥4000

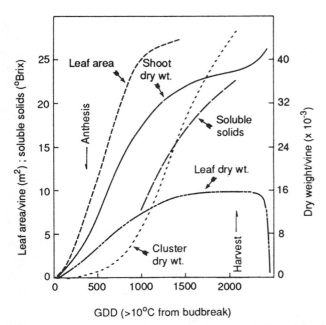

Figure 5.7 Increment of dry weight, leaf area, and soluble solids of 'Thompson seedless' as a function of degree-days (GDD) exceeding 10°C. Average bud break in the San Joaquin valley was March 9. (Data from Williams, 1987, graphed in Williams and Matthews, 1990, reproduced by permission.)

for bud break and leaf production are 3.5° and 7.1°C, respectively (Moncur *et al.*, 1989). Delay in grapevine leafing out permits potential shrub and tree supports to partially produce foliage before the vine establishes its canopy. Thus, the vine can position its leaves optimally for light exposure, relative to the foliage of the support plant. During the remainder of the growing season, lateral bud activation appears to be regulated by growth substances released from apical meristematic regions. This plasticity of growth permits the vine to respond to favorable environmental conditions, or foliage loss, throughout much of the year.

Temperature has marked and critical effects on both the duration and effectiveness of flowering and fruit set. Flowering typically does not occur until the average temperature reaches 20°C (18°C in cool regions). Low temperatures slow anthesis, as well as pollen release, germination, and growth. For example, pollen germination is low at 15°C but high at 30° to 35°C, and style penetration and fertilization may take 5 to 7 days at 15°C but only a few hours at 30°C (Staudt, 1982). If fertilization is delayed significantly, the ovules abort. Cold temperatures slowly reduce pollen viability.

Although pollen germination and tube growth are favored by warm temperatures, optimal fertilization and subsequent fruit set become progressively poorer as temperatures rise above 20°C. Ovule fertility, seed number per berry, and berry weight are greater at lower than

high temperatures. Even soil temperature can affect vine fertility. For example, cool soil temperatures tend to suppress bud break but enhance the number of berries produced per cluster (Kliewer, 1975). Fruit set seems proportional to the leaf area generated.

Temperature conditions affect the photosynthetic rate, but they are not known to dramatically influence overall annual vine growth under normal conditions. In fact, leaves partially adjust to seasonal temperature fluctuations by changing the optimal temperature range for photosynthesis. In midsummer, the optimal range generally varies between 25° and 32°C, but in the autumn it may decline to between 22° and 25°C (Stoev and Slavtcheva, 1982).

For centuries it has been known that temperature markedly affects berry ripening and quality. This is the basis of the degree-day concept of site selection. Temperature also differently affects specific reactions occurring during berry maturation. Consequently, fruit composition and potential wine quality can be significantly influenced.

Higher temperatures generally result in increased sugar levels but reduced malic acidity. Because taste, color, stability, and aging potential are all influenced by grape sugar and acid contents, temperature conditions throughout the season have a prominent effect in delimiting grape quality at harvest. Based on sugar and malic acid levels, the optimal temperature range for grape maturation lies between 20° and 25°C; for anthocyanin synthesis, slightly cooler temperatures may be preferable (Kliewer and Torres, 1972). Daytime temperatures appear to be more important than nighttime temperatures for pigment formation. Although moderate temperature conditions often favor fruit coloration, cool temperatures may limit the commercial cultivation of red cultivars owing to poor coloration.

Surprisingly little research has been conducted to confirm the generally held view that cool temperatures favor varietal aroma development in grapes. However, cool temperatures increase the frequency of vegetable odors in 'Cabernet Sauvignon,' while berry aroma formation is favored under warmer climatic conditions (Heymann and Noble, 1987). With 'Pinot noir,' however, several aroma compounds are reported to be produced in higher amounts in long, cool compared to early, hot seasons (Watson *et al.*, 1988).

CHILLING AND FROST INJURY

Although cool temperatures may be beneficial during grape maturation, prolonged exposure to temperatures below 10°C is reported to induce irreversible physiological damage, retarding ripening (Becker, 1985b). This type of damage, called **chilling injury**, is usually reversible when of short duration. Injury apparently results from the excessive gelling of semifluid cell membranes. This increases cell permeability, resulting in electrolyte loss and disruption of respiratory, photosynthetic, and other cellular functions (George and Lyons, 1979). Sensitive species are generally characterized by higher proportions of saturated fatty acids in their membranes. Saturated fatty acids maintain appropriate membrane fluidity at high temperatures but make the membrane too rigid at cool temperatures. In some plants, exposure to cool temperatures increases the proportion of linolenic acid (an unsaturated fatty acid) in cellular membranes.

Cool temperatures also increase the likelihood of dew formation by raising atmospheric relative humidity. Dew formation on the vine can increase the frequency and severity of most fungal infections.

Further cooling may result in ice crystal formation on and within plant tissues. As heat is lost from the tissues, water in the cell wall begins to form ice. Water then diffuses out of the cell, replacing the water molecules lost in ice crystal formation. In frost-sensitive plants, the dehydration induces protein and nucleic acid denaturation and subsequent aggregation. These changes can create irreversible damage, causing cell death. The degree of dehydration that results depends largely on the osmotic potential of the cells. Irreparable damage also can result from puncturing of cellular membranes by intracellularly formed ice crystals. Further damage may result from tissue deformation caused by the differential expansion of the phloem and xylem (Meiering *et al.*, 1980). The degree of damage often depends as much on the rate of cooling and subsequent thawing as on the minimum temperature reached. Rapid cooling tends to induce the more destructive intracellular ice formation, while slow cooling initially induces a less destructive, extracellular crystallization.

During the autumn, vine tissues progressively develop a degree of frost tolerance. Cold acclimation is directly associated with increased levels of soluble sugars (Wample and Bary, 1992) and partial dehydration of the tissues (Wolpert and Howell, 1985). Acclimation is especially marked in vascular tissues and dormant buds. These tissues eventually may withstand freezing down to −20°C or below.

The potential to tolerate freezing varies markedly between *Vitis* species (Table 5.3) and *V. vinifera* cultivars (Fig. 5.8). For example, 'Riesling,' 'Gewürztraminer,' and 'Pinot noir' are relatively winter hardy, being able to withstand short exposures to temperatures down to −26°C when fully acclimated. Varieties such as 'Cabernet Sauvignon,' 'Sémillon,' and 'Chenin blanc' possess moderate hardiness and suffer damage at −17° to −23°C. 'Grenache' is considered tender and may suffer severe damage even at −14°C. *Vitis labrusca* and French–American hybrid cultivars are relatively cold

Table 5.3 Cultivar Classification and Bud Survival after Minimum Temperatures of −20° to −30°C on December 24, 1980[a]

	Percent bud survival, 1981		
Cultivar classification	Primary	Secondary	Tertiary
Vitis labrusca	54.7	64.8	74.1
French–American hybrids	31.3	54.0	63.5
Vitis vinifera	11.3	14.7	23.8

[a] From Pool and Howard (1985), reproduced by permission.

hardy, and some *V. amurensis* hybrids can survive prolonged exposure to temperatures below −30°C.

Winter hardiness is a complex factor and inadequately represented by the minimum temperature the cultivar can survive. Cold hardiness is influenced by many genetic, environmental, and viticultural conditions. In addition, various parts of the grapevine show differential sensitivity to frost and winter damage. In the dormant

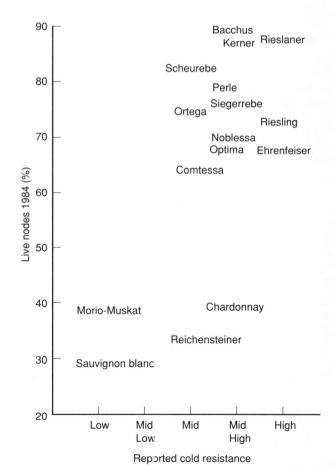

Figure 5.8 Effect of cold temperatures (−24°C) on bud survival of various cultivars compared with reported hardiness. (From Pool and Howard, 1985, reproduced by permission.)

bud, the most differentiated (primary) bud is the most sensitive, while the least differentiated (tertiary bud) is the most cold hardy. Buds and the root system are generally less winter hardy than old wood.

Although cold hardiness is primarily a physiological property, anatomical features may influence frost sensitivity. For example, hardy rootstocks generally have less bark tissue, containing relatively small phloem and ray cells, and possess woody tissues with narrow xylem vessels. Winter hardy rootstocks also may increase the resistance of the scion to cold damage. This may result from factors such as restrained scion vigor, modified contents of growth regulators, and earlier limitation of water availability.

In general, small vines have been considered more cold tolerant than large vines, except where vine size is restricted by stress factors such as disease, overcropping, and lime-induced chlorosis. Although this view has been challenged (Striegler and Howell, 1991), there is little doubt that late-season growth prolongs cellular activity and delays cold acclimation. This may develop because of growth regulator and/or nutrient influences. Reduced carbohydrate accumulation could limit supercooling of cellular fluids. During cold acclimation, starch is hydrolyzed to oligosaccharides (Fig. 5.9) and simple sugars. This change decreases the osmotic potential of the cytoplasm and reduces the freezing point.

The ability of surviving buds to replace winter-killed tissues is often critical to the commercial success of culti-

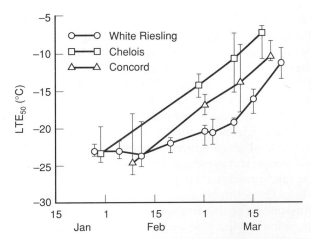

Figure 5.10 Deep supercooling median low temperature exotherm (LTE_{50}) of dormant vines and deacclimating the primary buds of *Vitis vinifera* ('Riesling'), *Vitis labrusca* ('Concord'), and French–American hybrid ('Chelois') cultivars in 1983. Vertical bars indicate the limits of LTE_{10} and LTE_{90}. (From Andrews *et al.*, 1984, reproduced by permission.)

vars in cool climates. The tendency of base buds on French–American hybrids to burst and develop flowering shoots often compensates for periodic severe winter damage. Even some *V. vinifera* cultivars, such as 'Chardonnay' and 'Riesling,' can withstand severe (80–90%) bud kill and produce a substantial crop from the activation of remaining buds (Pool and Howard, 1985). Nevertheless, retention during pruning of well-matured canes with superior cold-hardy buds is often crucial for commercial viability in cold viticultural regions.

The rates of the seasonal cold acclimation and deacclimation also can be critical to varietal success. With several cultivars such as 'Concord' (*V. labrusca*), 'Chelois' (French–American hybrid), and 'Riesling' (*V. vinifera*), winter hardiness is inversely proportional to the rate of deacclimation during the winter (Fig. 5.10). As cold acclimation is influenced, albeit slightly, by the rootstock, an appropriate rootstock choice might permit commercial cultivation of a particular variety (Miller *et al.*, 1988).

Because rapid temperature changes often are particularly destructive, the frequency of rapid temperature fluctuations can be more limiting than the lowest yearly temperature might suggest.

MINIMIZING FROST AND WINTER DAMAGE

Various techniques have been used to minimize the **radiative frost damage** that commonly develops on cool, still nights. One of the oldest protective measures involves the selection of sloped vineyard sites adjacent to open valleys, lakes, or rivers. The slope allows the cool air to flow down and away from the vineyard. In addition, air circulation created by the temperature differen-

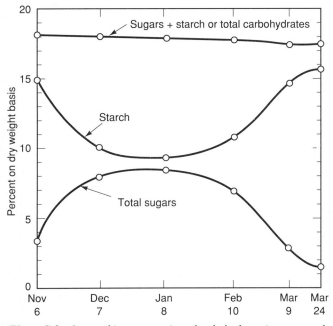

Figure 5.9 Seasonal interconversion of carbohydrates in cane wood, from starch to sugars and back, during fall cold acclimation and winter deacclimation, respectively. Total carbohydrate content remains almost constant. (From Winkler, 1934, reproduced by permission.)

tial between land and water can diminish the development of temperature inversions and radiative frost development (Fig. 5.11). Another procedure has involved the positioning of fruiting wood high on tall trunks, to place the buds above the coldest air. The lowest temperatures usually occur at or just above ground level (Fig. 5.12). Where early fall frosts are a prime concern, vines on slopes may be trained close to the ground, as is usually done in Champagne. In this case, heat radiation from the soil may combine with the flow of cold air to minimize the drop in air temperature and prevent frost damage.

Where practical, water sprinklers can reduce the likelihood of frost damage. Permanently raised sprinklers are commonly used. The heat of fusion released as the water freezes can protect underlying tissue from freezing. It is critical that sprinkling continue until the sun or air temperature melts the ice coating. Otherwise, heat is removed from the tissues as the ice begins to melt, causing frost injury. Water dispersion should be as uniform and constant as possible. Otherwise, parts of the plant may freeze as heat is lost to the surrounding air or ice.

Another technique of frost protection involves air mixing with fans or propellers. With sufficient agitation, the cold air at ground level is blended with warmer air above the vine canopy. Winds of about 3.5 km/hr (2 miles/hr) are often sufficient to prevent frost damage. Smudge pots may have a similar effect by increasing air turbulence; air pollution concerns, however, have made smudge pots largely a thing of the past.

A new method under investigation is the inoculation of vines with ice-nucleation-deficient strains of bacteria. This technique is based on the involvement of bacteria, notably *Pseudomonas syringae* and *Erwinia herbicola*, in the initiation of ice crystal formation on leaves at about −2° or −4°C. Crystallization may not commence until temperatures reach −8° to −11°C in the absence of the bacteria. Frost injury has been correlated directly with epiphytic bacterial populations under freezing conditions (Fig. 5.13). The population of ice-nucleating bacteria on leaf surfaces also generally increases in the autumn. Therefore, replacing them with ice-nucleation-deficient strains should reduce the likelihood of frost damage. While elimination might be produced by antibiotic application, biological control through the application of competitive ice-nucleation-deficient strains is preferable ecologically (Lindemann and Suslow, 1987).

Another new technique under investigation is the application of cryoprotectants. Cryoprotectants have been shown to lower the potential freezing temperature of dormant buds during the winter, and especially in the late winter when vines commence cold deacclimation (Himelrick *et al.*, 1991). Cryoprotectant application also may have potential in protecting vine tissues from frost damage after bud break. In addition, growth retardants applied about flowering also have been noted to enhance cold hardiness during the following winter (Shamtsyan *et al.*, 1989).

Where late spring frosts are frequent, delayed pruning often is recommended. By deferring pruning, the number of viable buds present at pruning is likely to be increased. The degree of pruning can then reflect the extent of winter injury. This may be assessed by opening a representative sampling of buds and looking for indications of

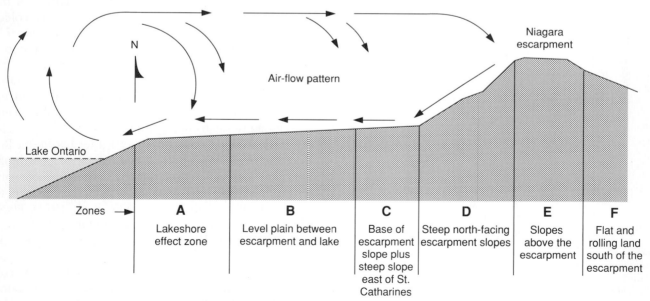

Figure 5.11 Influence of Lake Ontario on air circulation during frosty conditions over the Niagara peninsula up to the Niagara escarpment. (From Wiebe and Anderson, 1976, reproduced by permission).

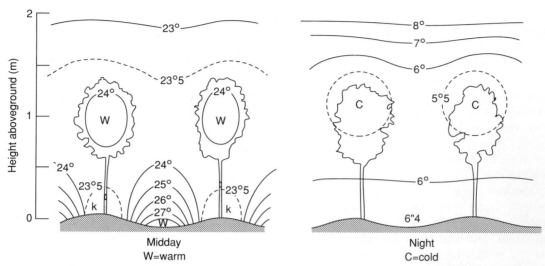

Figure 5.12 Temperatures at midday (left) and night (right) in a Rheinpfalz vineyard on September 17, 1933. (After Sonntag, 1934, reproduced by permission.)

bud viability. An alternate approach is double-pruning. In this procedure, more buds are left per vine than normally considered appropriate. After bud break, the actual degree of damage can be determined and the unnecessary buds removed.

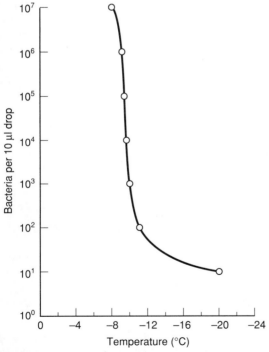

Figure 5.13 Effect of cell concentration on the nucleating activity of *Pseudomonas fluorescens* F-12. Data points represent the temperature at which 90% of the test drops were frozen. (From Maki and Willoughby, 1978, *J. Appl. Microbiol.*, reproduced by permission of the American Meteorological Society.)

Where trunk winterkill is a major problem, retaining multiple trunks is often practiced. Two-trunk vines are common in northeastern North America, and occasionally vines with up to five trunks of differing ages are developed. It is unlikely that all trunks will suffer damage equally in any one year. Additional trunks also limit the damage caused by crown gall or Eutypa dieback, both of which tend to be considerable problems in cold vineyard regions.

A valuable technique, especially for grafted vines, is hilling soil over the crown of the grapevine in the fall. A newer technique involves positioning empty tubes around the trunk (Pool and Howard, 1985). Earth pushed up around the tube provides insulation for the enclosed trunk. Although the graft union region is colder than when buried underground, it is considerably warmer than unhilled graft regions. Where climatic conditions are more severe, as in China, Russia, and the central United States, the whole vine may be laid down and buried for the winter. The difficulty and labor involved with this procedure, as well as the potential damage caused to the trunk(s), limit its use.

Cultural conditions also affect winter hardiness. Reduced nitrogen availability limits vegetative growth, favors cane maturation, and thereby should promote winter hardiness. Nevertheless, the actual effect appears to be minor or insignificant (Wample *et al.*, 1991). Phosphorus, and its proportion relative to nitrogen (Mozera and Ponomorev, 1969), as well as calcium application (Eifert *et al.*, 1986) have been reported to significantly increase winter bud survival. Ground covers tend to collect snow which can increase trunk and root insulation from rapid temperature fluctuations and severe cold.

Wind breaks also can favorably influence snow cover and distribution. Limiting irrigation, by restricting water uptake, can facilitate evacuation of the xylem that is a component of cold acclimation. Deacclimation is partially associated with refilling of the vessels with water.

Solar Radiation

As noted above, temperature plays the major role in regulating the yearly growth cycle of the grapevine. Nevertheless, it is solar radiation that generates climatic temperature cycling. The degree to which soil and vines receive light and heat radiation from the sun is largely a function of the angle at which the solar rays impact the earth's surface. The solar angle varies with the time of day, the season, and the inclination and orientation of the site. To comprehend these factors, it is necessary to understand certain aspects of the earth's rotation and movement around the sun and some principles of atmospheric and radiation science.

Solar radiation provides the energy that heats the earth. Because the distribution of the radiation is unequal, both geographically and temporally, the forces that generate the earth's climate are set in motion. The earth's two principal motions that generate global climatic change are its **rotation** and **revolution.** Rotation refers to the earth's spin about its axis. It creates the 24-hr day/night cycle and the arc through which the sun seems to pass during the day. Because the sun's position in the sky is constantly changing, the angle at which solar radiation impacts the earth is continually varying. This markedly influences the energy level reaching a point on the earth's surface throughout the day.

The other motion, revolution, is the movement of the earth on its orbit around the sun. The latter produces the cyclical changes in solar inclination and the earth's changing seasons (Fig. 5.14). During the longest day of the year (the summer solstice), the Northern Hemisphere is maximally exposed to the sun's energy, while the Southern Hemisphere is experiencing its shortest day. As the earth continues its revolution around the sun, the days in the Northern Hemisphere progressively become shorter as the earth's axis tilts away from the sun, and the sun appears to rise and set further south. During the winter solstice, the Northern Hemisphere is tilted at an angle of 23.5° away from the sun, while the Southern Hemisphere is tilted at 23.5° toward the sun.

The earth's revolution around the sun has several important effects on plant and animal life. Changing day length produces a yearly photoperiodic cycle that is used by many organisms to regulate yearly growth cycles. However, most *Vitis vinifera* cultivars are relatively insensitive to photoperiod, the yearly cycle being influenced primarily by annual temperature fluctuations.

In addition to the obvious effect on day length, the revolution of the earth around the sun significantly affects the intensity and spectral quality of sunlight. The intensity of solar radiation reaching the earth decreases as the sun's **altitude** becomes lower. This occurs for a number of reasons. One involves the dispersion of solar energy over an increasingly large surface area, especially at high latitudes (Fig. 5.15). At low latitudes (~20°), there is little change in the intensity of noon radiation received from spring through fall (~6%). At high latitudes (50°), however, there is a 28% variation in radiation intensity during the same period. Corresponding

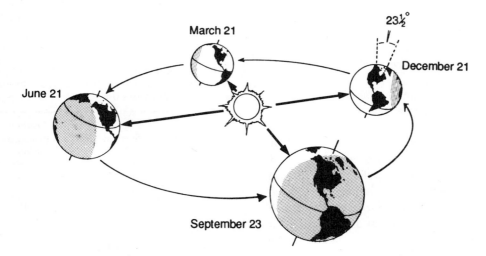

Figure 5.14 Path of the earth around the sun. The orbital characteristics include the tilt of the earth's axis (23.5°), the direction of the axis, and the ellipticity of the earth's orbit (exaggerated here). (From Pisias and Imbrie, 1986, 1987, reproduced by permission.)

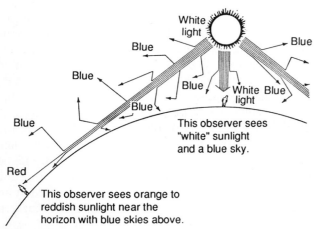

Figure 5.15 Direct sunlight consisting of all the visible wavelengths produces a white appearance, while the differential scattering and absorbency of light produces the blue of the sky and the orange/red color at sunset. (From Frederick K. Lutgens/Edward J. Tarbuck, THE ATMOSPHERE: An Introduction to Meteorology, 3e, © 1986, p. 29. Reprinted by permission of Prentice-Hall, Englewood Cliffs, New Jersey.)

winter/summer differences in energy imput are about 27 and 68%, respectively.

The seasonal variation in solar radiation input at different latitudes is considerably affected by the **azimuth**. The azimuth refers to the arching path through which the sun appears to move across the sky. As the summer passes and the earth's axis tilts away from the sun, the sun appears to rise closer to the equator, and its arc through the sky is progressively lower. This increases the atmospheric depth through which the incoming radiation must pass, both decreasing its intensity, and modifying its spectral quality. These changes result from increased absorption and reflection as the radiation interacts with atmospheric gases and particles. An example is the marked effect clouds (water droplets) have on the warmth of sunlight. Part of the benefit of sun-facing slopes at high latitudes is to offset some of the negative effects of low sun altitudes.

Because the various wavelengths of solar radiation are affected differentially by passage through the atmosphere, the longer the path, the greater the effect. These influences are particularly significant in the visible portion of the spectrum (Fig. 5.15). Scattering enriches diffuse radiation (skylight) in blue wavelengths. Although reducing the photosynthetic value of low-angle direct sunlight, scattering increases the relative importance of skylight. Atmospheric dispersion and absorption of short-wavelength infrared solar radiation diminishes the warming effect of sunlight at low sun angles.

The effects of atmospheric passage on the relative strengths of visible and short-wavelength infrared radiation in direct and diffuse light are given in Table 5.4.

These effects produce not only the blue sky, but also the diffuse radiation that comes from the sky. Because of the efficiency with which blue light (400 to 500 nm) is absorbed by chlorophyll, diffuse sunlight is often as photosynthetically active as direct sunlight. Diffuse radiation also penetrates more deeply into leaf canopies because of its hemispherical origin. As the two states of phytochrome differ in absorption of blue light, this might have important physiological consequences. The nondirectional origin of diffuse light under cloud cover also neutralizes the benefits normally gained from sun-facing slopes. Even under sunny conditions, diffuse radiation (skylight) contains little solar infrared radiation. Thus, diffuse radiation provides little solar heating. The coolness of tree shade in the summer is well known. Most direct solar heating comes from beam sunlight.

PHYSIOLOGICAL EFFECTS

Light has multiple effects on grapevine growth. The most significant is activating photosynthesis, the energy source for vine growth. Most of the radiant energy comes directly from beam radiation or skylight, with only about 3.5% coming from light reflected from the soil or adjacent vines (Smart, 1973). The proportion of light received directly by the horizontal (top) versus vertical (sides) of the vine changes (Fig. 5.16), depending on the shape and size of the vine and the angle of the leaves to the sun. However, the influence of vine shape and size on canopy shading is of greater significance.

When light directly impinges on a leaf, about 85 to 90% of the photosynthetically active radiation (PAR) is absorbed (Fig. 5.17). Of that spectral region (400–700 nm), the blue and red portions are absorbed. The green and far-red portions are either reflected or transmitted through the leaf, which gives leaves their green coloration. The efficiency of light absorption by leaves

Table 5.4 Contribution of Clear-Sky Diffuse Solar Radiation to Total Solar Radiation at Different Wavelengths at a Sun Altitude of 60°[a]

Wavelength (nm)	Spectral region	Fraction arriving as diffuse	Fraction arriving in direct beam
300	Ultraviolet	0.72	0.28
400	Blue	0.33	0.67
500	Green	0.19	0.81
600	Yellow	0.13	0.87
700	Red	0.09	0.91
1000	Far red	0.05	0.95
2000	Infrared	0.02	0.98

[a] Data from Schulze (1970), modified by Miller (1981), reproduced by permission.

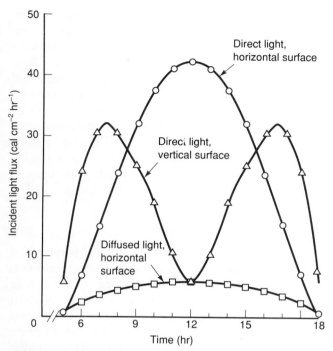

Figure 5.16 Flux densities of direct sunlight on horizontal and vertical surfaces, and of diffused skylight on a horizontal surface, December 22, 35° S. (From Smart, 1973, reproduced by permission.)

greatly reduces both the intensity and quality of the radiation transmitted to the inner canopy foliage. Second layer leaves under sunny conditions usually receive less than 10% of the exterior canopy radiation, and a third layer may receive about 1 to 2% of the surface radiation (Fig. 5.17). In addition, the selective removal of photosynthetically active radiation further reduces its photosynthetic value. The combined effect of the quantitative and qualitative changes in solar radiation is to limit the

effective photosynthetic canopy to the exterior and first inner leaf layers of a vine (Smart, 1985). About 80 to 90% of the photosynthate is produced by the exterior canopy. Inner leaves eventually turn yellow and drop prematurely.

The light intensity of direct sunlight can reach about 1000 W/m^2 [~10,000 footcandles (fc)]. Grapevine leaves can use little more than about one-quarter of that value (Kriedemann and Smart, 1971). Because the full intensity of sunlight cannot be used photosynthetically, most of the excess radiation energy absorbed is liberated as heat. If the leaves are fully exposed, and the roots have an ample water supply for rapid transpiration, increase in leaf temperature may be held to about 5°C. If the leaves are water stressed, or there is little air movement, heating may be sufficient to suppress photosynthesis. For example, the rate of photosynthesis at 35°C may be only 15% of that at 25°C (Fig. 5.18). The actual reduction depends on the variety, its water supply, light intensity, and adaptation to high light intensities. Nevertheless, surface leaves often do not photosynthesize maximally because of heating in full sunlight. Second layer leaves are less likely to experience photosynthetic limitation because of solar heating, but they are exposed to a light regime close to the **compensation point**. Diffuse skylight, or radiation reflected from the soil and adjacent vines, may augment the light transmitted through the foliage. Even temporary breaks in the canopy cover, called **sunflecks**, may significantly improve photosynthesis within the vine canopy. Nevertheless, most internal leaves contribute little to the photosynthate supply of the vine.

In addition to the direct and indirect effects on photosynthesis, shading affects cane maturation, inflorescence initiation, fruit maturation and aromatic character. It has long been known that sun exposure improves cane maturation. Whether this is due to the intensity or spectral quality of the sunlight appears to be unknown. It is also possible that sunlight may enhance cane maturation through an influence on heating and drying of the shoot surface.

Flower cluster initiation begins in early summer, and development stops only in the fall. Sunlight affects both inflorescence initiation and fruitfulness. Sun-exposed buds generally bear more flowers per cluster than buds developed under shaded conditions. As with other properties, there is considerable variation in the response of particular cultivars (Fig. 5.19).

In most regions, it is common to position shoots to increase fruit exposure to the sun. This long-standing practice favors fruit coloration and maturation. In contrast, training may be designed to protect the fruit from direct sun exposure in hot, arid environments. Sun exposure often raises the temperature of dark colored berries 5°C above the surrounding air (Fig. 5.20), and berry

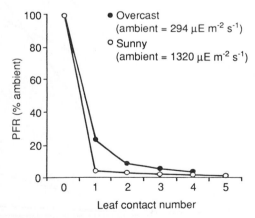

Figure 5.17 Attenuation of photosynthetically active radiation (PFR) by 'Merlot' canopies on sunny and overcast days in Bordeaux expressed as a percentage of above canopy ambient. (From Smart, 1985, reproduced by permission.)

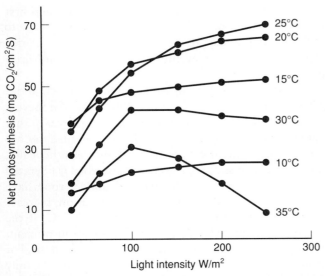

Figure 5.18 Influence of temperature on the photosynthetic capacity of grapevine leaves. (After Törökfalvy and Kriedemann, 1977, reproduced by permission.)

temperature can reach 15°C above ambient (Smart and Sinclair, 1976). Sun exposure also influences grape composition. Although the response is cultivar dependent, sun-exposed berries generally are lower in pH and in potassium and malic acid content. The berries also are generally smaller. Shaded grapes show the opposite tendencies. Sun-exposed berries typically show properties associated with higher quality, namely, higher °Brix, anthocyanin, and tartaric acid levels, as well as better pH and malic acid levels. Shading is associated with a greater vegetative character in the wines of some cultivars (Arnold and Bledsoe, 1990). Whether this is due to intensity, spectral quality, or thermal effects on fruit or canopy shading is unclear. Although sun exposure is usually beneficial, excessive exposure can produce sunburn or baking that generates atypical wine fragrances.

Wind

Unless strong winds are a characteristic feature of a region, wind seldom is considered in vineyard selection or row alignment. In regions such as the Valtellina area of Lombardy and the Margaret River area of Western Australia, however, strong winds can severely limit grape cultivation. Severe winds can cause both lingering physiological as well as extensive physical damage to the vine. The benefits of windbreaks on reduced soil erosion and evapotranspiration have made them a common landscape feature in the central plains of North America

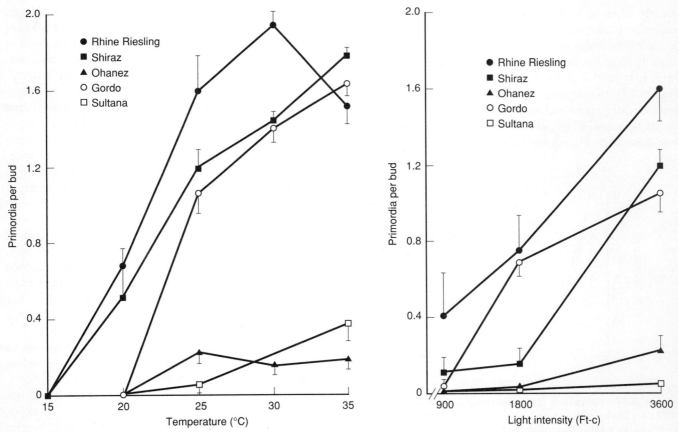

Figure 5.19 Effect of temperature and light intensity on the average number of bunch primordia per bud for the basal 12 buds of five grape varieties. (From Buttrose, 1970, reproduced by permission.)

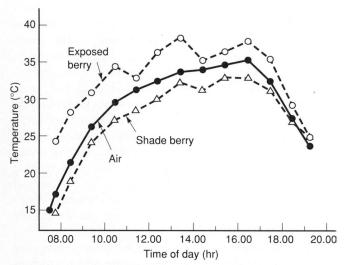

Figure 5.20 Temperature of exposed and shaded 'Carignan' berries, relative to the air temperature at cluster height, on February 10, 1972. (After Millar, 1972, reproduced by permission.)

and the southern Rhône Valley of France. Nevertheless, windbreaks are not a typical feature of vineyard regions. This may reflect not only the cost of construction or planting, but also a failure to realize their considerable

benefits. Instead, cultural practices such as hedging (Switzerland) and cane intertwining (Lombardy) have been more common. Where severe winds tend to occur primarily in the early spring, late pruning can provide up to a 2-week delay in bud break (Hamilton, 1988).

With appropriate design, shelter belts reduce both wind velocity and turbulence. If the windbreak is porous, the downwind influence can extend up to 25 times the height of the barrier (Fig. 5.21). Dense windbreaks are less effective in reducing wind speed and usually increase wind turbulence.

Not only do strong winds produce physical damage, but they also can reduce shoot length, leaf size, and stomal density, induce smaller and fewer clusters per vine, and generate lower °Brix levels (Kobriger *et al.*, 1984; Takahashi *et al.*, 1976). Several of these effects may result from stomatal closure, caused by an increased rate of water loss. In the leaves, carbon dioxide availability subsequently declines and oxygen content rises, both of which reduce photosynthesis and thus vine growth. In addition to the immediate effects of wind exposure, physiological effects may linger long after the wind has died down (Freeman *et al.*, 1982). Because the benefits of a windbreak increase with wind velocity, the value of

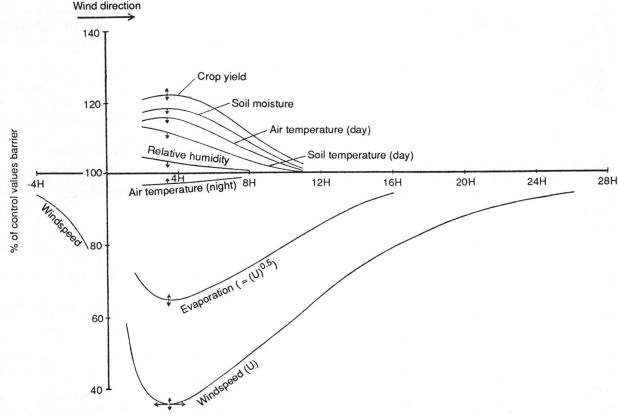

Figure 5.21 Diagrammatic representation of barrier effects on microclimate. Arrows indicate the directions in which values of different factors vary relative to unsheltered areas. *H* refers to barrier height. (From Marshall, 1967, reproduced by permission.)

shelter belts is directly proportional to average and peak wind speeds.

Occasionally, concern has been expressed about shading from windbreaks and the effect on photosynthesis. Generally, this is not a important factor as only the vines adjacent to the windbreak are likely to be affected. In addition, with a north/south alignment of the windbreak, the areas shaded in the morning and evening receive additional radiation reflected from the windbreak in the afternoon and morning, respectively. It also has been suggested that reduced evapotranspiration in the shaded area may permit prolonged photosynthetic activity owing to reduced water stress (Marshall, 1967).

Because of reduced wind velocity and air mixing, temperature inversions in sheltered vineyards may be more marked. Consequently, windbreaks may increase the frequency and severity of spring and fall frosts. This likelihood is reduced if the vineyard topography permits free flow of cool air out and away from the vineyard.

Another potential problem associated with windbreak use is higher atmospheric humidity. Although this could favor disease development, it seldom has this influence as most environments benefiting from wind protection are generally dry.

In addition to windbreaks and shelter belts, aligning vine rows relative to the prevailing wind can modify its effects. Positioning the rows at an oblique angle to the prevailing wind turns the outermost rows into a low windbreak. For example, vineyard rows aligned at angles other than the normally preferred north/south orientation may retain heat by limiting wind flow down the rows (Brandtner, 1974). Thus, in windy climates, slopes facing other than due south (in northern latitudes) may be preferable because of the protection provided from northerly winds.

In contrast, there may be advantages in planting the rows parallel to the prevailing wind under dry conditions (Hamilton, 1985). Winds moving directly down the rows produce less foliage drag and minimize water loss (Hicks, 1973). It is estimated that this could produce up to a 10 to 20% saving in water demand. Wind moving across rows creates turbulence and greater foliage movement than moving unimpeded down the rows. Turbulence may increase relative to the density of the grapevine canopy.

Water

Most of the water absorbed by roots is used as a coolant. Evaporation of water from the leaves helps to maintain normal physiology by minimizing overheating. When water loss continues to exceed replacement from the roots, the stomata close. As the tissues overheat, most metabolic activity slows. Physiological activity returns to normal only some time after the water deficit has been overcome. If the deficit persists, continued water loss through the cuticle results in cell plasmolysis and death. Mature vines seldom show marked wilting, owing to extensive root systems that can tap water meters below the soil surface. Nevertheless, physiological disruption occurs long before wilting. Reduced shoot growth is one of the most sensitive signs of grapevine water stress.

Where irrigation is permitted and economically feasible, the damage caused by naturally occurring water shortages can be prevented. Inexpensive and reliable sources of irrigation water are essential for commercial viticulture in most semiarid and arid climates. Water sprinkling systems may limit heat stress but are usually unnecessary or ineffectual.

Problems with excessive rainfall, fog, or high humidity are often more serious, or at least more difficult to control. Good soil porosity and drainage diminish problems associated with heavy rainfall. Basal leaf removal and divided-canopy training systems, with their larger surface area, can increase water evaporation. Angling vineyard rows to increase wind turbulence around the vines can further enhance surface drying of the foliage and fruit. Nevertheless, the enhancement of fungal diseases under foggy, humid conditions means that they are likely to be controlled only with fungicides. Planting new, more resistant cultivars is currently impractical because of Appellation Control restrictions (in Europe) or marketing concerns. Generally, areas of high humidity are avoided because of the severity of disease losses in such regions.

Water also has a major influence on global and regional vineyard distribution owing to the moderating effect of rivers, lakes, and oceans. The impact has been noted in de Castella's (1912) observation of the proximity of most vineyard regions to the cool edges of warm continents, and warm edges of cool continents.

Suggested Readings

Soil

Dubos, J. (1984). Importance du terroir comme facteur de differentiation qualitative des vins. *Bull. O.I.V.* 57, 420–434.

Hancock, J. M., and Price, M. (1990). Real chalk balances the water supply. *J. Wine Res.* 1, 45–60.

Morlat, R., Remoue, M., and Pinet, P. (1984). Influence de la densité de plantation et du mode d'entretien du sol sur l'enracinement d'un peuplement de vigne plante en sol favorable. *Agronomie* 4, 485–491.

Saayman, D. (1977). The effect of soil and climate on wine quality. *Int. Symp. Quality Vintage*, pp. 197–208. Oenol. Vitic. Res. Inst., Stellenbosch, South Africa.

Schreier, P. (1983). Correlation of wine composition with cultivar and site. *Centenary Grape Wine Symp.*, May 23, 1983, pp. 63–95. Roseworthy Agricultural College, Roseworthy, South Australia.

Seguin, G. (1986). "Terroirs" and pedology of wine growing. *Experientia* 42, 861–873.

Climate

Becker, N. (1985). Site selection for viticulture in cooler climates using local climatic information. *Proc. Int. Cool Climate Vitic. Enol. Symp.* (D. A. Heatherbell, P. B. Lombard, F. W. Bodyfelt, and S. F. Price, eds.), Agric. Exp. Station Tech. Publ. No. 7628, pp. 20–34. Oregon State Univ., Corvallis, Oregon.

Jones, H. G. (1983). "Plants and Microclimate. A Quantitative Approach to Environmental Plant Physiology." Cambridge Univ. Press, Cambridge.

Koblet, W. (1985). Influence of light and temperature on vine performance in cool climates and applications to vineyard management. *Proc. Int. Cool Climate Vitic. Enol. Symp.* (D. A. Heatherbell, P. B. Lombard, F. W. Bodyfelt, and S. F. Price, eds.), Agric. Exp. Station Tech. Publ. No. 7628, pp. 139–157. Oregon State Univ., Corvallis, Oregon.

Rosenberg, N. J., Blad, B. L., and Verma, S. B. (1983). "Microclimate. The Biological Environment," 2nd Ed. Wiley (Interscience), New York.

Schreier, P. (1983). Correlation of wine composition with cultivar and site. *Centenary Grape Wine Symp. May 23, 1983,* pp. 63–95. Roseworthy Agricultural College, Roseworthy, Australia.

Smart R. E. (1973). Sunlight interception by vineyards. *Am. J. Enol. Vitic.* 24, 141–147.

Smart, R. E. (1985). Some aspects of climate, canopy microclimate, vine physiology, and wine quality. *Proc. Int. Cool Climate Vitic. Enol. Symp.* (D. A. Heatherbell, P. B. Lombard, F. W. Bodyfelt, and S. F. Price, eds.), Agric. Exp. Station Tech. Publ. No. 7628, pp. 1–19. Oregon State Univ., Corvallis, Oregon.

Smart, R. E. (1985). Principles of grapevine canopy microclimate manipulation with implications for yield and quality. A review. *Am. J. Enol. Vitic.* 36, 230–239.

Wahl, K. (1988). Climate and soil effects on grapevine and wine: The situation on the northern borders of viticulture—The example Franconia. *Proc. 2nd Int. Symp. Cool Climate Vitic. Oenol., Jan. 11–15, 1988, Auckland, N.Z.* (R. E. Smart, S. B. Thornton, S. B. Rodriguez, and J. E. Young, eds.), pp. 1–15. New Zealand Soc. Vitic. Oenol., Auckland, New Zealand.

Climate Indices

Bentryn, G. (1988). World climate patterns and viticulture. *Proc. 2nd Int. Symp. Cool Climate Vitic. Oenol., Jan. 11–15, 1988, Auckland, N.Z.* (R. E. Smart, S. B. Thornton, S. B. Rodriguez, and J. E. Young, eds.), pp. 9–12. New Zealand Soc. Vitic. Oenol., Auckland, New Zealand.

Croser, B. J. (1987). Cool area winemaking. *Proc. 6th Aust. Wine Ind. Tech. Conf.* (T. Lee, ed.), pp. 34–39. Australian Industrial Publishers, Adelaide, Australia.

Jackson, D. I., and Cherry, N. J. (1988). Prediction of a district's grape-ripening capacity using a latitude–temperature index (LTI). *Am. J. Enol. Vitic.* 39, 19–28.

Kenny, G. J., and Harrison, P. A. (1992). The effects of climate variability and change on grape suitability in Europe. *J. Wine Res.* 3, 163–183.

Kirk, J. T. O. (1986). Application of a revised temperature summary system to Australian viticultural regions. *Aust. Grapegrower Winemaker* 268, 48, 51, and 52.

McIntyre, G. N., Kliewer, W. M., and Lider, L. A. (1987). Some limitations of the degree-day system as used in viticulture in California. *Am. J. Enol. Vitic.* 38, 128–132.

Prescott, J. A. (1965). The climatology of the vine (*Vitis vinifera* L.). The cool limits of cultivation. *Trans. R. Soc. South Aust.* 89, 5–22.

Cold Hardiness and Frost Protection

Kasimatis, A. N., Bearden, B. E., Sisson, R. L., and Bowers, K. (1982). "Frost Protection for North Coast Vineyards." Leaflet No. 2743. Coop. Extension, Div. Agric. Sci., Univ. California, Berkeley, Berkeley, California.

Landers, J. N., and White, K. (1967). Irrigation for frost protection. *In* "Irrigation of Agricultural Lands" (R. M. Hagan, H. R. Haise, and T. W. Edminster, eds.), Agronomy Monogr. No. 11, pp. 1037–1057. Am. Soc. Agronomy, Madison, Wisconsin.

Margaritis, A., and Bassi, A. S. (1991). Principles and biotechnological applications of bacterial ice nucleation. *Crit. Rev. Biotechnol.* 11, 277–295.

Pellet, H. M., and Carter, J. V. (1981). Effect of nutritional factors on cold hardiness of plants. *Hortic. Rev.* 30, 144–171.

Poghossian, K. S. (1985). Biological principles of grape cultivation in high-stemmed training systems under the continental climate conditions of southern USSR. *Proc. Int. Cool Climate Vitic. Enol. Symp.* (D. A. Heatherbell, P. B. Lombard, F. W. Bodyfelt, and S. F. Price, eds.), Agric. Exp. Station Tech. Publ. No. 7628, pp., 217–226. Oregon State Univ., Corvallis, Oregon.

Pool, R. M., and Howard, G. E. (1985). Managing vineyards to survive low temperatures with some potential varieties for hardiness. *Proc. Int. Cool Climate Vitic. Enol. Symp.* (D. A. Heatherbell, P. B. Lombard, F. W. Bodyfelt, and S. F. Price, eds.), Agric. Exp. Station Tech. Publ. No. 7628, pp. 184–197. Oregon State Univ., Corvallis, Oregon.

Rieger, M. (1989). Freeze protection for horticultural crops. *Hortic. Rev.* 11, 45–109.

Wind Protection

Hamilton, R. P. (1988). Wind effects on grapevines. *Proc. 2nd Int. Symp. Cool Climate Vitic. Oenol., Jan. 11–15, 1988, Auckland, N.Z.* (R. E. Smart, S. B. Thornton, S. B. Rodriguez, and J. E. Young, eds.), pp. 65–68. New Zealand Soc. Vitic. Oenol., Auckland, New Zealand.

Kobriger, J. M., Kliewer, W. M., and Lagier, S. T. (1984). Effects of wind on water relations of several grapevine cultivars. *Am. J. Enol. Vitic.* 35, 164–169.

Ludvigsen, K. (1989). Windbreaks—Some considerations. *Aust. Grapegrower Winemaker* 302, 20–22.

Rosenberg, N. J., Blad, B. L., and Verma, S. B. (1983). "Microclimate. The Biological Environment," 2nd Ed. Wiley (Interscience), New York.

van Eimern, J., Karschon, R., Razumova, L. A., and Robertson, G. W. (1964). "Windbreaks and Shelterbelts." World Meteorol. Organ. Tech. Note No. 59, Geneva, Switzerland.

References

Amerine, M. A., and Winkler, A. J. (1944). Composition and quality of musts and wines of California grapes. *Hilgardia* 15, 493–675.

Andrews, P. K., Sandidge, C. R., and Toyama, T. K. (1984). Deep supercooling of dormant and deacclimating *Vitis* buds. *Am. J. Enol. Vitic.* 35, 175–177.

Arnold, R. A., and Bledsoe, A. M. (1990). The effect of various leaf removal treatments on the aroma and flavor of Sauvignon blanc wine. *Am. J. Enol. Vitic.* **41**, 74–76.

Ballif, J. L., and Dutil, P. (1975). The warming of chalk soils by plastic films. Thermic measurements and evaluations. *Ann. Agron.* **26**, 159–167. (Reported in Rosenberg *et al.*, 1983, see suggested readings.)

Becker, N. (1985a). Site selection for viticulture in cooler climates using local climatic information. *Proc. Int. Cool Climate Vitic. Enol. Symp.* (D. A. Heatherbell, P. B. Lombard, F. W. Bodyfelt, and S. F. Price, eds.), Agric. Exp. Station Tech. Publ. No. 7628, pp. 20–34. Oregon State Univ., Corvallis, Oregon.

Becker, N. (1985b). Panel—Session I. *Proc. Int. Cool Climate Vitic. Enol. Symp.* (D. A. Heatherbell, P. B. Lombard, F. W. Bodyfelt, and S. F. Price, eds.), Agric. Exp. Station Tech. Publ. No. 7628, pp. 35–42. Oregon State Univ., Corvallis, Oregon.

Bentryn, G. (1988). World climate pattern and viticulture. *Proc. 2nd Int. Symp. Cool Climate Vitic. Oenol. Jan. 11–15, 1988, Auckland, N. Z.* (R. E. Smart, S. B. Thornton, S. B. Rodriguez, and J. E. Young, eds.), pp. 9–12. New Zealand Soc. Vitic. Oenol., Auckland, New Zealand.

Bootsma, A. (1976). Estimating minimum temperature and climatological freeze risk in hilly terrain. *Agric. Meteorol.* **16**, 425–443.

Bowers, S. A., and Hanks, R. J. (1965). Reflection of radiant energy from soils. *Soil Sci.* **100**, 130–138.

Brady, N. C. (1974). "Nature and Properties of Soil," 8th Ed. Macmillan, New York.

Brandtner, E. (1974). "Die Bewertung geländeklimatischer Verhältnisse in Weinbaulagen." Deutscher Wetterdienst, Zentralamt, Offenbach, Germany.

Bridley, S. F., Taylor, R. J., and Webber, T. J. (1965). The effects of irrigation and rolling on nocturnal air temperatures in vineyards. *Agric. Meteorol.* **24**, 373–383.

Büttner, K., and Sutter, E. (1935). Die Abkühlungsgrösse in den Dünen etc. *Strahlentherapie* **54**, 156–173.

Buttrose, M. S. (1970). Fruitfulness in grapevines: The response of different cultivars to light, temperature, and daylength. *Vitis* **9**, 121–125.

Chapman, H. D. (1965). Chemical factors of the soil as they affect microorganisms. *In* "Ecology of Soil-Borne Plant Pathogens" (K. F. Baker and W. C. Snyder, eds.), pp. 120–141. John Murray, London.

Coony, J. J., and Pehrson, J. E. (1955). "Avocado Irrigation," Leaflet No. 50. Calif. Agric. Extension Station, Oakland.

de Blij, H. J. (1983). "Wine: A Geographic Appreciation." Rowman and Allanheld, Totowa, New Jersey.

de Castella, F. (1912). The influence of geographical situation on Australian viticulture. *Victoria Geog. J.* **29**, 38–48.

Eifert, J., Szöke, L., Varnai, M., and Nagy, E. (1986). The effect of liming on the frost resistance of grape buds. *Szölötermesztés Boraszat.* **8**, 34–36 [*Vitis Abstr.* **26**, 99 (1987)].

Freeman, B. M., Kliewer, W. M., and Stern, P. (1982). Influence of windbreaks and climatic region on diurnal fluctuation of leaf water potential, stomatal conductance and leaf temperature of grapevines. *Am. J. Enol. Vitic.* **31**, 233–236.

Fregoni, M. (1977). Effects of soil and water on the quality of the harvest. *O.I.V. Int. Symp. Quality Vintage*, pp. 151–168. Oenol. Vitic. Res. Inst., Stellenbosch, South Africa.

Galet, P. (1979). "A Practical Ampelography—Grapevine Identification" (translated and adapted by L. T. Morton). Cornell Univ. Press, Ithaca, New York.

Geiger, R. (1966). "The Climate Near the Ground." Harvard Univ. Press, Cambridge, Massachusetts.

George, M. F., and Lyons, J. M. (1979). Non-freezing cold temperatures as a plant stress. *In* "Modification of the Aerial Environment of Plants" (B. J. Barfield and J. F. Gerber, eds.), Monogr. No. 2, pp. 85–96. Am. Soc. Agric. Engineer. St. Joseph, Missouri.

Hamilton, R. P. (1985). Severe wine grape losses in the Margaret River region of Western Australia. Research Report, Western Australian Dept. of Agriculture, South Perth, Western Australia.

Hamilton, R. P. (1988). Wind effects on grapevines. *Proc. 2nd Int. Symp. Cool Climate Vitic. Oenol., Jan. 11–15, 1988, Auckland, N.Z.* (R. E. Smart, S. B. Thornton, S. B. Rodriguez, and J. E. Young, eds.), pp. 65–68. New Zealand Soc. Vitic. Oenol., Auckland, New Zealand.

Heymann, H., and Noble, A. C. (1987). Descriptive analysis of Pinot noir wines from Carneros, Napa and Sonoma. *Am. J. Enol. Vitic.* **38**, 41–44.

Hicks, B. B. (1973). Eddy fluxes over a vineyard. *Agric. Meteorol.* **12**, 203–215.

Himelrick, D. G., Pool, R. M., and McInnis, P. J. (1991). Cryoprotectants influence freezing resistance of grapevine bud and leaf tissue. *HortScience* **26**, 406–407.

Jackson, D. I., and Cherry, N. J. (1988). Prediction of a district's grape-ripening capacity using a latitude–temperature index (LTI). *Am. J. Enol. Vitic.* **39**, 19–28.

Kliewer, W. M. (1975). Effect of root temperature on budbreak, shoot growth and fruit set of Cabernet Sauvignon grapevines. *Am. J. Enol. Vitic.* **26**, 82–84.

Kliewer, W. M., and Torres, R. E. (1972). Effect of controlled day and night temperatures on grape coloration. *Am. J. Enol. Vitic.* **23**, 71–76.

Kobriger, J. M., Kliewer, W. M., and Lagier, S. T. (1984). Effects of wind on water relations of several grapevine cultivars. *Am. J. Enol. Vitic.* **35**, 164–169.

Kriedemann, P. E., and Smart, R. E. (1971). Effects of irradiance, temperature and leaf water potential on photosynthesis of vine leaves. *Photosynthetica* **5**, 6–15.

Lindemann, J., and Suslow, T. V. (1987). Competition between ice nucleation-active wild type and ice nucleation-deficient deletion mutant strains of *Pseudomonas syringae* and *P. fluorescens* Biovar 1 and biological control of frost injury on strawberry blossoms. *Phytopathology* **77**, 882–886.

Luft, G., Morgenschweis, G., and Vogelbacher, A. (1983). Influence of large-scale changes of relief on runoff characteristics and their consequences for flood-control design. *In* "Scientific Procedures Applied to the Planning, Design and Management of Water Resources System. *Proc. Hamburg Symp. Aug. 1983.*" *International Association of Hydrological Sciences.* (IAHS) Publ. No. 147.

Lutgens, F. K., and Tarbuck, E. J. (1986). "The Atmosphere. An Introduction to Meteorology," 3rd Ed. Prentice-Hall, Englewood Cliffs, New Jersey.

McIntyre, G. N., Lider, L. A., and Ferrari, N. L. (1982). The chronological classification of grapevine phenology. *Am. J. Enol. Vitic.* **33**, 80–85.

McIntyre, G. N., Kliewer, W. M., and Lider, L. A. (1987). Some limitations of the degree-day system as used in viticulture in California. *Am. J. Enol. Vitic.* **38**, 128–132.

Maki, L. R., and Willoughby, K. J. (1978). Bacteria as biogenic sources of freezing nuclei. *J. Appl. Meteorol.* **17**, 1049–1053.

Marshall, J. K. (1967). The effect of shelter on the productivity of grasslands and field crops. *Field Crop Abstr.* **20**, 1–14.

Meiering, A. G., Paroschy, J. H., Peterson, R. L., Hostetter, G., and Neff, A. (1980). Mechanical freezing injury in grapevine trunks. *Am. J. Enol. Vitic.* **31**, 81–89.

Millar, A. A. (1972). Thermal regime of grapevines. *Am J. Enol. Vitic.* **23**, 173–176.

Miller, D. H. (1981). "Energy at the Surface of the Earth. An Introduction to the Energetics of Ecosystems." Academic Press, New York.

Miller, D. P., Howell, G. S., and Striegler, R. K. (1988). Cane and bud hardiness of own-rooted White Riesling and scions of White Riesling and Chardonnay grafted to selected rootstocks. *Am. J. Enol. Vitic.* 39, 60–66.

Moncur, M. W., Rattigan, K., Mackenzie, D. H., and McIntyre, G. N. (1989). Base temperatures for budbreak and leaf appearance of grapevines. *Am. J. Enol. Vitic.* 40, 21–26.

Morlat, R., Asselin, C., Pages, P., Leon, H., Robichet, J., Remoue, M., Salette, J., and Caille, M. (1983). Caractérisation integrée de quelques terroirs du val de Loire influence sur less qualité des vins. *Connaiss. Vigne Vin* 17, 219–246.

Mozera, L., and Ponomorev, B. F. (1969). *In* "High Stemmed Grape Culture" (I. Mikhailuk, M. Kyharski, and I. Milhaiake, eds.), pp. 23–24. Kishinev, Moldavia, 1978 (in Russian).

Nagarajah, S. (1987). Effects of soil texture on the rooting patterns of Thompson seedless vines on own roots and on Ramsey rootstock in irrigated vineyards. *Am. J. Enol. Vitic.* 38, 54–58.

Ough, C. S., and Nagaoka, R. (1984). Effect of cluster thinning and vineyard yields on grape and wine composition and wine quality of Cabernet Sauvignon. *Am. J. Enol. Vitic.* 35, 30–34.

Pagliai, M., La Marca, M., Lucamante, G., and Genovese, L. (1984). Effects of zero and conventional tillage on the length and irregularity of elongated pores in a clay loam soil under viticulture. *Soil Tillage Res.* 4, 433–444.

Pisias, N. G., and Imbrie, J. (1986, 1987). Orbital geometry, CO_2, and Pleistocene climate. *Oceanus* 29(4), 43–49.

Pool, R. M., and Howard, G. E. (1985). Managing vineyards to survive low temperatures with some potential varieties for hardiness. *Proc. Int. Cool Climate Vitic. Enol. Symp.* (D. A. Heatherbell, P. B. Lombard, F. W. Bodyfelt, and S. F. Price, eds.), Agric. Exp. Station Tech. Publ. No. 7628, pp. 184–197. Oregon State Univ., Corvallis, Oregon.

Pope, D. F., and Lloyd, P. S. (1974). Hemispherical photography, topography and plant distribution. *In* "Light as an Ecological Factor: II" (G. C. Evans, R. Brainbridge, and O. Rackham, eds.), pp. 385–408. Blackwell, Oxford.

Rankine, B. C., Fornachon, J. C. M., Boehm, E. W., and Cellier, K. M. (1971). Influence of grape variety, climate and soil on grape composition and on the composition and quality of table wines. *Vitis* 10, 33–50.

Schulze, R. (1970). "Strahlenklima der Erde." Steinkopf, Darmstadt, Germany.

Shamtsyan, S. M., Tsertsvadze, T. A., and Rapava, L. P. (1989). The effect of growth retardants on frost resistance of grapevines. *Soobshch. Akad. Nauk, Tbilisi* 133, 381–384 (in Russian).

Shulman, Y., Nir, G., Fanberstein, L., and Lavee, S. (1983). The effect of cyanamide on the release from dormancy of grapevine buds. *Sci. Hortic.* 19, 97–104.

Seguin, G. (1971). Influence des facteurs naturels sur les caractéres des vins. *In* "Sciences et Techniques de la Vigne" (J. Ribéreau-Gayon and E. Peynaud, eds.), Vol. 1, pp. 671–725. Dunod, Paris.

Seguin, G. (1986). "Terroirs" and pedology of wine growing. *Experientia* 42, 861–873.

Seguin, G., and Compagnon, J. (1970). Une cause du développement de la pourriture grise sur les sols gravelo-sableux du vignoble bordelais. *Connaiss. Vigne Vin* 4, 203–214.

Smart, R. E. (1973). Sunlight interception by vineyards. *Am. J. Enol. Vitic.* 24, 141–147.

Smart, R. E. (1985). Principles of grapevine canopy microclimate manipulation with implications for yield and quality. A review. *Am. J. Enol. Vitic.* 36, 230–239.

Smart, R. E., and Sinclair, T. R. (1976). Solar heating of grape berries and other spherical fruits. *Agric. Meteriol.* 17, 241–259.

Sonntag, K. (1934). Bericht über die Arbeiten des Kalmit-Observatoriums. D. Met. Jahrb. f. Bayern, Anhang D.

Soyer, J. P., Delas, J., Molot, C., and Andral, P. (1984). Techniques d'entretien du sol en vignoble bordelais. *Prog. Agric. Vitic.* 101, 315–320.

Staudt, G. (1982). Pollenkeimung und Pollenschlauchwachstum *in vivo* bei *Vitis* und die Abhähgigkeit von der Temperatur. *Vitis* 21, 205–216.

Stoev, K., and Slavtcheva, T. (1982). La photosynthèse nette chez la vigne (*V. vinifera* L.) et les facteurs écologiques. *Connaiss. Vigne Vin* 16, 171–185.

Striegler, R. K., and Howell, G. S. (1991). The influence of rootstock on the cold hardiness of Seyval grapevines. I. Primary and secondary effects on growth, canopy development, yield, fruit quality and cold hardiness. *Vitis* 30, 1–10.

Takahashi, K., Kuranaka, M., Miyagawa, A., and Takeshita, O. (1976). The effect of wind on grapevine growth: Windbreaks for vineyards. *Bull. Shimane Agric. Exp. Stn. (Shimane-ken Nogyo Shikenjo Kenkyu Hokoku)* 14, 39–83 (in Japanese).

Törökfalvy, E., and Kriedemann, P. (1977). Unpublished data, vineleaf photosynthesis. *O.I.V. Int. Symp. Quality Vintage*, pp. 67–87. Oenol. Vitic. Res. Inst., Stellenbosch, South Africa.

Verbrugghe, M., Guyot, G., Hanocq, J. F., and Ripoche, D. (1991). Influence de différents types de sol de la basse Vallée du Rhône sur les températures de surface de raisins et de feuilles de *Vitis vinifera*. *Rev. Fr. Oenol.* 128, 14–20.

Volk, O. H. (1934). Ein neuer für botanische Zwecke geeigneter Lichtmesser. *Ber. Dtsch. Bot. Ges.* 52, 195–202.

Wagner, R., and Simon, H. (1984). Studies on the microclimate of neighbouring vineyards under the same climatical conditions. *Bull. O.I.V.* 57, 573–583.

Wallace, P. (1972). Geology of wine. *24th Int. Geology Cong. (IGC)*, Sect. 6, 359–365.

Wample, R. L., and Bary, A. (1992). Harvest date as a factor in carbohydrate storage and cold hardiness of Cabernet Sauvignon grapevines. *J. Am. Soc. Hortic. Sci.* 117, 32–36.

Wample, R. L., Spayd, S. E., Evans, R. G., and Stevens, R. G. (1991). Nitrogen fertilization and factors influencing grapevine cold hardiness. *Proc. Int. Symp. Nitrogen Grapes Wine, Seattle, WA, June, 1991* (J. M. Rantz, ed.), pp. 120–125. Am. Soc. Enol. Vitic., Davis, California.

Watson, B., Lombard, P., Price, S., McDaniel, M., and Heatherbell, D. (1988). Evaluation of Pinot noir clones in Oregon. *Proc. 2nd Int. Symp. Cool Climate Vitic. Oenol., Jan. 11–15, 1988, Auckland, N.Z.* (R. E. Smart, S. B. Thornton, S. B. Rodriguez, and J. E. Young, eds.), pp. 276–278. New Zealand Soc. Vitic. Oenol., Auckland, New Zealand.

Wiebe, J., and Anderson, E. T. (1976). "Site Selection of Grapes in the Niagara Peninsula." Hort. Res. Inst. Ontario, Vineland Station, Ontario, Canada.

Williams, L. E. (1987). Growth of 'Thompson Seedless' grapevines: I. Leaf area development and dry weight distribution. *J. Am. Soc. Hortic. Sci.* 112, 325–330.

Williams, L. E., and Matthews, M. A. (1990). Grapevines. *In* "Irrigation of Agricultural Crops" (B. A. Stewart and D. R. Nielsen, eds.), Agronomy Ser. Monogr. No. 30, pp. 1019–1055. American Society of Agronomy, Madison, Wisconsin.

Winkler, A. L. (1934). Pruning *vinifera* grapevines. *Calif. Agric. Ext. Ser. Circ.* 89, 1–68.

Wolpert, J. A., and Howell, G. S. (1985). Cold acclimation of Concord grapevines. II. Natural acclimation pattern and tissue moisture decline in canes and primary buds of bearing vines. *Am. J. Enol. Vitic.* 36, 189–194.

6

Chemical Constituents of Grapes and Wine

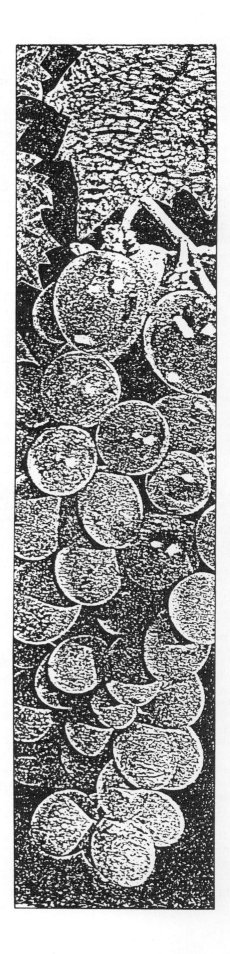

Introduction

The chemical understanding of grapes and wine has advanced greatly since the late 1960s. Although much still needs to be learned, the basic picture of what makes wine distinctive is beginning to emerge. Even the mysteries of the benefits of aging and barrel maturation are yielding their secrets. This knowledge is guiding vineyard and winery practice toward more consistent and better quality wines. Plant breeders are also using the information to streamline the production of new grape varieties.

Although significant progress has been made, there are still considerable limitations on the application of the data obtained. Some compounds cannot be detected by certain techniques, and some procedures produce artifacts. These problems are being minimized through improved instrumentation and the use of more than one analytical procedure. A more serious limitation lies in the sensory interpretation of the data. Perception is separated by many neural steps from sensation in the mouth or nose. In addition, compounds may interact in complicated ways to influence sensory stimulation. Therefore, it is often difficult to predict how chemical composition will affect sensory perception. For example, only rarely can a particular varietal aroma or wine bouquet be ascribed to one or a few volatile compounds. Distinctive fragrances usually arise from the relative concentrations of many aromatic compounds, not single varietally unique substances.

The rapid increase in the number of compounds found in wine has been spawned by developments in gas chromatography (GC), thin-layer chromatography (TLC), high-performance liquid chromatography (HPLC), droplet counter-current chromatography (DCCC), infrared spectroscopy, and nuclear magnetic resonance spectroscopy (NMR) (Brun *et al.*, 1986). Especially valuable has been the combination of gas chromatography with mass spectrometry (Sharpe and Chappell, 1991). Over 500 compounds have been isolated and identified from wines. Alone, over 160 esters have been distinguished. Most of the compounds occur at concentrations between 10^{-4} to 10^{-9} g/liter. At these levels, most are below the limit of human sensory perception. Therefore, most compounds isolated from wine *individually* play no role in the sensory characteristics of wine. In combination, though, they may be very significant.

The vast majority of chemicals found in wine are the metabolic by-products of yeast activity during fermentation. By comparison, the number of aromatic and sapid (taste and touch) substances derived from grapes are comparatively few. Nevertheless, the latter often are the compounds that make one wine distinct from another.

Other than alcohol, wines generally contain about 0.8 to 1.2 g of aromatic compounds per liter. This constitutes about 1% of the wine's ethanol content. The most common aromatic compounds are fusel alcohols, volatile acids, and fatty acid esters. Of these, fusel alcohols often constitute 50% of all volatile substances in wine. Although present in much smaller concentrations, carbonyls, phenols, lactones, terpenes, acetals, hydrocarbons, and sulfur and nitrogen compounds are more important qualitatively to the varietal and specific sensory features of the fragrance of a wine.

The taste and mouth-feel sensations of a wine are due primarily to the few compounds that occur individually at concentrations above 0.1 g/liter. These include water, alcohol (ethanol), fixed acids (primarily tartaric and malic or lactic acids), sugars (glucose and fructose), and glycerol. Tannins are important sapid substances in red wines, but they occur rarely in significant amounts in white wines without maturation in oak cooperage.

Overview of Chemical Functional Groups

Wine consists of two primary ingredients, water and ethanol. However, the basic flavor of wine depends on an additional 20 or more compounds. The subtle differences that distinguish one varietal wine from another depend on an even larger number of compounds. Because of the bewildering array of compounds found in grapes and wine, a brief overview is given before discussing each group in detail. Table 6.1 illustrates most of the groups.

Organic chemistry deals with compounds containing carbon and hydrogen atoms. Other elements may be, and usually are, involved. Carbon atoms may bond covalently to one another, or other elements, to form straight chains, branched chains, or ring structures. When only hydrogen and carbon atoms are involved, the molecule is classified as a **hydrocarbon.** If some carbons in a hydrocarbon chain are joined by double covalent bonds, the compounds are said to be **unsaturated.** In contrast, **saturated** hydrocarbons contain only single covalent bonds. Linear-chain compounds are called **aliphatic,** whereas those forming at least one circular linkage of six carbon atoms (a benzene ring) are called **aromatic.** The latter term comes from the fact that the first identified benzene derivatives were fragrant. However, not all odorous compounds are chemically aromatic, nor are all chemically aromatic compounds odorous. Because "aromatic" is commonly used to refer to the olfactory property of volatile compounds, the term aromatic will be restricted in this text to the fragrance, not the chemical nature, of compounds.

Most organic molecules contain more than just carbon and hydrogen atoms. The additional elements, along with double covalent bonds, give organic molecules most of their chemical (reactive) characteristics. The region of a molecule that generally gives a compound its primary reactive property is called the **functional group.** However, molecules may contain more than one functional group. This can complicate the classification and chemical properties shown by compounds. Compounds with more than one type of functional group express to varying degrees the properties of the different functional groups they possess. Not surprisingly, compounds may be classified differently by various authors. For example, methyl anthranilate may be classified as a benzoic acid derivative (and therefore a phenyl derivative), as an ester (because of the ester group), or as a nitrogen compound (because of the amine group).

methyl anthranilate

Most atoms in organic molecules are connected by covalent bonds, in which electrons are shared between the associated atoms. The sharing is equal between two

Table 6.1A   Some Important Functional and Chemical Groups in Grapes and Wine[a]

Compound class	General structure	Functional group	Example
Aliphatic compounds			
Alcohols	$R-OH$	$-OH$	Ethanol
Carbonyls			
Aldehydes	$R-\overset{\overset{O}{\|\|}}{C}-H$	$-\overset{\overset{O}{\|\|}}{C}-H$	Acetaldehyde
Ketones	$R_1-\overset{\overset{O}{\|\|}}{C}-R_2$	$-\overset{\overset{O}{\|\|}}{C}-$	Diacetyl
Carboxylic acids	$R-\overset{\overset{O}{\|\|}}{C}-OH$	$-\overset{\overset{O}{\|\|}}{C}-OH$	Acetic acid
Esters	$R_1-\overset{\overset{O}{\|\|}}{C}-O-R_2$	$-\overset{\overset{O}{\|\|}}{C}-O-$	Ethyl acetate
Amides	$R-\overset{\overset{O}{\|\|}}{C}-NH_2$	$-\overset{\overset{O}{\|\|}}{C}-NH_2$	Acetamide
Amines	$R-NH_2$	$-NH_2$	Histamine
α-Amino acids	$H_2N-\overset{\overset{H}{\|}}{\underset{\underset{R}{\|}}{C}}-\overset{\overset{O}{\|\|}}{C}-OH$	$H_2N-\overset{\overset{H}{\|}}{\underset{\underset{R}{\|}}{C}}-\overset{\overset{O}{\|\|}}{C}-OH$	Alanine
Acetals	$R_1-O-\overset{\overset{H}{\|}}{\underset{\underset{R}{\|}}{C}}-O-R_2$	$-O-\overset{\overset{H}{\|}}{\underset{\underset{R}{\|}}{C}}-O-$	Acetal
Terpenes			Linalool
Thiols	$R-SH$	$-SH$	Ethanethiol
Thioesters	$R_2-\overset{\overset{O}{\|\|}}{S}-OR_1$	$-\overset{\overset{O}{\|\|}}{S}-O-$	Methyl thiolacetate

Table 6.1B Some Important Cyclic Chemical Compounds in Grapes and Wine[a]

Compound class	General structure		Example
Cyclic compounds			
Phenolics			Vanillin
Lactones			3-Methyl-γ-octalactone
Pyrazines			2-Methoxy-3-isobutylpyrazine
Pyridines			2-Acetyltetrahydropyridine
C_{13} Norisoprenoids		(probable precursor)	Vitaspirane
Thiolanes			2-Methylthiolane-3-ol
Thiazoles			5-(2-Hydroxyethyl-4-methythiazole)

[a] The letter R usually designates the rest of the molecule. Complex molecules may have more than one type of R group, designated by different subscripts.

or more carbon atoms, or between carbon and hydrogen atoms. Because no electrical charge develops around those bonds, organic molecules tend to be **nonpolar** and **hydrophobic** (poorly water soluble). Water solubility is largely dependent on the attraction between electrically charged groups on a molecule and adjacent water molecules. For most organic compounds, solubility is dependent on the presence of atoms such as oxygen and nitrogen, which often form **polar** covalent bonds with hydrogen. Because the electrons are unequally shared between the atoms, weakly charged (polar) areas form

that can associate with the charged (polar) regions of water molecules. In some compounds, one element completely removes one or more electrons from another, forming an **ionic** bond. In water, the components of an ionic molecule tend to dissociate, forming soluble charged components termed **ions**. The presence of ionic bonds in an organic molecule often greatly enhances the water solubility of the compound.

The symbol **R** is used to represent hydrogen, hydrocarbon, or hydrocarbon derivative groups attached to a carbon chain. These are either **alkyls**, if based on an aliphatic carbon chain, or **phenyls,** if based on a benzene derivative.

In grapes and wine, most organic functional groups are based on the double covalent bond (especially benzene derivatives) or on bonds with oxygen, nitrogen, and sulfur. The most prominent of the functional groups are based on carbon–oxygen bonds. A common example is the **hydroxyl** group that provides the properties of an

$$-OH \qquad -O-R \qquad -C=O \qquad -\overset{\overset{\displaystyle OH}{|}}{C}=O$$

hydroxyl   alkoxy   carbonyl   carboxyl

alcohol to organic compounds. Substitution of the hydrogen of the hydroxyl group with a carbon atom yields the **alkoxy** group that characterizes **ethers.** Another common oxygen-based functional group is the **carbonyl** group. Depending on the location in the group, it characterizes **aldehydes** (terminal position) or **ketones** (internal position). Association with a hydroxyl group on the same carbon generates the **carboxyl** group. The latter gives organic molecules the properties of an acid.

When an organic alcohol (hydroxyl) group reacts with an acid (carboxyl) group, it produces the **ester** linkage.

$$-\overset{\overset{\displaystyle O}{\|}}{C}-O-C \qquad\qquad -O-\overset{\overset{\displaystyle H}{|}}{\underset{\underset{\displaystyle R}{|}}{C}}-O-$$

ester   acetal

The interaction of the hydroxyl group of alcohols with the carbonyl group of an aldehyde generates the **acetal** group of **acetals.** Oxygen can also be involved in the formation of carbon ring structures. An example is the **cyclic ester** group called **lactones.**

An important group of aromatic compounds in wines are the **terpenes.** Although not possessing a particular functional group, they are constructed from repeating groups of five carbon atoms in a distinctive arrangement called the **isoprene** unit.

isoprene

Additional functional groups are based on carbon–nitrogen linkages. The most important of these is the **amine** group. Alone, it forms compounds called **amines.** Associated with a carbonyl group on the same carbon atom forms the **amide** grouping. With a carboxyl group on an adjacent carbon atom, the amine group forms the **amino** grouping that characterizes **α-amino acids.** Nitro-

$$\overset{\overset{\displaystyle |}{}}{C}-\overset{\overset{\displaystyle |}{}}{N}- \qquad -\overset{\overset{\displaystyle H_2N}{|}}{C}=O \qquad -\overset{\overset{\displaystyle H_2N}{|}}{C}-\overset{\overset{\displaystyle O}{\|}}{C}-OH$$

amine   amide   amino acid

gen also may be part of a carbon ring structure. Important examples are **pyrazines** and **pyridines.**

Sulfur-based functional groups generate important aromatic compounds in wine. They are often structurally analogous to functional groups based on oxygen. Examples are the **thiol** group of mercaptans, **thioethers** and

$$-SH \qquad\qquad -S- \qquad\qquad -\overset{\overset{\displaystyle O}{\|}}{C}-S-$$

thiol, sulfhydryl   thioether   thioester

thioesters. Sulfur, alone or along with nitrogen, also may be involved in the formation of ring structures. Examples are **thiolanes** and **thiazoles,** respectively.

Chemical Constituents

Water

The water content of grapes and wine is seldom discussed as its presence is taken for granted. Nevertheless, as the predominant chemical constituent of grapes and wine, water plays critical roles in establishing the basic characteristics of wine. For example, compounds not at least slightly soluble in water rarely play a significant role in wine. Water also governs the basic flow characteristics of wine. Even the occurrence of *tears* in a glass of wine is partially dependent on the properties of water. In addition, the high specific heat of water slows the warming of a glass of wine. Water is also an essential component in many of the chemical reactions involved in grape growth, juice fermentation, and wine aging.

Sugars

Sugars are a category of carbohydrates distinguishable by the presence of several hydroxyl groups and an aldehyde or ketone group. Simple sugars may bond to-

$$-\overset{\overset{\displaystyle H}{|}}{C}=O \qquad\qquad\qquad -\overset{\overset{\displaystyle O}{\|}}{C}-$$

aldehyde   ketone

gether to form polymers, such as pectins, gums, starches, hemicelluloses, and cellulose, or with other compounds, such as lactones and anthocyanidins, to form glycosides. Only some of the simpler sugars taste sweet.

The principal grape sugars are **glucose** and **fructose.** They often occur in roughly equal proportions at maturity, whereas overmature grapes often have a higher proportion of fructose. Sugars other than glucose and fructose do occur, but in relatively insignificant amounts. **Sucrose** is rarely found in *Vitis vinifera* grapes, but it may constitute up to 10% of the sugar content in non-*V.*

vinifera cultivars. Sucrose, whether natural or added, is enzymatically split into glucose and fructose during fermentation.

Grape sugar content varies depending on the species, variety, maturity, and health of the fruit. Cultivars of *Vitis vinifera* generally reach a sugar concentration of 20% or more by maturity. Other winemaking species, such as *V. labrusca* and *V. rotundifolia*, seldom reach this level. Sugar commonly needs to be added to the juice of the latter species to permit the production of the standard 10 to 12% alcohol content of most wines.

In North America, sugar content (**total soluble solids**) is measured in °Brix. Brix is a good indicator of berry sugar content at levels above 18°, when sugars become the predominant soluble solids in grapes (Crippen and Morrison, 1986). °Brix is compared with other specific gravity measurements such as Oechsle and Baumé in Appendix 6.1.

Grape sugar content is critical to yeast growth and metabolism. *Saccharomyces cerevisiae*, the primary wine yeast, derives most of its metabolic energy from glucose and fructose. Because *Sacch. cerevisiae* has limited abilities to ferment other substances, it is important that most grape nutrients be in the form of glucose and fructose. Unfermented sugars are collectively termed **residual sugars**. In dry wines, the residual sugar content consists primarily of pentose sugars, such as **arabinose, rhamnose,** and **xylose,** and small amounts of unfermented glucose and fructose (~0.1–0.2 g/liter). These levels may increase slightly during maturation in oak cooperage via the breakdown of glycosides in the wood. Sugars also may be synthesized and released by yeast cells. Trehalose is commonly found in botrytized wines.

The residual sugar content in dry wine is generally below 1.5 g/liter. At this concentration, the presence of sugar is undetectable on the palate. The nutritive value also is insufficient to constitute a major threat to the microbial stability of bottled wine. At concentrations above 0.5 g/liter residual, yeast fermentable sugars, sugars increasingly pose a microbial hazard, especially above 1.2 g/liter (Amerine *et al.*, 1980) and if the acid and alcohol contents of the wine are low. Special procedures are required to prevent undesirable yeast and bacterial growth in wine containing high sugar contents.

When the residual sugar content rises above 0.2%, particulary sensitive individuals may begin to detect sweetness. Generally, though, the sugar content must be over 1% to possess distinct sweetness for most people. The detection of sweetness is markedly influenced by other wine constituents, notably ethanol, acids, and tannins. Conversely, detectable sweetness has a mitigating effect on the perception of sourness and bitterness. Very sweet table wines, such as sauternes, trockenbeerenausleses, and eisweins, may reach residual sugar contents well over 10%.

Although residual sugars are of obvious importance to the sweetness of wine, fermentable sugars are absolutely essential for fermentation. The single most significant by-product of fermentation is ethanol. In addition, sugars are metabolized to higher alcohols, fatty acid esters, and aldehydes, which give different wines much of their individual aromatic character. It is the abundance of fermentable sugars in grapes, in contrast to other fruits, that probably made wine the first fruit-based alcoholic beverage discovered.

Slow structural transformations in sugars may be involved in the darkening of dry white wines during aging. This is especially so in the production of brown melanoidin pigments found in sweet sherries, madeiras, and similar fortified wines. These may develop by Maillard reactions between reducing sugars (primarily glucose and fructose in grapes) and various nitrogenous compounds (such as amino acids), or by direct thermal degradation of carbohydrates.

Sugar concentration also can increase the volatility of aromatic compounds (Sorrentino *et al.*, 1986). The significance of sugars to the fragrance of fortified and sweet table wines is unknown.

Pectins, Gums, and Related Polysaccharides

Pectins, gums, and related substances typically are mucilaginous polymers of sugar acids that hold plant cells together. They commonly occur as complex branched chains. Being partially water soluble, they are extracted into the juice during crushing and pressing. Extraction is favored when whole or crushed grapes are heated to hasten anthocyanin liberation. During fermentation, the polysaccharides form complex colloids in the presence of alcohol and tend to precipitate. Consequently, grape pectins, gums, and glucosans seldom cause serious wine clouding or filtration problems. However, their presence can cause difficulty during the pressing of pulpy grapes, notably *V. labrusca* cultivars. The addition of pectinase following crushing significantly reduces the pectin content and their effect on pressing.

In contrast to the polysaccharides of healthy grapes, β-glucans produced in *Botrytis*-infected grapes can cause severe problems. They hinder juice and wine clarification by inhibiting the precipitation of other colloidal materials, such as tannins and proteins. In addition, β-glucans form a fibrous mat on filters, plugging the pores, whereas grape polysaccharides produce nonplugging, spherical colloids (Ribéreau-Gayon *et al.*, 1980). Because *Botrytis* glucans are localized under the grape skin, gentle harvesting and pressing can minimize their extraction.

The polysaccharide level in finished wine is generally low. The significance of polysaccharides to the sensory properties of wine has not been adequately studied, but it is commonly believed to be negligible.

Alcohols

Alcohols are organic compounds containing one or more hydroxyl groups (—OH). Simple alcohols contain a single hydroxyl group, while diols and polyols contain two or more hydroxyl groups, respectively. Phenols are six-carbon ring compounds containing one or more hydroxyl groups on the phenyl ring. The chemistry of phenols is sufficiently distinct that they are treated separately below.

ethanol (an alcohol)

glycerol (a triol)

phenol (a phenol)

ETHANOL

Ethanol is undisputably the most important alcohol in wine. Although small quantities are produced in grape cells during carbonic maceration, the primary source of ethanol is yeast fermentation. Ethanol is the principal organic by-product of fermentation.

Under standard fermentation conditions, ethanol can accumulate up to about 14 to 15%. Higher levels can be reached by the sequential addition of sugar during fermentation. Generally, though, ethanol concentrations above 14% in wine are the result of fortification. The prime factors controlling ethanol production are sugar content, temperature, and yeast strain.

Alcohol content is variously indicated in terms of percent by volume (vol %), percent by weight (wt %), grams per 100 ml, specific gravity, or proof. These expressions are compared in Appendix 6.2. In the text, percent by volume is implied unless otherwise indicated.

Besides its significant physiological and psychological effects, ethanol is crucial to the stability, aging, and sensory properties of wine. During fermentation, the increasing alcohol content increasingly limits the growth of most microorganisms. The relative alcohol insensitivity of *Saccharomyces cerevisiae* assures that it dominates the process of fermentation. Microbes that might produce off-odors are generally inhibited. The inhibitory

action of ethanol, combined with the acidity of the wine, permits wine to remain sound for years in the absence of air.

Ethanol acts as an important solvent in the extraction of pigments and tannins during red wine vinification. By affecting the metabolic activity of yeasts, ethanol also influences the types and amounts of aromatic compounds produced. Furthermore, ethanol acts as an essential reactant in the formation of volatile compounds and adds its own distinctive odor.

Ethanol has multiple effects on taste and mouth-feel. It directly enhances sweetness through its own sweet taste. It indirectly modifies the perception of acidity, making acidic wines appear less sour and more balanced. At high concentrations, alcohol produces a burning sensation and contributes to the feeling of weight or *body*, especially in dry wines. Ethanol also can increase the intensity of bitterness and decrease the sensation of astringency of tannins (Lea and Arnold, 1978).

The role of alcohol as a solvent has already been mentioned concerning pigment and tannin extraction from grapes. Ethanol also helps to dissolve volatile compounds produced during fermentation and those formed during maturation in wood cooperage. The dissolving action of the alcohol probably reduces the escape of aromatic compounds with carbon dioxide during fermentation. Conversely, at low concentrations (0.5–0.75%), alcohol enhances the release of certain aromatic compounds (Williams and Rosser, 1981). This may be important in the expression of the finish of a wine.

Ethanol plays several roles in the aging of wine. Along with other alcohols, ethanol slowly reacts with organic acids to produce esters. The ethanol concentration also influences the stability of esters. Furthermore, ethanol reacts slowly with aldehydes to produce acetals.

METHANOL

Methanol is not a major constituent in wines, nor is it considered important in flavor development. Within the usual range (0.1–0.2 g/liter), methanol has no direct sensory effect. Of the over 160 esters found in wine, few are associated with methanol.

The interest shown in methanol derives from its oxidation to formaldehyde and formic acid in the body. Both metabolites are toxic to the central nervous system. One of the first targets of formaldehyde is the optic nerve, damage to which causes blindness. Methanol never accumulates to toxic levels under legitimate winemaking procedures.

Methanol is predominately generated from the enzymatic breakdown of pectins. On degradation, methyl groups associated with pectin are released as methanol. The methanol content of fermented beverages is primarily a function of the pectin content of the fermentable substrate. Unlike most fruits, grapes are low in pectin. Thus, wine generally has the lowest methanol content of any fermented beverage. Pectolytic enzymes added to juice or wine to aid clarification inadvertently increase the methanol content of the wine. Adding distilled spirits to a wine also may slightly increase the methanol content. The concentration of ethanol and other flavor compounds achieved with distillation unintentionally augments the methanol content of the distillate.

HIGHER (FUSEL) ALCOHOLS

Alcohols with more than two carbon atoms are commonly called **higher,** or **fusel, alcohols.** They may be present in healthy grapes but seldom occur at significant levels. Hexanols are the major exception to this generalization, and they produce herbaceous odors in certain wines. Other potentially significant higher alcohols that survive fermentation unchanged are **2-ethyl-1-hexanol, benzyl alcohol, 2-phenylethanol, 3-octanol,** and **1-octen-3-ol.** However, most higher alcohols found in wine occur as by-products of yeast fermentation. Their synthesis closely parallels that of ethanol production (Fig. 6.1). They commonly account for about 50% of the aromatic constituents of wine, excluding ethanol.

Quantitatively, the most important higher alcohols are the straight-chain alcohols: **1-propanol, 2-methyl-1-propanol** (isobutyl alcohol), **2-methyl-1-butanol,** and **3-methyl-1-butanol** (isoamyl alcohol). **2-Phenylethanol** (phenethyl alcohol) is the most important phenol-derived higher alcohols.

Most straight-chain higher alcohols have a strong pungent smell. At low concentrations (~0.3 g/liter or less), they generally add an aspect of complexity to the bouquet. At higher levels, they increasingly overpower the fragrance. In distilled beverages, such as brandies and whiskeys, fusel alcohols give the beverage much of its distinctive aromatic character. Only in *porto* wine is a distinctive fusel character expected and appreciated. This property comes from the brandy added during port production.

The formation of higher alcohols during fermentation is markedly influenced by winery practices (see Sponholz, 1988). Synthesis is favored by the presence of oxygen, high fermentation temperatures, and the presence of suspended material in the fermenting must. Chaptalization and pressure-tank fermentation also tend to enhance synthesis of higher alcohols. Conversely, prefermentative clarification, the presence of sulfur dioxide, and low fermentation temperatures suppress production. Yeasts also vary in ability to produce higher alcohols.

Higher alcohols may originate from grape-derived aldehydes, by the reductive denitrification of amino acids, or via synthesis from sugars. The relative importance of those sources appears to vary with the specific fusel alcohol. Amino acid deamination is especially important in the generation of longer chain higher alcohols (Chen, 1978).

Additional higher alcohols come from the metabolic action of spoilage yeasts and bacteria. Wines showing a

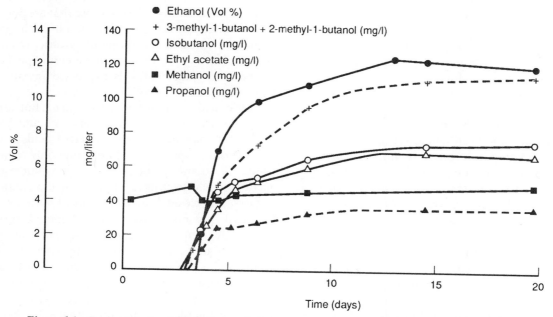

Figure 6.1 Production of various alcohols and ethyl acetate during fermentation. (From Rapp and Mandery, 1986, reproduced by permission.)

fusel off-odor usually have been infected at some stage by spoilage organisms. Pleasant smelling higher alcohols may be produced by microorganisms; for example the so-called mushroom alcohol (1-octen-3-ol) synthesized during "noble" rotting by *Botrytis cinerea* (see Rapp and Güntert, 1986).

Higher alcohols also play an indirect role in the development of an aged wine bouquet. By reacting with organic acids, they add to the number of esters found in wine. During fermentation, the production of esters occurs rapidly under the control of yeast enzymes. Esterification continues during aging, but at a much slower, nonenzymatic pace (Rapp and Güntert, 1986).

OTHER ALCOHOLS

In addition to the alcohols mentioned above, there are important terpene- and phenol-derived alcohols. Because they show stronger terpenic and phenolic characteristics than alcoholic properties, respectively, they are discussed under their predominant chemical associations.

DIOLS, POLYOLS, AND SUGAR ALCOHOLS

The most prominent diol in wine is **2,3-butanediol** (2,3-butylene glycol). It has little odor and only a mildly bittersweet taste. It appears to have little sensory significance in wine.

By far the most prominent of wine polyols is **glycerol**. In dry wine, glycerol is commonly the most abundant compound, after water and ethanol. Because of its high concentration, it has been assumed to be of sensory importance, notably in viscosity. Nevertheless, glycerol rarely reaches a concentration that can perceptibly affect viscosity (≥ 26 g/liter) (Noble and Bursick, 1984). Glycerol has a slight sweet taste but is unlikely to be noticeable in a sweet wine. It might play a minor role in dry wines, where the concentration of glycerol often surpasses the sensory threshold for sweetness (>5 g/liter).

Glycerol is an important nutrient source for growth of flor yeasts in sherry production. Certain spoilage bacteria also may metabolize glycerol.

Variety, maturity, and health all affect the amount of glycerol present in grapes. Infection by *Botrytis cinerea* produces juice with the highest glycerol contents. During fermentation, yeast strain, temperature, sulfur dioxide and pH level can influence the synthesis of glycerol.

Sugar alcohols, such as **alditol, arabitol, erythritol, mannitol, *myo*-inositol,** and **sorbitol,** commonly are found in small amounts in wine. Higher concentrations usually are the result of fungal infection of the grapes or bacterial growth in the wine. Sugar alcohols can be oxidized by some acetic acid bacteria to the respective sugars. The sensory and enological significance of this conversion is unknown. Combined, polyols and sugar

alcohols may have a slight effect on the sensation of *body*.

Acids

Acids are characterized by ionization and the release of hydrogen ions (H^+) into water. With organic com-

$$R - COOH \rightleftharpoons R - COO^- + H^+$$

$$\underset{\text{acid}}{\text{organic}} \qquad \underset{\text{radical}}{\text{carboxyl}} \quad \underset{\text{ion}}{\text{hydrogen}}$$

$$H_2CO_3 \rightleftharpoons HCO_3^- + H^+ \rightleftharpoons CO_3^{2-} + 2H^+$$

$$\underset{\text{acid}}{\text{carbonic}} \qquad \underset{\text{ion}}{\text{bicarbonate}} \qquad \underset{\text{ion}}{\text{carbonate}}$$

pounds, this property is primarily associated with the carboxyl group. The carboxyl group dissociates into a negatively charged carboxyl radical and a free, positively charged hydrogen ion. Inorganic acids, such as carbonic acid, dissociate into negatively charged ions and one or more positively charged hydrogen ions. The degree of dissociation in wine depends primarily on the pH and the characteristics of the particular acid.

The principal inorganic acids in wine are carbonic and sulfurous acids. Both also occur as dissolved gases, namely, CO_2 and SO_2, respectively. Because they are more important as gases in wine, and do not noticeably affect wine pH or perceptible acidity, discussion of these acids is incorporated under dissolved gases.

Acidity in wine is customarily divided into two categories: volatile and fixed. **Volatile acidity** refers to acids that can be readily removed by steam distillation, whereas **fixed acidity** refers to those that are poorly volatile. **Total acidity** is the combination of both categories. It may be expressed in terms of tartaric, malic, citric, lactic, sulfuric, or acetic acid equivalents (Appendix 6.3). In the text, total acidity is expressed in tartaric acid equivalents.

Acetic acid is the main volatile acid, but other carboxylic acids such as **formic, butyric,** and **propionic** acids also may be involved. Related acids possessing hydrocarbon chains ($>C_{10}$) are commonly termed **fatty acids.** All volatile carboxylic acids have intense odors: acetic acid is vinegarlike, propionic acid is characterized as fatty, butyric acid resembles rancid butter, and C_6 to C_{10} acids possess a goaty odor. Although these acids commonly occur in wine, they typically occur at detectable levels only in wines spoiled by microbes. Because acetic acid is the major volatile acid, volatile acidity is usually measured in terms of its presence alone.

Fixed acidity refers to all organic acids not included under the volatile category. Quantitatively, they control the pH of wine. In grapes, **tartaric** and **malic acid** often constitute more than 90% of the fixed acidity. The same two acids dominate the acid composition of wines. If the wine undergoes malolactic fermentation, malic acid is replaced by the smoother tasting lactic acid. Fermentation has little effect on total acidity, but it does increase the types of acids present. The increased complexity may play a minor role in the development of an aged bouquet.

Other organic acids, such as **citric, isocitric, fumaric, and α-ketoglutaric,** are TCA (tricarboxylic acid) cycle intermediates from grape or yeast metabolism (Fig. 7.12). They can be produced on the metabolic breakdown of sugars, amino acids, and fatty acids. Most of these acids are found in minor amounts and are generally not known to be of sensory significance in wine. A possible exception may be β-ketoglutarate; it can bind sulfur dioxide and reduce its free, active level in wine. Occasionally, citric acid may be added to acidify high pH wines. Tartaric acid also may be used for the same purpose.

Sugar acids, such as **gluconic, glucuronic,** and **galacturonic acids,** are associated with grape infection by *Botrytis cinerea.* Gluconic acid is so characteristic of this disease that its presence has been used as an indicator of the degree of infection (McCloskey, 1974). Nevertheless, these acids appear to be produced primarily by acetic acid bacteria that grow concurrently with *Botrytis* in the fruit (Sponholz and Dittrich, 1984).

Other than as a diagnostic tool, gluconic acid appears to have no enological significance. None of the sugar acids affects taste or odor. However, galacturonic acid may be involved in the browning of white wines (Jayaraman and van Buren, 1972). The production of brown pigments is catalyzed by copper and iron ions. Besides being formed by bacterial synthesis, galacturonic acid may be liberated through the enzymatic breakdown of pectins.

Phenolic acids are either synthesized in grapes or yeast cells via the shikimic acid pathway or are extracted from oak cooperage. They may add slightly to wine bitterness. Amino acids are usually excluded from a discussion of organic acids because of the dominating influence of the amino group on the molecule. They are correspondingly discussed under nitrogen compounds. Equally, vitamins and growth regulators with acidic groups are discussed under their main chemical headings (see section on Macromolecules and Growth Factors).

As a group, acids are almost as important to wines as alcohols. Acids not only produce a refreshing taste (or sourness if in excess), but they also modify the perception of other taste and mouth-feel sensations. This is especially noticeable in a reduction in perceived sweetness. In addition, the release of acids from cell vacuoles on crushing is probably instrumental in the initiation of acid hydrolysis of nonvolatile precursors in fruit (Winterhal-

ter *et al.,* 1990). Several important aroma compounds, such as monoterpenes, phenolics, C_{13} norisoprenoids, benzyl alcohol, and 2-phenylethanol may occur in the fruit of several grape varieties as acid-labile nonvolatile glycosides (Strauss *et al.,* 1987a).

The role of acids in maintaining a low pH is crucial to the color stability of red wines. As the pH rises, anthocyanins decolorize and may eventually turn blue. Acidity also affects the ionization of other phenolic compounds. The ionized (phenolate) form of phenols is more readily oxidized than is the nonionized form. Consequently, wines of high pH are much more likely to become oxidized and lose their fresh aroma and young color (Singleton, 1987).

Acids are involved in the precipitation of pectins and proteins that otherwise could cloud a finished wine. Conversely, acids can solubilize copper and iron, which can induce haziness (*casse*).

The low pH produced by wine acids has a beneficial antimicrobial effect. At low pH values, most bacteria do not grow. Low pH values also enhance the antimicrobial properties of fatty acids. Fatty acids are more toxic in the undissociated (nonionized) state found under acidic conditions (see Doores, 1983). However, by inhibiting yeast metabolism, carboxylic acids such as decanoic acid can favor the premature termination of fermentation (*sticking*).

At or just below the threshold levels, fatty acids contribute to the complexity of the bouquet of a wine. Acetic acid is considered sufficiently beneficial, at low concentrations to be permitted as an additive in wines. Above the recognition threshold, carboxylic acids have a negative influence on wine fragrance. The mild odors of lactic and succinic acids are generally considered inoffensive, as is the butterlike smell of sorbic acid.

During fermentation and aging, acids are involved in reactions leading to the formation of esters. These are often important to the fresh, fruity fragrance of wines. The influence of acidity, in favoring reductive reactions, favors the aging process and possibly the development of a desirable bottle bouquet. Low pH also facilitates the hydrolysis of disaccharides, such as trehalose, and various polysaccharides. This can slowly add fermentable sugars to wine.

ACETIC ACID

Small amounts of acetic acid are produced by yeasts during fermentation. At normal levels in wine (<300 mg/liter), acetic acid can be a desirable flavorant, adding to the complexity of the taste and odor of wine. It is equally important in the production of several acetate esters that can give wine a fruity character. Above 300 mg/liter, however, it progressively makes the wine sour and taints the fragrance. High levels of acetic acid are usually associated with contamination of grapes, juice, or wine with acetic or lactic acid bacteria.

MALIC ACID

Malic acid may constitute about half the total acidity of grapes and wine. Its concentration in the fruit tends to decrease as the grapes mature, especially during hot periods at the end of ripening. This can lead to the production of wine that tastes flat and is more liable to microbial spoilage. Conversely, under cool conditions, the malic acid level may remain high and give the resultant wine a sour taste. Therefore, malic acid content is one of the prime indicators used in determining harvest dates.

LACTIC ACID

A small amount of lactic acid is produced by yeast cells during fermentation. However, when lactic acid occurs as a major constituent in wine, it comes from the metabolic activity of bacteria. The bacteria most commonly involved are lactic acid bacteria. They produce an enzyme that decarboxylates malic acid to lactic acid. The process, called malolactic fermentation, is encouraged in red and some white wines produced in cool climates. The major benefit of malolactic fermentation is the conversion of the harsher tasting malic acid to the smoother tasting lactic acid. The predominance of L-lactic acid, one of the two stereoisomers of the acid, is usually an indicator of malolactic fermentation. In contrast, yeasts, and some bacteria synthesize equal quantities of both (L and D) forms of lactic acid during their metabolism of malic acid.

SUCCINIC ACID

Succinic acid is one of the more common by-products of yeast metabolism. It is resistant to microbial attack under anaerobic conditions and is particularly stable in wine. However, the bitter/salty taste of succinic acid limits its use in wine acidification.

TARTARIC ACID

Tartaric acid is the other major grape acid, along with malic acid. Unlike the case for malic acid, the concentration of tartaric acid does not drop markedly during grape ripening. In addition, tartaric acid is resistant to attack by most microorganisms. Tartaric acid is occasionally added to increase the acidity of low acid wines, but it carries the risk of increasing bitartrate instability.

Tartaric acid is commonly found associated with potassium as a tartrate salt in both grapes and wine. As wines age, tartrate salts crystallize and tend to precipitate. As chilling speeds the process, wines often are cooled before bottling to enhance early tartrate precipitation and avoid crystal formation in the bottle. Nevertheless, crystals may continue to form after bottling. This arises at least partially from the conversion of the natural (L form) of tartaric acid found in grapes to the D isomer. The calcium salt of both isomers is about one-eighth as soluble as the L-tartrate salt alone. Therefore, most wines form a salt deposit when aged sufficiently long.

Phenols and Related Phenol (Phenyl) Derivatives

Phenols are a large and complex group of compounds of particular importance to the characteristics and quality of red wines. They are also significant in white wines, but occur at much lower concentrations. Phenols, and related compounds, can affect the appearance, taste, mouth-feel, fragrance, and antimicrobial properties of wine. They may come from the fruit and vine stems, be produced by yeast metabolism, or be extracted from wood cooperage.

CHEMICAL GROUPS OF WINE PHENOLS AND RELATED COMPOUNDS

Chemically, phenols are cyclic benzene compounds possessing one or more hydroxyl groups associated directly with the ring structure. Although containing an alcohol group, they do not show the properties of an alcohol. Two distinct phenol groups occur in grapes and wine, namely, the **flavonoids** and the **nonflavonoids** (Table 6.2). In addition there are related (phenolic) compounds that do not possess one or more hydroxyl groups on the phenyl ring (and strictly are not phenols). In conformity with standard practice in the viticultural and enological literature, they are discussed here along with true phenols.

Flavonoids are characterized as molecules possessing two phenols joined by a pyran (oxygen-containing) carbon ring structure. The most common flavonoids in wine are **flavonols, catechins** (flavan-3-ols), and, in red wines, **anthocyanins.** Small amounts of free **leucoanthocyanins** (flavan-3,4-diols) also occur. Flavonoids may exist free, or polymerized to other flavonoids, sugars, nonflavonoids, or a combination of these. Those esterified to sugars or nonflavonoids are called, respectively, **glycosides** and **acyl** derivatives.

Polymerization of catechins and leucocyanidins produces a class of polymers called **procyanidins.** They may be classified based on the nature of the flavonoid monomers, their bonding, esterification to other compounds, or functional properties. The most common structural class in grapes and wine (type B) contains only single carbon bonds between the adjacent procyanidin subunits (Fig. 6.2). In some cultivars, structural differences exist between skin, stem, and seed procyanidins. There also are considerable differences in types and concentrations between cultivars (Kovač *et al.,* 1990). While procyanidins occur primarily as monomers in grapes (Dumon *et*

Table 6.2 Phenolic and Related Substances in Grapes and Wine[a]

General type	General structure	Examples	Major source[b]
Nonflavonoids			
Benzoic acid	COOH (benzene ring structure)	Benzoic acid	G, O
		Vanillic acid	O
		Gallic acid	G, O
		Protocatechuic acid	G, O
		Hydrolyzable tannins	G
Benzaldehyde	CHO (benzene ring structure)	Benzaldehyde	G, O, Y
		Vanillin	O
		Syringaldehyde	O
Cinnamic acid	CH=CHCOOH (benzene ring structure)	*P*-Coumaric acid	G, O
		Ferulic acid	G, O
		Chlorogenic acid	G
		Caffeic acid	G
Cinnamaldehyde	CH=CHCHO (benzene ring structure)	Coniferaldehyde	O
		Sinapaldehyde	O
Tyrosol	CH₂CH₂OH (benzene ring structure with OH)	Tyrosol	Y
Flavonoids			
Flavonols	(flavonol structure with R₁, R₂, OH, HO, O—Glucose)	Quercetin	G
		Kaempferol	G
		Myricetin	G

(*Continued*)

Table 6.2 (*Continued*)

General type	General structure	Examples	Major source[b]
Anthocyanins		Cyanin	G
		Delphinin	G
		Petunin	G
		Peonin	G
		Malvin	G
Flavan-3-ols		Catechin	G
		Epicatechin	G
		Gallocatechin	G
		Procyanidins	G
		Condensed tannins	G

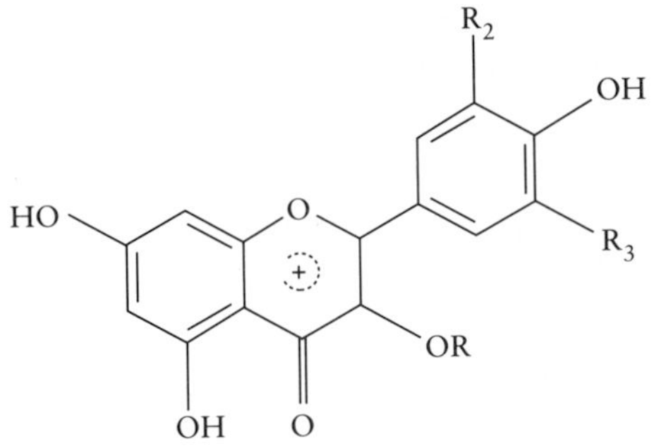

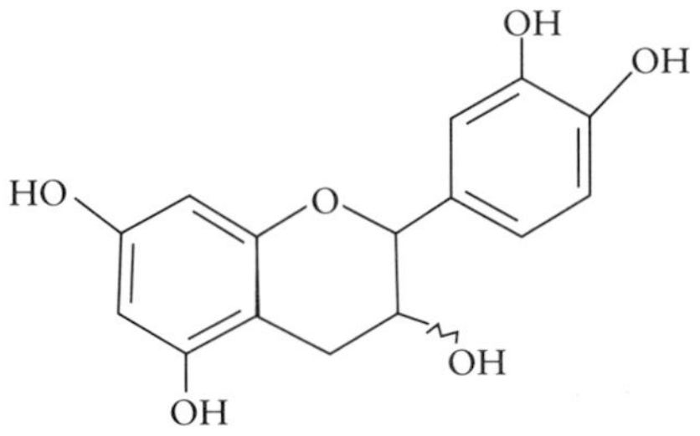

[a] Data from Amerine and Ough, 1980; Ribéreau-Gayon, 1964.
[b] G, Grape; O, oak; Y, yeast.

al., 1991), they tend to polymerize and predominate in wine as **condensed tannins,** mid-sized procyanidin polymers containing from two to five subunits. The presence of many unconjugated hydroxy–phenolic groups is thought to give procyanidins their distinctive protein-binding property. Larger polymers, being insoluble in aqueous solutions, do not react with or precipitate proteins.

Flavonoids are derived primarily from the skins, seeds, and stems (rachis and pedicels) of the fruit. Flavonols and anthocyanins come predominately from the skins, while catechins and leucoanthocyanins originate primarily from the seeds and stems. Flavonoids characterize red wines more than they do white wines. In red wines, they commonly constitute more than 85% of the phenol content (≥ 1000 mg/liter). In white wines, flavonoids typically comprise less than 20% of the total phenolic content (≤ 50 mg/liter).

The degree to which flavonoids are extracted is influenced by many factors. Extraction is ultimately limited by the amount present in the fruit. This content varies markedly from variety to variety (Table 6.3), with climatic conditions, and with fruit maturity. Traditional fermentation, probably owing to the longer maceration of the juice in contact with the seeds and skins, tends to extract more phenolic compounds than carbonic maceration or thermovinification. Phenol extraction also is markedly influenced by the pH, sulfur dioxide and ethanol contents of the juice, as well as the temperature, and duration of fermentation. Consequently, there is no simple means of predetermining the level of phenols extracted from grapes.

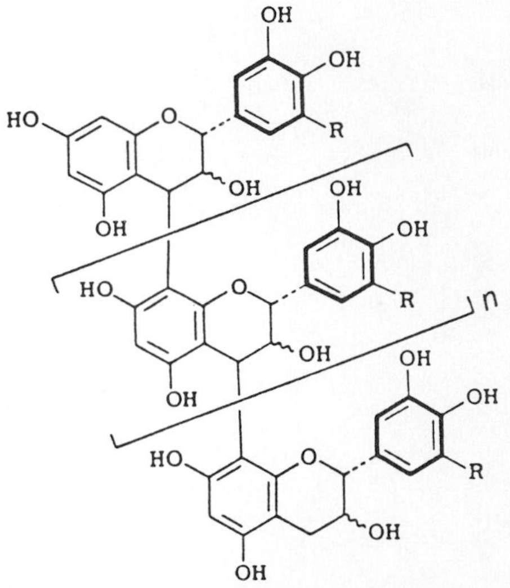

Figure 6.2 Structure of B-type procyanidin polymers (R = H, OH), with n = 0, 1, 2, 3, 4, 5, . . . (From Haslam, 1981, reproduced by permission.)

Table 6.3 Average Total Phenol and Anthocyanin Content of the Fruit of Different Grape Varieties in the South of France[a]

Variety	Level (mg/kg fresh weight)		% Individual monoglucoside anthocyanins		
	Total phenolics	Total anthocyanins	Delphinin	Petunin	Malvin and Peonin
'Colobel'	10,949	9967	20	21	28
'Pinot noir'	7722	631	4	12	77
'Alicante Bouschet'	7674	4893	7	10	55
'Cabernet Sauvignon'	6124	2339	17	8	48
'Syrah'	6071	2200	11	12	45
'Tempranillo'	5954	1493	25	16	41
'Gamay noir'	5354	844	1	3	64
'Malbec'	4613	1710	7	8	54
'Chardonnay'	4126	—	—	—	—
'Grenache'	3658	1222	7	10	63
'Carignan'	3582	1638	18	14	43
'Sauvignon blanc'	2446	—	—	—	—
'Villard blanc'	2280	—	—	—	—
'Cinsaut'	2154	575	6	9	51

[a] Data from Bourziex *et al.* (1983).

Due to the multiple factors affecting phenol extraction, it is not surprising that the phenol content shows greater variation than that of any other major wine constituent. In addition, there are greater quantitative and qualitative changes in the concentration and structure of phenolics during aging than in most other wine constituents.

Nonflavonoids are structurally simpler but their origin in wine more diverse. In wines not aged in oak, the primary nonflavonoids are derivatives of **hydroxycinnamic** and **hydroxybenzoic acids**. They are stored primarily in cell vacuoles of grape cells and are easily extracted on crushing. The most numerous are the hydroxycinnamic acid derivatives. They commonly occur esterified to sugars, organic acids, or various alcohols.

Wines matured in oak possess elevated levels of hydroxybenzoic acid derivatives, notably **ellagic acid**. Ellagic acid comes from the breakdown of **hydrolyzable tannins**, polymers of ellagic acid, or gallic and ellagic acids, with glucose (Fig. 6.3). Degradation of lignins in wood also liberates various **cinnamaldehyde** and **benzaldehyde** derivatives. In addition, small amounts of other nonflavonoid phenols, such as **esculin** and **scopoline**, are extracted from wood. Wines aged in chestnut cooperage extract slightly different, but related, hydrolyzable tannins and nonflavonoids.

Both flavonoid and nonflavonoid polymers are termed **tannins** because of their ability to tan leather. Nonflavonoid-based, hydrolyzable tannins separate readily into the component parts under acidic conditions. The low pH of wine weakens the bonding between the hydrogen and oxygen atoms of associated phenols. In contrast, flavonoid-based condensed tannins are relatively stable under similar conditions. They are held together by strong covalent bonds.

Yeast metabolism may provide additional nonflavonoid phenolics. The most prevalent is **tyrosol**. **Tryptophol** also is synthesized, but in smaller quantities.

Nonflavonoids of grape origin are initially synthesized from phenylalanine (Hrazdina *et al.*, 1984), while those of yeast origin are derived from acetic acid (Packter, 1980). Flavonoids are derived from the combination of derivatives synthesized from both phenylalanine (via the shikimic acid pathway) and acetic acid.

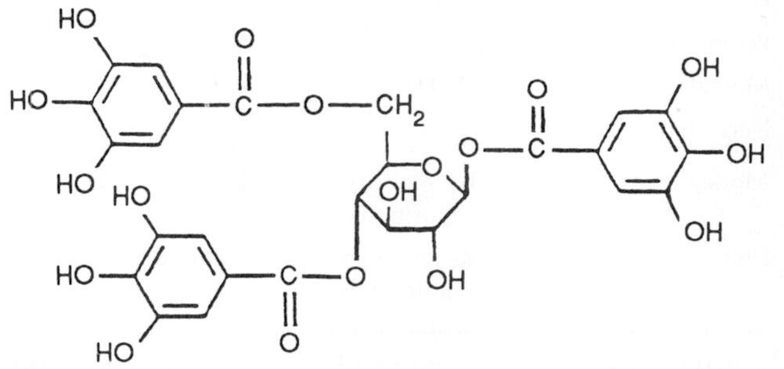

Figure 6.3 Structure of a hydrolyzable tannin. (From Bourzeix and Kovać, 1989, reproduced by permission.)

The phenolic content of wine increases during the early stages of fermentation, if the juice is in contact with the seeds and skins. Subsequently, the content begins to fall, as phenols bond and precipitate with proteins and cell remnants. During fining and maturation, the content continues to decline. Aging in wood cooperage results in a temporary increase in the phenolic content.

COLOR—RED WINES

Anthocyanins predominantly exist in grapes as glucosides, which form through the conjugation of the flavonoid component, called an **anthocyanidin**, with glucose. The sugar component increases the chemical stability and water solubility of the anthocyanidin. Each anthocyanin may be further complexed by acetic acid, coumaric acid, or caffeic acid bonding to the sugar component.

Anthocyanin classification is based primarily on the position of the hydroxyl and methyl groups on the B ring of the anthocyanidin molecule (Table 6.4). On this basis, grape anthocyanins are divided into five classes, namely, **cyanins, delphinins, malvins, peonins,** and **petunins.** The proportion and amount of each class vary widely among cultivars and growing conditions (Wenzel *et al.*, 1987). The proportion of anthocyanins markedly influences both hue and color stability. Both properties are directly affected by the hydroxylation pattern of the anthocyanidin B ring. Blueness increases with the number of free hydroxyl groups, while redness increases with the degree of methylation.

Sensitivity to oxidation in anthocyanins is influenced by the presence of *ortho*-diphenols (adjacent hydroxyl groups) on the B ring (Hrazdina *et al.*, 1970). The *o*-diphenol group is particularly sensitive to enzymatic and nonenzymatic oxidation. Except for laccase, most polyphenol oxidases oxidize only *o*-diphenol sites. As neither malvidin nor peonidin possesses ortho-positioned hydroxyl groups, they are the most resistant to oxidation. Resistance to oxidation also is affected by conjugation with sugar and other compounds (Robinson *et al.*, 1966). As the predominant anthocyanin in most red grapes is malvin—the reddest of anthocyanins—it commonly donates much of the color to young red wines.

Anthocyanins also are grouped based on the number of sugar molecules per anthocyanidin. In most grape species, both mono- and diglucosidic anthocyanins are produced. Diglucoside anthocyanins appear to be more stable than the monoglucoside counterparts, but they are more susceptible to browning (Robinson *et al.*, 1966). However, *Vitis vinifera* produces only monoglucosidic anthocyanins, as it lacks the dominant allele regulating the production of diglucosidic anthocyanins. Thus, first generation interspecies crosses with *V. vinifera* produce both mono- and diglucosidic anthocyanins. Complex hybrids involving several backcrosses to *V. vinifera* may produce only monoglucosidic anthocyanins as they may no longer possess the dominant allele for diglucoside synthesis (van Buren *et al.*, 1970). The use of diglucosidic anthocyanins to detect the illegal addition of juice from hybrid grapes in the production of Appellation Control (AC) red wines functions only because most red hybrid cultivars grown in Europe produce diglucosidic anthocyanins. The test would not work for 'Plantet' as it produces only monoglucosidic anthocyanins. This Seibel hybrid (#5455) is currently little grown in France.

In the fruit of red grape varieties, anthocyanins tend to exist in loose complexes, either with themselves or with other compounds. The anthocyanin conglomerates are held together by processes called **self-association** and **copigmentation** (see Somers and Vérette, 1988). Both lead to stacked molecular aggregates held together primarily by hydrophobic interactions between anthocyanidins and hydrophilic attractions of the glucose components. Self-association is more important at acidic pH levels, and copigmentation occurs more frequently at higher pH values. Increased free amino acid content in the fruit favors copigmentation. Flavonoid phenols, hydroxycinnamoyl esters, and polyphenols often are involved in these complexes (Somers, 1982). Both types of anthocyanin associations increase light absorption and color density.

Table 6.4 Anthocyanins Occurring in Wines[a]

Specific name	R_3	R_4	R_5
Cyanidin	OH	OH	
Peonidin	OCH$_3$	OH	
Delphinidin	OH	OH	OH
Petunidin	OCH$_3$ ·	OH	OH
Malvidin	OCH$_3$	OH	OCH$_3$

Derivatives	Structure
Monoglucoside	R_1 = glucose (bound at the glucose 1-position)
Diglucoside	R_1 and R_2 = glucose (bound at the glucose 1-position)

[a] After *Methods for Analysis of Musts and Wines*, M. A. Amerine and C. S. Ough, Copyright © 1980 John Wiley & Sons, Inc. Reprinted by permission of John Wiley & Sons, Inc.

Various factors may lead to disruption of the anthocyanin complexes. For example, heating grapes or must to improve color extraction (thermovinification) destabilizes the structure. This can lead to a serious loss of color during wine maturation, if insufficient tannins are extracted from the grapes or pomace. Alcohol also destabilizes the hydrogen bonding between anthocyanin aggregates. Consequently, must fermented in contact with the seeds and skins for only a few days may show a loss in color as fermentation continues (Fig. 6.4). The loss is primarily due to a decrease in light absorbency as the complexes dissociate, not to a decline in absolute anthocyanin content. Contained within the complexes are anthocyanins in quinoidal states (see Fig. 6.5), which lose their bluish color when freed into the acidic environment of wine.

In young red wine, anthocyanins occur predominantly in a dynamic equilibrium between five major molecular states, one bonded to sulfur dioxide and four free forms (Fig. 6.5). Most forms are colorless within the range of pH values found in wine. Color comes primarily from the small proportion of anthocyanin that exists in the red-colored **flavylium** state. This proportion depends on the pH and free sulfur dioxide content of the wine. Low

pH increases the concentration of the flavylium state, enhancing the red color of the wine. As the pH rises, the color density and proportion of anthocyanins in the flavylium state rapidly decrease. The blue/mauve cast of high pH wines comes from the slight increase in the proportion of **quinoidal** anthocyanins. However, the most common factor affecting color density is not pH, but the amount of free sulfur dioxide. Sulfur dioxide has the property of being an effective bleaching agent.

Early in the winemaking process, anthocyanin molecules tend to hydrolyze to anthocyanidins, losing their acetate, caffeate, or coumarate constituents along with the glucose component(s).

In addition to anthocyanins, flavonoid tannins are extracted from the grape skins, stems, and seeds (pomace) during fermentation. During the usual extended contact between the pomace and the fermenting juice, many phenolics are liberated. These compounds, mostly catechins and procyanidins, begin to polymerize with free anthocyanins and anthocyanidins. By the end of fermentation, about 25% of the anthocyanin content may have polymerized with tannins. This level may rise to 40% or more within about 1 year (Somers, 1982). Thereafter, polymerization continues at a slower pace until the level approaches 100% after several years (Fig. 6.6). Cultivar variability in the content of tannins able to react with anthocyanins has been suggested as one of the reasons for differences in color stability between red wines (McCloskey and Yengoyan, 1981). Lack of appropriate tannins appears to be involved in the color instability of muscadine wines (Sims and Morris, 1986). In addition, the absence of acylated anthocyanins and caffeoyl tartaric acid may limit color stability.

Polymerization is an important factor in stabilizing wine color by protecting the anthocyanidin molecule from oxidation or other chemical modifications. Polymerization makes anthocyanidins more resistant to decolorization by sulfur dioxide or high pH. In addition, more anthocyanin molecules are colored when covalently bonded with tannins. For example, about 60% of the polymerized anthocyanins are colored at a pH of 3.4, whereas only 20% of the equivalent free anthocyanin may be colored (Fig. 6.7). Polymerization increases the proportion of both the flavylium and quinoidal states. However, polymerization and oxidation change their tint to yellow-brown (Glories, 1984). Combined with the loss in red color, the yellow-brown of flavylium and quinoidal anthocyanin/tannin polymers results in the wine progressively taking on a brickish shade.

Because of the sensitivity of free anthocyanins to irreversible degradation, it is preferable that polymerization occur early during wine maturation. Initially, the procyanidin molecules found in wine are small and highly soluble, and form soluble complexes with anthocyanins.

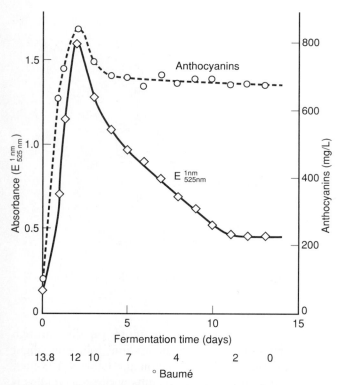

Figure 6.4 Changes in absorbance at 525 nm and in total anthocyanin content of the fermenting fluid during a traditional fermentation on skins. The must was pressed after 2 days. (From Somers, 1982, reproduced by permission.)

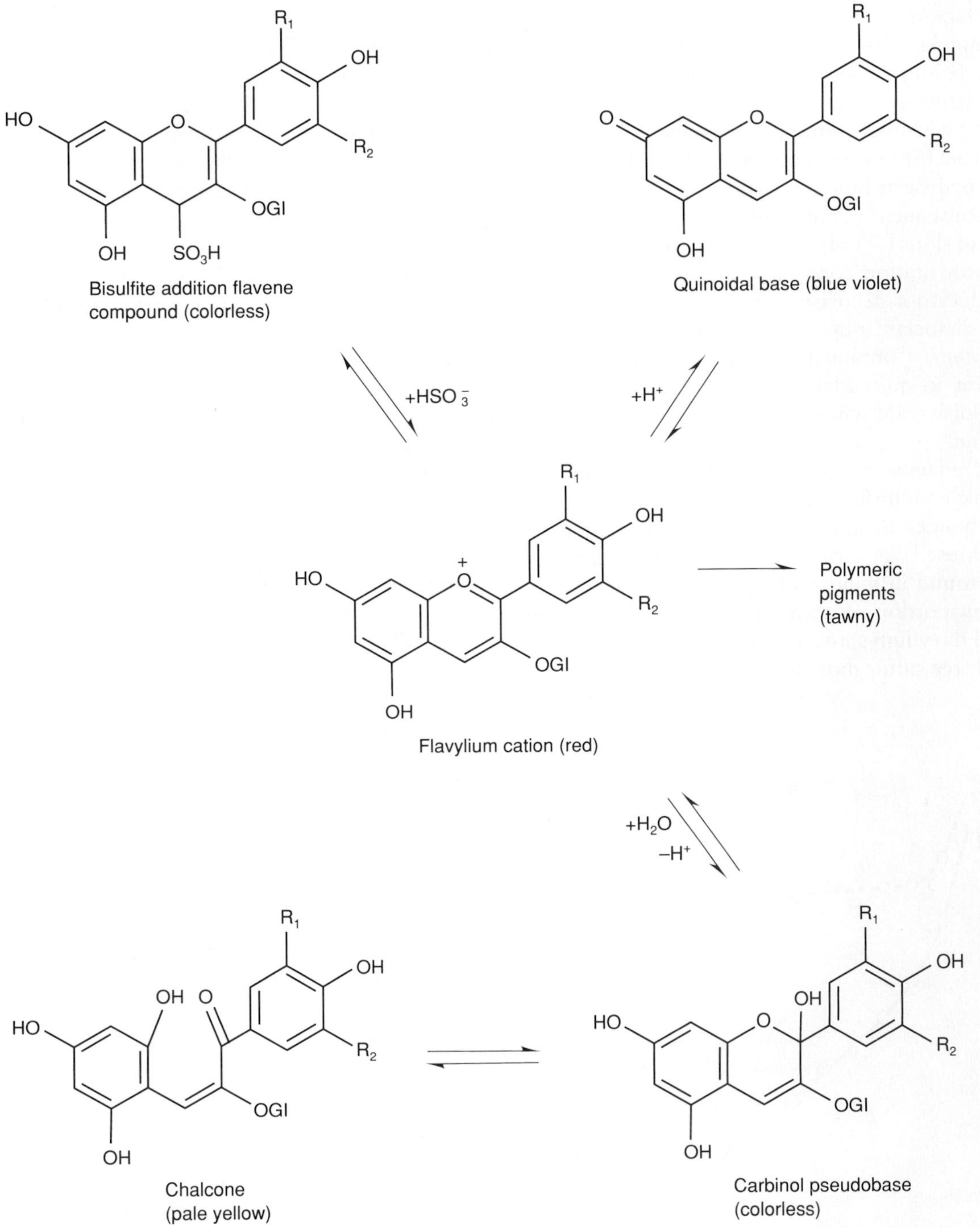

Figure 6.5 Equilibria between the different forms of anthocyanins in wine. Other molecular species are also present. Gl, glucose.

As aging progresses, procyanidins fuse into large condensed tannins. Reaction between anthocyanins and the large polymers often results in pigment precipitation and color loss. Procyanidin/anthocyanin polymerization occurs in the absence of oxygen through the enhanced concentration of the flavylium state at low pH values (Somers and Evans, 1986). However, polymerization is commonly believed to occur more rapidly through acetaldehyde-induced copigmentation, following slight wine oxidation (Ribéreau-Gayon and Glories, 1987).

Polymerization of anthocyanins with procyanidins is enhanced in the presence of acetaldehyde. This probably explains the color enhancing and stabilizing effect of exposing young red wines to small amounts of oxygen

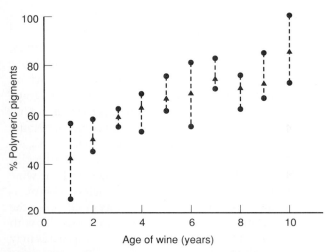

Figure 6.6 Increasing contributions of polymeric pigments to wine color density during the aging of 'Shiraz' wines. ▲, Mean values; •, extremes. (From Somers, 1982, reproduced by permission.)

(about 40 mg O_2 per year) before bottling (Ribéreau-Gayon *et al.*, 1983). Acetaldehyde is produced from ethanol by reaction with a strong oxidant synthesized during the oxidation of phenols. Acetaldehyde also may reverse the bleaching action of sulfur dioxide by removing it from its association with anthocyanins.

Alcohol and sugars also aid color enhancement. This is thought to result from their polymerization with anthocyanins. Free anthocyanins also may complex with peptides and polysaccharides.

As red wine ages, its coloration changes to reflect the increasing proportion of anthocyanin/tannin polymers. They give variously yellow, yellow-red, yellow-brown, red, and violet shades, depending on their chemical na-

ture (Table 6.5). Because most polymers have a brownish hue, the wine progressively takes on a more brickish shade. Color density also diminishes with time. This may result from the destruction of free anthocyanins, but most likely comes from the gradual formation and precipitation of pigment polymers.

COLOR — WHITE WINES

In contrast to red wines, comparatively little is known about the chemical nature and development of white wine color. Of the small amount of phenolic material found in white wines, most consists of readily soluble nonflavonoids such as caftaric acid (caffeoyl tartaric acid) and the related derivatives *p*-coumaric acid and ferulic acid (Lee and Jaworski, 1987). If the juice is treated with pectic enzymes, caftaric acid is largely hydrolyzed to its components, caffeic and tartaric acids (Singleton *et al.*, 1978). On crushing, caftaric acid and related compounds rapidly oxidize and associate with glutathione to form *s*-glutathionyl complexes. The latter may hydrolyze in wine to form other glutathione–cinnamate compounds. Unless further oxidized, notably by laccase, the glutathione–hydroxycinnamic acid complexes usually do not turn brown (Singleton *et al.*, 1985).

Because flavonols and other flavonoid phenols are extracted slowly, they are only found in significant quantities in juice macerated with the pomace. Those found are primarily catechins and catechin–gallate polymers (Lee

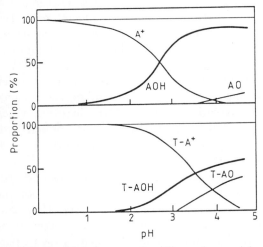

Figure 6.7 Equilibria between the different forms of free anthocyanins (A) and combined anthocyanins (T-A) extracted from wine +, red flavylium ion; OH, colorless carbinol pseudobase; O, blue violet quinoidal base. (From Ribéreau-Gayon and Glories, 1987, reproduced by permission.)

Table 6.5 Color and Molecular Weight of Several Wine Phenols[a,b]

Name[b]	Color	Molecular weight
A^+	Red	⎤
AOH	Noncolored	
AO	Violet	500
$AHSO_3$	Noncolored	⎦
P	Noncolored	600
T	Yellow	1000–2000
$T\text{-}A^+$	Red	⎤
T-AOH	Noncolored	
T-AO	Violet	1000–2000
$T\text{-}AHSO_3$	Noncolored	⎦
TC	Yellow-red	2000–3000
TtC	Yellow-brown	3000–5000
TP	Yellow	5000

[a] From Ribéreau-Gayon and Glories (1987), reproduced with permission.

[b] A, Anthocyanin; P, procyanidin; T, tannin; TC, condensed tannin; TtC, very condensed tannin; TP, tannin condensed with polysaccharides.

and Jaworski, 1987). It is believed that most of the yellow coloring in young white wine is derived from the limited extraction and oxidation of flavonols such as quercetin and kaempferol. Nonflavonoids and lignins extracted from oak cooperage during wine maturation also may add significantly to the color of white wines. The darkening that develops during aging may come from the oxidation of existing phenols or galacturonic acid, the formation of melanoidin compounds by Maillard reactions, or the caramelization of sugars.

TASTE AND MOUTH-FEEL

Flavonoid tannins have a marked influence on the taste and mouth-feel of red wines as they constitute the major group of phenolic compounds. In contrast, nonflavonoids constitute the major phenolic group in white wine.

Although abundant in red wines, anthocyanins directly contribute little to the taste of wine. Nevertheless, polymerization of anthocyanins with tannins is important to the retention of tannins in wine (Singleton and Trousdale, 1992). Thus, the absence of anthocyanins in white wine helps to explain the lower astringency of white wines given long skin contact or fermented with the skins.

Flavonoids are slightly bitter, but because of their low concentration relative to other flavonoids in red wines, they likely have a sensory impact only in white wine. For example, the bitter flavanone glycoside **naringin** could contribute to the bitterness of varietal wines such as 'Riesling' and 'Silvaner' (Drawert, 1979).

Catechins, and their polymers the procyanidins and condensed tannins, are the major sapid substances in red wine. As they occur at well above threshold levels, they constitute the predominant source of bitter and astringent sensations. The smaller catechins and procyanidins are relatively more bitter than astringent. Small condensed tannins are both bitter and astringent. The larger condensed tannins have little influence on taste. They appear to be too massive to react well with taste receptors or to precipitate proteins.

The small flavonoid content of white wines usually consists of flavan-3-ols (catechins) and flavan-3,4-diols (leucoanthocyanins). They give the wine *body* and can add to the perceived quality of the wine. The quality classification of German wines, customarily based on fruit ripeness, has been correlated with increasing flavonoid content (Dittrich *et al.*, 1974). Nevertheless, the role of flavonoids in browning and bitterness (~40 mg/liter) places an upper limit on their desirability.

Hydrolyzable tannins, extracted from wood cooperage, can play a significant role in the bitterness and astringency of wine. As a group, they tend to be more astringent than condensed tannins. Their significance is generally greater in white wines, which have milder flavors and generally lack significant amounts of flavonoid tannins. Cinnamaldehyde and benzaldehyde derivatives, extracted from oak cooperage, add to the cumulative bitterness of the nonflavonoid content of white wines. The derivatives of hydroxycinnamic acid found in grape juice are usually joined via ester linkages to tartaric acid. They often occur at concentrations that could contribute to bitterness in white wines (Ong and Nagel, 1978).

Most phenolic acids such as caftaric acid occur at concentrations below the detection thresholds in wine (Singleton and Noble, 1976). Nevertheless, combinations of phenolic acids have lower thresholds than the individual components. This property may increase with the alcoholic strength of wine. Thus, phenolic acids may contribute jointly to the bitterness and flavor of wine phenolics.

At the typical content of about 25 mg/liter, tyrosol could contribute to white wine bitterness. It may be more important in this regard in sparkling wines, where its level increases during the second fermentation.

Some phenol-related compounds may contribute to the peppery sensation associated with certain grape cultivars. 2-Phenylethanol and methyl anthranilate appear to have this effect. Other phenol derivatives generate a pungent mouth-feel or add to the varietal aroma of some cultivars. Besides the direct influence on bitterness and astringency, phenols have complex influences on the perception of sweetness and acidity. They also may have direct effects on the sensation of *body* and *balance*.

ODOR

Few volatile phenols or phenolic derivatives appear to come from grapes. **Acetovanillone,** which has a faintly vanillalike odor, is one exception. The most important, though, is **methyl anthranilate.** This phenol-derived ester is an important component of the characteristic aroma of some *V. labrusca* varieties. The roselike fragrance of **2-phenylethanol** often distinguishes varieties of *V. rotundifolia*. In addition, several volatile phenol derivatives, such as 2-phenylethanol, vanillin, and zingerone, occur in nonvolatile conjugated forms in several *V. vinifera* grape varieties. Release by enzymatic or acid hydrolysis could significantly influence the sensory impact of the compounds.

Instead of directly contributing to the varietal aroma of grapes, fruit phenolics appear to be more important as a source of hydroxycinnamic acid esters. They can be transformed into volatile phenols during fermentation. In this regard, **coumaric** and **ferulic acid esters** are particularly important. Fermentation and storage in oak are additional sources of the acids. The transformation of hydroxycinnamates into volatile phenols occurs under the action of enzymes derived from either yeasts or lactic acid bacteria. The derivatives, **4-ethyl guaiacol** and **4-vinyl guaiacol,** could add smoky, vanilla, and clovelike notes to wine. **Eugenol,** another clovelike guaiacyl compound, also may occur. At the usual concentration, eugenol is sufficient only to add a general spicy note. **Guaiacol** also may be generated, but the concentration is insufficient to influence the bouquet directly. At higher concentrations, guaiacol may be involved in an off-odor derived from contaminated stoppers. Guaiacol has a sweet, smoky odor (see Dubois, 1983).

Certain lactic acid bacteria generate volatile phenolics from nonphenolic compounds. An example is the synthesis of **catechol** from skikimic or quinic acids.

Besides 2-phenylethanol, the other major phenolic alcohol in wine is **tyrosol.** This yeast-synthesized phenol has a mild beeswax, honeylike odor. Whether it plays a role in the honeylike bouquet of certain wines, such as botrytized wines, is unknown.

Oak cooperage has already been mentioned as an important source of several volatile and nonvolatile phenolic acids. In addition, oak is the principal source of phenolic aldehydes (Chatonnet *et al.,* 1990), which are primarily derivatives of benzaldehyde and cinnamaldehyde. **Benzaldehyde** is the most prominent and possesses an almondlike odor. Benzaldehyde occurs in sufficient quantities in sherries to potentially participate in the nutlike bouquet of the wines. Benzaldehyde also may occur in wine following the oxidation of benzyl alcohol by enzymes produced in *Botrytis*-infected grapes (Goetghebeur *et al.,* 1992) or by some yeasts. Other important phenolic aldehydes are **vanillin** and **syringaldehyde,** which possess vanillalike fragrances. They form during the breakdown of lignins in wood. Another source of volatile phenolic aldehydes involves the heating of must or wine. For example, fructose is rapidly converted to 5-(hydroxymethyl)-2-furaldehyde during baking or, very slowly, during aging. 5-(Hydroxymethyl)-2-furaldehyde is considered to have a camomilelike odor. **Furfural** is commonly produced during distillation and the "toasting" of oak staves during barrel construction.

OXIDANT AND ANTIOXIDANT ACTION

Initially, it may seem contradictory that phenols are involved in both the activation and limitation of oxidation. This apparent anomaly results from the involvement of various phenols in the generation of peroxide and/or the reduction of oxygen to water. As a consequence, oxygen is consumed and unavailable to oxidize other wine constituents.

One of the initial reactions in must appears to involve the enzymatic oxidation of phenols, such as caftaric acid, to quinones. The latter may be reduced back to phenols through the coupled oxidation of procyanidins (Cheynier and Ricardo da Silva, 1991) or ascorbic acid. This

has the advantage of inducing the polymerization and early precipitation of readily oxidizable phenols during fermentation. Subsequently, nonenzymatic mechanisms become the predominant modes of oxidation in wine. Nonenzymatic oxidation primarily involves phenols containing two adjacently positioned phenol groups, termed *ortho*-diphenols (vicinal diphenols or catechols). In the presence of oxygen, *o*-diphenols are oxidized to *o*-quinones, and the oxygen probably reduced to hydrogen peroxide (Wildenradt and Singleton, 1974). Subsequently, the hydrogen peroxide oxidizes ethanol to acetaldehyde. The predominant substrate oxidized is thought to be ethanol because of its abundance in wine. Other readily oxidizable substrates in wine, besides phenols, are sulfur dioxide and amino acids.

Because quinones react with other phenols, they enhance phenol polymerization. By slow structural rearrangement, quinone/phenol dimers can generate new *o*-diphenol dimers (Fig. 6.8). The dimers subsequently may react with additional oxygen molecules, producing more peroxide. The *o*-quinone dimer so generated can react with additional phenols, producing even more complex polyphenols, while the additional peroxide can oxidize more ethanol to acetaldehyde. By structural rearrangements similar to those noted in Fig. 6.8, "nonoxidizable" phenols may continue to generate oxidizable *o*-diphenol sites. Because of the extensive degree of polymerization and rearrangement possible within complex-polyphenols, the ability of red wine slowly to assimilate oxygen is often considerable (Singleton, 1987).

Anthocyanin/tannin polymerization also is enhanced by the oxidative production of acetaldehyde. By reacting with anthocyanins, acetaldehyde forms a complex that facilitates bonding with catechin and procyanidins (Ribéreau-Gayon and Glories, 1987). This helps stabilize the color in red wine by limiting anthocyanin oxidation and decolorization by sulfur dioxide. In addition, it consumes acetaldehyde and restricts the development of a stale, oxidized odor.

The antioxidant (oxygen assimilating) action of phenols is improved if oxygen penetration is slow or infrequent. Under such conditions, the regeneration of new oxidizable phenols occurs sufficiently rapidly to assimilate the oxygen. With the rapid removal of oxygen, activation of less desirable oxidative reactions is limited. An additional advantage to the slight oxidation of red wines ($\sim$40 mg O_2/liter per year) is the precipitation of tannins through the generation of large condensed tannins (Ribéreau-Gayon *et al.,* 1983). This can reduce both wine bitterness and astringency.

Because of the efficient removal of oxygen, phenols can help maintain the low redox potential of a wine. This is considered beneficial in the development of an aged bouquet during long in-bottle maturation.

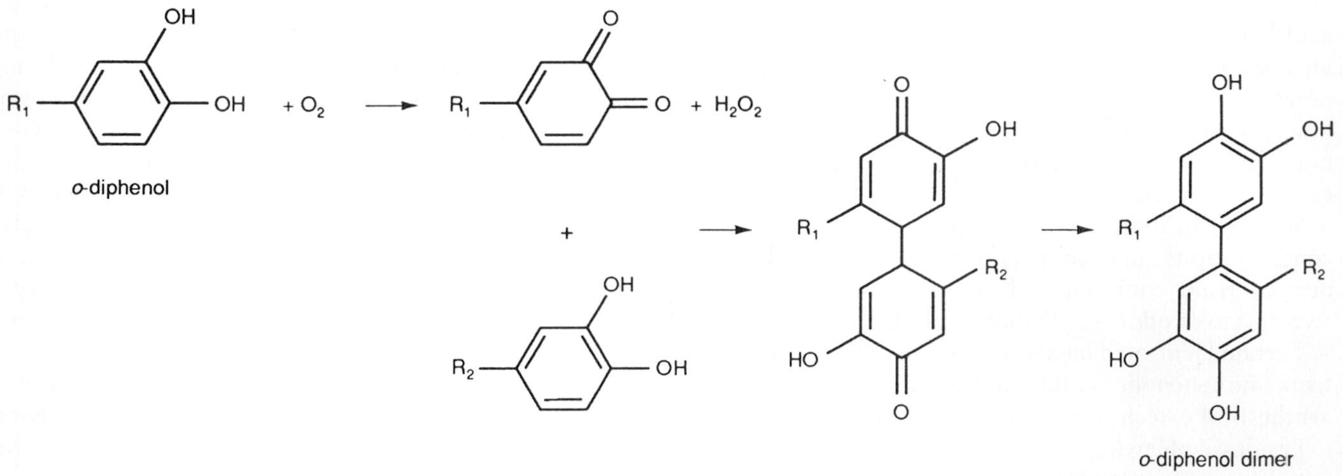

Figure 6.8   Generation of o-diphenol polymers following the oxidation of simpler o-diphenols to o-diquinones.

The low phenolic content of white wines makes them particularly susceptible to oxidative browning. Although the predominant phenol, caftaric acid, is readily oxidized by polyphenol oxidases, browning may be limited by the combination of the o-quinone by-products with glutathione (Singleton *et al.*, 1985). On glutathione depletion, o-quinones can activate the condensation of flavonoids (procyanidins) in the juice (Cheynier and Ricardo da Silva, 1991). If this occurs shortly after crushing, the brown polyphenols generated tend to precipitate, leaving the wine less susceptible to browning. This constitutes one of the primary advantages of permitting the limited juice oxidation that occurs normally during crushing.

Grape varieties have long been known to differ markedly in sensitivity to oxidation and oxidative browning. This is at least partially explained by the differing phenolic compositions and the diverse effects of various phenols to oxidative browning, both enhancing and inhibiting (Yokotsuka *et al.*, 1991). Grape cultivars also differ in the presence of phenol-oxidizing enzymes.

ANTIMICROBIAL ACTION

The action of wine against certain intestinal infections has been known for a long time—centuries before the source of the action was understood. Even now, the precise mechanisms by which wine, notably its phenols, exerts an antimicrobial action remain unclear. Even the particular phenols involved are unknown. Part of the problem arises from the diverse effects phenols have on living systems (see Scalbert, 1991) and the various abilities of different phenols to bind to substances. Owing to the binding of tannins with proteins, tannins can limit enzyme action by modifying enzyme solubility and structure. The latter probably results from restriction in the

movement of the catalytic site of the enzymes. Because bacteria and fungi digest complex nutrients exterior to their cytoplasm, inactivation of the digestive enzymes would be inhibitory, if not lethal. In addition, tannins can bind to membrane phospholipids and proteins, disrupting membrane function. Finally, phenols have strong chelating properties that can restrict microbial access to essential minerals.

Under normal fermentation conditions, phenols do not inhibit the growth or metabolism of yeasts and lactic acid bacteria. This may be partially due to their nutrient needs not requiring the action of exterior digestive enzymes. However, phenols can complicate the initiation of the second fermentation of sparkling wine. For this reason, red wines are seldom used in the production of sparkling wine.

CLARIFICATION

The effectiveness of tannins in precipitating proteins, and vice versa, is often used in wine clarification. Red wines may contain excessive amounts of tannins, which can make the wine overly astringent and generate large amounts of sediment. Conversely, white wines may contain excessively high levels of colloidal proteins, which can lead to haziness. Thus, proteins may be added as fining agents to remove the excess tannins in red wines, and tannins added to white wine to precipitate colloidal proteins.

Aldehydes and Ketones

Aldehydes are carbonyl compounds distinguished by the terminal location of the carbonyl functional group (—C=O) on the molecule. Ketones are related com-

acetaldehyde
(an aldehyde)

diacetyl
(an α-diketone)

phenyl acetaldehyde
(a phenolic aldehyde)

p-quinone
(a phenolic diketone)

pounds with the carbonyl group located on an internal carbon.

ALDEHYDES

Grapes produce few aldehydes important in the generation of varietal wine aromas. This may result from the reduction of aldehydes to alcohols during fermentation. Of those surviving fermentation, **hexanals** and **hexenals** (leaf aldehydes) appear to be the most significant. They may be involved in the grassy or herbaceous odor associated with certain grape varieties, such as 'Grenache' and 'Sauvignon blanc,' or with wines made from immature grapes. They appear to be formed during crushing of the fruit by the enzymatic oxidation of grape lipids. The dienal **2,4-hexadienal** also may be generated by the same process. However, most aldehydes in wine are produced during fermentation, processing, or exposure to oak cooperage.

Acetaldehyde is the major aldehyde found in wine. It often constitutes more than 90% of the aldehyde content of a wine. Above threshold values, it usually is considered an off-odor. Combined with other oxidized compounds, it contributes to the fragrance of sherries and other oxidized wines.

Acetaldehyde is one of the early metabolic by-products of fermentation (see Fig. 7.13). As fermentation approaches completion, acetaldehyde is transported back into yeast cells and reduced to ethanol. Thus, the acetaldehyde content usually falls to a low level by the end of fermentation. In fino sherries, most of the acetaldehyde accumulated is thought to be a by-product of the respiratory metabolism by film (*flor*) yeasts.

Another source of acetaldehyde in wine is the coupling of the autooxidation of *o*-diphenols and ethanol (Fig. 6.8). Other than generating the temporary stale odor in newly bottled wine ("bottle sickness"), acetaldehyde sel-

dom accumulates to detectable levels. It commonly reacts with sulfur dioxide or is consumed in the polymerization of acetaldehyde and procyanidins. In either case, the free acetaldehyde generally remains below the detection threshold.

Other aldehydes occasionally having a sensory impact in wine are **furfural** and **5-(hydroxymethyl)-2-furaldehyde,** synthesis of which involves sugar oxidation. Because furfural synthesis is activated by heat, furfurals primarily occur in wine heated during processing. They add to the "baked" fragrance of the wines. Phenolic aldehydes, such as **cinnamaldehyde** and **vanillin,** may accumulate in wines aged in oak. They are degradation products of lignins found in wood cooperage. Other phenolic aldehydes such as **benzaldehyde** may have diverse origins. Its bitter-almond odor is considered characteristic of certain wines, for example, those produced from 'Gamay' grapes. Benzaldehyde can also be produced from the oxidation of benzyl alcohol, used as a plasticizer in some epoxy resins, or through the metabolic action of some strains of yeasts and *Botrytis cinerea*.

KETONES

Few ketones are found in grapes, but those present usually survive fermentation. Examples are the norisoprenoid ketones **damascenone, α-ionone,** and **β-ionone.** The roselike scent of damascenone appears to play a contributing role in the aroma of several important grape varieties, such as 'Chardonnay' (Simpson and Miller, 1984) and 'Riesling' (Strauss *et al.*, 1987b). The violet, raspberry scent generated by **α-** and **β-ionones** may bestow part of the distinctive aroma of grape varieties characterized by these fragrances.

Many ketones are produced during fermentation, but few appear to have a sensory influence on wine. The major exception is **diacetyl** (biacetyl, or 2,3-butanedione). Diacetyl usually plays a minor role in wines, except when the content reaches levels sufficient to produce a buttery off-odor. This commonly occurs in association with spoilage induced by certain strains of lactic acid bacteria. Diacetyl also may be produced by yeasts, especially when fermentation temperatures are high. The odor of diacetyl is characterized variously as sweet, buttery, or butterscotchlike. Diacetyl occurs in fairly high concentrations in sherries, along with another ketone, acetoin. **Acetoin** (3-hydroxy-2-butanone) has a sugary, butterlike character. Its sensory significance in table wines, where it occurs at low concentrations, is doubtful.

Acetals

Acetals are formed when an aldehyde reacts with the hydroxyl groups of two alcohols. Acetals are produced on aging and during distillation. Acetals commonly have a vegetable-like character. Although over 20 acetals have been isolated from wines, their concentration and volatility seem to suggest that they have little sensory significance in table wines. Acetals may play a minor role in the bouquet of sherries and similar wines, where conditions are more favorable to their production.

$$R^1-\overset{\displaystyle O}{\underset{\displaystyle H}{C}} \;+\; 2\,R^2-OH \;\rightleftharpoons\; R^2-O-\overset{\displaystyle H}{\underset{\displaystyle R^1}{C}}-O-R^2$$

aldehyde   alcohols   acetal

Esters

Esters form as condensation products between the carboxyl group of an organic acid and the hydroxyl group of an alcohol or phenol. A prominent example is the formation of ethyl acetate from acetic acid and ethanol.

$$\underset{\text{acetic acid}}{H-\overset{H}{\underset{H}{C}}-\overset{O}{C}{\overset{\diagup}{\diagdown}}\,OH} \;+\; \underset{\text{ethanol}}{H-\overset{H}{\underset{HO}{C}}-\overset{H}{\underset{H}{C}}-H}$$

$$\rightleftharpoons\; \underset{\text{ethyl acetate}}{H-\overset{H}{\underset{H}{C}}-\overset{O}{\overset{\|}{C}}-O-\overset{H}{\underset{H}{C}}-\overset{H}{\underset{H}{C}}-H} \;+\; H_2O$$

Of all functional groups in wine, ester groups are the most frequently encountered. Over 160 specific esters have been identified. As most esters are found only in trace amounts, and have either low volatility or mild odors, their importance to fragrance development is probably negligible. However, the more common esters occur at or above the sensory threshold. Because some of these have fruity aspects, they are important in the generation of the bouquet of young white wines. Their importance to the fragrance of red wines is less well investigated.

CHEMICAL NATURE

Esters may be grouped into straight-chain (aliphatic) or cyclic (phenolic) classes. Most phenolic esters possess low thresholds of detection. However, because of their poor volatility and presence in trace amounts, they generally do not significantly influence wine fragrance. The primary exception is **methyl anthranilate,** which gives a general "grapy" aroma to several *Vitis labrusca* cultivars, for example 'Concord.'

Aliphatic esters comprise the larger ester group in wine. They are subdivided into monocarboxylic acid

esters (those containing a single carboxyl group in the parent acid), di- or tricarboxylic acid esters (those containing, respectively, two or three carboxyl groups in the parent acid), and hydroxy and oxo acid esters (those, respectively, containing a hydroxyl or ketone group in the parent acid). Of the three subgroups, only the first is believed to be of aromatic significance.

Of the monocarboxylic acid esters, the most important are those based on ethanol and saturated carboxylic acids, such as **hexanoic** (caproic), **octanoic** (caprylic), and **decanoic** (capric) acids, and those based on acetic acid and higher alcohols, such as isoamyl and isobutyl alcohols. These are often considered to give wine much of its vinous fragrance. The low molecular weight esters, often called "fruit" esters, have distinctly fruitlike fragrances. For example, **isoamyl acetate** (3-methylbutyl acetate) is bananalike, while **benzyl acetate** is applelike. As the length of the hydrocarbon chain increases, the odor shifts from being fruity to soaplike and, finally, lardlike with C_{16} and C_{18} fatty acids. The presence of certain of these esters, for example, hexyl acetate and ethyl octanoate, have occasionally been considered indicators of red wine quality (Marais *et al.*, 1979).

The di- and tricarboxylic acid esters generally occur at concentrations up to 1 mg/liter and above, especially ethyl lactate following malolactic fermentation. Nevertheless, because of their weak odors, they do not appear to be of aromatic significance. Examples are esters based on malic, tartaric, succinic, and lactic acids.

Hydroxy and oxo acid esters have low volatility and, correspondingly, appear to play little sensory role. The major esters of this group are associated with lactic acid.

Ethyl and methyl esters of amino acids occur in the milligram per liter range, but their sensory significance is unknown. In contrast, the high concentration of phenolic acid esters, such as caffeoyltartrate (caftaric acid) in 'Riesling' wines, may explain their typical slight bitterness (Ong and Nagel, 1978).

ORIGIN

Esters are produced in grapes, but the amount and sensory importance are often negligible. The prime exceptions are the phenolic ester methyl anthranilate in some *Vitis labrusca* cultivars and possibly isoamyl acetate in 'Pinotage' (Marais *et al.*, 1979). Synthesis of ethyl 9-hydroxynonanoate by *Botrytis cinerea* may contribute to the distinctive aroma of botrytized wines (Masuda *et al.*, 1984).

Most esters are produced by yeasts after cell division essentially has ceased. Subsequent synthesis and hydrolytic breakdown continue nonenzymatically, based on the chemical composition and storage conditions of the wine (Rapp and Güntert, 1986).

At the end of fermentation, fruit esters are generally in excess of the equilibrium constant. As a result, many acetate esters hydrolyze back to the component alcohols and acetic acid. Hydrolysis is favored at elevated temperatures and low pH levels (Ramey and Ough, 1980). For wines that derive much of their fragrance from fruit esters, aging can result in a fading of the bouquet. In contrast, fusel alcohol esters are generally retained within yeast cells, rather than being released into the fermenting must. As their concentration in wine is commonly below the equilibrium constant at the end of fermentation, there is a slow synthesis of fusel alcohol esters during aging. Esters or dicarboxylic acids also increase during aging. Nonenzymatic synthesis appears higher in sherries than in table wines (Shinohara *et al.*, 1979).

Ester formation during fermentation is influenced by many factors. In certain instances, the ability of the must to support ester formation declines as the grapes reach maturity (see Fig. 7.21). Variation in the esterase activity of yeast strains is another important factor affecting wine ester content. Low fermentation temperatures (~10°C) favor synthesis of fruity esters, such as isoamyl, isobutyl, and hexyl acetates, while higher temperatures (15° to 20°C) favor the production of higher molecular weight esters, such as ethyl octanoate, ethyl decanoate, and phenethyl acetate (Killian and Ough, 1979). Higher temperatures also tend to suppress ester accumulation by favoring hydrolysis. Both low SO_2 levels and juice clarification favor ester synthesis and retention. Intercellular grape fermentation (carbonic maceration) and the absence of oxygen during yeast fermentation further enhance ester formation.

Of all the esters, ethyl acetate has been the most investigated. In sound wines, the concentration of ethyl acetate generally is below 50 to 100 mg/liter. At low levels (<50 mg/liter) it may be pleasant and add to general fragrance complexity, while above 150 mg/liter it is likely to donate a sour-vinegar off-odor (Amerine and Roessler, 1983). The development of undesirable levels of ethyl acetate is usually associated with grape, must, or wine contamination with acetic acid bacteria. The bacteria not only directly synthesize ethyl acetate, but the acetic acid the bacteria produce can react nonenzymatically with ethanol to form ethyl acetate. Ethyl acetate can seriously flaw the fragrance of a wine long before the acetic acid level reaches a concentration sufficient to make the wine undrinkable.

Lactones and Other Oxygen Heterocycles

Lactones are a special subgroup of esters formed by internal esterification between carboxyl and hydroxyl groups. The result is the formation of a **cyclic ester**. Like other esters, lactones exist in equilibrium with the reactants, in this case a hydroxy acid:

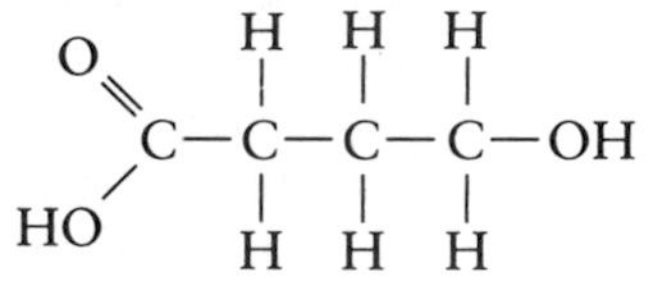

hydroxy acid

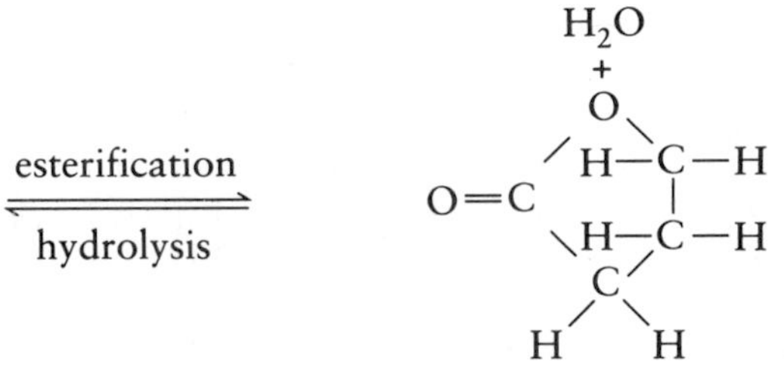

lactone

Most lactones in wines are four-carbon esterified rings, as illustrated above. Most also are gamma-lactones (γ-lactones); that is, the hydroxyl group involved in esterification is at carbon position 4 along the chain.

Lactones may come from grapes, may be synthesized during fermentation and aging, or may be extracted from oak cooperage. Lactones derived from grapes generally are not involved in the development of varietal odors. The lactone **2-vinyl-2-methyltetrahydrofuran-5-one** is one exception. It is associated with the distinctive aroma of 'Riesling' and 'Muscat' varieties (Schreier and Drawert, 1974). Because lactone formation is enhanced during heating, some of the raisined character of sunburned grapes may come from lactones, such as **2-pentenoic acid-γ-lactone**. Sotolon (4,5-dimethyltetrahydro-2,3-furandione) is characteristic of *Botrytis*-infected grapes and wine. It commonly occurs at levels above the sensory threshold (>5 ppb) in botrytized wines (Masuda *et al.*, 1984). Sotolon also may be a significant fragrance component of sherries (Martin *et al.*, 1992).

Most lactones in wine appear to be produced during fermentation. They are apparently derived from amino or organic acids, notably glutamic acid and succinic acid. **Solerone** (4-acetyl-4-hydroxybutyric acid-γ-lactone) and **pantolactone** (2,4-dihydroxy-3,3-dimethylbutyric acid-γ-lactone) are examples. Although solerone has been considered to possess a "bottle-aged" fragrance, it apparently does not accumulate to levels above the detection threshold in either table wines or sherries (Martin *et al.*, 1991).

Oak is an additional source of lactones (Chatonnet *et al.*, 1990). The most important of these are the so-called oak lactones, namely, isomers of 3-methyl-γ-octalactone (3-methyl-4-hydroxyoctanoic acid-γ-lactone). They have an oaky or coconutlike fragrance. Several **γ-nonalactones** also have an oaky aspect and may be partially involved in development of an oaky fragrance.

In the popular wine literature, allowing a wine to "breathe" is commonly recommended to improve wine flavor. To date, no solid evidence has been presented to support this belief. If true, the lactone **5-(3-methylthio-proionyl)dihydro-2(3H)-furanone** may be involved (Muller *et al.*, 1973). The susceptibility of the thiol group of the compound to oxidation makes it highly labile, and would result in the rapid dissipation of its unpleasant odor. The occurrence of this highly labile lactone in wine has yet to be confirmed.

Among other oxygen heterocyclic compounds, the spiroether **vitispirane** has been the most extensively investigated (see Etiévant, 1991). It may be derived from several compounds, such as free or glucosidically bound 3-hydroxytheaspirane and megasigma-3,6,9-triols. Vitispirane is slowly generated during aging, reaching concentrations of 20 to 100 ppb. It consists of two isomers, each of which have qualitatively different odors. The cis isomer has a chrysanthemum, flower–fruity odor, whereas *trans*-vitispirane has a heavier exotic fruit scent. Other authors have considered vitispirane to have a camphoraceous or eucalyptus odor. However, its sensory significance remains in doubt as its concentration is close to the perception threshold.

Terpenes and Oxygenated Derivatives

Terpenes are an important group of aromatic compounds characterizing the odor of many flowers, fruits, seeds, leaves, woods, and roots. As such, terpenes are often important in the fragrance of herb-flavored wines, such as vermouth, and fruit-flavored wines. In addition, they also characterize the aromas of several grape varieties.

Chemically, terpenes are grouped together because of their distinctive carbon skeleton. This consists of a basic five-carbon **isoprene** (2-methyl-1,3-butadiene) structure. Terpenes generally are composed of two, three, four, or six isoprene units. These are called **monoterpenes, sesquiterpenes, diterpenes,** and **triterpenes,** respectively.

Terpenes may contain a variety of functional groups. Many important terpenes contain hydroxyl groups, making them terpene alcohols. Other terpenes are ketones. Terpene oxides are terpenes having an oxygen-containing ring structure as well as the basic isoprenoid structure. In other words, they contain a cyclic ether (C—O—C) bond.

Unlike many of the aromatic constituents of wine, terpenes are primarily derived from grapes (see Strauss *et*

isoprene   isoprene skeleton

terpene skeleton

al., 1986). They exist in grapes in three forms (Fig. 6.9). Most are found as free monoterpene alcohols or oxides. As such, they are volatile and may contribute to the fragrance of a wine. A variable proportion of terpenes also exist complexed with glycosides or occur as di- or triols. Neither of the latter groups are aromatic.

Considerable interest has been shown lately in augmenting the release of free (volatile) monoterpenes from glycosidic linkages by the addition of several enzyme preparations (Günata *et al.*, 1990). Although most studied in relation to white grapes, nonspecific hydrolytic glycosidases may be useful in increasing the desirable flavor of wines made from red grapes, for example, 'Shiraz' (Abbott *et al.*, 1991). The chemical nature of the flavor compound(s) involved in the latter case is unknown.

Although essentially unaffected by crushing, maceration, or fermentation, the terpene content varies considerably from cultivar to cultivar. Thus, monoterpene content can be used to determine the varietal origin of certain wines (Rapp and Mandery, 1986). Geographical origin appears not to modify monoterpene content significantly. The varieties most easily characterized by their terpene contents are members of the 'Muscat' and 'Riesling' cultivar families. Other grape cultivars may possess terpenes but are less dependent on them for much of their varietal distinctiveness (Strauss *et al.*, 1987c).

Although the terpene content of healthy grapes is generally stable, infection by *Botrytis cinerea* markedly diminishes their content. The loss of aromatic terpenes undoubtedly plays a major role in the destruction of varietal aromas in botrytized wines (Bock *et al.*, 1988).

During aging, the types and proportions of terpenes found in wine change (Rapp and Güntert, 1986). Some increase may result from the release of free terpenes from

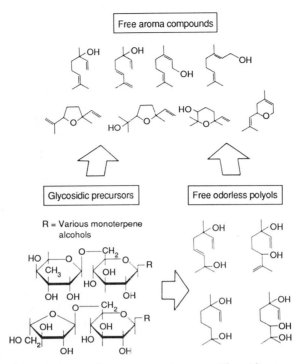

Figure 6.9 Categories of monoterpenes in grapes. Glycosidic precursors and free odorless polyols comprise a reserve of odorless precursors in the fruit. Only free aroma compounds make a direct contribution to fruit character. (From Williams *et al.*, 1987, reproduced by permission.)

glucosidically bound forms. Losses also may occur owing to oxidation or other transformations. Although the latter occur more slowly than the changes in fruit esters mentioned above, a marked loss of aroma can result over several years. This is particularly significant in 'Muscat' varieties, which depend on monoterpene alcohols for much of their fragrance. Most monoterpene alcohols are replaced by terpene oxides that have sensory thresholds about 10 times higher than the precursors. Correspondingly, the importance of terpenes to wine fragrance usually drops during aging. Nevertheless, additive or synergistic effects of the various terpenes make prediction of sensory influences difficult. Changes in terpene content also affect odor quality. For example, the muscaty, irislike odor of linalool is progressively replaced by the musty, pinelike scent of α-terpineol.

During aging, additional structural changes can affect wine bouquet. Some terpenes become cyclic and form lactones, for example, 2-vinyl-2-methyltetrahydrofuran-5-one from linalool oxides. Other terpenes may transform into ketones, such as α- and β-ionone, or spiroethers like vitispirane.

Changes in terpene content have little effect on the fragrance of red wines, where their occurrence is minimal. The major exception is the variety 'Black Muscat.'

Although most terpenes have pleasant odors, some are unpleasant. For example, *Penicillium roquefortii* growing on cork produces musky-smelling sesquiterpenes (Heimann *et al.*, 1983), and species of *Streptomyces* may synthesize earthy-smelling sesquiterpenes on cork or cooperage wood. If these compounds were to contaminate wine, they would seriously flaw its fragrance.

Of nonaromatic terpenes, the most important is the triterpene oleanolic acid. It is a primary constituent of the waxy covering on grapes and can act as a precursor for sterol synthesis by yeasts during fermentation.

Nitrogen-Containing Compounds

Many nitrogen-containing compounds are found in grapes and wine. These include inorganic forms, such as ammonia and nitrates, and many diverse organic forms including amines, amides, amino acids, pyrazines, nitrogen bases, pyrimidines, proteins, and nucleic acids. While complex organic nitrogen compounds (pyrimidines, proteins, and nucleic acids) are essential for the growth and metabolism of grape and yeast cells, they seldom are involved directly in the sensory properties of wine. Occasionally, colloidal proteins cause haziness in wine.

Amines are organic compounds associated with an ammonia group. Several simple volatile amines have been found in grapes and wine, including **ethylamine, phenethylamine, methylamine,** and **isopentylamine.** The concentration of volatile amines tends to decline as they are metabolized by yeasts during fermentation. Reten-

$$NH_3 \qquad\qquad R-NH_2$$

ammonia   amines

$$CH_3CH_2NH_2 \qquad\qquad C_6H_5CH_2CH_2NH_2$$

ethylamine   phenethylamine

tion is favored at both high and low fermentation temperatures. Subsequently, the level may increase owing to release from autolyzing yeasts. Their importance to flavor development is uncertain (see Etiévant, 1991). In beer, volatile amines are known to produce harsh tastes. While the higher flavor of red wines probably precludes a similar effect in red wines, white wines could be affected.

Wine also contains small amounts of nonvolatile amines. The most well-studied is **histamine,** but other physiologically active amines such as **tyramine** and **phenethylamine** (volatile) also occur. At sufficiently high concentrations, the compounds can induce headaches, hypertension, and allergic reactions. However, they do not appear to occur in wines at levels necessary to induce these effects (Radler and Fäth, 1991).

Polyamines, such as **putrescine** and **cadaverine,** may be present as a result of bacterial contamination. They appear to have no sensory importance in wine.

Amides are amines with a carbonyl group associated with the ammonia-containing carbon. Although a number of amides may occur in wines, none appear to significantly affect wine flavor.

Urea is a simple nitrogen compound related to the amides. It consists of two ammonia groups attached to a

amides   urea

ethyl carbamate

common carbonyl. Urea is produced in wine as a by-product of yeast metabolism and has been added to juice to promote yeast growth. Until relatively recently, its presence in wine was not considered to be of significance. However, urea may be important in the synthesis of ethyl carbamate (urethane) in wine and other fermented beverages (Ough *et al.,* 1990). As ethyl carbamate is a potential carcinogen, minimizing its production is especially important in wine exposed to heating (Stevens and Ough, 1993). Although the addition of acid urease can reduce the urea content of wines (Kodama *et al.,* 1991), other precursors of ethyl carbamate appear to occur in grapes (Tegmo-Larsson and Henick-Kling, 1990).

Yeasts may be involved in ethyl carbamate synthesis via the production of ethanol (a constituent in ethyl carbamate formation), through the production of carbamyl phosphate (Sponholz *et al.,* 1991), and by the synthesis and degradation of urea. The timing and degree of aeration during fermentation can significantly influence the production and degradation of urea (Henschke and Ough, 1991). The metabolic activity of lactic acid bacteria also may enhance the potential for ethyl carbamate production in wine.

α-amino acids   pyrazines   pyridines

Amino acids are another class of amine derivatives. They contain a carboxyl group attached to the amine-containing carbon. Amino acids are most important as the subunits of enzymes and other proteins. In addition, amino acids may act both as nitrogen and energy sources for yeast metabolism, and may be involved in the synthesis of flavor components. For example, amino acids may be metabolized to organic acids, higher alcohols, aldehydes, phenols, and lactones. During distillation, heating may convert amino acids into aromatic pyrazines. Amino acids also are associated with the caramelization of sugars during the production of baked sherries and madeira. Although some amino acids have bitter, sweet, or sour tastes, they are unlikely to have an appreciable influence on taste because of the low concentration in finished wine.

Pyrazines, cyclic nitrogen-containing compounds, contribute significantly to the flavor of many natural and baked foods. They are also important to the varietal aroma of several grape varieties. **2-Methoxy-3-isobutylpyrazine** plays a major role in the green-pepper odor often noticeable in 'Sauvignon blanc' (Allen *et al.,* 1991) and 'Cabernet Sauvignon' (Boison and Tomlinson, 1990) wines.

Another group of cyclic nitrogen compounds are the pyridines. Thus far, their involvement appears to be restricted to the production of mousy off-odors in wine. The odor has been associated with **2-acetyltetrahydropyridines** (Heresztyn, 1986).

Discussion of the presence and significance of proteins and nucleic acids is given below.

Hydrogen Sulfide and Organosulfur Compounds

Hydrogen sulfide and sulfur-containing organic compounds generally occur only in trace amounts in finished wines. Nevertheless, their presence can be significant as their sensory thresholds are equally low, often a few parts per trillion. At or above the recognition thresholds, the compounds produce odors that are almost uniformly unpleasant to nauseating. Thus, preventing their occurrence at detectable levels is of major concern in wine-making.

Generally, **hydrogen sulfide** (H_2S) is the most common, volatile, sulfur metabolic by-product found in wine. It is recognized by its rotten egg (*böcker*) odor. At near threshold levels, hydrogen sulfide is part of the yeasty odor of newly fermented wines. It is primarily produced by yeasts through the reduction of elemental sulfur found on grapes at harvest time (Schütz and Kunkee, 1977). Smaller amounts may be derived from the metabolism of sulfur-containing amino acids, notably cystine (Henschke and Jiranek, 1991).

Organic sulfur compounds found in wine consist of a wide variety of straight-chain and cyclic molecules. They may be produced during the metabolism of sulfur-containing amino acids, peptides, and proteins. Figure 6.10 shows the close association between fermentation and the production of organosulfur compounds.

The autolysis of dead and dying yeast cells has been implicated in the production of organosulfur off-odors. Exposure to light also may activate the production of organosulfur compounds. The *goût de la lumière* off-odor in champagne, caused by organic sulfides, has this origin (Charpentier and Maujean, 1981). In beer, a "sun-struck" off-odor is induced by **prenyl mercaptan** (3-methyl-2-butene-1-thiol).

Structurally, the simplest organosulfur compounds are the **mercaptans.** They are hydrocarbon chains attached to a sulfhydryl (—SH) group. A significant member of this group is **ethanethiol.** It produces an onion, rubberlike odor at threshold levels. At higher levels, it has a fecal odor. A related thiol compound, **2-mercaptoetha-**

nol, is implicated in the production of a barnyardlike (*böxer*) odor (Rapp *et al.*, 1985).

Thioethers are organosulfur compounds characterized by the presence of one or more sulfur atoms bonded between two carbon atoms. For example, **dimethyl sulfide** (DMS) has the structural formula CH_3—S—CH_3. At above the threshold level, dimethyl sulfide has a shrimplike odor. At low levels, it apparently has asparagus, corn, and molasses aspects. Other thioethers occasionally found in wine are dimethyl and diethyl disulfides.

Thiolanes are ring structures containing etherlike sulfur bonds. An example is **2-methylthiolane-3-ol.** It has a faint onionlike smell.

Thiazoles are additional cyclic compounds that contain both sulfur and nitrogen as part of the ring. The medicinal, peanutlike smell of **5-(2-hydroxyethyl)-4-methylthiazole** has been detected in wine and grape distillates. It is not known whether the compound occurs at levels sufficient to directly influence the fragrance of wine.

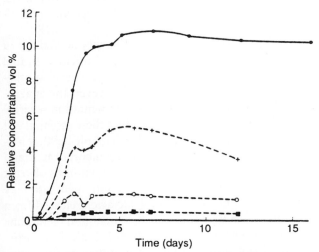

Thioesters are formed between a carboxyl-containing thiol compound and an alcohol. The most important of these may be **ethyl 3-mercaptopropionate,** a proposed source of the "foxy" (fox-den) odor of some *Vitis labrusca* varieties (Kolor, 1983).

Hydrocarbons and Derivatives

Hydrocarbons are compounds composed solely of carbon and hydrogen atoms. Because of their poor solubility in water, they usually remain associated with cellular grape debris and are lost before or during clarification. Thus, they usually do not directly influence the sensory characteristics of wine. Nevertheless, hydrocarbon degradation products may be involved in the formation of important volatile compounds. For example, several important aromatic compounds such as **damascenone, vitispirane,** and **TDN** (1,1,6-trimethyl-1,2-dihydronaphthalene) are derived from the hydrolysis and degradation of megastigmane derivatives in grapes (Sefton *et al.*, 1989).

Possibly the most significant aromatic hydrocarbon present or subsequently generated in wine is the norisoprenoid TDN. Its precursor, probably 2,6,10,10-tetramethyl-1-oxaspiro[4,5]dec-6-ene-2,8-diol, appears to accumulate simultaneously with the increase in sugar content during ripening of 'Riesling' grapes (Winterhalter, 1991). In wine, the concentration of TDN has been

Figure 6.10 Production of organosulfur compounds during fermentation. ●, Ethanol (%); +, 3-methylthio-1-propanol; ○, (3-methylthiopropyl)acetamide; ■, 2-methylthiolane-3-ol. (Data from Tucknott, 1977, in Rapp and Mandery, 1986, reproduced by permission.)

noted to rise from undetectable to about 40 ppb after several years (Rapp and Güntert, 1986). The sensory threshold is about 20 ppb. It has a smoky, kerosene, "bottle-aged" fragrance, which may be desirable at low concentrations. Its production appears to be enhanced by sun exposure of the fruit (Marais *et al.*, 1992).

A cyclic hydrocarbon occasionally found in wine is **styrene**. Although styrene is synthesized by yeast cells from 2-phenylethanol, amounts sufficient to give a plastic taint usually come from storage in plastic cooperage or transport containers (Hamatschek, 1982).

Additional hydrocarbon taints may come from microbially contaminated corks. For example, **methyl tetrahydronaphthalene** has been implicated in one of the types of corky off-odors occasionally found in wine (Dubois and Rigaud, 1981).

Macromolecules and Growth Factors

Macromolecules are the polymers that constitute the structural and major regulatory molecules of cells. These include carbohydrates, proteins, nucleic acids, and some lipids. The specific roles of macromolecules in the growth and reproduction of grape, yeast, and bacterial cells are beyond the scope of this text. However, without them life as we know it would not exist. Few occur in wine in significant quantities. Owing to poor solubility, macromolecules usually remain and precipitate with cellular debris during clarification or fermentation.

CARBOHYDRATES

The major carbohydrate polymers of plant cells are cellulose, hemicelluloses, pectins, and starch. They function primarily as structural elements in cell walls or as forms of energy storage. Cellulose is too insoluble to be extracted into wine and remains with the pomace. Hemicelluloses are poorly soluble and, if extracted, precipitate during fermentation. Pectins either precipitate or are enzymatically degraded to soluble galacturonic acid during fermentation. Correspondingly, they seldom occur in finished wine. Starch, the major storage carbohydrate of plants, is not found in significant quantities in mature grapes. Grapes are atypical in using soluble sugars as the storage carbohydrate.

Mannans and glucans, the major carbohydrate polymers of yeast cell walls, are either insoluble or precipitate during fermentation. Related, but smaller, polysaccharides are soluble and may be released in significant amounts (>400 mg/liter) in wine. Most are mannans combined with proteins. In combination with cell wall degradation products released during yeast autolysis, they are important in producing a stable effervescence in sparkling wines.

The glucans and chitins of most fungal cell walls are too insoluble to be incorporated into wine. However, the extracellular glucans produced by *Botrytis cinerea* can cause serious enological problems. High molecular weight forms can induce severe plugging during filtration, while the low molecular weight forms can inhibit yeast metabolism (see Ribéreau-Gayon, 1988).

Yeast cell walls may be added to fermenting juice to prevent the premature termination of fermentation. They appear to act by removing toxic carboxylic acids and as sources of vital nutrients (see Munoz and Ingledew, 1990).

LIPIDS

Plant lipids consist of two major groups. One group is based on fatty acids esterified with a polyol, usually glycerol, and the other is based primarily on isoprene subunits. The first includes phospholipids, fats, oils, waxes, glycolipids, and sulfolipids; the second, steroids. All lipids are vital to the structure and function of plant and yeast cells. However, only oils, waxes, and steroids directly influence wine quality.

Plant oils generally do not occur in wine. Their presence would likely indicate the use of excessive pressure during grape crushing, which could rupture grape seeds and release the oils. On oxidation, the oils could generate a rancid taint. Modern crushers have almost eliminated this source of wine contamination. Fruit, and especially leaves accidentally macerated with grapes, may release small amounts of oil constituents such as **linoleic** and **linolenic acids** into the juice (Roufet *et al.*, 1986). Grape lipoxygenases activated on crushing can rapidly oxidize these compounds and release aromatic C_6 aldehydes and alcohols, such as ***trans*-2-hexenal** and ***cis*-2-hexenol** (Iglesias *et al.*, 1991). Because these compounds generate both herbaceous odors and bitter tastes, they are occasionally termed "leaf" aldehydes and alcohols.

Both the growth and continued metabolic activity of yeast cells require the presence of sterols and unsaturated fatty acids. In the presence of oxygen, yeast cells synthesize their own lipid requirements. However, the anaerobic conditions that develop during fermentation severely restrict the ability of yeasts to produce these essential cell constituents. **Oleanolic acid** (oxytriterpenic acid), a major component of grape wax, can be incorporated and used under anaerobic conditions in the synthesis of yeast sterols. The unsaturated fatty acid requirement also may be satisfied by linoleic and linolenic acids released from grape cells. Both types of lipids help maintain membrane function and enhance yeast tolerance to alcohol during and after fermentation. Extraction of the lipids is significantly improved by leaving the juice in contact with the skins for several hours after crushing.

PROTEINS

During ripening, the soluble protein content of grapes increases; the degree of enrichment being cultivar dependent. After crushing, the soluble protein content may increase by a further 50% during cold settling (Tyson *et al.*, 1982). The addition of bentonite reverses this trend. During fermentation, the soluble protein content may increase, decrease, or fluctuate markedly, depending on the cultivar. At the end of fermentation, many proteins are precipitated with tannins, especially in red wines.

In most wines, soluble proteins are considered undesirable, as they can induce haze formation. In sparkling wines, however, soluble proteins help to stabilize the effervescence (Maujean *et al.*, 1990).

In juice, the most important group of enzymatic proteins are hydrolases activated on release from grape cells during crushing or pressing. The most well-known class are the **phenol oxidases.** They activate the browning of juice in the presence of oxygen. The principal grape enzymes of this class are polyphenol oxidases (*o*-diphenol oxidases). Because the enzymes tend to remain with the cellular debris of grapes, enzyme activity is largely restricted to the period before pressing. The addition of sulfur dioxide further limits enzyme action. The most serious group of phenol oxidases are the laccases produced in *Botrytis*-infected grapes. Not only are laccases readily soluble and relatively insensitive to SO_2 inhibition, but they also oxidize monophenols, *o*-, *m*-, and *p*-diphenols, *o*-triphenols, anthocyanins, catechins, procyanidins, and 2-*S*-glutathionylcaftaric acid (see Macheix *et al.*, 1991). The solubility of laccases means that they are not effectively removed by most clarification techniques including bentonite addition. One of the potential advantages of ultrafiltration is its removal of laccase. The presence of laccase can be measured with the syringaldazine test developed by Dubourdieu *et al.* (1984) (see Zoecklein *et al.*, 1990).

Pectinases are important in fruit softening and maceration following crushing. Tissue disintegration eases juice release and flavor liberation from the pomace. Correspondingly, the juice may be left in contact with the skins and pulp for several hours before pressing. Commercial pectinases may be added to enhance tissue breakdown and reduce the high pectin levels that characterize certain grape varieties.

Lipoxygenases are being viewed with increasing interest owing to their ability to oxidize fruit and leaf oils, notably linoleic and linolenic acids. This action could generate several aromatic C_6 and C_9 aldehydes including hexanals, hexenals, nonenals, nonadienal, and the corresponding alcohols. Although C_6 ("leaf") aldehydes and alcohols can give wine a grassy off-odor, lipoxygenases also generate aromatic compounds from carotenoids and release oleanolic acid from grape skins.

Proteases (protein-hydrolyzing enzymes) have been detected in grape must, but their significance, if any, is unknown. By releasing amino acids, they may increase nitrogen availability to yeasts.

NUCLEIC ACIDS

Nucleic acids are long polymers of nucleotides that function in the storage, transmission, and translation of genetic information in cells. The molecular weight of nucleic acids is so great that they are not released in significant amounts from grape cells on crushing. Although degradation products of nucleic acids are readily soluble and easily assimilated, yeasts can synthesize their own nucleotides. Nucleotides and nucleic acids are not known to influence wine quality directly.

VITAMINS

Vitamins encompass a series of diverse chemicals involved in the regulation of cellular activity. They are found in small quantities in grape cells, juice, and wine. The concentration of vitamins generally falls during fermentation and aging. For example, **ascorbic acid** (vitamin C) is oxidized rapidly following crushing; **thiamine** (vitamin B_1) is degraded by reaction with SO_2, exposure to heat, or absorption to bentonite; and **riboflavin** (vitamin B_2) is oxidized on exposure to light. The only vitamin to increase notably during fermentation is **p-aminobenzoic acid** (PABA).

Vitamin levels in wine are inadequate to be of major significance in human nutrition, but they usually are ample for microbial growth. **Biotin** (vitamin H) and **nicotinic acid** (niacin) contents are adequate for most yeast strains, as are the vitamin and growth factor levels required by lactic acid bacteria.

Dissolved Gases

Wines contain varying levels of several gases. All, except nitrogen gas, can have marked effects on the sensory properties of wine. Nitrogen gas is chemically inert and poorly soluble in wine.

CARBON DIOXIDE

Carbon dioxide in wine comes primarily from the metabolic activity of yeasts. Additional small amounts may be generated by lactic acid bacteria; minute amounts may arise from the breakdown of amino acids and phenols during aging.

Most of the carbon dioxide produced by yeast action escapes during fermentation. Nevertheless, wine remains supersaturated with carbon dioxide at the end of fermentation. During maturation, much of the carbon dioxide escapes, and the CO_2 concentration falls to about 2 g/liter (saturation) by bottling. At this level, carbon dioxide has no sensory effect. If wine is bottled while still supersaturated, bubbles are likely to form in the glass when the wine is poured. At above 5 g/liter, carbon dioxide begins to produce a prickling sensation on the tongue (Amerine and Roessler, 1983).

Usually, refermentation is the primary source of detectable effervescence in still wines, and it may be associated with off-odors and haziness. This potential fault can be turned to advantage if induced by desirable strains of yeasts and cloudiness prevented. The result is sparkling wine. However, wines may receive effervescence more inexpensively, by carbonation.

OXYGEN

Before crushing, grapes contain very low levels of oxygen. Crushing results in the rapid uptake of about 6 ml (9 mg) of O_2 per liter at 20°C. Use of crushers employing minimal agitation limits oxygen uptake. Slight juice aeration often is preferred as it favors complete fermentation. The oxygen is used by yeasts to synthesize essential compounds, such as unsaturated fatty acids, sterols, and nicotinic acid. It also limits browning by converting caftaric acid to a less oxidizable complex with glutathione and promoting the early oxidation and precipitation of readily oxidized phenols. Juice from white grapes is occasionally purposely hyperaerated to encourage the latter process.

Because oxygen is rapidly consumed in various oxidative reactions, the greater portion of fermentation occurs in the absence of oxygen. Protection from oxygen is provided by the generation of carbon dioxide, which rapidly blankets the fermenting juice. Once fermentation is complete, however, the wine must be protected from exposure to oxygen.

Much of the oxygen absorbed by must or wine is consumed in oxidative reactions with phenols. Oxygen consumption is rapid in red wine, often being complete within 6 days at 30°C (Singleton, 1987). Consumption is complete at lower temperatures, but takes longer.

The slow or periodic incorporation of small amounts of oxygen ($\sim$40 ml/liter) is thought beneficial during the maturation of red wines (Ribéreau-Gayon *et al.*, 1983). Oxygen aids color stabilization and reduces the bitterness and astringency of tannins in wine. In contrast, oxygen uptake by white wine generally is detrimental. White wines are less able to bind the acetaldehyde generated on oxidation, and mask the stale oxidized odor that develops. White wines also do not derive color stabilization or bitterness reduction from slight aeration.

Although red wines benefit from limited aeration during the early stages of maturation, excessive oxygen exposure produces an oxidized (aldehyde) odor and browning. Oxygen exposure also may favor the growth of spoilage organisms. Fortified wines that develop a complex, aged, oxidized odor avoid the microbial spoilage associated with oxygen exposure owing to the high alcohol content. Sherries begin to take on an oxidized bouquet following exposure to about 60 ml O_2/liter (Singleton *et al.*, 1979). Table 6.6 illustrates one view on the relative need of different wines for limited oxygen exposure during maturation.

SULFUR DIOXIDE

Sulfur dioxide is a normal constituent of wine, accumulating to levels between 12 and 64 mg/liter due to yeast metabolism (Larue *et al.*, 1985). Major factors influencing biosynthesis of sulfur dioxide are the yeast strain, fermentation temperature, and sulfur content of the grapes (Würdig, 1985). However, SO_2 levels above 30 mg/liter usually result from its addition during or after vinification.

Table 6.6 Relative Need for Oxygen during the Maturation and Processing of Certain Types of Wines[a]

Type of wine	Oxygen demand	Typical period between production and bottling (years)	Typical aging potential (years)
Table wines			
White	None	0.3–1	1–3
Rosé	None	0.5–1	1–2
Light red	Slight	0.5–2	2–4
Deep red	Slight to moderate	2–4	5–40+
Fortified wines			
Flor sherries	Considerable	3–7	0.5–1
Oloroso sherries	Extensive	4–10	2–4
Tawny ports	Considerable	10–40+	1–2

[a] After Somers (1983), reproduced by permission.

Although burning elemental sulfur may have been used for vessel fumigation during ancient Roman times (Roberts and McWeeny, 1972), its modern use is reported to have begun about the sixteenth century (Anonymous, 1986). It appears to have been used first as sulfur wicks in the disinfection of wooden cooperage. Up to 60 mg of SO_2 per liter can apparently be absorbed by wine from the sulfur dioxide left in barrels after fumigation (Amerine *et al.*, 1980). In contrast to inadvertent absorbtion, the intentional addition of sulfur dioxide to wine began early in the twentieth century (Somers and Wescombe, 1982). Sulfur dioxide is commonly incorporated directly as a gas from cylinders, although potassium metabisulfite ($K_2S_2O_5$) is also frequently used. Bisulfite salts rapidly ionize under the acidic conditions of must or wine, releasing sulfur dioxide.

In wine, sulfur dioxide can exist in a variety of "free" and "bound" forms. Only a small portion of the total sulfur dioxide content exists as free dissolved gas (Fig. 6.11). An additional small fraction exists as free sulfite ions (SO_3^{2-}). However, most of the free ionic sulfur dioxide occurs as bisulfite ions (HSO_3^-). The remainder of the free sulfur dioxide exists as undissociated sulfurous acid (H_2SO_3). Because sulfur dioxide binds readily with several constituents in wine, it often occurs as hydroxysulfonates. Much of this is reversibly associated with acetaldehyde. Bisulfite addition products also form with anthocyanins, tannins, pyruvic acid, α-ketoglutarate, sugars, and sugar acids. Binding of sulfur dioxide greatly reduces the active (free) concentration of sulfur dioxide.

$$SO_2 + H_2O \rightleftharpoons H_2SO_3 \rightleftharpoons HSO_3^- + H^+ \rightleftharpoons SO_3^{2-} + 2H^+$$

Because sulfur dioxide exists in many interconvertible states in wine, the active level of sulfur dioxide is influenced by many factors. Of these, the pH of the wine and the concentration of binding compounds are the most significant. For example, acetaldehyde synthesis during fermentation lowers the level of free SO_2, while metabolism of acetaldehyde by lactic acid bacteria releases the SO_2 from its bound association.

$$\text{acetaldehyde} + \text{sulfur dioxide} \rightleftharpoons \text{acetaldehyde hydroxysulfonate}$$

One of the consequences of the release of bound sulfur dioxide can be the slowing or inhibition of malolactic fermentation. In addition, wines can lose color intensity, notably owing to the bleaching of anthocyanins.

Although undesirable in excess, sulfur dioxide has many valuable attributes. These include antimicrobial and antioxidant properties, as well as the potential to bleach pigments and suppress oxidized odors in wines. The relative value of SO_2 addition often depends as much on when it is added as on how much is added.

As with other aspects of sulfur dioxide activity, the antimicrobial property of sulfur dioxide is largely dependent on the free component. Because bound forms are weakly antimicrobial, it is common to consider only the free sulfur dioxide content when assessing the antimicrobial action of sulfur dioxide. Of the various free forms of sulfur dioxide, molecular SO_2 is most readily absorbed by cells. However, as yeast cytoplasm has a pH of 6, the active form(s) presumably are either sulfite or bisulfite (see Beech and Thomas, 1985).

The antimicrobial action of sulfur dioxide probably involves its reaction with the disulfide bonds of proteins. Because disulfide bonds help maintain the structure of many enzymes and regulatory proteins, sulfonate formation can disrupt enzyme and protein function. The binding of sulfur dioxide with nucleic acids and lipids also may cause genetic and membrane dysfunction, respectively. Additional antimicrobial activity may result from the destruction of thiamine.

Although sulfur dioxide has wide-spectrum antimicrobial activity, wine yeasts are relatively insensitive to its action. About 1.5 mg/liter molecular SO_2 inhibits most spoilage yeasts and bacteria (Sudraud and Chauvet, 1985). Because dormant yeast cells are more sensitive to sulfur dioxide than metabolically active cells, it often selectively restricts the growth of indigenous wine yeasts derived from grape skins and winery equipment. Thus, added yeast strains generally act as the major agents of fermentation.

In addition to its antimicrobial activity, sulfur dioxide is an effective antioxidant. It suppresses the activity of several oxidases and nonenzymatic oxidative reactions. Although the mechanism of enzymatic disruption is unclear, it likely involves the direct reaction between sulfite ions and enzymes. Sulfites also may act reductively, by converting oxidation products back to the reduced

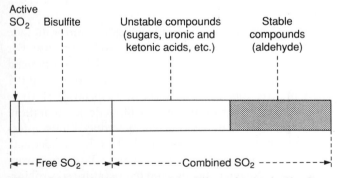

Figure 6.11 Relative amounts of the different forms of sulfur dioxide in wine. (From *Knowing and Making Wine*, E. Peynaud, Copyright © 1984 John Wiley & Sons, Inc. Reprinted by permission of John Wiley & Sons, Inc.)

forms, for example, quinones to phenols. Sulfites can additionally suppress nonenzymatic browning (Maillard) reactions between sugars and amino acids. This involves the binding of sulfites with the carbonyl group of sugars. A similar reaction with quinones can limit the participation of quinones in further oxidative reactions (see Taylor *et al.*, 1986). Sulfur dioxide is also important in the antioxidant action of ascorbic acid. Sulfites mediate the rapid reduction of hydrogen peroxide, produced as a by-product of the action of ascorbic acid, to water.

One negative property of sulfur dioxide, relative to oxidation, is the restriction of the enzymatic reaction between caftaric acid and glutathione. As a result, readily oxidizable phenols may remain in wine and enhance the tendency of white wines to brown during aging (Singleton *et al.*, 1985).

The addition of small amounts of sulfur dioxide to wine can produce a fresher odor by forming nonvolatile sulfonates with acetaldehyde. Also, by reacting with phenols, sulfur dioxide can reduce the nonenzymatic generation of acetaldehyde. However, its addition to juice can enhance the synthesis of acetaldehyde during fermentation. The formation of acetaldehyde hydroxysulfonate shifts the equilibrium between acetaldehyde in fermenting juice and yeast cells. As a result, yeast cells may produce and secrete more acetaldehyde. Thus, during fermentation, sulfur dioxide increases the absolute while reducing the volatile concentration of acetaldehyde.

Sulfur dioxide can have a further beneficial action in white wines. By bleaching brown pigments, SO_2 causes the wine to develop a paler color. However, the same action can result in undesirable color loss in red wines. By destroying hydrogen peroxide produced during the autooxidation of phenols, SO_2 limits the formation of acetaldehyde and the generation of color-stabilizing anthocyanin/tannin polymers. This is not compensated during maturation by the recoloration of anthocyanins released from association with sulfur dioxide. Free anthocyanins are much more liable to nonreversible decolorization than polymerized forms. In addition, delaying anthocyanin/tannin copolymerization restricts its occurrence. Tannins self-polymerize and become progressively unable to bond with anthocyanins.

Not only is sulfur dioxide the most important antimicrobial and antioxidant wine additive, it is also the primary sterilant of winery equipment. Nevertheless, excessive use can be detrimental. The corrosive action of SO_2 can solubilize metal ions from unprotected metal surfaces. Sulfur dioxide also may react with oak constituents, forming lignosulfurous acid. It has been proposed that, on decomposition, lignosulfurous acid may release hydrogen sulfide that reacts with pyrazines in the wood, forming musty-smelling thiopyrazines (Tanner and Zanier, 1980). Extraction of the latter would taint the wine.

At sulfur dioxide levels between 15 and 40 mg/liter free SO_2, most individuals begin to detect a distinctive burned-match odor (Amerine and Roessler, 1983). Possibly owing to habituation, consumers seldom seem to object to such levels.

Consumption of up to 400 mg SO_2 (free and bound) per day for several weeks has no adverse effects on some people (Hötzel *et al.*, 1969). Most commercial wines contain considerably less than 10 mg/liter. The Joint FAO/WHO Expert Committee on Food Additives (1974) established an acceptable daily sulfite intake of 0.7 mg/kg body weight (0.32 mg/lb). Although SO_2 can precipitate an asthma attack in sensitive individuals, most wines possess insufficient freely volatile sulfur dioxide to induce an attack. Nevertheless, for a small portion of asthmatics, all forms of sulfur dioxide are potentially dangerous. Air passage constriction may result from the translocation of sulfites, absorbed from the digestive tract, to the lungs. In addition, people expressing a rare genetic disease (sulfituria) are unable to produce **sulfite oxidase.** As a consequence, they cannot detoxify sulfites to sulfates. In normal individuals, sulfites are converted rapidly to sulfates that are effectively removed from the blood and excreted by the kidneys. While sulfite oxidase deficient individuals are at risk from sulfites in wine, many foods are a more common risk factor. It is estimated that humans release about 2.4 g of sulfate in urine per day (Institute of Food Technologists Expert Panel on Food Safety and Nutrition, 1975), most of which comes from the metabolism of sulfur-containing amino acids in food. Individuals deficient in sulfite oxidase must avoid foods containing significant levels of sulfur-containing amino acids. Sulfites released from food intake normally exceed the amount of sulfite derived from wine.

On the positive side, the presence of sulfur dioxide suppresses the toxicity or mutagenicity of several compounds found in foods and beverages (see Taylor *et al.*, 1986).

As a result of the few disadvantages of sulfur dioxide addition to wine, the current trend is to limit its use. For example, Somers and Wescombe (1982) recommend that sulfur dioxide addition to red wines not exceed 0.5 to 1 mg free molecular SO_2 per liter (spectral) or 10 to 15 mg/liter (aspiration) after malolactic fermentation. The need for sulfur dioxide can be reduced most effectively by practices that promote the health and adequate acidity of grapes delivered to the winery. Protection from undesired oxygen exposure and temperature control further limit the need for the preservation provided by sulfur dioxide. Sulfur addition usually is delayed if malolactic fermentation is desired.

Minerals

Many mineral elements are found in grapes and wine. In most situations, the mineral concentration reflects uptake by the grapevine. However, high levels of elemental sulfur may arise from fungicides applied to the vines for disease control. Elevated calcium levels are usually found in wines stored in unlined cement tanks. Augmented chlorine and sodium contents may originate from the use of ion-exchange columns, and high levels of copper and iron can result from contact with corroded winery equipment. Greater than trace amounts of lead in wine may be associated with vines located near highways (Médina *et al.*, 1977), corroded lead capsules (Sneyd, 1988), or prolonged storage in lead crystal decanters (Falcone, 1991). Atypically high aluminum contents may come from the use of bentonite for wine clarification (McKinnon *et al.*, 1992).

At naturally occurring levels, many minerals are important cofactors in vitamins and enzymes. However, heavy metals such as lead, mercury, cadmium, and selenium are toxic. If present in the fruit, heavy metals usually precipitate during fermentation (von Hellmuth *et al.*, 1985). Thus, their occurrence in wine at above trace amounts usually indicates contamination after fermentation. At higher than normal levels, minerals such as iron and copper can also be undesirable. They can catalyze oxidative reactions, modify taste characteristics, or induce haziness (*casse*).

Copper ions can slowly associate with dissolved proteins and induce copper *casse*. In the presence of high levels of both phosphate and iron, a ferric-derived *casse* may develop. Under appropriate conditions, iron can also react with tannic acid, giving rise to blue *casse*. High calcium levels can delay tartrate precipitation and augment crystal formation after bottling.

Oxidative reactions may be catalyzed in the absence of oxygen by copper and, to a lesser extent, iron. An example is the oxidization of ascorbic acid to dehydroascorbic acid and its cleavage into oxalic and threonic acids. In the presence of oxygen, iron is an important ion catalyzing oxidative reactions involved in browning.

Although copper and iron can induce metallic/astringent tastes, this occurs only at concentrations higher than usually found in wine (Amerine and Roessler, 1983). High sulfate contents can give wines a slightly salty–bitter taste.

Chemical Nature of Varietal Aromas

There has long been an interest in the origin of grape aromas. Knowledge of the chemical nature of aromas could directly benefit grape growers and winemakers. It might permit a more precise determination of a desirable harvest date. It also could allow an assessment of how various viticultural and vinicultural practices influence one of the most important determinants of wine quality, fragrance. In addition, such knowledge could streamline the production of new grape varieties by permitting the breeder to select early for lines showing particular fragrance characteristics. An objective measure of the varietal origin also would be of particular interest to those charged with enforcing Appellation Control laws. In Europe, regulations stipulate the varieties permitted in the production of Appellation Control wines. In certain instances, differences are sufficiently marked to permit such a distinction (Rapp and Mandery, 1986). Chemical changes occurring during aging may limit the technique to young wines from a common vintage.

Determination of the chemical nature of a varietal aroma is fraught with difficulties. The first step usually involves the separation of grape and wine volatile components with gas chromatography. The column may be split to divert a fraction of each compound for sniffing, while the remainder undergoes physicochemical analysis. The study is easier if the crucial compound(s) occurs in volatile form in both grapes and wine. However, aroma compounds in wine may often exist in nonvolatile forms in the grapes. They may be released only on crushing (e.g., C_{18} fatty acids into "leaf" aldehydes and alcohols), through yeast activity (e.g., phenol into vinyl guaiacol), or during aging (e.g., linalool to α-terpineol). In addition, varietal aromas may originate from a particular combination of compounds, not a single unique compound. Extracting procedures can also influence the stability and isolation of potentially important compounds. When compounds of likely importance are isolated, identification and quantification of the compounds are required. Only by comparing the concentration found in wine with its sensory threshold can the relative importance of a compound be assessed. Because several hundred volatile compounds may occur in a wine, multivariate analysis may facilitate the detection of compounds that deserve more detailed investigation.

Even with the highly precise analytical tools currently available, great difficulty can be encountered in the detection of certain groups of aromatic compounds (e.g., aldehydes bound to sulfur dioxide). However, the situation is made even more demanding when the significant compounds are labile and/or occur in trace amounts. For example, the occurrence of 2-methoxy-3-isobutylpyrazine has only recently been detected in wines made from 'Sauvignon blanc' and 'Cabernet Sauvignon' grapes. The compound both is highly labile and may occur in wine at up to 35 ppt. It has a detection threshold of about 2 ppt.

In relation to the comparative importance in aroma production, volatile ingredients have been classified as impact, **contributing,** or **insignificant** compounds. Impact compounds are those that have a marked and distinctive effect on wine fragrance. They generally give wines their varietal distinctness. Although usually desirable, they may impart notoriety to wines, for example, the "foxy" aroma of certain *Vitis labrusca* varieties.

Contributing compounds are those that contribute to the overall complexity of the fragrance of a wine. For example, C_{10} and C_{12} esters of unsaturated fatty acids add to the fruity odor of 'Concord' fruit and wine (Schreier, 1982). Contributing compounds are also important to the development of an aged bouquet. They are equally responsible for the basic wine bouquet generated by yeast metabolism during fermentation.

Although most grape varieties do not develop distinctive varietal aromas, and many that do appear not to show unique aroma compounds, there is a growing list that do. For example, the foxy character of some *Vitis labrusca* cultivars has been ascribed to **ethyl 3-mercaptopropionate** (Kolor, 1983) or, more recently, to **N-(N-hydroxy-N-methyl-γ-aminobutyryl)glycin** (Boison and Tomlinson, 1988) and **o-aminoacetophenone** (Acree *et al.*, 1990). The latter has also been isolated from some *V. vinifera* wines (Rapp *et al.*, 1993). Other *V. labrusca* cultivars may derive much of their distinctive aroma from **furaneol** (2,5-dimethyl-4-hydroxy-2,3-dihydro-3-furanone) (Schreier and Paroschy, 1981) or from **methyl anthranilate** and **damascenone** (Acree *et al.*, 1981). The bell-pepper character of some 'Cabernet Sauvignon' (Boison and Tomlinson, 1990) and 'Sauvignon blanc' wines (Lacey *et al.*, 1991) is due primarily to the presence of **2-methoxy-3-isobutylpyrazine. Isopropyl** and **sec-butyl methoxypyrazines** are also present in 'Sauvignon blanc' wines, but at lower concentrations. The source of the desirable black currant fragrance of many 'Cabernet Sauvignon' wines is unknown, but the property appears not to come from any chemical similarity to the natural black currant aroma. The latter is based on a complex terpene composition (Marriott, 1986).

The spicy character of 'Gewürztraminer' wines appears to be associated with the production of **4-vinyl guaiacol** in association with several terpenes (Versini, 1985). A guavalike feature, occasionally associated with 'Chenin blanc' and 'Colombard' wines, has been attributed to the presence of the mercaptan **4-methyl-4-mercaptopantan-2-one** (du Plessis and Augustyn, 1981). 'Muscat' varieties are distinguished by the prominence of monoterpene alcohols in their varietal aromas. Similar monoterpene alcohols are important, but at lower concentrations, in the aromas of cultivars similar or related to 'Riesling'. The relative and absolute concentrations of these compounds, and their respective sensory thresholds, distinguish the varieties in each group from one another.

In some instances, compounds that are varietally distinctive also occur as by-products of fermentation. For example, the important impact compound in muscadine varieties, 2-phenylethanol, may also be produced by yeasts (Lamikanra, 1987). Equally, isoamyl acetate, a distinctive flavorant of 'Pinotage' wines, also may be produced by yeasts (van Wyk *et al.*, 1979).

For most grape varieties, including the well-known cultivars 'Chardonnay' (Simpson and Miller, 1984) and 'Pinot noir' (Brander *et al.*, 1980), no distinctive aroma compounds have been isolated. Nevertheless, some markedly aromatic compounds are found in higher than usual concentrations in these varieties. For example, damascenone tends to be prominent in the aroma profile of 'Riesling' and 'Chardonnay' (Strauss *et al.*, 1987b; Simpson and Miller, 1984), β-ionone in 'Muscat' cultivars (Etiévant *et al.*, 1983), and α-ionone and benzaldehyde in 'Pinot noir' and 'Gamay,' respectively (Dubois, 1983).

The absence of unique aroma compounds characterizing varietally distinctive wines may simply reflect their not having been isolated or identified as yet. It is also possible that no major impact compound exists in some aromatically distinctive cultivars. Varietally distinctive odors may arise from quantitative, rather than qualitative, aromatic differences. These may be grape-derived, produced by yeast metabolism, or formed during maturation. For example, 'Colombard' wines are reported to derive most of their distinctive "aroma" from products formed during fermentation (Marais, 1986).

Appendix 6.1

Conversion Table for Various Hydrometer Scales Used to Measure Sugar Content of Must[a]

°Brix (Balling)	Specific gravity at 20°C	Oechsle[b]	Baumé[c]	°Brix (Balling)	Specific gravity at 20°C	Oechsle[b]	Baumé[c]
0.0	1.00000	0.0	0.00	21.2	1.08823	88	11.8
0.2	1.00078	0	0.1	21.4	1.08913	89	11.9
0.4	1.00155	1	0.2	21.6	1.09003	90	12.0
0.6	1.00233	2	0.3	21.8	1.09093	91	12.1
0.8	1.00311	3	0.45	22.0	1.09183	92	12.2
1.0	1.00389	4	0.55	22.2	1.09273	93	12.3
2.0	1.00779	8	1.1	22.4	1.09364	94	12.45
3.0	1.01172	12	1.7	22.6	1.09454	95	12.55
4.0	1.01567	15	2.2	22.8	1.09545	95	12.7
5.0	1.01965	20	2.8	23.0	1.09636	96	12.8
6.0	1.02366	24	3.3	23.2	1.09727	97	12.9
7.0	1.02770	28	3.9	23.4	1.09818	98	13.0
8.0	1.03176	32	4.4	23.6	1.09909	99	13.1
9.0	1.03586	36	5.0	23.8	1.10000	100	13.2
10.0	1.03998	40	5.6	24.0	1.10092	101	13.3
11.0	1.04413	44	6.1	24.2	1.10193	102	13.45
12.0	1.04831	48	6.7	24.4	1.10275	103	13.55
13.0	1.05252	53	7.2	24.6	1.10367	104	13.7
14.0	1.05667	57	7.8	24.8	1.10459	105	13.8
15.0	1.06104	61	8.3	25.0	1.10551	106	13.9
16.0	1.06534	65	8.9	25.2	1.10643	106	14.0
17.0	1.06968	70	9.4	25.4	1.10736	107	14.1
17.4	1.07142	71	9.7	25.6	1.10828	108	14.2
18.0	1.07404	74	10.0	25.8	1.10921	109	14.3
18.4	1.07580	76	10.2	26.0	1.11014	110	14.45
19.0	1.07844	78	10.55	26.2	1.11106	111	14.55
19.2	1.07932	79	10.65	26.4	1.11200	112	14.65
19.4	1.08021	80	10.8	26.6	1.11293	113	14.85
19.6	1.08110	81	10.9	26.8	1.11386	114	14.9
19.8	1.08198	82	11.0	27.0	1.11480	115	15.0
20.0	1.08287	83	11.1	27.2	1.11573	116	15.1
20.2	1.08376	84	11.2	27.4	1.11667	117	15.2
20.4	1.08465	85	11.35	27.6	1.11761	118	15.3
20.6	1.08554	86	11.45	27.8	1.11855	119	15.45
20.8	1.08644	86	11.55	30.0	1.12898	129	16.57
21.0	1.08733	87	11.7				

[a] After *Methods for Analysis of Musts and Wines*, M. A. Amerine and C. S. Ough, Copyright © 1980 John Wiley & Sons, Inc. Reprinted by permission of John Wiley & Sons, Inc.

[b] The approximate sugar content on the Oechsle scale is given by dividing the Oechsle by 4 and subtracting 2.5 from the result. Thus, for a must of 80 Oechsle, 80/4 − 2.5 = 17.5. The approximate prospective percentage alcohol by volume is derived by multiplying the Oechsle by 0.125.

[c] Sugar content on the Baumé scale is approximated by the use of the equation °Brix = 1.8 × Baumé. Reducing sugar is about 2.0 less than the °Brix. The °Brix × 0.52 gives the approximate prospective alcohol production.

Appendix 6.2

Conversion Table for Various Measures of Ethanol Content at 20°C[a]

Alcohol (vol %)	Alcohol (wt %)	Alcohol (g/100 ml)	Proof	Specific gravity	Alcohol (vol %)	Alcohol (wt %)	Alcohol (g/100 ml)	Proof	Specific gravity
0.00	0.00	0.00	0.0	1.00000	11.50	9.27	9.13	23.0	0.98471
0.50	0.40	0.40	1.0	0.99925	12.00	9.679	9.52	24.0	0.98412
1.00	0.795	0.79	2.0	0.99851	12.50	10.08	9.92	25.0	0.98354
1.50	1.19	1.19	3.0	0.99777	13.00	10.487	10.31	26.0	0.98297
2.00	1.593	1.59	4.0	0.99704	13.50	10.90	10.71	27.0	0.98239
2.50	1.99	1.98	5.0	0.99633	14.00	11.317	11.11	28.0	0.98182
3.00	2.392	2.38	6.0	0.99560	14.50	11.72	11.51	29.0	0.98127
3.50	2.80	2.78	7.0	0.99490	15.00	12.138	11.90	30.0	0.98071
4.00	3.194	3.18	8.0	0.99419	15.5	12.54	12.30	31.0	0.98015
4.50	3.60	3.58	9.0	0.99360	16.00	12.961	12.69	32.0	0.98960
5.00	3.998	3.97	10.0	0.99281	16.50	13.37	13.09	33.0	0.97904
5.50	4.40	4.37	11.0	0.99214	17.00	13.786	13.49	34.0	0.97850
6.00	4.804	4.76	12.0	0.99149	17.50	14.19	13.89	35.0	0.97797
6.50	5.21	5.16	13.0	0.99084	18.00	14.612	14.28	36.0	0.97743
7.00	5.612	5.56	14.0	0.99020	18.50	15.02	14.68	37.0	0.97690
7.50	6.02	5.96	15.0	0.98956	19.00	15.440	15.08	38.0	0.97638
8.00	6.422	6.36	16.0	0.98894	19.50	15.84	15.47	39.0	0.97585
8.50	6.83	6.75	17.0	0.98832	20.00	16.269	15.87	40.0	0.97532
9.00	7.234	7.14	18.0	0.89771	20.50	16.67	16.26	41.0	0.97479
9.50	7.64	7.54	19.0	0.89711	21.00	17.100	16.66	42.0	0.97425
10.00	8.047	7.93	20.0	0.98650	21.50	17.51	17.06	43.0	0.97372
10.50	8.45	8.33	21.0	0.98590	22.00	17.993	17.46	44.0	0.97318
11.00	8.862	8.73	22.0	0.98530	22.50	18.34	17.86	45.0	0.97262

[a] Reprinted from Official Methods of Analysis, 11th edition, Appendix 6.2, table 8, 1970. Copyright 1970 by AOAC International.

Suggested Readings

General Review Articles and Books

Etiévant, P. X. (1991). Wine. *In* "Volatile Compounds in Foods and Beverages" (H. Maarse, ed.), pp. 483–586. Dekker, New York.

Nykänen, L., and Suomalainen, H. (1983). "Aroma of Beer, Wine and Distilled Alcoholic Beverages." Reidel, Dordrecht, The Netherlands.

Rapp, A. (1988). Wine aroma substances from gas chromatographic analysis. *In* "Wine Analysis" (H. F. Linskens and J. F. Jackson, eds.), pp. 29–66. Springer-Verlag, Berlin.

Rapp, A., and Mandery, H. (1986). Wine aroma. *Experientia* **42**, 873–880.

Schreier, P. (1979). Flavor composition of wines: A review. *Crit. Rev. Food Sci. Nutr.* **12**, 59–111.

Schreier, P. (1984). Formation of wine aroma. *Proc. Alko Symp. Flavour Res. Alcoholic Beverages, Helsinki 1984* (L. Nykänen and P. Lehtonen, eds.), Vol. 3, pp. 9–37. Foundation Biotech. Indust. Ferm., Kauppakirjapino Oy, Helsinki, Finland.

Usseglio-Tomasset, L. (1989). "Chimie Oenologique." Techinque & Documentation, Lavoisier, Paris.

Williams, A. A. (1982). Recent developments in the field of wine flavour research. *J. Inst. Brew.* **88**, 43–53.

Würdig, G., and Woller, R. (1988). "Chemie des Weines." Verlag Eugen Ulmer, Stuttgart.

Analytical Techniques

Linskens, H. F., and Jackson, J. F. (eds.) (1988). "Wine Analysis." Springer-Verlag, Berlin.

Maarse, H. (1991). Introduction. *In* "Volatile Compounds in Foods and Beverages" (H. Maarse, ed.), pp. 1–39. Dekker, New York.

Ough, C. S. and Amerine, M. A., (1988). "Methods for Analysis of Musts and Wines." Wiley, New York.

Sharpe, F. R., and Chappell, C. G. (1991). An introduction to mass spectrometry and its application in the analysis of beer, wine, whiskey, and food. *J. Inst. Brew.* **96**, 381–394.

Zoecklein, B., Fugelsang, K. C., Gump, B. H., and Nury, F. S. (1990).

Appendix 6.3

Interconversion of Acidity Units

A. Conversion of total (titratable) acidity expressed in terms of one acid to another[a,b]

Initial units	Expressed as					
	Tartaric	Malic	Citric	Lactic	Sulfuric	Acetic
Tartaric	1.000	0.893	0.853	1.200	0.653	0.800
Malic	1.119	1.000	0.955	1.343	0.731	0.896
Citric	1.172	1.047	1.000	1.406	0.766	0.667
Lactic	0.833	0.744	0.711	1.000	0.544	0.677
Sulfuric	1.531	1.367	1.306	1.837	1.000	1.225
Acetic	1.250	1.117	1.067	1.500	0.817	1.000

[a] After *Methods for Analysis of Musts and Wines*, M. A. Amerine and C. S. Ough, Copyright © 1980 John Wiley & Sons, Inc. Reprinted by permission of John Wiley & Sons, Inc.

[b] For example, 5 g/liter (sulfuric) would be equivalent to 7.65 g/liter (tartaric) [5 × 1.531].

B. Conversion of acidity expressed in terms of milliequivalents (meq) per liter to grams per liter is achieved by multiplying the meq/liter value by the millinormality (mN) of the acid involved. The millinormalities of tartaric, malic, citric, lactic, sulfuric, and acetic acids are 0.075, 0.067, 0.064, 0.090, 0.049, and 0.060, respectively.
For example, 100 meq/liter (sulfuric) is equivalent to 4.9 g/liter [100 × 0.049].

"Production Wine Analysis." Van Nostrand-Rheinhold, New York.

Acids and Amino Acids

Ough, C. S. (1988). Acids and amino acids in grapes and wine. *In* "Wine Analysis" (H. F. Linskens and J. F. Jackson, eds.), pp. 92–146. Springer-Verlag, Berlin.

Amines

Ough, C. S., and Daudt, C. E. (1981). Quantitative determination of volatile amines in grapes and wines. I. Effect of fermentation and storage temperature on amine concentrations. *Am. J. Enol. Vitic.* **32**, 185–188.

Esters

Nykänen, L. (1986). Formation and occurrence of flavor compounds in wine and distilled alcoholic beverages. *Am. J. Enol. Vitic.* **37**, 89–96.

Ramey, D. D., and Ough, C. S. (1980). Volatile ester hydrolysis or formation during storage of model solutions and wines. *J. Agric. Food Chem.* **28**, 928–934.

Lactones

Muller, C. J., Kepner, R. E., and Webb, A. D. (1973). Lactones in wines—A review. *Am. J. Enol. Vitic.* **24**, 5–8.

Organosulfur Compounds

Rapp, A., Güntert, M., and Almy, J. (1985). Identification and significance of several sulfur-containing compounds in wine. *Am. J. Enol. Vitic.* **36**, 219–221.

Phenolics

Bourzeix, M., and Kovać, V. (1989). Mise au point: Procyanidines ou proanthocyanidols? *Bull. O.I.V.* **62**, 167–175.

Bourzeix, M., Weyland, D., Heredia, N., and Desfeux, N. (1986). Étude des catéchines et des procyanidols de la grappe de raisin, du vin et d'autres dérivés de la vigne. *Bull. O.I.V.* **59**, 1171–1254.

Haslam, E., Lilley, T. H., Warminski, E., Liao, H., Cai, Y., Martin, R., Gaffney, S. H., Goulding, P. N., and Luck, G. (1992). Polyphenol complexation. A study in molecular recognition. *In* "Phenolic Compounds in Food and Their Effects on Health. I. Analysis, Occurrence, and Chemistry" (C.-T. Ho, C. Y. Lee, and M.-T. Huang, eds.), ACS Symp. Ser. No. 506, pp. 8–50. American Chemical Society, Washington, D.C.

Macheix, J. J., Sapis, J. C., and Fleuriet, A. (1991). Phenolic compounds and polyphenoloxidases in relation to browning grapes and wines. *Crit. Rev. Food Sci. Nutr.* **30**, 441–486.

Ribéreau-Gayon, P., and Glories, Y. (1987). Phenolics in grapes and wines. *Proc. 6th Aust. Wine Ind. Tech. Conf.* (T. Lee, ed.), pp. 247–256. Australian Industrial Publ., Adelaide, Australia.

Singleton, V. L. (1987). Oxygen with phenols and related reactions in musts, wines, and model systems: Observation and practical implications. *Am. J. Enol. Vitic.* **38**, 69–77.

Singleton, V. L. (1988). Wine phenols. *In* "Wine Analysis" (H. F. Linskens, and J. F. Jackson, eds.), pp. 173–218. Springer-Verlag, Berlin.

Somers, T. C. (1987). Assessment of phenolic components in viticulture and oenology. *Proc. 6th Aust. Wine Ind. Tech. Conf.* (T. Lee, ed.), pp. 257–260. Australian Industrial Publishers, Adelaide, Australia.

Somers, T. C., and Vérette, E. (1988). Phenolic composition of natural wine types. *In* "Wine Analysis" (H. F. Linskens and J. F. Jackson, eds.), pp. 219–257. Springer-Verlag, Berlin.

Sulfur Dioxide

Larue, F., Park, M. K., and Caruana, C. (1985). Quelques observations sur les conditions de la formation d'anhydride sulfureux en vinification. *Connaiss. Vigne Vin* **19**, 241–248.

Ough, C. S. (1983). Sulfur dioxide and sulfites. *In* "Antimicrobials in Foods" (A. L. Branen and P. M. Davidson, eds.), pp. 177–203. Dekker, New York.

Rose, A. H. (1993). Sulphur dioxide and other preservatives. *J. Wine Res.* **4**, 43–47.

Somers, T. C., and Wescombe, L. F. (1982). Red wine quality: The critical role of SO$_2$ during vinification and conservation. *Aust. Grapegrower Winemaker Tech. Issue* **220**, 1–7.

Taylor, S. L., Higley, N. A., and Bush, R. K. (1986). Sulfites in foods: Uses, analytical methods, residues, fate, exposure assessment, metabolism, toxicity, hypersensitivity. *Adv. Food Res.* **30**, 1–76.

Terpenes

Rapp, A., and Güntert, M. (1986). Changes in aroma substances during the storage of white wines in bottles. *In* "The Shelf Life of Foods and Beverages" (G. Charalambous, ed.), pp. 141–167. Elsevier, Amsterdam.

Rapp, A., Mandery, H., and Güntert, M. (1984). Terpene compounds in wine. *Proc. Alko Symp. Flavour Res. Alcoholic Beverages. Helsinki 1984* (L. Nykänen and P. Lehtonen, eds.), Vol. 3, pp. 255–274. Foundation Biotech. Indust. Ferm., Kauppakirjapino Oy, Helsinki, Finland.

Strauss, C. R., Wilson, B., Gooley, P. R., and Williams, P. J. (1986). The role of monoterpenes in grape and wine flavor—A review. *In* "Biogeneration of Aroma Compounds" (T. H. Parliment and R. B. Croteau, eds.), ACS Symp. Ser. No. 317, pp. 222–242. American Chemical Society, Washington, D.C.

Williams, P. J., Strauss, C. R., Aryan, A. P., and Wilson, B. (1987). Grape flavour—A review of some pre- and postharvest influences. *Proc. 6th Aust. Wine Ind. Tech. Conf.* (T. Lee, ed.), pp. 111–116. Australian Industrial Publ., Adelaide, Australia.

Varietal Aromas

Strauss, C. R., Wilson, B., and Williams, P. J. (1987). Flavour of non-muscat varieties. *Proc. 6th Aust. Wine Ind. Tech. Conf.* (T. Lee, ed.), pp. 117–120. Australian Industrial Publ., Adelaide, Australia.

Williams, P. J., Strauss, C. R., Aryan, A. P., and Wilson, B. (1987). Grape flavour—A review of some pre and postharvest influences. *Proc. 6th Aust. Wine Ind. Tech. Conf.* (T. Lee, ed.), pp. 111–116. Australian Industrial Publ., Adelaide, Australia.

References

Abbott, N. A., Coombe, B. G., and Williams, P. J. (1991). The contribution of hydrolyzed flavor precursors to quality differences in Shiraz juice and wines: An investigation by sensory descriptive analysis. *Am. J. Enol. Vitic.* **42**, 167–174.

Acree, T. E., Braell, P. A., and Butts, R. M. (1981). The presence of damascenone in cultivars of *Vitis vinifera* (Linnaeus), *rotundifolia* (Michaux), and *labruscana* (Bailey). *J. Agric. Food. Chem.* **29**, 688–690.

Acree, T. E., Lavin, E. H., Nishida, R., and Watanabe, S. (1990). *o*-Aminoacetophenone, the "foxy" smelling component of Labruscana grapes. Wöhrmann Symposium, Wädenswil, Switzerland.

Allen, M. S., Lacey, J. J., Harris, R. L. N., and Brown, W. B. (1991). Contribution of methoxypyrazines to Sauvignon blanc wine aroma. *Am. J. Enol. Vitic.* **42**, 109–112.

Amerine, M. A., and Ough, C. S. (1980). "Methods for Analysis of Musts and Wines." Wiley, New York.

Amerine, M. A., and Roessler, E. B. (1983). "Wines. Their Sensory Evaluation." Freeman, New York.

Amerine, M. A., Berg, H. W., Kunkee, R. E., Ough, C. S., Singleton, V. L., and Webb, A. D. (1980). "The Technology of Wine Making." AVI Publ., Westport, Connecticut.

Anonymous. (1986). The history of wine: Sulfurous acid—used in wineries for 500 years. *Ger. Wine Rev.* **2**, 16–18.

AOAC. (1970). "Official Methods of Analysis," 11th Ed. Association of Official Analytical Chemists, Washington, D.C.

Beech, F. W., and Thomas, S. (1985). Action antimicrobienne de l'anhydride sulfureux. *Bull. O.I.V.* **58**, 564–579.

Bock, G., Benda, I., and Schreier, P. (1988). Microbial transformation of geraniol and nerol by *Botrytis cinerea*. *Appl. Microbiol. Biotechnol.* **27**, 351–357.

Boison, J., and Tomlinson, R. H. (1988). An investigation of the volatile composition of *Vitis labrusca* grape must and wines. II. The identification of *N*-(*N*-hydroxy-*N*-methyl-γ-aminobutyryl)glycin in native North American grape varieties. *Can. J. Spectrosc.* **33**, 35–38.

Boison, J. O. K., and Tomlinson, R. H. (1990). New sensitive method for the examination of the volatile flavor fraction of Cabernet Sauvignon wines. *J. Chromatogr.* **522**, 315–328.

Bourzeix, M., and Kovać, V. (1989). Mise au point: Procyanidines ou proanthocyanidols? *Bull. O.I.V.* **62**, 167–175.

Bourzeix, M., Heredia, N., and Kovać, V. (1983). Richesse de différents cépages en composés phénoliques totaux et en anthocyanes. *Prog. Agric. Vitic.* **100**, 421–428.

Brander, C. F., Kepner, R. E., and Webb, A. D. (1980). Identification of some volatile compounds of wine of *Vitis vinifera* cultivar Pinot noir. *Am. J. Enol. Vitic.* **31**, 69–75.

Brun, S., Cabanis, J. C., and Mestres, J. P. (1986). Analytical chemistry. *Experientia* **42**, 893–904.

Charpentier, N., and Maujean, A. (1981). Sunlight flavours in champagne wines. *Flavour '81, Proc. 3rd Weurman Symp., Munich, Apr. 28–30, 1981*, pp. 609–615. de Gruyter, Berlin.

Chatonnet, P., Boidron, J. N., and Pons, M. (1990). Élevage des vins rouges en fûts de chêne: Évolution de certains composés volatils et leur impact arômatique. *Sci. Aliments* **10**, 565–587.

Chen, E. C.-H. (1978). The relative contribution of Ehrlich and biosynthetic pathways to the formation of fusel alcohols. *J. Am. Soc. Brew. Chem.* **35**, 39–43.

Cheynier, V., and Ricardo da Silva, J. M. R. (1991). Oxidation of grape procyanidins in model solutions containing *trans*-caffeoyltartaric acid and polyphenol oxidase. *J. Agric. Food Chem.* **39**, 1047–1049.

Crippen, D. D., Jr., and Morrison, J. C. (1986). The effects of sun exposure on the phenolic content of Cabernet Sauvignon berries during development. *Am. J. Enol. Vitic.* **37**, 243–247.

Dittrich, H. H., Sponholz, W. R., and Kast, W. (1974). Vergleichende Untersuchungen von Mosten und Weinen aus gesunden und aus *Botrytis*-infizierten Traubenbeeren. I. Säurestoffwechsel, Zuckerstoffwechsel produckte. Leucoanthocyangehalte. *Vitis*13, 36–49.

Doores, S. (1983). Organic acids. *In* "Antimicrobials in Foods" (A. L. Branden and P. M. Davidson, eds.), pp. 75–99. Dekker, New York.

Drawert, F. (1970). Causes déterinant l'amertume de certains vins blancs. *Bull. O.I.V.* **43**, 19–27.

Dubois, P. (1983). Volatile phenols in wines. *In* "Flavour of Distilled Beverages" (J. R. Piggott, ed.), pp. 110–119. Ellis Horwood, Chichester.

Dubois, P., and Rigaud, J. (1981). A propos de goût de bouchon. *Vignes Vins* **301**, 48–49.

Dubourdieu, D., Grassin, C., Deruche, C., and Ribéreau-Gayon, P. (1984). Mise au point d'une mesure rapide de l'activité laccase dans le moûts et dans les vins par la méthode à la syringaldazine. Application à l'appréciation de l'état sanitaire des vendages. *Connaiss. Vigne Vin* **18**, 237–252.

Dumon, M. C., Michaud, J., and Masquelier, J. (1991). Dosage des procyanidols des pépins de raisin de cépages rouges et blancs du Bordelais. *Bull. O.I.V.* **64**, 533–542.

du Plessis, C. S., and Augustyn, O. P. H. (1981). Initial study on the guava aroma of Chenin blanc and Colombar wines. *S. Afr. J. Enol. Vitic.* **2**, 101–103.

Etiévant, P. X. (1991). Wine. *In* "Volatile Compounds in Foods and B everages" (H. Maarse, ed.), pp. 483–546. Dekker, New York.

Etiévant, P. X., Issanchou, S. N., and Bayonove, C. L. (1983). The flavour of Muscat wine: The sensory contribution of some volatile compounds. *J. Sci. Food Agric.* **34**, 497–504.

Falcone, F. (1991). Migration of lead into alcoholic beverages during storage in lead crystal decanters. *J. Food Protect.* **54**, 378–380.

Glories, Y. (1984). La couleur des vins rouges. Part 2. Mesure, origine et interpretation. *Connaiss. Vigne Vin* **18**, 253–271.

Goetghebeur, M., Nicolas, M., Blaise, A., Galzy, P., and Brun, S. (1992). Étude sur le rôle et l'origine de la benzyl alcool oxydase responsable du goût d'amande amère des vins. *Bull. O.I.V.* **65**, 345–360.

Günata, Y. Z., Bayonove, C. L., Tapiero, C., and Cardonnier, R. E. (1990). Hydrolysis of grape monoterpenyl β-D-glucosides by various β-glucosidases. *J. Agric. Food Chem.* **38**, 1232–1236.

Hamatschek, J. (1982). Aromastoffe im Wein und deren Herkunft. *Dragoco Rep.* (*Ger. Ed.*) **27**, 59–71.

Haslam, E. (1981). Vegetable tannins. *In* "The Biochemistry of Plants" (E. E. Conn, eds.), Vol. 7, pp. 527–556. Academic Press, New York.

Heimann, W., Rapp, A., Völter, J., and Knipser, W. (1983). Beitrag zur Entstehung des Korktons in Wein. *Dtsch. Lebensm.-Rundsch.* **79**, 103–107.

Henschke, P. A., and Jiranek, V. (1991). Hydrogen sulfide formation during fermentation: Effect of nitrogen composition in model grape musts. *Proc. Int. Symp. Nitrogen Grapes Wine, Seattle, WA, June, 1991* (J. M. Rantz, ed.), pp. 177–184. Am. Soc. Enol. Vitic., Davis, California.

Henschke, P. A., and Ough, C. S. (1991). Urea accumulation in fermenting grape juice. *Am. J. Enol. Vitic.* **42**, 317–321.

Heresztyn, T. (1986). Formation of substituted tetrahydopyridines by species of *Brettanomyces* and *Lactobacillus* isolated from mousy wines. *Am. J. Enol. Vitic.* **37**, 127–131.

Hötzel, D., Muskat, E., Bitsch, I., Aign, W., Althoff, J.-D., and Cremer, H. D. (1969). Thiamin-Mangel und Unbedenklichkeit von Sulfit für den Menschen. *Int. Z. Vitaminforsch.* **39**, 372–383.

Hrazdina, G., Borzell, A. J., and Robinson, W. B. (1970). Studies on the stability of the anthocyanidin-3,5-diglucosides. *Am. J. Enol. Vitic.* **21**, 201–204.

Hrazdina, G., Parsons, G. F., and Mattick, L. R. (1984). Physiological and biochemical events during development and maturation of grape berries. *Am. J. Enol. Vitic.* **35**, 220–227.

Iglesias, J. L. M., Dabiila, F. H., Marino, J. I. M., De Miguel Gorrdillo, C., and Exposito, J. M. (1991). Biochemical aspects of the lipids of *Vitis vinifera* grapes (Macebeo var.). Part 1. Linoleic and linolenic acids as aromatic precursors. *Nahrung* **35**, 705–710.

Institute of Food Technologists Expert Panel on Food Safety and Nutrition. (1975). Sulfites as food additives. *Food Technol.* **29**, 117–120.

Jayaraman, A., and van Buren, J. P. (1972). Browning of galacturonic acid in a model system stimulation fruit beverages and white wine. *J. Agric. Food Chem.* **20**, 122–124.

Joint FAO/WHO Expert Committee on Food Additives. (1974). Toxicological evaluation of certain food additives with a review of general principles and of specifications. 17th Report Food and Agricultural Organization, Rome.

Killian, E., and Ough, C. S. (1979). Fermentation esters—Formation and retention as affected by fermentation temperature. *Am. J. Enol. Vitic.* **30**, 301–305.

Kodama, S., Suzuki, T., Fujinawa, S., De la Teja, P., and Yotsuzuka, F. (1991). Prevention of ethyl carbamate formation in wine by urea degradation using acid urease. *Proc. Int. Symp. Nitrogen Grapes Wine, Seattle, WA, June, 1991* (J. M. Rantz, ed.), pp. 270–273. Am. Soc. Enol. Vitic., Davis, California.

Kolor, M. K. (1983). Identification of an important new flavor compound in Concord grape: Ethyl 3-mercaptopropionate. *J. Agric. Food Chem.* **31**, 1125–1127.

Kovać, V., Bourzeix, M., Heredia, N., and Ramos, T. (1990). Études des catéchines et proanthocyanidols de raisins et vins blancs. *Rev. Fr. Oenol.* **125**, 7–15.

Lacey, M. J., Allen, M. S., Harris, R. L. N., and Brown, W. V. (1991). Methoxypyrazines in Sauvignon blanc grapes and wines. *Am. J. Enol. Vitic.* **42**, 103–108.

Lamikanra, O. (1987). Aroma constituents of Muscadine wines. *J. Food Qual.* **10**, 57–66.

Larue, F., Park, M. K., and Caruana, C. (1985). Quelques observations sur les conditions de la formation d'anhydride sulfureux en vinification. *Connaiss. Vigne Vin* **19**, 241–248.

Lea, A. G. H., and Arnold, G. M. (1978). The phenolics of ciders: Bitterness and astringency. *J. Sci. Food Agric.* **29**, 478–483.

Lee, C. Y., and Jaworski, A. W. (1987). Phenolic compounds in white grapes grown in New York. *Am. J. Enol. Vitic.* **38**, 277–281.

McCloskey, L. P. (1974). Gluconic acid in California wines. *Am. J. Enol. Vitic.* **25**, 198–201.

McCloskey, L. P., and Yengoyan, L. S. (1981). Analysis of anthocyanins in *Vitis vinifera* wines and red color versus aging by HPLC and spectrophotometry. *Am. J. Enol. Vitic.* **32**, 257–261.

Macheix, J. J., Sapis, J. C., and Fleuriet, A. (1991). Phenolic compounds and polyphenoloxidases in relation to browning grapes and wines. *Crit. Rev. Food Sci. Nutr.* **30**, 441–486.

McKinnon, A. J., Cattrall, R. W., and Schollary, G. R. (1992). Aluminum in wine—Its measurement and identification of major sources. *Am. J. Enol. Vitic.* **43**, 166–170.

Marais, J. (1986). Effect of storage time and temperature of the volatile composition and quality of South African *Vitis vinifera* L. cv. Colombar wines. *In* "The Shelf Life of Foods and Beverages" (G. Charalambous, ed.), pp. 169–185. Elsevier, Amsterdam.

Marais, J., van Rooyen, P. C., and du Plessis, C. S. (1979). Objective quality rating of Pinotage wine. *Vitis* **18**, 31–39.

Marais, J., van Wyk, C. J., and Rapp, A. (1992). Effect of sunlight and shade on norisoprenoid levels in maturing Weisser Riesling and Chenin blanc grapes and Weisser Riesling wines. *S. Afr. J. Enol. Vitic.* **13**, 23–32.

Marriott, R. J. (1986). Biogenesis of blackcurrant (*Ribes nigrum*) aroma. *In* "Biogeneration of Aromas" (T. H. Parliment and R. Crouteau, eds.), ACS Symp. Ser No. 317, pp. 184–192. American Chemical Society, Washington, D.C.

Martin, B., Etiévant, P. X., and Le Quéré, J.-L. (1991). More clues of the occurrence and flavor impact of solerone in wine. *J. Agric. Food Chem.* **39**, 1501–1503.

Martin, B., Etiévant, P. X., Le Quéré, J. L., and Schlich, P. (1992). More clues about sensory impact of sotolon in some flor sherry wines. *J. Agric. Food Chem.* **40**, 475–478.

Masuda, J., Okawa, E., Nishimura, K., and Yunome, H. (1984). Identification of 4,5-dimethyl-3-hydroxy-2(5*H*)-furanone (sotolon) and ethyl 9-hydroxynonanoate in botrytised wine and evaluation of the roles of compounds characteristic of it. *Agric. Biol. Chem.* **48**, 2707–2710.

Maujean, A., Poinsaut, P., Dantan, H., Brissonnet, F., and Cossiez, E. (1990). Étude de la tenue et de la qualité de mousse des vins effervescents. II. Mise au point d'une technique de mesure de la moussabilité, de la tenue et de la stabilité de la mousse des vins effervescents. *Bull. O.I.V.* **63**, 405–427.

Médina, B., Guimberteau, G., and Sudraud, P. (1977). Dosage de plomb dans les vins. Une cause d'enrichissement: Les capsules de surbouchage. *Connaiss. Vigne Vin* **2**, 183–193.

Muller, C. J., Kepner, R. E., and Webb, A. D. (1973). Lactones in wines—A review. *Am. J. Enol. Vitic.* **24**, 5–8.

Munoz, E., and Ingledew, W. M. (1990). Yeast hulls in wine fermentations. A review. *J. Wine Res.* **1**, 197–209.

Noble, A. C., and Bursick, G. F. (1984). The contribution of glycerol to perceived viscosity and sweetness in white wine. *Am. J. Enol. Vitic.* **35**, 110–112.

Ong, B. Y., and Nagel, C. W. (1978). High-pressure liquid chromatographic analysis of hydroxycinnamic acid tartaric acid esters and their glucose esters in *Vitis vinifera*. *J. Chromatogr.* **157**, 345–355.

Ough, C. S., Stevens, D., Sendovski, T., Huang, Z., and An, A. (1990). Factors contributing to urea formation in commercially fermented wines. *Am. J. Enol. Vitic.* **41**, 68–73.

Packter, N. M. (1980). Biosynthesis of acetate-derived phenols (polyketides). *In* "The Biochemistry of Plants" (P. K. Strumpf, ed.), Vol. 4, pp. 535–570, Academic Press, New York.

Peynaud, E. (1984). "Knowing and Making Wine." Wiley, New York.

Radler, F., and Fäth, K. P. (1991). Histamine and other biogenic amines in wines. *Proc. Int. Symp. Nitrogen Grapes Wine, Seattle, WA, June, 1991* (J. M. Rantz, ed.), pp. 185–195. Am. Soc. Enol. Vitic., Davis, California.

Ramey, D. D., and Ough, C. S. (1980). Volatile ester hydrolysis or formation during storage of model solutions and wines. *J. Agric. Food Chem.* **28**, 928–934.

Rapp, A., and Güntert, M. (1986). Changes in aroma substances during the storage of white wines in bottles. *In* "The Shelf Life of Foods and Beverages" (G. Charalambous, ed.), pp. 141–167. Elsevier, Amsterdam.

Rapp, A., and Mandery, H. (1986). Wine aroma. *Experientia* **42**, 873–880.

Rapp, A., Güntert, M., and Almy, J. (1985). Identification and significance of several sulfur-containing compounds in wine. *Am. J. Enol. Vitic.* **36**, 219–221.

Rapp, A., Versini, G., and Ullemeyer, H. (1993). 2-Aminoacetophenon: Verursachende komponente der "untypischen alterungsnate" ("Naphthalinton," "Hybridlon") bei Wein. *Vitis* **32**, 61–62.

Ribéreau-Gayon, J., Ribéreau-Gayon, P., and Seguin, G. (1980). *Botrytis cinerea* in Enology. *In* "The Biology of Botrytis" (J. R. Coley-Smith, K. Verhoeff, and W. R. Jarvis, eds.), pp. 251–274. Academic Press, London.

Ribéreau-Gayon, P. (1964). Les composés phénoliques du raisin et du vin. I. II. III. *Ann. Physiol. Veg.* **6**, 119–147, 211–242, 259–282.

Ribéreau-Gayon, P. (1988). *Botrytis*: Advantages and disadvantages for producing quality wines. *Proc. 2nd Int. Symp. Cool Climate Vitic. Oenol., Auckland, N.Z.* (R. E. Smart, S. B. Thornton, S. B.

Rodriguez, and J. E. Young, eds.), pp. 319–323. New Zealand Soc. Vitic. Oenol., Auckland, New Zealand.

Ribéreau-Gayon, P., and Glories, Y. (1987). Phenolics in grapes and wines. *Proc. 6th Aust. Wine Ind. Tech. Conf.* (T. Lee, ed.), pp. 247–256. Australian Industrial Publ., Adelaide, Australia.

Ribéreau-Gayon, P., Pontallier, P., and Glories, Y. (1983). Some interpretations of colour changes in young red wines during their conservation. *J. Sci. Food Agric.* **34**, 505–616.

Roberts, A. C., and McWeeny, D. J. (1972). The uses of sulphur dioxide in the food industry—A review. *J. Food Technol.* **7**, 221–238.

Robinson, W. B., Weirs, L. D., Bertino, J. J., and Mattick, L. R. (1966). The relation of anthocyanin composition to color stability of New York State wines. *Am. J. Enol. Vitic.* **17**, 178–184.

Roufet, M., Bayonove, C. L., and Cordonnier, R. E. (1986). Changes in fatty acids from grape lipidic fractions during crushing exposed to air. *Am. J. Enol. Vitic.* **37**, 202–205.

Scalbert, A. (1991). Antimicrobial properties of tannins. *Phytochemistry* **30**, 3875–3883.

Schreier, P. (1982). Volatile constituents in different grape species. *Grape Wine Centennial Symp. Proc. 1980*, pp. 317–321. Univ. of California, Davis, California.

Schreier, P., and Drawert, F. (1974). Gaschromatographisch—massenspektometrische Untersuchung flüchtiger Inhaltsstoffe des Weines. V. Alkohole, Hydroxy-Ester, Lactone und andere polare Komponenten des Weinaromas. *Chem. Mikrobiol. Technol. Lebensm.* **3**, 154–160.

Schreier, P., and Paroschy, J. H. (1981). Volatile constituents from Concord, Niagara (*Vitis labrusca*) and Elvira (*V. labrusca* × *V. riparia*) grapes. *Can. Inst. Food Sci. Technol. J.* **14**, 112–118.

Schütz, M., and Kunkee, R. E. (1977). Formation of hydrogen sulfide from elemental sulfur during fermentation by wine yeast. *Am. J. Enol. Vitic.* **28**, 137–144.

Sefton, M. A., Skouroumounis, G. K., Massey-Westropp, R. A., and Williams, P. J. (1989). Norisoprenoids in *Vitis vinifera* white wine grapes and the identification of a precursor of damascenone in these fruits. *Aust. J. Chem.* **42**, 2071–2084.

Sharpe, F. R., and Chappell C. G. (1991). An introduction to mass spectrometry and its application in the analysis of beer, wine, whiskey, and food. *J. Inst. Brew.* **96**, 381–394.

Shinohara, T., Shimizu, J., and Shimazu, Y. (1979). Esterification rates of main organic acids in wines. *Agric. Biol. Chem.* **43**, 2351–2358.

Simpson, R. F., and Miller, G. C. (1984). Aroma composition of Chardonnay wine. *Vitis* **23**, 143–158.

Sims, C. A., and Morris, J. R. (1986). Effects of acetaldehyde and tannins on the color and chemical age of red Muscadine (*Vitis rotundifolia*) wine. *Am. J. Enol. Vitic.* **37**, 164–165.

Singleton, V. L. (1987). Oxygen with phenols and related reactions in must, wines and model systems: Observations and practical implications. *Am. J. Enol. Vitic.* **38**, 69–77.

Singleton, V. L., and Noble, A. C. (1976). Wine flavour and phenolic substances. *In* "Phenolic, Sulfur and Nitrogen Compounds in Food Flavors" (G. Charalambous and G. Katz, eds.), Am. ACS Symp. Ser. No. 26, pp. 47–70. American Chemical Society, Washington, D.C.

Singleton, V. L., and Trousdale, E. (1992). Anthocyanin–tannin interactions explaining differences in polymeric phenols between white and red wines. *Am. J. Enol. Vitic.* **43**, 63–70.

Singleton, V. L., Timberlake, C. F., and Lea, A. G. H. (1978). The phenolic cinnamates of white grapes and wine. *J. Sci. Food Agric.* **29**, 403–410.

Singleton, V. L., Trousdale, E., and Zaya, J. (1979). Oxidation of wines. 1. Young white wines periodically exposed to air. *Am. J. Enol. Vitic.* **30**, 49–54.

Singleton, V. L., Salgues, M., Zaya, J., and Trousdale, E. (1985). Caftaric acid disappearance and conversion to products of enzymic oxidation in grape must and wine. *Am. J. Enol. Vitic.* **36**, 50–56.

Sneyd, T. N. (1988). Tin lead capsules. *Aust. Wine Res. Inst. Techn. Rev.* **56**, 1.

Somers, T. C. (1982). Pigment phenomena—From grapes to wine. *Grape Wine Centennial Symp. Proc. 1980* pp. 254–257. University of California, Davis, California.

Somers, T. C. (1983). Influence du facteur temps de conservation. *Bull. O.I.V.* **57**, 172–188.

Somers, T. C., and Wescombe, L. G. (1982). Red wine quality: The critical role of SO_2 during vinification and conservation. *Aust. Grapegrower Winemaker Tech. Issue* **220**, 1–7.

Somers, T. C., and Evans, M. E. (1986). Evolution of red wines. I. Ambient influences on colour composition during early maturation. *Vitis* **25**, 31–39.

Somers, T. C., and Vérette, E. (1988). Phenolic composition of natural wine types. *In* "Wine Analysis" (H. F. Linskens and J. F., Jackson eds.), pp. 219–257.

Sorrentino, F., Voilley, A., and Richon, D. (1986). Activity coefficients of aroma compounds in model food systems. *AIChE J.* **32**, 1988–1993.

Sponholz, W. R. (1988). Alcohols derived from sugars and other sources and fullbodiedness of wines. *In* "Wine Analysis" (H. F. Liskens and J. F. Jackson, eds.), pp. 147–172. Springer-Verlag, Berlin.

Sponholz, W. R., and Dittrich, H. H. (1984). Galacturonic, glucuronic, 2- and 5-oxo-gluconic acids in wines, sherries fruit and dessert wines. *Vitis* **23**, 214–224.

Sponholz, W. R., Kürbel, H., and Dittrich, H. H. (1991). Beiträge zur Bildung von Ethylcarbamat in Wein. *Wein-Wiss.* **46**, 11–17.

Stevens, D. F., and Ough, C. S. (1993). Ethyl carbamate formation: Reaction of urea and citrulline with ethanol in wine under low to normal temperature conditions. *Am. J. Enol. Vitic.* **44**, 309–312.

Strauss, C. R., Wilson, B., Gooley, P. R., and Williams, P. J. (1986). The role of monoterpenes in grape and wine flavor—A review. *In* "Biogeneration of Aroma Compounds" (T. H. Parliment and R. B. Croteau, eds.), ACS Symp. Ser. No. 317, pp. 222–242. American Chemical Society, Washington, D.C.

Strauss, C. R., Gooley, P. R., Wilson, B., and Williams, P. J. (1987a). Application of droplet countercurrent chromatography to the analysis of conjugated forms of terpenoids, phenols, and other constituents of grape juice. *J. Agric. Food Chem.* **35**, 519–524.

Strauss, C. R., Wilson, B., Anderson, R., and Williams, P. J. (1987b). Development of precursors of C_{13} nor-isoprenoid flavorants in Riesling grapes. *Am. J. Enol. Vitic.* **38**, 23–27.

Strauss, C. R., Wilson, B., and Williams, P. J. (1987c). Flavour of non-muscat varieties. *Proc. 6th Austr. Wine Ind. Tech. Conf.* (T. Lee, ed.), pp. 117–120. Australian Industrial Publ., Adelaide, Australia.

Sudraud, P., and Chauvet, S. (1985). Activité antilevure de l'anhydride sulfureux moléculaire. *Connaiss. Vigne Vin* **19**, 31–40.

Tanner, H., and Zanier, C. (1980). Der Kork als Flaschenverschluss aus der Sicht des Chemikers. *Mitt. Geb. Lebensmittelunters. Hyg.* **71**, 62–68.

Taylor, S. L., Higley, N. A., and Bush, R. K. (1986). Sulfites in foods: Uses, analytical methods, residues, fate, exposure assessment, metabolism, toxicity, and hypersensitivity. *Adv. Food Res.* **30**, 1–75.

Tegmo-Larsson, I.-M., and Henick-Kling, T. (1990). Ethyl carbamate precursors in grape juice and the efficiency of acid urease on their removal. *Am. J. Enol. Vitic.* **41**, 189–192.

Tucknott, O. G. (1977). The mousy taint in fermented beverages. Ph. D. Thesis. Univ. of Bristol, Bristol.

Tyson, P. J., Luis, E. S., and Lee, T. H. (1982). Soluble protein levels in grapes and wine. *Grape Wine Centennial Symp. Proc. 1980,* pp. 287–290. Univ. of California, Davis, California.

van Buren, J. P., Bertino, J. J., Einset, J., Remaily, G. W., and Robinson, W. B. (1970). A comparative study of the anthocyanin pigment composition in wines derived from hybrid grapes. *Am. J. Enol. Vitic.* **21**, 117–130.

van Wyk, C. J., Augustyn, O. P. H., de Wet, P., and Joubert, W. A. (1979). Isoamyl acetate—A key fermentation volatile of wines of *Vitis vinifera* cv Pinotage. *Am. J. Enol. Vitic.* **30**, 167–173.

Versini, G. (1985). Sull'aromna del vino «Traminer aromatico» 0 «Gewürztraminer». *Vignevini* **12**, 57–65.

von Hellmuth, K. H., Fischer, E., and Rapp. A. (1985). Über das Verhalten von Spurenelementen und Radionukliden in Traubenmost bei der Gärung und beim Weinausbau. *Dtsch. Lebensm.-Rundsch.* **81**, 171–176.

Wenzel, K., Dittrich. H. H., and Heimfarth, M. (1987). Die Zusammensetzung der Anthocyane in den Beeren verschiedener Rebsorten. *Vitis* **26**, 65–78.

Wildenradt, H. L., and Singleton, V. L. (1974). The production of aldehydes as a result of oxidation of polyphenolic compounds and its relation to wine aging. *Am. J. Enol. Vitic.* **25**, 119–126.

Williams, A. A., and Rosser, P. R. (1981). Aroma enhancing effects of ethanol. *Chem. Senses* **6**, 149–153.

Williams, P. J., Strauss, C. R., Aryan, A. P., and Wilson, B. (1987). Grape flavour—A review of some pre and postharvest influences. *Proc. 6th Aust. Wine Ind. Tech. Conf.* (T. Lee, ed.), pp. 111–116. Australian Industrial Publ., Adelaide, Australia.

Winterhalter, P. (1991). 1,1,6-Trimethyl-1,2-dihydronaphthalene (TDN) formation in wine. 1. Studies on the hydrolysis of 2,6,10,10-tetramethyl-1-oxaspiro[4.5]dec-6-ene-2,8-diol rationalizing the origin of TDN and related C_{13} norisoprenoids in Riesling wine. *J. Agric. Food Chem.* **39**, 1825–1829.

Winterhalter, P., Sefton, M. A., and Williams, P. J. (1990). Volatile C_{13} norisoprenoid compounds in Riesling wine are generated from multiple precursors. *Am. J. Enol. Vitic.* **41**, 277–283.

Würdig, G. (1985). Levures produisant du SO_2. *Bull. O.I.V.* **58**, 582–589.

Yokotsuka, K., Shimizu, T., and Shimizu, T. (1991). Polyphenoloxidase from six mature grape varieties on their activities towards various phenols. *J. Ferment. Bioeng.* **71**, 156–162.

7

Fermentation

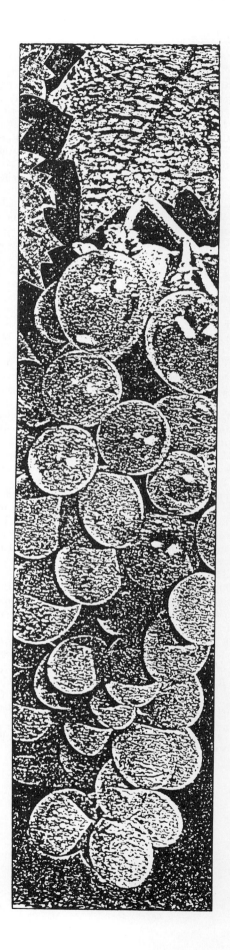

Introduction

The theory and practice of enology have developed enormously since its simple beginnings some six thousand years ago. Although advancements occurred sporadically, the pace of change quickened dramatically in the seventeenth century, reflecting parallel developments in science and technology. Improvements in glass production and the use of cork favored the development of wine styles that benefited from long aging. Sparkling wine also became possible. The research by Pasteur on problems in the wine industry during the 1860s led to solutions to several wine "diseases", and the foundation of our understanding of the nature of fermentation. Subsequent work has perfected wine making skills to their current high standards. Future study should result in premium wines showing more consistently the quality characteristics consumers deserve. In addition, distinctive features, based on varietal, regional, or stylistic differences, hopefully will become more discernible and controlled. Dr. Richard Peterson, a highly respected winemaker in California, has commented that Mother

Nature is a "nasty old lady, who must be controlled." Modern enological and viticultural science is increasingly providing the means by which many of the vicissitudes of Mother Nature can be moderated or controlled.

Basic Procedures of Wine Production

Vinification formally begins when the grapes, or juice, reach the winery. The basic steps in the production of table wines are outlined in Fig. 7.1.

The first step involves removing the stems, leaves, and any other extraneous material (Fig. 7.1). The fruit is then crushed to release the juice and begin the process of **maceration,** which facilitates the extraction of compounds in the seeds and skins. Initially, maceration is activated by the action of hydrolytic enzymes released from cells ruptured during crushing. Enzymatic maceration releases flavor ingredients from the skins, seeds, and pulp and promotes the syntheses of additional flavor compounds. Enzymes also may hydrolyze macromolecules into forms readily utilized by yeast and bacterial cells. In addition, the cytotoxic action of pectic enzymes on undamaged grape cells effects the release of cell contents into the **must** (grape macerate).

For white wines, maceration either is kept to an absolute minimum or is kept brief, seldom lasting more than a few hours. The juice that runs feely from the crushed grapes (**free-run**) is usually combined with that released on gentle pressing. The combined fractions are then fermented. Subsequent pressings are usually fermented separately.

With red wines, maceration is prolonged and occurs simultaneously with alcoholic fermentation. The alcohol generated by yeast action enhances the extraction of anthocyanin and is crucial to the uptake of tannins from the seeds and skins (**pomace**). The phenolic compounds solubilized give red wines their basic properties of appearance, taste, and flavor. They are also required to give red wines their aging and mellowing characteristics. In addition, ethanol is important in the release of aromatic ingredients from the pulp and skins. After partial or complete fermentation, the **free-run** is allowed to flow away under gravity. Subsequent pressing extracts most of the remaining juice (**press fractions**). Press fractions are commonly added to the free-run in proportions determined by the style of wine desired.

Rosé wines are made from red grapes subjected to a comparatively short maceration on the skins, the duration depending primarily on the intensity of rosé color desired. Owing to the short duration of maceration, it may end before significant alcoholic fermentation has occurred.

Fermentation may start spontaneously due to indigenous yeasts derived from the grape, or picked up from crushing equipment. More commonly, however, the juice or must is inoculated with a yeast strain of known characteristics. Yeasts not only produce alcohol, but also generate the basic bouquet and flavor of wines.

On completing **alcoholic** fermentation, the wine may be encouraged to undergo a second **malolactic** fermentation. Malolactic fermentation is particularly valuable in cool climatic regions, where the acidity reduction improves the taste of the wine. Although most red wines benefit from malolactic fermentation, fewer white wines profit from its occurrence. The milder fragrance of most white wines makes them less able to mask potentially undesirable flavor changes induced by malolactic fermentation. In warm viticultural regions, malolactic fermentation is often unneeded and undesirable; its development usually is discouraged by practices such as the addition of sulfur dioxide, early clarification, and storage under cool conditions.

Newly fermented wine often is protected from or given only limited exposure to air. This is designed to restrict oxidation and microbial spoilage, while the processes of maturation permits the loss of yeasty odors, the dissi-

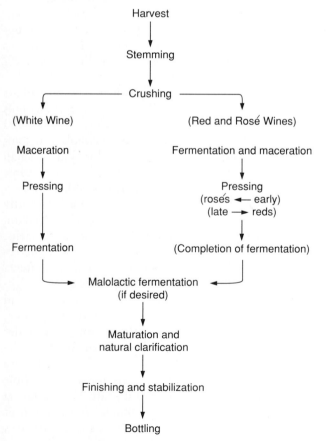

Figure 7.1   Flow diagram of winemaking.

pation of excess carbon dioxide, and the precipitation of suspended particular matter. Changes in aroma, and the development of an aged bouquet, also commence. Where maturing wines are exposed to air, exposure is usually restricted to that which occurs during racking. Such exposure to air can help oxidize hydrogen sulfide and favors color stability in red wines.

After several weeks or months, the wine is racked. **Racking** separates the wine from solids that settle out during spontaneous or induced clarification. Sediment consists primarily of yeast and bacterial cells, grape cell remains, and precipitated tannins, proteins, and potassium tartrate crystals. If left in contact with wine, they may lead to the production of off-odors, and some may favor microbial spoilage. Racking from small cooperage usually results in the absorption of limited amounts of oxygen.

Prior to bottling, the wine may be **fined** to remove traces of dissolved proteins and other materials that can lead to the development of haziness, especially on exposure to heat. Fining also is used to soften the taste of the wine by removing excess tannins. Wines are commonly chilled and filtered to further enhance clarification and stability.

At bottling, wines are generally given a small dose of sulfur dioxide to prevent oxidation and microbial spoilage. Sweet wines usually are sterile filtered as a further precaution against microbial spoilage.

Newly bottled wines are normally aged at the winery for several months to years before distribution to wholesalers. This period permits wines blended shortly before bottling time to "harmonize". In addition, it allows acetaldehyde that may be produced as a result of slight aeration during bottling to be converted to nonaromatic compounds. As a consequence, "bottle sickness" induced by acetaldehyde usually dissipates before the wine reaches the consumer.

Prefermentation Practices

Stemming and crushing are commonly conducted as soon as possible after harvesting. During the harvest, some grapes are unavoidably broken and their juice released, while others may be bruised. Thus, oxidative browning often begins before the grapes reach the winery and crushing begins. The juice also becomes "field-inoculated" with the yeast and bacterial flora present on the grape surface. If the berries are harvested during the heat of the day, undesirable microbial contamination can rapidly develop. To minimize this occurrence, grapes may be sulfited on harvest, and they are picked during cool parts of the day.

Left in containers, harvested fruit quickly warm owing to endogenous metabolic activity. This can aggravate contamination by speeding microbial activity. In addition, warming may necessitate cooling to bring the temperature down to an acceptable prefermentation value.

Stemming

The modern trend is to separate the processes of stemming and crushing. Removal of the stem, leaves, and grape stalks before crushing has several advantages. Notably, it minimizes the excessive uptake of phenols and lipids from vine parts. Extraction of stem phenols is of potential value only when dealing with red grape varieties low in phenol content. Stem phenols generally produce more astringent and bitter tastes than phenols released by the seeds and skins.

In the past, stems were often left with the must throughout fermentation, especially in the production of red wines. Presence of the stems made pressing easier, presumably by creating drainage channels along which the wine could escape. Modern improvements in press design have made stem retention unnecessary. The higher tannin contents derived from a prolonged contact with the stems gave red wines made during poor vintage years extra "body" and improved color density by stabilizing the limited anthocyanin content (see Chapter 6).

In addition to facilitating phenol extraction and pressing, maceration with the stems may increase the fermentation rate. This appears to be due to the increased uptake of oleanolic acid (Bréchot et al., 1971). This is especially valuable under cool cellar conditions by favoring complete fermentation.

Leaf removal before crushing is beneficial as it limits the production of C_6 ("leaf") aldehydes and alcohols generated during the enzymatic oxidation of linoleic and linolenic acids. The aldehydes and alcohols can taint wine with a grassy to herbaceous odor, but in small amounts can contribute to the typical aroma of some wines. High leaf content also may result in the considerable uptake of quercitin. If the wine is bottled shortly after fermentation, quercitin can lead to the production of a yellowish haze in white wines (Somers and Ziemelis, 1985). When the wines are matured sufficiently, quercetin precipitates before bottling. High flavonol contents also can produce bitterness in white wines.

For convenience and efficiency, stemming and crushing often are performed by the same machine. Stemmers usually contain an outer perforated cylinder that permits berries to pass through but prevents the passage of stems, stalks, and leaves (Fig. 7.2). Often there are a series of spirally arranged arms, possessing flexible paddle ends, situated on a central shaft. Shaft rotation draws grape clusters into the stemmer, forces the fruit through the perforations, and expels the stems and leaves out the end. When crusher–stemmers are working optimally, the fruit is removed largely unbroken. Expelling

Figure 7.2 Internal view of a crusher–stemmer. (Photograph courtesy of the Wine Institute.)

the stems and leaves in a dry state avoids juice loss and facilitates their disposal. The stems may be chopped for subsequent soil incorporation.

Crushing

Crushing typically follows immediately on stemming because stemming unavoidably crushes some of the fruit. The juice so released is highly susceptible to oxidative browning and microbial contamination. Crushing the fruit without delay permits fermentation to commence almost immediately, limits microbial contamination, and permits better control of oxidation.

Crushing is accomplished by any one of a number of procedures. Those generally preferred involve pressing the fruit against a perforated wall or passing the fruit through a set of rollers. In the former, the berries are broken, and the juice, pulp, seeds, and skins pass through openings to be collected and pumped to a retaining tank or vat. In the latter process, berries are crushed between a pair of rollers turning in opposite directions. The rollers usually have spiral ribbing or contain grooves with interconnecting profiles to draw the grapes down and through the rollers. Spacing between the rollers usually can be adjusted to accommodate the variation in berry

size found among different cultivars. It is important to avoid crushing the seeds to preclude contaminating the must with seed oils, the oxidation of which could produce rancid odors. Crushed seeds also provide an additional and undesirable source of bitter tannins.

Crushing also can be achieved using centrifugal force. In centrifugal crushers, the fruit is flung against the sides of the crusher. Because they tend to turn the fruit into a pulpy slurry, centrifugal crushers generally are undesirable. Clarification of the juice is difficult, and seeds are commonly broken.

Although grapes are customarily crushed prior to vinification, there are a few exceptions. Juice for sparkling wine production is commonly obtained by pressing intact grapes. Special presses extract the juice with a minimum of pigment and tannin extraction. The absence of pigments and tannins is particularly important where white sparkling wines are made from red-skinned grapes.

Botrytized grapes also may be pressed, rather than crushed. The gentler separation of the juice minimizes the incorporation of fungal dextran polymers (β-glucans) into the juice that can plug filters used in clarification. In the production of the famous botrytized wine Tokaji Eszencia, juice is derived solely from the liquid that drains freely from heavily infected grapes; no pressure other than the weight of the fruit promotes juice release.

In the production of wines employing carbonic maceration, such as *vino novello* and beaujolais, it is essential that most of the fruit initially remain uncrushed. Only within intact berries does the internal grape fermentation occur that develops the characteristic fragrance shown by the wines. After a variable period of autofermentation, berries that have not broken under their own weight are pressed to release the juice. Fermentation is completed by yeast action.

Supraextraction

An alternative to crushing being investigated in France is supraextraction (Defranoux *et al.*, 1989). It involves cooling the grapes to −4°C, followed by warming to about 10°C before pressing. Freezing causes both grape cell rupture and skin splitting that facilitate juice escape during pressing. While increasing the extraction of sugars and phenolics, supraextraction reduces total acidity and raises the pH. The latter may result from induced crystallization of tartaric acid.

Maceration: White Wines

Maceration refers to the breakdown of grape solids following crushing of the grapes. The rupture and release of enzymes from grape cells facilitates the liberation and solubilization of compounds bound in cells of the skin, flesh, and seeds. While maceration is always involved in the initial phase of red wine fermentation, until recently the trend has been to limit maceration in white wine production. However, there is a shift back to limited maceration for white wines, along with slight juice oxidation before fermentation.

The major factors influencing the extraction and types of compounds released during maceration are the temperature and duration of the process. Extraction may be a linear function of the temperature and length of skin contact. For example, cool temperatures and short duration minimize flavonoid uptake (Fig. 7.3), and thereby limit wine bitterness and astringency. Occasionally, the concentration of extracted compounds decreases with prolonged maceration, presumably owing to precipitation or degradation. Extraction also varies markedly with the class of compounds involved. For example, flavonoid phenols from the skin are more rapidly solubilized than nonflavonoids from seeds (Fig. 7.3).

In addition to phenolic compounds, the concentration of many nutrients and flavorants in juice and wine is influenced by maceration. For example, amino acid, fatty acid, and higher alcohol contents rise, while total acidity falls (Ramey *et al.*, 1986; Soufleros and Bertrand, 1988). The decline in acidity appears to be due to the extraction of potassium that induces tartrate salt forma-

tion. Other changes result from indirect effects on yeast metabolism. For example, increased amino acid availability has been correlated with a reduction in the production of hydrogen sulfide (Vos and Gray, 1979).

Occasionally, a short exposure (15 min) to high temperatures (70°C) greatly increases the release of volatile compounds, such as monoterpenes (Marais, 1987). Although the concentrations of most monoterpenes increase on short high-temperature maceration, not all follow this trend. For example, the concentration of geraniol decreases.

Maceration temperature also may affect the subsequent production of flavor compounds during fermentation. Production of volatile esters may increase with a rise in maceration temperature up to 15°C, whereas it decreases at higher temperatures. Synthesis of most alcohols is reduced following maceration at warm temperatures (Fig. 7.4), except for methanol. Synthesis of methanol is probably spurred by the increased action of grape pectinases, which release methyl groups from pectins.

The sensory influence of maceration can be affected by the degree of simultaneous oxidation. Oxidation speeds phenol polymerization, the products of which can cause browning and increase bitterness and astringency. However, polymerization also aids early tannin precipitation, leaving the wine less sensitive to subsequent in-bottle oxidation.

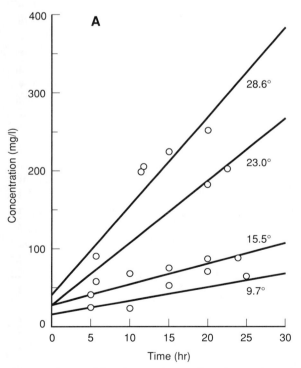
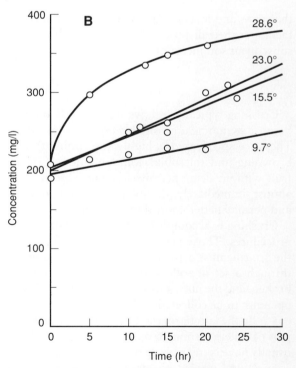

Figure 7.3 Flavonoid (A) and nonflavonoid (B) phenol content in 'Chardonnay' must during skin contact. Temperatures are in °C. (From Ramey *et al.*, 1986, reproduced by permission).

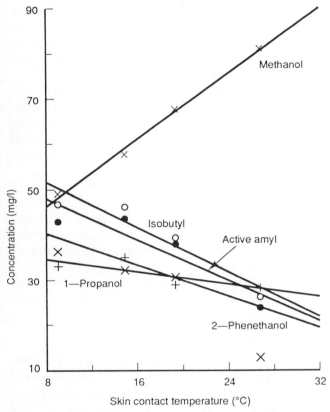

Figure 7.4 Concentration of various alcohols in 'Chardonnay' wine as a function of skin contact temperature. (From Ramey *et al.*, 1986, reproduced by permission.)

Maceration directly and indirectly improves juice fermentability (Ollivier *et al.*, 1987). Part of this effect is due to the release of particulate matter into the juice. Particulate matter is well known to increase microbial growth. The solids provide a surface for yeast and bacterial growth, adsorption of nutrients, binding of toxic carboxylic acids, and the escape of carbon dioxide. The latter is thought to increase must agitation and, thereby, promote more uniform nutrient distribution. Skin contact facilitates the extraction of unsaturated lipids, such as oleanolic, linolenic, and linoleic acids. The lipids are important in permitting yeast cells to synthesize essential steroids and build cell membranes under anaerobic fermentation conditions. The small amounts of oxygen absorbed during crushing, and during other prefermentation cellar activities, likely activate the synthesis of sterols by yeast cells.

Although the oxygen absorbed during crushing ($\sim$6 mg O_2/liter at 20°C) benefits certain wines, most winemakers prefer to avoid contact between the juice and oxygen during maceration.

Sulfur dioxide is added to the juice, depending on the health of the crop and the maceration temperature. Even small amounts of moldy fruit can significantly increase microbial contamination. The laccase concentration tends to rise with the degree of grape infection. Sulfur dioxide also may be added to restrict the growth of indigenous grape flora in the juice at warm maceration temperatures. In contrast, juice from healthy grapes, chilled and macerated at cool temperatures, seldom requires the addition of sulfur dioxide. The rapid disruption of cell membrane function by sulfur dioxide may be useful in speeding the release of grape constituents. In addition, sulfur dioxide can inhibit the action of grape polyphenol oxidases, delay the inception of alcoholic fermentation, and retard the onset of malolactic fermentation. Whether these effects are desirable depends on grape maturity, the cultivar involved, and the wine style desired. Addition of sulfur dioxide, at commonly used concentrations ($\sim$50 mg/liter), does not markedly affect the residual SO_2 content of the wine.

Minimal maceration at cool temperatures often leads to the production of young, fresh, fruity wines. Longer, warmer maceration typically produces a wine deeper in color and of fuller flavor. The latter may age more quickly and develop a more complex character than wines produced with minimal skin contact (Ramey *et al.*, 1986). Thus, varietal characteristics (Singleton *et al.*, 1980), fruit quality, equipment availability, and market response all influence the decision of the winemaker on whether and how to conduct maceration.

A new means of complementing or replacing maceration is called **cell-cracking** (Bach *et al.*, 1990). Cell-cracking involves forcing must through narrow gaps separating steel balls positioned in a small bore.

MACERATION: RED WINES

In red wine production, maceration studies have focused primarily on the extraction of pigments and tannins. Both the style and consumer acceptance of the wine can be dramatically altered by the duration and conditions of maceration. Thus, maceration provides one of the primary means by which winemakers can adjust the character of a wine. Short macerations (<24 hr) commonly produce a rosé wine. For early consumption, red musts are commonly pressed after 3 to 5 days. This provides good coloration but avoids the undue extraction of tannins. Wines for long aging may be macerated on the seeds and skins from 5 days to as long as 3 weeks. Long maceration may result in a decline in free anthocyanin content (see Fig. 7.9), but enhance color stability and aging potential.

Because of the importance of phenol solubilization during alcoholic fermentation, little attention has been given to the extraction of aromatic compounds. In one of the few studies on the subject, the "berry" aspect of 'Cabernet Sauvignon' was increased on prolonged maceration, while the less desirable canned bean/asparagus aspects were diminished (Schmidt and Noble, 1983).

Dejuicing

Dejuicers are especially useful when dealing with large volumes of must. The capacity of presses can be used more economically to extract only the remaining juice (**pressings**).

Dejuicers often consist of a tank sealed at the exit by a perforated basket. Gravity forces much of the juice from crushed grapes pumped into the tank into the basket, from which the juice flows into a receiving tank (sump). Carbon dioxide pressure may be used to speed the separation. When drainage is complete, the basket is raised to ease pomace discharge for transport to a press.

Dejuicers of simpler design may consist of a sloped central cylinder containing perforations that permit juice escape, but retain most of the pomace. The crushed grapes are moved up the cylinder by rotation of a central screw. The dejuiced grapes are dumped into a hopper for loading in a press. The upward flow of the crush in the dejuicer supplies the gravitational force needed to speed juice release.

Pressing

If the crush has not been previously dejuiced, the must may be allowed to rest in the press for several minutes during which juice runs out under its own weight.

One of the first major advances in press design involved the use of hydraulic force. It replaced muscle power with mechanical force. The use of a removable bottom permitted easier pomace discharge. Previously, presses had to be dismantled, or the pomace shoveled out, at the end of each press cycle. Both tasks were unpleasant, time consuming, and labor intensive.

Increasing drainage surface area has been one of the modern goals of press design. This not only speeds juice release, but also reduces the flow path for juice escape. Increasing the area over which the pressure is applied also has been a major design improvement. By reducing the force required for juice extraction, the presses diminish the release of grape tannins, and pigments.

Placing the press on its side (horizontally) permitted additional improvements. Because the length (former height) of the press could be increased considerably, the surface area over which juice could escape was greatly increased. A horizontal orientation also permitted a section of the press to be hinged, providing access for convenient filling and emptying. By suspending the press on heavy gears, the press could easily be rotated for pomace **crumbling** (tumbling) and inverted for emptying. Crumbling breaks the compacted pomace produced during

pressing and helps entrapped juice escape on subsequent pressing. Previously, chains or manual mixing were used to acheive crumbling. This had the disadvantages of both crushing seeds and increasing juice clouding, owing to the greater release of solids into the juice.

Another major innovation was the development of the continuous screw press. By permitting uninterrupted operation, such equipment avoids time-consuming filling and emptying cycles. This is especially valuable when large volumes of must, or wine, need to be pressed in a short period.

Because presses produce juice and wine fractions of differing physicochemical properties, winemakers can influence the character by the choice of press. The degree of fining and blending of the various press fractions provides additional means of adjusting the final character of the wine.

Brief descriptions of the three major types of presses in current use are given below. Figure 7.5 compares vertical, horizontal, and pneumatic presses.

HORIZONTAL PRESSES

A well-known press of horizontal design is that produced by Vaslin (Fig. 7.6). Both crushed and uncrushed grapes, as well as fermented juice, are effectively pressed in Vaslin-type presses.

Loading occurs through an opening in the upper, raised end of the press. Pressing is conducted by moving one or both end plates inward. The rate at which pressure is applied can be modified to suit the needs of the grape variety involved and characteristics of the press fraction desired. Fluid escape occurs between the slats of the pressing cylinder. Chains and/or rotation of the press break the pomace cake between successive pressings. Once pressing is complete, inversion of the press places the exit port downward for convenient dumping of the pomace.

The primary drawback to horizontal presses is the progressive reduction in drainage surface area during pressing. Consequently, the force required to maintain a rapid discharge increases during pressing.

PNEUMATIC (TANK OR MEMBRANE) PRESSES

Pneumatic presses, such as those produced by Willmes, Diemme, and Bucher, come in forms that effectively press crushed or uncrushed grapes as well as fermented must. The press is filled through an elongated opening in the top. Once filled and closed, the press is inverted to allow the free-run juice or wine to escape. Gas forced into the press between the sack and the cylinder wall compresses the grape mass against perforated plates that project into the central cavity (Fig. 7.7). Alternately, grapes or wine are placed between the cylinder wall and a central sack. Gas forced into the sack forces

Press type	Vertical	Horizontal	Pneumatic
Size of the basket (cm)	113 x 90	215 x 73	215 x 73
Volume (m³)	0.9	0.9	0.9
Pressure area (m²)	1	0.42	4.95
Pressure per 1 cm² (MPa)	1.25 – 1.6	1.2	0.6
Pressure over the whole area (MPa)	12,500 – 16,000	5000	29,700
Pressure per 1 dm of pomace (MPa)	13.9 – 17.8	5.6	33.0
Average size of the cake (cm) at one half of the original volume	113 x 18	73 x 43	215 x 239 x 3.3
Shape of the cake			
Flowing out of the must (time)	long	short	very short
Time of one pressing (min)	100 – 120	100 – 120	50 – 90
Number of pressings	2	1	1
Total time of pressing (hr)	3 – 4	2	1

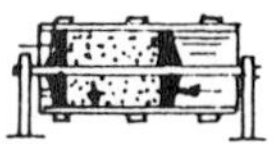

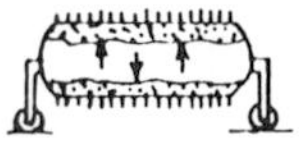

Figure 7.5 Comparison of various types of presses. (From Farkaš, 1988, reproduced by permission.)

the grape mass against the slatted sides of the press. In either case, a fairly constant surface for drainage is maintained throughout pressing. Crumbling of the pomace cake is achieved by rotating the pressing cylinder. Opening of the filling trap and inversion discharge the pomace.

Small volume presses (5 to 22 hl) also are being constructed by producers such as Willmes. These presses are or particular value when small lots of high quality juice or wine need to be kept separate.

Both horizontal and pneumatic presses yield high quality pressings. The pressings are relatively low in suspended solids, and press operation neither crushes the seeds nor extracts high amounts of tannins. A common drawback involves the time associated with the repeated filling and emptying, and fairly fixed press cycle (~1–2 hr).

CONTINUOUS SCREW PRESS

Continuous-type presses have the advantage of running uninterruptedly. While working best with fermented

must, they can be adjusted to handle crushed, nonpulpy grapes. They do not function adequately with uncrushed grapes.

Crushed grapes as well as fermenting or fermented must are pumped into the press via a hopper at one end of the press (Fig. 7.8). A fixed helical screw forces the material into a pressing chamber whose perforated wall allows the juice or wine to escape. Pressed pomace accumulates at the end of the pressing cylinder, where it is periodically discharged through a port opening.

The primary disadvantage of the continuous press is the poorer quality of the released juice or wine. This is particularly noticeable in older models, where separation of different press fractions was not possible. Newer models permit such separations. The first fractions (closest to the intake) possess characteristics similar to free-run material. Fractions obtained nearer the end of the pressing cylinder progressively resemble the first, second, and third pressings of conventional presses. Slower pressing

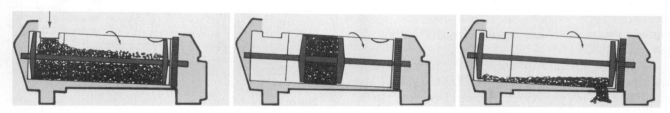

Figure 7.6 Schematic diagram of the operation of a horizontal press. (Courtesy of CMMC, Chalonnes-sur-Loire, France.)

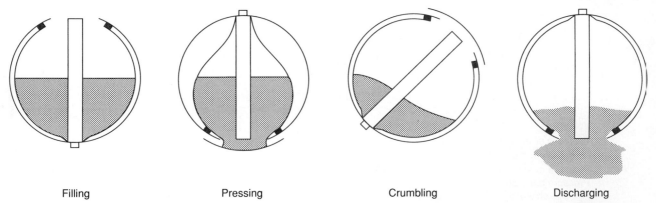

| Filling | Pressing | Crumbling | Discharging |

Figure 7.7   Schematic diagram of the operation of a pneumatic press. Note the centrally located perforated plates for drainage and the inward moving bladder membrane. (Courtesy of WILLMES.)

decreases the incorporation of suspended solids that diminish juice or wine quality, but also reduces the principal advantage of continuous-type presses, namely, speed.

Because different presses and press fractions produce fluids of distinct physicochemical properties, they can influence the sensory properties of the wine produced. For example, the production of fruit esters during fermentation tends to be lower in juice derived from continuous screw presses. The amount of tannins, pigments, and particulate matter also can vary considerably between fractions.

Pressing aids, such as cellulose or rice hulls, may be added to improve extraction. Occasionally, though, the addition has been noticed to influence fragrance development in the wine. Addition of pectinase to the crush also has a marked effect in improving juice release, especially with slip-skin (*V. labrusca*) or other pulpy cultivars.

Must Clarification

White must typically is clarified before fermentation to favor the retention of a fruity character. Loss of fruitiness may result from the excessive production of fusel alcohols, associated with juice containing high amounts of suspended solids. The largest particles in the solids fraction seem to be the most active in inducing higher alcohol synthesis (Klingshirn *et al.*, 1987). In addition, much of the polyphenol oxidase activity is associated with the particulate material. Thus, early removal of suspended material is important in minimizing enzyme-catalyzed oxidation. High levels of suspended solids also are reported to increase hydrogen sulfide production (Singleton *et al.*, 1975).

While high amounts of suspended solids in the juice are generally undesirable, highly clarified juices are also unsuitable, owing to increased susceptibility to "stick" during fermentation. For example, filtration and centrifugation can remove more than 90% of the higher fatty acids from must (Bertrand and Miele, 1984). Because suspended solids favor early malolactic fermentation, retention of small amounts of suspended solids in the juice can be beneficial. Although concentrations of suspended solids between 0.1 and 0.5% seem desirable for several white grapes (Groat and Ough, 1978), the optimal value can vary with the cultivar and the style of wine wanted. Within the range noted above, juice fermentation usually goes to completion and is associated with the production of desirable amounts of fruit esters and higher alcohols. The precise reasons for these benefits are poorly understood but may involve factors such as the adsorption of toxic carboxylic acids produced during fermentation and the availability of essential nutrients.

White juice commonly is allowed to settle spontaneously for several hours (~12 hr) before racking. Bentonite may be added to facilitate settling and subsequent protein stability. When used, bentonite is commonly added after an initial period of spontaneous settling. This avoids the production of voluminous amounts of loose sediment, and the associated loss of juice. Occasionally, some of the precipitate may be left with the juice during alcoholic fermentation. This permits vital nutrients, such as sterols and unsaturated fatty acids, to remain available. Although bentonite is commonly used, its effects on

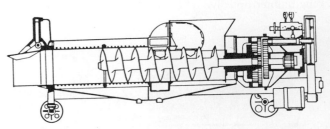

Figure 7.8   Schematic representation of a continuous press with hydraulic control. (Courtesy of Diemme.)

wine quality are still contentious (Groat and Ough, 1978).

To speed clarification, the juice may be centrifuged. By removing only suspended particles, centrifugation affects the chemical composition of the juice the least of any clarification technique. Although centrifugation equipment is expensive, minimal juice loss and speed have made it particularly popular.

Filtration with diatomaceous earth also may be used to clarify must prior to fermentation.

Adjustments to Juice and Must

ACIDITY AND pH

Juice and must failing to possess the desired acidity and pH may be adjusted before fermentation. Acidification of low acid juice or must often occurs before fermentation because it limits the growth of spoilage microorganisms and may be illegal after fermentation in some jurisdictions. In contrast, deacidification typically occurs after fermentation, when its effect on acidity is known. Deacidification can be based on actual rather than projected need. Flavor production also is generally better in musts fermented at a low pH. Finally, postfermentative deacidification permits the process to be delayed until spring, when other winery activities are less urgent.

One of the oldest procedures of adjusting the pH of juice, known as plastering, is rarely used today. The addition of gypsum acts by converting some of the potassium bitartrate to the free acid form:

$$CaSO_4 + 2\ KH(C_4H_4O_6) \longrightarrow K_2SO_4 + H_2(C_4H_4O_6) \\ + Ca(C_4H_4O_6)$$

Gypsum   + Potassium → Potassium + Tartaric + Calcium
(Calcium sulfate)   bitartrate   sulfate   acid   tartrate

The procedure has fallen out of favor not only because it increases the sulfur content of wine, but also because organic acids are readily available, inexpensive, and do not markedly effect the chemistry of wine.

Currently, the high pH usually associated with low total acidity is corrected by the addition of organic acids (**acidification**) (Buechsensteing and Ough, 1979). Tartaric acid often is preferred because of its relative insensitivity to microbial decomposition and its ability to increase pH by inducing the precipitation of excess potassium as a bitartrate salt. Citric acid may be substituted because of its iron stabilization properties, but it is susceptible to microbial degradation.

Deacidification of excessively acidic juice low in pH may involve blending with juice of lower acidity but higher pH. Alternately, some of the acid may be neutralized by the addition of calcium carbonate, potassium carbonate, or Acidex.

The most difficult situation occurs when juice shows both high acidity and high pH. This situation is particulary common in cool climatic regions where grapes may possess both high malic acid and potassium contents. Nagel *et al.* (1988) suggest adding tartaric acid to adjust the malic/tartaric ratio to unity. This is followed by precipitation of the excess potassium with Acidex and acidification with tartaric acid to a desirable pH/acidity.

Amelioration is a means of deacidification involving the dilution of juice acidity by the addition of water. Because dilution also reduces juice sugar content, sugar addition is required to readjust the °Brix upward. Although amelioration is illegal in most countries, it has the advantage that it little affects juice pH. This results because of the dicarboxylic nature of tartaric acid and its low dissociation constant. The dilution of H^+ that results from the addition of water is counterbalanced by the increased dissociation of tartaric acid. Thus, acidity falls but pH is only slightly affected. While reduced color, body, and flavor are usually undesirable consequences of the use of amelioration, the effects may not all be undesirable in intensely flavored varieties, such as *V. labrusca* cultivars. However, the greatest disadvantage of amelioration is its consumer image. The addition of sugar, and especially water, is commonly viewed as unscrupulous behavior.

Acidity and pH adjustment of wine are discussed in Chapter 8.

SUGAR CONTENT

The sugar content (**total soluble solids**) of juice is commonly measured with a hydrometer in units variously called **Brix, Balling, Baumé,** and **Oechsle** (Appendix 6.1). Because sugars constitute the major component of grape soluble solids when over 18° Brix (Crippen and Morrison, 1986), °Brix is a fairly accurate indicator of the capacity of the juice to support alcohol production. More precise measurements of sugar content are available, but the hydrometer determinations are usually adequate early in the winemaking process. In the field, refractometer readings are often used to assess grape sugar content.

As fermentation progresses, hydrometer readings become imprecise measures of sugar content. This results because the alcohol produced during fermentation independently affects specific gravity, the property measured by the hydrometer. Although specific gravity is an adequate indicator of the termination of fermentation in dry table wines, correction tables are necessary for use with sweet fortified wines (Amerine and Ough, 1980).

Following the completion of fermentation, precise chemical analysis of the residual sugar content of the wine usually is required (Zoecklein *et al.*, 1990). Even small amounts of residual sugars can affect the microbial stability of the wine and, therefore, how the wine should be treated up to and during bottling.

When juice °Brix is insufficient to generate the desired alcohol content, chaptalization may be used. **Chaptalization** usually involves the addition of a concentrated solution of sugar to the juice or must. It was first advocated by Dr. Chaptal in 1801 to improve the stability and character of wines produced from immature or rain-swollen grapes. The increased alcohol content generated by the added sugar improved both features.

Chaptalization is typically illegal in regions or countries where warmer growing conditions obviate its need, but it often is permissible in areas where cool climates may prevent full ripening of the grapes. Where permissible, chaptalization usually occurs under strict governmental regulation.

Although many factors influence the conversion of sugars to alcohol (Jones and Ough, 1985), 17 g of sucrose (i.e., cane sugar) typically yields about 10 g of ethanol. The sugar is first dissolved in grape juice and added near the end of the exponential phase of yeast growth (commonly 2 to 4 days after the commencement of fermentation) (Ribéreau-Gayon *et al.*, 1987). By this time, yeast multiplication is essentially complete, and the sugar does not disrupt fermentation. Simultaneous aeration of the fermenting juice or must is recommended.

In addition to elevating the alcohol content, chaptalization slightly augments the production of certain compounds in wine, for example, glycerol, succinic acid, and 2,3-butanediol. Synthesis of some aromatically important esters also may be increased, while that of others is decreased (Fig. 7.21). However, these influences do not make up for the lack of varietal character found in immature grapes or grapes diluted by rains. In some varietal wines, such as 'Riesling,' chaptalization can diminish the "green" or "unripe" taste derived from immature fruit (Bach and Hess, 1986).

Various techniques are under investigation to improve the character of wines produced in poor vintages without the addition of sugar. **Reverse osmosis** is one such technique (Duitschaever *et al.*, 1991). Although first designed as an economical means of obtaining fresh water from salt water, reverse osmosis has found many applications in other industries, from sewage treatment to fruit juice concentration. It is the latter application that has attracted the attention of enologists. In addition to offsetting some of the problems of poor vintages, reverse osmosis can concentrate fruit flavors in the juice.

Reverse osmosis operates by forcing water out of the juice through a membrane that retains most of the sugars and flavoring components. The principles of the operation are discussed more fully in Chapter 8.

Although effective, reverse osmosis has its limitations and drawbacks. Presently, it concentrates the juice only up to about 30° Brix. During concentration, acids may accumulate to a degree requiring deacidification. More significantly, important aroma components may be lost. Small, highly volatile, water-soluble compounds such as esters and aldehydes are the most likely to be lost. The addition of untreated juice to the concentrated juice can partially alleviate this problem. Concentration of volatiles removed with water and reintroduction into the treated juice constitute another possible solution. Development of filters with improved selective permeability may eliminate this problem.

Cryoextraction is another technique being investigated to overcome deficiencies in sugar and flavor content (Chauvet *et al.*, 1986). As with reverse osmosis, cryoextraction can be used with immature grapes or berries swollen with water after rains. It also may be used to augment the sugar and flavor content of grapes in the production of sweet table wines. Cryoextraction is the technical equivalent of *eiswein* production, except that overmature grapes are not used. Cryoextraction involves freezing and subsequent crushing and pressing of the frozen grapes.

As water in the grapes forms ice, dissolved substances become increasingly concentrated in the remaining liquid juice. Because berries of greater maturity (sugar content) freeze more slowly than immature grapes, preferential extraction of juice from the more mature grapes can be achieved. Although temperatures down to −15°C increase solute concentration, temperatures between −5° and −10°C are generally sufficient to remove unwanted water. Cryoextraction appears not to produce undesirable sensory consequences.

Another technique under investigation is the Entropie concentrator (Froment, 1991). It involves juice concentration under vacuum and at moderate temperatures (~20°C).

Brix adjustment usually is designed to generate a higher alcohol content in the wine. There is, however, a growing market for low alcohol wines. Reduced alcohol contents are usually produced following alcoholic fermentation by dealcoholization. A new technique offers the possibility of diminishing the capacity of juice to support alcohol production (Villettaz, 1987). The process involves the action of two enzymes, glucose oxidase and peroxidase. Glucose oxidase converts glucose to gluconic acid, a nutrient that yeasts cannot ferment. Hydrogen peroxide, produced as a by-product of glucose oxidation, is destroyed by peroxidase. The two reactions are as follows:

$$2 \text{ Glucose} + 2 \text{ H}_2\text{O} + \text{O}_2 \xrightarrow{\text{glucose oxidase}} 2 \text{ gluconic acid} + 2 \text{ H}_2\text{O}_2$$

$$2 \text{ H}_2\text{O}_2 \xrightarrow{\text{peroxidase}} 2 \text{ H}_2\text{O} + \text{O}_2$$

With glucose oxidase, alcohol production can be reduced by about half, equivalent to the proportional concentration of glucose in the juice. Thus, ethanol production is dependent on the remaining fructose content of the juice.

Because a steady supply of oxygen is required for enzymatic dealcoholization, the juice becomes oxidized and turns brown. Although much of these colored, oxidized compounds formed precipitate during fermentation, the wine is still left with a distinct golden color. The effects of this, and other factors, on the sensory quality of the wine have yet to be fully assessed.

COLOR EXTRACTION: THERMOVINIFICATION

The grapes of several red varieties seldom produce a dark red wine using standard vinification techniques, for example, 'Pinot noir.' Standard procedures may extract only about 30% of the anthocyanin content of the grapes. Poor color also may result from the action of fungal polyphenol oxidases such as laccase. **Thermovinification** is one technique of improving the color of red wines.

Thermovinification involves the heating of intact or crushed grapes to between 50° and 80°C. Some versions involve rapid heating of whole grapes with steam or boiling water. Such treatments are typically short (~1 min) and heat the outer pigment-containing layers of the fruit to about 80°C. Other procedures involve heating some or all of the pomace, or both the pomace and juice. The juice and pomace are typically heated to about 70°C for 30 min. Where this treatment damages subtle varietal aromas, temperatures as low as 50°C may be used. For especially delicate varieties such as 'Pinot noir,' heating may be as low as 32°C for 12 hr (Cuénat **et al.,** 1991). Only mold-free grapes can safely be treated at temperatures below 60°C, as laccase activity increases up to this temperature. Heating may be conducted with or without continuous stirring. Subsequent vinification may be conducted in the presence or absence of the seeds and skins. Each variation influences the attributes of the wine generated from the must.

The heat dramatically increases anthocyanin extraction (Fig. 7.9), and temperatures above 60°C inactivate laccases. Thermovinification is used primarily to produce wines designed for early consumption. In addition to generating a rich red color, thermovinification improves juice fermentability (both alcoholic and malolactic), produces wines low in astringency, and reduces varietal aroma. Although not normally considered beneficial, diminished varietal aroma can be desirable with strongly flavored cultivars. Low astringency contributes to a soft mouth-feel, appropriate for wines designed for early consumption. Rapid completion of fermentation has numerous benefits, including the

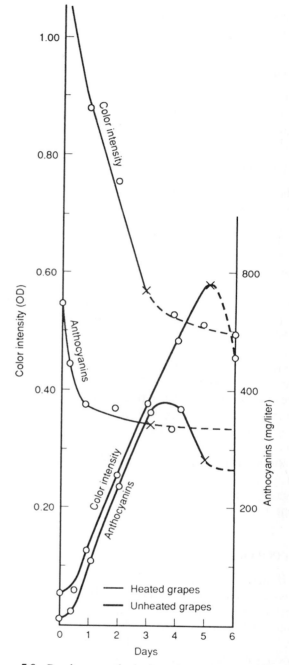

Figure 7.9 Development of color intensity (OD) and level of anthocyanins (mg/liter) during fermentation; × indicates the end of alcoholic fermentation. (After Ribéreau-Gayon *et al.,* 1976, reproduced by permission.)

liberation of fermentors for additional fermentations. However, it increases the need for temperature control during fermentation.

Occasionally, thermovinification generates undesirable bluish colors and "cooked" flavors. These usually can be avoided by appropriate adjustments to the technique. Difficulties with clarification and filtering also may be experienced, but they often can be corrected by the addition of pectinase. Some proprietary pectinase combinations are reported to enhance color extraction as well.

OTHER ADJUSTMENTS

The addition of nitrogen (usually as an ammonium salt) is uncommon, but can improve the fermentation of highly clarified white juice.

BLENDING

For white wine production, it is common to combine free-run juice with that from the first pressing. Occasionally, the second pressing also is added. Other pressings usually are too tannic and difficult to clarify to be used in making white wine. However, several finings and centrifugations may permit later pressings to be incorporated with the other fractions. Alternately, late pressings and the pomace may be fermented to obtain alcohol for distillation, or they may be sold for vinegar production.

For grape varieties such as 'Riesling,' pressings may contain two to five times the concentration of terpenes found in the free-run juice (Marais and van Wyk, 1986). Because of the importance of terpenes to the distinctive aroma of the cultivars, the addition of pressings improves wine quality. The distribution of individual monoterpenes within the fruit varies with the cultivar, grape maturity, and the free versus bound state of the terpenes (Park *et al.*, 1991). Thus, the addition of pressings may affect both the quantitative and qualitative aspects of wine aroma.

DECOLORIZATION

If necessary, decolorization is normally conducted after fermentation. However, addition of the enzyme **anthocyanase** permits decolorization before fermentation. Anthocyanase removes the sugar component from anthocyanins, converting them to anthocyanidins. The lower solubility of anthocyanidins promotes their precipitation during fermentation. Because anthocyanase is inactivated by sulfur dioxide, ethanol, and high temperatures, treatment normally follows juice clarification. At this point, the free sulfur dioxide content is likely to be low and little alcohol will have been produced by the indigenous yeast inoculum.

Alcoholic Fermentation

Fermentors

Fermentors come in a wide variety of shapes, sizes, and technical designs. Most differ little in design from those used centuries ago. However, some are highly complex and designed for specific functions.

Most fermentors are straight sided or have the form of slightly inverted cones. **Tanks** are differentiated from vats by being closed; **vats** have open tops. Tanks commonly double as storage cooperage, while vats can be used only as fermentors.

BATCH-TYPE FERMENTORS

During the fermentation of red wines, carbon dioxide released by yeast metabolism becomes entrapped in the pomace. This causes the pomace to rise to the top, where it forms a **cap**. However, the entrapped carbon dioxide prevents contact between the pomace and most of the juice, retarding the extraction of anthocyanins and other compounds from the skins and pulp.

With vats, periodic submerging (**punching down**) of the cap into the fermenting must is adequate. In addition to improving color removal, punching down can aerate the fermenting must, limit the growth of spoilage organisms in the cap, and help equalize the temperature throughout the fermenting must. Before our own era, there were no simple, convenient methods of achieving the benefits of punching down other than manually.

With modern developments, tanks have almost completely replaced vats for the fermentation of all type of wine. By having a closed top, tanks prevent must exposure to air-borne contaminants and oxygen. Thus, tanks can act as fermentors during the harvest period and as storage cooperage during the rest of the year.

Since the 1950s, there has been a move away from wooden tanks (oak, chestnut, redwood, etc.) to more impervious and inert materials. Cement is favored in some regions, but stainless steel is probably the most generally preferred material. Fiberglass tanks are becoming more popular because of their light weight and lower cost. Nevertheless, stainless steel has one distinct advantage over other materials, namely rapid heat transfer. This property can be used to cool the fermenting juice. This can be achieved by flushing water over the sides of the tank, with evaporating water acting as the coolant. However, double-jacketed tanks, circulating a coolant between the inner and outer walls, can provide more versatile temperature regulation of the fermenting juice.

Although cement is a poor insulator, the rate of heat transfer of cement is usually insufficient to prevent excessive heat buildup during fermentation. Thus, the juice may be passed through a heat exchanger to achieve temperature control. Cement is also more difficult to surface-sterilize. While an epoxy coating helps, it requires frequent maintenance. However, cement tanks do have the initial advantage of being less expensive to construct than equivalent stainless steel tanks.

Fermentors for white wine production are generally of simple design. The primary technical requirement is for efficient temperature control. If not initially cool, the juice may be chilled before yeast inoculation, and the ferment is maintained at cool temperatures (8° to 15°C) throughout vinification.

Fermentors for red wine production may be even simpler in design. Cap formation and a higher phenol content give fermenting red wines considerable protection from oxidation. Also, if the cellar is cool, and the fermentor volume relatively low (50 to 100 hl), cooling during fermentation may be unnecessary. Red wines generally are fermented at, or allowed to warm to, 25° to 28°C. Punching down slightly cools the fermenting must.

The shift to tanks for red must fermentation has demanded means of replacing manual punching down. This requirement has spawned an incredible array of solutions. One of the more novel is the *pileage* fermentor with mechanical cap punchers to simulate the action of manual punching down (Anonymous, 1983). Other solutions include pumping the must over the cap for about an hour several times a day. This may be combined with passing the wine through cooling coils for temperature control. If the headspace over the cap does not contain an inert gas (N_2 or CO_2), pumping over produces some aeration. By submerging the cap, growth of potential spoilage organisms in the cap is limited. **Pumping over** may be manual or automated, as well as periodic or continuous. Other devices designed to mix the pomace with the must consist of large mechanical stirrers.

Autofermentors are specifically designed to facilitate the extraction of pigments from the pomace. They generally possess two superimposed fermentation chambers. The lower, main chamber contains two traps into the upper smaller chamber. An elongated, perforated cylinder descends from one trap into the main chamber. As fermentation progresses, carbon dioxide accumulation increases the must volume. At a certain point, the pressure forces this trap open and a portion of the carbon dioxide and fermenting juice escapes into the upper chamber. Here, the carbon dioxide escapes and the fermenting juice cools slightly. The weight of the juice forces the other trap to open downward. The flush of juice back in the main chamber ruptures the cap and temporarily disperses it into the fermenting must. Perforations in the cylinder prevent pomace from escaping with the juice into the upper chamber.

The cap in autofermentors is normally submerged. A simpler system of achieving a submerged cap involves a grill located below the surface of the must. Although autofermentors avoid the need of punching down, additional agitation is still required to achieve adequate extraction of color from the cap.

Fermentors of modern design typically include some system to ease pomace discharge. For this purpose, sloped bottoms with trap doors are often used. Removable bottoms are another solution.

CONTINUOUS FERMENTATION AND RELATED PROCEDURES

Most fermentors for winemaking are of the **batch** type. In other words, separate lots (batches) are individually fermented to completion. In most industrial fermentations, continuous fermentation is the norm. In continuous fermentation, substrate is added constantly, or at frequent intervals. Equivalent volumes of the fermenting liquid are removed to maintain a constant volume. Continuous fermentors may remain in uninterrupted operation for weeks or months. For the industrial production of single metabolic products, synthesized primarily during a particular phase of colony growth, continuous fermentation has many economic advantages. The technique is less compatible with wine production, however, especially with high quality wines showing subtle and complex associations of hundreds of compounds.

Despite the potential advantages of continuous fermentation, it is rarely used in the wine industry. Because of their design, and expense, continuous fermentors are economically feasible only if used year-round. This in turn demands a constant supply of must. With the seasonal character of the grape harvest, this requires the storage of must under sterile, nonoxidizing conditions. These requirements demand more sophisticated storage than would be needed to store the corresponding volume of wine. Thus, technical and financial concerns generally outweigh the benefits of product uniformity and the easier alcoholic and malolactic fermentations achieved by the use of continuous fermentors.

A new technique under investigation is the repeated use of yeast in fermentation. After each fermentation, the yeast is removed and used to inoculate successive fermentations. Removal is achieved either by filtration, centrifugation, or spontaneous sedimentation. In addition to cost saving, there are further benefits to what is called **cell-recycle batch fermentation** (Rosini, 1986). The duration of fermentation is considerably reduced, and the conversion of sugars to ethanol is slightly improved. There also is a reduction in the synthesis of sulfur dioxide by yeast cells, but an increase in volatile acidity. Although yeast multiplication continues at progressively reduced rates, continual monitoring for contamination by undesirable yeasts and bacteria is necessary. Periodic assessment to determine that the genetic character of the yeast population has not changed is required.

FERMENTOR SIZE

Optimal fermentor size has more to do with the volume of juice or must typically fermented than almost any other factor. When the volumes are large, immense fermentors are both needed and economically appropriate. When modest volumes need to be fermented, suitably small fermentors are required. Specially designed tanks of 50 to 60 hl capacity are produced by several European manufacturers that mix the pomace and fermenting juice automatically (Rieger, 1993). They have been particularly valuable for poorly colored varieties such as 'Pinot noir.' Traditional use of carbonic maceration (see Chapter 9) also requires the use of special, shallow, small volume fermentors.

Small must volumes may result from limited land holdings or when wine is made from small lots of special quality fruit. The latter situation can develop when grape harvesting occurs at different times or states of maturity. This is particularly striking with the higher level Prädikat wines of Germany. Separate fermentation is essential to maintain the individuality of the different grape lots. In these situations, fermentation often is conducted in barrels (~225 liters), puncheons (~500 liters), or fuders (~1000 liters).

In addition to permitting the separation of small lots of juice or must, "in-barrel" fermentation has a number of potential advantages. Because cooling during fermentation occurs only by heat radiation through the sides of the cooperage, fermentation may occur at temperatures higher than currently recommended, especially for white wines. As a result, fruit-smelling acetate esters may dissipate more readily along with escaping carbon dioxide (Fig. 7.25) and varietal aromas achieve a clearer expression in the wine.

As the wine usually is left on the lees (dregs) longer than in larger fermentors, there is an increased likelihood of malolactic fermentation. There also is an increased risk of off-odor production which can be countered by using yeast strains with a low potential for hydrogen sulfide synthesis. In addition, periodic mixing of the lees and wine provides aeration that further decreases the generation of sulfide odors. Furthermore, yeast viability is enhanced by slight aeration. This may permit the enzymatic oxidation of lipid precursors needed to maintain membrane function. Enhanced yeast viability is viewed as contributing to a better integration of oak flavors and tannins from the cooperage wood and to improved mouth-feel.

On the negative side, more effort is involved in sterilizing, cleaning, topping, and maintaining small wood fermentors. There also is increased risk of oxidation and acetic acid bacterial activity during maturation (Stuckey *et al.*, 1991). In small amounts, acetaldehyde and acetic acid and the uptake of oak flavors can increase wine complexity, but excess amounts can mar wine flavor. Barrel reuse increases the risks of microbial spoilage.

Another complicating factor with the use of small wooden cooperage is the loss of water and alcohol by evaporation. Depending on the relative humidity of the cellar, the wine may either lose more water or alcohol. While this results in a loss in wine volume, evaporation

from barrel surfaces also influences the concentration of most wine flavorants.

For many premium wines, fermentors commonly range in size from 50 to 100 hl. Such a volume appears to strike an appropriate balance between economics and ease of operation and the desire to maintain individuality. For standard quality wines, the economics of size shift the balance toward fewer but larger fermentors. In this case, fermentors in capacities from 200 to more than 2000 hl (~50,000 gal) become preferable. Computers have proved useful in monitoring and regulating the course of fermentation in mammoth fermentors.

Associated with increased fermentor volume are temperature control problems. In large fermentors, passive heat dissipation via the surface is insufficient to prevent excessive heat buildup during fermentation. As a result, overheating of the must is likely, and the fermentation will "stick." However, the economics of size permit the use of sophisticated cooling systems to maintain a favorable fermentation temperature.

Fermentation

Fermentation is an energy-releasing form of metabolism where both the substrate (initial electron donor) and by-product (final electron acceptor) are organic compounds. It differs fundamentally from respiration in not requiring the involvement of molecular oxygen. Although many fermentative pathways exist, *Saccharomyces cerevisiae* possesses the most common, **alcoholic fermentation.** In it, ethanol acts as the final electron acceptor (by-product), while glucose is the preferred electron donor (substrate). Although *Sacch. cerevisiae* possesses the ability to respire, it preferentially ferments even in the presence of oxygen.

Although most organisms are able to ferment sugars, they do so only when oxygen is lacking. This partially results from the toxic action of the usual end products of fermentation, lactic acid or ethanol. In addition, fermentation is an inherently inefficient mode of energy release. Its converts only about 6 to 8% of the chemical bond energy of glucose into readily available metabolic energy, (ATP, adenosine triphosphate). Much of the eneregy remains bound in the terminal by-product of electron acceptance, ethanol.

The two main organisms involved in vinification, *Saccharomyces cerevisiae* and *Leuconostoc oenos,* are somewhat unusual in selectively employing fermentative metabolism. They are also atypical in withstanding moderately high ethanol concentrations.

The combined properties of alcohol tolerance and preferential alcoholic fermentation endows *Sacch. cerevisiae* with the ability to dominate rapidly in grape must in the absence of oxygen. The mechanism(s) by which

Sacch. cerevisiae avoids ethanol toxicity is incompletely understood but may involve the rapid diffusion or export of alcohol out of the cell.

Leuconostoc oenos is less well adapted to growing in grape juice or must than *Sacch. cerevisiae*. It typically grows slowly in wine after *Sacch. cerevisiae* has completed alcoholic fermentation and has become inactive. In most ecological habitats, production of lactic acid by lactic acid bacteria lowers the pH and excludes competitive bacteria. Lactic acid bacteria are one of the few acid-tolerant bacterial groups. However, the high acidity of juice and wine often retards or inhibits their growth. Thus, the metabolic conversion of malic to lactic acid, a weaker acid, has the result of increasing the pH and favoring bacterial growth. Malolactic fermentation also makes excessively acidic wines more acceptable to the human palate and may improve microbial stability by removing residual fermentable substrates.

As noted above, wine is usually **batch fermented.** Thus, nutrient availability is maximal at the beginning of fermentation, and progressively declines thereafter. By the end of fermentation, most sugars have been metabolized, leaving the wine "dry."

Batch fermentations generally show a growth pattern consisting of four phases: lag, log, stationary, and decline. Immediately following inoculation, cells need to adjust to the new environment. Because some cells do not make the adjustment successfully, the number of new cells produced approximates the number that die. Thus, there is no net increase in the number of viable cells. This is called the **lag** phase.

Once adjustment is complete, most cells begin to multiply at a steady rate until conditions become unfavorable. Because of the unicellular nature of most microbes, the growth curve approximates an exponential equation. This phase is appropriately called the **exponential** or **log** (logarithmic) phase. During this period, the population of viable cells rapidly increases to its maximum value.

Under batch conditions, the nutrient content progressively falls and toxic metabolic by-products accumulate. Thus, after a period of rapid growth, the rate of cell division (growth) declines and approaches the rate at which cells die. The culture is now said to have moved into the **stationary** phase. As nutrient conditions continue to deteriorate and the concentration of toxic metabolites escalates, more cells die than divide. At this point, the culture enters a **decline** phase. Because most viable cells are not replaced, the colony eventually perishes, or the remaining cells become dormant.

Although basically similar, the population growth pattern displayed by yeast growth in grape juice shows several variations from the norm (Fig. 7.10). Typically the lag phase is short or undetectable; the exponential growth phase is relatively short (seldom amounting to

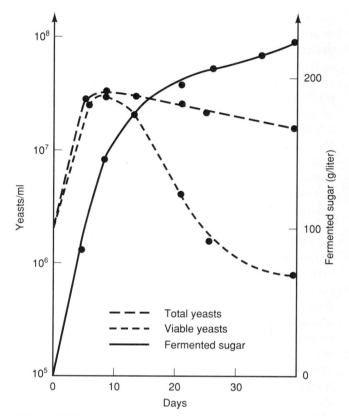

Figure 7.10 Growth cycle of yeasts and fermentation kinetics in grape must with a high sugar content (320 g/liter). (From Ribéreau-Gayon *et al.*, 1976, reproduced by permission.)

more than eight divisions); the stationary phase may be short and commence long before nutrients become limiting; and the decline phase is atypically long and stabilizes at a high population level. As much as 40% of the sugar metabolized to alcohol occurs during the decline phase (Ribéreau-Gayon, 1985).

The brevity or apparent absence of a lag phase in yeast growth may result from the preadapted state of the cells initiating fermentation. Active dry yeast commonly used for juice or must inoculation comes from cultures grown exponentially in aerated media. Thus, the cells possess the enzymatic capacity necessary to commence exponential growth almost immediately. Equally, the indigenous yeast population on grapes may require little enzymatic adaptation to commence rapid cell growth. However, the absence of a noticeable lag period with spontaneous fermentation also may be an artifact. Indigenous yeasts are commonly bathed in juice released from broken grapes and may pass through the lag phase before fermentation officially begins in the winery. In addition, yeasts growing on berry skins may exist under minimal, but concentrated, nutrient conditions.

Although physiological adjustment to growth in juice appears to be minimal, a lag phase may be observed when conditions are less than optimal. Conditions such as low temperatures ($\leq 10°C$) and low oxygen content in the juice may damage many yeast cells. Active dry yeast cells are often "leaky" and initially may lose vital nutrients (Kraus *et al.*, 1981). In addition, low juice pH can prolong the lag period, probably through enhancement of the antimicrobial effects of sulfur dioxide (Ough, 1966a). High °Brix values or ethanol contents (second fermentation in sparkling wine production) also inhibit yeast growth and the rate of fermentation (Ough, 1966a,b).

The early termination of exponential growth may be partially explained by the large inoculum supplied from grapes or added via inoculation. Juice or must often contains about 10^5 to 10^6 cells/ml at the onset of fermentation, which rises to a maximum of little more than 10^8 viable cells/ml. Considerably more cell divisions may occur when the cell concentration is initially low or reduced during clarification. Ethanol accumulation and sensitivity may partially explain the fact that viable cell counts seldom reach more than 10^8 cells/ml. However, other factors appear to be more important, as populations reaching 10^6 to 10^8 can be reached in juice initially fortified to 8% ethanol. These factors probably include the inability of yeasts to synthesize essential sterols and long-chain unsaturated fatty acids and the buildup of toxic mid-sized carboxylic acids. The latter are by-products of yeast metabolism and can affect the function of the yeast cell membrane. Sterols and unsaturated fatty acids needed for the synthesis and maintenance of cell membranes are not synthesized under the anaerobic conditions of fermentation. One factor not involved is an obvious lack of fermentable nutrients. The stationary phase usually begins when no more than half of the sugars have been metabolized. The remaining sugars are slowly metabolized during the stationary and decline phases, which may constitute up to 80% of the fermentation period (Ribéreau-Gayon, 1985).

Initiation of the decline phase probably results from increasing membrane dysfunction becoming progressively lethal to cellular function. Membrane disruption results from the combined effects of ethanol and carboxylic acid toxicity and a shortage in sterol precursors. The absence of oxygen also may be a factor. Molecular oxygen is required for the synthesis of nicotinic acid, a vital component of the electron carriers NAD^+ and $NADP^+$. However, why the decline initially stabilizes at about 10^5 to 10^6 cells/ml is unknown. These cells die slowly over the next few months.

Another distinction from most industrial fermentations is the nonpure status of wine fermentation. The normal procedure in most fermentations is to sterilize the medium before inoculation. Except for continuous fermentation, grape juice or must is not sterilized. In Europe, the traditional procedure has been to allow indigenous yeasts to conduct the fermentation. However, this is changing with a shift toward induced fermentation with selected yeast strains (Barre and Vezinhet, 1984). Induced fermentation is standard throughout much of the rest of the world. Although sulfur dioxide may be added to inhibit the growth of indigenous (wild) yeasts, it may be ineffectual in this regard (see Martínez *et al.*, 1989).

Until recently, there was no adequate means of determining whether indigenous strains of *Sacch. cerevisiae* were controlled by sulfur dioxide. With techniques such as mitochondrial DNA sequencing (Dubourdieu *et al.*, 1987) and gene marker analysis (Petering *et al.*, 1991), it is possible to identify the strain(s) conducting fermentation. Although species occurring on grapes or winery equipment occasionally may dominate the fermentation of inoculated juice (Bouix *et al.*, 1981), inoculated yeasts appear to be the primary if not only yeasts detectable at the end of most fermentations (Querol *et al.*, 1992).

Vinification of red must routinely occurs in the presence of high concentrations of indigenous yeasts, regardless of yeast inoculation. White grapes, which are pressed shortly after crushing, cold settled, and quickly clarified, usually contain a diminished indigenous yeast population. Nevertheless, white juice may still possess a significant indigenous yeast population (Fig. 7.17). Although indigenous yeasts may be present and viable, whether they are metabolically active during fermentation remains unclear.

Biochemistry of Alcoholic Fermentation

Glucose and fructose are metabolized to ethanol primarily via glycolysis (the glycolytic or Embden Meyerhof pathway) (Fig. 7.11). Although the primary by-product is ethanol, additional yeast metabolites generate most of the aromatic compounds found in wine. Yeast action also may influence the development of the varietal aroma of wine by hydrolyzing nonvolatile aroma precursors or releasing aromatic terpenes, phenols, or norisoprenoids (Laffort *et al.*, 1989). In addition, the changing physicochemical conditions produced in the fermenting juice progressively modify yeast metabolism. This is reflected in the changing phases of colony growth noted above, and related adjustments in the nutrient and energy status of the cells, and the substances released and absorbed throughout fermentation. Thus, much of the fragrance of wine can be interpreted in terms of the changing primary and intermediary metabolism of yeast cells during fermentation.

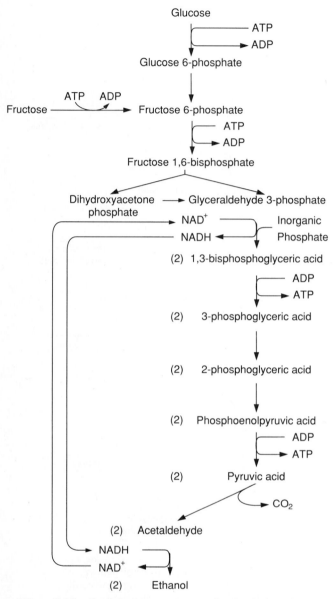

Figure 7.11 Alcoholic fermentation via the glycolytic pathway.

ENERGY BALANCE AND THE SYNTHESIS OF METABOLIC INTERMEDIATES

At different phases in colony growth, yeasts require differing amounts of ATP and reducing power in the form of NADH or NADPH. These energy forms are required to activate cellular functions and maintain an acceptable ionic and redox balance in the cell. Redox balance refers primarily to the equilibrium between the oxidized and reduced forms of the two major pyridine nucleotides (NAD$^+$ and NADH, and NADP$^+$ and NADPH, respectively).

As noted in Fig. 7.11, glucose and fructose are oxidized to pyruvate primarily via glycolysis. During the process, electrons are transferred to NAD$^+$, reducing it to NADH. Pyruvate subsequently may be decarboxylated to acetaldehyde, which is reduced to ethanol on the transfer of electrons from NADH. In the process NADH is reoxidized to NAD$^+$.

The release of energy from glucose and fructose and its storage in ATP and NADH are much less efficient via fermentation than by respiration. Most of the chemical energy initially associated with glucose or fructose remains bound in the end product, ethanol. Furthermore, the energy trapped in NADH is used to reduce acetaldehyde to ethanol. The latter process is necessary to maintain an acceptable redox balance. Cells contain only a limited supply of NAD$^+$, and they are unable to transfer the energy stored in NADH to the more abundant adenosine diphosphate (ADP), generating ATP. Under the anaerobic conditions of fermentation, regeneration of oxidized NAD$^+$ requires the reduction of an organic molecule, in this case acetaldehyde to ethanol. Without the regeneration of NAD$^+$, fermentation itself would cease. Thus, alcoholic fermentation generates only about two molecules of ATP per sugar molecule, in contrast to the potential 24 to 34 ATPs produced via respiration. Most of the ethanol produced during fermentation escapes from the cell and accumulates in the surrounding juice.

The low respiratory capacity of *Sacch. cerevisiae* reflects its limited ability to produce the requisite enzymes. The high proportion of glycolytic enzymes in yeast cytoplasm, about 50% of the soluble protein content, indicates the importance of fermentation to wine yeasts (Hess *et al.*, 1969). Yeasts correspondingly show high rates of glycolysis, usually about 200 to 300 μmol glucose/min/g cell weight (de Deken, 1966). It is estimated that about 85% of the sugars incorporated by *Sacch. cerevisiae* are used in energy production, and about 15% for biosynthetic reactions. Specific values vary depending on the prevailing conditions.

Although most fermentable sugars in juice are metabolized via glycolysis, some are channeled through the pentose phosphate pathway (PPP) (upper right, Fig. 7.12). The diversion is important to the production of pentose sugars needed for nucleic acid synthesis. The PPP also generates NADPH required to activate various cellular functions. Amino acid availability in the juice can decrease activity of the PPP as NADPH is primarily required for amino acid biosynthesis (Gancelos and Serrano, 1989). Intermediates of the PPP not required for biosynthesis are normally directed through phosphoglycerate to pyruvate.

In alcoholic sugar fermentation, yeasts do not generate

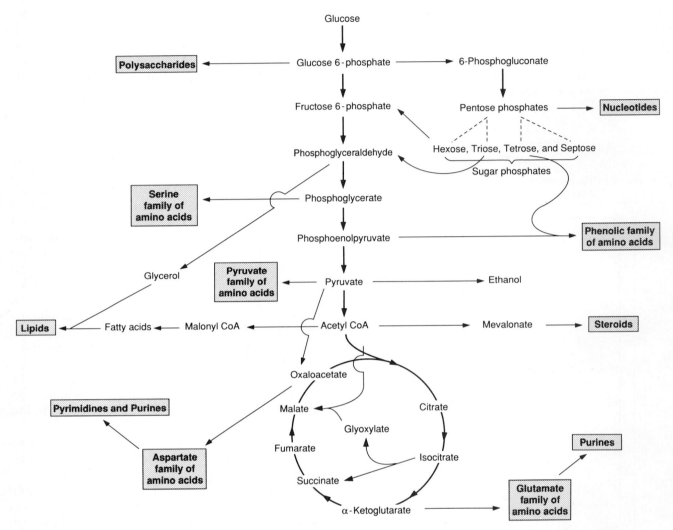

Figure 7.12 Core reactions of metabolism showing the main energy-yielding pathways (bold arrows) and the major biosynthetic products derived from central metabolism (boxes). The central pathway is the Embden Meyerhof pathway of glycolysis, the top right shows a highly schematic pentose phosphate pathway (PPP), and the bottom is the TCA (tricarboxylic acid) cycle. Each pathway has been simplified for clarity by the omission of several intermediates. The directions of the reactions are shown as being unidirectional although several are reversible. Energy transformations and the loss or addition of carbon dioxide are not shown.

a net supply of NADH. However, yeasts need reducing power for growth and reproduction during the early stages of juice fermentation. Some of the reducing power comes from the operation of the pentose phosphate pathway and the oxidation of pyruvic acid to acetic acid. Additional supplies come from NADH generated in glycolysis. This diversion of NADH to biosynthetic functions means that acetaldehyde cannot be reduced to ethanol. The changing needs of yeasts for reducing power during fermentation probably explains why compounds, such as acetaldehyde and acetic acid, are initially released into the juice during fermentation but subsequently reincorporated (Figs. 7.13 and 14). Early in fermentation, growth and cell division require considerable supplies of reducing power; in the decline phase, NADH and NADPH probably accumulate. The associated redox disruption would suppress sugar fermentation by diminishing the supply of the requisite NAD$^+$ (Fig. 7.11). The incorporation and reduction of compounds such as acetaldehyde and acetic acid help to balance the redox potential and permit fermentation to continue.

In addition to generating metabolic energy, yeasts

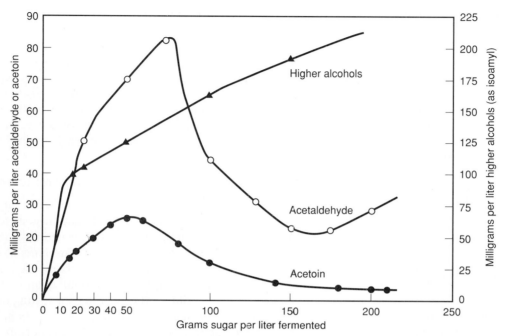

Figure 7.13 Formation of acetaldehyde, acetoin, and higher alcohols during alcoholic fermentation. (From Amerine and Joslyn, 1970, reproduced by permission.)

need to synthesize metabolic intermediates for cell growth and maintenance. The metabolic intermediates generally are synthesized from components of the glycolytic, PPP, and TCA (tricarboxylic acid) cycle pathways of central metabolism (Fig. 7.12).

In fermentation, the main function of the TCA cycle is to produce metabolic intermediates for the synthesis of several amino acids. However, normal operation would produce an excess of NADH, disrupting the redox bal-

ance of the cell. This can be prevented if the NADH produced during the oxidation of citrate to succinate (the right-hand side of the TCA cycle) is used to reduce oxaloacetate to succinate (the left-hand side of the TCA cycle). This can generate TCA cycle intermediates without a change in redox balance. Alternately, NADH generated in glycolysis may be oxidized in the reduction of oxaloacetate to succinate, rather than in the reduction of acetaldehyde to ethanol. In both cases an excess of suc-

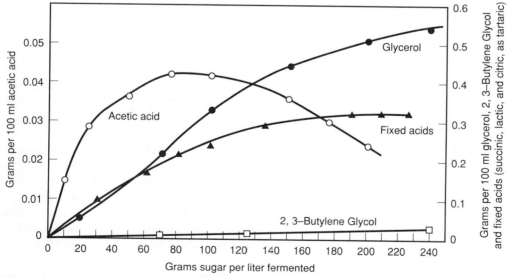

Figure 7.14 Formation of acetic acid, glycerol, 2,3-butylene glycol, and fixed acids during alcoholic fermentation. (From Amerine and Joslyn, 1970, reproduced by permission.)

cinate is generated. This probably explains why succinate is one of the major by-products of fermentation.

Replacement of TCA cycle intermediates lost to biosynthesis, or secreted as succinate, probably comes from pyruvate. Pyruvate may be directly channeled through acetate, carboxylated to oxaloacetate, or indirectly routed via the glyoxylate pathway. The involvement of biotin in the carboxylation of pyruvate to oxaloacetate may partially explain its requirement by yeast cells.

The accumulation of another major by-product of fermentation, glycerol, also is probably explained in terms of maintaining a favorable redox balance. The reduction of dihydroxyacetone phosphate to glycerol 3-phosphate can oxidize the NADH generated in the oxidation of glyceraldehyde 3-phosphate in glycolysis (Fig. 7.15). However, coupling of the two reactions does not generate net ATP production. This is in contrast to the net production of 2 ATP molecules during fermentation to ethanol.

The increased production of glycerol in the presence of sulfur dioxide is probably explained by the need to regenerate NAD^+. The binding of sulfur dioxide with acetaldehyde inhibits its reduction to ethanol, the usual means of NAD^+ regeneration during alcoholic fermentation.

Throughout fermentation, yeast cells adjust continuously to the changing conditions in the juice to produce adequate levels of ATP, maintain favorable redox and ionic balances, and synthesize necessary metabolic intermediates. Consequently, the concentration of yeast by-products in the juice changes continually during fermentation (Figs. 7.13 and 7.14). Because several of the products are aromatic, for example, acetic acid, acetoin, and succinic acid, their presence can affect bouquet development.

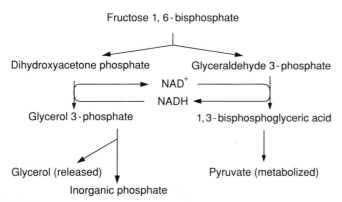

Figure 7.15 Simplified pathway showing how NADH derived from the oxidation of glyceraldehyde 3-phosphate to 1,3-bisphosphoglyceric acid is used in the reduction of dihydroxyacetone phosphate to glycerol. As a consequence NADH is unavailable to reduce acetaldehyde to ethanol.

Although different strains of *Sacch. cerevisiae* possess similar enzymes, the relative proportions and catalytic activities may vary. The differences probably depend on the precise functioning of the regulatory systems of the cells and the number of copies of each gene. Thus, no two strains are likely to respond identically to the same set of environmental conditions. This variability in response undoubtedly explains much of the subtle, and not so subtle, differences between fermentations conducted by different yeast strains.

INFLUENCE ON GRAPE CONSTITUENTS

Yeasts have their major effect on the sugar content of the juice or must. If fermentation goes to completion, only minute amounts of fermentable sugars remain (preferably ≤2 g/liter). Small amounts of nonfermentable sugars, such as arabinose, rhamnose, and xylose, also remain (~0.2 g/liter). The small quantities of sugars are imperceptible and leave the wine tasting dry.

Yeasts may increase the pH by metabolizing malic acid to lactic acid. However, the proportion converted is highly variable in *Saccharomyces cerevisiae* and can differ among strains from 3 to 45% (Rankine, 1966). In addition, some *Sacch. cerevisiae* strains liberate significant amounts of malic acid during fermentation (Farris *et al.*, 1989). In contrast, *Schizosaccharomyces pombe* completely decarboxylates malic acid to lactic acid. It has been little used in juice deacidification as the sensory impact on wine is generally negative. Undesirable flavors, such as hydrogen sulfide, often mask the fragrance of the wine. Delaying inoculation until after *Sacch. cerevisiae* has been active for a few days, or has completed fermentation, apparently reduces the negative impact of *Schizosacch. pombe* on wine quality (Carre *et al.*, 1983).

During fermentation, the release of alcohols and other organic solvents helps extract compounds from seeds and skins. Quantitatively, the most significant of these are the anthocyanins and tannins found in red wines. The extraction of tannins is especially dependent on the solubilizing action of ethanol. Anthocyanin extraction often reaches a maximum after 3 to 5 days, when the alcohol content produced during fermentation has reached about 5 to 7% (Somers and Pocock, 1986). As the alcohol concentration continues to rise, color intensity may begin to fall. This can result from the coprecipitation of anthocyanins with grape and yeasts cells, to which they bind. Nevertheless, the primary reason for color loss is the disruption of weak anthocyanin complexes present in the juice. Freed anthocyanins may change into uncolored states in wine (see Chapter 6). Although removal of tannic compounds occurs more slowly, tannin content often reaches higher values than that of anthocyanins. Extraction of tannins from the stems (rachis) may reach a plateau after about 7 days. Tannins from the seeds are the slowest to be liberated. Accumulation of seed tannins may still be active after several weeks (Siegrist, 1985).

Ethanol also extracts various aromatic compounds from grape cells. Regrettably, little is known about the details of the effects. Conversely, ethanol decreases the solubility of other grape constituents, notably pectins and other carbohydrate polymers. The pectin content may fall by upward of 70% during fermentation.

The metabolic action of yeasts, besides producing many of the most important wine volatiles, notably higher alcohols, fatty acids, and esters, also degrades some grape aromatics, notably aldehydes. This potentially could limit the expression of the herbaceous odor generated by C_6 aldehydes and alcohols produced during the grape crush.

Yeasts

Classification and Life Cycle

Yeasts are classified taxonomically among the fungi. However, the unicellular habit, the possession of a chemically distinct cell wall, the budding or fission form of cell division, and the presence of a single nucleus per cell make yeasts a unique fungal group. Although characterized by a distinctive set of properties, yeasts are not a single, evolutionarily related group. The yeastlike growth habit has evolved independently in at least three major fungal taxa.

The members of only two yeastlike groups occur in wine. These are the ascomycete and imperfect yeasts. Most imperfect yeasts are derived from ascomycete yeasts that have lost the ability to undergo sexual reproduction. Under appropriate conditions, the cells of ascomycete yeasts differentiate into asci. For *Saccharomyces cerevisiae*, this often means culturing on acetate-containing media. Asci are the structures in which haploid spores are produced through meiosis and cytoplasmic separation. In *Sacch. cerevisiae*, four haploid spores are produced (Fig. 7.16). On breakdown of the ascal

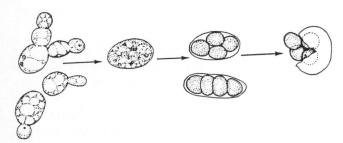

Figure 7.16 Phases of yeast sporulation. From left to right: budding vegetative cells, cessation of budding, development of ascospores, and spore release from the ascus. (From Ribéreau-Gayon *et al.*, 1975, reproduced by permission).

wall, spores may germinate to produce new vegetative cells. Cells of opposite mating type typically fuse shortly after germination to reestablish the diploid state. The diploid cells divide by budding until appropriate conditions induce ascal development and spore production. Although wine strains of *Sacch. cerevisiae* possess the potential for sexual reproduction, the property is rarely expressed in must or wine. Ascal development is suppressed by high concentrations of carbon dioxide and either glucose or ethanol.

Until the late 1970s, yeast classification was based primarily on physiological properties and the few morphological traits readily observable under the light microscope. To these have now been added protein serology, amino acid sequence analysis, nucleotide sequence analysis, and DNA–DNA hybridization. The new procedures are likely to permit taxonomists to develop more stable classification based on evolutionary relationships.

In the most recent major taxonomic treatment of yeasts (Kreger-van Rij, 1984), many named species of *Saccharomyces* have been reduced to synonyms of *Sacch. cerevisiae* or other genera. This does not deny the reality of the differences used to distinguish the former species, but rather indicates that they were either minor and/or genetically unstable. Most of the former species are now viewed as physiological variants of recognized species. For example, *Saccharomyces fermentari* and *Sacch. rosei* are considered to be strains of *Torulaspora rosei*. The taxonomic treatment presented in Kreger-van Rij (1984) is followed in the text.

A listing of currently accepted names and synonyms of some of the more commonly found yeasts on grapes or in wine is given in Appendix 7.1. Differences between some of the physiological races of *Sacch. cerevisiae* are noted in Appendix 7.2.

Ecology

Saccharomyces cerevisiae is undoubtedly the most important yeast to mankind. In its various forms, it functions as the wine yeast, brewer's yeast, distiller's yeast, and baker's yeast. Laboratory strains are extensively used in industry and in fundamental studies on genetics, biochemistry, and molecular biology. For all the importance of *Sacch. cerevisiae*, its original habitat in nature is uncertain (Phaff, 1986). Strains similar to those used in winemaking are rarely if ever isolated from natural sources. *Saccharomyces cerevisiae* var. tetrasporus isolated from oak tree exudate may be the ancestral form. Although *Sacch. cerevisiae* is occasionally isolated in nature from the intestinal tract of fruit flies (*Drosophila* spp.), the importance of insects in the dispersal of *Saccharomyces* is unclear (Phaff, 1986; Wolf and Benda, 1965). *Saccharomyces cerevisiae* is usually absent or rare on grapes in all but long-established vineyards. Even in such vineyards, the presence of wine yeasts on grapes becomes detectable only as they mature. By this time, about half of the isolated yeasts may be strains of *Sacch. cerevisiae*. They are generally less numerous on grapes in cool viticultural regions.

Yeasts seldom occur in significant numbers on grape leaves or stems. They may be found on fruit pedicels but occur more frequently on the callused terminal ends, the receptacle. Yeasts are most frequently found on fruit surfaces, around stomata, or next to cracks in the cuticle, where they routinely form small colonies (Belin, 1972). They presumably grow on nutrients seeping out of openings in the fruit, receptacle, and pedicel.

Yeasts are not held by the plates of wax that cover much of the berry surface and produce its mattelike **bloom.** In fact, yeasts cease to grow where they come in contact with the waxy plates of the cuticle.

Commonly, yeast cells are in a dormant or slowly reproducing state on fruit surfaces. This undoubtedly results from the predominantly dry state of the cuticle. Microbes require at least a thin coating of water to be metabolically active. This occurs only during rainy spells, or when fog or dew condenses on grape surfaces.

As noted above, grapes appear not to have been the natural (ancestral) habitat of *Saccharomyces cerevisiae*. It is suspected that wineries, or the pomace spread as a vineyard fertilizer, are the primary sources. Inoculation of grape surfaces also may be favored by wind and insect dispersal.

Next to *Sacch. cerevisiae*, the most frequently occurring yeast species on mature grapes tends to be *Kloeckera apiculata*. In warm regions, the perfect state of *Kloeckera* (*Hanseniaspora*) tends to replace the asexual form. Other yeasts occasionally isolated from grapes include species of *Brettanomyces, Candida, Debaryomyces, Hansenula, Kluyveromyces, Metschnikowia, Nadsonia, Pichia, Saccharomycodes*, and *Torulopsis* (see Lafon-Lafourcade, 1983). Filamentous fungi such as *Aspergillus, Botrytis, Penicillium, Plasmopara*, and *Uncinula* are rarely isolated except from diseased or damaged fruit. Similarly, acetic acid bacteria are usually found in significant numbers only on damaged and diseased fruit.

Under most conditions, yeasts other than *Saccharomyces cerevisiae* are generally believed to be of little significance in winemaking. The acidic, highly osmotic conditions of grape juice restrict the growth of most yeasts, fungi, and bacteria. Also, the rapid development of anaerobic conditions and the production of ethanol further limit the growth of competitive species.

Nevertheless, other yeasts may occur at concentrations as high as those of *Sacch. cerevisiae* (Fig. 7.17). At present, it is not established how commonly or for how long other yeast species are metabolically active during alcoholic fermentation. Thus, their significance in vini-

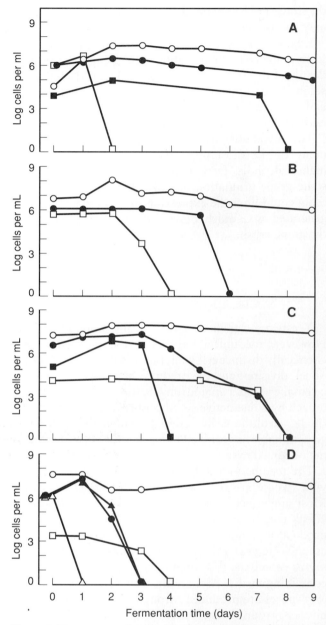

Figure 7.17 Yeast numbers during fermentation of white (A, B) and red (C, D) wines. ○, *Saccharomyces cerevisiae*; ●, *Kloeckera apiculata*; □, *Candida stellata*; ■, *C. pulcherrima*; ▲, *C. colliculosa*; △, *Hansenula anomala*. (From Heard and Fleet, 1985, reproduced by permission.)

fication remains unclear. However, in some instances they maintain or increase their population numbers during fermentation (Mora and Mulet, 1991). Only with diseased or damaged grapes are most of the microbial grape flora known to play an important, albeit negative, role in vinification.

In addition to the fruit, the juice or must may become inoculated from winery equipment (notably crushers, presses, and sumps) and the winery itself. This is especially true in old wineries, where the equipment and buildings are thinly covered with wine yeasts. In addition, *Hansenula anomala* and *Pichia membranae-faciens* commonly occur in wineries, with *Brettanomyces* and *Aureobasidium pullulans* often being isolated from the walls of most cellars. The hygienic operation of modern wineries greatly limits juice and must inoculation, or contamination, from the winery and its equipment.

Succession during Fermentation

In spontaneous fermentation, there is an early and rapid succession of yeast species (Fig. 7.17). Fermentation initially may involve the action of species such as *Kloeckera apiculata* and *Candida stellata*. *Kloeckera apiculata* commonly grows in the must, but it subsequently passes into a decline phase and soon becomes a minor component of the yeast population (Fleet *et al.,* 1984). However, during the first few days, the species could add significantly to the production of compounds such as acetic acid, glycerol, and various esters. *Kloeckera apiculata* also may unfavorably influence the ability of some *Sacch. cerevisiae* strains to complete fermentation (Velázquez *et al.,* 1991). More significant, though, may be the growth of *C. stellata*. The occasional ability of *C. stellata* to persist in fermenting juice (Fleet *et al.,* 1984) and produce high concentrations of acetic acid may affect the sensory characteristics of wine. Although *K. apiculata* and *C. stellata* may ferment only up to about 6 and 10% alcohol, respectively, they are able to survive much higher alcohol concentrations (Gao and Fleet, 1988). Most other members of the indigenous grape flora, other than *Sacch. cerevisiae*, either are slow growing or are inhibited by sulfur dioxide, low pH, high ethanol contents, or oxygen deficiency. Consequently, species of *Candida*, *Pichia*, *Cryptococcus*, and *Rhodotorula* that may be found initially in must probably do not contribute significantly to fermentation. Their populations seldom rise above 10^4 cells/ml, and they usually disappear quickly from fermenting juice and must. Most bacteria which could grow during fermentation are inhibited by *Sacch. cerevisiae*, with the occasional exception of lactic acid bacteria. Therefore, strains of *Sacch. cerevisiae* generally come to dominate and complete fermentation, even when their initial presence in the must is rare (less than 1 in 5000 colonies isolated) (Holloway *et al.,* 1990).

In inoculated fermentations, *Sacch. cerevisiae* is usually added to achieve a population of about 10^5 to 10^6 cells/ml in the must. The indigenous *Sacch. cerevisiae* grape flora often constitutes an additional 10^4 to 10^5 cells/ml. The addition of sulfur dioxide to a concentration of about 1.5 mg/liter free molecular SO_2 has generally been thought to inhibit, if not kill, most of the indigenous yeast population of the grapes (Sudraud and

Chauvet, 1985). However, as the content of free SO_2 rapidly falls during maceration and fermentation, indigenous strains may grow and multiply. A study using mitochondrial DNA restriction analysis has shown that although the inoculated strains may soon become dominant, other *Sacch. cerevisiae* may remain active for several days before being suppressed (Querol *et al.,* 1992). Thus, indigenous strains of *Sacch. cerevisiae* may play a significant sensory role during the early stages of fermentation.

Indigenous yeast populations could have a much greater influence on alcoholic fermentation if it contained "killer" yeast strains. These could replace sensitive inoculated strains even if the killer strain occurred at a concentration of 0.1% (Jacobs and van Vuuren, 1991). Use of a commercial strain containing both killer plasmids should prevent elimination of the inoculated strain. Possession of a killer plasmid protects the cell from the effect of the toxic protein it produces (see below).

In *fino* sherry production, species other than *Sacch. cerevisiae* often appear involved in the development of a *flor* character. For example, *Zygosaccharomyces fermentati* (*Sacch. fermentati*) often is the dominant species covering the wine in sherry barrels. Alternately, various strains of *Sacch. cerevisiae* (*Sacch. cerevisiae* f. *bayanus* and *Sacch. cerevisiae* f. *prostoserdovii*) may dominate. Other film-forming yeasts frequently isolated from *flor* include species of *Pichia*, *Hansenula*, and *Candida*.

There has been much discussion over the years concerning the relative merits of spontaneous versus induced fermentations. A definitive resolution of the issue is unlikely, as it partially depends on stylistic preferences. Spontaneous fermentations may vary from year to year, location to location, and with the source and age of the fermentation cooperage. Conversely, induced fermentations are not strictly "pure culture" fermentations, since indigenous strains usually are present. Occasionally, but not consistently, spontaneous fermentations generate higher concentrations of volatile acidity than induced fermentations. Nevertheless, those who favor spontaneous fermentation believe that indigenous yeasts donate a desired subtle character (Mateo *et al.,* 1991) and possibly provide some of the distinctive character of a regional wine.

An alternative to either spontaneous or induced fermentations with a single yeast strain is inoculation with a mixed culture of local and commercial yeast strains (Moreno *et al.,* 1991). The various species appear to diminish the differences in the metabolic properties of one another, producing a more uniform and regionally distinctive character.

In only a few instances is intentional inoculation essential to winemaking. Intentional inoculation is needed to achieve a rapid initiation of fermentation after thermovinification, and it is essential for the fermentation of pasteurized juice. In addition, yeast inoculation is necessary to restart "stuck" fermentations and promote fermentation of juice containing significant numbers of moldy grapes. Moldy grapes generally possess various inhibitors, such as high concentrations of acetic acid, that slow yeast growth and metabolism. Finally, inoculation is required to assure the initiation of the second fermentation in sparkling wine production. However, the predominant reason for using specific yeast strains is to avoid the production of undesirable flavors occasionally associated with spontaneous fermentation. In addition, distinctive flavor characteristics generated by a particular yeast strain may favor selective use of that strain.

Desirable Yeast Characteristics and Breeding

Saccharomyces cerevisiae is amazingly suited to its role as the predominant fermenter of grape must. It is remarkably tolerant to high sugar, ethanol, and sulfur dioxide concentrations. Consequently, it typically metabolizes the fermentable sugars in must completely. Because *Sacch. cerevisiae* has a low respiratory potential, sugars are primarily converted to ethanol and other flavor products, rather than cell mass. It also grows and ferments rapidly at the low pH values that typify grape must. In addition, strains may possess valuable properties such as fermentation at low temperatures and/or high pressures, may synthesize little SO_2, H_2S, acetic acid, or urea, may show killer factor resistance, may selectively ferment fructose or glucose, and may flocculate rapidly and completely after fermentation. For specific wine types, specific combinations of properties may be desired. For example, the strains used to induce the second fermentation in sparkling wines customarily show the following properties: they commence fermentation at cool temperatures and in the presence of at least 10% alcohol, are relatively insensitive to high carbon dioxide pressures, and flocculate well.

Important as these major physiological features are, subtle variations between yeast strains in the synthesis of aromatic and sapid substances can be equally important (Fig. 7.18). Strains often differ significantly in the production of ethanol, higher alcohols, glycerol, acetic acid, and esters, as well as most minor metabolic by-products. In addition, strains possessing wine stabilizing properties are known. While these tend to produce significant amounts of sulfur dioxide, the SO_2 is produced gradually during fermentation and occurs only in a bound form (Suzzi *et al.*, 1985).

In general, grape cultivars possessing mild varietal aromas are those likely to benefit most from the use of aromatic yeast strains. In such grape varieties, most of the fragrances of the wine comes from yeast by-products.

Genetic Modification

In contrast to most industrial fermentations, winemaking has made little use of genetically engineered microbes. This is probably explained by the complex chemical origin of wine quality, which makes delineating specific and detectable improvements difficult. In addition, other factors, such as grape variety, fruit maturity, and fermentation temperature, are generally thought to be of greater importance. Generally, more is known about the negative influences of certain yeast properties than about their positive sensory effects.

In genetic improvement, features controlled by one or a few genes are the most easily influenced. For example, inactivating the gene that encodes sulfite reductase limits the conversion of sulfite to H_2S. Improvement in other properties, such as flocculation, have been more difficult.

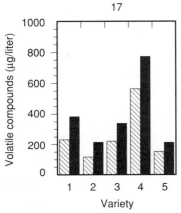 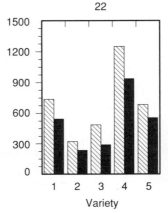 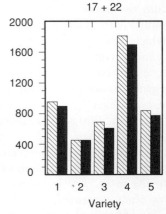

Figure 7.18 Concentration of the volatile components hexyl acetate (17), hexanol (22), and the sum (17 + 22) in wines of five cultivars (1, 'Chenin blanc'; 2, 'Sémillon'; 3, 'Muscat of Alexandria'; 4, 'Cape Riesling'; 5, 'Colombar') with the yeast strains WE 14 (hatched bars) and WE 452 (solid bars). (After Houtman and du Plessis, 1985, reproduced by permission.)

Expression of the flocculation property is regulated by several genes, epistatic (modifier) genes, and possibly cytoplasmic genetic factors.

Many important enological properties are under multigenic control. Alcohol tolerance and the ability to ferment steadily and cleanly at low temperatures are undoubtedly multigenic: individual genes alone having slight effects. Genetic improvement is possible but will probably take considerable time to achieve.

Additional problems arise from the inability to predict accurately the consequences of changing the direction of metabolic pathways. Modifying the regulation of one pathway can have important and unexpected consequences on others (Guerzoni *et al.*, 1985). For example, increasing ester production could have unsuspected, and undesired, effects on alcohol tolerance. Unforeseen metabolic disruptions are less likely when the compound concerned is at the end of a metabolic pathway. For example, terpene synthesis has been incorporated by transfer of farnesyl diphosphate synthetase from a laboratory strain into a wine strain of *Sacch. cerevisiae* (Javelot *et al.*, 1991). Use of the modified yeast in fermentation gives the wine a Muscat-like fragrance.

Many techniques are available to the researcher interested in improving wine yeasts. The simplest and most direct approaches involve simple selection. This is much facilitated if a selective culture medium can be devised to permit only cells containing the desired trait to grow. Otherwise, cells need to be isolated and individually studied for the presence of the desired trait.

Initially, considerable improvement may be obtained employing simple selection. However, further progress usually requires modifying the genetic makeup of the yeast. This can involve standard procedures such as hybridization, backcrossing, and mutagenesis, or newer techniques such as transformation, somatic fusion, and genetic engineering.

One of the difficulties in breeding *Sacch. cerevisiae* is its **diploid** nature, that is, each cell contains two copies of most genes. Although typical in plants and animals, diploidy is uncommon in microbes. The diploid state complicates improvement by permitting the masking of potentially desirable **recessive** genes. In **haploid** organisms, recessive genes are expressed directly as the cells contain only one copy of each gene. Thus, elimination of undesirable traits is both easier and faster in haploid organisms. In diploid organisms, two out of three individuals carrying a recessive gene will not express its presence.

Additional difficulties include the low frequency of sexual reproduction, poor spore germination, and a rapid return to the diploid state (Bakalinsky and Snow, 1990a). Many strains undergo meiosis only infrequently, a precondition for sexual reproduction. Of the haploid spores produced, most do not germinate. Spore sterility is suspected to be due to the high frequency of aneuploidy (unequal numbers of similar chromosomes) in wine yeasts. Although often inducing spore sterility, aneuploidy appears to be beneficial to vegetative growth under the conditions of continuous culture. Of the spores that germinate, most rapidly become diploid by mating with neighboring cells, thus precluding further mating. Alternately, a proportion of the cells derived from a single spore switch mating type. Thus, after a few divisions, daughter cells of a single spore may begin to fuse to form diploid cells. As a consequence, matings need to take place almost immediately after spore germination. However, a technique developed by Bakalinsky and Snow (1990b) can ease crossing by making wine yeasts heterothallic. Introduction of the *HO* gene prevents switching of the mating-type gene and increases the frequency and ease of achieving designed crosses. It also facilitates the elimination of recessive lethal genes that impair spore germination. The mating-type switch is regulated by a single gene with the recessive allele conferring heterothallism and the dominant allele conferring homothallism.

When single traits are to be incorporated into a desirable strain, **backcross breeding** is the preferred technique. Figure 7.19 illustrates a situation in which a dominant flocculant gene is transferred from a donor strain to a valuable nonflocculant recipient strain. Hybridized cells containing the flocculant gene are backcrossed to the recipient strain for several generations. Combined with strong selection for flocculation, backcrossing rapidly eliminates undesired donor genes incorporated unintentionally from the donor strain in the original cross.

Where traits not found in existing *Sacch. cerevisiae* strains are desired, incorportion involves procedures beyond those permitted by conventional genetics. If the feature is found in another yeast species, **somatic fusion** may permit incorporation. Somatic fusion requires enzymatic cell wall dissolution and subsequent mixing of the protoplasts generated in the presence of chemicals such as polyethylene glycol. Polyethylene glycol enhances cell fusion by promoting the physical union of the membranes of cells that fuse.

One of the major problems with somatic fusion is the frequent instability of the association. Fused cells often revert to one of the original species. Also undesirable is the incorporation of genes that interfere with the expression of desirable wine yeast traits.

Potentially less disruptive is the incorporation of only one or several genes from a donor organism. The procedure, called **transformation,** has the added advantage that the donor and recipient need not be closely related. In transformation, yeast protoplasts are bathed in a solu-

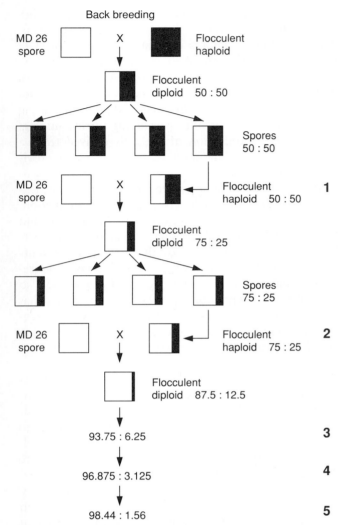

Figure 7.19 Use of backcrossing to eliminate undesired genetic information following the crossing of two yeast strains. In the example, a nonflocculant (recipient) strain is crossed with a flocculent (donor) strain to obtain the flocculant trait. While some haploid progeny will express the flocculant property, the hybrid cells possess only 50% of the genes from the recipient (MD 26) parent. Haploid flocculant cells of the hybrid strain are backcrossed to spores of the parental MD 26 strain. This reduces the proportion of genes from the flocculant parent to 25%. Haploid flocculant progeny from this backcross are again backcrossed (2) to parental MD 26 spores. Several similar backcrossings of flocculant progeny to MD 26 (3, 4, and 5) essentially eliminate all but the desired flocculant genes derived from the flocculant strain. (From Thornton, 1985, reproduced by permission.)

tion containing DNA from the donor organism. Incorporation requires uptake of the DNA containing the gene and its insertion into a yeast or plasmid chromosome. Plasmids are circular, cytoplasmic DNA segments partially controlling their own replication. Although frequently found in *Saccharomyces cerevisiae,* plasmids are not essential to yeast existence.

Using transformation, the malolactic gene from *Leu-*

conostoc oenos has been transferred into *Sacch. cerevisiae* (Snow, 1985). As the gene functions in *Sacch. cerevisiae* at about 1% of its normal value, the example is only illustrative of the potential of the technique. Incorporation of genes improving malate uptake, or improving expression of the malolactic gene, may produce a commercial strain capable of inducing both malolactic and alcoholic fermentations. Incorporation of desired genes into the naturally occurring yeast **2μm plasmid** as a vector may ease the incorporation, expression, and maintenance of foreign genes.

A requirement for all useful strains, new and old, is genetic stability. Although a property of most yeast strains, genetic stability is not a property of all. For example, flocculant strains often lose the ability to form large clumps of cells and settle out as a powdery sediment. Loss of a genetic property is thought to be due to factors such as aneuploidy or mutation.

Environmental Factors Affecting Fermentation

CARBON AND ENERGY SOURCES

The major carbon and energy sources for fermentation are glucose and fructose. Other nutrients may be utilized, but they either are present in small amounts (amino acids), are poorly incorporated into the cell (glycerol), or can only be respired (acetic acid and ethanol). Sucrose can be readily fermented, but it is seldom present in significant amounts in grapes. It may be added, however, in the techniques of chaptalization and amelioration. Sucrose is enzymatically split into its component monosaccharides, glucose and fructose, by one of several invertases. Hydrolysis usually occurs external to the cell membrane by an invertase located between the cell wall and plasma membrane (periplasm). Most other grape sugars are not fermented by *Sacch. cerevisiae* but may be used by several spoilage yeasts and bacteria.

At maturity, the sugar concentration of most wine grapes ranges between 20 and 25%. At this concentration, the osmotic effect of sugar can delay the onset of fermentation. Yeast cells may be partially plasmolyzed, inducing a noticeable lag period (Nishino *et al.,* 1985). Cell viability may be decreased, cell division limited, and sensitivity to alcohol toxicity increased. At sugar concentrations about 25 to 30%, the likelihood of fermentation terminating prematurely increases considerably. Strains of *Sacch. cerevisiae* differ greatly in sensitivity to sugar concentration.

The nature for the remarkable tolerance of wine yeasts to the plasmolytic action of sugar is unclear, but the property appears related to increased synthesis of, or reduced permeability of the cell membrane to, glycerol (see Brewster *et al.,* 1993). These responses to an in-

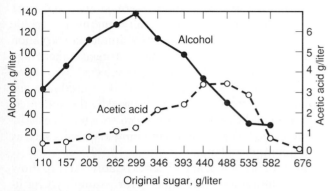

Figure 7.20 Effect of sugar concentration on alcohol and volatile acid production. (After unpublished data of C. von der Heide from Schanderl, 1959, reproduced by permission.) A rough approximation of degrees Brix can be obtained by dividing the sugar concentration by 10.

creased osmolarity of the environment permit glycerol to act as an osmoticum, equilibrating the osmotic potential of the cytoplasm with the surrounding juice.

Sugar content affects the synthesis of several important aromatic compounds. High sugar concentrations increase the production of acetic acid (Fig. 7.20) and its esters. However, as indicated in Fig. 7.21, the effect of total soluble solids on esterification is not solely due to sugar content. For example, synthesis of isoamyl and 2-phenethyl acetate rises with increasing maturity (°Brix) but decreases in juice from immature grapes augmented with sugar to achieve the same degrees Brix.

Over a wide range of sugar concentrations, ethanol production is directly related to juice sugar content. However, above about 30% sugar, ethanol production per gram sugar begins to decline (Fig. 7.20). In some strains of *Saccharomyces cerevisiae* sugar content also

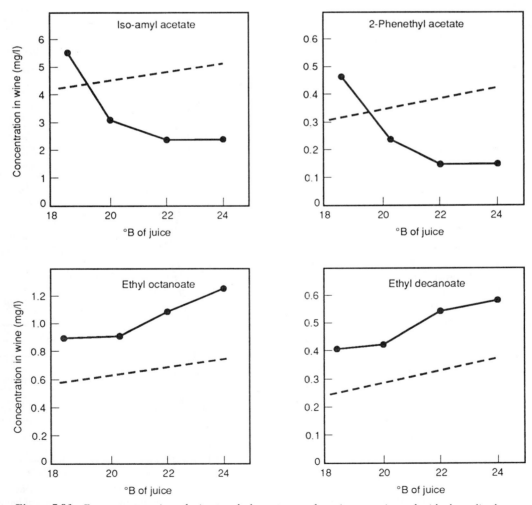

Figure 7.21 Ester concentration of wines made from grapes of varying maturity and with chaptalized must: wines for musts of four degrees of maturity (—); average slope from wines produced from chaptalized must (– – –). (From Houtman *et al.*, 1980, reproduced by permission.)

directly affects the synthesis of acetic acid (Henschke and Dixon, 1990).

ALCOHOLS

All alcohols are toxic to varying degrees. Because *Sacch. cerevisiae* shows considerable ethanol tolerance, much effort has been spent attempting to understand the nature of this tolerance, and why it breaks down at high concentrations. Although alcohol buildup eventually inhibits fermentation, it begins to affect yeast action at much lower concentrations. For example, suppression of sugar uptake can begin at about 2% ethanol (Dittrich, 1977), and rise quickly with increasing temperature. Disruption of the translocation of ammonium and several amino acids also occurs as the alcohol content increases (see Casey and Ingledew, 1986). Although higher (fusel) alcohols are more inhibitory than ethanol, their occurrence at substantially lower concentrations limits their toxicity.

Although most strains of *Sacch. cerevisiae* can ferment in the presence of up to 13 to 15% ethanol, there is wide variation in this ability. Cessation of growth routinely occurs at concentrations well below those that inhibit fermentation. It is generally believed that one of the major toxic effects of alcohol is disruption of the semifluid nature of the cell membrane. This destroys the ability of the yeast to control cell functions and can lead to nutrient loss and cell death.

Ethanol, usually in the form of brandy, may be used to arrest microbial activity. This property is used selectively in sherry and port production. In port, ethanol is added early in fermentation to retain much of the sugar content of the must. This, correspondingly, retains the concentration of most volatile compounds found in the fermenting must. Early cessation of fermentation probably leaves the wine higher in acetic acid, acetaldehyde, and acetoin content, but lower in glycerol, fixed acids, and higher alcohols (Figs. 7.13 and 7.14). In sherry production, the addition of wine spirits at the end of fermentation inhibits the growth of acetic acid bacteria (>15% ethanol) and flor yeasts (>18% ethanol).

The accumulation of alcohol during fermentation has a very important dissolving action on phenolic compounds. Most of the distinctive taste and color of red wines depends on the extraction of flavanols and anthocyanins by ethanol during vinification. Extraction is further enhanced in the presence of sulfur dioxide (Oszmianski *et al.,* 1988).

NITROGENOUS COMPOUNDS

Under most circumstances, juice and must contain nitrogen sufficient for several fermentations. However, there are several conditions in which nitrogen can become limiting. For example, preformative clarification diminishes juice nitrogen content; if the reduction

is sufficiently marked, it can slow fermentation and cause it to become stuck. This may result from the irreversible inactivation of sugar transport by ammonia starvation (Lagunas, 1986). In addition, juice nitrogen content may be reduced by 33 to 80% in grapes infected by *Botrytis cinerea* (Rapp and Reuther, 1971). In sparkling wine production, nitrogen deficiency caused by the initial fermentation and clarification of the *cuveé* wines is usually counteracted by the addition of ammonium salts such as ammonium phosphate.

The juice of some grape varieties is more likely to show nitrogen limitation than others, for example, 'Chardonnay' and 'Colombard.' This is especially true when the juice has been given undue centrifugation or filtration.

Nitrogen demand appears to be greatest during the exponential phase of growth during fermentation, and nitrogen is incorporated most rapidly during that period (Fig. 7.22). The inorganic nitrogen source preferentially incorporated is ammonia. The oxidized state of ammo-

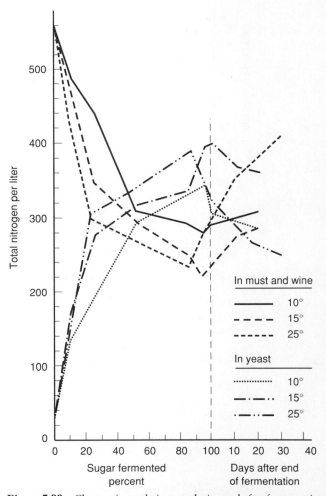

Figure 7.22 Changes in total nitrogen during and after fermentation in musts and in yeasts fermenting them. (After Nilov and Valuiko, 1958, in Amerine *et al.,* 1980, reproduced by permission.)

nia permits its direct incorporation into organic compounds. Although ammonia is potentially capable of repressing the uptake of amino acids, its concentration in the juice is insufficient for that to occur. *Saccharomyces cerevisiae* has several amino acid transport systems (Cartwright *et al.*, 1989). One is nonspecific and directs the uptake of all amino acids except proline. The other systems are more selective, transporting only particular groups of amino acids. Certain amino acids are preferentially incorporated, such as phenylalanine, leucine, isoleucine, and tryptophan, whereas others, such as alanine, arginine, and proline, are poorly assimilated (Ough *et al.*, 1991). Urea can be readily incorporated into yeast cells, but it is no longer recommended as a form of nitrogen supplement. Urea has been implicated in the production of the carcinogen ethyl carbamate (Ough *et al.*, 1990). Amines and peptides also may be incorporated as nitrogen sources, but protein nitrogen is unavailable. Wine yeasts are capable of neither transporting proteins across the cell membrane nor enzymatically degrading them to amino acids outside the cell.

Nitrogen is required for the synthesis of proteins, pyrimidine nucleotides, and nucleic acids. Yeast cells generally synthesize their amino acid and nucleotide requirements from inorganic nitrogen and sugar. Consequently, most yeast strains do not require these metabolites in the must, but they can be assimilated from the medium when available. Assimilation of metabolites avoids the diversion of metabolic intermediates and energy to amino acid biosynthesis.

Nitrogen content can influence the synthesis of aromatic compounds during fermentation. Most noticeable is the reduction in fusel alcohol content by ammonia and urea. This effect can be reversed by the assimilation of certain amino acids from the juice or must. These opposing effects appear to result from the use of fusel alcohols in the biosynthesis of amino acids from ammonia and urea, and their release on deamination of amino acids assimilated from the must. The sensory impact of these changes, if any, is unknown.

During fermentation, and especially after, there is a slow release of nitrogen compounds back into the wine, probably owing to autolysis of dead yeast cells (Fig. 7.22). This release may activate subsequent microbial activity. For this reason, the first racking often occurs shortly after fermentation. When malolactic fermentation is desired, however, racking is delayed until the bacterial conversion of malic to lactic acid is complete. Racking also may be delayed if lees contact with the wine is desired.

LIPIDS

Lipids function as the basic constituents of cellular membranes (phospholipids and sterols), in energy storage (oils), as pigments (carotenoids), and as regulator molecules complexed with proteins (lipoproteins) and carbohydrates (glycolipids).

Yeasts synthesize their own lipid requirements when grown aerobically, but they are unable to generate long-chain unsaturated fatty acids and sterols under anaerobic conditions. This is less significant in red wine vinification as adequate supplies of precursors may be obtained during fermentation on the skins. Anaerobic inhibition of lipid synthesis, however, can result in sluggish fermentation of highly clarified white juice. Clarification can remove more than 90% of the fatty acid content. This is particularly marked with the unsaturated fatty acids—oleic, linoleic, and linolenic acids (Bertrand and Miele, 1984). In addition, sterols such as ergosterol and oleanolic acid are probably removed from the juice. Nevertheless, yeasts typically possess sufficient reserves of these vital compounds to initiate fermentation and complete several cell divisions. Eventually, though, the accumulated deficit in sterols and unsaturated fatty acids can enhance the ethanol-induced reduction in glucose uptake and result in a stuck fermentation.

Wine yeasts are particularly sensitive to the toxicity of mid-chain carboxylic acids such as octanoic and decanoic acid. As by-products of yeast metabolism, they accumulate during fermentation. They increase the ethanol-induced leakage of nutrients such as amino acids from the cells (Sá Correia *et al.*, 1989). The toxic effects are limited or reversed by the addition of ergosterol and long-chain unsaturated fatty acids (Fig. 7.23) or by the addition of various absorptive substances, such as activated charcoal, bentonite, silica gel, or yeast hulls ("ghosts"). The latter consist primarily of cell wall remnants. By removing octanoic and decanoic acids via absorption from the juice, their potential to disrupt yeast membranes is reduced. In addition, yeast hulls can be sources of sterols and unsaturated fatty acids (see Munoz and Ingledew, 1990).

Another possible solution to problems associated with low sterol and unsaturated fatty acid content is to use yeasts high in sterol content. Sterol synthesis is commonly suppressed in the presence of glucose, but some strains do not show repression. Growing yeasts under highly aerobic conditions, as in active dry yeast production, generates cells high in unsaturated fatty acids, and with up to three times the sterol content of cells grown semiaerobically (Tyagi, 1984). Musts inoculated with strains possessing high sterol contents frequently ferment more sugar than strains possessing low sterol contents.

PHENOLS

The phenolic content of must can have various effects on the course of fermentation. For example, the phenols in red grapes, notably anthocyanins, can stimulate fermentation, while the procyanidins of white grapes can be slightly inhibitory (Cantarelli, 1989). Phenolic com-

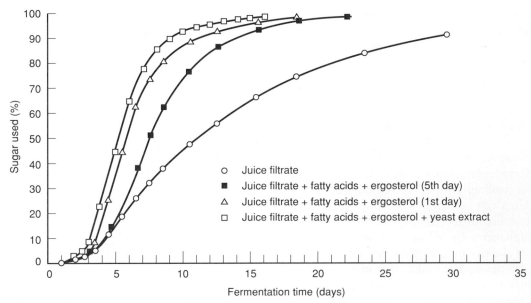

Figure 7.23 Fermentation curves of juice filtrates with additions of yeast extract, unsaturated fatty acids, and ergosterol at inoculation and on the fifth day thereafter. (From Houtman and du Plessis, 1986, reproduced by permission.)

pounds also are a determining factor in the activation of film formation important in *fino* sherry production (Cantarelli, 1989). Certain phenols, notably the esters of gallic acid, are toxic, but others such as chlorogenic and isochlorogenic acids may stimulate fermentation. The primary situation where phenols commonly suppress yeast metabolism is during the second fermentation in sparkling wine production. This is one of the reasons why few red sparkling wines are produced using traditional techniques.

In addition to affecting fermentation, phenols also may be modified by yeast action. For example, ferulic and *p*-coumaric acids are decarboxylated to aromatic vinyl phenols (4-vinyl guaiacol and 4-vinyl phenol, respectively) by some strains of *Sacch. cerevisiae* (Chatonnet *et al.*, 1989). Other phenolic constituents in the must may influence the conversion.

SULFUR DIOXIDE

In most cases, the primary reason for adding sulfur dioxide is to restrict or prevent the growth of undesirable microbes. Sulfur dioxide has a distinct advantage over most other antimicrobial agents, because of the relative insensitivity of wine yeasts to its action. In contrast, sulfur dioxide is toxic, or inhibitory, to most bacteria and other yeasts at low concentrations. Actively growing yeasts are even more tolerant than their dormant counterparts. The toxicity of sulfur dioxide also is aided by the low pH of grape must (Fig. 7.24).

At currently used concentrations (usually less than 50 ppm in healthy grapes), sulfur dioxide is unlikely to affect the rate of alcoholic fermentation. However, sulfur dioxide can slow the onset of fermentation. The

presence of 15 to 20 ppm can reduce the viability of a yeast inoculum from 10^6 to 10^4 cells/ml or less (Lehmann, 1987). Sulfur dioxide also can significantly influence yeast metabolism. Sulfur dioxide readily binds with several carbonyl compounds produced by yeasts, notably acetaldehyde, pyruvic acid, and α-ketoglutaric acid. Binding with sulfur dioxide increases the production and subsequent release of carbonyl compounds into the juice or must. Thus, their concentration in the finished wine correlates well with the concentration of sulfur dioxide initially added to the must. Sulfur dioxide also favors glycerol synthesis, while acetic acid production tends to decline. Fixed acidity generally does not change, par-

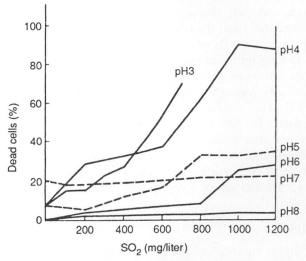

Figure 7.24 Effect of pH on the lethal dose of SO_2 for wine yeasts. (From Farkaš, 1988, reproduced by permission.)

tially because sulfur dioxide suppresses the metabolism of both lactic acid and acetic acid bacteria.

The binding of sulfur dioxide to carbonyl compounds inadvertently increases the amount of sulfur dioxide needed to suppress the action of spoilage organisms. Bound sulfur dioxide is much less antimicrobial than molecular SO_2.

Although sulfur dioxide is the best wine antimicrobial agent currently available, it does not control certain spoilage yeasts. For example, many strains of the yeasts *Saccharomycodes ludwigii*, *Zygosaccharomyces bailii*, and *Brettanomyces* spp. are particularly tolerant to sulfur dioxide (Hammond and Carr, 1976).

When present, elemental sulfur can be assimilated and used in the synthesis of sulfur-containing amino acids and coenzymes. It also may be oxidized to sulfate and sulfur dioxide, or reduced to hydrogen sulfide. Reduction of sulfur to hydrogen sulfide may be a means, albeit aromatically unpleasant, of maintaining a favorable redox balance in yeast cells under anaerobic conditions.

OXYGEN AND AERATION

The process of fermentation itself requires no oxygen. Even in the presence of ample oxygen, *Sacch. cerevisiae* preferentially ferments sugars. Nevertheless, trace amounts of oxygen favor fermentation by permitting the direct oxidation of precursors in the biosynthesis of sterols and long-chain unsaturated fatty acids. Production and proper functioning of cell membranes in wine yeasts require sterols (ergosterol and lanosterol) as well as C_{16} and C_{18} fatty acids. Some of the latter, such as linoleic and linolenic acids, are unsaturated. Molecular oxygen also is required for the synthesis of the vitamin nicotinic acid.

Stemming and crushing usually induce sufficient oxidation of the juice for yeast growth. The juice quickly becomes saturated with oxygen during the processes. The capacity of juice to absorb oxygen depends partially on the duration of skin contact (maceration). With the increase in extraction of phenols, oxygen consumption is increased. In addition, this speeds the removal of free oxygen, and the shift from oxidative to reductive conditions. Aeration beyond that which occurs coincidental to stemming and crushing (**hyperoxidation**) is variously viewed as being potentially beneficial (Cheynier *et al.*, 1991) or detrimental (Dubourdieu and Lavigne, 1990). The initial browning commonly associated with crushing is permissible since the color compounds commonly precipitate during fermentation or clarification. It also gives white wine a degree of resistance to subsequent oxidative browning by removing readily oxidizable phenols early in vinification.

During the fermentation of red wine, additional oxygen may be absorbed during pumping over. The resultant incorporaton of about 10 mg O_2 per liter often speeds the process of fermentation. This is more marked when aeration occurs at the end of the exponential phase of yeast growth (Sablayrolles and Barre, 1986). The yeast population is increased and average cell viability is extended. Aeration also increases the production of acetaldehyde, thus favoring color stability by assisting the early formation of polymers of anthocyanins and tannins.

With white juice fermentations, winemakers typically try to avoid oxidation as it increases the synthesis of fusel alcohols and acetaldehyde. The higher concentrations of acetic acid noted may be due to the activation of acetic acid bacteria in the juice. In addition, semiaerobic conditions can depress the synthesis of esters (Nykänen, 1986). However, the effects may be reversed with short aerations at the beginning or a few days after the commencement of fermentation (Bertrand and Torres-Alegre, 1984). The effect of aeration on hydrogen sulfide production is another example of the complexity of the effects of introducing oxygen. Aeration may remove/oxidize H_2S but may also enhance its synthesis (Houtman and de Plessis, 1981). The timing and extent of aeration also can influence urea accumulation and, thereby, the potential to produce ethyl carbamate (Henschke and Ough, 1991).

Reactivation of fermentation in stuck wines usually requires aeration. *Cuvée* wines also may be aerated slightly prior to the second fermentation in sparkling wine production.

Following fermentation, limited aeration (~40 mg O_2/liter, or four saturations) may benefit the maturation of red wines. In contrast, most white wines are painstakingly protected from air contact following fermentation. The exception to this practice occurs when the wines are given extended contact with the lees (occasionally up to 10 months). In that case, limited oxidation occurs when the wine is stirred with the lees.

Oxygen absorption is influenced by many factors, including clarification, skin contact, phenol concentration, sulfur dioxide content, presence of polyphenol oxidases, sugar proportion, temperature, pumping over, rate of fermentation, and, of course, protection from air before, during, and after fermentation.

CARBON DIOXIDE AND PRESSURE

During fermentation, large volumes of carbon dioxide gas are produced, about 260 ml/gram glucose. This equates to over 50 times the volume of the juice fermented. The escape of carbon dioxide is estimated to remove about 20% of the heat generated during fermentation. Some of that heat discharge is probably due to the evaporative heat loss connected with the accompanying escape of water vapor.

Carried off with the carbon dioxide are various volatile compounds. Ethanol loss is estimated to be about 1 to 1.5% of the ethanol produced (Williams and Boulton, 1983) but varies with temperature and sugar utilization. Higher alcohols and monoterpenes are lost to about the same degree (~1%). In contrast, significant losses of both ethyl and acetate esters can occur. Depending on the grape variety, and especially the fermentation temperature, upward of 25% of these aromatically important compounds may be lost (Miller *et al.*, 1987). On average, more acetate esters are lost than ethyl esters. This loss could diminish significantly the fruity character of the resulting wine. Figure 7.25 illustrates the loss of some of these compounds.

Loss of volatiles from fermenting juice is a function of both the relative rates of synthesis and destruction and the relative solubility of compounds in the increasingly alcoholic juice. Loss is further affected by fermentor size and shape; for example, the increased surface area and low liquid pressures of small fermentors favor volatility. In addition, while the reduction of vapor pressure at low temperatures tends to diminish loss, the slower release of carbon dioxide could partially offset this by favoring the absorption and liberation of volatile compounds.

The action of carbon dioxide production causes the development of strong convection currents within the juice. This helps to equilibrate the nutrient and temperature status throughout the juice. However, the presence of a floating or submerged cap, as with red wines, disrupts equilibration (Fig. 7.28).

In vats, and in most tanks, the carbon dioxide produced during fermentation escapes into the surrounding air. When the gas is trapped, the pressure rapidly rises. At pressures above 700 kPa (~7 atm), yeast growth ceases. Low pH and high alcohol contents increase yeast sensitivity to high pressures (Kunkee and Ough, 1966). This has a significant influence in the production of sparkling wine, where the finished product may develop pressures upward of 600 kPa. Nevertheless, yeast fermentation ability may not be inhibited completely until about 3000 kPa. The major inhibitory effects of carbon dioxide buildup seem to result from changes in yeast membrane composition and permeability. In addition, carbon dioxide buildup may affect the balance between carboxylation and decarboxylation reactions (Table 7.1). The effect of pressure on the synthesis of aromatic compounds during vinification seems not to have been investigated.

The pressure created by trapping the carbon dioxide produced during fermentation has occasionally been used to encourage a more constant rate of fermentation. It also has been used to induce the premature termination of fermentation, to give the wine a sweet finish. However, care must be used with this technique as spoilage yeasts, such as *Torulopsis* and *Kloeckera,* are less sensitive to high pressures than is *Sacch. cerevisiae.* The production of acetic acid by spoilage yeasts can give a

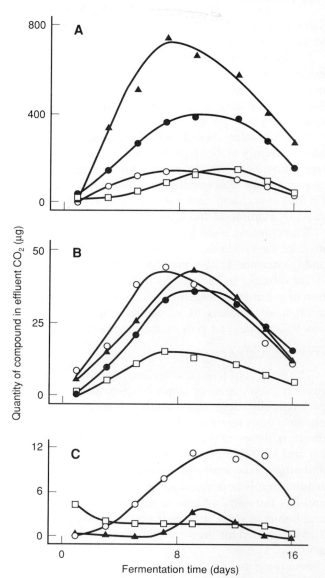

Figure 7.25 Yeast aromatics released with CO_2 during fermentation at 15°C. (A) ▲, Isoamyl acetate; ○, ethyl *n*-hexanoate; ■, ethyl *n*-octanoate; ●, isoamyl alcohol. (B) □, Isobutyl acetate; ○, hexyl acetate; ●, ethyl *n*-butanoate; △, isobutanol. (C) ○, Ethyl *n*-decanoate; □, 1-hexanol; ▲, 2-phenylethanol. (From Miller *et al.,* 1987, reproduced by permission.)

vinegary taint. Caution also needs to be taken as lactic acid bacteria (*Lactobacillus*) are little affected by pressures that restrict the growth of wine yeasts (Dittrich, 1977).

pH

The pH range normally found in juice or must has little effect on the rate of fermentation, or on the synthesis and release of aromatic compounds by yeasts. Only at abnormally low pH values (<3.0) is fermentation somewhat inhibited. However, low pH may assist the uptake of some amino acids by providing protons required for transport across the cell membrane (Cartwright *et al.,*

Table 7.1 Possible Effects of Carbon Dioxide on Key Enzymes of *Saccharomyces cerevisiae*[a]

Reaction	Comment
Pyruvate → acetaldehyde → ethanol (CO_2)	Reduced production of ethanol
Pyruvate (ATP → ADP, CO_2) → oxaloacetate → amino acids	Stimulation, less available pyruvate for ethanol production
Acetyl-CoA → malonyl-CoA → fatty acids (CO_2)	Stimulation, less available pyruvate for ethanol production
Pyruvate → malate (CO_2)	Stimulation, less available pyruvate but malate enzyme level is not high
Phosphoenolpyruvate (ADP → ATP, CO_2) → oxaloacetate	Stimulation, less available pyruvate but enzyme is repressed by glucose
6-Phosphogluconate ($NADP^+$ → $NADPH$, CO_2) → ribulose 5-phosphate	Reduced production of biosynthetic precursors, thus cell yield will decrease; will reduce rate of production of ethanol

[a] From Jones *et al.* (1981), reproduced by permission.

1989). Although the growth rate of yeasts at a pH of 3.0 may be about half that at pH 4.0, this appears to be of little practical significance in winemaking.

The most important effects of pH on fermentation are indirect, such as noted above concerning the antibiotic action of sulfur dioxide. Low pH also prevents many potentially competitive organisms from growing in must. In addition, pH affects the survival of some fermentation by-products in wine. The most well-known effect concerns the hydrolysis of ethyl and acetate esters, where hydrolytic breakdown occurs more rapidly at low pH values.

VITAMINS

Vitamins play a crucial role in the regulation of yeast metabolism as coenzymes or enzyme precursors (Table 7.2). Although vitamins are not metabolized as energy sources, their concentrations decrease markedly during fermentation (see Amerine and Joslyn, 1970). Nevertheless, yeast requirements typically are satisfied by either biosynthesis or assimilation from the juice. However, certain conditions may reduce the availability or concentration of vitamins in the juice. Fatty acids produced during fermentation can inhibit the uptake of thiamine,

oversulfiting degrades thiamine, and infection of the grapes by fungi lowers the vitamin content. Under such conditions, vitamin supplements may improve or be required to reinitiate fermentation.

Adequate concentrations of thiamine reduce the synthesis of carbonyl compounds that bind to sulfur dioxide, thereby diminishing the amount of SO_2 needed to control spoilage organisms adequately. In addition to limiting carbonyl synthesis, thiamine also reduces the concentration and relative proportions of higher alcohols produced during fermentation.

Although seldom a problem, deficiencies in pyridoxine and pantothenic acid can disrupt yeast metabolism, resulting in increased hydrogen sulfide synthesis.

INORGANIC ELEMENTS

Inorganic elements often are essential components in the active sites of many enzymes, and they help in regulating cellular metabolism and maintaining cytoplasmic pH and ionic balance (Table 7.3). Although inorganic elements are normally assumed to be in adequate supply in juice, there is difficulty in assessing accurately yeast requirements and available ion concentrations in grape juice and must. Not only do organic compounds such as

Table 7.2 Role of Vitamins in Yeast Metabolism[a]

Vitamin	Active form	Metabolic role	Optimum conc. (mg/liter)
Biotin	Biotin	All carboxylation and decarboxylation reactions	0.005–0.5
Pantothenate	Coenzyme A	Keto acid oxidation reactions; fatty acid, amino acid, carbohydrate, and choline metabolism	0.2–2.0
Thiamine (B$_1$)	Thiamine-pyrophosphate	Fermentative decarboxylation of pyruvate; oxo acid oxidation and decarboxylation	0.1–1.0
Pyridoxine	Pyridoxal phosphate	Amino acid metabolism; deamination, decarboxylation, and racemization reactions	0.1–1.0
p-Aminobenzoic acid and folic acid	Tetrahydro-folate	Transamination; ergosterol synthesis; transfer of one-carbon units	0.5–5.0
Niacin (nicotinic acid)	NAD$^+$, NADP$^+$	Dehydrogenation reactions	0.1–1.0
Riboflavin (B$_2$)	FMN, FAD	Dehydrogenation reactions and some amino acid oxidations	0.2–0.25

[a] From Jones et al. (1981), reproduced by permission.

amino acids sequester elements, thereby reducing their effective concentration, but ions can antagonize uptake of one another. Occasionally, as in the case of potentially toxic aluminum ions, this may be beneficial.

The abundance of potassium ions probably makes K$^+$ the most significant metallic ion in juice and must. High potassium content may interefere with the efficient uptake of amino acids, such as glycine. Under anaerobic conditions, potassium excretion may be necessary to maintain an acceptable ionic balance, owing to the simultaneous incorporation of protons (H$^+$) with glycine (Cartwright et al., 1989). High potassium concentrations also can generate tartrate instability, which is associated with high juice and wine pH. High pH can lead to microbial instability, increase the tendency of white wines to brown, and induce color instability in red wines.

TEMPERATURE

Temperature is one of the most influential factors affecting fermentation. Not only does temperature both directly and indirectly influence yeast metabolism, but it is one of the factors over which the winemaker has the greatest control.

At the upper and lower limits, temperature can cause cell death. However, inhibitory effects are experienced well within the extremes. The inhibitory effects of high

Table 7.3 Major Inorganic Elements Required for Yeast Growth and Metabolism[a]

Ion	Role	Concentration (μM)[b]
K$^+$	Enhances tolerance to toxic ions; involved in control of intercellular pH; K$^+$ excretion is used to counterbalance uptake of essential ions, e.g., Zn^{2+}, Co^{2+}; K$^+$ stabilizes optimum pH for fermentation	20×10^3
Mg^{2+}	Levels regulated by divalent cation transport system; Mg^{2+} seems to buffer cell against adverse environmental effects and is involved in activating sugar uptake	5×10^3
Ca^{2+}	Actively taken up by cells during growth and incorporated into cell wall proteins; Ca^{2+} buffers cells against adverse environments; Ca^{2+} counteracts Mg^{2+} inhibition and stimulates effect of suboptimal concentrations of Mg^{2+}	1.5×10^3
Zn^{2+}	Essential for glycolysis and for synthesis of some vitamins; uptake is reduced below pH 5, and two K$^+$ ions are excreted for each Zn^{2+} taken up	50
Mn^{2+}	Implicated in regulating the effects of Zn^{2+}; Mn^{2+} stimulates synthesis of proteins	15
Fe^{2+}, Fe^{3+}	Present in active site of many yeast proteins	10
Na$^+$	Passively diffuses into cells; stimulates uptake of some sugars	0.25
Cl$^-$	Acts as counterion to movement of some positive ions	0.1
Mo^{2+}, Co^{2+}, B^{2+}	Stimulates growth at low concentrations	0.5

[b] For growth of 25 g cells/liter.

temperatures are increased by other growth-limiting factors, such as the presence of ethanol and certain C_8 to C_{10} carboxylic acids. At low temperatures, yeasts tend to be less sensitive to the toxic effects of high alcohol concentrations. This may be due to the higher proportion of unsaturated fatty acid residues in the plasma membrane (Rose, 1989). This influence may help to explain the higher maximum viable cell count of fermentations conducted at cooler temperatures (Ough, 1966a). It also may explain the increased yeast viability associated with periodic aeration of barrels of wine left on the lees. Oxygen could permit the desaturation of precursors in the production of needed unsaturated fatty acids.

The growth rate of yeast cells is also strongly affected by the fermentation temperature. This is particularly marked during the exponential growth phase. For example, cell division may occur about every 12 hr at 10°C, every 5 hr at 20°C, and every 3 hr at 30°C (Ough, 1966a). The rate of cell division also is affected by pH and °Brix value of the juice.

At warm temperatures (>20°C), yeast cells experience a rapid decline in viability at the end of fermentation. At cooler temperatures, cell growth is retarded, but viability is enhanced. Cool temperatures also exaggerate the lag phase of fermentation. For this reason, winemakers may warm white juice to 20°C before adding the yeast inoculum. Once fermentation has commenced, the juice may subsequently be cooled to a more desirable fermentation temperature, commonly between 10° and 15°C.

The temperature at which active dry yeast is rehydrated prior to inoculation can affect the onset of fermentation. Temperatures between 38° and 40°C appear optimum for rehydration (Kraus *et al.*, 1981). Active dry yeast often contains about 20 to 30 × 10⁹ cells/g. A rapid onset of fermentation limits oxidation and the growth of spoilage organisms.

In addition to the above-mentioned effects on yeast growth and survival, temperature has many subtle, and not so subtle, effects on yeast metabolism. One of the most marked is the influence of temperature on the rate of fermentation. Cool temperatures dramatically slow the rate of fermentation and may cause premature termination. Excessively high temperatures may disrupt enzyme and membrane function and also result in stuck fermentation. Although quick onset and completion of fermentation have advantages, the preferred temperature for vinification is often less than the optimum for ethanol production. Because yeast strains differ in response to temperature, the optimum temperature for vinification can vary widely.

The modern preference in most wine producing regions is to conduct white wine fermentations between 8° and 15°C. Some European wineries still prefer to ferment between 20° and 25°C. Most New World winemakers favor cool temperatures because they give fresher more fruity wines. Fruitiness in wine is a highly valued characteristic throughout much of the world. Important in this regard is the increased synthesis of fruit esters, such as isoamyl, isobutyl, and hexyl acetates. The esters are both synthesized and retained to a greater degree at cool temperatures (Fig. 7.26A). Other esters, such as ethyl octanoate and ethyl decanoate, are produced optimally at 15°C, whereas 2-phenethyl acetate achieves its highest concentration at 20°C (Fig. 7.26B). Greater production of ethanol and higher alcohols also may be observed at cool fermentation temperatures. In addition, cooler fermentation temperatures reduce the liberation of yeast colloids and thereby facilitate rapid clarification.

Red wines are typically fermented at temperatures higher than those for white wines. Temperatures between 24° and 27°C are commonly considered optimal. However, such temperatures are not universally preferred. For example, wines from 'Pinotage' are reported as being better when fermented at 15°C (du Plessis, 1983). The warmer temperatures preferred for red wine vinification probably are related more to the effect on phenol extraction than on fermentation rate. Temperature and alcohol are the major factors influencing pigment and tannin extraction from seeds and skins. Both groups of compounds dominate the characteristics of young red wines. The potentially undesirable consequences of higher fermentation temperatures, such as the production of increased amounts of acetic acid, acetaldehyde, and acetoin and lower concentrations of some esters, probably are less noticeable against the more intense fragrance of red wine. The greater synthesis of glycerol at higher temperatures is often considered to give red wines a smoother mouth-feel. Data from Noble and Bursick (1984) are in conflict with this common view.

Other important influences arise from factors not directly related to the effect of temperature on fermentation. For example, temperature affects the rate of ethanol loss during vinification (Williams and Boulton, 1983). Nevertheless, losses of hydrophobic low molecular weight compounds such as esters are more marked, and such losses have a greater potential impact on the sensory quality of the wine produced.

During fermentation, much of the chemical energy stored in grape sugars is released as heat. It is estimated that the release is equivalent to about 23.5 kcal/mol glucose (see Williams, 1982). This is sufficient for juice with a reading of 23° Brix to increase in temperature by about 30°C during the course of fermentation. If this were to occur, the yeast cells would die before completing fermentation. In practice, such temperature increases are not realized. Because the heat is liberated over several days to weeks, some of the heat is lost with escaping

A

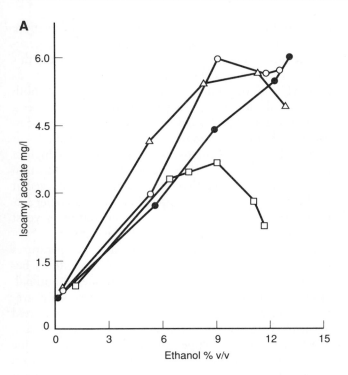

B

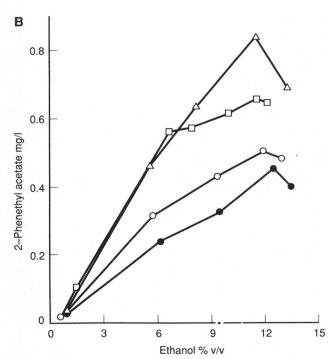

Figure 7.26 Effect of temperature and progress of fermentation on isoamyl acetate (A) and 2-phenethyl acetate (B) content. ●, 10°C; ○, 15°C; △, 20°C; and □, 30°C. (From Killian and Ough, 1979, reproduced by permission.)

control measures are not implemented in large fermentors.

Important in heat buildup is the initial juice or must temperature. This sets the potential rate at which the temperature rises. The higher the juice temperature, the greater the initial rate of fermentation and heat release, and the sooner a lethal temperature may be reached. Thus, a cool temperature at the beginning of fermentation can limit the degree of temperature control required.

Also important to temperature control is the size and shape of the fermentor, and the presence or absence of a cap. The rate of heat lost is directly related to the surface area to volume ratio of the fermentor. By retaining heat, the volume of juice fermented can significantly affect the rate of fermentation: the larger the fermentor, the greater the retention of heat and the subsequent likelihood of overheating. This feature is illustrated in Fig. 7.27.

The tumultuous release of carbon dioxide during fermentation may be sufficient to maintain a uniform temperature throughout the vat or tank. This is usual for fermenting white and rosé juice, where vertical and lateral variation in temperature is seldom more than 1°C. At cold fermentation temperatures, however, turbulence may be insufficient to equilibrate the temperature throughout the fermentor, and temperature strata may develop.

With red wines, cap formation can disrupt effective circulation and mixing of the must. The maximum cap-

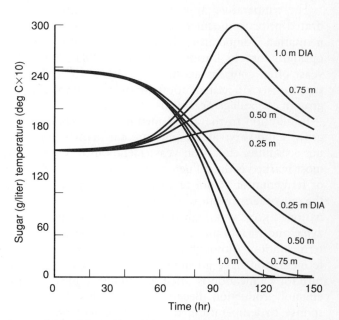

Figure 7.27 Effect of barrel diameter on fermentation rate and temperature rise during fermentation. (From Boulton, 1979, by permission.) Although the data are not presented in terms of cooperage capacity, barrels possessing maximum diameters of 0.5, 0.75, and 1.0 m, respectively, could have capacities ranging from 75 to 150, 225 to 500, and 500 to 1200 liters depending on barrel height and stave length.

carbon dioxide and water vapor. Heat also radiates through the surfaces of the fermentor into the cellar environment. Nevertheless, the rise in temperature can easily reach levels critical to yeast survival if temperature

to-liquid temperature difference is often about 10°C (Fig. 7.28). Punching down induces only a transitory temperature equilibration between the cap and the juice. In contrast, little temperature variation exists within the main volume of the must. Because high cap temperatures are a common feature of red wine fermentations, it has been suggested that red wine vinification consists of two simultaneous but different phases, namely, a liquid phase, where the temperature is cooler and readily controlled, and a largely uncontrolled high-temperature phase in the cap (Vannobel, 1986). Because the rate of fermentation is much more rapid in the cap, the alcohol content rises quickly to above 10%. The higher temperatures found in the cap, plus the association of alcohol, probably increase the speed and efficiency of phenol extraction from the skins trapped in the cap.

Temperature regulation is achieved by a variety of techniques. Appropriate timing of the harvest can provide fruit at a desired temperature for the initiation of fermentation. Relatively small fermentation cooperage and vinification in cool cellars have been used for centuries to achieve a degree of temperature control. Carbonic maceration (see Chapter 9) slows the rate of fermentation and correspondingly diminishes the peak temperature reached during fermentation. However, maintenance of fermentation temperatures within a narrow range requires direct cooling in all but small barrels (~225 liters).

Where heat transfer through the fermentor wall is sufficiently rapid, cooling the fermentor surface with water, or by passing a coolant through an insulating jacket, can be effective. Where thermal conductance is insufficient, fermenting must may be pumped through external heat exchangers, or cooling coils may be inserted directly into the fermenting must. In special fermentors, carbon dioxide is trapped and the pressure buildup used to slow fermentation and heat accumulation.

PESTICIDE RESIDUES

Under most situations, no more than trace amounts of pesticide residues are found in juice or must. At such concentrations, they have little or no perceptible effect on fermentation, or on the sensory qualities of the wine. Used properly, pesticides help the fruit reach maximum quality. When used in excess or applied just before harvest, however, pesticides may negatively affect winemaking.

Various factors influence the pesticide content on or within fruit. For example, heavy rains or sprinkler irrigation may wash contact pesticides off the fruit. Rains have less of an effect on systemic pesticides that are absorbed into plant tissues. Ultraviolet radiation in sunlight can degrade some pesticides and decrease the residual levels. Microbial decomposition is also likely.

Crushing, and especially maceration, can influence the incorporation of crop protection chemicals into must. The long maceration used in red wine production can increase the extraction of contact fungicides. Maceration generally has little effect on the content of systemic pesticides as they are already present in the juices before crushing.

Clarification, either by cold settling or by centrifugation, significantly reduces the concentration of contact fungicides, such as elemental sulfur, but has less effect on systemic pesticide residues (Fig. 7.29). The persistence of pesticide residues, once dissolved, depends largely on their stability under the physicochemical conditions found in juice and wine. For example, more than 70% of dichlofluanid (Euparen) residues may be degraded under the acidic conditions of juice and wine (Wenzel *et al.*, 1980).

Of pesticide residues, fungicides not surprisingly have the greatest effect on the growth and fermentability of yeasts. Newer fungicides, such as metalaxyl (Ridomil) and cymoxanil (Curzate), do not appear to affect fermentation. In contrast, triadimefon (Bayleton) can depress fermentation, presumably by disrupting sterol metabolism. The older, broad-spectrum fungicides such as

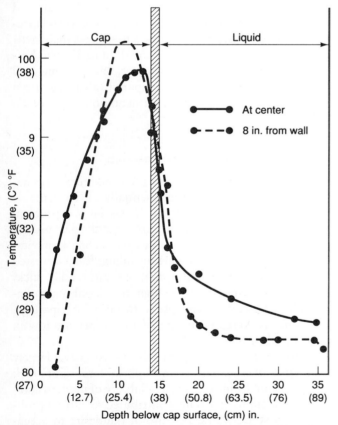

Figure 7.28 Vertical temperature profile through cap and liquid at 40 hr. Cross-hatching indicates that the boundary between the cap and liquid is not sharply defined. (From Guymon and Crowell, 1977, reproduced by permission.)

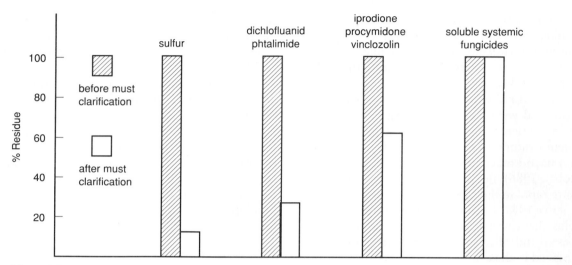

Figure 7.29   Influence of must settling on the elimination of different fungicides employed in viticulture. (After Gnaegi *et al.*, 1983, reproduced by permission.)

dinocap, captan, mancozeb, and maneb generally are toxic to yeasts. Fungicides such as copper sulfate and elemental sulfur seldom have a significant effect, except at abnormally high concentrations (Conner, 1983). This tolerance is probably related to the relative insensitivity of wine yeasts to other sulfur compounds, such as sulfur dioxide.

Fungicides can have several direct and indirect effects on fermentation. Delaying the start of fermentation is probably the most common. As the depression primarily affects the lag phase, subsequent fermentation is unaffected. This may explain why the addition of 5 to 10 g/hl active dry yeast avoids most fungicide-induced suppression of fermentation (Lemperle, 1988). In addition, the increased yeast population reduces the amount of fungicide available to react with each cell. Occasionally, the onset of fermentation occurs normally, but the rate is depressed (Gnaegi *et al.*, 1983). Such suppression may result in stuck fermentation.

Fungicides may affect the sensory qualities of wine by influencing the relative activities of various yeast biosynthetic pathways. Elemental sulfur augments the synthesis of hydrogen sulfide in some yeast strains, as may Bordeaux mixture, folpet, and zineb. While hydrogen sulfide favors the subsequent production of mercaptans, residual copper can limit mercaptan synthesis by forming insoluble cupric sulfide with H_2S. Vinclozolin (Ronilan) and iprodione (Rovral) occasionally appear to affect fermentation and induce the development of off-flavors (Romao and Belchior, 1982).

Fungicides also may have a selective effect on the yeast strains or species present during vinification. For example, captan may favor the growth of *Torulopsis bacillaris* by inhibiting the growth of most other yeast species (Minárik and Rágala, 1966).

Several herbicides (2,4-dichlorophenoxyacetic acid [2,4-D] and simazine) and insecticides (lindane, dieldrin, methiocarb, and rotenone) have been found to be nontoxic to wine yeasts (Conner, 1983).

Much less is known about the possible effects of pesticide residues on the action of bacteria, notably lactic acid and acetic acid bacteria. Vinclozolin (Ronilan) and iprodione (Rovral) have been reported to depress malolactic fermentation but increase the growth of acetic acid bacteria (San Romáo and Coste Belchior, 1982). Cymoxanil (Curzate) and dichlofluanid (Euparen) also have been reported to inhibit malolactic fermentation (Haag *et al.*, 1988).

Stuck Fermentation

Stuck fermentation refers to the premature termination of fermentation before essentially all fermentable sugars have been metabolized. It has been a problem since time immemorial. Its occurrence in the past usually has been associated with overheating during fermentation. In the absence of adequate cooling, fruit harvested and fermented under hot conditions can readily stick. The resulting wines often are high in residual sugar content and particularly susceptible to microbial spoilage. Instability is increased further if the grapes are low in acidity and/or high in pH.

Improvements in temperature control have largely eliminated overheating as an important cause of stuck fermentation. Ironically, the availability of modern cooling equipment has partially contributed to current problems with stuck wines. Because of the desire to accentuate the fresh, fruity character of white wines, excessively cool fermentation temperatures may be used. A similar desire to enhance freshness occasionally has resulted in undue juice clarification. The resultant loss in sterols, unsaturated fatty acids, and nitrogenous nutrients can increase yeast sensitivity to the combined toxicity of ethanol and saturated mid-chain carboxylic acids, notably octanoic and decanoic acids and their esters. Crushing, pressing, and other preformentative activities scrupulously conducted in the absence of oxygen heighten these effects. Molecular oxygen is required by yeasts for the biosynthesis of sterols and long-chain unsaturated fatty acids essential for cell membrane synthesis and function.

Juice from overmature or botrytized grapes generally has very high sugar contents. The osmotic effect of high sugar concentrations can partially plasmolyze yeast cells, resulting in slow and/or incomplete fermentation. Botrytized juice also contains lower than usual concentrations of available nitrogen as well as vitamins, owing to assimilation by *Botrytis cinerea*. Nutrient depletion adds to the combined inhibitory effects of high sugar content and the toxicity of ethanol and C_8 and C_{10} saturated carboxylic acids. In addition, polysaccharides synthesized by *B. cinerea* may have inhibitory effects on yeast fermentation (Ribéreau-Gayon *et al.*, 1979). Further, the toxicity of acetic acid typically found in botrytized grapes may contribute to poor fermentability. The presence of 10^5 to 10^6 acetic acid bacteria/ml can be lethal to *Sacch. cerevisiae* (Grossman and Becker, 1984). Depending on the style of wine desired, the retention of significant concentrations of residual sugar may or may not be desirable.

For the production of low alcohol wines possessing high residual sugar contents, a stuck fermentation may be purposely induced by chilling and clarification to remove the yeasts.

"Killer" yeasts can generate off-flavors and disrupt fermentation. To counter the effects of killer strains, several workers have incorporated one or both primary killer traits into commercial wine yeast strains (Boone *et al.*, 1990; Sulo *et al.*, 1992). Killer properties have been found in several naturally occurring *Sacch. cerevisiae* strains, as well as species of *Hansenula, Pichia, Torulopsis, Candida,* and *Kluyveromyces.*

Expression of the killer property occurs variously in different yeasts. A cytoplasmic double-stranded RNA virus encodes the property in *Sacch. cerevisiae,* whereas two linear DNA molecules control the property in *Kluyveromyces lactis.* In other genera, chromosomal genes may be involved. The toxic principle is associated with the production and release of a protein or glycoprotein. The toxin attaches to the wall, and subsequently the membrane, of sensitive yeast cells. The attachment creates pores in the membrane, destroying the ability of the cell to control ion flow and resulting in cell death.

Most killer strains of *Sacch. cerevisiae* produce one of two types of killer proteins (K_1 and K_2), though others are suspected. Killer cells are immune to the effects of their own toxin but may be sensitive to those produced by other strains. Eleven different killer (K) factors have been identified, most of which do not affect *Sacch. cerevisiae.* Killer toxins commonly affect only related yeast strains.

The killer proteins produced by *Sacch. cerevisiae* act optimally at a pH of 4 to 5, a range above that normally found in wine. However, both toxins are stable within wine pH values. Therefore, the toxins appear to be at least partially active during fermentation. Cells appear to be most sensitive to the toxin during the exponential phase of cell growth. The importance of killer toxins in winemaking probably will depend on juice pH, the addition of protein-binding substances such as bentonite or yeast hulls, the ability of killer strains to ferment, and the degree to which the yeast population multiplies during fermentation.

Control of stuck or sluggish fermentation in any particular case will depend on knowing the precise cause (Munoz and Ingledew, 1990). Temperature regulation has eliminated overheating as the major cause of stuck fermentations. Use of commercial strains carrying both K_1 and K_2 factors and addition of materials to reduce the active concentration of killer toxins or fungicides are additional solutions to particular problems of stuck or sluggish fermentations. The likelihood of stuck fermentation also may be minimized by reduced preformentative clarification, limited oxidation during crushing and pressing, the addition of ergosterol and/or long-chain unsaturated fatty acids (i.e., oleic, linoleic, or linolenic acids), the addition of ammonium salts, and/or the addition of yeast ghosts or other absorptive materials. The addition of nutrients and absorptive substances appears to have optimal effects when applied midway in or near the end of the exponential phase of yeast growth.

Malolactic Fermentation

It is unlikely that the value of any winemaking process other than malolactic fermentation is associated with such diversity of opinion. The diversity is not surprising since malolactic fermentation to varying degrees can improve or reduce wine quality. In addition, conditions where its occurrence would be beneficial discourage the processs. Conversely, situations where malolactic fermentation is either unnecessary or undesirable tend to promote its development.

The principal effect of malolactic fermentation is a reduction in acidity. In wines of excessive acidity, the reduction is desirable. Thus, winemakers in most cool wine-producing regions view malolactic fermentation positively, especially for red wines. In contrast, wines produced in warm regions may be low in acidity or high in pH. Malolactic fermentation can aggravate a difficult situation, leaving the wine "flat" tasting and microbially unstable. Consequently, the issue of malolactic fermentation elicits markedly diverse responses from winemakers in different parts of the world.

Lactic Acid Bacteria

Lactic acid bacteria are characterized by a unique set of properties. Their name refers to one of these, the production of large amounts of lactic acid. Depending on the genus or species, lactic acid bacteria may ferment sugars solely to lactic acid or to lactic acid, ethanol, and carbon dioxide. The former mechanism is called **homofermentation,** and the latter is termed **heterofermentation.** Bacteria capable of both types of fermentation grow in wine. Homofermentation potentially yields two ATPs per glucose (similar to yeast fermentation), while heterofermentation yields but one.

The most beneficial number of the group is one of the heterofermentative species, *Leuconostoc oenos.* It is probably the most frequently occurring species of lactic acid bacteria in wine and commonly is the only species inducing malolactic fermentation in wines of low pH (≤3.5). Spoilage forms are generally members of the genera *Lactobacillus* and *Pediococcus. Lactobacillus* contains both homo- and heterofermentative species, whereas *Pediococcus* is strictly homofermentative.

Although the bacteria are classified primarily on the basis of sugar metabolism, whether sugar metabolism is important to their growth in wine is unclear. Even dry wines possess sufficient residual sugars (mostly pentoses) to support considerable bacterial growth. However, sugars are not necessarily metabolized in wine. Occasionally, the concentrations of glucose and fructose even increase during malolactic fermentation (Davis *et al.,* 1986). This increase apparently is unrelated to malolactic fermentation, as it also occurs in its absence.

Lactic acid bacteria are further distinguished by their limited biosynthetic abilities. They require a complex series of nutrients, including B vitamins, purine and pyrimidine bases, and several amino acids. Indicative of the limited synthetic capabilities is the inability of lactic acid bacteria to produce heme proteins. As a consequence, they produce neither cytochromes nor the enzyme catalase. Without cytochromes, they cannot respire. Consequently, energy metabolism is strictly fermentative.

Most bacteria incapable of synthesizing heme molecules are strict anaerobes, that is, are unable to grow in the presence of oxygen. Oxygen reacts with certain cytoplasmic components, notably flavoproteins, to produce the toxic oxygen radicals **superoxide** and **peroxide.** In aerobic organisms, superoxide dismutase and catalase rapidly inactivate these toxic radicals. Lactic acid bacteria produce neither enzyme. They escape the toxicity of the oxygen radicals by accumulating large quantities of Mn^{2+} ions, or by producing a peroxidase. Manganese detoxifies superoxide by converting it to oxygen. The rapid action of manganese also prevents the synthesis of hydrogen peroxide from superoxide. Peroxidase reduces organic compounds in the presence of peroxide, oxidizing the peroxide to water. Lactic acid bacteria are the only bacterial group that are both strictly fermentative and aerotolerant.

Although fermentative metabolism is inefficient in terms of energy production, the generation of large amounts of acidic wastes quickly lowers the pH of most substrates. The resulting low pH inhibits the growth of most other bacteria. Lactic acid bacteria are one of the few bacterial groups capable of growing below a pH of 5. This property has preadapted lactic acid bacteria to grow in acidic environments such as wine.

Although lactic acid bacteria grow under acidic conditions, growth is poor in must and wine. For example, species of *Lactobacillus* and *Pediococcus* commonly cease growing below pH 3.5. Even *Leuconostoc oenos,* the primary malolactic bacterium, is inhibited below pH 3.0 to 2.9. *Leuconostic oenos* grows optimally within a pH range of 4.5 to 5.5. Thus, the major benefit of malolactic fermentation for the bacteria surprisingly may be acid reduction. By metabolizing a dicarboxylic acid (malic acid) to a monocarboxylic acid (lactic acid), acidity is reduced and pH increased. The increased pH provides conditions more favorable to bacterial growth.

L-malic acid

L-lactic acid

The enzyme involved in malolactic fermentation by lactic acid bacteria, the **malolactic enzyme,** is unique. Unlike other conversions of malic to lactic acid, the enzymatic reaction directly decarboxylates L-malic acid to L-lactic acid. Because the reaction is not associated directly with energy capture, the conversion can only dubiously be called a fermentation. However, some ATP is generated through the joint export of lactic acid and hydrogen ions (protons) from the cell (Cox and Henick-

Kling, 1990). The pH differential produced across the cell membrane is sufficient to drive the oxidative phosphorylation of ADP to ATP. Release of a small amount of reducing energy also may occur via the oxidation of malic acid to pyruvic acid, some of which is subsequently reduced to lactic acid.

The primary energy source for the growth of lactic acid bacteria in wine is still unclear. The situation is complicated by the marked influence pH has on the abilities of the bacteria to ferment sugars and by the considerable variability between strains. *Leuconostoc oenos* appears to show little ability to ferment sugars, at least below pH 3.5 (Davis *et al.*, 1986). However, a more recent report suggests that fructose and glucose may be the energy source for *Leuco. oenos* (Arnick *et al.*, 1992). Species of *Lactobacillus* and *Pediococcus*,, which generally grow only above pH 3.5, appear to ferment hexoses and pentoses in wine. Fumaric and citric acids are metabolized by several strains of *Leuco. oenos*, but not by most lactobacilli and pediococci. The metabolism of citric acid is apparently associated with the accumulation of acetoin and diacetyl (Shimazu *et al.*, 1985). Amino acids, notably arginine, also may act as energy sources (Feuillat *et al.*, 1985).

All lactic acid bacteria growing in wine assimilate acetaldehyde and other carbonyl compounds. The metabolism of carbonyl compounds may retard malolactic fermentation by liberating sulfur dioxide bound to carbonyl compounds.

As noted with anaerobic yeast metabolism, fermentation can result in the generation of an excess of reduced NAD^+ (NADH). To maintain an acceptable redox balance, the bacteria must regenerate NAD^+. How lactic acid bacteria accomplish this in wine is unclear. Some species reduce fructose to mannitol, presumably for this purpose. This may explain the common occurrence of mannitol in wine associated with malolactic fermentation. Some strains also regenerate NAD^+ with flavoproteins and oxygen. This reaction might explain the reported improvement in malolactic fermentation in the presence of small amounts of oxygen.

Besides the important physiological differences, anatomical features distinguish the various genera of lactic acid bacteria (Fig. 7.30). *Leuconostoc* usually consists of spherical to lens-shaped cells, commonly occurring in pairs or chains but occasionally singly. *Pediococcus* species usually occur as packets of four spherical cells. *Lactobacillus* produces long, slender, occasionally bent, rod-shaped cells commonly occurring in chains. Some of the lactic acid bacteria that may occur in wine are listed in Table 7.4.

Effects of Malolactic Fermentation

Malolactic fermentation has three distinct, but interrelated, effects on wine quality. It reduces acidity, influences microbial stability, and may affect the sensory characteristics of the wine.

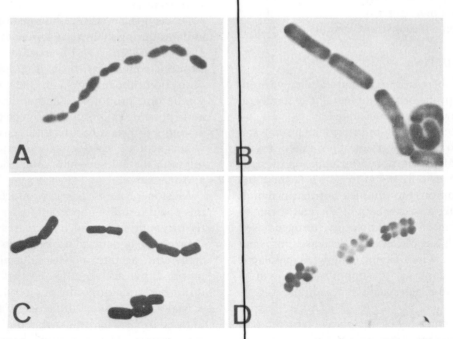

Figure 7.30 Micrographs of important members of the Lactobacillaceae found in wine. (A) *Leuconostoc oenos* (×6000). (B) *Lactobacillus casei* (×8500). (C) *Lactobacillus brevis* (×5500) (D) *Pediococcus cerevisiae* (×5000). (From Radler, 1972, reproduced by permission.) Cell shape and grouping may depend on the medium in which the bacteria grow.

Table 7.4　Lactic Acid Bacteria Occurring in Wine

Genus	Species
Leuconostoc	Leuco. oenos
Pediococcus	P. pentosaceus, P. damnosus, (P. cerevisiae), P. parvulus
Lactobacillus	L. plantarum, L. brevis, L. cellobiosis, L. buchneri, L. casei, L. hilgardii, L. trichodes, L. mesenteroides

ACIDITY

Deacidification, and the corresponding rise in pH, are the most consistent effects of malolactic fermentation. Reduction in acidity increases the smoothness and drinkability of red wines, but excess reduction generates a "flat" taste. The desirability of deacidification depends primarily on the pH and acidity of the grapes. In general, the higher the acidity and lower the pH, the greater is the benefit; conversely, the lower the acidity and higher the pH, the greater is the likelihood of undesirable effects. The higher the proportion of tartaric acid, the less likely malolactic fermentation will significantly affect the acidity and pH of the wine.

As the pH of a wine changes, so too does the relative proportion of the various colored and uncolored forms of anthocyanin pigments. The metabolism of carbonyl compounds (notably acetaldehyde) by lactic acid bacteria, and the accompanying release of SO_2, also may result in some pigment bleaching. In general, color loss associated with malolactic fermentation is significant only in pale-colored wines or those initially of high pH.

MICROBIAL STABILITY

For many years, increased microbial stability was considered one of the prime benefits of malolactic fermentation. This view is now being questioned.

Winemakers assumed that improved microbial stability resulted from the metabolism of residual nutrients left after alcoholic fermentation. Consumption of malic and citric acids leaves only the more stable tartaric and lactic acids. In addition, the complex nutrient demands of lactic acid bacteria were thought to reduce significantly the concentrations of amino acids, nitrogen bases, and vitamins. While such levels may decrease, this is not a consistent finding. Wines having completed malolactic fermentation may continue to support the growth of *Leuconostoc oenos* or lactobacilli and pediococci (Costello *et al.*, 1983).

Contrary to common belief, malolactic fermentation occasionally may *decrease* microbial stability. This can occur when the wine initially has marginal to high pH. The resultant rise in pH can favor the subsequent growth

of spoilage forms of lactic acid bacteria. Spoilage organisms generally do not grow in wines at a pH below 3.5, but their ability to grow increases rapidly as the pH rises from 3.5 to 4.0 and above.

Thus, the stabilizing action of malolactic fermentation may come more from preservation practices employed after its completion, rather than from the consumption of residual nutrients by the bacteria. The early onset and completion of malolactic fermentation permits the prompt addition of sulfur dioxide, storage at cool temperatures, and clarification. Early completion of malolactic fermentation also avoids its possible occurrence after bottling.

FLAVOR MODIFICATION

The greatest contention concerning the relative merits of malolactic fermentation revolves around flavor modification. Several studies have shown that perceptible wine quality may be little influenced by malolactic fermentation (see Davis *et al.*, 1985). When differences are detected, it may reflect only the influences of reduced acidity (Càstino *et al.*, 1975). However, similar wines occasionally can be distinguished organoleptically, based on the strains of bacteria that induced the malolactic fermentation (McDaniel *et al.*, 1987). Because of the considerable variability between strains and species of lactic acid bacteria, and the marked influence of pH and temperature on their metabolism, it is not surprising that the reported sensory influences of malolactic fermentation are so inconsistent. Table 7.5 lists some of the substrates metabolized and by-products liberated by lactic acid bacteria.

Diacetyl (biacetyl, 2,3-butanedione) commonly accumulates during malolactic fermentation (Rankine *et al.*, 1969). Between 1 and 4 mg/liter, diacetyl often adds desirable complexity to the fragrance; however, at concentrations above 5 to 7 mg/liter, its buttery character can become pronounced and undesirable. Other flavorants potentially produced by lactic acid bacteria in amounts sufficient to affect the sensory character of the wine include acetaldehyde, acetic acid, acetoin, 2-butanol, diethyl succinate, ethyl acetate, ethyl lactate, and 1-hexanol.

Most lactic acid bacteria produce esterases. Although this could result in important losses in the fruity character of young wines, such decreases generally are small. Conversely, nonenzymatic synthesis of some esters, especially ethyl acetate, increases during malolactic fermentation. Ethyl acetate also may accumulate because of direct bacterial biosynthesis.

Malolactic fermentation below pH 3.5 is generally induced by *Leuconostoc oenos* and is less likely to generate off-odors. The undesirable buttery, cheesy, or milky odors are usually connected with malolactic fermentation induced by pediococci or lactobacilli above pH 3.5.

Table 7.5 Substances and Fermentation Products of Lactic Acid Bacteria[a]

Substrate	Products
Acids	
L-Malate	L-Lactate, CO_2, succinate, acetate
Citrate; pyruvate	Lactate, acetate, CO_2, acetoin, diacetyl
Gluconate	Lactate, acetate, CO_2
2-Oxoglutarate	4-Hydroxybutyrate, CO_2, succinate
Tartrate	Lactate, acetate, CO_2, succinate
Sorbate	2,4-Hexadien-1-ol (sorbic alcohol)
Chlorogenate	Ethylcatechol, dihydroshikimate
Sugars	
Glucose	Lactate, ethanol, acetate, CO_2
Fructose	Lactate, ethanol, acetate, CO_2, mannitol
Arabinose, xylose, or ribose	Lactate, acetate
Polyols	
Mannitol	(Probably as from glucose)
2,3-Butanediol	2-Butanol
Glycerol	1,3-Propanediol
Amino acids	
Arginine	Ornithine, CO_2, NH_4
Histidine	Histamine, CO_2
Phenylalanine	2-Phenylethylamine, CO_2
Tyrosine	Tyramine, CO_2
Ornithine	Putrescine, CO_2
Lysine	Cadaverine, CO_2
Serine	Ethanolamine, CO_2
Glutamine	Aminobutyrate, CO_2
Unknown substrates (probably sugars)	Propanol, isopropanol, isobutanol, 2-methyl-1-butanol, 3-methyl-1-butanol, ethyl acetate, acetaldehyde, *n*-hexanol, *n*-octanol, glycerol, 2,3-butanediol, erythritol, arabitol, dextran, diacetyl

[a] After Radler (1986), reproduced by permission.

AMINE PRODUCTION

Lactic acid bacteria, notably pediococci, produce amines through the decarboxylation of amino acids. Amine synthesis appears to be important only in wines of high pH. Although some amines can induce blood vessel constriction, headaches, and other associated effects, the amine contents in wine appear to be insufficient to induce these physiological effects in humans (Radler and Fäth, 1991).

Origin and Growth of Lactic Acid Bacteria

The ancestral habitat of *Leuconostoc oenos* is apparently unknown as it rarely occurs on grape and leaf surfaces, and then only in low numbers. No other species of *Leuconostoc* grows in wine. Species of *Pediococcus* and *Lactobacillus* generally occur more frequently on grapes than does *Leuconostoc*. Their numbers may occur in the range of 10^3 to 10^4 cells/ml shortly after crushing (Costello *et al.*, 1983). The population size de-

pends largely on the maturity and health of the fruit, higher numbers frequently occurring on mature and/or mold-infected fruit. Nevertheless, the relatively low numbers indicate that grapes are unlikely to be the primary habitat of the species found.

Although malolactic fermentation may be induced by bacteria growing indigenously on grape surfaces, strains also may originate from winery equipment. Stemmers, crushers, presses, fermentors, etc., all may harbor populations of lactic acid bacteria. The relative importance of grape versus winery sources has yet to be established. Nevertheless, an increasing number of winemakers inoculate the wines with carefully selected strains of lactic acid bacteria when malolactic fermentation is desired.

Unlike yeast growth during alcoholic fermentation, no consistent bacterial growth sequence develops in the must or wine during malolactic fermentation. A pattern occasionally found in spontaneous malolactic fermentations is shown in Fig. 7.31. Significant modifications often occur owing to factors such as pH, total acidity, malic acid content, temperature, and the duration of contact with skin and lees.

In most spontaneous fermentations, cells lyse rapidly as alcoholic fermentation begins, dropping the bacterial population from about 1×10^3 to about 1 cell/ml. Most species of lactic acid bacteria initially found die out during alcoholic fermentation. Wines with pH values higher than 3.5 may show temporary growth of some species, such as *Lactobacillus plantarum*. Occasionally when sulfiting is low and the pH above 3.5, *Leuco. oenos* may induce a malolactic fermentation coincident with alcoholic fermentation.

The usual initial decline in the population of lactic acid

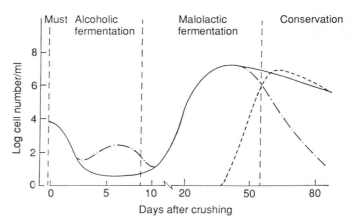

Figure 7.31 Diagrammatic representation of the growth of indigenous lactic acid bacteria during the vinification of a red wine. Population development of *Leuconostoc oenos* below (—) and above (—·—) pH 3.5, before and during malolactic fermentation, and as affected by the growth of other species (- - -) during the later stages of malolactic fermentation or wine maturation. (From Wibowo *et al.*, 1985, reproduced by permission.)

bacteria has been variously ascribed to sulfur dioxide toxicity, acidity, the synthesis of ethanol and toxic carboxylic acids, or the increasingly nutrient-poor status of fermenting must. All these factors may be involved to some degree. At the end of alcoholic fermentation, a lag period generally ensues before the bacterial populations begin to rise. This phase may be of short duration, or it may last several months. Once growth initiates, the bacterial population may rise to 10^6 to 10^8 cells/ml. In most wines of low pH, only *Leuconostoc oenos* grows. However, there may be variations in the proportion of different strains of *Leuco. oenos* throughout malolactic fermentation. At high pH values, species of both *Lactobacillus* and, especially, *Pediococcus* predominate. Depending on the strain or species involved, decarboxylation of malic acid may occur simultaneously with bacterial multiplication or only after cell growth has ceased.

At the end of exponential growth, the cell population enters a prolonged decline phase. The slope of the decline can be dramatically changed by cellar practices. For example, storage of the wine above 20° to 25°C or the addition of sulfur dioxide results in the rapid death of *Leuco. oenos*. If the pH is above 3.5, and other conditions favorable, previously inactive strains of *Lactobacillus* or *Pediococcus* may begin to multiply. Their growth often produces a corresponding decline in the population of *Leuco. oenos*. The nature of this apparently competitive antibiosis is unknown.

Factors Affecting Malolactic Fermentation

PHYSICOCHEMICAL FACTORS

pH The initial pH of juice or wine strongly influences not only if and when malolactic fermentation will occur, but how and what species will conduct the process (Bousbouras and Kunkee, 1971). Low pH not only slows the rate, but can delay its initiation. Below pH 3.5 *Leuconostoc oenos* is the predominant species inducing malolactic fermentation, while above pH 3.5 *Pediococcus* and *Lactobacillus* spp. become increasingly prevalent (Costello *et al.*, 1983). Some of the inhibitory effects observed at low pH values are probably indirect, acting by increasing the ethanol sensitivity of the bacterial membrane and the proportion of the free molecular component of sulfur dioxide.

In addition, pH modifies significantly the metabolic activity of lactic acid bacteria. For example, the bacteria ferment sugars much more effectively at higher pH values. Similarly, the synthesis of acetic acid increases, while diacetyl production decreases as the pH increases (see Wibowo *et al.*, 1985). Metabolism of malic and tartaric acid also is influenced by pH with decarboxylation of malic acid being favored at low pH while the

potential for tartaric acid degradation increases above pH 3.5. The latter results primarily from the growth of *Lactobacillus brevis* and *L. plantarum* above pH 3.5 (Radler and Yannissis, 1972).

Temperature The effect of temperature on malolactic fermentation has long been known, as the process often took place only in the spring when cellars began to warm. To speed development, cellars may be heated in the fall to maintain a wine temperature above 20°C. Although temperature directly affects bacterial growth rates, its most significant effect is on the rate of malic acid decarboxylation. While the growth curve and maximum population generated are roughly similar within a range of 20° to 35°C, maximal decarboxylation of malic acid happens between 20° and 25°C (Ribéreau-Gayon *et al.*, 1975). The decarboxylation of malic to lactic acid is slow at 15° and 30°C. At temperatures below 10°C, essentially no decarboxylation occurs. Most strains of *Leuconostoc oenos* grow very slowly or not at all below 15°C. However, cool temperatures maintain cell viability. Thus, wines cooled after malolactic fermentation commonly retain a high population of viable bacteria for months. Temperatures around 25°C favor rapid decline in the *Leuco. oenos* population (Lafon-Lafourcade *et al.*, 1983) but may promote the growth of pediococci and lactobacilli.

Cellar Practices Many cellar practices can affect when, and if, malolactic fermentation occurs. Maceration on the skins commonly increases the frequency and speed of malolactic fermentation (Guilloux-Benatier *et al.*, 1989). The precise factors are unknown but may involve the action of phenols as electron acceptors in the oxidation of sugars in fermentation (Whiting, 1975). The usual long maceration of red wines helps to explain why malolactic fermentation develops more commonly in red wines than in white wines. The higher pH of most red wines is undoubtedly involved as well.

Clarification can directly reduce the population of lactic acid bacteria by inducing coprecipitation with yeasts or grape residues. Racking, fining, centrifugation, and other similar practices also remove nutrients or limit their uptake into wine as a consequence of yeast autolysis.

CHEMICAL FACTORS

Carbohydrates and Polyols The chemical composition of juice or wine has a profound influence on the outcome of malolactic fermentation. Carbohydrates and polyols constitute the most potentially significant group of fermentable compounds (Davis *et al.*, 1986). Most dry wines contain between 1 and 3 g/liter of residual hexoses and pentoses. There are also variable amounts of di- and trisaccharides, sugar alcohols, glycosides, glycerol, and other polyols. There is considerable heterogeneity in the ability of strains and species of lactic acid bacteria to use these nutrients. Ethanol and pH also influence the ability to ferment carbohydrates. Consequently, few generalizations about carbohydrate use appear possible. The major exception may be the poor utilization of most polyols. For example, few lactobacilli metabolize glycerol, the most prevalent polyol in wine. Although uncommon, glycerol metabolism can lead to the production of the bitter-tasting compound acrolein (Meyrath and Lüthi, 1969).

Occasionally, significant increases in the concentration of glucose and fructose have been noted following malolactic fermentation. The increase appears to be coincidental and not causally related. The sugars may arise from the breakdown of various complex sugars, such as trehalose, or from the hydrolysis of phenolic glycosides.

Organic Acids Although malic acid is the most important acid metabolized by lactic acid bacteria, other organic acids are also metabolized. For example, citric acid metabolism by *Leuco. oenos* has been correlated with the synthesis of acetic acid, diacetyl, and acetoin (Shimazu *et al.*, 1985). Few other species appear to metabolize citric acid. Gluconic acid, common in botrytized wines, is metabolized by most lactic acid bacteria, except the pediococci. Some lactic acid bacteria, notably lactobacilli, have been reported to degrade tartaric acid (Radler and Yannissis, 1972) and have been associated with a wine "disease" called *tourne* (see Chapter 8).

Formerly, the addition of fumaric acid was proposed as a practical inhibitor of malolactic fermentation. However, because the suppression caused by fumaric acid decreases dramatically at pH values above 3.5 (Pilone *et al.*, 1974) it becomes progressively ineffective under conditions where protection is increasingly required.

Because lactic acid bacteria can metabolize sorbic acid, and thus generate the production of a strong geranium-like off-odor, sorbic acid is no longer used as a yeast inhibitor in sweet wines.

The importance of organic acid metabolism to the energy budget of lactic acid bacteria is still unclear. However, the association of organic acid metabolism with proton transport across the cell membrane can generate a chemiosmotic potential that can activate the phosphorylation of ADP to ATP. A small proportion of malic acid may be metabolized by pathways other than decarboxylation to lactic acid. These could generate a small amount of reducing energy (NADH).

The ability of lactic acid bacteria to metabolize and/or tolerate fatty acids is largely unknown. However, some related carboxylic acid by-products of yeast metabolism are toxic to lactic acid bacteria, notably decanoic and octanoic acids. The addition of yeast hulls helps to limit the toxicity of carboxylic acids by removing the acids from the wine by absorption.

Nitrogen-Containing Compounds Although lactic acid bacteria have complex nitrogen growth requirements, few generalizations about their influence on the nitrogen composition of wines can be made. For example, the concentration of individual amino acids in wine may increase, decrease, or remain stable during malolactic fermentation. Studies of this aspect are made complex by the potential release of proteases that could liberate amino acids from soluble proteins. In addition, amino acids are released via yeast autolysis during maturation of the wine. Only the concentration of arginine appears to change consistently, through its bioconversion to ornithine.

Most of the reduction in amino acid content is probably associated with uptake and incorporation into proteins. Further losses may result from decarboxylation to amines, as in the production of histamine from histidine. It is not established that amino acids are used as energy sources, but this appears possible (Feuillat *et al.*, 1985).

Ethanol Ethanol inhibits the growth of lactic acid bacteria but is an even more active inhibitor of malolactic fermentation (Fig. 7.32). Of lactic acid bacteria, species of *Lactobacillus* are the most ethanol tolerant. For example, *L. trichodes* can grow in wines at up to 20% ethanol (Vaughn, 1955). Nevertheless, a few strains of *Leuconostoc oenos* grow in culture media at up to 15% alcohol. Alcohol tolerance appears to decrease both with higher temperatures and lower pH values. However, at low concentrations (1.5%), ethanol appears to favor bacterial growth (King and Beelman, 1986).

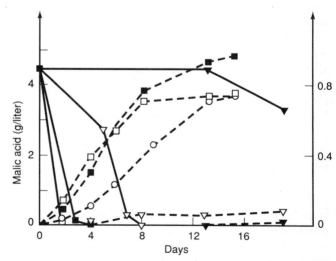

Figure 7.32 Influence of ethanol on the growth (− −, OD_{620}nM) and malolactic activity (—, g malic acid/liter) of *Leuconostoc oenos*. Ethanol at □, 0%; ■, 5%; ○, 8%; ▽, 11%; ▼, 13%. (After Guilloux-Benatier, 1987, reproduced by permission.)

The toxic mechanism of ethanol is unknown, but it probably involves changes in the semifluid nature of the cell membrane. Reduction in the concentration of neutral lipids and increases in the proportion of glycolipids have been correlated with high alcohol concentrations (Desens and Lonvaud-Funel, 1988). The membrane dysfunction so produced could explain growth disruption, reduced viability, and poor malolactic fermentation at high alcohol contents. Up to 80% reduction in the rate of malic acid decarboxylation has been reported with an increase in alcohol content from 11 to 13% (Lafon-Lafourcade, 1975).

Other Organic Compounds During malolactic fermentation, the bacteria assimilate a wide range of compounds. Degradation of organic compounds may affect the process of malolactic fermentation itself. For example, the metabolism of carbonyl compounds bound to sulfur dioxide can release sufficient SO_2 to slow or terminate malolactic fermentation. In most cases, the role of assimilated organic compounds in bacterial metabolism is unknown.

Many strains of lactic acid bacteria produce esterases. Nevertheless, the concentrations of most esters, such as 2-phenethyl acetate and ethyl hexanoate, change little during malolactic fermentation. Reductions that do occur appear insufficient to cause noticeable losses in wine fruitiness. Other esters, such as ethyl acetate, ethyl lactate, and diethyl succinate, may increase. While the latter two are unlikely to have sensory significance, owing to low volatility, an increased ethyl acetate content could donate a vinegary aspect.

Various phenolic acids, such as ferulic, quinic, and shikimic acids, as well as their esters, are metabolized by some lactic acid bacteria. Species of *Lactobacillus* are notable in this regard. Their metabolism can generate volatile phenolics, such as ethyl guaiacol and ethyl phenol. These generally have spicy, medicinal, creosotelike odors (Whiting, 1975). As these fragrances do not generally characterize malolactic fermentations, the contents of volatile phenolics are probably too low to be perceptible.

At concentrations usually found in wine, phenols do not seriously retard malolactic fermentation. Nevertheless, leaving the stems to ferment with the grapes can noticeably slow its initiation (Feuillat *et al.*, 1985).

Gases Sulfur dioxide has the greatest influence on lactic acid bacteria. The effect is complex because of the differing concentrations and toxicities of the many forms of sulfur dioxide in wine. Free forms of sulfur dioxide are more toxic than bound forms (Fig. 7.33), and of the free forms, molecular SO_2 is the most antimicrobial. By affecting the relative proportions of these forms, pH has a marked influence on the toxicity of sulfur dioxide. Tem-

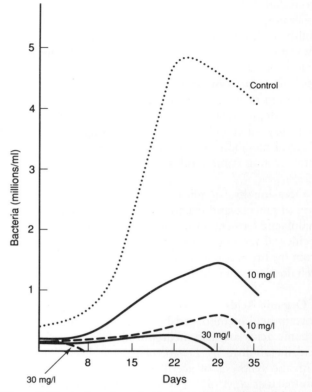

Figure 7.33 Action of two levels (10 and 30 mg/liter) of free SO_2 (– – –) and acetaldehyde bisulfite (▬) on the growth of *Leuconostoc gracile*. Control, (· · · ·). (From Lafon-Lafourcade, 1975, reproduced by permission).

perature also dramatically influences the sensitivity of the bacteria to sulfur dioxide (Lafon-Lafourcade, 1981).

Different species and strains vary considerably in sensitivity to sulfur dioxide. In general, strains of *Leuconostoc oenos* appears to be the most sensitive. Because of the greater tolerance of pediococci and lactobacilli to SO_2, they may be unintentionally selected in wines of high pH at customarily used amounts of sulfur dioxide.

Although lactic acid bacteria are strictly fermentative, small amounts of oxygen may favor malolactic fermentation. As the bacteria neither respire nor require steroids or unsaturated fatty acids for growth, the oxygen may act by maintaining a favorable redox balance through reaction with flavoproteins.

Somewhat controversial is the value of carbon dioxide. One potential mechanism of involvement could entail the production of oxaloacetate via the carboxylation of pyruvate. This could both help maintain a desirable redox balance and favor the biosynthesis of amino acids.

BIOLOGICAL FACTORS

Biological interactions in malolactic fermentation are as intricate as the complex interplay of physical and chemical factors noted above. Although mutually beneficial effects may occur, most interactions suppress growth.

Yeast Interactions Occasionally, alcoholic and malolactic fermentation occur simultaneously. Infrequently, lactic acid bacteria may exert an inhibitory effect on yeast growth, causing stuck fermentations. More commonly, yeast metabolism inhibits bacterial growth (Edwards *et al.*, 1990). Consequently, malolactic fermentation generally develops days, weeks, or months after alcoholic fermentation is complete. Contamination of wine with spoilage yeasts, such as *Pichia, Candida,* and *Saccharomycodes,* also may inhibit the growth of lactic acid bacteria.

Various explanations have been offered for the inhibitory action associated with yeast growth on lactic acid bacteria. The addition of sulfur dioxide, or SO_2 production by yeasts, may be the major factor. Depletion of arginine and other amino acids during the early stages of alcoholic fermentation is another possibility. Certain yeast strains also induce coprecipitation with the bacteria, delaying malolactic fermentation. The increasing ethanol content and the accumulation of toxic carboxylic acids such as octanoic and decanoic acid in the fermenting juice are undoubtedly involved (Lafon-Lafourcade *et al.*, 1984). Finally, proteins with antibacterial activities (e.g., lysozyme) have been isolated from some strains of *Sacch. cerevisiae* (Dick *et al.*, 1992).

After alcoholic fermentation, yeast cells die and begin to undergo autolysis. The associated release of nutrients may explain the initiation of bacterial growth. Thus, wines may be left in contact with the lees for several weeks or months to encourage the development of malolactic fermentation. While nutrient release is undoubtedly involved, maintenance of a high CO_2 concentration by the lees may be more important in some instances (Mayer, 1974).

Surprisingly, malolactic fermentation may be stimulated in botrytized juice (San Romáo *et al.*, 1985). This occurs in spite of the well-known reduction in available nitrogen content of grapes infected with *Botrytis cinerea.* Promotion of the growth of lactic acid bacteria may result from the metabolism of the higher concentrations of acetic acid found in botrytized grapes and/or the removal of toxic carboxylic acids by adsorption to *Botrytis*-synthesized polysaccharides. The presence of glycerol can favor the development of the wine disease called mannitic fermentation.

Bacterial Interactions The growth of acetic acid bacteria often favors malolactic fermentation during alcoholic fermentation. This may result indirectly from the suppression of yeast growth, thus resulting in higher nutrient levels and lower concentrations of alcohol and toxic carboxylic acids.

Some intraspecies competition between *Leuconostoc oenos* strains is suggested by changes in the relative numbers of different strains throughout malolactic fermentation. More importantly, there may be antagonism between different species of bacteria. Antagonism is rare below pH 3.5, where typically only *Leuco. oenos* grows, but above pH 3.5 pediococci and lactobacilli progressively have a selective advantage over *Leuco. oenos.* Antagonism appears most noticeably when the decline of *Leuco. oenos* is mirrored by an equivalent rise in the numbers of lactobacilli and pediococci (Fig. 7.31). The mechanism of this apparent antagonism is unknown.

Viral Interactions Bacterial viruses (**phages**) can severely disrupt malolactic fermentation by attacking *Leuconostoc oenos* cells. *Leuconostoc oenos* also may possess viral infections in the inactive **prophage** state (Cavin *et al.*, 1991). Because the prophage state tends to be unstable at high pH values, the lytic phase may be reestablished as the pH rises during malolactic fermentation and kill the bacterial host cell. Malolactic fermentation may continue, but less predictably, under the action of *Lactobacillus* or *Pediococcus* species.

Leuconostoc oenos is most sensitive to phage infection during exponential growth. The virus attaches to the bacterium and injects its DNA into the cell. In the cytoplasm, the viral DNA begins to replicate and establish control over bacterial functions. After multiple copies are made of both the DNA and the viral coat, these components self-assemble into virus particles. At this point, the cell bursts open (lyses), releasing new virus particles. These subsequently may initiate further cycles of infection and lysis.

The main defense against viral disruption of malolactic fermentation involves massive inoculation with *Leuconostoc.* Large populations reduce the number of cell divisions and, thereby, the sensitivity of the bacteria to infection. Use of a mixed culture, containing several resistant strains, also is recommended. Mixed cultures minimize the likelihood that all strains will be equally sensitive to the races of phage present. Infection of one strain might delay, but not inhibit, malolactic fermentation. Besides delaying or inhibiting malolactic fermentation, phage infection can leave the wine open to the undesirable development of pediococci and lactobacilli. Specificity among phage strains means that those attacking *Leuco. oenos* are unlikely to attack pediococci or lactobacilli and vice versa.

Control

The occurrence of spontaneous malolactic fermentation has been notoriously difficult to control. Malolactic fermentation often takes months to begin, even under conditions chosen to stimulate development. It has also happened when not desired. In attempts to obtain more control over the process, several procedures are employed, either to assure or prevent the development of malolactic fermentation.

INOCULATION

To assure the development of malolactic fermentation, winemakers commonly inoculate the wine with one or more strains of *Leuconostoc oenos*. By maintaining conditions favorable to bacterial activity (warm temperatures, minimal or no sulfiting, and delayed racking), winemakers usually are able to promote rapid completion.

Inoculation often involves cell populations reactivated from lyophilized or frozen concentrates. Reactivation is required before addition to fermenting must or wine, as direct addition often results in a rapid loss in viability. In addition, proper reactivation is critical to maintaining the ability of the cells to induce malolactic fermentation (Hayman and Monk, 1982). Reactivation commonly takes place in nonsulfited, diluted grape juice (1:1 with water), adjusted to a pH of 3.6 or above. Yeast extract may be added at a concentration of 0.05% w/v. Reactivation generally requires at least 24 hr. Inoculation customarily aims at achieving a population of about 10^6 to 10^7 cells/ml. This usually assures the quick onset of malolactic fermentation and the dominance of the inoculated strain over indigenous bacteria.

When simultaneous alcoholic and malolactic fermentation is desired, the reactivation medium is inoculated with both yeasts and bacteria. Occasionally, however, this is associated with the production of high concentrations of acetic acid. Thus, Prahl *et al.* (1988) suggest the use of *Lactobacillus plantarum* because it is insensitive to low pH, rapidly decarboxylates malic acid, does not generate acetic acid, and rapidly dies out as the alcohol content rises.

Different researchers hold various views on when inoculation should occur. The view in Bordeaux has been to recommend inoculation after the completion of alcoholic fermentation, to avoid the risk of off-odor production (Lafon-Lafourcade, 1980). Early malolactic fermentation also occasionally results in stuck fermentation. In contrast, delayed inoculation avoids much of the toxicity of carboxylic acids, as their concentration declines after alcoholic fermentation. Nevertheless, exposure to the highest levels of ethanol occur with late inoculation.

American experience has not confirmed French findings concerning the development of high concentrations of acetic acid, yeast antagonism, or stuck fermentation associated with the early growth of lactic acid bacteria (Beelman and Kunkee, 1985). Simultaneous inoculation with both yeast and bacteria has commonly resulted in the concurrent development of alcoholic and malolactic fermentations. As a result, the wine can be racked off the lees, cooled, and sulfited immediately after alcoholic fer-

mentation. The bacteria suffer neither from a shortage of nutrients nor from the toxicity of alcohol.

An alternate procedure involves inoculation midway through alcoholic fermentation. Theoretically, this should avoid the toxicity of high concentrations of ethanol and sulfur dioxide. Sulfur dioxide added during the grape crush is largely inactivated by binding with carbonyl and other compounds in the juice. However, the most intense level of yeast-induced antagonism by metabolites such as decanoic acid may be encountered with this method.

Since the 1980s, there has been growing interest in the application of modern fermentation technologies to malolactic fermentation. This has involved the use of both immobilized bacterial cells and enzymes. Such techniques offer the possibility of faster decarboxylation of malic acid, better control over the degree of the conversion (which is especially valuable for wines of pH 3.5 and above), and a reduction in the production of undesirable flavor compounds.

Immobilization involves coating a dense population of active cells with a gel, such as alginate, carrageenan, or polyacrylamide (see McCord and Ryu, 1985). Extrusion of the gel/cell mixture forms small beads, about 2 to 3 mm in diameter. The gel helps prevent cell division and infiltration into the wine. In many industrial uses, division of the entrapped cells is unnecessary. This is equally valid with malolactic fermentation, where stationary phase cells actively decarboxylate malic acid. A support system (reactor) generally holds the beads in a rigid framework through which the wine is slowly pumped. Rapid conversion of malic acid has been achieved by this means (Crapisi *et al.*, 1987). One of the benefits of the system is the reduced bacterial sensitivity to pH, ethanol, and sulfur dioxide. Improvement in the useful life of the reactor may bring its application into common practice.

Theoretically, immobilizing the malolactic enzyme into an inert permeable medium would be even simpler, as it avoids the need to maintain living cells. Enzyme immobilization also would avoid any flavor modification as only malic acid decarboxylation would be induced. However, a considerable obstacle to successful application is the instability of the crucial enzyme cofactor NAD^+ at low pH values.

INHIBITION

Inhibition of malolactic fermentation involves the reverse of all the factors that might favor its occurrence. Thus, storage of the wine at or below 10°C, maintenance of a total sulfur dioxide content above 50 mg/liter, racking early and frequently, early wine clarification, acidification of high pH musts or wines, and minimal maceration favor inhibition of malolactic fermentation. Fumaric acid may be added, but only after alcoholic fermentation, as yeast cells can readily metabolize fumaric acid. Owing to its poor solubility in wine, about 1.5 to 2 g/liter fumaric acid is often required. An alternate antimicrobial agent, nisin, appears to show considerable promise in preventing malolactic fermentation (Radler, 1990).

Pasteurization or sterile filtration associated with sterile bottling should assure the absence of malolactic fermentation after bottling. In-bottle malolactic fermentation may produce clouding, *pétillance* (slight sparkle), and the generation of off-odors.

Appendix 7.1

Partial Synonymy of Several Important Wine Yeasts[a]

Brettanomyces intermedius (Krumbholz & Tauschanoff) van der Walt & van Kerken (1971)
Perfect state: *Dekkera intermedia* van der Walt (1964)
Synonyms: *Mycotorula intermedia* Krumbholz & Tauschanoff (1933),
Brettanomyces vini Peynaud & Domercq (1956)

Candida stellata (Kroemer & Krumbholz) Meyer & Yarrow (1978)
Synonyms: *Brettanomyces italicus* Verona & Florenzano (1947),
Torulopsis bacillaris (Kroemer & Krumbholz) Lodder (1932),
Torulopsis stellata (Kroemer & Krumbholz) Lodder (1932),
Saccharomyces bacillaris Kroemer & Krumbholz (1931)

Candida vini (Desmazières *ex* Lodder) van Uden & Buckley (1970)
Synonyms: *Candida mycoderma* (Reess) Lodder & Kreger-van Rij (1952),
Mycoderma cerevisiae Desmazières (1823) *ex* Leberle (1909),
Mycoderma vini Desmazières (1823) *ex* Lodder (1934),
Saccharomyces mycoderma Reess (1870)

Dekkera intermedia van der Walt (1964)
Imperfect state: *Brettanomyces intermedius* (Krumbholz & Tauschanoff) van der Walt & van Kerken (1971)

Hanseniaspora uvarum (Niehaus) Shehata, Mrak, & Phaff comb. nov.
Imperfect state: *Kloeckera apiculata* (Reess emend. Klöcker) Janke (1928)
Synonyms: *Kloeckera lodderi* van Uden & Assis-Lopes (1953)

Hansenula anomala (Hansen) H. & P. Sydow (1919)
Imperfect state: *Candida pelliculosa* Redaelli
Synonyms: *Saccharomyces anomalus* Hansen (1891)

Kloeckera apiculata (Reess emend. Klöcker) Janke (1928)
Perfect state: *Hanseniaspora uvarum* (Niehaus) Shehata, Mrak, & Phaff
Synonyms: *Saccharomyces apiculatus* Reess (1870)

Metschnikowia pulcherrima Pitt & Miller (1968)
Imperfect state: *Candida pulcherrima* (Lindner) Windisch (1940)
Synonyms: *Torula pulcherrima* Linder (1901),
Torulopsis pulcherrima (Linder) Sacc. (1906),
Saccharomyces pulcherrima (Linder) Beijerinck (1912)

Pichia fermentans Lodder (1932)
Imperfect state: *Candida lambica* (Lindner & Genoud) van Uden & Buckley (1970)
Synonyms: *Saccharomyces dombrowskii* Sacchetti (1933),
Pichia dombrowskii Sacchetti

Pichia membranaefaciens Hansen (1904)
Imperfect state: *Candida valida* (Leberle) van Uden & Buckley (1970)
Synonyms: *Saccharomyces membranaefaciens* Hansen (1888)

Saccharomyces cerevisiae Meyen *ex* Hansen (1883)
Synonyms: *Saccharomyces aceti* Santa María (1959),
Saccharomyces bayanus Saccardo (1895),
Saccharomyces beticus Marcilla ex Santa Maria (1970),
Saccharomyces capensis van der Walt & Tscheuschner (1956),
Saccharomyces carlsbergensis Hansen (1908),
Saccharomyces chevalieri Guilliermond (1914),
Saccharomyces coreanus Saito (1910),
Saccharomyces diastaticus Andrews & Gilliland *ex* van der Walt (1965),
Saccharomyces ellipsoideus Meyen *ex* Hansen (1883),
Saccharomyces fructuum Lodder & Kreger-van Rij (1952),
Saccharomyces globosus Osterwalder (1924),

(continued)

Appendix 7.1

Partial Synonymy of Several Important Wine Yeasts[a] (*continued*)

    Saccharomyces inusitatus van der Walt (1965),
    Saccharomyces italicus nom. nud. (Castelli 1938),
    Saccharomyces norbensis Santa María (1959),
    Saccharomyces oleaceus Santa María (1959),
    Saccharomyces oleaginosus Santa María (1959),
    Saccharomyces oviformis Osterwalder (1924),
    Saccharomyces prostoserdovii Kudriavzev (1960),
    Saccharomyces steineri Lodder & Kreger-van Rij (1952),
    Saccharomyces uvarum Beijerinck (1898),
    Saccharomyces vini Meyer *ex* Kudriavzev (1960)

Saccharomycodes ludwigii Hansen (1904)
  Synonyms: *Saccharomyces ludwigii* Hansen (1889)

Torulaspora delbrueckii (Lindner) Lindner (1904)
  Imperfect state: *Candida colliculosa* (Hartmann) Meyer & Yarrow
  Synonyms: *Saccharomyces delbrueckii* Lindner (1985),
    Saccharomyces fermentati (Saito) Lodder and Kreger-van Rij (1952),
    Saccharomyces rosei (Guilliermond) Lodder & Kreger-van Rij (1952)

Zygosaccharomyces bailii (Lindner) Guilliermond (1912)
  Synonyms: *Saccharomyces acidifaciens* (Nickerson) Lodder & Kreger-van Rij (1952),
    Saccharomyces bailii Lindner (1895)

Zygosaccharomyces bisporus (Naganishi) Lodder & Kreger-van Rij (1952)
  Synonyms: *Saccharomyces bisporus* Naganishi (1917)

Zygosaccharomyces florentinus Castelli *ex* Kudriavzev (1960)
  Synonyms: *Saccharomyces florentinus* (Castelli *ex* Kudriavzev) Lodder & Kreger-van Rij (1952)

Zygosaccharomyces rouxii (Boutroux) Yarrow (1977)
  Imperfect state: *Candida mogii* Vidal-Leiria
  Synonyms: *Saccharomyces rouxii* Boutroux (1884),
    Zygosaccharomyces barkeri Saccardo & Sydow (1902)

[a] Data from Kreger-van Rij (1984), reproduced by permission.

Appendix 7.2

Physiological Races of *Saccharomyces cerevisiae* Previously Given Species Status[a]

	Fermentation substrate					
	Galactose	Sucrose	Maltose	Raffinose	Melibiose	Starch
Saccharomyces aceti	−	−	−	−	−	−
Sacch. bayanus	−	+	+	+	−	−
Sacch. capensis	−	+	−	+	−	−
Sacch. cerevisiae	+	+	+	+	−	−
Sacch. chevalieri	+	+	−	+	−	−
Sacch. coreanus	+	+	−	+	+	−
Sacch. diastaticus	+	+	+	+	−	+
Sacch. globosus	+	−	−	−	−	−
Sacch. heterogenicus	−	+	+	−	−	−
Sacch. hienipiensis	−	−	+	−	+	−
Sacch. inusitatus	−	+	+	+	+	−
Sacch. norbensis	−	−	−	−	+	−
Sacch. oleaceus	+	−	−	+	+	−
Sacch. oleanginosus	+	−	+	+	+	−
Sacch. prostoserdovii	−	−	+	−	−	−
Sacch. steineri	+	+	+	−	−	−
Sacch. uvarum	+	+	+	+	+	−

[a] From van der Walt (1970), reproduced by permission.

Suggested Readings

General Texts

Amerine, M. A., Berg, H. W., Kunkee, R. E., Ough, C. S., Singleton, V. L., and Webb, A. D. (1980). "The Technology of Wine Making," 4th Ed. AVI Publ., Westport, Connecticut.

Dittrich, H. H. (1987). "Mikrobiologie des Weines: Handbuch der Lebensmitteltechnologie," 2nd Ed. Ulmer, Stuttgart, Germany.

Farkaš, J. (1988). "Technology and Biochemistry of Wine," Vols. 1 and 2. Gordon & Breach, New York.

Fleet, G. H. (ed.) (1992). "Wine Microbiology and Biotechnology." Harwood Academic Publ., New York.

Margalit, Y. (1992). "Winery Technology and Operations: A Handbook for Small Wineries." The Wine Appreciation Guild, San Francisco, California.

Ough, C. S. (1992). "Winemaking Basics." Food Products Press, Binghamton, New York.

Ribéreau-Gayon, J., Peynaud, E., Ribéreau-Gayon, P., and Sudraud, P. (eds.) (1972–1977). "Traité d'Oenologie: Sciences et Techniques du Vin," Vols. 1–4. Dunod, Paris.

Troost, R. (1988). "Technologie des Weines: Handbuch der Lebensmitteltechnologie," 2nd Ed. Ulmer, Stuttgart.

Vine, R. P. (1981). "Commercial Winemaking." AVI Publ. Westport, Connecticut.

Wurdig, G., and Woller, R. (eds.) (1988). "Chimie des Weines." Ulmer, Stuttgart, Germany.

Maceration/Skin Contact

Kinzer, G., and Schreier, P. (1980). Influence of different pressing systems on the composition of volatile constituents in unfermented grape musts and wines. *Am. J. Enol. Vitic.* **31,** 7–13.

Long, Z. R., and Lindblom, B. (1987). Juice oxidation in California Chardonnay. *Proc. 6th Aust. Wine Ind. Tech. Conf.* (T. Lee, ed.), pp. 267–271. Australian Industrial Publ., Adelaide, Australia.

Marais, J., and van Wyk, C. J. (1986). Effect of grape maturity and juice treatments on terpene concentrations and wine quality of *Vitis vinifera* L. cv. Weisser Riesling and Bukettraube. *S. Afr. J. Enol. Vitic.* **7,** 26–35.

Ramey, D., Bertrand, A., Ough, C. S., Singleton, V. L., and Sanders, E. (1986). Effects of skin contact temperature on Chardonnay must and wine composition. *Am. J. Enol. Vitic.* **37,** 99–106.

Adjustments before Fermentation

Beelman, R. B., and Gallander, J. F. (1979). Wine deacidification. *Adv. Food Res.* **25,** 1–53.

Chauvet, S., Sudraud, P., and Jouan, T. (1986). La cryoextraction sélective des moûts. *Rev. Oenologues* **39,** 17–22.

Giesbrecht, R., and Fuleki, T. (1985). Ultrafiltration and reverse osmosis: Revolutionary technologies in the wine cellar. *East. Grape Grow. Winery News* **10**(6), 20–24.

Ribéreau-Gayon, J., Peynaud, E., Ribéreau-Gayon, P., and Sudraud, P. (1977). Vinification avec chauffage de la vendange. *In* "Traité d'Oenologie: Sciences et Techniques du Vin" (J. Ribéreau-Gayon, E. Peynaud, P. Ribéreau-Gayon, and P. Sudraud, eds.), Vol. 4, p.p. 315–360. Dunod, Paris.

General Fermentation Reviews

Dubourdieu, D. (1986). Wine technology: Current trends. *Experientia* **42,** 914–921.

Fleet, G. H. (1990). Growth of yeasts during wine fermentations. *J. Wine Res.* **1,** 211–223.

Goswell, R. W. (1986). Microbiology of table wines. *Dev. Food Microbiol.* **2,** 21–65.

Lafon-Lafourcade, S. (1983). Wine and Brandy. *In* "Biotechnology, Volume 5: Food and Feed Production with Microorganisms" (G. Reed, ed.), pp. 81–163. Verlag Chemie, Weinheim, Germany.

Lafon-Lafourcade, S. (1986). Applied microbiology. *Experientia* **42,** 904–914.

Reed, G., and Nagodawithana, T. W. (1988). Technology of yeast usage in winemaking. *Am. J. Enol. Vitic.* **39,** 83–90.

Ribéreau-Gayon, P. (1985). New developments in wine microbiology. *Am. J. Enol. Vitic.* **36,** 1–9.

Biochemistry of Yeast Fermentation

Bertrand, A. (1983). Volatiles from grape must fermentation. *In* "Flavour of Distilled Beverages: Origin and Development" (J. R. Piggott, ed.), pp. 93–109. Ellis Horwood, Chichester.

Gancelos, C., and Serrano, R. (1989). Energy-yielding metabolism. *In* "The Yeasts" (A. H. Rose and J. S. Harrison, eds.), Vol. 3, 2nd Ed., pp. 205–259. Academic Press, New York.

Guerzoni, M. E., Marchetti, R., and Giudici, P. (1985). Modifications de composants aromatiques des vins obtenus par fermentation avec des mutants de *Saccharomyces cerevisiae. Bull. O.I.V.* **58,** 230–233.

Lagunas, R. (1986). Misconceptions about the energy metabolism of *Saccharomyces cerevisiae. Yeast* **2,** 221–228.

Radler, F. (1986). Microbial biochemistry. *Experientia* **42,** 884–893.

Yeast Classification

Kreger-van Rij, N. J. W. (ed.) (1984). "The Yeasts: A Taxonomic Study," 3rd Ed. Elsevier, Amsterdam.

Yeast Genetics

Pretorius, I. S., and van der Westhuizen, T. J. (1991). The impact of yeast genetics and recombinant DNA technology on the wine industry—A review. *S. Afr. J. Enol. Vitic.* **12,** 3–31.

Subden, R. E. (1988). Current developments in wine yeasts. *Crit. Rev. Biotechnol.* **5,** 49–65.

Thornton, R. J., and Rodriguez, S. B. (1987). Genetics of wine microorganisms: potentials and problems. *Proc. 6th Aust. Wine Ind. Tech. Conf.* (T. Lee, ed.), pp. 98–102. Australian Industrial Publ., Adelaide, Australia.

Tuite, M. F. (1992). Strategies for the genetic manipulation of *Saccharomyces cerevisiae. Crit. Rev. Biotechnol.* **12,** 157–188.

Environmental Factors

Bisson, L. F. (1991). Influence of nitrogen on yeast and fermentation. *Proc. Int. Symp. Nitrogen Grapes Wine, Seattle, WA, June, 1991* (J. M. Rantz, ed.), pp. 78–89. Am. Soc. Enol. Vitic., Davis, California.

Guymon, J. F., and Crowell, E. A. (1977). The nature and cause of cap–liquid temperature differences during wine fermentation. *Am. J. Enol. Vitic.* **28,** 74–78.

Houtman, A. C., and du Plessis, C. S. (1985). Influence du cépage et de la souche de levure. *Bull. O.I.V.* **58,** 236–246.

Houtman, A. C., Marais, J., and du Plessis, C. S. (1980). Factors affecting the reproducibility of fermentation of grape juice and of the aroma composition of wines. I. Grape maturity, sugar, inoculum concentration, aeration, juice turbidity and ergosterol. *Vitis* **19**, 37–84.

Jones, R. P., Pamment, N., and Greenfield, P. F. (1981). Alcohol fermentation by yeasts—the effect of environmental and other variables. *Process Biochem.* **16**(3), 42–49.

Lemperle, E. (1988). Fungicide residues in musts and wines. *Proc. 2nd Int. Symp. Cool Climate Vitic. Oenol., Jan. 11–15, 1988, Auckland, N.Z.,* (R. E. Smart, S. B. Thornton, S. B. Rodriguez, and J. E. Young, eds.), pp 211–218. New Zealand Soc. Vitic. Oenol, Auckland, New Zealand.

Miller, G. C., Amon, J. M., and Simpson, R. F. (1987). Loss of aroma compounds in carbon dioxide effluent during white wine fermentation. *Food Technol. Aust.* **39**, 246–253.

Nishino, H., Miyazakim S., and Tohjo, K. (1985). Effect of osmotic pressure on the growth rate and fermentation activity of wine yeasts. *Am. J. Enol. Vitic.* **36**, 170–174.

Ough, C. S., and Amerine, M. A. (1966). Effects of temperature on wine making. *Calif. Agric. Exp. Stn. Bull.* No. 827.

Sablayrolles, J. M., and Barre, P. (1987). Evaluation des besoins en oxygène de fermentations alcooliques en conditions oenologiques simulées. *Rev. Fr. Oenol.* **107**, 34–38.

White, B. B., and Ough, C. S. (1973). Oxygen uptake studies on grape juice. *Am. J. Enol. Vitic.* **24**, 148–152.

Stuck and Sluggish Fermentation

Houtman, A. C., and du Plessis, C. S. (1986). Nutritional deficiencies of clarified white grape juices and their correction in relation to fermentation. *S. Afr. J. Enol. Vitic.* **7**, 39–46.

Jacobs, C. J., and van Vuren, H. J. J. (1991). Effects of different killer yeasts on wine fermentations. *Am. J. Enol. Vitic.* **42**, 295–300.

Larue, F., and Lafon-Lafourcade, S. (1989). Survival factors in wine fermentation. *In* "Alcohol Toxicity in Yeasts and Bacteria" (N. van Uden, ed.), pp. 193–215. CRC Press, Boca Raton, Florida.

Munoz, E., and Ingledew, W. M. (1990). Yeast hulls in wine fermentation. A review. *J. Wine Res.* **1**, 197–209.

Malolactic Fermentation

Beelman, R. B., and Kunkee, R. E. (1985). Inducing simultaneous malolactic–alcoholic fermentation in red table wines. *In* "Malolactic Fermentation" (T. H. Lee, ed.), pp. 97–112. Aust. Wine Res. Inst., Urrbrae, South Australia.

Davis, C. R., Wibowo, D., Eschenbruch, R., Lee, T. H., and Fleet, G. H. (1985). Practical implications of malolactic fermentation: A review. *Am. J. Enol. Vitic.* **36**, 290–301.

Davis, C. R., Wibowo, D., Fleet, G. H., and Lee, T. H. (1988). Properties of wine lactic acid bacteria: Their potential enological significance. *Am. J. Enol. Vitic.* **39**, 137–142.

Eggenberger, W. (1988). Malolactic fermentation of wines in cool climates. *In Proc. 2nd Int. Symp. Cool Climate Vitic. Oenol., Jan. 11–15, 1988, Auckland, N.Z.* (R. E. Smart, S. B. Thornton, S. B. Rodriguez, and J. E. Young, eds.), pp. 232–237. New Zealand Soc. Vitic. Oenol, Auckland, New Zealand.

Guilloux-Benatier, M. (1987). Les souches de bactéries lactiques et les divers essais d'ensemencement de la fermentation malolactique en France. *Bull. O.I.V.* **60**, 624–642.

Kunkee, R. B. (1991). Some roles of malic acid in the malolactic fermentation in winemaking. *FEMS Microbiol. Rev.* **88**, 55–72.

Radler, F. (1986). Microbial biochemistry. *Experientia* **42**, 884–893.

van Vuuren, H. J. J., and Dicks, L. M. T. (1993). *Leuconostoc oenos:* A review. *Am. J. Enol. Vitic.* **44**, 99–112.

Wibowo, D., Eschenbruch, R., Davis, C. R., Fleet, G. H., and Lee, T. H. (1985). Occurrence and growth of lactic acid bacteria in wine: A review. *Am. J. Enol. Vitic.* **36**, 302–313.

References

Amerine, M. A., and Joslyn, M. A. (1970). "Table Wines: The Technology of Their Production," 2nd Ed. Univ. of California Press, Berkeley.

Amerine, M. A., and Ough, C. S. (1980). "Methods for Analysis of Musts and Wines." Wiley, New York.

Amerine, M. A., Berg, H. W., Kunkee, R. E., Ough, C. S., Singleton, V. L., and Webb, A. D. (1980). "The Technology of Wine Making." AVI Publ., Westport, Connecticut.

Anonymous (1983). Steel feet punch the cap at Buena Vista. *Wines Vines* **64**(2), 52.

Arnick, K. S., Kriger, S. A., and Henick-Kling, T. (1992). Utilization of sugars by six strains of *Leuconostoc oenos. Am. J. Enol. Vitic.* **43**, 112 (Abstract).

Bach, H. P., and Hess, K. H. (1986). Der Einfluss der Alkoholerhöhung auf Weininhaltsstoffe und Geschmack. *Weinwirsch.-Tech.* **122**, 437–440.

Bach, H. P., Schneider, P., Bamberger, U., and Wintrich, K. H. (1990). Cell-Cracking und sein Einfluss auf die Qualität von Weissweinen. *Weinwirtsch. Tech.* **8**, 26–34.

Bakalinsky, A. L., and Snow, R. (1990a). The chromosomal constitution of wine strains of *Saccharomyces cerevisiae. Yeast* **6**, 367–382.

Bakalinsky, A. L., and Snow, R. (1990b). Conversion of wine strains of *Saccharomyces cerevisiae* to heterothallism. *Appl. Environ. Microbiol.* **56**, 849–857.

Barre, P., and Vezinhet, F. (1984). Evolution towards fermentation with pure culture yeasts in wine making. *Microbiol. Sci.* **1**, 159–163.

Beelman, R. B., and Kunkee, R. E. (1985). Inducing simultaneous malolactic–alcoholic fermentation in red table wines. *In* "Malolactic Fermentation" (T. H. Lee, ed.), pp. 97–111. Aust. Wine Res. Inst., Urrbrae, South Australia.

Belin, J. M. (1972). Recherches sur la répartition des levures à la surface de la grappe de raisin. *Vitis* **11**, 135–145.

Bertrand, A., and Miele, A. (1984). Influence de la clarification du moût de raisin sur sa teneur en acides gras. *Connaiss. Vigne Vin* **18**, 293–297.

Bertrand, A., and Torres-Alegre, V. (1984). Influence of oxygen added to grape must on the synthesis of secondary products of the alcoholic fermentation. *Sci. Aliments* **4**, 45–64.

Boone, C., Sdicu, A.-M., Wagner, J., Degré, R., Sanchez, C., and Bussey, H. (1990). Integration of the yeast K_1 killer toxin gene into the genome of marked wine yeasts and its effect on vinification. *Am. J. Enol. Vitic.* **41**, 37–42.

Bouix, M., Leveau, J. Y., and Cuinier, C. (1981). Applications de l'électrophocell des fractions exocellulaires de levures au contrôle de l'efficacité d'un levurage en vinification. *In* "Current Developments in Yeast Research." (G. G. Stewart and I. Russell, eds.). Pergamon Press.

Boulton, R., (1979). The heat transfer characteristics of wine fermentors. *Am. J. Enol. Vitic.* **30**, 152–156.

Bousbouras, G. E., and Kunkee, R. E. (1971). Effect of pH on malolactic fermentation in wine. *Am. J. Enol. Vitic.* **22**, 121–126.

Bréchot, P., Chauvet, J., Dupuy, P., Croson, M., and Rabatu, A. (1971). Acide olénolique, facteur de croissance anaérobie de la levure de vin. *C. R. Acad. Sci. Ser. D* **272**, 890–893.

Brewster, J. L., de Valoir, T., Dwyer, N. D., Winter, E., and Gustin,

M. C. (1993). An osmosensing signal transduction pathway in yeast. *Science* 259, 1760–1763.

Buechsensteing, J., and Ough, C. S. (1979). Comparison of citric, dimalic, and fumaric acids as wine acidulants. *Am. J. Enol. Vitic.* 30, 93–97.

Cantarelli, C. (1989). Phenolics and yeast: Remarks concerning fermented beverages. *Yeast* 5, S53–61.

Carre, E., Lafon-Lafourcade, S., and Bertrand, A. (1983). Désacidification biologique des vins blancs secs par fermentation de l'acide malique par les levures. *Connaiss. Vigne Vin* 17, 43–53.

Cartwright, C. P., Rose, A. H., Calderbank, J., and Keenan, M. J. (1989). Solute transport. *In* "The Yeasts" (A. H. Rose and J. S. Harrison, eds.), Vol. 3, 2nd Ed., pp. 5–56. Academic Press, New York.

Casey, G. P., and Ingledew, W. M. (1986). Ethanol tolerance in yeasts. *Crit. Rev. Microbiol.* 13, 219–280.

Càstino, M. L., Usseglio-Tomasset, L., and Gandini, A. (1975). Factors which affect the spontaneous initiation of the malolactic fermentation in wines: The possibility of transmission by inoculation and its effect on organoleptic qualities. *In* "Lactic Acid Bacteria in Beverages and Food" (J. G. Carr, C. V. Cutting, and G. C. Whiting, eds.), pp. 139–148. Academic Press, London.

Cavin, J. F., Drici, F. Z., Prevost, H., and Divies, C. (1991). Prophage curing in *Leuconostoc oenos* by mitomycin C induction. *Am. J. Enol. Vitic.* 42, 163–166.

Chatonnet, P., Dubourdieu, P., and Boidron, J. N. (1989). Incidence de certains facteurs sur la décarboxylation des acides phénols par la levure. *Connaiss. Vigne Vin* 23, 59–63.

Chauvet, S., Sudraud, P., and Jouan, T. (1986). La cryoextraction sélective des moûts. *Rev. Oenologues* 39, 17–22.

Cheynier, V., Souquet, J.-M., Samson, A., and Moutounet, M. (1991). Hyperoxidation: Influence of various oxygen supply levels on oxidation kinetics of phenolic compounds and wine quality. *Vitis* 30, 107–115.

Conner, A. J. (1983). The comparative toxicity of vineyard pesticides to wine yeasts. *Am. J. Enol. Vitic.* 34, 278–279.

Costello, P. J., Morrison, R. H., Lee, R. H., and Fleet, G. H. (1983). Numbers and species of lactic acid bacteria in wines during vinification. *Food Technol. Aust.* 35, 14–18.

Cox, D. J., and Henick-Kling, T. (1990). A comparison of lactic acid bacteria for energy-yielding (ATP) malolactic enzyme systems. *Am. J. Enol. Vitic.* 41, 215–218.

Crapisi, A., Spettoli, P., Nuti, M. P., and Zamorani, A. (1987). Comparative traits of *Lactobacillus brevis*, *Lact. fructivorans* and *Leuconostoc oenos* immobilized cells for the control of malolactic fermentation in wine. *J. Appl. Bacteriol.* 63, 513–521.

Crippen, D. D., Jr., and Morrison, J. C. (1986). The effects of sun exposure on the phenolic content of Cabernet Sauvignon berries during development. *Am. J. Enol. Vitic.* 37, 243–247.

Cuénat, P., Zufferey, E., Kobel, D., Bregy, C. A., and Crettenand, J. (1991). Le cuvage du Pinot noir. Rôle des températures. *Rev. Suisse Vitic. Arboric. Hortic.* 23, 267–272.

Davis, C. R., Wibowo, D., Eschenbruch, R., Lee, T. H., and Fleet, G. H. (1985). Practical implications of malolactic fermentation: A review. *Am. J. Enol. Vitic.* 36, 290–301.

Davis, C. R., Wibowo, D. J., Lee, T. H., and Fleet, G. H. (1986). Growth and metabolism of lactic acid bacteria during and after malolactic fermentation of wines at different pH. *Appl. Environ. Microbiol.* 51, 539–545.

de Deken, R. H. (1966). The Crabtree effect: A regulatory system in yeasts. *J. Gen. Microbiol.* 44, 149–156.

Defranoux, C., Gineys, D., and Joseph, P. (1989). Le potentiel aromatique du Chardonnay. Essai d'utilisation des techniques de cryoextraction et de supraextraction (procédé Kreyer). *Rev. Oenologues* 55, 27–29.

Desens, C., and Lonvaud-Funel, A. (1988). Étude de la constitution lipidique des membranes de bactéries lactiques utilisées en vinification. *Connaiss. Vigne Vin.* 22, 25–32.

Dick, K. I., Molan, P. C., and Eschenbruch, R. (1992). The isolation from *Saccharomyces cerevisiae* of two antibacterial cationic proteins that inhibit malolactic bacteria. *Vitis* 31, 105–116.

Dittrich, H. H. (1977). "Mikrobiologie des Weines: Handbuch der Getränketechnologie." Ulmer, Stuttgart.

Dubourdieu, D., and Lavigne, V. (1990). Incidence de l'hyperoxidation sur la composition chimique et les qualitiés organoleptiques des vins blancs du Bordelais. *Rev. Fr. Oenol.* 30, 58–61.

Dubourdieu, D., Sokol, A., Zucca, J., Thalouarn, P., Dattee, A., and Aigle, M. (1987). Identification des souches de levures isolées de vins par l'analyse de leur ADN mitochondrial. *Connais. Vigne Vin.* 21, 267–278.

Duitschaever, C. L., Alba, J., Buteau, C., and Allen, B. (1991). Riesling wines made from must concentrated by reverse osmosis. I. Experimental conditions and composition of musts and wines. *Am. J. Enol. Vitic.* 42, 19–25.

du Plessis, C. S. (1983). Influence de la température d'élaboration et de conservation. *Bull. O.I.V.* 524, 104–115.

Edwards, C. G., Beelman, R. B., Bartley, C. E., and McConnell, A. L. (1990). Production of decanoic acid and other volatile compounds and the growth of yeast and malolactic bacteria during vinification. *Am. J. Enol. Vitic.* 41, 48–56.

Farkaš, J. (1988). "Technology and Biochemistry of Wine," Vols. 1 and 2. Gordon & Breach, New York.

Farris, G. A., Fatichenti, F., and Deiana, P. (1989). Incidence de la température et du pH sur la production d'acide malique par *Saccharomyces cerevisiae*. *J. Int. Sci. Vigne Vin* 23, 89–93.

Feuillat, M., Guillox-Benatier, and Gerbaux, V. (1985). Essais d'activation de la fermentation malolactique dans les vins. *Sci. Aliments* 5, 103–122.

Fleet, G. H., Lafon-Lafourcade, S., and Ribéreau-Gayon, P. (1984). Evolution of yeasts and lactic acid bacteria during fermentation and storage of Bordeaux wines. *Appl. Environ. Microbiol.* 48, 1034–1038.

Froment, T. (1991). Auto-enrichissement des moûts de raisin: Le concentrateur Entropie (type M.T.A.) par évaporation sous vide à très basse température (20°C). *Rev. Oenologues* 17, 7–11.

Gancelos, C., and Serrano, R. (1989). Energy-yielding metabolism. *In* "The Yeasts" (A. H. Rose and J. S. Harrison, eds.), Vol. 3, 2nd Ed., pp. 205–259. Academic Press, New York.

Gao, C., and Fleet, G. H. (1988). The effects of temperature and pH on the ethanol tolerance of the wine yeasts, *Saccharomyces cerevisiae*, *Candida stellata* and *Kloeckera apiculata*. *J. Appl. Bacteriol.* 65, 405–410.

Gnaegi, F., Aerny, J., Bolay, A., and Crettenand, J. (1983). Influence des traitements viticoles antifongiques sur la vinification et la qualité du vin. *Rev. Suisse Vitic. Arboric. Hortic.* 15, 243–250.

Groat, M. L., and Ough, C. S. (1978). Effect of particulate matter on fermentation rates and wine quality. *Am. J. Enol. Vitic.* 29, 112–119.

Grossman, M. K., and Becker, R. (1984). Investigations on bacterial inhibition of wine fermentation. *Kellerwirtschaft* 10, 272–275.

Guerzoni, M. E., Marchetti, R., and Giudici, P. (1985). Modifications de composants aromatiques des vins obtenus par fermentation avec des mutants de *Saccharomyces cerevisiae*. *Bull. O.I.V.* 58, 230–233.

Guilloux-Benatier, M. (1987). Les souches de bactéries lactiques et les divers essais d'ensemencement de la fermentation malolactique in France. *Bull. O.I.V.* 60, 624–642.

Guilloux-Benatier, M., Le Fur, Y., and Feuillat, M. (1989). Influence de la macération pelliculaire sur la fermentiscibilité malolactique des vins blancs de Bourgogne. *Rev. Fr. Oenol.* 29, 29–34.

Guymon, J. F., and Crowell, E. A. (1977). The nature and cause of cap–liquid temperature differences during wine fermentation. *Am. J. Enol. Vitic.* **28**, 74–78.

Haag, B., Krieger, S., and Hammes, W. P. (1988). Hemmung der Startezkulturen zur Einlertung des biologischen Säurenabbaus durch Spritzmittelrückstände. *Wein-Wiss.* **43**, 261–278.

Hammond, S. M., and Carr, J. C. (1976). The antimicrobial activity of SO₂—with particular reference to fermented and non-fermented fruit juices. *In* "Inhibition and Inactivation of Vegetative Microbes" (F. A. Skinner and W. B. Hugo, eds.), pp. 89–110. Academic Press, London.

Hayman, D. C., and Monk, P. R. (1982). Starter culture preparation for the induction of malolactic fermentation in wine. *Food Technol. Aust.* **34**, 14, and 16–18.

Heard, G. M., and Fleet, G. H. (1985). Growth of natural yeast flora during the fermentation of inoculated wines. *Appl. Environ. Microbiol.* **50**, 727–728.

Henschke, P. A., and Dixon, G. D. (1990). Effect of yeast strain on acetic acid accumulation during fermentation of *Botrytis* affected grape juice. *Proc. 7th Aust. Wine Ind. Tech. Conf.* (P. J. Williams, D. Davidson, and T. H. Lee, eds.), pp. 242–244. Australian Industrial Publishers, Adelaide, Australia.

Henschke, P. A., and Ough, C. S. (1991). Urea accumulation in fermenting grape juice. *Am. J. Enol. Vitic.* **42**, 317–321.

Hess, B., Boiteux, A., and Krüger, J. (1969). Cooperation of glycolytic enzymes. *Adv. Enzyme Regulation* **7**, 149–167.

Holloway, P., Subden, R. E., and Lachance, M. A. (1990). The yeasts in a Riesling must from the Niagara grape-growing region of Ontario. *Can. Inst. Food Sci. Technol. J.* **23**, 212–216.

Houtman, A. C., and du Plessis, C. S. (1981). The effect of juice clarity and several conditions promoting yeast growth on fermentation rate, the production of aroma components and wine quality. *S. Afr. J. Enol. Vitic.* **2**, 71–81.

Houtman, A. C., and du Plessis, C. S. (1985). Influence du cépage et de la souche de levure. *Bull. O.I.V.* **58**, 236–246.

Houtman, A. C., and du Plessis, C. S. (1986). Nutritional deficiencies of clarified white grape juices and their correction in relation to fermentation. *S. Afr. J. Enol. Vitic.* **7**, 39–46.

Houtman, A. C., Marais, J., and du Plessis, C. S. (1980). Factors affecting the reproducibility of fermentation of grape juice and of the aroma composition of wines. I. Grape maturity, sugar, inoculum concentration, aeration, juice turbidity and ergosterol. *Vitis* **19**, 37–84.

Jacobs, C. J., and van Vuuren, H. J. J. (1991). Effects of different killer yeasts on wine fermentations. *Am. J. Enol. Vitic.* **42**, 295–300.

Javelot, C., Girard, P., Colonna-Ceccaldi, B., and Valdescu, B. (1991). Introduction of terpene-producing ability in a wine strain of *Saccharomyces cerevisiae*. *J. Biotechnol.* **21**, 239–252.

Jones, R. P., and Ough, C. S. (1985). Variations in the percent ethanol (v/v) per Brix conversions of wines from different climatic regions. *Am. J. Enol. Vitic.* **36**, 268–270.

Jones, R. P., Pamment, N., and Greenfield, P. F. (1981). Alcohol fermentation by yeasts—The effect of environmental and other variables. *Process Biochem.* **16**, 42–49.

Killian, E., and Ough, C. S. (1979). Fermentation esters—Formation and retention as affected by fermentation temperature. *Am. J. Enol. Vitic.* **30**, 301–305.

King, S. W., and Beelman, R. B. (1986). Metabolic interactions between *Saccharomyces cerevisiae* and *Leuconostoc oenos* in a model grape juice/wine system. *Am. J. Enol. Vitic.* **37**, 53–60.

Klingshirn, L. M., Liu, J. R., and Gallander, J. F. (1987). Higher alcohol formation in wines as related to the particle size profiles of juice insoluble solids. *Am. J. Enol. Vitic.* **38**, 207–209.

Kraus, J. K., Scoop, R., and Chen, S. L. (1981). Effect of rehydration on dry wine yeast activity. *Am. J. Enol. Vitic.* **32**, 132–134.

Kreger-van Rij, N. J. W. (ed). (1984). "The Yeasts: A Taxonomic Study," 3rd Ed. Elsevier, Amsterdam.

Kunkee, R. E., and Ough, C. S. (1966). Multiplication and fermentation of *Saccharomyces cerevisiae* under carbon dioxide pressure in wine. *Appl. Microbiol.* **14**, 643–648.

Laffort, J.-F., Romat, H., and Darriet, Ph. (1989). Les levures et l'expression aromatique des vins blancs. *Rev. Oenologues* **53**, 9–12.

Lafon-Lafourcade, S. (1975). Factors of the malolactic fermentation of wines. *In* "Lactic Acid Bacteria in Food and Beverages" (J. G. Carr, C. V. Cutting, and G. C. Whiting, eds.), pp. 43–53. Academic Press, London.

Lafon-Lafourcade, S. (1980). Les origins microbiologiques de l'acidité volatile des vins. *Microbiol. Ind. Aliment. Ann. Congr. Int.* **2**, 33–48.

Lafon-Lafourcade, S. (1981). Connaissances actuelles, dans la maîtrise de la fermentation malolactique dans les moûts et les vins. *In* "Actualités Oenologiques et Viticoles" (P. Ribéreau-Gayon and P. Sudraud, eds.), (pp. 243–251.) Dunod, Paris.

Lafon-Lafourcade, S. (1983). Wine and brandy. *In* "Biotechnology. Volume 5. Food and Feed Production with Microorganisms" (G. Reed, ed.), pp. 81–163. Verlag Chemie, Weinheim.

Lafon-Lafourcade, S., Carre, E., and Ribéreau-Gayon, P. (1983). Occurrence of lactic acid bacteria during different stages of vinification and conservation of wines. *Appl. Environ. Microbiol.* **46**, 874–880.

Lafon-Lafourcade, S., Geneix, C., and Ribéreau-Gayon, P. (1984). Inhibition of alcoholic fermentation of grape must by fatty acids produced by yeasts and their elimination by yeast ghosts. *Appl. Environ. Microbiol.* **47**, 1246–1249.

Lagunas, R. (1986). Misconceptions about the energy metabolism of *Saccharomyces cerevisiae*. *Yeast* **2**, 221–228.

Lehmann, F. L. (1987). Secondary fermentations retarded by high levels of free sulfur dioxide. *Aust. N.Z. Wine Ind. J.* **2**, 52–53.

Lemperle, E. (1988). Fungicide residues in musts and wines. *Proc. 2nd Int. Symp. Cool Climate Vitic. Oenol., Jan. 11–15, 1988, Auckland, N.Z.* (R. E. Smart, S. B. Thornton, S. B. Rodriguez, and J. E. Young, eds.), (pp. 211–218). New Zealand Soc. Vitic. Oenol, Auckland, New Zealand.

McCord, J. D., and Ryu, D. D. Y. (1985). Development of malolactic fermentation process using immobilized whole cells and enzymes. *Am. J. Enol. Vitic.* **36**, 214–218.

McDaniel, M., Henderson, L. A., Watson, B. T., Jr., and Heatherbell, D. (1987). Sensory panel training and screening for descriptive analysis of the aroma of Pinot noir wine fermented by several strains of malolactic bacteria. *J. Sens. Stud.* **2**, 149–167.

Marais, J. (1987). Terpene concentrations and wine quality of *Vitis vinifera* L. cv. Gewürztraminer as affected by grape maturity and cellar practices. *Vitis* **26**, 231–245.

Marais, J., and van Wyk, C. J. (1986). Effect of grape maturity and juice treatments on terpene concentrations and wine quality of *Vitis vinifera* L. cv. Weisser Riesling and Bukettraube. *S. Afr. J. Enol. Vitic.* **7**, 26–35.

Martínez, J., Millán, C., and Ortega, J. M. (1989). Growth of natural flora during the fermentation of inoculated musts from 'Pedro Ximenez' grapes. *S. Afr. J. Enol. Vitic.* **10**, 31–35.

Mateo, J. J., Jimenez, M., Huerta, T., and Pastor, A. (1991). Contribution of different yeasts isolated from musts of Monastrell grapes to the aroma of wine. *Int. J. Food Microbiol.* **14**, 153–160.

Mayer, K. (1974). Mikrobiologisch und kellertechnisch wichtige neue Erkenntnisse in bezug auf den biologischen Säureabbau. *Schweiz. Z. Obst. Weinbau.* **110**, 291–297.

Meyrath, J., and Lüthi, H. R. (1969). On the metabolism of hexoses and pentoses by *Leuconostoc* isolated from wines and fruit. *Lebensm. Wiss. Technol.* **2**, 22–27.

Miller, G. C., Amon, J. M., and Simpson, R. F. (1987). Loss of aroma compounds in carbon dioxide effluent during white wine fermentation. *Food Technol. Aust.* **39**, 246–253.

Minárik, E., and Rágala, P. (1966). Einfluss einiger Fungizide auf die Hefeflora bei der spotanen Mostgärung. *Mitt. Rebe Wein, Obstbau Fruechteverwert.* (*Klosterneuburg*) **16**, 107–114.

Mora, J., and Mulet, A. (1991). Effects of some treatments of grape juice on the population and growth of yeast species during fermentation. *Am. J. Enol. Vitic.* **42**, 133–136.

Moreno, J. J., Millán, C., Ortega, J. M., and Medina, M. (1991). Analytical differentiation of wine fermentations using pure and mixed yeast cultures. *J. Ind. Microbiol.* **7**, 181–190.

Munoz, E., and Ingledew, W. M. (1990). Yeast hulls in wine fermentations. A review. *J. Wine Res.* **1**, 197–209.

Nagel, C. W., Weller, K., and Filiatreau, D. (1988). Adjustment of high pH–high TA musts and wines. *Proc. 2nd Int. Symp. Cool Climate Vitic. Oenol., Jan. 11–15, 1988, Auckland, N.Z.*, (R. E. Smart, S. B. Thornton, S. B. Rodriguez, and J. E. Young, eds.), pp. 222–224. New Zealand Soc. Vitic. Oenol, Auckland, New Zealand.

Nilov, V. I., and Valuiko, G. G. (1958). Changes in nitrogen during fermentation. *Vinodel. Vinograd. SSSR* **18**, 4–7 (in Russian).

Nishino, H., Miyazakim, S., and Tohjo, K. (1985). Effect of osmotic pressure on the growth rate and fermentation activity of wine yeasts. *Am. J. Enol. Vitic.* **36**, 170–174.

Noble, A. C., and Bursick, G. F. (1984). The contribution of glycerol to perceived viscosity and sweetness in white wine. *Am. J. Enol. Vitic.* **35**, 110–112.

Nykänen, L. (1986). Formation and occurrence of flavor compounds in wine and distilled alcoholic beverages. *Am. J. Enol. Vitic.* **37**, 89–96.

Ollivier, Ch., Stonestreet, Th., Larue, F., and Dubourdieu, D. (1987). Incidence de la composition colloidale des moûts blancs sur leur fermentescibilité. *Connaiss. Vigne Vin.* **21**, 59–70.

Oszmianski, J., Ramos, T., and Bourzeix, M. (1988). Fractionation of phenolic compounds in red wine. *Am. J. Enol. Vitic.* **39**, 259–262.

Prahl, C., Lonvaud-Funel, A., Korsgaard, S., Morrison, E., and Joyeux, A. (1988). Étude d'un nouveau procédé de déclenchement de la fermentation malolactique. *Connaiss. Vigne Vin* **22**, 197–207.

Querol, A., Barrio, E., Huerta, T., and Ramón, D. (1992). Molecular monitoring of wine fermentations conducted by active dry yeast strains. *Appl. Environ. Microbiol.* **58**, 2948–2953.

Radler, F. (1972). Problematik des bakteriellen Säureabbaus. *Weinberg Keller* **19**, 357–370.

Radler, F. (1986). Microbial biochemistry. *Experientia* **42**, 884–893.

Radler, F. (1990). Possible use of nisin in winemaking. II. Experiments to control lactic acid bacteria in the production of wine. *Am. J. Enol. Vitic.* **41**, 7–11.

Radler, F., and Fäth, K. P. (1991). Histamine and other biogenic amines in wines. *Proc. Int. Symp. Nitrogen Grapes Wine, Seattle, WA, June, 1991* (J. M. Rantz, ed.), pp. 185–195. Am. Soc. Enol. Vitic., Davis, California.

Radler, F., and Yannissis, C. (1972). Weinsäureabbau bei Milchsäurebakterien. *Arch. Mikrobiol.* **82**, 219–239.

Ramey, D., Bertrand, A., Ough, C. S., Singleton, V. L., and Sanders, E. (1986). Effects of skin contact temperature on Chardonnay must and wine composition. *Am. J. Enol. Vitic.* **37**, 99–106.

Rankine, B. C. (1966). Decomposition of L-malic acid by wine yeasts. *J. Sci. Food Agric.* **17**, 312–316.

Ough, C. S. (1966a). Fermentation rates of grape juice. II. Effects of initial °Brix, pH, and fermentation temperature. *Am. J. Enol. Vitic.* **17**, 20–26.

Ough, C. S. (1966b). Fermentation rates of grape juice. III. Effects of initial ethyl alcohol, pH, and fermentation temperature. *Am. J. Enol. Vitic.* **17**, 74–81.

Ough, C. S., Stevens, D., Sendovski, T., Huang, Z., and An, A. (1990). Factors contributing to urea formation in commercially fermented wines. *Am. J. Enol. Vitic.* **41**, 68–73.

Ough, C. S., Huang, Z., An, D., and Stevers, D. (1991). Amino acid uptake by four commercial yeasts at two different temperatures of growth and fermentation: Effects on urea excretion and reabsorption. *Am. J. Enol. Vitic.* **42**, 26–40.

Park, S. K., Morrison, J. C., Adams, D. O., and Noble, A. C. (1991). Distribution of free and glycosidically bound monoterpenes in the skin and mesocarp of Muscat of Alexandria grape during development. *J. Agric. Food Chem.* **39**, 514–518.

Petering, J. E., Henschke, P. A., and Langridge, P. (1991). The *Escherichia coli* β-glucuronidase gene as a marker for *Saccharomyces* yeast strain identification. *Am. J. Enol. Vitic.* **42**, 6–12.

Phaff, H. J. (1986). Ecology of yeasts with actual and potential value in biotechnology. *Microb. Ecol.* **12**, 31–42.

Pilone, G. J., Rankine, B. C., and Pilone, A. (1974). Inhibiting malolactic fermentation in Australian dry red wines by adding fumaric acid. *Am. J. Enol. Vitic.* **25**, 99–107.

Rankine, B. C., Fornachon, J. C. M., and Bridson, D. A. (1969). Diacetyl in Australian dry red wines and its significance in wine quality. *Vitis* **8**, 129–134.

Rapp, A., and Reuther, K. H. (1971). Der Gehalt an freien Aminotäuren in Trabenmosten von gesunden und edelfaulen Beeren verichiedener Rebsorten. *Vitis* **10**, 51–58.

Ribéreau-Gayon, J., Peynaud, E., Ribéreau-Gayon, P., and Sudraud, P. (eds.) (1975). "Traité d'Oenologie: Sciences et Techniques du Vin," Vol. 2. Dunod, Paris.

Ribéreau-Gayon, J., Peynaud, E., Ribéreau-Gayon, P., and Sudraud, P. (eds.) (1976). "Traité d'Oenologie: Sciences et Techniques du Vin," Vol. 3. Dunod, Paris.

Ribéreau-Gayon, P. (1985). New developments in wine microbiology. *Am. J. Enol. Vitic.* **36**, 1–9.

Ribéreau-Gayon, P., Lafon-Lafourcade, S., Dubourdieu, D., Lucmaret, V., and Larue, F. (1979). Métabolism de *Saccharomyces cerevisiae* dans le moût de raisins parasités par *Botrytis cinerea*: Inhibition de la fermentation, formation d'acide acétique et de glycerol. *C. R. Acad. Sci. Paris, Ser. D* **289**, 441–444.

Ribéreau-Gayon, P., Larue, F., and Chaumet, P. (1987). The effect of addition of sucrose and aeration to grape must on growth and metabolic activity of *Saccharomyces cerevisiae*. *Vitis* **26**, 208–214.

Rieger, T. (1993). Rotary fermenters stir more interest at U.S. wineries. *Vineyard Winery Management* **19**(1), 40–44.

Rose, A. H. (1989). Influence of the environment on microbial lipid composition. *In* "Microbial Lipids" (C. Ratledge and S. G. Wilkinson, eds.), Vol. 2, pp. 255–278. Academic Press, London.

Rosini, G. (1986). Wine-making by cell-recycle-batch fermentation process. *Appl. Microbiol. Biotechnol.* **24**, 140–143.

Sablayrolles, J. M., and Barre, P. (1986). Evaluation of oxygen requirement of alcoholic fermentations in simulated enological conditions. *Sci. Aliments* **6**, 373–383.

Sá Correia, I., Salgueiro, S. P., Viegas, C. A., and Novais, J. M. (1989). Leakage induced by ethanol, octanoic and decanoic acids in *Saccharomyces cerevisiae*. *Yeast* **5**, S124–129.

San Romáo, M. V., and Coste Belchior, A. P. (1982). Study of the influence of some antifungal products on the microbiological flora of grapes and musts. *Cienc. Tec. Vitivinic.* **1**, 101–112 (in Portugese).

San Romáo, M. V., Coste Belchior, A. P., Silva Alemáo, M., and Gonçalves Bento, A. (1985). Observations sur le métabolisme des bacteries lactiques dans les moûts de raisins altéres. *Connaiss. Vigne Vin* **19**, 109–116.

Schanderl, H. (1959). "Die Mikrobiologie des Mostes und Weines," 2nd Ed. Ulmer, Stuttgart.

Schmidt, J. O., and Noble, A. C. (1983). Investigation of the effect of skin contact time on wine flavor. *Am. J. Enol. Vitic.* **34**, 135–138.

Shimazu, Y., Uehara, M., and Watanbe, M. (1985). Transformation of citric acid to acetic acid, acetoin and diacetyl by wine making lactic acid bacteria. *Agric. Biol. Chem.* **49**, 2147–2157.

Siegrist, J. (1985). Les tanins et les anthocyanes du Pinot at les phénomènes de macération. *Rev. Oenologues* **11**(38), 11–13.

Singleton, V. L., Sieberhagen, H. A., de Wet, P., and van Wyk, C. J. (1975). Composition and sensory qualities of wines prepared from white grapes by fermentation with and without grape solids. *Am. J. Enol. Vitic.* **26**, 62–69.

Singleton, V. L., Zaya, J., and Trousdale, E. (1980). White table wine quality and polyphenol composition as affected by must SO_2 content and pomace contact time. *Am. J. Enol. Vitic.* **31**, 14–20.

Snow, R. (1985). Genetic engineering of a yeast strain for malolactic fermentation of wine. *Food Technol.* **39**, 98–101.

Somers, T. C., and Pocock, K. F. (1986). Phenolic harvest criteria for red vinification. *Aust. Grapegrower Winemaker* **256**, 24–30.

Somers, T. C., and Ziemelis, G. (1985). Flavonol haze in white wines. *Vitis* **24**, 43–50.

Soufleros, E., and Bertrand, A. (1988). Les acides gras libres du vin. Observations sur leur origine. *Connaiss. Vigne Vin* **22**, 251–260.

Stuckey, W., Iland, P., Henschke, P. A., and Gawel, R. (1991). The effect of lees contact time on Chardonnay wine composition. *Proc. Int. Symp. Nitrogen Grapes Wine, Seattle, WA, June, 1991* (J. M. Rantz, ed.), pp. 315–319. Am. Soc. Enol. Vitic., Davis, California.

Sudraud, P., and Chauvet, S. (1985). Activité antilevure de l'anhydride sulfureux moléculaire. *Connaiss. Vigne Vin.* **19**, 31–40.

Sulo, P., Michačáková, S., and Reiser, V. (1992). Construction and properties of K_1 type killer wine yeast. *Biotechnol. Lett.* **14**, 55–60.

Suzzi, G., Romano, P., and Zambonelli, C. (1985). *Saccharomyces* strain selection in minimizing SO_2 requirement during vinification. *Am. J. Enol. Vitic.* **36**, 199–202.

Thornton, R. J. (1985). The introduction of flocculation into a homothallic wine yeast. A practical example of the modification of winemaking properties by the use of genetic techniques. *Am. J. Enol. Vitic.* **36**, 47–49.

Tyagi, R. D. (1984). Participation of oxygen in ethanol fermentation. *Process Biochem.* **19**(4), 136–141.

van der Walt, J. P. (1970). The genus *Saccharomyces* (Meyer) Reess. *In* "The Yeasts, A Taxonomic Study" (J. Lodder, ed.), 2nd Ed. pp. 555–718. North-Holland Publ., Amsterdam.

Vannobel, C. (1986). Réfexions sur l'évolution récente de la technologie en matière de maîtrise des températures de fermentation. *Prog. Agric. Vitic.* **21**, 488–494.

Vaughn, R. H. (1955). Bacterial spoilage of wines with special reference to California conditions. *Adv. Food Res.* **7**, 67–109.

Veláquez, J. B., Longo, E., Sieiro, C., Cansado, J., Calo, P., and Villa, T. G. (1991). Improvement of the alcoholic fermentation of grape juice with mixed cultures of *Saccharomyces cerevisiae* wild strains. Negative effect of *Kloeckera apiculata*. *World J. Microbiol. Biotechnol.* **7**, 485–489.

Villettaz, J. C. (1987). A new method for the production of low alcohol wines and better balanced wines. *Proc. 6th Aust. Wine Ind. Tech. Conf.* (T. Lee, ed.), pp. 125–128. Australian Industrial Publ., Adelaide, Australia.

Vos, P. J. A., and Gray, R. S. (1979). The origin and control of hydrogen sulfide during fermentation of grape must. *Am. J. Enol. Vitic.* **30**, 187–197.

Wenzel, K., Dittrich, H. H., Seyffardt, H. P., and Bohnert, J. (1980). Shwefelrüchstände auf Trauben und Most und ihr Einfluss auf die H_2S Bildung. *Wein-Wiss.* **6**, 414–420.

Whiting, G. C. (1975). Some biochemical and flavour aspects of lactic acid bacteria in ciders and other alcoholic beverages. *In* "Lactic Acid Bacteria in Food and Beverages" (J. G. Carr, C. V. Cutting, and G. C. Whiting, eds.), pp. 68–85. Academic Press, London.

Wibowo, D., Eschenbruch, R., Davis, C. R., Fleet, G. H., and Lee, T. H. (1985). Occurrence and growth of lactic acid bacteria in wine: A review. *Am. J. Enol. Vitic.* **36**, 302–313.

Williams, L. A. (1982). Heat release in alcoholic fermentation: A critical reappraisal. *Am. J. Enol. Vitic.* **33**, 149–153.

Williams, L. A., and Boulton, R. (1983). Modelling and prediction of evaporative ethanol loss during wine fermentations. *Am. J. Enol. Vitic.* **34**, 234–242.

Wolf, E., and Benda, I. (1965). Qualitat und Resistenz. III. Das Futterwahlvermögen von *Drosophila melanogaster* gegenüber natürlichen Weinhefe-Arten und -Rassen. *Biol. Zentralbl.* **84**, 1–8.

Zoecklein, B., Fugelsang, K. C., Gump, B. H., and Nury, F. S. (1990). "Production Wine Analysis." Van Nostrand-Rheinhold, New York.

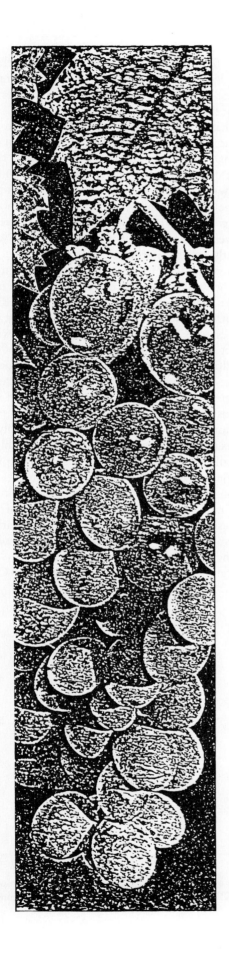

8

Postfermentation Treatments and Related Topics

All wines undergo a period of adjustment (maturation) before bottling. Maturation involves the precipitation of particulate and colloidal material from the wine as well as a complex range of physical, chemical, and biological changes that tend to maintain and/or improve the sensory characteristics of the wine. These processes occur spontaneously, but can be facilitated by the winemaker. Although wines usually benefit from the judicious intervention of the winemaker, undue use can alter the natural character of the wine.

Wine Adjustments

Adjustments attempt to correct deficiencies found in the grapes and sensory imbalances that develop during fermentation. In certain jurisdictions, acidity and sweetness adjustments are permitted only before fermentation. This is regrettable, as it is impossible to predict precisely the course of fermentation. Judicious adjustment after vinification can improve the finished wine, without compromising regional or varietal characteristics.

Acidity and pH Adjustment

Theoretically, acidity and pH adjustment could occur at almost any stage during vinification. Nevertheless, postfermentative correction is probably optimal. During fermentation, deacidification often occurs spontaneously owing to acid precipitation and yeast or bacterial metabolism. In addition, some strains of *Saccharomyces cerevisiae* synthesize significant amounts of malic acid during fermentation (Farris *et al.,* 1989). Thus, the extent and type of deacidification needed are difficult to assess before the end of fermentation. However, if the juice is above pH 3.4, some lowering of the pH before fermentation may be advisable to favorably influence fermentation and avoid large adjustments following fermentation.

No precise recommendations for optimal acidity are possible as they reflect stylistic and regional preferences. More fundamentally, acidity and pH are complexly interrelated. The major fixed acids in grapes (tartaric and malic) occur in a dynamic equilibrium of various ionized and nonionized states (Fig. 8.1). Both tartaric and malic acids may exist as undissociated (nonionized) acids, in half-ionized forms (one ionized carboxyl group), in fully ionized states (both carboxyl groups ionized), as half-salts (one carboxyl group associated with a cation), as full salts (both carboxyl groups bound to cations), or as double salts with other acid molecules and cations. The proportion of interconvertible states depends largely on the concentration of the acids and potassium ions. Because of the complexity of the equilibria, it is presently impossible to predict precisely the consequences of changing any one of the factors affecting acidity. A range

between 0.55 and 0.85% total acidity is generally considered desirable. Red wines are customarily preferred at the lower end of the range, while white wines are preferred at the upper end.

Another important aspect of acidity is pH. It represents the proportion of H^+ to OH^- ions in aqueous solutions. The higher the proportion of H^+, the lower the pH; conversely, the higher the proportion of OH^-, the higher the pH. Wines vary considerably in pH, with values below 3.1 being sensed as sour while those above 3.7 taste "flat." White wines are commonly preferred at the lower end of the pH range, while red wines are frequently favored in the mid range.

Relatively low pH values are preferred for many reasons. They give wines a fresh taste, improve microbial stability, reduce browning, diminish the need for SO_2 addition, and enhance the production and stability of "fruit" esters. Concentrations of monoterpenes also may be affected. For example, the concentration of geraniol, citronellol, and nerol may rise at low pH, while those of linalool, α-terpineol, and hotrienol fall. In red wines, color intensity and hue are better at lower pH values.

Because of the importance of pH, choice of acidity correction procedure is influenced considerably by how it affects pH. Because tartaric acid is more highly ionized than malic acid within the usual range of wine pH values, adjusting the concentration of tartaric acid has a greater effect on pH than an equivalent change in the concentration of malic acid (Fig. 8.2). Thus, where correction in pH is desired, adjusting the concentration of tartaric acid is preferable. Where changes in pH should be minimized, adjustment in the malic acid concentration is favored. Acidification before fermentation is discussed in Chapter 7.

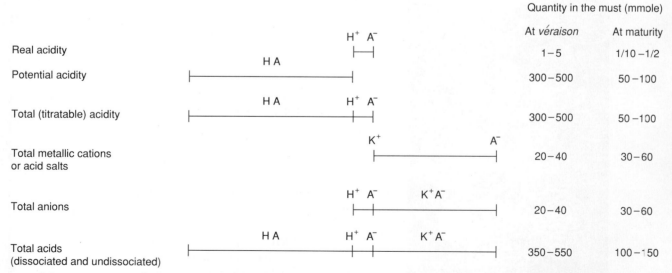

Figure 8.1 Different notions characterizing the acid–base equilibrium of a solution. The values given correspond to those found in must. A^-, Acid anion; H^+, hydrogen ion; HA, undissociated acid; K^+, potassium ion; K^+A^-, undissociated potassium salt. (After Champagnol, 1986, reproduced by permission.)

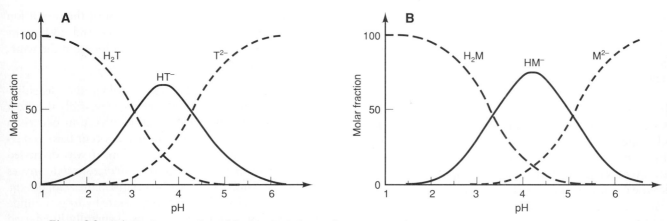

Figure 8.2 Relative concentrations of the three main forms of (A) tartaric and (B) malic acids as a function of pH. H_2, Full acid (unionized); H, half-acid (half-ionized); $^{2-}$, fully ionized acid. (After Champagnol, 1986, reproduced by permission.)

DEACIDIFICATION

Wine may be deacidified by either physicochemical or biological means. Physicochemical deacidification involves either acid precipitation or column ion-exchange. Biological deacidification usually involves malolactic fermentation (see Chapter 7).

Precipitation Precipitation primarily entails the neutralization of tartaric acid; malic acid is less involved owing to the higher solubility of its salts in wine. Neutralization can result as cations (positively charged ions) of an added inorganic salt exchange with hydrogen ion(s) of the organic acid. Salt formation can reduce the solubility of an acid, inducing crystallization and precipitation. Removal of the precipitated salt by racking, filtration, or centrifugation makes the reaction nonreversible.

To induce neutralization and precipitation of tartaric acid, finely ground calcium carbonate may be added to the wine. The reaction is as follows:

Deacidification with calcium carbonate is probably the most common procedure, as use of potassium carbonate is prohibited in several countries and potassium tartrate tends to be more expensive. Although widely used, calcium carbonate has a number of disadvantages. Its primary drawback is the slow rate at which calcium tartrate precipitates. In addition, formation of the soluble salt of calcium malate may produce a salty taste. Furthermore, if tartrate removal is excessive, the resultant increase in pH may leave the wine tasting "flat" and susceptible to microbial spoilage.

Some of the disadvantages of calcium carbonate addition may be avoided by the use of "double-salt" deacidifiction. The name refers to the belief that the technique functioned primarily by the formation of an insoluble double salt between malic acid, tartaric acid, and calcium. It now appears that very little of the hypothesized double salt forms. Instead, the procedure seems to both speed the precipitation of calcium tartrate and facilitate

$$2 \,\,\overset{\displaystyle O}{\underset{\displaystyle HO}{>}}C-\underset{\underset{\displaystyle H}{|}}{\overset{\overset{\displaystyle OH}{|}}{C}}-\underset{\underset{\displaystyle OH}{|}}{\overset{\overset{\displaystyle H}{|}}{C}}-C\overset{\displaystyle OH}{\underset{\displaystyle O}{<}} + CaCO_3 \rightarrow \,\,\overset{\displaystyle O}{\underset{\displaystyle HO}{>}}C-\underset{\underset{\displaystyle H}{|}}{\overset{\overset{\displaystyle OH}{|}}{C}}-\underset{\underset{\displaystyle OH}{|}}{\overset{\overset{\displaystyle H}{|}}{C}}-C\overset{\displaystyle O^- \,\,Ca^{2+} \,\,^-O}{\underset{\displaystyle O}{<}} \,\,\overset{\displaystyle}{>}C-\underset{\underset{\displaystyle OH}{|}}{\overset{\overset{\displaystyle H}{|}}{C}}-\underset{\underset{\displaystyle H}{|}}{\overset{\overset{\displaystyle OH}{|}}{C}}-C\overset{\displaystyle O}{\underset{\displaystyle OH}{<}} + CO_2 + H_2O$$

Neutralization also can result from the formation of insoluble half-salts. Deacidification with potassium tartrate acts in this manner:

$$\overset{\displaystyle O}{\underset{\displaystyle K^+{}^-O}{>}}C-\underset{\underset{\displaystyle H}{|}}{\overset{\overset{\displaystyle OH}{|}}{C}}-\underset{\underset{\displaystyle OH}{|}}{\overset{\overset{\displaystyle H}{|}}{C}}-C\overset{\displaystyle O^-K^+}{\underset{\displaystyle O}{<}} + 2 \,\,\overset{\displaystyle O}{\underset{\displaystyle HO}{>}}C-\underset{\underset{\displaystyle H}{|}}{\overset{\overset{\displaystyle OH}{|}}{C}}-\underset{\underset{\displaystyle OH}{|}}{\overset{\overset{\displaystyle H}{|}}{C}}-C\overset{\displaystyle OH}{\underset{\displaystyle O}{<}} \longrightarrow 2 \,\,\overset{\displaystyle O}{\underset{\displaystyle HO}{>}}C-\underset{\underset{\displaystyle H}{|}}{\overset{\overset{\displaystyle OH}{|}}{C}}-\underset{\underset{\displaystyle OH}{|}}{\overset{\overset{\displaystyle H}{|}}{C}}-C\overset{\displaystyle O^-K^+}{\underset{\displaystyle O}{<}} + \,\,\overset{\displaystyle O}{\underset{\displaystyle HO}{>}}C-\underset{\underset{\displaystyle H}{|}}{\overset{\overset{\displaystyle OH}{|}}{C}}-\underset{\underset{\displaystyle OH}{|}}{\overset{\overset{\displaystyle H}{|}}{C}}-C\overset{\displaystyle OH}{\underset{\displaystyle O}{<}}$$

the partial precipitation of calcium malate (Cole and Boulton, 1989).

The major difference between the single- and double-salt procedures is that the latter involves the addition of calcium carbonate to only a small proportion (~10%) of the wine to be deacidified. Sufficient calcium carbonate is added to raise the pH above 5.1 This assures adequate dissociation of both malic and tartaric acids (see Fig. 8.2) and induces the rapid formation and precipitation of the salts. A patented modification of the double-salt procedure (Acidex) incorporates 1% calcium malate–tartrate with the calcium carbonate. The double salt possibly acts as seed crystals, promoting the rapid growth of salt crystals.

In double-salt procedures, the remainder of the wine is slowly blended into the treated portion with vigorous stirring. Subsequent crystal removal occurs by filtration, centrifugation, or settling. Stabilization of the treated wine may take 3 months, during which residual salts are allowed to precipitate before bottling.

Although precipitation works well with wines of medium to high total acidity (6–9 g/ml) and medium to low pH (<3.5), it can result in an excessive pH rise in wines showing both high acidity (>9 g/ml) and high pH (>3.5). This situation is most common in cool climatic regions, where malic acid constitutes the major acid and the potassium content is high. In this situation, column ion-exchange may be used (Bonorden *et al.*, 1986) or tartaric acid added to the wine before addition of calcium carbonate in double-salt deacidification (Nagel *et al.*, 1988). Precipitation following neutralization removes the excess potassium with acid salts, and the additional tartaric acid lowers the pH to an acceptable value.

Because "protective" colloids can significantly affect the precipitation of acid salts, it is important to conduct deacidification trials on small samples of the wine to be treated. This will establish the amount of calcium carbonate or Acidex required for the desired degree of deacidification.

Column Ion-Exchange   Ion exchange involves passing the wine through a resin-containing column. During passage, ions in the wine exchange with those in the column. The types of ions exchanged can be adjusted by the type of resin used and the ions present for exchange on the resin.

For deacidification, the column is packed with an anion-exchange resin. Tartrate ions are commonly exchanged with hydroxyl ions (OH^-), thus removing tartrate from the wine. The hydroxyl ions released from the resin associate with hydrogen ions, forming water. Alternately, malate may be removed by exchange with a tartrate-charged resin. The excess tartaric acid may be subsequently removed by neutralization and precipitation.

Currently, the major limiting factor in the use of ion exchange other than legal restrictions and cost, is its tendency to remove flavorants and color from the wine, reducing wine quality.

Biological   Deacidification   Biological deacidification, via malolactic fermentation, is possibly the most common means by which acidity correction occurs in wine. As malolactic fermentation can occur before, during, and after alcoholic fermentation, it was discussed previously in Chapter 7. The yeast *Schizosaccharomyces pombe* also effectively decarboxylates malic acid, but the strong tendency of the yeast to generate hydrogen sulfide and other off-odors makes its use generally ill-advised (see Beelman and Gallander, 1979). However, as some strains of *Saccharomyces cerevisiae* can degrade nearly 50% of the malic acid content in juice (Rankine, 1966), their use in fermentation could achieve partial deacidification.

ACIDIFICATION

When wines are too low in acidity or pH, tartaric acid is commonly added to correct the fault. The advantages of tartaric acid as an acidulant include its high microbial stability and a dissociation constant (K_a, commonly expressed as the negative logarithm pK_a) that allows it to have a marked effect in lowering the pH. The main disadvantage is crystal formation if the wine has a high potassium content. Citric acid addition does not have this problem and can assist in preventing ferric casse. However, the ease with which citric acid can be metabolized means that it is microbially unstable. Alternately, ion exchange may be used to lower pH by exchanging H^+ for the Ca^{2+} or K^+ of tartrate and malate salts.

Sweetening

In the past, stable naturally sweet wines were rare. Most of the sweet wines of antiquity probably contained boiled-down must or honey. Stabilization by the addition of distilled alcohol is a comparatively recent development. The stable naturally sweet wines of the past few centuries seem to have been produced from highly botrytized grapes (see Chapter 9). In contrast, modern technology can produce a wide range of sweet wines without recourse to botrytization, baking, or fortification.

Wines may be sweetened with sugar, for example, sparkling wines. However, most still wines possessing a sweet finish are sweetened by the addition of partially fermented or unfermented grape juice, termed **sweet reserve** (*süssreserve*). The base wine is typically fermented dry and sweetened just before bottling. To avoid microbial contamination, both wine and sweet reserve are sterilized by filtration, or pasteurized, and the blend is bottled under aseptic conditions using sterile bottles and corks.

Various techniques are used in preparing and preserving sweet reserve. One procedure involves separating a small portion of the juice to produce the sweet reserve with the same varietal, vintage, and geographical origin as the wine it sweetens. If the sweet reserve is partially fermented, yeast activity may be terminated prematurely by chilling, filtration, or centrifugation, or by trapping the carbon dioxide released during fermentation. The pressure buildup stops fermentation. If the sweet reserve is stored as unfermented juice, microbial activity is restricted after clarification by cooling to $-2°C$, applying CO_2 pressure, pasteurizing, or sulfiting to above 100 ppm of free SO_2. In the latter case, desulfiting before use is conducted by flash heating and sparging with nitrogen gas. If desired, reverse osmosis or cryoextraction can concentrate the juice. Heat and vacuum concentration are additional possibilities but are likely to result in greater flavor modification and fragrance loss.

Dealcoholization

In recent years, there has been an increasing market for low alcohol and dealcoholized wines. Techniques have been available for the production of such wines since the early 1900s. Until comparatively recently, the process required heating and the evaporation of 50 to 70% of the wine to reduce the alcohol content to 4 g/liter. With the advent of vacuum distillation, the temperature required could be reduced, thus avoiding the baked or cooked character associated with heating. Nevertheless, important volatiles were lost with the alcohol. Most can be retrieved and added back to the wine, but the final product was still lacking in some of the original character. Modern strip column distillation techniques apparently require a loss of only about 20 to 30% of the wine to reach 9 g ethanol/liter (Ireton, 1990; Duerr and Cuénat, 1988). The use of dialysis apparently results in little flavor loss (Wucherpfennig *et al.*, 1986). Nevertheless, the most widely used dealcoholization technique currently involves reverse osmosis.

Reverse osmosis derives its name from the reversal of the normal flow of water in osmosis. Osmosis is the diffusion of water across a differentially permeable membrane, from a region of higher to lower concentration. If sufficient pressure is exerted on the more concentrated solution, diffusion of water and other permeable substances is reversed, with net movement occurring across the membrane into the dilute solution. The technique is widely used in the selective concentration or elimination of low molecular weight compounds from juices. The limitation of its use in viniculture to dealcoholization probably results from the current unavailability of membranes with appropriate permeability characteristics.

With the use of a proper support system, and sufficient pressure, reverse osmosis can reduce the alcohol content of wine to almost any degree desired. However, as water is removed along with the ethanol, water must be added back to the concentrated wine or added to the wine before use of reverse osmosis. This creates legal problems where the addition of water to wine is prohibited. Depending on the particular membrane used, compounds such as esters, aldehydes, and organic acids may be lost. This can result in significant fragrance loss in the dealcoholized wine. The isolation of the compounds and their addition to the wine generally has not proved satisfactory. However, addition of grape juice or concentrate may provide some of the former fragrance of the wine. Addition of organic acids lost during reverse osmosis may further enhance the flavor of the wine.

The problem of adding water has been circumvented by an ingenious system involving double reverse osmosis (Bui *et al.*, 1986). It produces alcohol-reduced and alcohol-enriched wines simultaneously. By interconnecting both systems, no water needs to be added to the alcohol-reduced concentrate, nor is there the legal problem of producing an alcohol "distillate." The system cannot produce completely dealcoholized wines, however.

Where the addition of water is permissible in the production of low alcohol beverages, dilution is the simplest means of dealcoholization. Flavor enhancement, as with wine coolers, offsets flavor dilution.

Color Adjustment

Heating newly fermented wine to 35 to 40°C for 24 to 48 hr before pressing also has been advocated where immature or slightly molded grapes are involved (Ribéreau-Gayon and Ribéreau-Gayon, 1980). The process is reported to improve flavor, color, and tannic structure. However, excessive heating can induce the loss of color, make the wine aggressively tannic, and require inoculation to achieve malolactic fermentation.

Both red and white wines may be partially or completely decolored by the membrane technique called **ultrafiltration.** Depending on the permeability characteristics of the membrane, ultrafiltration selectively retains macromolecules based on molecular size. With membranes of lower cutoff values ($\sim$500 daltons), ultrafiltration also can remove the "pinking" produced by small procyanidin molecules. Use of filters with even lower cutoff values can produce blush or white wines from red or rosé wines. The major factor limiting the more widespread use of ultrafiltration in enology is its potential to remove important flavorants along with macromolecules.

An alternative technique for removing brown and pink pigments involves the addition of polyvinylpolypyrrolidone (PVPP). By binding tannins into large macromolecular complexes, PVPP greatly facilitates their removal by filtration or centrifugation.

Other means of color removal have involved the addition of casein or special preparations of activated charcoal. With activated charcoal, the simultaneous removal of aromatic compounds and the occasional donation of off-odors have limited its use.

Blending

Blending of wine from different varieties is a long established procedure in many wine regions. The mixed varietal planting of the past had a similar result but precluded the selective blending of individual varietal wines after fermentation. Separating lots of wine based on varietal, vineyard, or maturity differences is more difficult and expensive, but it permits blending based on the wishes of the cellarmaster.

The art of the blender is especially important in the production of fortified and sparkling wines. Without blending, the creation and maintenance of the brand name products that characterize fortified and sparkling wines would be impossible. Computer-aided systems have been proposed to facilitate this important activity (Datta and Nakai, 1992).

The production of standard-quality table wines also is largely dependent on blending. Consistency of character typically is more important than the vintage, varietal, or vineyard origin of the wine. The skill of the blender is often amazing, given the diversity of wines often involved in the formation of each blend.

Blending also is used in the production of many premium table wines. In this case, however, the wines usually come from the same geographical region and may be from a single vineyard and/or vintage. Limitations on blending are usually precisely articulated in the Appellation Control laws affecting the region concerned; frequently the more prestigious the region, the more restrictive the legislation.

Currently, there is little to guide blenders other than past experience. Few studies have investigated the scientific basis of the blender's art. Nevertheless, several studies have proposed methods of predicting color based on the pigmentation characteristics of potential base wines. Color is very important because of its strong biasing influence on quality perception. Blending diagrams may be developed using colorimeter readings (Little and Liaw, 1974). The diagrams are based on the reflectance of the wines in the red, green, and blue portions of the visible spectrum. Based on the results, the selection and proportions of each lot required to achieve a desired color may be established. A more complex system based on a Scheffé design has been proposed by Negueruela *et al.* (1990).

One of the advantages of blending is an improvement in fragrance and flavor. In a classic study by Singleton and Ough (1962), pairs of wines ranked similarly, but noted to be distinctly different, were reassessed along with a 50 : 50 blend of both wines. In no case was the blend ranked more poorly than the lower ranked of the two source wines. More importantly, about 20% of the blends were ranked higher than the unblended source wines (Singleton and Ough, 1962).

Although the origin of the improved sensory quality of blended wines is unknown, it probably relates to the increased flavor subtlety of the blend. Human taste and odor perception commonly respond in a nonlinear manner to increasing or decreasing concentrations of flavorants. Therefore, dilution of a flavor ingredient during blending need not necessarily diminish perception. Thus, a blend may express the flavors of all the wines used. In addition, dilution may reduce the concentration of undesirable odors to below the detection threshold. Finally, the qualitative perception of some off-odors improves with dilution (see Chapter 11).

When blending should occur depends largely on the type and style of wine involved. In sherry production, blending occurs periodically throughout the long maturation period. With sparkling wines, blending is commonly done in the spring following the harvest. At this point the unique features of the wines are beginning to be expressed. Red wines also are typically blended in the spring following the vintage. Important in deliberation is not only the proportional amounts of each wine to be blended, but also the amount of pressings to use. Wines from poor vintages customarily benefit more from the addition of extra press wine than do wines from good vintages. Press fractions contain a higher proportion of pigment and tannins than the free-run. After blending, the wine is often aged for several weeks, months, or years before bottling. The addition of pressings also can provide extra "body" and color to white wines.

Wines may not be blended for a number of reasons. Wines produced from grapes of especially high quality are usually kept and bottled separately to retain the distinctive characteristics undiluted. For wines produced from famous vineyard sites, blending with wine from other sites, regardless of quality, would prohibit the owner from using the name of the site. This could markedly reduce the market value of the wine. With famous sites, uniqueness of origin can be more important than perceptible quality.

Stabilization and Clarification

Stabilization and clarification involve procedures designed to produce a permanently clear wine with no flavor faults. Because the procedures can themselves create problems, it is essential that they be used judiciously and only to the degree necessary to solve specific problems.

Stabilization

TARTRATE AND OTHER CRYSTALLINE SALTS

Tartrate stabilization is one of the facets of wine technology most influenced by consumer perception. The presence of even a few tartrate crystals is inordinately feared, or at least misinterpreted, by many wine consumers. As a consequence, considerable effort and expense are expended in avoiding the formation of crystalline deposits in bottled wine. Stabilization is normally achieved by enhancing crystallization, followed by removal. However, it also may be attained by delaying or inhibiting crystallization.

Potassium Bitartrate Instability Juice is typically supersaturated with potassium bitartrate at crushing. As the alcohol content rises during fermentation, the solubility of bitartrate decreases. This induces the slow precipitation of potassiumn bitartrate (cream of tartar). Given sufficient time, the salt crystals usually precipitate spontaneously. In northern regions, the low temperatures found in unheated cellars may produce adequately rapid precipitation, but spontaneous precipitation is seldom satisfactory in warmer areas. Early bottling aggravates the problem. Where spontaneous precipitation is inadequate, refrigeration often achieves rapid and satisfactory bitartrate stability.

Because the rate of bitartrate crystallization is directly dependent on the degree of supersaturation, wines only mildly unstable may be insufficiently stabilized by cold treatment. In addition, protective colloids may retard crystallization such as mannoproteins (Lubbers *et al.*, 1993). If the colloids precipitate after bottling, the released bitartrates are free to crystallize. Because of the incomplete understanding of tartrate crystallization in wine, and the importance of the process, it is still under active investigation.

At the simplest, potassium bitartrate exists in a dynamic equilibrium between various ionized and salt states:

Under supersaturated conditions, crystals form and eventually reach a critical mass that induces precipitation. Crystallization continues until an equilibrium develops. If sufficient crystallization and removal occur before bottling, bitartrate stability is achieved. Because chilling decreases solubility it speeds crystallization.

Theoretically, chilling should establish bitartrate stability. However, charged particles in wine can interfere with crystal initiation and growth. For example, positively charged bitartrate crystals attract negatively charged colloids onto their surfaces, blocking growth. The charge on the crystals is created by the tendency of more potassium than bitartrate ions to associate with the crystals early in growth (Rodriguez-Clemente and Correa-Gorospe, 1988). Crystal growth also may be delayed by the binding of bitartrate ions to positively charged proteins. This reduces the amount of free bitartrate and, thereby, the rate of crystallization. Because both bitartrate and potassium ions may bind with tannins, crystallization tends to be delayed more in red than in white wines. The binding of potassium with sulfites is another source of delayed bitartrate stabilization.

For cold stabilization, table wines are routinely chilled to near the freezing point of the wine. Five days is usually sufficient at $-5.5°C$, but 2 weeks may be necessary at $-3.9°C$. Fortified wines are customarily chilled to between $-7.2°$ and $-9.4°C$, depending on the alcoholic strength. The stabilization temperature can be estimated using the formula empirically established by Perin (1977):

$$\text{Temperature } (-°C) = (\% \text{ alcohol} \div 2) - 1$$

Direct seeding with potassium bitartrate crystals is occasionally employed to stimulate crystal growth and deposition. Another technique involves filters incorporating seed crystals. The chilled wine is agitated and then passed through the filter, where crystal growth is encouraged by the dense concentration of crystal nuclei. The filter acts as a support medium for the seed nuclei.

At the end of conventional chilling, the wine is filtered or centrifuged to remove the crystals. Crystal removal is performed before the wine warms to ambient temperatures.

Because of the expense of refrigeration, various procedures have been developed to determine the need for cold stabilization. At present, none of the techniques appears to be sufficiently adequate. Potassium conductivity, while valuable, is too complex for regular use in most wineries. Thus, emperical freeze tests are still the most commonly used means of assessing bitartrate stability. For details on the various tests, the reader is directed to Goswell (1981) or Zoecklein *et al.* (1990).

Reverse osmosis is an alternative technique to chilling, agitation, and the addition of nucleation crystals. With the removal of water, the increased bitartrate concentration augments crystallization and precipitation. After crystal removal, the water is added back to reestablish the original balance of the wine. Electrodialysis is another membrane technique occasionally used for bitartrate stabilization.

Another technique particularly useful with wines having high potassium contents is column ion-exchange. Passing the wine through a column packed with sodium-containing resin exchanges sodium for potassium. Sodium bitartrate is more soluble than the potassium salt and is therefore much less likely to precipitate. Although effective, ion exchange is not the method of choice. Not only is it prohibited in certain jurisdictions, for example, EEC countries, but it also increases the sodium content of the wine. The high potassium, low sodium content of wine is one of its healthful properties.

If the wine is expected to be consumed shortly after bottling, treatment with metatartaric acid is an inexpensive means of establishing short-term tartrate stability. Metatartaric acid is produced by the formation of ester bonds between hydroxyl and acid groups of tartaric acid. The polymer is generated during prolonged heating of tartaric acid at 170°C. When added to wine, metatartaric acid restricts potassium bitartrate crystallization and interferes with the growth of calcium tartrate crystals. As metatartaric acid slowly hydrolyzes back to tartaric acid, the effect is temporary. At storage temperatures between 12° and 18°C, it may be effective for about 1 year. Because hydrolysis is temperature dependent, the stabilizing action of metatartaric acid quickly disappears above 20°C. The metatartaric acid is added just before bottling.

Calcium Tartrate Instability Instability caused by calcium tartrate is more difficult to control than that induced by potassium bitartrate. Fortunately, it is less common. Calcium-induced problems usually arise from the excessive use of calcium carbonate in deacidification, but they also may come from cement cooperage, filter pads, and fining agents.

Calcium tartrate stabilization is more difficult because precipitation is not activated by chilling. It may take many months for stability to develop spontaneously. Seeding with calcium tartrate crystals, simultaneously with calcium carbonate in deacidification, greatly enhances precipitation (Neradt, 1984). Because the formation of crystal nuclei requires more free energy than crystal growth, seeding circumvents the major limiting factor in the development of stability. A racemic mixture of calcium tartrate seed nuclei, containing both L and D isomers of tartaric acid, is preferred. The racemic mix-

ture is about one-eighth as soluble as the naturally occurring L-tartrate salt. The slow conversion of the L-form to the D-form during aging is one of the major causes of the calcium tartrate instability in bottled wine. Because premature clarification removes "seed" crystals that promote crystallization, filtration of wines deacidified with calcium carbonate should be delayed until stability has been achieved. Crystal growth and precipitation occur optimally between 5° and 10°C. Protective colloids such as soluble proteins and tannins can restrict crystal nucleation, but they do not inhibit crystal growth (Postel, 1983).

Calcium content may be directly reduced through ion exchange. Because of the efficiency of ion removal, typically only part of the wine needs to be treated. The treated sample is then blended back into the main volume of the wine. Treating only a small portion of the wine minimizes the flavor loss often associated with the ion-exchange technique.

Other treatments showing promise are the addition of pectic and alginic acid colloids to restrict crystallization and keep calcium tartrate in solution (Wucherpfennig et al., 1984).

Other Calcium Salt Instabilities Occasionally, crystals of calcium oxalate form in wine. The development occurs late, commonly after bottling. The redox potential of most wines stabilizes the complex formed between oxalic acid and metal ions such as iron. However, as the redox potential of the wine may eventually rise during aging, ferrous oxalate changes into the unstable ferric form. On dissociation, oxalic acid may bond with calcium, forming calcium oxalate crystals.

Oxalic acid is commonly derived from grape must, but small amounts may form from iron-induced structural changes in tartaric acid. Oxalic acid can be removed by blue fining early in maturation (Amerine et al., 1980), but avoiding the development of high calcium levels in the wine is preferable.

Other potential troublesome sources of crystals are saccharic and mucic acids. Both are produced by the fungus *Botrytis cinerea* during grape infection and form insoluble calcium salts. The addition of calcium carbonate for bitartrate stability can induce their crystallization, precipitation, and separation before bottling.

PROTEIN STABILIZATION

Although less commonly a cause of wine rejection than crystal formation, protein haze can cause considerable economic loss in bottle returns. Protein haze results from the clumping of dissolved proteins into light-dispersing particles. Heating accelerates the process, as does reaction with tannins and heavy metals.

The majority of proteins suspended in wine have an

isoelectric point (**p*I***) above the pH range of wine. The isoelectric point is the pH at which a protein is electrically neutral. Consequently, most soluble proteins in wine possess a net positive charge, generated by ionization of the amino groups of proteins to $-NH_3^+$ groups. The similar charge on proteins slows clumping, while Brownian movement and association (hydration) with water delay settling. However, adsorption, denaturation, or neutralization by cation exchange with fining agents can induce proteins to coalesce and produce a haze.

The issue of which proteins are primarily involved in haze production is still poorly understood. In addition, protein instability is only weakly correlated with total protein content. Recent studies suggest that proteins ranging between 12,000 and 30,000 daltons, and having a low p*I* values (4.0 to 6.0), may constitute the most significant unstable protein fraction (Hsu and Heatherbell, 1987). Some glycoproteins also appear to be involved. Smaller proteins may become involved by reacting with tannins, pectins, and metal ions.

A number of procedures have been developed to achieve protein stability. The most common involves the addition of bentonite (Fig. 8.3). Because of the abundance of soluble cations associated with bentonite, extensive exchange of ions can occur with ionized protein amino groups. By weakening the association with water, cations make the proteins more liable to coalesce and precipitate. Flocculation and precipitation are further enhanced by adsorption onto the negatively charged plates of bentonite. Sodium bentonite is preferred as it

separates more readily into individual silicate plates in wine. This gives it the largest surface area of any clay and, thereby, the greatest potential for cation exchange and protein adsorption.

Although negatively charged proteins bind less effectively to bentonite, they also less commonly induce haze formation. They generally adhere only to the edges of the clay plates, where positive charges tend to occur. All proteins, including those with a neutral charge, may associate with bentonite through weak hydrogen bonds.

Other fining agents, such as tannins, are occasionally used in lieu of bentonite. Addition of tannins is ill-advised with white wine, as it can leave an off-odor and generate an astringent mouth-feel. However, when immobilized in porous silicon dioxide, tannic acid causes minimal flavor modification or wine loss (Weetal *et al.*, 1984). Kieselsol, a colloidal suspension of silicon dioxide, has occasionally been used to remove proteins. Wines also may be protein stabilized through heat treatment, followed by filtration or centrifugation.

In a recent comparison of various protein stabilization tests, Dubourdieu *et al.* (1988) recommend exposure of wine samples to 80°C for 30 min. On cooling, the sample is observed for signs of haziness.

Recently, ultrafiltration has been investigated as an alternative to bentonite or other types of fining (Hsu *et al.*, 1987). It has the advantage of minimizing wine loss caused by sediment formation. Ultrafiltration also eliminates the need for a final "polishing," centrifugation, or filtration. Although generally applicable for white wines, ultrafiltration results in excessive color and flavor loss in red wines.

POLYSACCHARIDE REMOVAL AND STABILITY

Pectinaceous and other mucilaginous polysaccharides can cause difficulty during filtration, as well as induce haze formation. The polysaccharides act as "protective" colloids by binding with other colloidal materials, slowing or preventing their precipitation. Multiple hydrogen bonds formed between water molecules and the polysaccharides help them remain in suspension.

Pectin levels can be reduced by the addition of pectinase. Pectinase is an mixture of enzymes that breaks the pectin polymer down into simpler, noncolloidal, galacturonic acid subunits. Other grape-derived polysaccharides, such as arabinans, galactans, and arabinogalactans, have little effect on haziness or filtration and do not require specific removal. The same is true for the mannans produced by yeasts. In contrast, β-glucans present in botrytized juice can cause serious filtration problems even at low concentrations (Villettaz *et al.*, 1984). This is especially serious in highly alcoholic wines, where ethanol induces aggregation of the glucans. A Kieselsol/gelatin mixture is apparently effective in re-

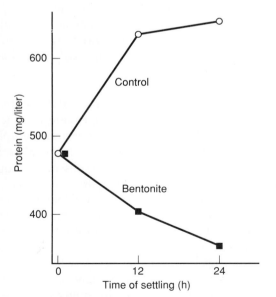

Figure 8.3 Total soluble protein in 'Gewürztraminer' must settled at 20°C for 24 hr in the presence or absence of 0.5 g/liter bentonite. (From Tyson *et al.*, 1982, reproduced by permission.)

moving these mucilaginous polymers. Alternately, the wine may be treated with a formulation of β-glucanases (Villettaz *et al.*, 1984). The enzymes hydrolzye the polymer, destroying both its protective colloidal action and filter plugging property.

TANNIN REMOVAL AND OXIDATIVE CASSE

Tannins may be directly and indirectly involved in haze development. On exposure to oxygen, tannins oxidize and polymerize into brown, light-diffracting colloids, causing **oxidative casse**. Polyphenol oxidases released from crushed grape cells speed the reactions, but slower nonenzymatic autooxidation reactions continue after enzyme inactivation (Fig. 6.8). Depending on the timing and degree of oxidation, tannin oxidation can result in a loss in color intensity or shift in hue, and it may enhance long-term color stability. Addition of sulfur dioxide limits oxidation through its joint antioxidant and antienzymatic properties. Moldy fruit contaminated with fungal polyphenol oxidases (laccases) are particularly susceptible to oxidative casse. Because laccases are poorly inactivated by sulfur dioxide, pasteurization may be the only convenient means of protecting the juice from excessive oxidation. Grapes free of fungal infection rarely develop oxidative casse. As the casse usually develops early during maturation and precipitates before bottling, it does not cause in-bottle clouding.

Chilling wine to achieve bitartrate stability may induce a protein/tannin haze. Filtration must occur while the wine is still cold, as the association between protein and tannin dissociates on warming. The removal of protein–tannin complexes enhances both tannin and protein stability.

Tannin stability is normally achieved by adding fining agents such as gelatin, egg albumin, and casein. The positive charge of the proteins attracts tannins, which are negatively charged. The interaction produces large protein–tannin complexes that settle out quickly and can be eliminated effectively by racking. The complexes may be removed earlier by filtration or centrifugation if required. The removal of excess tannins diminishes a major source of astringency, generates a smoother mouth-feel, reduces the likelihood of oxidative casse, and limits the formation of sediment following bottling.

With white wines, addition of PVPP (polyvinylpolylpyrrolidone) is a particularly effective means of removing tannins. Ultrafiltration also may be used to remove undesired tannins and other polyphenolic compounds from white wine. Ultrafiltration is seldom used with red wines, as the filter simultaneously may remove important flavorants and anthocyanins.

Additional, though infrequent, sources of phenolic instability include oak chips or shavings used to rapidly develop oak character (Pocock *et al.*, 1984) and the accidental incorporation of excessive amounts of leaf material in the grape crush (Somers and Ziemelis, 1985). Both can generate in-bottle precipitation if the wine is bottled early, but they can be avoided by permitting sufficient time for spontaneous precipitation during maturation. The instability associated with oak chip use results from the overextraction of ellagic acid. The **phenolic deposit** produced consists of a fine precipitate composed of off-white to fawn-colored ellagic acid crystals. The **flavonol haze** associated with the presence of leaf material during the crushing of white grapes is produced by the formation of fine yellow quercetin crystals.

Many premium-quality red wines develop a **tannin sediment** during prolonged bottle aging. This potential source of haziness is typically not viewed as a fault by wine connoisseurs. Individuals who customarily consume aged wines know its origin and often consider it a sign of quality.

METAL CASSE STABILIZATION

A number of heavy metals form insoluble salts and induce additional forms of haziness (*casse*). Although occurring much less frequently than in the past, metal casse is still of concern to winemakers.

The most important metal ions involved in casse formation are iron (Fe^{3+}, Fe^{2+}) and copper (Cu^{2+}, Cu^{+}). They may be derived from grapes, soil contamination, fungicidal residues, or winery equipment. Most metal ions so derived are lost during fermentation by coprecipitation with yeast cells. Troublesome concentrations of metal contaminants usually are associated with pickup subsequent to vinification.

Corroded stainless steel, improperly soldered joints, unprotected copper or bronze piping, and tap fixtures are the prime sources of contamination. Additional sources may be fining and decoloring agents, such as gelatin, isinglass, and activated charcoal, as well as cement cooperage.

Ferric Casse   Two forms of iron casse are known, white and blue (Fig. 8.4). **White casse** is most frequent in white wine and forms when soluble ferrous phosphate, $Fe_3(PO_4)_2$, is oxidized to insoluble ferric phosphate, $FePO_4$. The white haziness that results may be due to solely ferric phosphate or to a complex between it and soluble proteins. In red wines, the oxidation of ferrous ions (Fe^{2+}) to the ferric state (Fe^{3+}) can result in the formation of **blue casse**. Ferric ions form insoluble particles with anthocyanins and tannins. The oxidation of ferrous to ferric ions usually occurs when the wine is exposed to air. Sufficient oxygen may be absorbed during bottling to induce clouding in an unstable wine.

Ferric casse development is dependent on both the metallic content of the wine and its redox potential. Its

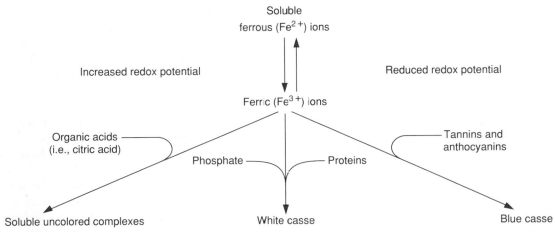

Figure 8.4 Iron-induced casse formation.

occurrence also is affected by pH, temperature, and the level of certain acids. White casse forms only below pH 3.6, and is generally suppressed at cool temperatures, whereas blue casse is accentuated at cold temperatures. The frequency of white casse increases sharply as the iron concentration rises above 15 to 20 mg/liter. Critical iron concentrations for blue casse production are more difficult to predict, as its occurrence is influenced markedly by the phosphate content of the wine and traces of copper (1 mg/liter). In addition, citric acid can chelate ferric and ferrous ions and reduce the effective (free) concentration in wine. Citric acid may be added to wine (~120 mg/liter) to limit the occurrence of ferric casse (Amerine and Joslyn, 1970).

Wines may be directly stabilized against ferric casse by iron removal. For example, the addition of phytates such as calcium phytate selectively removes iron ions. EDTA (ethylenediaminetetraacetic acid), pectinic acid, and alginic acid can be used to remove both iron and copper ions. Removal with ferrocyanide is probably the most efficient method, as it precipitates most metal ions, including iron, copper, lead, zinc, and magnesium. Addition as potassium ferrocyanide is known as **blue fining**. Blue fining is prohibited in many countries and is strictly controlled where it is permitted. As a buffered complex with bentonite, potassium ferrocyanide is sold as a proprietary formulation called Cufex. The Cufex formulation reduces the likelihood of toxic cyanide residues remaining in the wine after treatment. In either case, insoluble metal ferrocyanide complexes are removed by filtration.

Ferric casse may be equally controlled by the addition of agents that limit the flocculation of insoluble ferric complexes. Gum arabic acts in this manner. It functions as a protective colloid, restricting haze formation. Because gum arabic limits the clarification of colloidal ma-

terial, it can only be safely applied after the wine has received all other stabilization procedures.

Copper Casse While iron casse forms as a result of exposure to oxygen, copper casse forms only in the absence of oxygen. It develops only after bottling and is associated with the decrease in redox potential that accompanies aging. Exposure to light speeds the reduction of copper, critical in casse development. Sulfur dioxide is important, if not essential, as a sulfur source in copper casse formation. In a series of incompletely understood reactions, involving the generation of hydrogen sulfide, cupric and cuprous sulfides may form. The sulfides produce a fine, reddish brown deposit, or they flocculate with proteins to form a reddish haze. Copper casse is particularly a problem in white wines, but it also can cause haziness in rosé wines. Wines with copper contents greater than 0.5 mg/liter are particularly susceptible to copper casse (Langhans and Schlotter, 1985).

MASQUE

Occasionally, a deposit called *masque* forms on the inner surface of bottles of sparkling wine. It results from the deposition of material formed by the interaction of albumin, used as a fining agent, and fatty acids (Maujean *et al.*, 1978). Riddling and disgorging used to remove yeast sediment does not remove *masque*. *Masque* affects only traditionally produced (*méthode champenoise*) sparkling wines. With this technique, the bottle used for the second fermentation is the same as that in which the wine is sold.

MICROBIAL STABILIZATION

Microbial stability is not necessarily synonymous with microbial sterility. At bottling, wines may contain a considerable number of viable, but dormant, microorganisms. Under most situations, they cause no stability or sensory problems.

The simplest procedure conferring limited microbial stability is racking. Racking removes cells that have fallen out of the wine by flocculation or coprecipitation with tannins and proteins. The sediment includes both viable and nonviable microorganisms. The latter slowly undergo autolysis and release nutrients back into the wine. Cold temperatures help to maintain microbial viability but retard or prevent growth.

For long-term microbial stability, especially with sweet wines, the addition of antimicrobial compounds or sterilization is required. The antimicrobial agent most frequently used is sulfur dioxide. It may be added at various times during wine production, but almost always after fermentation. Concentrations of 0.8 to 1.5 mg/liter molecular sulfur dioxide inhibit the growth of most yeasts and bacteria. The total sulfur dioxide content required to maintain a desirable level of molecular SO_2 depends on the pH of the wine and the number of sulfur-binding compounds (see Chapter 6).

At present, no other permitted wine additive possesses the wide-spectrum antimicrobial activity of sulfur dioxide. Sorbic acid is an effective inhibitor of several yeasts but not others, such as *Zygosaccharomyces bailii* (Rankine and Pilone, 1973). In addition, sorbic acid can be metabolized by lactic acid bacteria, generating a strong geranium-like odor. Benzoic acid and sodium benzoate were once employed as yeast inhibitors, but their general ineffectiveness and taste modification have essentially eliminated their use. If used just before bottling, dimethyl dicarbonate (DMDC) can effectively sterilize the wine without producing sensory defects or leaving a residue. DMDC decomposes rapidly to carbon dioxide and methanol (Calisto, 1990). In the absence of sulfur dioxide or DMDC, bottled wines can be securely stabilized against microbial growth only by physical means, namely, pasteurization and filter sterilization.

Pasteurization is the older of the two techniques. It has the advantage of inactivating polyphenol oxidases (laccases), which are little affected by the concentrations of sulfur dioxide commonly used. It also facilitates protein and copper casse stabilization by denaturing and precipitating colloidal proteins. Although pasteurization may generate increased amounts of "protective" colloids, cause slight decolorization, and modify wine fragrance, it does not influence the aging process of phenols in wine (Somers and Evans, 1986).

Because the low pH and ethanol content of wine markedly depress the thermal resistance of yeasts and bacteria, heating for shorter periods or at lower temperatures than typical is recommended by Barillère *et al.* (1983). Barillère *et al.* indicate that about 3 min at 60°C should be sufficient for a wine at 11% ethanol. Flash

pasteurization at 80°C usually requires only a few seconds. Sulfur dioxide reduces still further the need for heating, as high temperatures markedly increase the proportion of free SO_2 in wine. Although pasteurization destroys most microbes, it does not inactivate the endospores of *Bacillus* species that occasionally cause wine spoilage.

Partially because of the complexities of establishing the most appropriate time and temperature conditions for pasteurization, membrane filters have replaced pasteurization in most situations. Filters also result in few physical or chemical disruptions to the sensory characteristics of wine. Membrane filters with a pore size of 0.45 μm or less are adequate for juice and wine filtration.

Wine sterilization requires the simultaneous use of measures to avoid recontamination. This involves sterilizing all parts of the bottling line and the use of sterile bottles and corks.

Sulfur dioxide is commonly added before wines are pasteurized or sterile filtered to confer protection against oxidation.

Fining

Fining is commonly used to accelerate the spontaneous precipitation of suspended material in wine. Fining agents bind to or adsorb particulate matter. The aggregates formed are generally sufficiently large to precipitate quickly. If not, removal can be achieved by centrifugation or filtration. In addition to facilitating clarification, fining may help stabilize wines against haze formation by removing compounds involved in haze production.

Because fining is an aid to, not a replacement for, spontaneous stabilization, it should be used only to the degree necessary. It is important to avoid sensory disruption by minimizing changes to the chemical and physical balance of the wine. Figure 8.5 illustrates the potential of fining agents to produce aroma changes through the selective removal of aromatic compounds. Fining also should be conducted as quickly as possible to avoid oxidation. Tests to determine the need for fining are discussed in most enological references, for example, Zoecklein *et al.* (1990). A brief description of the primary fining agents is given below.

ACTIVATED CHARCOAL (CARBON)

Activated charcoal is used primarily to decolorize wine and remove off-odors. Different preparations may be required for each application. The large surface area (between 500 and 1500 m^2/g) and electrical charge allow activated charcoal to adsorb a wide range of compounds effectively. Although this is useful in removing mercaptan off-odors, it simultaneously removes desir-

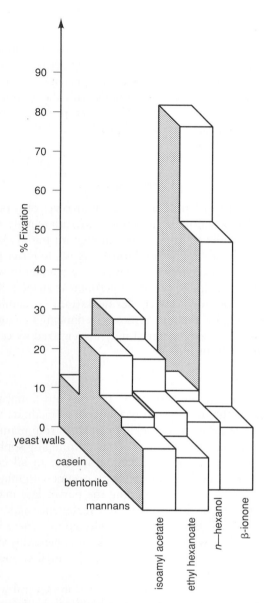

Figure 8.5 Percent removal of several aromatic compounds during fining with bentonite or 1% casein, mannans, or yeast cell walls. (Reprinted with permission from Voilley *et al.*, 1990. Copyright 1990 American Chemical Society.)

able flavor compounds. It also may give the treated wine an atypical odor. Furthermore, activated charcoal has an oxidizing property. As a consequence, ascorbic acid may be incorporated along with the activated charcoal. Although valuable in some situations, activated charcoal must be used with caution.

ALBUMIN

Egg white has long been used in fining wines. The active ingredient is the protein albumin. As with other protein fining agents, albumin is primarily employed to

remove excess tannins. Peptide linkages of the albumin form hydrogen bonds with hydroxyl groups on tannins. The opposing charges on the molecules favor the formation of large protein–tannin aggregates.

Egg albumin is presently available in pure form. Use of pure albumin avoids the necessity of adding sodium chloride to whipped egg whites to solubilize the albumins.

BENTONITE

Bentonite is a form of montmorillonite clay widely used as a fining agent. It is used in clarifying juice and wines, in removing heat-unstable proteins, and in deterring the development of copper casse. Depending on the objectives of the winemaker, the ability of bentonite to induce partial decolorization and remove nutrients such as amino acids is either an advantage or disadvantage. Together with other fining agents, such as tannins and casein, bentonite can speed the settling of particulate matter. It also can correct for the addition of excessive amounts of proteinaceous fining agents, by inducing their precipitation. Because bentonite settles out well and is easily filtered, it is one of the few fining agents that does not itself create a stability or clarification problem. Bentonite also has comparatively little effect on the sensory properties of the treated wine. The major drawbacks are promotion of color loss from red wines and a tendency to produce voluminous sediment. The latter can cause considerable wine loss during racking.

The bentonite often preferred in the United States is Wyoming bentonite. Because the predominant cation is monovalent (sodium), the particles swell readily in water and separate into separate sheets of aluminum silicate. The sheets are about 1 nm thick and 500 nm wide. The separation of the sheets provides an immense surface area over which cation exchange, adsorption, and hydrogen bonding can occur. When fully expanded, sodium bentonite has a surface area of about 700 to 800 m^2/g. Calcium bentonite is less commonly used as it tends to clump on swelling and provides less surface area for fining. Nevertheless, it has the advantage of producing a heavier sediment that is easier to remove and does not liberate sodium ions into the wine.

The net negative charge of bentonite attracts positively charged proteins. The latter are neutralized by cation exchange with the clay, and adsorption to the clay results in flocculation and settling as clay–protein complexes.

CASEIN

Casein is the major protein found in milk. In association with sodium or potassium ions, it forms a soluble caseinate salt that easily dissolves in wine. In wine, the salt dissociates and insoluble caseinate is released. The caseinate adsorbs and removes negatively charged particles as it settles out. Casein finds its primary use as a decolorant in white wines. It also has some deodorizing properties.

GELATIN

Gelatin is a soluble albuminlike protein derived from the tissues of animals after prolonged boiling. As a result the product loses some of its gelling properties but becomes a more effective fining agent.

Gelatin is employed primarily to remove excess tannins from wines. When gelatin is added to white wine, there is a risk of leaving a gelatin-derived haze. This may be avoided by the simultaneous addition of flavorless tannins, Kieselsol, or other protein-binding agents. These materials assist in the formation of the fine meshwork of gelatin fibers that removes tannins and other negatively charged particles. Excessive fining with gelatin can result in undesirable color removal in red wine.

KIESELSOL

Kieselsol is an aqueous suspension of silicon dioxide. Because it is available in both positively and negatively charged forms, Kieselsol can be formulated to adsorb and remove both positively and negatively charged colloidal material. It is commonly used to remove bitter polyphenolic compounds from white wine. Combined with gelatin, it is effective in clarifying wines containing mucilaginous protective colloids. Kieselsol tends to produce a less voluminous sediment than bentonite and removes little color from red wines.

ISINGLASS

Isinglass is derived from proteins extracted from the air bladder of fish, notably sturgeons. Similar to most other proteinaceous fining agents, isinglass is primarily used to remove tannins. Because it is less subject to overfining, isinglass requires less added tannin than gelatin to function in fining white wine. Regrettably, it produces a voluminous sediment that tends to plug filters.

POLYVINYLPOLYPYRROLIDONE

Polyvinylpolypyrrolidone (PVPP) is a resinous polymer that acts analogously to proteins in binding tannins. It is very efficient in removing brown pigments from white wines. It functions well at cool temperatures and precipitates spontaneously. PVPP can be isolated from the sediment, purified, and reused.

TANNIN

Insect galls on oak leaves are the typical source of tannins used in fining. Tannins are commonly used in combination with gelatin. The tannin/gelatin mixture forms a delicate meshwork that sweeps colloidal

proteins out of wine. Tannins in the mesh join with soluble proteins in the wine to form both weak and strong chemical bonds. Nonionized carboxyl and hydroxyl groups of the tannins establish weak hydrogen bonds with peptide linkages of the proteins, while quinone groups form covalent bonds with the amino and sulfur groups of proteins. The latter produce strong, stable links between soluble proteins and the tannin/protein meshwork.

Clarification

In contrast to fining, clarification involves only physical means of removing suspended particulate matter. As such, it usually follows fining, though an initial clarification often occurs before fermenting white juice. Juice clarification often improves flavor development in white wine, and it helps to prevent microbial spoilage following fermentation. After fermentation, preliminary clarification by racking removes sedimented material. Subsequently, finer material may be removed by centrifugation or filtration.

RACKING

Until the twentieth century, racking and fining were essentially the only methods available to facilitate wine clarification. Presently, racking can vary from manually decanting wine from barrel to barrel to highly sophisticated, automated, tank-to-tank transfers. In all cases, separation from the sediment (**lees**) occurs with minimal agitation to avoid resuspending the particulate matter. Decanting stops when unavoidable turbulence makes the wine cloudy. The residue may be filtered to retrieve wine otherwise lost with the lees. Racking is generally more effective in clarifying wine matured in small cooperage than in large tanks.

The first racking is conventionally done several weeks after alcoholic fermentation. Racking may be delayed if malolactic fermentation is desired until the process has come to completion. Racking also may be delayed to permit prolonged lees contact, which is considered important to flavor development in some wines. The delay permits yeast autolysis to occur and favors the release of cell-bound compounds, such as ethyl octanoate, ethyl decanoate, amino acids, and cell wall mannoproteins. Prolonged lees contact is often associated with periodic manual stirring. This may provide sufficient aeration to limit the production of reduced-sulfur off-odors.

By the first racking, most of the yeast, bacterial, and grape cell fragments have settled out of the wine. Subsequent rackings remove most of the residual microbial population, along with precipitated tannins, pigments, and crystalline material. Later rackings also remove sediment induced by fining. If sufficient time is provided, racking and fining can produce stable, crystal clear wines. However, the trend to early bottling, within a few weeks or months of fermentation, provides insufficient time for racking and spontaneous precipitation to generate a stable, clear wine. Consequently, centrifugation and filtration are used to achieve the required level of clarity.

In addition to aiding clarification, racking plays several additional valuable roles in wine maturation. By removing microbial cells and other sources of nutrients, racking enhances microbial stability. The transfer process also disrupts stratification that may develop within the wine. This is particularly important in large storage tanks, where stratification can lead to variation in redox potential and rates of aging throughout the wine. Racking also removes the primary source of such reduced-sulfur taints as hydrogen sulfide and mercaptans, which may form under low redox potentials that develop in thick layers of lees.

Aeration and liberation of carbon dioxide result as a consequence of racking. Modest oxygen uptake during racking assists color stability in red wine, but its value in white wine maturation is more controversial. As noted earlier, slight aeration appears to be beneficial in white wines matured on the lees but is avoided otherwise. Oxygen exposure can be minimized with modern automatic pumping systems using carbon dioxide or nitrogen sparging. The turbulence generated during pumping and filling helps to release carbon dioxide found in its supersaturated state following fermentation. The escape is essential for the wine to lose its slight *pétillance* before bottling. If necessary, adjustment with additional sulfur dioxide occurs during racking.

The number of rackings recommended varies considerably from region to region, depending on empirically established norms. The size of the cooperage also is a determining factor. The larger the storage vessel, the more frequent the required racking. This is necessary to avoid the development of a thick layer that is conducive to off-odor production.

The method of racking depends largely on the size of the cooperage and the economics of manual versus mechanical transfer. Manual draining by gravity, or with a simple hand pump, is adequate where volumes are small and labor costs low. For most large wineries, however, manual racking would be prohibitively expensive, in terms of both labor and time. Mechanical pumping is the only reasonable option. Also, if aeration and sulfiting are deemed desirable, they can be controlled more precisely through mechanical rather than manual racking.

CENTRIFUGATION

Centrifugation employs rotation at high speeds to expedite settling. It is equivalent to spontaneous sedimentation, but occurs within minutes rather than months. It often replaces racking when early bottling is desired. Centrifugation also is useful when the wine is heavily laden with particulate matter. Centrifugation is much more efficient at removing large amounts of particulates compared to plate filters. Highly turbid wines are prone to off-odor development if permitted to clarify spontaneously.

Blanketing the wine with inert gases has minimized a former liability of centrifugation, namely, oxidation. Automation combined with continuous centrifugation has improved the efficiency and economy of the process to such an extent that centrifugation is often the preferred clarification technique.

FILTRATION

Filtration involves the physical retention of material on, or within, a fibrous or porous material. Depending on the pore size, filtration removes coarse particles with diameters larger than 100 μm down to molecules and ions with diameters less than 10^{-3} μm. However, the greater the retentive property of the filter, the greater is the likelihood of rapid plugging. As a consequence, filtration typically follows preliminary clarification by fining, centrifugation, or racking. This is especially important when using membrane sterilization or ultrafiltration.

With the development of new filters and support systems, filtration is currently classified into four categories. Conventional filtration employs depth-type fibrous filters that remove particles down to about 1 μm in diameter. Other filtration techniques involve membranes containing crevices, channels, or pores that cross the membrane. Depending on the size range of the membrane perforations, the sieving action is termed microfiltration, ultrafiltration, or reverse osmosis/dialysis. Ultrafiltration and microfiltration usually are differentiated on the basis of nominal pore size (0.2–0.05 and 1.0–0.1 μm, respectively). Microfiltration is used primarily to remove fine particles or in sterilization. Ultrafiltration may be used to remove macromolecules and colloidal materials. Reverse osmosis and dialysis are utilized to remove or concentrate low molecular weight molecules or ions. Dialysis involves the same principle (diffusion) as reverse osmosis discussed earlier in the chapter, but it does not use pressure to reverse the direction of diffusion. Electrodialysis uses an electrical differential across the membrane to influence the diffusion of charged particles.

Filtration primarily acts by blocking the passage of material larger than the maximum pore size of the filter (Fig. 8.6). However, as material smaller than the smallest perforations are often retained by a filter, other principles are involved. Surface adsorption by electrical at-

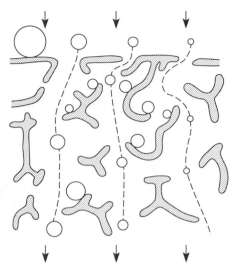

Figure 8.6 Mechanism of membrane filtration. Particles larger than the diameter of the filter pores become trapped at the surface while smaller particles pass through or adhere to the filter matrix. (After Helmcke, 1954, from Brock, 1983, reproduced by permission.)

traction can be more important than physical blockage at the lower limit of filtration. Adsorption is generally important with depth filters but less significant with membrane filters. Conversely, capillary forces may facilitate the movement of material through filters. With depth filters, microbial growth within the filter can result in "grow through." Thus, it is important that depth filters be frequently cleaned and sterilized, or replaced periodically.

Depth Filters   Depth filters are composed of randomly overlapping fibers of relatively inert material. They may be purchased preformed (**filter pads**) or produced during the filtration process (**filter beds**). Although asbestos fibers were effective and widely used in the past, concern about potential health hazards led to their being abandoned. Less inert cellulose fibers have replaced asbestos in most situations.

Filter pads come in a broad range of porosities, permitting differential flow rates and selective particle removal. **Tight filters** remove smaller particles but retain most of the material on the filter surface. Correspondingly, they tend to plug quickly. In contrast, **loose filters** retain most of the material within the tortuous channels of the pad, plug less quickly, but remove only larger particles. Tight filters are commonly used just before bottling for the final **polishing filtration.**

Filter beds develop as a **filter aid** suspended in the wine is progressively deposited on a internal framework during filtration. Filter beds may be employed prior to polishing, sterilization, or ultrafiltration. The filter aid most commonly used has been **diatomaceous earth.** It consists

of the remains of countless generations of diatoms, microscopic, unicellular algae that construct their cell walls primarily out of silicon dioxide. Depending on the filtration rate and the particle size to be removed, different formulations of diatomaceous earths are available. Diatomaceous earth is often added to the wine just prior to filtration at about 1 to 1.5 g/liter. Loose cellulose fibers, treated to have a positive charge, may be added to facilitate the adsorption of colloidal materials. **Perlite,** the pulverized remains of heat-treated volcanic glass, has a very fine structure and is occasionally used instead of diatomaceous earth.

A screen of cloth, plastic, or stainless steel is covered with a **precoat** of filter aid. The plates are inserted into the framework of a **filter press,** and the press is closed. Filter aid is continuously added and mixed with the wine being filtered. Pressure forces the wine through the filter bed (Fig. 8.7). Porous metal or plastic sheets may support the filter and provide channels through which the filtered wine can escape. During filtration, the depth of the filter bed grows. The continual addition of filter aid is essential to maintaining a high flow rate at low pressures. Without additional filter aid, the bed would soon become plugged. The use of higher pressures to maintain a high flow rate tends to compact the filter material, aggravating plugging. Choosing the correct grade of filter aid is essential as the particulate size distribution affects flow and plugging rates, and consequently filtration efficiency. After a period of operation, filtration is temporarily halted to allow the removal of the accumulated filter aid and trapped particulate matter.

Filter beds are usually associated with **plate-and-frame, recessed-plate,** or **leaf** press construction. Plate-and-frame presses consist of alternating precoated plates and frames that provide space for cake development. Recessed-plate presses are similar, but each plate serves

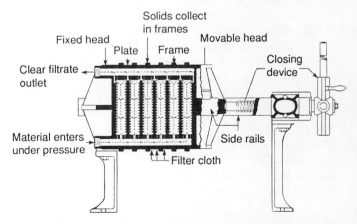

Figure 8.7   Sectional view of a filter press showing the arrangement of plates, filter cloth, and frames, as well as the flow of material. (Courtesy of T. Shriver & Co., division of Eimco Inc.)

both plate and frame functions. However, filter beds can be constructed quite differently. The **rotary vacuum drum** is a prime example (Fig. 8.8). It consists of a large, perforated, cloth-covered hollow drum. The drum is precoated with 5 to 10 cm of filter aid, usually diatomaceous earth. Part of the drum is immersed in the wine being filtered. Filter aid is added and kept uniformly dispersed as the wine is drawn through the filter bed into the drum. Shaving off the accumulated filter aid and particulate matter occurs automatically as the drum rotates. The rotary vacuum drum works particularly well with wines highly charged with particulate matter or mucilaginous colloids. Other than high purchase and operation costs, wine oxidation is the major potential drawback of rotary vacuum drum filtration. Extensive aeration is a serious risk as part of the drum is raised out of the wine during rotation. Thus, a blanketing atmosphere devoid of oxygen is typically used to avoid oxidation.

Because filter aids and pads occasionally can be sources of metal and calcium contamination, the material is commonly treated with a tartaric acid wash. In addition, the first lot filtered is often kept aside, at least initially, to determine that no metal contamination or earthy, paperlike off-odors have been acquired.

Membrane Filters   Membrane filters are constructed out of a wide range of synthetic materials, including cellulose acetate, cellulose nitrate (collodion), polyamide (nylon), polycarbonate, polypropylene, and polytetrafluoroethylene (PTFE or Teflon). With the exception of polycarbonate filters, most form a complex network of fine interconnected channels. Polycarbonate (Nuclepore) filters contain cylindrical pores of uniform diameter that pass directly through the filter. Because polycarbonate filters have a small filtration area, they are seldom used for wine filtration. By comparison, most other membrane filters contain 50 to 85% filtering surface. Thus, they have a higher flow rate for the same cutoff point (rated pore size). The applicability of a new inorganic membrane filter (Anapore) with uniform capillary pores and higher flow rates has yet to be assessed.

Because of the small pore diameter, membrane filters have a slower flow rate than depth filters. They are also more liable to plug, since most of the filtering action occurs at the surface. To circumvent rapid plugging, filter holders may be employed that direct the fluid flow parallel to, rather than straight through, the filter (Fig. 8.9). The former system is called **tangential** or **cross-flow filtration.** The conventional, perpendicular flow is called **dead-end filtration.** In tangential filtration, the flow of the fluid tends to prevent suspended material from accumulating on the membrane and causing plugging. Back flushing further increases the functional life of membrane filters. Tangential filtration partially eliminates the need for prefiltering, which is normally required with membrane filters.

Another development in tangential filtration is the synthesis of cylinders of complex internal structure. The external portion exposed to the wine contains pores that become progressively smaller toward the center. The central region contains pores of constant diameter to assure retention of particles or molecules below a certain

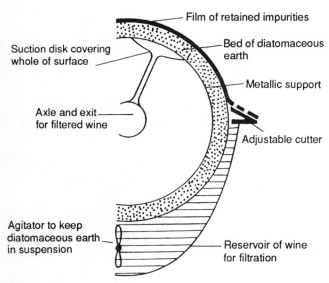

Figure 8.8   Cross-sectional view of a rotary vacuum drum filtration apparatus. (After Ribéreau-Gayon *et al.*, 1977, reproduced by permission.)

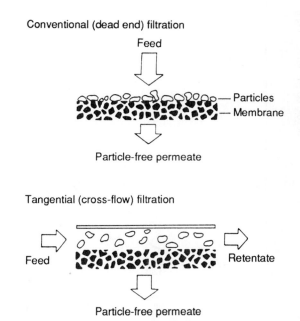

Figure 8.9   Distinction between conventional filtration and tangential (cross-flow) filtration. The washing action of the fluid passing tangentially across the surface of the membrane keeps the filter from becoming clogged. (From Brock, 1983, reproduced by permission.)

size or molecular weight. Such **cartridge filters** contain some of the nonplugging features of depth filters and offer the control of particle size retention of conventional membrane filters. These polypropylene filters are resistant to most chemical reagents. This allows them to be cleansed and used repeatedly. Such new developments may reduce, if not eliminate, the need to conduct filterability tests prior to sterile filtration. Filterability tests are fully discussed by Peleg *et al.* (1979) and de la Garza and Boulden (1984).

Microfiltration is extensively used to sterilize wines. Microfiltration avoids possible flavor modification occasionally associated with pasteurization.

Ultrafiltration has been used to a limited extent in protein stabilization in wine. While effectively removing most colloidal material from wine, ultrafiltration also can remove important pigments and tannins from red wines (Fig. 8.10). Its use with white wines appears not to produce unacceptable flavor loss (Flores *et al.*, 1991). The benefits and liabilities of ultrafiltration are still under active investigation.

Aging

The tendency of wine to improve during aging is one of its most fascinating properties. Regrettably, most wines improve only for a few months to years before showing irreversible loss in quality. In contrast, red wines produced from varieties such as 'Cabernet Sauvignon,' 'Tempranillo,' 'Nebbiolo,' 'Pinot noir,' and 'Syrah' may continue to improve in flavor and subtlety for decades. In addition, white wines produced from varieties such as 'Riesling,' 'Chardonnay,' 'Sauvignon blanc,' and 'Viura' also show excellent aging potential.

Quality loss is commonly explained as a dissipation of the fresh, fruity bouquet, along with any aroma donated by the grape. Wines noted for continued improvement typically show similar aromatic losses, but they gain in **aged bouquet.** Aging is considered desirable when the aged bouquet, subtle flavor, and smooth texture more than compensate for the fading varietal and fruity character of the young wine.

Only since the early 1980s have sufficiently precise

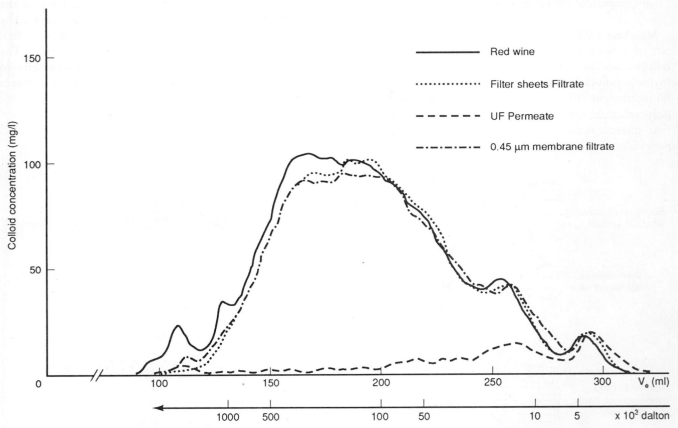

Figure 8.10 Removal of colloids from red wine during different filtration procedures, expressed as elution volumes (V_e, in ml) and molecular mass (daltons) of the various colloidal fractions. UF, Ultrafiltration. (From Cattaruzza *et al.*, 1987, reproduced by permission.)

analytical tools become available to begin unraveling the nature of wine aging. However, as aging has been studied in only a few wines, caution must be used in extrapolating the findings to other wines. Currently, more is known about why most wines decline in quality after several years than why a few wines retain or improve in character for decades.

Knowledge of how wines age, and how the effects of aging might be directed, is important to all involved or interested in wine. At the simplest, quality loss can adversely affect shelf life and the financial return to the producer. At the other end of the spectrum, the prestige connected with long aging potential adds greatly to the desirability and value of a wine to connoisseurs. It also permits consumers to participate through the conditions and duration of aging they permit. Because the factors affecting aging are poorly understood, a mystique is often associated with vineyards and varieties making wines that age well.

Aging is occasionally considered to consist of two phases. The first, called **maturation,** refers to changes that occur between alcoholic fermentation and bottling. Although maturation often lasts from 6 to 24 months, it may continue for decades. During maturation, the wine may undergo malolactic fermentation, be stored in oak cooperage, be racked several times, and be treated to one or more clarification techniques. During racking and clarification, wines may absorb about 40 ml O_2 per year, an amount insufficient to give the wine a noticeable oxidized character. Only in some fortified wines is obvious oxidation an important component of maturation.

The second phase of aging commences with bottling. Because this phase occurs essentially in the absence of oxygen, it has been called **reductive aging.** This contrasts with **oxidative aging,** an alternative term for maturation occasionally used for the aging of some fortified wines.

Effects of Aging

Many age-related changes in wine chemistry have been noted. As with other aspects of wine chemistry, determining the significance of the changes is more difficult than simply detecting them. To establish significance, it is necessary to show that the changes detectably influence sensory perception. Because most chemicals occur at concentrations below the sensory threshold, most changes affect neither wine flavor nor the development of an aged bouquet.

APPEARANCE

One of the most obvious changes to occur during aging is a color shift toward brown. Red wines initially deepen in color after fermentation, but subsequently become lighter and take on a brickish hue. Decreased color intensity and browning are indicated, respectively, by a drop in optical density and a shift in the absorption spectrum toward the blue (Fig. 8.11). Such changes are often measured as a ratio of the optical density measurements at 520 and 420 nm. High 520/420 nm values indicate a red color, while low values indicate a brown color. In contrast, white wines darken in color and take on yellow, gold, and eventually brown shades during aging.

Although little studied in white wine, color change has been extensively investigated in red wines. Nevertheless, no consensus has been reached about the relative importance of the mechanisms involved. While small amounts of acetaldehyde produced during limited aeration enhance the polymerization of anthocyanins and flavonoid tannins (Ribéreau-Gayon and Glories, 1987), polymerization also occurs directly under anaerobic, acid-catalyzed conditions (Somers and Evans, 1986). Because it occurs throughout the wine, direct polymerization may be more significant. Since temperature markedly affects the rate of direct polymerization, mild heating has been recommended as an alternative to aeration for color stabilization (Somers and Pocock, 1990). Aeration has the risk of activating dormant acetic acid bacteria and increasing volatile acidity (acetic acid and its esters).

The origin of the shift in color in white wines is poorly understood. It probably involves structural modification in existing pigments or their formation by one or more of the following processes: the slow oxidation of grape and oak phenols and related compounds, metal ion-induced modifications in galacturonic acid, Maillard reactions between sugars and amino acids, and sugar caramelization.

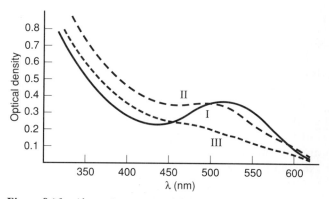

Figure 8.11 Absorption spectra of three red wines of different ages. I, One year old; II, 10 years old; III, 50 years old. (From Ribéreau-Gayon, 1986, reproduced by permission.)

TASTE AND MOUTH-FEEL SENSATIONS

During aging, glucose and fructose may react with other compounds and undergo structural rearrangement. Nevertheless, these reactions do not appear to occur to a degree sufficient to affect perceptible sweetness. In contrast, aging can induce small losses in total acidity. The mellowing of taste that may result probably comes from the formation of esters and the precipitation of acids. Esterification involves the removal of carboxyl groups involved in the sensation of sourness. Slow deacidification also can result from the isomerization of the natural L to the D form of tartaric acid. The racemic mixture so generated is less soluble than the L form, and this is one of the origins of tartrate instability in wine. Isomerization also results in forming racemic mixtures of free L- and D-amino acids (Chaves das Neves *et al.*, 1990). The potential significance of the toxicity of the D-amino acids is unknown.

Important changes also occur in the bitter and astringent sensations of red wines. The best understood of these reactions is the polymerization of tannins with anthocyanins. Their eventual precipitation results in a decline in bitterness and astringency. However, the initial effect may be an increase in astringency. This results from the greater astringency of medium-sized tannin polymers. The reaction of both hydrolyzable and condensed tannins with proteins leads to additional loss in bitter, astringent compounds. The hydrolytic breakdown of hydrolyzable tannins further reduces astringency but may increase bitterness.

FRAGRANCE

Whereas most studies on aging in red wines have concentrated on color stability, most research on aging in white wines has focused on fragrance modification. Flavor loss in most white wines is associated with changes in the ester content. Other sources of reduced fragrance involve structural rearrangements in terpenes.

Loss of Aroma and Fermentation Bouquet Esters produced during fermentation generate much of the fresh, fruity character of young white wines. Esters formed from acetic acid and higher alcohols, such as isoamyl and isobutyl acetates, are particularly important in this regard. Because yeasts produce and release more of these esters than the equilibrium in wine permits, the esters tend to hydrolyze back to the corresponding acids and alcohols. Thus, the fruity aspect donated by the esters fades with time.

Concurrent with the hydrolytic breakdown of acetate esters is the synthesis of certain ethyl esters. These form slowly and nonenzymatically between ethanol and the primary fixed acids of wine. Synthesis of diethyl succinate is particularly marked (Fig. 8.12). However, ethyl

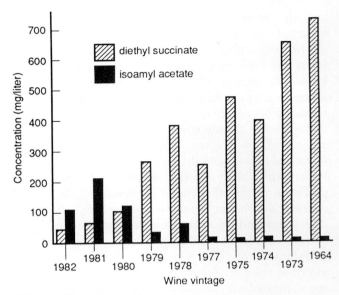

Figure 8.12 Examples of the influence of wine age on the concentration of esters, namely, acetate esters (isoamyl acetate, cross-hatching) and ethanol esters (dithyl succinate, hatching). (Data from Rapp and Güntert, 1986, Elsevier Science Publishers.)

esters probably play little role in bouquet development as they have low volatilities and nondistinctive odors. Their production may be more important in the mellowing wine acidity noted above (Edwards *et al.*, 1985).

A third class of esters is based on ethanol and carboxylic acids. The level of these esters remains stable or increases slightly during aging. Those with shorter hydrocarbon chains, such as butanoate, hexanoate, and octanoate ethyl esters, are somewhat fruity in character. As the hydrocarbon chain becomes longer, the odor shifts to being soapy and finally lardlike. Although important to wine fragrance, the significance of carboxylic acid esters to the development of an aged bouquet is unclear.

Another group of important flavorants that change during aging are the terpenes. The importance of terpenes to the aroma of Muscat and similar varieties has already been mentioned (Chapter 6). As with esters, changes in terpenes affect both the quantitative and qualitative aspects of wine fragrance.

The total concentration of monoterpene alcohols falls markedly during aging. The decline in geraniol, linalool, and citronellol is especially marked and can result in noticeable losses in floral character. For example, the linalool content of 'Riesling' wines can fall by 80% within 3 years. In contrast, the concentrations of linalool oxides, nerol oxide, hotrienol, and α-terpineol increase. Rapp and Güntert (1986) have proposed a reaction scheme for changes in monoterpene alcohols during aging. Most compounds generated have higher perception thresholds than do the terpene progenitors. The oxide terpene derivatives also have qualitatively different odors. For example, α-terpineol has a musty, pinelike odor, whereas its precursor linalool has a floral aspect. As a group, the four isomeric linalool oxides have an eucalyptus aspect. Terpene-related, heterocyclic oxygen compounds also develop during aging, but the sensory significance of most of these is unknown.

Some of the loss in aromatic terpenes associated with aging may be offset by the slow release of monoterpenes chemically bound in nonaromatic forms. A variable portion of the monoterpene content of wines (and grapes) occurs as nonvolatile glycosides. Terpene glycosides tend to hydrolyze slowly under acidic wine conditions (Gunata *et al.*, 1986). In addition, minor quantities of some monoterpene oxides, such as 2,6,6-trimethyl-2-vinyltetrahydropyran and the 2,2-dimethyl-5-(1-methylpropenyl)tetrahydrofuran isomers, may participate in the cineolelike fragrance of aged 'Riesling' wines (Simpson and Miller, 1983).

Origin of Bottle-Aged Bouquet Presently, three groups of compounds appear to be significantly involved in the generation of a bottle-aged bouquet. These include compounds derived from norisoprenoid precursors and related diterpenes, residual carbohydrates, and reduced-sulfur compounds.

Of isoprenoid degradation products, vitispirane and 1,1,6-trimethyl-1,2-dihydronaphthalene (TDN) appear to be the most potentially important. The two isomers of vitispirane have qualitatively different odors. However, as they occur in concentrations at or below the detection thresholds, the sensory significance of vitispiranes is doubtful. In contrast, the concentration of TDN seems sufficient to play a meaningful role in bouquet development in wines from varieties such as 'Riesling.' TDN also has been found in some red wines. Because TDN has a kerosenelike odor, accumulation considerably above the threshold value may be undesirable. The presence of glycosides and aglycones, precursors of TDN and vitispirane, has been suggested as an indicator for timing grape harvest to achieve an optimal aging potential (Winterhalter *et al.*, 1990). Other isoprenoid degradation products, such as theaspirane, ionene (a 1,1,6-trimethyl-1,2,3,4-tetrahydronaphthalene isomer), and damascenone, appear little involved in the development of bottle bouquet. For example, the concentration of the roselike damascenone declines during aging.

Carbohydrate degradation occurs rapidly during the baking of wines such as Madeira and baked sherries. Equivalent acid-catalyzed dehydration reactions occur much more slowly at cellar temperatures. For example, the caramellike 2-furfural shows a marked increase during aging (Table 8.1). The sensory significance of other degradation products, such as 2-acetylfuran, ethyl 2-furoate, 5-(hydroxymethyl)-2-furaldehyde, 2-formylpyrrole, and levulinic acid, is unknown. The fruity, slightly pungent ethyl ether, 2-(ethoxymethyl) furan, has been found to form on aging of 'Sangiovese' wine (Bertuccioli and Viani, 1976). This suggests that etherification of Maillard-generated alcohols may play a role in development of an aged bouquet.

The concentration of a number of reduced-sulfur compounds may change during aging. Of these, only one occasionally has been correlated with the development of a desirable aged bouquet, namely, dimethyl sulfide. Addition of 20 mg/liter dimethyl sulfide to wines containing 8 to 15 mg/liter dimethyl sulfide enhanced the flavor score of the wines (Spedding and Raut, 1982). Higher concentrations ($\geq$40 mg/liter) were considered detrimental. By itself, dimethyl sulfide has a shrimplike odor.

ADDITIONAL CHANGES

A number of age-related changes develop in sherries that generally do not occur in table wines. Notable are increases in the concentrations of aldehydes and acetals,

Table 8.1 Changes in Aroma Composition from Carbohydrate Decomposition during Aging of a 'Riesling' Wine[a, b]

Substance from carbohydrate degradation	1982	1978	1973	1964	1976 (frozen)	1976 (cellar stored)
2-Furfural	4.1	13.9	39.1	44.6	2.2	27.1
2-Acetylfuran	—	—	0.5	0.6	0.1	0.5
Furan-2-carbonic acid ethyl ester	0.4	0.6	2.4	2.8	0.7	2.0
2-Formylpyrrole	—	2.4	7.5	5.2	0.4	1.9
5-Hydroxymethylfurfural (HMF)	—	—	1.0	2.2	—	0.5

The Year column header spans: 1982, 1978, 1973, 1964, 1976 (frozen), 1976 (cellar stored).

[a] From Rapp and Güntert (1986), reproduced by permission.
[b] Relative peak height on gas chromatogram (mm).

which develop under the oxidizing conditions prevalent during sherry maturation. In table wines, the aldehyde concentration generally declines after bottling. Correspondingly, aldehyde-derived acetals do not accumulate during aging. An exception involves Tokaji Aszú wines that are often exposed to oxidizing conditions during their maturation (Schreier *et al.*, 1976).

Wines stored in oak slowly amass products extracted from the wood. A discussion of these products, and their significance, is given below.

Structural rearrangements of the main, fixed acids in wine occur during aging. For example, tartaric acid may give rise to oxalic acid and citric acid to citramalic acid via decarboxylation. The sensory significance of such changes is unknown.

During aging, a marked increase occurs in the concentration of abscisic acid. Although abscisic acid is important as a growth regulator in higher plants, its generation in wine is probably purely coincidental. Its production has no known sensory significance.

The concentration of several groups of compounds are little affected by aging, notably higher alcohols and lactones.

Factors Affecting Aging

Several years ago there was active interest in accelerating the aging of wine (Singleton, 1962). Subsequently, interest in the topic has waned. Accelerated aging simulates some of the changes of prolonged aging but generally does not generate the desired subtle complexity of spontaneously bottle-aged wines.

OXYGEN

Some wines are much more sensitive to oxidation than others. Currently, a precise explanation of these differences is not possible. However, variations in the concentration and types of phenols and polyphenol oxidases are undoubtedly important. For example, the most prevalent phenol in white wines is caftaric acid, an ester between caffeic and tartaric acids. On hydrolysis, caffeic acid (an *o*-diphenol) could significantly increase oxidative browning (Cillers and Singleton, 1990). In addition, red grape varieties particularly susceptible to oxidation, such as 'Grenache,' also contain high concentrations of caftaric acid and derivatives. Another important factor affecting the oxidation potential of wine is the pH. As the pH rises, the proportion of phenols in the highly reactive phenolate state increases, enhancing potential oxidation. Finally, differences in anthocyanin composition, the respective reactivity of the various anthocyanins, and their polymerization with tannins greatly affect their susceptibility to oxidation.

Because oxygen prevents the development of a reductive bottle bouquet, considerable effort is expended limiting its access to bottled wine. In addition, oxygen can destroy or mask fruity and varietal odors and generate an oxidized (aldehyde) odor. To assure that bottled wine remains under reductive, anaerobic conditions, good quality closures (corks or pilfer-proof screw caps) should be used and oxygen exposure during bottling minimized.

TEMPERATURE

To avoid loosening the seal between the cork and the bottle, wine needs to be stored under relatively stable temperature conditions. Rapid temperature changes put pressure on the cork/neck seal by generating sudden fluctuations in wine volume, which, if repeated frequently, will loosen the seal between the cork and the glass. This could result in the slow seepage of oxygen into the wine. An even more marked expansion of the wine occurs if the wine freezes. Freezing can produce sufficient volume increase to force the cork out of the bottle.

Temperature also directly influences the rate and direction of wine aging. Because the aging process is primarily physicochemical, heat both speeds and activates the reactions involved. Thus, cool storage (<10°C) tends to maintain the fresh, fruity character of most young wines. For example, fragrant acetate esters, such as isoamyl and hexyl acetates, hydrolyze slowly at 0°C but rapidly at 30°C (Marais, 1986). In contrast, formation of less aromatic ethyl esters is rapid at 30°C but negligible at 0°C. Temperature also has a marked effect on age-induced changes in the concentration and types of monoterpene alcohols found in some wines (Rapp and Güntert, 1986).

Heating favors the degradation of carbohydrates into furfurals and pyrroles. Whether a similar activation affects the conversion of norisoprenoid precursors to spiroesters such as vitispirane and theaspirane, or to hydrocarbons such as TDN and ionene, is unknown.

For most wines, prolonged exposure to high temperatures (≥40°C) rapidly results in quality deterioration. Carbohydrates in the wine undergo Maillard and thermal degradation reactions, turning brown and producing a baked (madeirized) flavor. The wines also tend to develop a heavy sediment. Even temperatures as low as 30°C can produce detectable losses in fragrance within several months, as well as cause the dissipation of a fresh, fruity aroma from 'Colombard' white wines within 2 years (Marais, 1986).

LIGHT

Exposure to sunlight is not known to directly affect the aging process in wine, but it can cause heating that can speed aging reactions. In addition, exposure to near-ultraviolet and blue radiation can activate detrimental oxidative reactions. The best known example of this is the shrimplike/skunky odor of the Champagne fault called **light-struck** (*goût de la lumière*) (Carpentier and Maujean, 1981). Light activates the synthesis of several sulfur compounds, including dimethyl sulfide, dimethyl disulfide, and methyl mercaptan. Synthesis is considered to involve the activation of riboflavin or its derivatives. Light also facilitates the production of copper casse. As a consequence, wine should be stored in darkness whenever possible. Bottling in glass opaque to ultraviolet and blue radiation also is advisable.

VIBRATION

Wines are normally stored in areas free of vibration. Although vibration is commonly believed to disrupt or accelerate aging, there appears to be no evidence to support this view. Old claims concerning the beneficial aspects of vibration probably refer to facilitated clarification rather than aging.

Rejuvenation of Old Wines

A process has been proposed to rejuvenate wines having lost their fresh character (Cuénat and Kobel, 1987). It involves the dilution of the affected wine with water. The water, along with the chemicals presumably involved in the development of the undesired aged character, is subsequently removed by reverse osmosis. Use of ultrafiltration may be similarly beneficial in rejuvenating old wines.

Oak and Cooperage

Oak has been used in Europe for the transport and maturation of wine for over 2000 years. Many types of wood, other than oak, have been used over this period, but they have generally been limited to the construction of large storage tanks. Oak has been used similarly, but small barrels for fermentation, maturation, and transport have been predominantly constructed from one of the several white oak species. White oak has the anatomical and chemical properties needed for tight cooperage. More recently, bottles have replaced oak as the primary transport container. Consumer preference for light, young, fruity wines also has led to a reduction in the use of wood cooperage. Currently, oak cooperage is reserved primarily for the production of premium white and red wines, and some fortified wines. The flavor and slight oxidation given by barrel maturation donate characteristics usually appreciated in premium wines.

Oak Species and Wood Properties

Not only does white oak possess the properties required for tight cooperage, but its traditional use has led to a habituation to, and appreciation of, its subtle frag-

rance. Other woods either have undesirable structural or aromatic characteristics or have been studied insufficiently to establish their acceptability.

Quercus alba, Q. robur, and *Q. sessilis* are the species most commonly used. *Quercus alba* and a series of some six related white oak species constitute the oaks employed in the construction of American oak cooperage. *Quercus alba* provides about 45% of the white oak lumber produced in North America. It has the widest distribution of all American white oak species and has the size and structure preferred for select oak lumber.

In Europe, *Q. robur* (*Q. pedunculata*) and *Q. sessilis* (*Q. petraea* or *Q. sessiliflora*) are the primary white oaks employed in cooperage production. Both species grow throughout much of Europe. The proportion of each species in any location depends on the prevailing soil and climatic conditions. *Quercus robur* does better on deep, rich, moist soils, while *Q. sessilis* does well on drier, shallow, hillside soils and accepts more shading and variation in soil pH.

Staves produced from different American white oak species are almost indistinguishable to the naked eye. The same is true for the two important white oak species in Europe. They may be differentiated only with difficulty, and then with certainty solely under the microscope. Both genetic and environmental factors often blur the few anatomical features that distinguish the wood of each species (Fletcher, 1978).

In North America, most of the oak used in barrel construction comes from Kentucky, Missouri, Arkansas, and Michigan, and there has been little tendency to separate or distinguish between oak coming from different states or diverse sites. In contrast, identification of oak origin is common in Europe. Geographical designation may indicate the country, region (i.e., Slovonian, Limousin), political district (i.e., Vosges, Allier), or forest (i.e., Nevers, Tronçais) of the wood (Fig. 8.13).

Conditions affecting growth also affect the anatomy and chemistry of the wood. Slow growth generally results in the development of a less dense heartwood owing to the higher proportion of large-diameter vessels produced in the spring. Subsequently, the large-diameter spring vessels accumulate more phenolics than do smaller summer-produced vessels. Phenol deposition generally occurs 10 to 15 years after vessel formation, when the sapwood differentiates into heartwood. Phenolics not only resist wood rotting, but also contribute to the flavors extracted by wine during in-barrel maturation.

Owing to the higher proportion of large-diameter vessels, slow-grown wood is softer. The lower percentage of cell wall material in the wood makes it more pliable than oak that grew rapidly and has more summer wood. In France, the properties of slowly grown *Q. sessilis,* found in forests such as Nevers and Allier, are commonly pre-

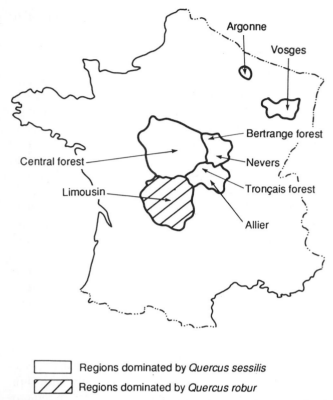

Regions dominated by *Quercus sessilis*

Regions dominated by *Quercus robur*

Figure 8.13 Location of the major oak forest in France. Outlined areas are dominated by *Quercus sessilis* and hatched areas by *Quercus robur*. (Modified from Seguin-Moreau, reproduced by permission.)

ferred for wine maturation. For brandies, the denser, more phenolic *Q. robur* found in the Limousin region is preferred. The properties and origin of the wood preferred depend largely on the desired balance between the varietal and oak character of the finished wine.

Besides growth rate-induced variations, differences in wood anatomy and chemistry also occur throughout the tree. Higher tannin levels occur in the heartwood at the base of the tree than near the crown. Higher tannin levels also develop in heartwood close to the sapwood. This reflects increased phenol deposition as the tree ages.

Although chemical variations arise from differences in growth rate and tree age, more variation arises from genetic dissimilarities between species. The most significant differences may occur between *Quercus alba* (and related species) and the European species *Q. sessilis* and *Q. robur*. American oak appears to possess about 40% of the extractable phenolics of European oak (Singleton *et al.*, 1971). However, not all European oak samples possess high phenol levels (Hoey and Codrington, 1987). *Quercus sessilis* generally contains considerably less extractable phenolics than *Q. robur*. Winemakers desiring higher tannin and phenol levels may choose oak such as that from Limousin, while those preferring less may

prefer American oak or a more mild European oak, such as found in Germany.

Although significant differences exist in levels of extractable tannins in American and European oaks, the intensity of oak flavor is similar, albeit different in character (Singleton, 1974). For example, American oak possesses markedly higher levels of isomers of volatile norisoprenoids (3,4-dihydro-3-oxoactinidol and oxoedulan derivatives) than Vosges oak (Sefton *et al.*, 1990). Oak species also differ in "oak" lactone content (isomers of β-methyl-γ-octalactone), and probably in sesquiterpene, hydrocarbon, and fatty acid concentrations. Significant differences exist even between the two European species and the same species obtained from different sites (Fig. 8.14). Additional sources of flavor variation come from the method of seasoning and the degree of "toasting" (see below).

Habituation to the flavor characteristic of locally or readily available oak supplies probably explain much of the preferential use of one oak over another. Europeans have developed historical associations with oak derived from particular regions. For example, Spanish vintners customarily prefer American oak cooperage, while French producers favor oak derived from their own extensive forests. In California, the choice of oak may depend on winemaker preferences, the intensity of the wine fragrance, and the style of wine desired. Although matching oak to the wine remains primarily an art, progress in oak chemistry soon may facilitate these decisions.

PRIMACY OF OAK

Cooperage requires very specific features of the wood used in barrel construction. The wood must be straight-

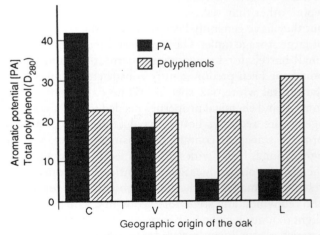

Figure 8.14 Aromatic (solid) and polyphenolic (hatched) profiles of European oak from different regions in France (C, Central group; V, Vosges group; B, Burgundy group; L, Limousin group). Aromatic potential (PA) is based on the concentrations of β-methyloctalactone isomers and eugenol content. (After Chatonnet, 1991, reproduced by permission.)

grained, that is, possess vessels and fibers running parallel to the length of the trunk, with no undulating growth patterns or vessel intertwining. In addition, the wood needs to possess both strength and resilience. Structurally, the wood must be free of faults that could make the cooperage leaky. The wood also must be free of pronounced or undesirable odors that could taint the wine. In all these aspects, *Quercus alba, Q. robur,* and *Q. sessilis* excel. The trees also grow large, straight, and tall. This minimizes wood loss during stave production (see below). Furthermore, white oaks exhibit qualities of two features, namely, rays and tyloses, that cause them to be eminently suited for tight cooperage.

All trees produce **rays,** collections of elongated cells positioned radially along the trunk axis. Rays function in conducting water and nutrients radially between the bark and wood tissues of the trunk. In oak, the rays are unusually large, being upward of 15 to 35 cells across and 100 or more cells high. In cross section, the rays resemble elongated lenses. Because staves are split (or sawn) along the radius of the log, the broad surface of the stave runs roughly parallel to the rays. This radial surface becomes the inner and outer surfaces of the cooperage. The high proportion of ray tissue (~28%) and their positioning parallel to the cooperage circumference make rays a major barrier to wine and air diffusion. Wine diffusing into ray cells is deflected along the stave width. Continued lateral flow is limited by the nonalignment of the rays of adjacent staves. Wine would have to navigate around five or more large rays to diffuse through the sides of a barrel. In practice, penetration of about 6 mm is typical (Singleton, 1974).

Positioning the radial axis of the wood along the circumference of the barrel has benefits in the construction of tight cooperage. The large number of rays permits only minor swelling around the circumference of the barrel. The swelling (~4%) is sufficient, though, to help compress the staves together and seal the joints. Positioning the radial plane of the wood outward also directs the axis of greatest wood expansion (tangential plane) inward. In this arrangement, wood expansion of about 7% (Peck, 1957) does not influence barrel tightness. The negligible longitudinal expansion of the staves has no effect on barrel tightness or strength.

The high proportion of rays gives oak much of its flexibility and resilience. Otherwise, the staves would be too tough to be bent without cracking and form the bulging shape (bilge) of the barrel. The bilge permits full barrels weighing several hundred kilograms to be easily rolled.

As a group, oaks produce especially large-diameter xylem vessels in the spring. These are large enough to be easily seen with the naked eye. The vessels allow the rapid flow of water and nutrients up the tree early in the season. However, the vessels also could make barrels excessively porous, permitting wine to seep out through the ends of the staves. In white oaks, the vessels become tightly plugged as the sapwood differentiates into heartwood. Plugging results from cellular ingrowths (**tyloses**) into the empty vessels (Fig. 8.15). Tylose production is so extensive that the vessels become essentially impervious to the movement of liquids or gases. Consequently, only the heartwood is used in the construction of tight barrels.

The combined effects of the rays, tyloses, and placement of the radial plane of the stave toward the exterior severely limit diffusion of air or wine through the wood. With proper construction and presoaking, oak cooperage is essentially an impervious, airtight container.

As sapwood completes its maturation into heartwood, deposition of tannins and phenols kill any remaining living wood cells. The phenolics render the heartwood resistant to decay, and, because the heartwood contains only dead cells, the lower moisture content makes the lumber less liable to crack or bend during drying. These features give heartwood the final properties required for superior quality cooperage wood.

Barrel Production

STAVES

For stave production, trees between 45 and 60 cm in diameter are preferred. Larger trees tend to be reserved for the production of head staves (*headings*). After felling, the trees may be cut into sections (bolts) equivalent to the stave length desired and split into quarters. The staves are then split or cut from the quarter-cut sections.

Splitting is preferred to sawing as it results in a natural fracturing of the wood along the plane of vessel elongation. Sawing cuts across the ends of some vessels, augmenting wood roughness and increasing potential permeability. Even with splitting, some staves are less than optimally tight. This results from most staves being removed obliquely rather than parallel to the radial axis of the wood. The procedure minimizes wood loss but may increase permeability. If all staves were removed parallel to the tree's radius, wedges of wood would be lost between each stave. Heading pieces are cut out similarly.

Stave length, width, and thickness depend on the volume of wine to be held and the rate of wine maturation desired. To accelerate maturation, barrels constructed of thinner (~1.8 cm), *Château*-style staves may be preferred. For barrels with a capacity of 225 liters, staves and headings are roughly 2.7 cm thick.

Once cut, the staves and heading pieces are stacked to dry and season. Natural seasoning for about 3 years is preferred, especially in Europe. Some winemakers consider that naturally dried oak gives a more pleasant

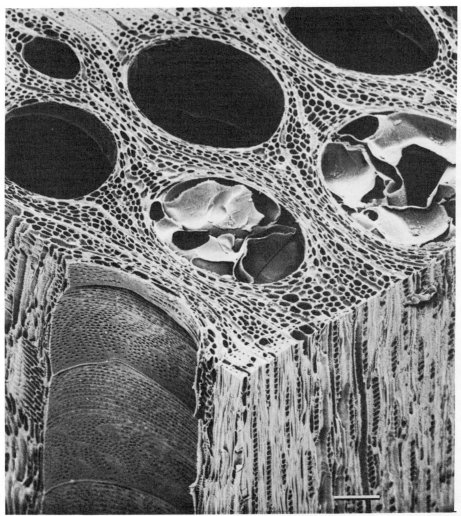

Figure 8.15 Scanning electron micrograph of large vessels of white oak heartwood plugged with tyloses. (Photograph courtesy of Dr. W. Côte, and from Hoadley, 1980).

woody, vanillalike character, whereas kiln drying produces a more aggressive, "green," occasionally resinous aspect (Pontallier *et al.*, 1982). In the United States, kiln drying between 50° and 60°C is commonly employed to rapidly bring the wood down to the desired moisture content of 14 to 18%. The effect of drying method depends considerably on the species involved (Chatonnet, 1991). For example, *Q. sessilis* releases more tannins following kiln drying than does *Q. robur*. In contrast, kiln drying decreases the production and/or release of "oak" lactones and eugenol from oak. The main exceptin is the increased uptake of *trans*-methyloctalactone from kiln-dried *Q. sessilis*.

The effects created by the natural drying (seasoning) may be largely indirect. During seasoning, fungi and bacteria may produce and liberate aromatic aldehydes and lactones from wood lignins (Chen and Chang, 1985). For example, the polyphenol oxidase produced by the wood-rot fungus *Coriolus versicolor* can degrade lignins as well as induce polymerization of phenolic compounds. Oxidation of the phenols so released could favor cellulose hydrolysis (see Evans, 1987). Kiln drying would markedly restrict but not prevent microbial action. Microbial growth could also modify other cell wall constituents. Sugars liberated by cell wall degradation could increase the furfural content generated during barrel toasting. In addition, the leaching and degradation of phenolic compounds by rain, oxygen, and ultraviolet radiation may be significant. The conversion of the bitter esculin to its less bitter aglycone, esculetin, may be another example of how wood character improves with outdoor seasoning.

BARREL ASSEMBLY

In barrel construction, the first step involves checking the staves for knots, cracks, or other structural faults.

Once the appropriate number have been assembled, the staves are *dressed*. This involves the selective shaving of wood in preparation for raising the barrel. The first of the dressing procedures (*listing*) tapers the ends of the staves to give them their basic shape (Fig. 8.16A). The amount of listing required depends on the desired *height* of the barrel, that is, the length of the staves relative to the maximal circumference of the barrel. Subsequently, small amounts of wood may be chiseled from the ends (*backing*) and center (*hollowing*) of the staves to facilitate bending (Fig. 8.16B,C). The staves are now ready for *jointing*, where a bevel is planed along the long axis of the staves (Fig. 8.16D). Jointing requires considerable skill as the angle changes along the length of the stave. The bevel depends on the barrel *height* and is maximal at the center (*bilge*) and least at the ends (*heads*). Jointing precision determines the tightness between adjoining staves.

The curved shape of the barrel provides much of its strength. This comes from the engineering principle called the *double arch*. The sloping sides also provide a point on which the barrel can be pivoted and rolled with comparative ease.

Once dressed, the staves are *raised*. This involves plac-

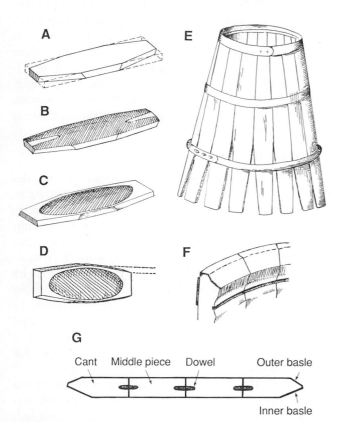

Figure 8.16 Stave preparation and barrel construction. (A) Listing; (B) backing; (C) hollowing; (D) jointing; (E) raising; (F) chimed, howelled, and crozed stave ends; (G) head cross section. (After Kilby, 1971, reproduced by permission.)

ing the staves together in an upright circle, with the aid of several hoops, including a trussing hoop (*runner*) (Fig. 8.16E). The latter is forced down and begins to bend the staves into the curved barrel outline. At this point, the barrel is inverted and placed about 5 cm above an open fire (*brazier*) for softening. Periodically, the inner and outer surfaces are moistened with water. Occasionally, staves may be steamed prior to or instead of firing. On sufficient softening of the lignins in the wood, the staves are slowly and periodically pulled together with a windlass (capstan). Positioning temporary hoops holds the staves in place until additional heating (~10–15 min) *sets* the staves in their curved shape. The firing helps shrink the innermost wood fibers, releasing tension caused by bending. Additional heating (toasting) produces the major sensory benefit by inducing important pyrolytic chemical changes in the wood. The heating may be performed closed, with a metal cover over the barrel top, or open. Closed firing requires more frequent moistenings, but produces a more uniform heating of the inner surfaces of the barrel (Chatonnet, 1991). Not only does moistening limit the rate of heating, but it also produces steam that facilitates the hydrolytic breakdown of hemicellulose, lignins, and tannins. Inner surface temperatures of the barrel typically reach 200°C and above. Carbonization (charring) of the wood begins about 250°C.

The degree and desirability of pyrolysis depend on the style and characteristics of the wine desired. Light toasting sufficient to ease bending of the staves (~5 min), produces few pyrolytic by-products and leaves the wood with a natural woody aspect (Fig. 8.17). Medium toasting (~15 min) generates many phenolic and furanilic aldehydes which donate a vanillin, roasted character. Prolonged exposure (~25 min) chars the innermost layers of the staves and destroys or limits the synthesis of phenolic and furanilic aldehydes, which are replaced by volatile phenols which give the wood a smoky, spicy aspect. Among the many pyrolytic compounds derived from tannins and hemicelluloses are aromatic aldehydes, furfurals, furans, oxygen heterocycles, pyrazines, pyridines, and pyrans, while lignin decomposition products include volatile phenols such as guaiacol, 4-methyl guaiacol, vanillin, syringaldehyde, and coniferaldehyde.

A new technique involves prolonged exposure to a small fire. This produces greater heat penetration and, presumably, the accumulation of additional volatile compounds. It also produces fewer dark-colored pyrolytic breakdown products on the inner surfaces of the staves (Hoey and Codrington, 1987).

After setting, the cooper puts a bevel on the inner surface of the stave ends and cuts the grooves that will receive the headpieces in a process called *chiming*. The first task involves smoothing the ends of the staves (the

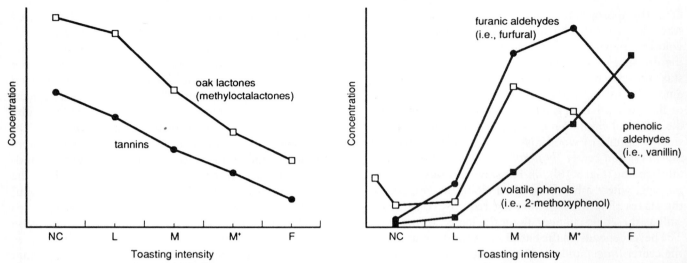

Figure 8.17 Effect of the degree of toasting on the concentration of several compounds extracted from oak barrels constructed from *Quercus sessilis.* NC, not heated; L, light toast (5 min); M, medium toast (10 min); M⁺, medium strong toast (15 min); F, charring (20 min). (After Chatonnet, 1989, reproduced by permission.)

chime). Shaving the inner edge produces the bevel. Cutting a concave groove slightly below the chime produces the *howel.* A deeper cut into the howel (the *croze*) produces a slot into which the headings fit (Fig. 8.16F).

The outer surface of the barrel is planed to give it a smooth surface, while the inner surface is left rough. The rough inner surface aids wine clarification by providing increased surface area for the deposition of suspended particulate matter.

Next, a bung hole is bored and enlarged with a special auger to receive a tapered wooden peg. A tap hole also may be bored near the end of the middle head stave.

If temporary hoops were employed, they are replaced with permanent hoops. For 225 liter barrels, this usually consists of two chime hoops located just below the heads of the barrel, two bilge hoops positioned one-third in from the ends, and a set of quarter hoops placed about one-fourth of the way in from the heads. So positioned, the hoops limit the wear on the staves during rolling. At this point, the heading pieces are produced.

The head consists of several heading pieces. In contrast to the staves, the joints between the heading are straight, not beveled. Dowels between each heading piece keep them in alignment. Caulking with river rushes, called *flags,* may be used to prevent leakage.

The circular shape of the head is now sawn, in preparation for *cutting the head.* Cutting the head involves shaving two bevels, called *basles,* on the upper and lower surfaces of the head (Fig. 8.16G).

The bottom head is inserted first. Removal of the bottom head hoop allows the head to be forced into the croze. After repositioning the head hoop, the barrel is inverted to permit removal of the opposite head hoop. A

heading vice may be screwed into the head and the head lowered sideways into the barrel. The head is pulled up into its groove with the vice. Alternately, a piece of iron forced in a joint between two staves levers the head into position. Positioning the wood grain of the two heads perpendicular to one another limits the pressure, that develops during wood swelling, from acting in the same direction, thus minimizing the likelihood of leakage.

The final task involves hammering the hoops tight. This forces the staves together and closes most cracks. After soaking for a day in water, a well-made barrel becomes leakproof.

COOPERAGE SIZE

Cooperage is produced in a wide range of sizes, depending on the intended use. In the past, large straight-sided tanks and vats were constructed with capacity greater than 5 to 10 hl. They also acted as storage cooperage after fermentation. The large size minimized oxidation and eased cleansing. However, wooden fermentors and storage tanks largely have been replaced by more durable and easily cleaned tanks constucted from stainless steel, epoxy-lined carbon steel, fiberglass, or cement. Currently, wooden cooperage is restricted largely to maturing wine in which an oak character is desired. In addition, small cooperage may be used for in-barrel fermentation.

Many of the benefits of barrel use come from the relatively large surface area of the barrels. Although surface area increases logarithmically with decreasing volume (Singleton, 1974), other factors place practical limits on minimum size. Production economy favors larger size, while ease of movement and earlier maturation of the wine favors small size. A compromise between the opposing factors has led to the widespread adoption of barrels with a capacity between 200 and 250 liters. Individual regions in Europe often use barrels of a particular capacity. For example, the Bordeaux *barrique* is 225 liters, the Chablis *feuillette,* 132 liters, the Rhine *doppelohm,* 300 liters, the sherry *butt,* 490.7 liters, and the port *pipe,* 522.6 liters. Premium white wines commonly receive about 3 to 6 months of maturation in oak, while red wines often receive between 18 and 24 months of oak maturation before bottling.

While much current literature focuses on the value of maturing wine in small oak cooperage (225 liters), many fine wines are aged in large oak cooperage (>1000 liters). This is especially true in European regions other than France. Well-sealed large tanks permit lower rates of oxidation and donate only small amounts of tannins and oak flavor to wine. In contrast, the Bordeaux *barrique* is estimated to permit the ingress of about 2 to 5 mg O_2/liter/year through the wood (Ribéreau-Gayon *et al.,* 1976). Large oak cooperage also can be used for decades, while small cooperage usually is replaced after several years. In addition, stratification of wine at different redox potentials may develop in large cooperage. Until the sensory effects of maturation in large oak cooperage are better understood, it would be unwise for winemakers automatically to reject this form of maturation; it has been used for centuries with favorable results.

CONDITIONING AND CARE

As with most winemaking practices, opinions differ considerably on how to condition new barrels. Furthermore, the need appears to vary with the source and seasoning of the staves. Minimal treatment usually involves rinsing and presoaking with warm water to swell the wood and seal the joints. In-barrel fermentation is also a well-established conditioning procedure. During the latter, the most readily extractable tannins and phenols dissolve, precipitate, and are lost with the lees. However, as the desirable flavors in oak dissolve more slowly, they are not unduly removed by in-barrel fermentation. Although an effective conditioning procedure, the technique is laborious, and barrels require cleansing before subsequent reuse in wine maturation. Consequently, barrels are conditioned more commonly with a solution of 1% sodium or potassium carbonate. The alkaline solution accelerates both phennol extraction and oxidation. Subsequently, the barrels require a thorough rinsing with a 5% solution of citric acid, and finally a water wash.

In-barrel maturation preferably follows an initial clarification of the wine. This minimizes both the adherence of material to the inner surfaces of the barrel and the excessive accumulation of lees. In addition, barrels commonly are racked several times a year to avoid the buildup of a thick layer of lees. These actions decrese the difficult and unpleasant task of barrel cleaning. In addition, potential contamination of the wood with spoilage microorganisms is minimized. However, tradition or personal preference may reseult in the wine being left on the lees for several months. Some vintners believe that the yeast and tartrate coating that develops slows the release of oak flavors.

After use, barrels require cleansing and disinfection before reuse. Where little precipitate has formed, rinsing with water under high pressure is usually sufficient for cleansing. Where a thick layer of tartrates has built up, barrels may require treatment with 0.1 to 1% sodium or potassium carbonate, followed by a thorough rinsing with water. Burning a sulfur wick in the barrel usually provides sufficient disinfection of the inner surfaces.

Barrels should be refilled with wine as soon as possible after cleansing and disinfection. If left empty for more than a few days, barrels should be thoroughly drained, sulfited, and tightly bunged. Barrels stored empty for more than 2 months should be filled with an acidified sulfur solution at 200 ppm SO_2. The sulfur dioxide inhibits the growth of most microbes, and the water prevents wood shrinkage and cracking. Before barrel reuse, the residual sulfur dioxide is removed with several water rinses.

Treating the outer surfaces of the cooperage with 1% rotenone in boiled linseed oil usually controls oak boring insects, but not fungal growth. Mold growth over the external surface of the barrel may mar the appearance but does not affect barrel strength or influence the sensory properties of the wine contained.

USEFEUL LIFE SPAN

For certain types of wine, legislation specifies the rate at which used barrels must be replaced. In most regions, however, oak use is left to the discretion of the winemaker. Thus, the frequency of reuse depends on economics and intensity of oak character desired. Making these decisions will be facilitated when more is known about the dynamics and sensory impact of flavorant extraction from oak.

Figure 8.18 shows differences in the total and nonflavonoid phenol extraction from American and French oak. Not surprisingly, the differences are most marked during the first fill. The differences are also more striking with French than with American oak. Subsequently, the differences become less marked. As the rate of nonflavonoid extraction does not drop as rapidly as that of total phenolics, the proportional extraction of nonflavonoids increases with each filling. Currently, the sensory significance of this change is unknown.

For aromatic compounds extracted from oak, those

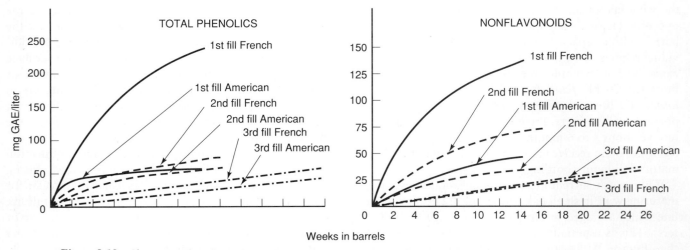

Figure 8.18 Changes in phenolics over time for French and American oak barrels. GAE, gallic acid equivalents. (From Rous and Alderson, 1983, reproduced by permission.)

such as furfurals, "oak" lactones, and phenolic aldehydes become progressively exhausted with barrel reuse. In contrast, extraction of several volatile phenols with less pleasant odors increases (Fig. 8.19). These changes probably reflect chemical extraction in the first instance and degradative synthesis in the second.

Because wine readily extracts material from oak staves only to a depth of 6 to 8 mm, shaving the innermost layers permits renewed access to oak flavorants. However, as the effect of toasting decreases rapidly away from the innermost surfaces, shaving would expose wood with different chemical characteristics. Refiring the exposed wood is possible, but the effects of such treatment appear not to have been studied.

To gain some control over differences produced by barrel-to-barrel variation (both new and used), it is common to maintain a constant proportion of new to used barrels (shaved and/or unshaved). Blending wine from different barrels before bottling helps to generate a relatively constant oak character in the wine.

Although phenol extraction apparently does not affect the internal structure of the wood, and presumably strength, after 80 years of contact (Peuch, 1984), repeated shaving would seriously weaken barrel strength.

Chemical Composition of Oak

The major chemical constituents of oak are not markedly different from those of other hardwoods. As wood consists primarily of dead cells, the major chemical components are cell wall constituents. In addition, heartwood contains infiltration substances deposited during the differentiation from sapwood.

CELL WALL CONSTITUENTS

Wood cell walls contain primarily cellulose, hemicelluloses, and lignins. Cellulose occurs in long fibers, locally grouped together to form small bundles called **micelles.** The fibers are deposited in different planes, forming complex interlacings resembling the plies of a tire. These are immersed in a matrix of hemicellulose and lignin polymers. In oak, hemicelluloses are predominantly polymers of xylose. Lignins are complex phenolic polymers.

Cellulose gives wood much of its strength and resilience, while lignins limit water permeability and provide much of the structural strength of the wood. Hemicelluloses act as binding substances, along with pectins, to hold the cellulose and lignins together.

Because of the high resistance of cellulose to both enzymatic and nonenzymatic degradation, cellulose is probably not involved in the development of oak flavor. However, hemicelluloses slowly hydrolyze on exposure to the acidic conditions of wine, releasing both sugars and acetyl groups. The latter may be converted to acetic acid during maturation. Hydrolysis is significantly increased during firing of the staves during barrel assembly. Heating also converts some of the sugars to furan aldehydes, such as furfural and 5-(hydroxymethyl)-2-furaldehyde.

Lignins are complex, branched-chain, phenylpropanoid polymers. In hardwoods, such as oak, the phenylpropanoid units contain either hydroxyl or methoxyl groups. These, respectively, form coniferyl and sinapyl alcohols that polymerize into guaiacol and syringyl lignins. Most lignin polymers contain both alcohols. A small proportion of lignins, called "native" lignins, are

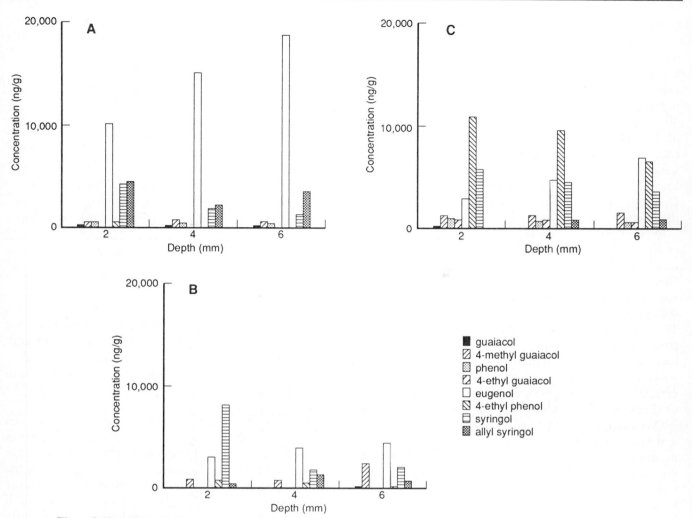

Figure 8.19 Effect of white wine maturation on the level of several volatile phenols present in oak staves at different depths. Oak cooperage used to mature a single wine, A; three wines, B; and five wines, C. (After Chatonnet, 1991, reproduced by permission.)

ethanol soluble and dissolve readily into wine. Lignins may undergo ethanolysis and be subsequently oxidized to aromatic compounds. Pyrolysis enhances degradation, leading to the production of aromatic phenolic aldehydes.

CELL LUMEN CONSTITUENTS

When wood cells die, the cytoplasm degrades, leaving only a central cavity (**lumen**) formed by the remaining cell wall. Later, when the sapwood matures into heartwood, phenolic compounds may be deposited in the lumen. Of these, tannins are the most common in oak heartwood. Most oak tannins, in contrast to those of grapes, are hydrolyzable tannins, although some condensed tannins occur. Oak hydrolyzable tannins are primarily polymers of ellagic acid, the dimer of gallic acid, and glucose (Quinn and Singleton, 1985). As noted

above, the tannin content of oak varies considerably, depending on the species, conditions of growth, age of the tree, and position of the wood in the tree.

Other phenolic compounds extracted from oak heartwood include cinnamic acid derivatives, namely, *p*-coumaric and ferulic acids. Yeasts and lactic acid bacteria may convert these to aromatic phenols, such as 4-ethyl phenol, 4-vinyl guaiacol, and 4-ethyl guaiacol (Dubois, 1983). Lyoniresinol is also apparently an important phenolic compound derived from oak (Moutounet *et al.*, 1989).

Additional components found in small quantities in oak heartwood include resins, sterols, lactones, and fats. Some of these are structurally modified during toasting and increase the concentration of *cis*-β-methyl-γ-octalactone and octanoic through octadecanoic fatty acids (Chatonnet, 1991).

COMPOUNDS EXTRACTED FROM OAK

The solubility of oak constituents, and their degradation by-products, varies widely. Compounds extracted in small amounts may affect only the bouquet, while those extracted in larger amounts may influence all the sensory perceptions of wine. For example, wine dissolves about 30% of the tannins in the innermost few millimeters of the oak staves. This is sufficient to affect the color, taste, mouth-feel, and fragrance of wine. In contrast, wine extracts only about 2% of oak lignins. Lignin breakdown products typically affect only the fragrance. Currently, over 200 volatile compounds have been identified from oak.

Quantitatively, phenolics are the most important group of oak extractives. Of these, about two-thirds are nonflavonoids. The hydrolyzable tannins (ellagitannins) comprise the most significant subgroup of oak nonflavonoid phenols. Lignin degradation products form the second most important group of extracted phenolics. Their extraction depends largely on the alcoholic strength and acidity of the wine. Both factors also are involved in the degradation of tannins and lignins to simpler, more soluble compounds. Toasting eases the extraction of ellagitannins by enhancing the hydrolysis of ellagic acid glycosides.

Oak tannins can add significantly to the astringency and bitter taste of wine. Consequently, white wines are usually matured in oak for shorter periods than red wines, and in barrels conditioned to release fewer extractable tannins. For red wines, the influence of oak on taste depends on the flavor intensity of the wine, with light wines being negatively influenced while full-flavored wines are little influenced. Oak tannins also participate in stabilizing the color of red wines.

Lignin breakdown products add significantly to the development of an oak bouquet. Lignin degradation involves the action of both alcohol and oxygen. It is believed that ethanol reacts with certain lignins, forming ethanol–lignins. As the complexes break down, the lignin monomers (coniferyl and sinapyl alcohols) are released along with the ethanol. The phenolic alcohols slowly oxidize under the acidic conditions of wine to form sinapaldehyde and syringaldehyde, and coniferaldehyde and vanillin, respectively (Puech, 1987). Above threshold values, the oxidized compounds donate woody, vanillalike odors. Toasting, especially at about 200°C, markedly augments their synthesis (Table 8.2). The lower amounts extracted from charred wood probably result from carbonization. Lignin degradation also may generate phenolic acids, such as vanillic and syringic acids, and the coumarin derivatives scopoletin and escutelin. The presence of scopoletin, along with "oak" lactones, is so characteristic that those compounds are considered diagnostic of oak maturation.

Table 8.2 Aromatic Aldehydes Produced by Toasting or Charring Oak Chips[a,b]

Product (ppm)	Toasting temperature			
	100°C	150°C	200°C	Charred
Vanillin	1.1	3.8	13.5	2.8
Propiovanillone	0.6	1.1	1.4	0.9
Syringaldehyde	0.1	3.8	32.0	9.2
Acetosyringone	—	0.025	1.5	0.6
Coniferylaldehyde	Trace	4.3	24.0	4.8
Vanillic acid	—	1.8	6.1	1.1
Sinapaldehyde	Trace	6.5	60.0	9.0

[a] Oak chips (2%, w/v, in ethanol) were toasted at various temperatures. Charring occurs above 250°C.
[b] From Nishimura *et al.* (1983), reproduced by permission.

Oak-derived phenols may be modified further by yeast and bacterial metabolism. The changes can influence both volatility and odor quality. For example, the reduction of furfurals to the corresponding alcohols results in a quality shift from almondlike to hay/verbenalike (Chatonnet, 1991).

Various phenolic and nonphenolic acids have been implicated in the synthesis of esters, acetals, and lactones in wine matured in oak (Nykänen, 1986). The acids can lower wine pH and increase acidity. By increasing the proportion of colored anthocyanins, the acids enhance color intensity. The most prevalent acid is acetic acid. It may be formed during the degradation of hemicelluloses or from the oxidation of acetaldehyde, but the most significant source is probably the metabolism of acetic acid bacteria. Oxygen uptake during cellar activities can activate the growth and metabolism of acetic acid bacteria, which generate acetic acid during the metabolism of ethanol and several sugars.

Lignins, tannins, and inorganic salts also influence the poorly understood phenomenon of ethanol/water interactions. Such interactions are believed to mellow the alcoholic taste of wine and distilled spirits (Nishimura *et al.*, 1983).

Although small amounts of sugars accumulate during the hydrolysis of hemicelluloses, they are insufficient to affect the taste of the wine. The simultaneous pyrolytic conversion of some of the sugars to furfurals appears to be the most significant sensory effect of sugar liberation.

"Oak" lactones occur in many wines matured in oak. Although present in oak, and formed on toasting, oak lactones are extracted slowly by wine. Consequently, it may take more than 1 year for the coconutlike fragrance of oak lactones to affect wine fragrance.

While oak maturation increases the concentration of

many important sensory compounds, it also reduces the concentration of others. For example, the concentration of dimethyl sulfide and dimethyl disulfide decrease during barrel maturation (Nishimura *et al.*, 1983). Methionyl acetate also decreases in quantity, but only in the joint presence of oxygen. The green bean/green chili aspect of some 'Cabernet Sauvignon' wines may dissipate when the wines are matured in oak barrels (Aiken and Noble, 1984).

Aeration

Slight oxidation is commonly viewed as a important consequence of maturation in oak. Wine placed in well-made barrels, bunged tight and rotated so that wine covers the bung, receives oxygen exposure only during cellaring procedures such as racking. Air does not usually diffuse into tight barrels in significant quantities. The water and alcohol lost through the surfaces of the barrel are not replaced. Instead, a partial vacuum develops in the space (*ullage*) vacated by the alcohol and water. As barrels may differ markedly in tightness, the negative pressures observed over barreled wine also may vary considerably (Fig. 8.20). Variation in oxygen penetration may explain the barrel-to-barrel diversity in maturation rates often noted by winemakers.

Evaporative wine loss is more marked in barrels left with the bungs upright, but sampling the wine to check its development is much easier. The bung up position also permits frequent topping to fill the ullage as it develops. Coincidentally, both procedures increase the

wine's exposure to oxygen. During normal racking, topping, and sampling, a wine may absorb about 30 to 40 ml O_2/liter per year. It is estimated that about 2 to 5 ml O_2/liter is absorbed through the wood, between 15 to 25 ml O_2/liter during topping, sampling, and through the bung, and up to about 6 ml O_2/liter during each racking (Ribéreau-Gayon *et al.*, 1976). In red wine, the absorbed oxygen is estimated to be consumed within about 6 days at 30°C (Singleton, 1987), which probably is equivalent to about 15 days at 15°C.

The addition of oak chips or shavings to wine has been investigated as an economical alternative to barrel aging. However, the sensory effects appear to differ from those obtained during barrel aging. This may result from the absence of heat-induced hydrolysis, aeration and/or the microbial modification of oak constituents.

In-Barrel Fermentation

Most vinifications take place in large tanks or vats. They permit more uniform fermentation and are easier and more economical to maintain. Fermentation in-barrel, or other small-volume cooperage, is commonly used only with modest quantities of must of unique qualities or origin. Although temperatures during in-barrel fermentation often are higher than those in cooled tank fermentors, the surface area to volume ratio is sufficient to avoid overheating and stuck fermentation. The likelihood of stuck fermentation also may be diminished by the uptake of sterols extracted from the wood (Chen, 1970). Sterols are required for the proper maintenance of yeast membranes during and after fermentation. Sugars released during maturation (Nykänen *et al.*, 1985) also may favor malolactic fermentation, by providing nutrients for the growth of lactic acid bacteria.

In addition to maintaining the individuality of small lots of juice, some winemakers prefer in-barrel fermentation for its effect on wine development. Wine fermented and matured in new oak incorporates less phenolic material than the same wine matured in equivalent barrels after fermentation. This partially results from the coprecipitation of tannins extracted during and shortly after fermentation with yeast cells. The early extraction and oxidation of tannins help consume oxygen, minimizing wine oxidation. Phenols also reduce the accumulation of volatile reduced-sulfur compounds (Nishimura *et al.*, 1983). As the more desirable oak flavors dissolve more slowly than oak tannins, the wine retains proportionally more oak flavor and less harsh tannins. Other differences have been noted, but the sensory significance is uncertain. For example, the reduction of furfural and 5-(hydroxymethyl)-2-furaldehyde to the corresponding alcohols is enhanced (Marsal and Sarre, 1987), as well as the reduction of ferulic and *p*-coumaric acids to 4-vinyl

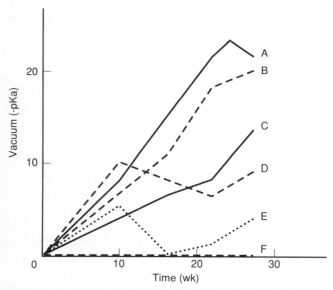

Figure 8.20 Development of a partial vacuum in six wine barrels (A–F) during undisturbed aging of 'Cabernet Sauvignon' wine. (From Peterson, 1976, reproduced by permission.)

guaiacol, 4-ethyl phenol (Dubois, 1983). The latter possess spicy, smoky odors. Another significant change is the metabolic conversion of phenolic aldehydes, notably vanillin, to less aromatic phenolic alcohols. In-barrel fermentation increases the level of "oak" lactones (β-methyl-γ-octalactones), nitrogen compounds, and polysaccharides, primarily those derived from yeast mannoproteins. The latter might enhance the smooth mouth-feel of the wine. In contrast, soluble protein levels may decrease.

Disadvantages of Oak Cooperage

For premium wines, fermentation or aging in oak is often worth the effort and expense involved. For most wines, however, exposure to oak is neither cost effective nor necessarily beneficial.

Oak barrels are costly to purchase and expensive to maintain, and new barrels need conditioning before use. The tartrates and tannins that build up on the inside the barrel during wine maturation are both difficult and unpleasant to remove. When not containing wine, barrels must be protected from drying and microbial contamination. Off-flavors produced by bacteria and fungi growing on internal surfaces can taint wine subsequently stored in the cooperage with "corky" off-odors (Amon *et al.*, 1987).

Because the rate of maturation varies from barrel to barrel, frequent and time-consuming barrel sampling is necessary to assess the progress of the wine. Racking is more labor intensive and inefficient than the equivalent in large cooperage. In addition, considerable economic loss can result from the evaporation of wine from the barrel. Up to 2 to 5% of the volume may be lost per year in this way (Swan, 1986). Depending on the relative humidity of the winery, wine may either increase or decrease in alcoholic strength (Guymon and Crowell, 1977). High relative humidity suppresses water evaporatoin but has no effect on alcohol loss. Consequently, the alcoholic strength of wine decreases in humid cellars. Under dry conditions, water evaporates more rapidly than ethanol, increasing the alcoholic strength. In addition to water and ethanol, acetaldehyde, acetal, acetic acid, and ethyl acetate are lost by evaporation from barrel surfaces (Hasuo and Yoshizawa, 1986). Relative humidity also influences the types and amounts of phenols extracted. Low relative humidity decreases total phenolic uptake but increases vanillin synthesis (Hasuo *et al.*, 1983).

Other Cooperage Materials

For both fermentation and storage, cooperage conspicuous from material other than wood has many advan-

tages. In addition, other materials are less expensive to maintain. Stainless steel often is preferred, but fiberglass and cement also are widely used. Because all are impervious to oxygen, wine oxidation is minimized. This preserves the fresh, fruity character important to most wines designed for early consumption. Stainless steel and fiberglass have the additional benefits of permitting construction in a variety of shapes more difficult to produce with wood or cement. Modern construction materials also facilitate cleaning and dry storage. Gas impermeability permits partial filling, as the ullage can be filled with carbon dioxide or nitrogen to limit oxidation. Furthermore, modern construction materials do not modify the fragrance of wines.

Stainless steel is generally the preferred modern cooperage material because of its strength and inertness. The inertness avoids the need for, and maintenance of, coatings of paraffin wax, glass, or epoxy resin. These are required for cement tanks as excessive amounts of calcium can seep into wine matured in unprotected tanks. In addition, the acidic nature of wine tends to corrode the cement. Stainless steel possesses heat transfer properties permitting comparatively easy temperature control during fermentation. Temperature contol often is obtained with coolant circulated within an insulated, double-lined jacket. Installation is rapid and subsequent movement of the cooperage is possible.

For wine production and storage, stainless steels high in chromium and nickel content are required. Contents of between 17 and 18% chromium by weight provide an adequate surface layer of insoluble chromium oxide. It is the chromium oxide that provides most of the anticorrosive properties of stainless steel. Nickel is present in amounts that may vary between 8 and 14%. It facilitates soldering and further enhances corrosion resistance. When the stainless steel is exposed to wine for short periods, molybdenum may be omitted from the steel. However, for prolonged contact or exposure to sulfited wines, molybdenum is required at a concentration of about 2 to 3%. Titanium may be incorporated as it increases the level of carbon permitted in the finished steel. Titanium also reduces the risk of corrosion next to soldered joints. It often is added at about 0.5%.

With stainless steel, it is important to avoid introducing scratches on the inner surfaces of the tank. Even rinsing with hard water containing minute rust particles can cause damage to the polished surfaces.

Fiberglass tanks also have become common replacements for wooden cooperage. Fiberglass has the advantages of being less expensive and lighter than stainless steel. However, it possesses less strength, is less conductive to heat, is more porous, and tends to be more difficult to clean (has a rougher surface) than stainless steel. In addition, residual styrene may diffuse into the wine from the polyester resin binding the glass fibers. At concentrations above 100 μg/liter, styrene may taint the wine with a plastic odor (Anonymous, 1991).

Stainless steel, resin-coated regular steels, and fiberglass have permitted the production of an extensive array of cooperage. While the containers may be used for wine maturation, most are designed to facilitate emptying and cleansing after fermentation. Thus, they typically possess a slanted floor and exit ports at or near the base. The position of the port (horizontal or vertical) is largely a function of whether cleaning occurs automatically or manually. Other designs may pivot tanks to facilitate emptying or a fixed helical blade may mix the must during fermentation and aid subsequent discharge.

Cork and Other Bottle Closures

Cork

Cork remains the bottle closure of choice after over 400 years of use in Europe. However, the use of cork in the preservation of wine predates its use as a bottle closure by some 2000 years. The ancient Greeks and Romans frequently used cork to stopper wine amphoras (Tchernia, 1986). The resin-coated stoppers were often covered by a cap of volcanic-tuff cement (*pozzolana*). The oldest known use of cork as a wine seal comes from an Etruscan amphora (sixth century B.C.) unearthed in Tuscany (Joncheray, 1976). The subsequent decline in the use of cork reflected the decline in the use of amphoras following the collapse of the Roman Empire. The major reemergence of cork as a closure for wine containers began about the mid-seventeenth century, coincident with the beginnings of industrial-scale glass bottle manufacture in England. Nevertheless, cork appears to have been used as a bottle closure as far back as the end of the fifteenth century (McKearin, 1973).

Cork is a tissue produced by a special layer of cells, the **cork cambium,** located in the outer bark of plants. The cambium produces cork (**phellem**) to the outside and a thinner layer of cells (**phelloderm**) to the inside. Together, these tissues constitute the outer bark. The inner bark, or **phloem,** consists of cells primarily involved in conducting organic nutrients throughout the plant.

In the majority of woody plants, the cork layer is relatively thin. In only a few species of oak is a deep, relatively uniform layer of cork produced. Of these, only the cork oak, *Quercus suber,* produces cork in commercial quantities. Not only does *Q. suber* produce a thick layer of cork, but the cork can be repeatedly harvested without damaging the tree.

CORK OAK

Quercus suber grows in a narrow region bordering the western Mediterranean (Fig. 8.21), with most commercial stands located in Portugal. Portugal produces about two-thirds of the world's supply ($\sim$200 million kg), with the remainder coming from Spain, Algeria, Morocco, Italy, France, and Greece. Dry, upland sites on rocky soils provide the better areas for cork production. Here, the bark is firmer and more resilient. On rich lowland soils, trees produce a thicker but more spongy layer of less valuable cork.

The cork oak grows about 16 m high and has a trunk diameter of 20 to 60 cm at breast height. Typically, the tree begins to branch out about 4 to 5 m above the ground. Thus, the trunk provides large, clear sections of bark. The lower branches of older trees also may yield bark sections of sufficient size to be of commercial value. Trees may live about 500 years, but the most productive period occurs between the first and second century of growth.

CULTURE AND HARVEST

Most commercial oak stands are of natural origin. However, selection and planting of superior seedlings are occurring in both existing stands and reforestation areas. Pruning helps to shape the trees for optimal production of quality cork.

When trees reach a diameter of over 4 cm, or are about 20 to 30 years old, the cork is stripped from the trunk for the first time. This stimulates the growth of new cork and the tree in general. The initial or **virgin cork** is not used for the production of bottle closures as its structure is too irregular and porous (Fig. 8.22).

On removal of virgin cork, the exposed tissues turn red, and, by an unknown mechanism, cork production is

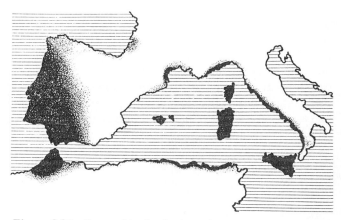

Figure 8.21 Geographic distribution of *Quercus suber,* the cork oak. (From *Subériculture* by J. V. Vieira Natividade, 1956, reproduced by permission of Éditions de l'École nationale du Génie rural des Eaux et des Forêts, Nancy, France.)

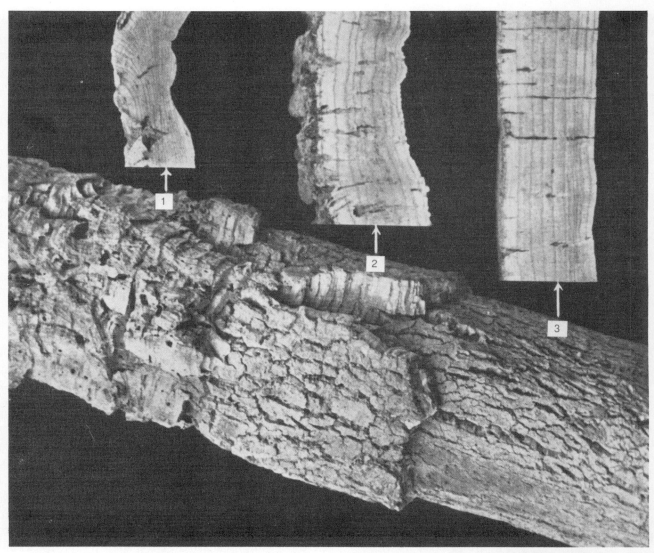

Figure 8.22 Cork oak showing the appearance of virgin (1), second (2), and reproduction cork (3). (From *Subériculture* by J. V. Vieira Natividade, 1956, reproduced by permission of Éditions de l'École nationale du Génie rural des Eaux et des Forêts, Nancy, France.)

spurred and becomes more uniform. Within 7 to 10 years, the tree may again be ready for stripping. This cork, called **second cork,** also lacks the qualities necessary for bottle closures. The third and all subsequent strippings are referred to as **reproduction cork** and yield cork suitable for bottle closures. Stripping occurs about every 9 years, except in mountainous regions where growth is slow and stripping may occur only every 12 to 18 years.

Stripping usually takes place in early summer, when the trees are actively growing. At this time, the soft, newly formed cork ruptures along a line just outside the cork cambium, easing bark removal. Stripping begins with the production of two circumferential cuts with a special ax, one around the base of the tree and the other just below the lowest branches. A subsequent vertical cut

down the trunk connects the initial cuts. A few additional cuts may be made to ease removal and handling. The cork is pried off with the wedge-shaped end of the ax handle. Workers remove the cork from branches if the diameter is sufficient to yield useful slabs. Because of the damage caused by improper stripping, skilled workers are used to remove the cork. Deep cuts damage the inner bark, causing permanent scarring that makes subsequent cork removal difficult.

Yield varies widely from tree to tree. Young trees often yield only about 15 kg of cork, while large trees can produce upward of 200 kg. Bark thickness varies from about 1.5 to 6.5 cm. For wine cork production, the bark needs to be sufficiently thick to yield sheets of usable cork 2.5 cm thick.

On harvest, laborers bundle and stack the slabs outdoors to season for several days or weeks. Subsequently, the bundles are immersed in tanks of boiling water for about 30 min. Boiling extracts water-soluble compounds and softens the bark for easier flattening and removal of the outer portion of the cork (**hard back**). The structure of the hard back layer, which can vary from 1.5 to 3 cm in thickness, is too irregular, stiff, and fractured to be of value in the production of cork closures. After removing the hard back, workers trim and sort the slabs into rough grades based on thickness and surface quality. After grading, the cork is bundled and seasoned for 1 to 3 years.

CELLULAR STRUCTURE

As cork cells enlarge and mature, they elongate along the radial axis of the trunk. During this period, they produce a chemically distinct cell wall composed primarily of fatty material. By maturity, the cytoplasm has disintegrated, leaving only the empty cell wall. Because the typical intercellular cytoplasmic connections (**plasmodesmata**) become plugged during cell development, each cell acts as a sealed unit. Cork tissue also lacks intercellular spaces, each cell abutting tightly against its neighbors. Cork tissue, especially the valuable reproduction cork, is relatively homogeneous in texture. The only significant disruptions to homogeneity result from the presence of lenticels and the sporadic occurrence of fissures. Lenticels are columns of thin-walled cells containing large, irregular, intercellular spaces. They allow gas exchange between the internal tissues of the tree and the exterior. All cork contains lenticels, but the best cork contains small, narrow lenticels and possesses few fissures.

Individual cork cells resemble elongated prisms in cross section (Fig. 8.23A). They possess from 4 to 9 sides, with most cells showing a total of 7 to 9 sides. The cells formed in the spring are larger and thinner walled than those produced later in the season. The weak, early cells often collapse under the pressure exerted by the subsequent growth of smaller, tougher cork cells. Even the latter frequently show corrugations in the side walls (Fig. 8.23B). This is caused by their being pressed against the existing outer layers of cork. The highly collapsed, corrugated nature of the early cork accounts for the bands that distinguish each year's growth.

Cork that grows slowly has a smaller proportion of spongy, early cork cells. The cork compresses less readily but is more elastic. The elastic, resilient nature of cork is one of the most important properties. Consequently, slow-grown cork is more valuable than rapidly grown cork for bottle closure.

The cell wall of cork tissue shows several unique features. The most notable is the presence of approximately 50 alternating layers of wax and suberin (Sitte,

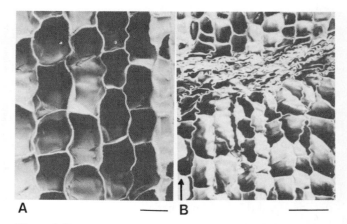

Figure 8.23 Scanning electron micrograph of reproduction cork. (A) Radial section; (B) transverse section showing the corrugated appearance and the collapse of cork cells formed in the spring. Bar, 20μm. (From Pereira *et al.*, 1987, reproduced by permission.)

1961). Both compounds are complex polymers, highly impermeable to gases and liquids and resistant to the action of acids. About 37% of the wall of reproduction cork consists of suberin (Asensio and Seoane, 1987), a polymeric ester of long-chain ω-hydroxy acids, with only a few free acids such as phloionic and phloionolic acids (Agulló and Seoane, 1981). The wax component, which makes up about 5% of the wall mass, consists primarily of cerin and friedlin, smaller amounts of betulin, and probably fatty acids, such as betulic, cerolic, oxyarachidic, phellonic, oleic, and linoleic acids (Lefebvre, 1988). It is believed that the platelike layers of wax and suberin permit the sliding of the layers past one another. Because the wall folds and becomes corrugated under compression, the pressure is absorbed and wall cracking minimized. Realignment of the wall layers probably explains the elastic return on pressure release.

About 28% of the cell wall constituents of reproduction cork are lignins, complex water-insoluble polymers composed of phenylpropanoid monomers. Cellulose and related hemicelluloses constitute about 13% of the wall mass. The wall also contains phenolic compounds, such as catechol, orcinol, gallic acid, and tannic acid.

PHYSICOCHEMICAL PROPERTIES

The physicochemical properties of cork ideally suit its use as a bottle closure. These include compressibility and resilience, chemical inertness, imperviousness to liquids, and a high coefficient of friction.

Cork is one of the few substances, natural or synthetic, that can be compressed without showing marked lateral expansion. In addition, cork shows remarkable resilience on the release of pressure. Cork returns almost immediately to 85% of its original dimensions, and within the next few hours it regains about 98% of the original volume (Fig. 8.24). These properties are undoubtedly related to the distinctive wall structure and the sealed nature of the cells. The latter gives the cells the property of minute air cushions. Some air is expelled in the early stages of compression, but then gas release ceases until high pressures are reached (Gibson *et al.*, 1981). At that point the cells rupture, and collapse of the cell wall occurs, destroying its resilient properties.

The ability of cork to spring back to the original shape gives cork much of its sealing properties. Its resilience exerts pressure on the neck of the bottle, providing a tight seal for years. Eventually, however, the resilience declines, and the cork may become loose in the neck.

The moisture content of cork significantly influences resilience and compressibility. Within a moisture range of 5 to 12%, cork remains sufficiently supple for insertion. Most corks are held at the lower end of the moisture range, between 5 and 7%, to help prevent mold growth during storage. Use of drier corks also avoids the extrusion of fluid which could contaminate the wine during insertion.

Cork generally resists microbial attack because of its hydrophobic nature and negligible nutrient status. Low moisture content further limits microbial growth. Microbial growth is limited largely to nutrients supplied by dust contaminants or by wine that seeps into crevices in the cork.

The chemical inertness of cork protects it during prolonged contact with wine. It also means that few breakdown products form and diffuse into the wine. Boiling cork after harvest extracts most compounds that

might unfavorably affect the sensory properties of a wine.

Impermeability to liquids comes from the tightly packed nature of cork tissue, which provides few channels through which fluids can pass. Cork presents a penetration barrier of about 500 cells/cm. In addition, the waxy nature of the wall restricts diffusion across or between the cells. Although gases, water vapor and fat-soluble compounds can diffuse into cork, significant movement of these substances through corks inserted in wine bottles appears limited. This may be due to the pressure created within the cork cells by compression in the neck of the bottle (Casey, 1993). It is estimated that oxygen penetrates cork-sealed bottles of wine at a rate of about 0.1 ml/liter per year (Singleton, 1976). In addition, up to about 0.5 ml of oxygen may be absorbed during the first few years of storage, associated with wine seepage caused by temperature fluctuations (Ribéreau-Gayon *et al.*, 1977). Nevertheless, it is unlikely that this level of oxygen uptake significantly affects the sensory properties of wine, even over many years.

Because cork is permeable to water vapor, it will dry if the stopper does not remain in continuous contact with the wine. If the cork dries, it shrinks, becoming less resilent and loses some of its sealing properties.

The high coefficient of friction of cork allows it to hold tightly, even to smooth surfaces. The cut surfaces of cork cells form microscopic suction cups that adhere tightly to glass. There is also an inelastic loss of energy during compression that increases the friction between the cork and the glass (Gibson *et al.*, 1981). Such properties combined with resilience allow cork to establish a long-lasting, tight seal with the glass in about 8 to 24 hr.

With time, cork loses its resilience, strength, and sealing properties. The reasons for these changes are unclear, but may relate to the slow escape of gas from the cork cells. Because of the loss of the desirable qualities of the cork, wines aged in-bottle for long periods may be recorked every 25 to 30 years. Recorking also permits topping up bottles that have developed a large headspace.

PRODUCTION OF STOPPERS

To make the cork more pliable and easy to cut, the slabs are boiled in water or steamed for about 20 min. The treatment also cleans the cork, removes residual water-soluble compounds, and helps kill microorganisms on or within the cork. The cork may be used immediately but is customarily stored for about 2 to 3 weeks, while the moisture content falls to about 10 to 12%.

Once ready for use, the slabs are cut into strips equivalent in width to the length of the stoppers desired (usually 24 to 30 mm). By placing the strips on their side, corks may be punched out parallel to the annual growth rings

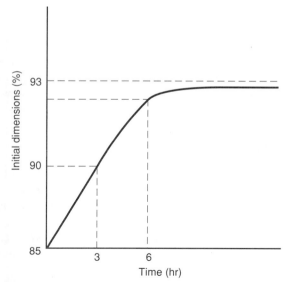

Figure 8.24   Elastic rebound of cork tissue. About 85% of the elastic return is instantaneous. (After Riboulet and Alegoët, 1986, reproduced by permission.)

of the bark (Fig. 8.25). This positions the lenticels and fissures in the cork at right angles to the length of the stopper. Thus, the major sources of porosity (lenticels and fissures) are positioned such that their ends abut against the neck of the bottle.

Skilled workers, rather than machines, punch out the stoppers. This permits a better positioning of the cutter to avoid structural faults that occur in even the best cork. Nevertheless, only 20 to 40% of a strip of cork can be used to produce stoppers.

If desired, the corners of the stopper may be trimmed at a 45° angle. This process, called **chamfering,** is most commonly used for corks destined to stopper fortified wines. Chambering eases reinsertion of the cork after bottle opening. Chamfered corks are routinely bonded to plastic tops to improve the grip on the cork. Such corks are called **T-corks.** Chamfered corks are seldom used to seal bottles of table wine as chamfering reduces the surface area over which the cork adheres to the glass.

After being punched out, corks are treated to a series of washes. The corks are initially rinsed in water to remove debris and cork dust, followed by a soak in a solution of calcium hypochlorite [$Ca(OCl)_2$]. The latter both bleaches and surface sterilizes the cork. Subsequently, the corks are placed in a bath of dilute oxalic acid. The oxalic acid reacts with residual calcium hypochlorite, liberating the chlorine as a gas and precipitating calcium as the oxalate salt. Oxalic acid also binds with iron particles that may have been left by the cork cutter. If not removed, iron may react with tannins in the cork,

forming black ferric tannate spots. After the oxalate wash, the corks are given a final water rinse.

Both the hypochlorite and oxalate treatments need to be kept to a minimum since hypochlorite can corrode the cell wall material of cork and oxalic acid can form crystalline deposits. The latter can cause leaks by reducing adherence of the cork to the glass.

Alternative bleaching and sterilizing agents are desirable as chlorine can generate one of the most common sources of cork off-odors, namely, 2,4,6-trichloroanisole (2,4,6-TCA). It also may generate polychlorinated dibenzo-*p*-dioxins and dibenzofurans (PCDDs/PCDFs). Production is, however, minimal, and uptake of PCDDs/PCDFs by wine is less than 0.2% of the current estimated daily uptake of the toxins from all sources (Frommberger, 1991).

The primary alternative bleaching and surface sterilizing agent is peracetic acid (CH_3CO_3H). On reacting with organic matter, it releases acetic acid and hydrogen peroxide. Hydrogen peroxide is reduced to water by its bleaching and sterilizing action, while acetic acid is eliminated either by rinsing or by evaporation on drying.

On request, producers give corks a surface coloration. Although coloring is less frequent than in the past, there is still a demand for corks given a "traditional" color. For example, Spanish wines may be stoppered with light brown corks, French wines occasionally are sealed with rose-tint stoppers, and Italian wines closed with white corks.

To bring the moisture content quickly down to about 5 to 8%, the corks are placed in centrifugal driers or in hot-air tunnels. The corks are next sorted into quality grades. The various grades reflect the presence and size of lenticels, fissures, and other surface structural faults. Sorting is essential as even the best quality bark may contain faults invisible until the corks have been punched out.

Lower-grade corks are usually treated to a process called *colmatage.* The process fills the surface cavities with a mixture of fine cork particles and glue. This normally occurs in a rotary drum in which the stoppers, cork particles, and glue are tumbled together. After an appropriate period, the corks are removed and rolled to produce a smooth surface. Colmatage improves not only the appearance but also the sealing qualities of the cork. The treatment is justifiable with corks slightly marred with surface imperfections, but not with corks so structurally flawed that their sealing properties are severely impaired. The cork/glue mixture does not possess the elastic, cohesive, and structural properties of natural cork.

Corks may subsequently be stamped, or burnt, with marks indicating the winery, producer, wine type, and vintage. Corks may be coated with paraffin or silicone to ease insertion into the bottle. Such coatings also reduce

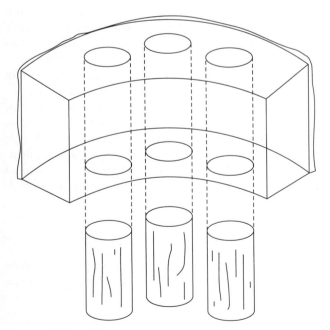

Figure 8.25 Positioning of the cutter for the removal of corks from strips of cork tissue. (Redrawn from Riboulet and Alegoët, 1986, reproduced by permission.)

penetration of wine into the cork and the extraction of potential taints. Silicone generally is preferred because its inorganic nature makes it resistant to microbial attack; however, silicone is unacceptable for champagne corks as it can disrupt effervescence production.

Finally, the corks are bundled in large sacks (~10,000 each) or sealed in plastic bags (500 to 1000 each). Packaging in jute sacks is still relatively common, despite a number of disadvantages. It not only makes stacking difficult, but also can result in a loss of quality. Baling exposes the corks to contamination from dust, microorganisms, and volatile chemicals. Absorption of naphthalene, from containers transported in proximity to baled corks, has been implicated in wine showing a musty taint (Strauss *et al.*, 1985). Guaiacol and other aromatic substances also may be adsorbed from the surrounding environment. In addition, exposure to high humidity can promote microbial growth, while low humidity results in a loss of suppleness.

Storage in polyethylene bags avoids many of the problems. Because bags are commonly packaged in carton boxes, the corks are easily stacked for storage and shipment. Holding corks at a moisture content of 5 to 7% usually prevents microbial activity. Nevertheless, sulfur dioxide is usually added to further protect the corks from microbial growth. As polyethylene is slowly permeable to sulfur dioxide, the inhibition of microbial activity seldom lasts more than 6 months to 1 year. The alternative of soaking the corks in a concentrated solution of sulfur dioxide prior to storage is not recommended; it leads to the breakdown of suberin in the surface layers of the cork and can generate sulfur taints with other cork constituents.

Because sulfur dioxide, hypochlorite, and peracetic acid disinfect only the outer layers of the cork, microorganisms found deep in the lenticels or that have grown into the cork tissue may remain unaffected. Other treatments offer the possibility of completely sterilizing corks. The most effective to date is exposure to gamma radiation. Microwaving also appears effective.

In most instances, cork sterilization is unnecessary as most wines and bottles are not free of viable microorganisms. In addition, sterilization does not destroy existing off-odors in the cork. However, because sweet wines are particularly susceptible to microbial deterioration, sterilized corks are used in sealing these wines. Sweet wines are typically sterile filtered into sterilized bottles to avoid microbial growth following filling.

AGGLOMERATE CORK

Agglomerate cork is synthesized from granules of natural cork, bound together with a glue such as polyurethane. The granules are produced by grinding strips of bark left from punching out natural cork stoppers. The granules are mixed with glue and extruded into long tubular molds. The molds are heated to 95° to 105°C to set the glue. The cylinders of agglomerate cork are then removed and cut into stoppers of the desired lengths. Subsequent treatment is equivalent to that of natural cork. The diameter of agglomerate cork stoppers is slightly less than that of natural cork equivalents, to adjust for the lower resilience. Lower resilience means that agglomerate corks stoppers are less easily compressed for bottle insertion and spring back less than natural cork equivalents.

The use of agglomerate cork to seal table wines has expanded in the past few years. However, agglomerate cork has been used extensively in closing sparkling wine for years. Originally, corks used for stoppering sparkling wines were produced from strips of natural cork. However, because of the larger stopper diameter required for adequte sealing (31 versus 24 mm), thicker slabs of cork were needed. Because of the cost, corks composed of two, three, or more layers of cork laminated together became common. The disks (*rondelles*) could be punched out of cork of normal thickness at right angles to the bark surface. Following the development of better quality agglomerate cork, agglomerate cork replaced all but the inner two disks of natural cork. The corks are positioned before insertion to have the natural cork in contact with the wine.

Rondelles are produced by the same technique used for metal crown caps. Cork of high quality, possessing a fine texture and free of tough spots or other faults, is steamed and softened. After a short seasoning, the cork is trimmed and its rough side pressed into a board containing short tacks. So attached, the upper surface of the cork is shaved off in preparation for punching out the *rondelles*. Strips are usually cut about 6 to 8 mm thick, though for crown caps the strips are usually about 2 mm thick. *Rondelles* are punched out of the cork strips.

To produce finished corks, two *rondelles* are commonly glued together with a casein paste and bonded to one end of a core of agglomerate cork. The corks are trimmed and fine sanded to obtain the standard dimensions of 31 by 47 mm. The familiar mushroom shape of the cork is produced by compressing the top of the cork just after insertion.

Cork Faults

Cork can be both a direct and indirect source of problems in bottled wine. Corks can cause leakage, generate deposits, and be a source of off-odors.

LEAKAGE

Leakage around, or rarely through, the cork may have many causes. Incorrect bore size or inperfection in the glass surface may leave or create gaps between the cork and the neck. Improper alignment or compression of the cork during insertion can create structural faults in an initially flawless cork. Laying the bottle on its side immediately after filling and corking, or rapid temperature changes during storage or shipment, also can induce leakage. Direct leakage due to the cork is typically caused by structural imperfections, though improper sizing during manufacture also may be a cause.

Cork may show a wide variety of structural imperfections and mechanical faults. Corks containing numerous flaws are customarily avoided by evading poor regions of the bark during manufacture, or they are rejected during subsequent grading. However, quality control systems cannot adequately check every specimen, and some flawed corks are found in all grades.

Lenticels can be the source of large cavernous or conical-shaped crevices in cork. Growth irregularities in the bark or wood may cause splits or fissures in the cork. Less frequently, holes are produced by cork-boring insects, including ants and several moth and beetle larvae. Some of the larvae also may produce holes in corks after insertion. The old practice of dipping the bottle neck in beeswax helped prevent the deposition of insect eggs on the cork. Plastic and metal capsules serve a similar function. Most cork-boring insects are larvae of the common cork moth (*Nemapogon cloacellus*), the grain moth (*Nemapogon granellus*), the wine moth (*Oenophila flavum*), the cave moth (*Dryadaula pactolia*), the seed moth (*Hofmannphila pseudopretella*), or the glue moth (*Eudrosis lacteella*) (see Vieira-Natividade, 1956).

Mechanical faults may result from the development of fibrous tissue in the cork ("woody" cork), the growth of fungi ("marbled" cork), or insufficient suberification ("green" cork).

DEPOSITS

Cork can occasionally be a source of wine deposits. The most common originates from lenticular dust. Dust also is generated during cork production, but most is removed prior to shipment. Mechanical agitation, though, may loosen more lenticular dust during transport. Coating corks with paraffin or silicone helps to limit but may not prevent dust release. If the coating is defective or nonuniform, it can itself release particulate material. Some coating materials, such as polyvinylidene chloride (PVDC), may become unstable and flake off in contact with highly alcoholic wines. Calcium oxalate crystals, formed after the oxalate wash to degrade residual hypochlorite, occasionally can generate visible deposits in wine.

Defective *colmatage* also can produce wine deposits. Sediment can come both from the filling material and from dust released from exposed lenticels or fissures.

Improper corking, by physically damaging the cork, can release cork particles into the wine.

TAINTS

Cork can give rise to several off-odors. Although the odors are often referred to as *corky*, this designation gives a false impression. Sound cork, containing no faults, donates little if any odor, even though cork may release several volatile compounds.

Improper treatment or storage near volatile chemicals can result in the adsorption of off-odors that can subsequently taint wine. Cork-derived taints, which are due primarily to microbial growth, are covered later in the chapter.

Several authors have estimated that up to 2 to 6% of all bottled wine may be adversely affected by cork-derived off-odors (Heimann *et al.*, 1983; Carey, 1988). If true, wine producers and retailers are fortunate that the majority of consumers cannot detect or do not object to these faults. One large retailer in England claims a complaint rate of about 1 bottle in 50,000 (Young, 1982). Whether bottles are returned or not, faulty wines undoubtedly result in a loss of repeat sales.

The ability of cork to adsorb highly aromatic compounds from the surroundings has been mentioned above. More common, though, is the formation of off-odors associated with bleaching and surface disinfection.

2,4,6-Trichloroanisole (2,4,6-TCA) is the most well-known example of a processing-generated cork taint (Simpson, 1990). 2,4,6-TCA produces a marked musty to moldy odor even at a few parts per trillion. It can originate from the hypochlorite [$NaOCl$ or $Ca(OCl)_2$] used in bleaching corks. If corks are steeped overly long in bleach, chloride ions diffuse into the cork through lenticels and fissures. Here, chlorine can react with phenolic compounds, producing chlorophenols. These can be methylated by microorganisms in the cork to 2,4,6-TCA and subsequently escape into and taint the wine. The absorption of water vapor can desorb volatile compounds from the cork, permitting them to diffuse into the wine (Casey, 1990). Use of substitute bleaching agents, such as sodium peroxide and peracetic acid, avoids the production of 2,4,6-TCA, but may themselves leave undesirable residues (Fabre, 1989; Puerto, 1992).

Another source of 2,4,6-TCA probably comes from cork oak trees, oak cooperage, or wooden structures in wineries or storehouses treated with pentachlorophenol (PCP) to control insects and/or wood rot. Because several fungi such as *Penicillium* can metabolize the pesticide to 2,4,6-TCA (Maujean *et al.*, 1985), significant levels of 2,4,6-TCA may occur in corks not exposed to bleaching. PCP also can directly contaminate corks and wine.

cork
(source)

fungal
metabolism
(source)

lignins
phenols + Ca (OCl)$_2$ →

2,4,6-TCP
2,4,6-trichlorophenol

→ microbial methylation →

2,4,6-TCA
2,4,6-trichloroanisole

← microbial methylation ←

PCP
(fungicide)
pentachlorophenol

Although microbial methylation of PCP is apparently involved in most of the production of 2,4,6-TCA, some white rot fungi (basidiomycetes) can directly incorporate chlorine into a benzene ring. Subsequent metabolism could produce 2,4,6-TCA without treatment of the cork with hypochlorite (Maarse *et al.*, 1988).

In addition to chlorophenols, various chloropyrazines may generate corky, moldy odors. Boiling cork slabs after harvesting produces pyrazines from the interaction of sugars and amino acids in the bark. Most pyrazines readily diffuse into the water and are discarded, but some can migrate back into the bark if the slabs are left in the water after boiling. Later, when the corks are bleached with hypochlorite, chloropyrazines may form.

Aqueous solutions of sulfur dioxide, often used to surface sterilize corks, have been implicated in the production of sulfurous off-odors. Sulfur dioxide can react with lignins in the cork to generate lignosulfurous acid. When leached into wine, lignosulfurous acid may decompose and generate hydrogen sulfide. Reaction with pyrazines could produce musty-smelling thiopyrazines. An additional source of pyrazines, above that produced by boiling cork slabs, comes from mold growth on the cork. Some pyrazines have strong, moldy, earthy odors.

Additional cork-associated off-odors are discussed below. Several of the compounds involved may be synthesized by microorganisms on substrates other than cork (i.e., oak staves). In addition, synthesis is not associated with any particular treatment of the cork, as is the case with 2,4,6-TCA. Because cork taints may come from a variety of sources, the combined effect of several compounds may be distinct from the odors of each individual component.

Cork Alternatives

Alternatives to cork have been sought for years. This has been spurred more recently by both the rising costs and demand for premium quality cork. In addition, some wine producers believe that the frequency of cork-related faults is on the rise. Industry estimates suggest that between 0.5 and 6% of wines may become tainted by contact with cork closures (Anonymous, 1992).

The most successful alternative has been the pilfer-proof, roll-on metal screw cap (Fig. 8.26). Part of its success comes from its not attempting to reproduce the appearance and characteristics of natural cork with synthetic materials. The roll-on (RO) closure is generally superior to cork in retaining sulfur dioxide and minimizing oxidation, especially in bottles stored upright (Bach, 1982). The primary disadvantage of the RO closure has been its association with inexpensive wines. Most producers of premium wines are unwilling to risk the stigma of sealing their wines with RO closures. As a consequence, most cork substitutes have attempted to reproduce the appearance and physical properties of cork.

Ethylene vinyl acetate is the major alternative that possesses most of the basic appearance and features of cork (Anonymous, 1992). When injected with air and a hardener, liquid ethylene vinyl acetate forms millions of microscopic gas pockets before hardening. The inflated plastic develops a resilience that permits it to return almost immediately to its original diameter (97%). The plastic stopper regains more than 99% of its initial volume within 1 hour (Anonymous, 1983).

The first plastic stoppers used successfully were made of polyethylene. The ease of resealing with polyethlyene stoppers made them especially valuable for sparkling wines. However, their primary disadvantage was slow oxygen permeability. For wines with a rapid turnover this was of little concern, but for premium wines oxygen permeability prevented adoption of the stoppers. Also, the consumer association between the plastic cork and inexpensive carbonated wines has prevented improvements from being accepted by wine producers.

The crown cap with a cork or polyethylene liner, formerly so familiar owing to its use with soft drinks, has seen little application as a wine closure. The one excep-

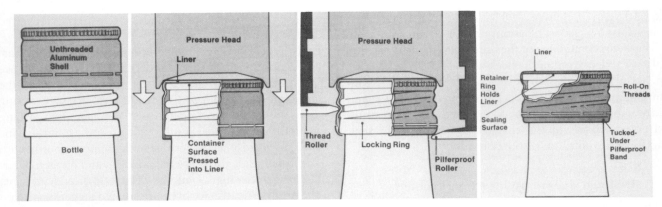

Figure 8.26 Insertion and formation of an Alcoa RO closure. (Courtesy of Alcoa Closure Systems International, Indianapolis, IN.)

tion is in the second in-bottle fermentation of sparkling wines. Here, consumer image is not a problem as the bottle cap is replaced by a traditional cork closure when the wine is disgorged and prepared for shipment.

Another solution to stoppering bottles has been to avoid the use of bottles altogether. Use of cans and bag-in-box packaging are current, if not elegant, solutions.

Cork Insertion

Various mechanical devices are available to insert corks into wine bottles by compressing the cork to a diameter smaller than the bore of the bottle. These function essentialy as well with natural as with the ethylene vinyl acetate corks. On compression, a plunger forces the cork down into the neck of the bottle.

Better corking machines apply a uniform pressure over the full length of the cork. This minimizes the production of creases, folds, or puckering, which can result in leakage. Standard 24 mm wide corks are compressed to about 14 to 15 mm prior to insertion. The plunger is adjusted to assure that, regardless of cork length, the top of the cork rests at, or just below, the lip of the bottle.

Most 750 ml wine bottles are produced to the CETIE (Centre Technique International de l'Embouteillage) standard, with an inner neck diameter (bore) of 18.5 ± 0.5 mm at the mouth and no more than 21 mm 4.5 cm below the mouth. An older standard, retained for most sparkling wines, has a bore diameter of 17.5 ± 0.5 mm at the lip. Because most wine corks have a 24 mm diameter, they remain compressed by about 6 mm after insertion. This level of compression is sufficient to generate a pressure of about 1 to 1.5 kg/cm^2 against the glass within 24 hr (Lefebvre, 1981). For moist corks of superior quality, a pressure of about 3 kg/cm^2 may be produced. Sweet wines or wines containing greater than 1 g/liter CO_2 typically use corks that remain compressed about 7 to 8 mm against the glass. Sparkling wines com-

monly use special 30 to 31 mm diameter corks to maintain 12 mm of compression against the neck.

Oversized corks can cause leaks as easily as those that are too narrow. In the former situation, creases may occur during insertion; in the latter case, the seal may be too weak.

Because the diameter of the bore commonly increases down the length of the neck. The desirable cork length depends partially on the extent of the increase. In the CETIE standard, the maximum bore diameter at a depth of 4.5 cm is 21 mm. This means that both medium (44/45 mm) and long (49/50 mm) corks of 25 mm diameter may be held by no more than 4 mm of compression at depths below 4.5 cm. This contrasts with about 6.5 mm of compression at the bottle mouth. Deeper than 4.5 cm, the large inner neck diameter may result in long corks being held only weakly at the bottom. Thus, cork diameter is generally more important to a good seal than length. The advantages of long corks for wines benefiting from long aging come from factors other than surface contact of the cork with the glass.

Cork is noted for its chemical inertness in contact with wine. Nevertheless, prolonged exposure slowly reduces the structural integrity of the cork. Because of the low permeability of cork to liquids, corrosion progresses slowly up the cork from where it contacts the wine. In addition, water uptake makes the cork more supple, and it loses its hold on the glass. As the rate of weakening is most rapid in wines of high sugar and alcohol contents, long corks are especially useful for such wines. Because corrosion affects the substance of the cork, denser cork is preferred where long storage of wine is anticipated. It is estimated that loosening of the cohesive attachment of the cork to the glass can progress at a rate of about 1.5 mm per year (Guimberteau *et al.*, 1977). The effect is reported to diminish the pressure exerted by the cork against the glass from its initial value of about 100–300 kPa to 80–100 kPa after 2 years, and about 50 kPa after 10 years (Lefebvre, 1981). This explains why fine wines are often recorked about every 25 years.

The slight cone shape of the bore (18.5 → 21 mm), the bulge or indentation in the neck about 1.5 cm below the lip, and the compression of the cork are all important in limiting movement of the cork in the neck. This is especially important if the wine is exposed to temperature extremes, where volume changes can weaken the seal and force the cork out of the bottle.

Leakage Caused by Insertion Problems

During insertion, air tends to be trapped underneath the cork in the neck of the bottle. As cork does not regain its full resilience for several hours, the stopper does not immediately exert its maximum force against the neck. If the bottle is laid on its side or turned upside down shortly after corking, the pressure exerted by the trapped air can force wine out between the cork and the neck of the bottle. To avoid this, wines often are set upright for several hours after corking. During the interval, the entrapped air can escape before the seal becomes firmly established. This is important as air consists of about 78% nitrogen, a gas poorly soluble in wine. If the pressure is not released, it continues to act on the wine and cork when the bottle is placed on its side. Oxygen, the other main atmospheric gas, dissolves quickly in the wine and ceases to exert pressure.

Nitrogen contained in the headspace can augment leakage induced by extremes of temperature. As shown in Fig. 8.27, temperature significantly influences the volume of wine. Temperature–volume changes modify the pressure exerted on the headspace volume and, indirectly, on the cork. Leakage becomes likely at internal pressures above 200 kPa (about twice atmospheric pressure), when the net outward pressure equals that exerted by the cork against the glass.

To reduce leakage, most bottles are placed under partial vacuum or flushed with carbon dioxide before corking. The creation of a partial vacuum eliminates the development of a positive headspace pressure during cork insertion. Any vacuum that remains after corking soon dissipates as carbon dioxide escapes from the wine into the headspace. Alternately, flushing the bottle with carbon dioxide displaces the nitrogen and oxygen of air. Because of the high solubility of carbon dioxide in wine, the positive headspace pressure created on cork insertion rapidly dissipates. The uptake of about 20 mg of carbon dioxide in this manner has no sensory effect on the wine.

The use of corking under vacuum or carbon dioxide has benefits beyond reducing the likelihood of leakage. It limits the development of "bottle sickness" by removing the 4 to 5 mg of oxygen otherwise absorbed from the trapped air. Both procedures also avoid lowering the content of free sulfur dioxide by removing oxygen that can react with SO_2. In addition, the procedures permit bottles to be packed and stored in an inverted position directly after corking.

Variation in bottle capacity may be an additional source of leakage problems. If the bottle has a capacity smaller than specified, very little headspace volume may remain after filling. Even with medium length corks (44/45 mm), bottles may possess a headspace volume of as little as 1.5 ml after corking. Since the wine in a 750 ml bottle can expand by about 0.23 ml/°C, a rise in temperature can quickly result in a significant increase in the pressure exerted on the cork. The effect is increased if the headspace gas contains nitrogen or if the wine is sweet or is supersaturated with carbon dioxide (Levreau *et al.*, 1977).

Sugar content can augment leakage by increasing the capillary action between the cork and the glass. Sugar also increases, by about 10%, the rate at which wine volume changes with temperature (Levreau *et al.*, 1977).

Finally, the moisture content of cork can influence the likelihood of leakage. At moisture contents between 5 and 7%, cork has sufficient suppleness not to crumble on compression. It also rebounds sufficiently slowly to allow pressurized headspace gases to escape before a tight seal develops. At lower moisture levels, compression and insertion are likely to rupture or crease the cork. At high moisture levels, more gas is likely to be trapped in the headspace (Levreau *et al.*, 1977).

Bottles and Other Containers

Over the years, wine has been stored and transported in a wide variety of containers. The first extensively used

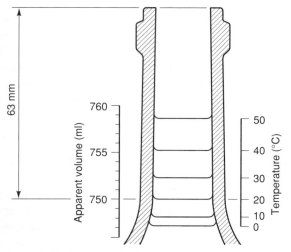

Figure 8.27 Variations in the level of fill as a function of temperature. (From Lefebvre, 1981, reproduced by permission.)

storage vessels were probably made from animal skins. In most areas, these were replaced by clay amphoras. Use of amphoras in Egypt was well-established at least by 1400 B.C. (Lesko, 1977), and they apparently were being used to store wine in Iran between 3500 and 2900 B.C. (Badler *et al.*, 1990). Amphoras continued to be used in the Middle East up until at least A.D. 625 (Bass, 1971). They fell into disuse throughout much of the western Mediterranean after the decline of the Roman Empire. North of the Alps, wood cooperage became the primary wine storage and transport vessel from the beginning of the Christian era. It held its preeminence in Europe until the twentieth century, when the glass bottle became the primary storage and transport vessel.

Glass bottles were used to a limited extent during Roman times, but only sporadically. This undoubtedly resulted from their fragility and costly production. Developments in glass technology in Italy during the 1500s reintroduced bottles as wine containers. However, because of their fragility, the bottles had to be covered with reeds to protect them from breaking. Further technological developments in England during the 1600s permitted the production of strong glass bottles. Initially, expense limited their widespread adoption, and bottles were used primarily as decanters to transport wine from barrel to table. Use for long distance transport occurred but was limited. Intentional bottle aging of wine appears to have developed with sweet wines, notably *porto* and Tokaji wines. Storage and transport in bottles became necessary only with the development of sparkling wine. Subsequently, bottles slowly began to supplant barrels for the transport and aging of wine.

Bottles are not without disadvantages, however. Because of the variety of shapes, sizes, and colors, collecting for reuse tends to be cost effective only with that portion of the market using standard bottle shapes and colors. Bottles also create a considerable disposal problem if not recycled in some other product. Once a bottle is opened, wine conservation is difficult, and noticeable loss in quality is usually evident within 1 day. Although decanting systems can replace the volume of consumed wine with nitrogen gas, they are practical only in large establishments and are of little value to the average consumer. As a consequence, new containers such as bag-in-box packages have begun to replace the bottle for many standard-quality wines.

Glass Bottles

Glass has many advantages over other materials in bottle manufacture. The chemcial inertness of glass is especially critical in aging premium wines for prolonged periods. Glass also is impermeable to gases and resists all but rough handling. Although trace amounts of sodium, chromium, and nickel may dissolve into wine, the levels involved generally are minuscule. However, changes in glass formulation may increase the amount of chromium extracted from 1 to 4 μg/liter (Médina, 1981). The transparency of glass to near-ultraviolet and blue radiation is a disadvantage but can be countered by the incorporation of several metal oxides. The primary disadvantages of glass are its weight and the energy required in its manufacture.

PRODUCTION

Bottle glass is formed by heating sand (largely silicon dioxide) to about 1500°C in the presence of soda (sodium carbonate), lime (calcium oxide), and small amounts of magnesium and aluminum oxides. The first three materials often contribute about 95% of the mass. Clear glass is typically produced by adding sufficient magnesium to decolorize the iron oxide contaminants that commonly occur in sand. Small amounts iron, manganese, nickel, and chromium oxides may be added to give the glass a desired color. The yellow to green color of most wine bottles comes primarily from ferric and ferrous oxides (Fe_2O_3 and FeO). The specific shade is influenced by the redox potential, degree of hydration, presence of other metals, and the chemical nature of the glass. Amber is often generated by maintaining the reducing action of sulfur during glass fusion, brown by the addition of manganese and nickel, and emerald green with the incorporation of chromic oxide (CrO_3). Adding chromic oxide under oxidizing conditions or vanadium pentoxide (V_2O_5) greatly improves ultraviolet absorption (Harding, 1972).

In bottle manufacture, an appropriate amount of molten glass, the *parison*, is placed in the upper end of a rough (*blank*) bottle mold (Fig. 8.28). Compressed air forces the molten glass down to the bottom of the mold, where the two portions of the neck mold occur. Air pressure, from the center of the neck mold, blows most of the glass back into the configuration of the blank mold and establishes the finished shape of the neck. The blank mold, containing the outer half of the neck mold, is opened and removed. The outer stiff layer of glass, next to the mold, maintains the shape of the bottle during its transfer to the finishing (*blow*) mold. The inner half of the neck mold (*neck ring*) is used to raise and invert the bottle, to position it between the halves of the blow mold. The bottle is released by the neck ring just before closing the halves of the blow mold.

The glass is reheated to bring it back to a moldable temperature. Compressed air, blown in via the neck, drives the glass against the sides of the blow mold. At the same time, a vacuum is created at the base of the mold, removing trapped air. These actions give the bottle its final dimensions. After a short cooling period, to assure

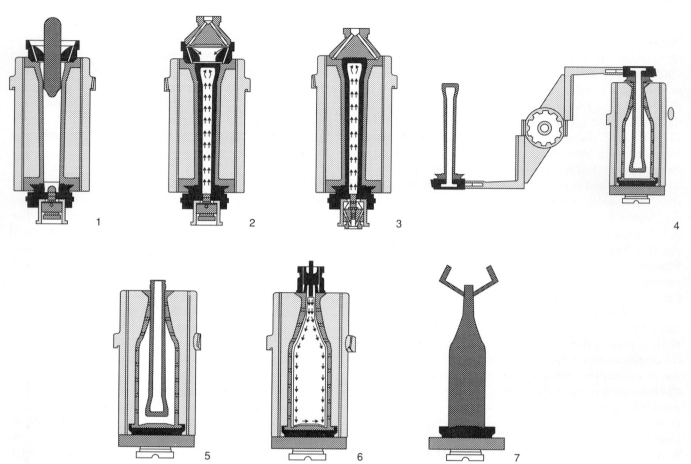

Figure 8.28 Blowing of glass wine bottles. 1, Molten glass (parison) added to the rough mold; 2, air pressure forces the parison to the base of the mold; 3, air pressure from the mold base forces the still molten glass into the rough mold shape; 4, removal of the rough mold and transferral of the bottle into the finishing mold; 5, reheating of the glass; 6, air pressure forces the molten glass into the shape of the finishing mold; 7, removal of the finished bottle. (After Riboulet and Alegoët, 1986, reproduced by permission.)

retention of the final shape, the bottle is removed from the blow mold.

During production, various parts of the bottle cool at different rates. This creates structural heterogeneity that makes the glass fragile. To remove the structural tensions, the bottle is *annealed* by heating to about 550°C. After sufficient annealing, the glass is slowly cooled through the annealing range, then rapidly cooled to ambient temperatures. During annealing, sulfur is typically burnt to produce a thin layer of sodium sulfate on the inner surface of the glass. The associated diffusion of sodium ions to the glass surface increases its chemical durability. Alternately, a thin coating of titanium or other ions may be added. Both procedures harden the surface of the glass and minimize lines of weakness.

SHAPE AND COLOR

Bottles of particular shapes and colors have come to be traditionally associated with wines from several European regions. These often have been adopted in the New World for wines of similar style, or to imply character similarity. Unique bottle shapes and markings also are used in increasing consumer awareness and recognition of particular wines.

In general, bottle shape and color are more important in marketing than to aging, storage, or transport. Tradition and image probably explain why bottles that filter out little ulraviolet radiation are still in common use. White wines are the most susceptible to light-induced damage (Macpherson, 1982), but they are often the least protected, being commonly sold in clear or light-colored bottles. To partially offset light-activated spoilage, white wines often receive higher doses of sulfur dioxide before bottling than red wines.

In contrast, bottle neck design depends primarily on pragmatic issues, such as the type of closure. Still table wines can use plain necks, as they require only a simple cork or RO closure. Sparkling wines, however, require a special lip to which the restraining wire mesh for the stopper is secured.

Bottles of differing filling heights, and permissible capacity variations, are available. Bottles having lower filling heights and smaller volume variation leave more headspace between the wine and the cork. By providing more space for temperature-induced wine expansion, the bottles are less susceptible to leakage.

FILLING

After a hot water rinse, bottles are steam cleaned and allowed to drip dry before filling. Although cleaning is always essential, actual sterilization is necessary only for wines that have been sterile filtered, or when used bottles are being recycled. Because the microbial contamination of new bottles is generally minimal and consists largely of organisms unable to grow in wine, sterilization is usually unnecessary. In addition, the microbial population of the wine may be considerably higher than that in the bottle.

Various types of automatic bottling machines are available. Some operate by siphoning or gravity feed, while others use pressure or vacuum. Siphoning and gravity feeding are the simplest but slowest, while the pressure and vacuum fillers are more appropriate for rapid, automated filling lines.

Regardless of the machine, precautions must be taken against the buildup of microbes within the equipment. Microbial contamination has been known to cause the spoilage of thousands of bottles of wine. Precaution also is taken to minimize oxidation during filling. This is best achieved by flushing the bottles before filling with carbon dioxide gas. Alternately, the headspace may be flushed with carbon dioxide, or filling and cork insertion may occur under vacuum. Bottle sickness, a mild form of oxidation that may follow bottling, is due primarily to oxygen left in the headspace. The quantity contained in headspace air can amount to eight times the oxygen absorbed during filling.

Bag-in-Box Containers

Bag-in-box technology has progressed dramatically from its initial start as a means of marketing battery fluid. Subsequent developments have found wide application in the wine industry. Because the bag collapses as wine is removed, its volume is not replaced with air. This permits wine to be periodically removed over several weeks to months without noticeable loss in quality. This convenience is credited with expanding the wine market in regions without a strong association with wine. In some countries, such as Australia, more than half the wine sold is by the "box" (Anderson, 1987). Bag-in-box packaging also is ideally suited for "house wine" sold in restaurants, especially the larger 10 to 20 liter sizes.

In its modern form, bag-in-box packaging protects

wine from oxidation for 9 months or longer. This is usually adequate for wines that require no additional aging and have a high rate of sale.

For protection and ease of stacking and storage, the bag is housed in a corrugated or solid fiber box. A handle permits easy transportation. The large surface area of the box provides ample space for marketing information and attracting consumer attention.

At present, no single membrane possesses all the features necessary for a collapsible wine bag. The solution has been to use a two-layered membrane (Webb, 1987). The inner layer, comprising a low density polyester or ethyl vinyl acetate film, provides the necessary protection against flex cracking. The outer layer, typically a metallized polyester laminated to polyethylene, slows the loss of sulfur dioxide and aromatic compounds and the inward diffusion of oxygen. Most modern plastic films are inert to wine and do not affect the sensory characteristics of the wine. Eliminating amide additives has removed a former source of off-odors.

Taps come in a variety of styles. Each has its own potential problems, such as tendency to leak, relative permeability to oxygen, and expense. Currently, the tap is considered the prime source of oxidized wine in bag-in-box packaging (Armstrong, 1987).

To minimize oxidation, the bags are placed under vacuum before filling, and the headspace is charged with inert gas after filling. To protect the wine against microbial spoilage, and to limit oxidation, the wine is usually adjusted to a final level of 50 mg/liter sulfur dioxide before filling.

Wine Spoilage

Cork-Related Problems

The difficulties associated with off-odor identification are well illustrated with cork-related faults. The origins of cork-related problems are diverse and their chemical nature often unknown. Even experienced tasters have great difficulty in correctly naming most faults. A difficult situation is made more complex if a wine is affected by more than one fault. Wines may be affected by several, distinct, off-odor compounds. Combinations of off-odors may influence both the quantitative and qualitative perception of the faults. Correspondingly, most off-odors can be identified with confidence only with sophisticated analytic equipment. Although much has been learned to date, considerably more needs to be known to limit the occurrence of off-odors.

Although several cork-derived taints have a "musty" or "moldy" odor, others do not. Therefore, cork-related taints are usually grouped relative to presumed origin.

Some off-odors come from the adsorption of highly aromatic compounds in the environment. Other taints originate during cork production, notably 2,4,6-trichloroanisole (2,4,6-TCA). Even some glues used in producing sparkling wine corks can generate off-odors, for example, butyl acrylate (Brun, 1980). Finally, microbial infestation can be an important source of cork-related taints. As several of these problems have already been discussed, consideration will be limited to microbially induced spoilage.

In the study by Simpson (1990) of tainted wines, the most frequent off-odor was 2,4,6-TCA (86%), followed by 1-octen-3-one (73%), 2-methylisoborneol (41%), guaiacol (30%), 1-octen-3-ol (19%), and geosmin (14%). Each compound has its own somewhat distinctive odor. In about 50% of the wines, 2,4,6-TCA was considered the most intense off-odor present.

Although 2,4,6-TCA appears to be the most common source of musty taints in wine, its musty, chlorophenol odor is considered to be different from the putrid, butyric smell of the so-called true corky taint of wine (Schanderl, 1971). Apparently, occurrence of the latter is now rare and its origin uncertain. The "yellow stain" fault of cork bark is often associated with the true corky taint. While some researchers believe that the tree pathogen *Armellaria mellea* is the likely causal agent (Pes and Vodret, 1971), others feel that *Streptomyces* is responsible (Lefebvre *et al.*, 1983). In the latter case, the contaminated cork contained several compounds well-known to possess moldy, smoky, earthy odors, for example, guaiacol, 1-octen-3-ol, 2,5-dimethylpyrazine, 2,4,6-TCA, and sesquiterpenes.

Moldy odors may originate from the growth of bacteria, filamentous fungi, and occasionally yeasts on cork before harvesting, during seasoning, while stored in plastic containers, or after insertion into the bottle. Cork also can be a source of yeasts and bacteria which can grow in and spoil wine.

Because nonselective media are commonly used when isolating organisms from cork, it is difficult to know whether most of the microbes isolated are members of an indigenous cork flora or contaminants. Regardless of origin, several fungal and bacterial isolates produce moldy/musty smelling compounds. As microorganisms require moist conditions for growth, control is commonly attained by keeping the moisture content of corks below 8%. Sulfur dioxide, added to plastic storage bags, further minimizes microbial growth. Sterilization by exposure to gamma radiation has been recommended to prevent microbial tainting during storage (Borges, 1985). Valuable as these procedures are, none address the problem of off-odors produced in the bark before harvesting or subsequent seasoning.

Fungi, such as *Penicillium roquefortii, P. citrinum,* and *Aspergillus versicolor,* produce a series of musty/corky smelling compounds, including 3-octanol. In addition, *P. roquefortii* and bacteria of the genus *Streptomyces* produce musty-smelling sesquiterpenes (Heimann *et al.,* 1983). The role of *Penicillium* in the synthesis of 2,4,6-TCA has already been mentioned above. Additional corky taints are suspected to come from the growth of fungi such as *Trichoderma harzianum* on the cork (Brezovesik, 1986). Species of *Streptomyces* can taint cork by metabolizing vanillin, derived from cork lignins, to guaiacol (Lefebvre *et al.,* 1983). Guaiacol possesses a sweet burnt odor.

Most of the fungi isolated from cork can be found in bark taken directly from the tree. However, *P. roquefortii* appears to originate as a winery contaminant. It is seldom isolated from corks prior to delivery and storage in wine cellars. It is one of the few organisms that can occasionally grow through cork in bottled wine (Moreau, 1978). Other commonly isolated fungi, such as *Penicillium glabrum, P. spinulosum,* and *Aspergillus conicus,* cannot grow in contact with wine. Thus, the lower two-thirds of a cork seldom yields viable fungi after a few months in the bottle (Moreau, 1978).

Among the genera of fungi growing on or in cork, few are known toxin producers. The major potentially toxigenic species is *P. roquefortii* (Leistner and Eckardt, 1979). Whether the strains of *P. roquefortii* that grow on cork are toxigenic and whether the toxin, if produced, seeps into wine appear not to have been investigated. The major aflatoxin-producing fungus, *Aspergillus flavus,* has not been isolated from cork.

Fungal growth on the upper surface of corks in bottled wine is favored by the moisture retained by unperforated lead capsules. Although growth rarely progresses through the cork, the production of organic acids can speed capsule corrosion. Corrosion eventually can result in contamination of the neck and upper cork surface with soluble lead salts.

Yeasts have seldom been implicated in cork-derived taints. Exceptions, however, involve *Rhodotorula* and *Candida.* Both have been isolated from corks of tainted champagne (Bureau *et al.,* 1974).

Although the importance of cork as a source of spoilage yeasts and bacteria has been little investigated, the presence of the microorganisms seems both highly variable and generally low. The most frequently encountered bacterial genus is *Bacillus.* This is not surprising because of its production of highly resistant dormant structures called **endospores.** Several *Bacillus* species have been associated with spoilage. *Bacillus polymyxa* has been implicated in metabolism of glycerol to acrolein, a bitter-tasting compound (Vaughn, 1955), and *B. megaterium* occasionally produces an unsightly deposit in brandy (Murrell and Rankine, 1979).

In addition to cork-related taints, cork can donate a slightly woody character to wine. This is derived from naturally occurring aromatics extracted from the cork. Over eighty volatile compounds have been isolated from cork (Boidron *et al.*, 1984). Many of the compounds are similar to those isolated from oak wood used in barrel maturation. This is not surprising as cork comes from the same genus, *Quercus*. Nevertheless, the donation of a woody odor from cork is infrequent; after all, one of the major advantages of cork is the relative absence of extractable odors.

Yeast-Induced Spoilage

A wide variety of yeast species have been implicated in wine spoilage. Occasionally, this is associated with haziness and the deposition of sediment. The number of cells required to generate haziness varies with the species. With *Brettanomyces* a distinct cloudiness is reported to develop at less than 10^2 cells/ml (Edelényi, 1966), while with most yeasts a slight haze begins to develop only at 10^5 cells/ml (Hammond, 1976).

Zygosaccharomyces bailii can generate both flocculant and granular deposits (Rankine and Pilone, 1973). It can grow in bottling equipment and thereby contaminate thousands of bottles. White and rosé wines tend to be more susceptible to attack than red wines. *Zygosaccharomyces bailii* is especially difficult to control owing to its high resistance to yeast inhibitors. For example, the yeast can grow in wine supplemented at 200 mg/liter with either sulfur dioxide, sorbic acid, or diethyl dicarbonate (Rankine and Pilone, 1973). Even 1 cell per 10 liters may be sufficient to induce spoilage (Davenport, 1982). Along with osmophilic *Kluyveromyces*, *Zygosacch. bailii* can contaminate sweet reserve.

Many yeasts form filmlike growths on the surface of wine under aerobic conditions. These include strains of *Saccharomyces cerevisiae* (*Sacch. bayanus* and *Sacch. prostoserdovii*) and *Zygosaccharomyces fermentati*. These are involved in producing the *flor* character of sherries and similarly matured wines. Species of *Candida*, *Pichia*, *Hansenula*, and *Brettanomyces* also may occur as minor members in film growths on *flor* sherries without apparent harm. Alone, however, they usually induce spoilage.

Of spoilage yeasts, species of *Brettanomyces* (imperfect state of *Dekkera*) are probably the most serious (Larue *et al.*, 1991). Both *B. intermedius* and *B. lambicus* produce 2-acetyltetrahydropyridines, compounds that possess mousy odors (Heresztyn, 1986a). *Brettanomyces* species also synthesize volatile phenolic compounds, including 4-ethyl guaiacol, 4-vinyl guaiacol, phenol, and syringol (Heresztyn, 1986b). These can give wine smoky, spicy, medicinal, and woody taints. In addition, certain

Brettanomyces species generate apple/cider odors, produce high levels of acetic acid, and cause haziness.

Some filamentous fungi, such as *Aureobasidium pullans* and *Exophiala jeanselmei* var. *heteromorpha*, can grow in must and under suitably aerobic conditions in wine. They often metabolize tartaric acid and can seriously reduce total acidity (Poulard *et al.*, 1983). Must or wine so affected is highly susceptible to further microbial spoilage.

Bacteria-Induced Spoilage

LACTIC ACID BACTERIA

Lactic acid bacteria have already been discussed in connection with malolactic fermentation. Although lactic acid bacteria are often beneficial, certain strains are spoilage organisms. Spoilage is largely limited to wines stored under warm conditions, in the presence of insufficient sulfur dioxide, and at pH values higher than 3.5. None of the spoilage problems are induced exclusively by a single species, and the frequency of spoilage strains varies considerably among species.

Tourne is a spoilage problem caused primarily by *Lactobacillus brevis*, though a few strains of *Leuconostoc oenos* may be involved. The primary action is the fermentation of tartaric acid to oxaloacetic acid. Depending on the strain, oxaloacetate is subsequently metabolized to lactic acid, succinic acid, or acetic acid and carbon dioxide. Associated with the rise in pH is the development of a "flat" taste. Affected red wines usually turn a dull red-brown color, become cloudy, and develop a viscous deposit. Some forms produce an abundance of carbon dioxide, giving the wine an effervescent aspect. In addition to an increase in volatile acidity, other off-odors may develop. These often are characterized as being sauerkrautlike or mousy.

Amertume is associated with the growth of a few strains of *Lactobacillus brevis* and *L. buchneri*. The strains are characterized by the ability to oxidize glycerol to acrolein, or reduce it to 1,3-propanediol:

$$\text{Glycerol} \xrightarrow{\quad H_2O \quad} \text{3-hydroxypropanal}$$

$$\begin{array}{c} \xrightarrow{\; H_2O \;} \text{acrolein} \\[1em] \xrightarrow[\text{NADH} \quad \text{NAD}^+]{} \text{1,3-propanediol} \end{array}$$

Acrolein possesses a bitter taste, giving the spoilage its French name. Alternative metabolic routing of glycerol may increase the concentrations of aromatic compounds, such as 2,3-butanediol and acetic acid. As a result of glycerol metabolism, the concentration of glycerol may decrease by 80 to 90%. In addition, there is a marked accumulation of carbon dioxide and often a doubling of the volatile acidity (Siegrist *et al.*, 1983).

Some heterofermentative strains of lactic acid bacteria, notably *Leuconostoc oenos*, induce what is called **mannitic fermentation.** This form of spoilage is characterized by the production of both mannitol and acetic acid.

Ropiness is associated with the synthesis of profuse amounts of mucilaginous polysaccharides (β-1,3-glucans). The bacteria typically involved are strains of *Leuconostoc oenos* and *Pediococcus*. The polysaccharides hold the bacteria together in long silky chains that appear as floating threads in affected wine. When dispersed, the polysaccharides give the wine an oily look and viscous texture. Although visually unappealing, ropiness is not consistently associated with off-odors and tastes.

The presence of mousy taints has already been mentioned relative to *tourne* and the growth of species of *Brettanomyces*. Mousiness is also associated with the growth of some strains of *Lactobacillus brevis, L. cellobiosis,* and *L. hilgardii* in must and wine. The strains synthesize 2-acetyltetrahydropyridines and 2-propionyltetrahydropyridines in the presence of ethanol and 1-propanol, respectively. Both compounds have distinctly mouselike odors.

Lactic acid bacteria also may produce a geranium-like taint in the presence of sorbic acid (Crowell and Guymon, 1975). Some strains of *Lactobacillus,* and most strains of *Leuconostoc oenos*, metabolize sorbic acid to *p*-sorbic alcohol (2,4-hexadienol). Under the acidic conditions of wine, 2,4-hexadienol isomerizes to *s*-sorbic alcohol (1,3-hexadienol). It in turn reacts with ethanol to form an ester, 2-ethoxyhexa-3,4-diene. It is the latter compound that generates a strong geranium-like odor.

High volatile acidity is a common feature of most forms of spoilage induced by lactic acid bacteria. Acetic acid can be produced by the metabolism of citric, malic, tartaric, and gluconic acids, as well as hexoses, pentoses, and glycerol. The level of acetic acid synthesis depends on the strain and conditions involved. However, production is very limited in the absence of suitable reducible substances. Correspondingly, acetic acid production is rare under strictly anaerobic conditions.

ACETIC ACID BACTERIA

Acetic acid bacteria were first recognized as causing wine spoilage in the nineteenth century. Their ability to oxidize ethanol to acetic acid both induces wine spoilage and is vital to commercial vinegar production. Although acetic acid synthesis during vinegar production has been extensively investigated, the action of acetic acid bacteria on grapes, and in must and wine, has escaped modern scrutiny until relatively recently. That acetic acid bacteria could remain viable in wine for years under anaerobic conditions was unexpected. They were thought to be strict aerobes and unable to grow or survive in the absence of oxygen. However, the ability of acetic acid bacteria to use hydrogen acceptors other than molecular oxygen suggests that they may show limited metabolism under anaerobic conditions. Thus, the role of acetic acid bacteria in all phases of winemaking is coming under renewed investigation.

As presently recognized, acetic acid bacteria form a distinct family of gram-negative, rod-shaped bacteria characterized by the ability to oxidize ethanol to acetic acid. For years, molecular oxygen was thought to be the only acceptable terminal electron acceptor for respiration. It is now known that quinones can substitute for oxygen (Aldercreutz, 1986). Thus, acetic acid bacteria may grow in barreled or bottled wine, if acceptable electron acceptors are present. Of even greater practical significance is their ability to grow with the traces of oxygen absorbed by wine during clarification and maturation (Joyeux *et al.*, 1984).

The metabolism of sugar by acetic acid bacteria is atypical in many ways. For example, the pentose phosphate pathway is used exclusively for sugar oxidation to pyruvate, whereas pyruvate oxidation to acetate is by decarboxylation to acetaldehyde, rather than to acetyl-CoA. Sugars also may be oxidized to gluconic and mono- and diketogluconic acids, rather than being metabolized to pyruvic acid (Eschenbruch and Dittrich, 1986). While this property is most common in *Gluconobacter oxydans*, some strains of *Acetobacter* also possess the ability to generate gluconic acid from glucose.

In addition to oxidizing ethanol to acetic acid, acetic acid bacteria oxidize other alcohols to the corresponding acids. In addition, they may oxidize polyols to ketones, for example, glycerol to dihydroxyacetone.

Only two genera of acetic acid bacteria are recognized, namely, *Acetobacter* and *Gluconobacter*. They can be distinguished both metabolically and by the position of the flagella. Members of the genus *Acetobacter* have the ability to "overoxidize" ethanol. That is, they may oxidize ethanol past acetic acid to carbon dioxide and water, via the TCA cycle. Under the alcoholic conditions of wine, though, ethanol overoxidation is suppressed. In contrast, *Gluconobacter* lacks a functional TCA cycle and cannot oxidize ethanol past acetic acid. The genus *Gluconobacter* is further characterized by a greater ability to use sugars compared to *Acetobacter*. Motile forms of the two genera may be distinguished by the flagellar attachment. *Gluconobacter* has polar flagellation (insertion at the end of the rod-shaped cells), whereas *Acetobacter* has a more uniform (peritrichous) distribution over the cell surface. Of the species of these genera, only *Acetobacter aceti, A. pasteurianus,* and *Gluconobacter oxydans* are commonly found on grapes or in wine.

Although all three species occur on grapes, and in must and wine, the frequency of occurrence differs markedly. Probably because of the greater ability of *Gluconobacter oxydans* to metabolize sugars, it is the predominant species on grape surfaces. On healthy fruit the bacteria commonly occur at about 10^2 cells/g, but on diseased or damaged fruit they occur at up to 10^6 cells/g (Joyeux *et al.*, 1984). *Acetobacter pasteurianus* is typically present in small numbers, and *A. aceti* is rarely found.

During fermentation, the number of viable bacteria tends to decrease. Nevertheless, the number generally does not fall below 10^2 to 10^3 cells/ml. The most marked change is in the relative proportion of the species present. The population of *Gluconobacter oxydans* often falls during fermentation, being replaced by *Acetobacter pasteurianus.* Subsequently, the population of *A. pasteurianus* may rise or fall during maturation, while *A. aceti* tends to become the dominant species. *Gluconobacter oxydans* tends to disappear entirely during maturation (Fig. 8.29).

Although the viable population of acetic acid bacteria tends to decline during maturation, racking can induce temporary increases. The increase is probably due to the slight incorporation of oxygen during racking. Oxygen can participate directly in bacterial respiration, but it also may indirectly generate electron acceptors for respiration, such as quinones. Small amounts of oxygen may seep into wine through the cooperage or incompletely sealed bung holes.

Spoilage can result from the activity of the bacteria at

any stage of wine production. For example, moldy grapes typically have a high population of acetic acid bacteria. By-products of metabolism, such as acetic acid and ethyl acetate, are retained throughout fermentation and can taint the resultant wine.

Spoilage by acetic acid bacteria during fermentation is currently rare, largely because most present winemaking practices restrict contact with air. Improved forms of pumping over and cooling have eliminated the major sources of must oxidation during fermentation. Also, a better understanding of stuck fermentation can limit its incidence, permitting the earlier application of techniques that reduce the likelihood of oxidation and microbial spoilage.

Although wine maturation occurs largely under anaerobic conditions, its duration often makes the maturation period the most susceptible to wine spoilage. Wood cooperage can be a major source of microbial contamination, if improperly stored, cleansed, and disinfected before use. Wood cooperage also may permit the slow incorporation of oxygen as noted above.

Alone, the levels of sulfur dioxide commonly maintained in maturing wine are generally insufficient to inhibit the growth of acetic acid bacteria. Therefore, combinations of techniques such as maintaining or achieving low pH values, minimizing oxygen incorporation, and storing at cool temperatures, along with sulfur dioxide, appear to be the most effective means of limiting the activity of acetic acid bacteria. Spoilage of bottled wine by acetic acid bacteria presumably is limited to situations where failure of the closure permits seepage of oxygen into the bottle.

The most well-known and serious consequence of spoilage by acetic acid bacteria is the production of high levels of acetic acid. The recognition threshold for acetic acid is approximately 0.7 g/liter (Amerine and Roessler, 1983). At twice this value, it gives wine an unacceptable vinegary odor and taste. Acetic acid production appears to be associated more with the stationary and decline phases of colony growth rather than the log phase of active cell division (Kösebalaban and Özilgen, 1992).

While wines mildly contaminated with acetic acid may be improved by blending with unaffected wine, seriously spoiled wines are fit only for distillation or conversion to wine vinegar.

Under aerobic conditions, acetic acid bacteria do not synthesize noticeable amounts of esters. While ethyl acetate production is increased at low oxygen levels, most of the ethyl acetate generated during acetic spoilage appears to form as a result of nonenzymatic esterification or the activity of other contaminant microorganisms. As a consequence, the strong sour vinegary odor of ethyl acetate is not consistently associated with spoilage by acetic acid bacteria (Eschenbruch and Dittrich, 1986).

Another important aromatic compound sporadically

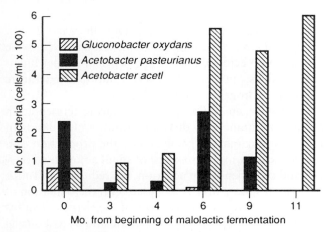

Figure 8.29 Evolution of acetic acid bacteria during malolactic fermentation and maturation in barrel of 'Cabernet Sauvignon' wine. (Data from Joyeux *et al.*, 1984.)

associated with spoilage by acetic acid bacteria is acetaldehyde. Under most circumstances, acetaldehyde is rapidly metabolized to acetic acid and seldom accumulates. However, the enzyme that oxidizes acetaldehyde to acetic acid is sensitive to denaturation by ethanol (Muraoka et al., 1983). As a result, acetaldehyde may build up in highly alcoholic wines. Low oxygen tensions also favor the synthesis of acetaldehyde from lactic acid.

Spoilage by acetic acid bacteria generally does not produce a fusel taint as the organisms oxidize higher alcohols to the corresponding acids. In oxidizing polyols, acetic acid bacteria generate either ketones or sugars. For example, glycerol and sorbitol are metabolized to dihydroxyacetone and sorbose, respectively. The conversion of glycerol to dihydroxyacetone may affect the sensory properties of wine as it has a sweet fragrance and cooling mouth-feel. Dihydroxyacetone also may react with several amino acids to generate a "crustlike" aroma. Whether the latter reaction occurs in wine is unknown.

The oxidation of organic acids under acidic conditions appears weak. Its only significance in wine spoilage by acetic acid bacteria may be the oxidation of lactic acid to acetaldehyde and acetoin.

Some strains of acetic acid bacteria produce one or more types of polysaccharides from glucose. Such production in grapes may account for some of the difficulties in filtering wines made with infected berries.

In addition to acetic acid, acetic acid bacteria may generate considerable quantities of gluconic and mono- and diketogluconic acids from glucose in grapes. These compounds occur in association with most fungal infections and may be used as indicators of the degree of infection.

OTHER BACTERIA-INDUCED SPOILAGE

Other than that caused by lactic acid and acetic acid bacteria, bacterial spoilage is rare. Nevertheless, its development can have serious financial consequences. The main genus associated with the spoilage problems is *Bacillus*. Members of the genus are gram-positive, rod-shaped bacteria that commonly produce long-lived, highly resistant endospores. Most species are aerobic, but some are facultatively anaerobic, as well as being acid and alcohol tolerant.

Bacillus polymyxa has been associated with the fermentation of glycerol to acrolein. Other species, including *B. circulans, B. coagulans, B. pantothenticus* and *B. subtilis,* have been isolated from spoiled fortified dessert wines (Gini and Vaughn, 1962). *Bacillus megaterium* may even produce sediment in bottled brandy.

Although unlikely to grow directly in wine, species of *Streptomyces* may be involved in tainting wine. The growth of *Streptomyces* has already been mentioned in regard to the production of off-odors in cork. *Streptomyces* also may grow on the surfaces of unfilled cooperage. Their ability to produce earthy, musty odors is well known. Synthesis by *Streptomyces* of sesquiterpenols, such as geosmin and 2-methylisoborneol, is believed to be the source of the earthy odor of soil.

Sulfur Off-Odors

In minute quantities, reduced-sulfur compounds may be important to the desirable and characteristic fragrance of wines. They also can be the source of revolting off-odors. The same compounds may induce both responses, depending on the concentration.

One of the most well-known reduced-sulfur compounds is hydrogen sulfide (H_2S). At much above the sensory threshold, hydrogen sulfide produces a rotten-egg odor. The origin of hydrogen sulfide can be very diverse because of its pivotal role in the metabolism of sulfur compounds. Hydrogen sulfide may be generated from sulfate found in grape tissue, sulfite derived from sulfur dioxide, amino acid degradation, and elemental sulfur fungicidal sprays.

During fermentation, organic sulfur compounds do not appear to be significant sources of hydrogen sulfide. The primary source of hydrogen sulfide during vinification is elemental sulfur, especially in colloidal form (Schütz and Kunkee, 1977). Elemental sulfur is derived mainly from its use in the control of powdery mildew. Its conversion to H_2S appears to be nonenzymatic, resulting from the reduction of sulfur particles touching yeast cells (Wainwright, 1970). Reducing substances in yeast cells are presumed to be involved, but yeast strain selection is unlikely to affect H_2S production via this nonenzymatic route. Nevertheless, strain selection can be of value in minimizing the enzymatic reduction of sulfite to hydrogen sulfide. A small proportion of yeast strains are deficient in sulfite reductase and are unable to generate hydrogen sulfide from sulfites (Zambonelli et al., 1984). Deficiencies in the vitamins pantothenate and pyridoxine or the amino acid methionine, presence of copper fungicide residues, and increased fermentation temperature augment hydrogen sulfide production.

The hydrogen sulfide produced during fermentation declines spontaneously during maturation. Oxidation of H_2S to sulfur may be catalyzed by the production of a strong oxidant in the presence of small amounts of oxygen and o-diphenols. In addition, the reduction of sulfur dioxide to sulfate may be coupled with the oxidation of hydrogen sulfide to sulfur. Most of the elemental sulfur produced would probably precipitate and be lost during racking. If high concentrations of hydrogen sulfide develop, however, the H_2S generally persists at perceptible concentrations.

Production of organosulfur compounds may occur during both fermentation and aging and tends to be more serious, as the malodorous compounds formed, such as mercaptans, are more difficult to remove. Several volatile organosulfur compounds also may form during fermentation. High levels of soluble solids and sulfur dioxide promote production of the compounds, especially methylmercaptopropanol (Lavigne *et al.*, 1992). Cultivar and vineyard conditions also play significant roles in the production of volatile organosulfur compounds.

The synthesis of mercaptans during maturation (Fig. 8.30) is favored by the development of low redox potentials in the lees. Under these conditions, elemental sulfur may be reduced and released as hydrogen sulfide. Iron ions are suspected to be involved in the generation of H_2S. Whether the hydrogen sulfide so produced is involved in the formation of mercaptans is uncertain. Reaction of H_2S with acetaldehyde to generate ethanethiol apparently does not occur under wine conditions (Bobet, 1987). However, ethanethiol forms slowly from diethyl disulfide and sulfite in winelike solutions (Bobet *et al.*, 1990). Sulfur-containing amino acids also may be an important source of organosulfur compounds. Methionine can be the origin of methyl mercaptan, while cysteine is a source of dimethyl sulfide (de Mora *et al.*, 1986). Dimethyl sulfide also might form from the use of dimethyl sulfoxide by acetic acid bacteria as an electron acceptor in anaerobic respiration. However, the precise origin of most volatile organosulfur compounds in wine is unknown, as are the precise conditions that lead to their formation.

The genesis of sulfide off-odors in bottled wine, notably champagne, appears clearer. The odor compounds are produced in a complex series of reactions involving methionine, cysteine, riboflavin, and light (Maujean and Seguin, 1983). When photoactivated by light, riboflavin catalyzes the degradation of sulfur-containing amino

acids. Various free radicals are formed, some of which combine to form methanethiol and dimethyl disulfide:

$$CH_3{-}S \cdot + H \cdot \longrightarrow CH_3 {-} SH \text{ methanethiol}$$

$$2\,CH_3{-}S \cdot \longrightarrow CH_3{-}S{-}S{-}CH_3 \text{ dimethyl disulfide}$$

Hydrogen sulfide also is generated in the process. Together, these compounds give rise to the light-struck (*goût de la lumière*) fault of champagne. The significance and origin of the associated high concentrations of dimethyl sulfide commonly found in champagne are unclear. They occur regardless of light exposure.

Goût de la lumière is induced by exposure to violet and near-ultraviolet radiation. These regions correspond to the visible and near-visible peaks in the absorption spectrum of riboflavin (440 and 370 nm, respectively). Only antiultraviolet and amber glass exclude both regions of the electromagnetic spectrum. Wines high in sulfur-containing amino acids and riboflavin are those most likely to be at risk. The ability of tannins to bind riboflavin, and limit its photoactivation, probably explains why red wines are much less sensitive to light-induced taints than are white wines.

Selective removal of mercaptans from wine is difficult. Activated charcoal absorbs mercaptans but reduces wine quality by removing other volatile compounds. Where legally permissible, addition of trace amounts of silver chloride (Schneyder, 1965) or copper sulfate (Petrich, 1982) can precipitate mercaptans. Ascorbic acid may be used in conjunction with copper sulfate to aid mercaptan removal. Early removal of the metal thiol precipitates is important as the reactions are reversible.

Various suggestions have been made as to how to reduce the production of volatile organosulfur compounds. Some of the newer suggestions include the use of SO_2 at concentrations below 50 mg/liter, maintenance of low levels of soluble solids during fermentation (Lavigne *et al.*, 1992), and centrifugation and slight oxidation after the first racking (Cuénat *et al.*, 1990).

Additional Spoilage Problems

In addition to the light-induced off-odor noted above, light also can cause a fault in Asti Spumante (Di Stefano and Ciolfi, 1985). Exposure to sunlight accelerates the oxidation of aromatic terpenes that give the wine its Muscat aroma. Changes in the composition of esters also can occur. In addition, novel terpenes are generated, notably 3-ethoxy-3,7-dimethyl-1-octen-7-ol. The changes produce an unpleasant taste and a fragrance resembling aromatic herbs and preserved vegetables.

Oxidation is a serious, and still too common, source of wine spoilage. Oxidation causes a broad range of sensory changes including an oxidized odor, browning, loss

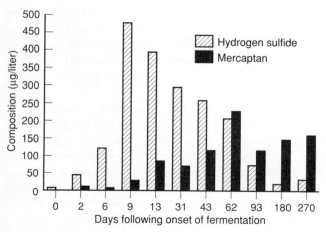

Figure 8.30 Development of reduced-sulfur compounds in wine during fermentation and maturation. The first racking occurred on day 93. (Data from Cantarelli, 1964).

of color in red wines, activation of spoilage by bacteria and yeasts, development of ferric casse, and the precipitation of tannins. These topics have been discussed earlier.

Exposure to heat is a problem receiving renewed attention because of increasing concern about quality loss during wine transport (Anonymous, 1984). Temperature extremes can induce serious spoilage problems during transit, as well as during storage in warehouses and on store shelves. Direct exposure to sunlight often compounds the effects of warm store temperatures.

Temperature directly affects the rate of reactions involved in maturation, for example the acceleration of hydrolysis of aromatic esters and the loss of terpene fragrances at high temperatures. Both changes can result in a decrease in the fruity, varietal character of wines. The synthesis of furfurals from sugars is believed to be partially involved in the development of a baked character. Exposure to temperatures over 50°C quickly accelerates the generation of caramelization products that affect both the bouquet and color of wine. Browning in red wines results from the destruction of free anthocyanins and the generation of polyphenolic pigments. Precipitation of the pigments can lead to sediment formation.

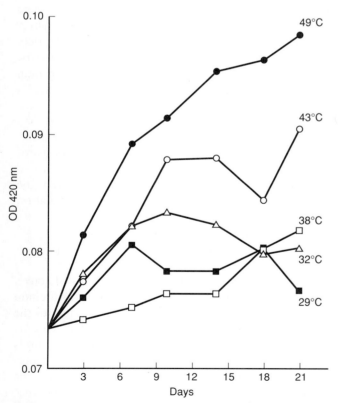

Figure 8.31 Changes in optical density (OD) of white wine with low SO_2 concentrations at five temperatures over a 21-day period. (From Ough, 1985, reproduced by permission.)

Important changes can occur even within a few days at elevated temperatures (Fig. 8.31). Consequently, temperature control is critical during all stages of wine transport and storage. Sulfur dioxide (100 mg/liter) apparently minimizes some of the effects (Ough, 1985). This is apparently unrelated to its antioxidant property.

By affecting wine volume, temperature can loosen the seal of a cork, leading to leakage, oxidation, and possibly microbial spoilage of bottled wine.

Accidental Contaminants

Intentional adulteration of wine has probably occurred since time immemorial. Lead salts were occasionally added during the Middle Ages to "sweeten" highly acidic wine. Laws forbidding such additives were enacted in Europe at least by 1327, when an edict by William, Count of Hennegau, Holland, outlawed the addition of mercury, lead sulfate, and similar compounds to wine (Beckmann, 1846). Concern about nontoxic, accidental contamination is more recent. Only since the 1970s have experimental tools become available that can begin to identify many of the contaminants that can periodically be found in wine.

Typically, heavy metals occur in wine at well below toxic levels. When metals occur at above trace amounts, they usually arise from accidental contamination after fermentation. For example, activated charcoal can be an unsuspected source of chromium, calcium, and magnesium, diatomaceous earth a source of iron contamination, and bentonite a source of high aluminum values.

Tin-lined lead capsules may be an additional source of metal contamination. Corrosion of the capsule can produce a deposit of soluble lead salts on the neck of the bottle. Although the salts do not diffuse through the cork into the wine (Gulson et al., 1992), failure to adequately cleanse the neck can contaminate the wine during pouring (Sneyd, 1988). Nevertheless, structural faults in cork cannot be ruled out as an occasional source of lead contamination in wine. When wine bottles show obvious corrosion of a lead capsule, the bottle mouth needs to be thoroughly cleaned and puncture of the cork with a corkscrew avoided.

Off-odors can arise from a wide variety of unsuspected sources, including blanketing gases, fortifying brandies, oil and refrigerant leaks, absorption of volatile pollutants from the wax coating of cement fermentation and storage tanks, the leaching of chemicals from epoxy paints and resins used on winery equipment, and chlorophenol-treated cooperage (see Strauss et al., 1985).

Naphthalene, absorbed by cork from the environment, has been mentioned previously as a source of wine contamination. Naphthalene also may come from fiberglass used in the construction of holding tanks, resin used to seal tank joints, compounds used to adhere polyvinyl chloride inlays in bottle caps, and from adsorption onto wax used to coat cement tanks. In the latter instance, the naphthalene may come from oil or solvent spills in or near fermentation/storage tanks (Strauss *et al.*, 1985). Naphthalene gives wine a musty note at levels above 0.02 mg/liter.

Another off-odor occasionally associated with the use of epoxy resins is benzaldehyde. It can produce a bitter almond odor when the concentration reaches 2 to 3 mg/liter (Brun, 1984). Benzaldehyde is derived from the oxidation of benzyl alcohol, which is occasionally used as a plasticizer in epoxy resins or found as a contaminant in liquid gelatin (Delfini, 1987). Oxidation of benzyl alcohol to benzaldehyde can be preformed by several microbes, including *Botrytis cinerea* and the yeasts *Schizosaccharomyces pombe* and *Zygosaccharomyces bailii* (Delfini *et al.*, 1991). The latter two organisms also may synthesize benzyl alcohol and benzoic acid. Various ketone notes also may be associated with epoxy resin contaminants. Improperly formulated resins also may be a source of methyl isobutyl ketone, methyl ethyl ketone, acetone, toluene, and xylene (Brun, 1984). In addition, trace amounts of the hardener methylenedianiline, and its monomers bisphenol and epichlorhydrin, may migrate into wine stored in cement and steel tanks coated with some epoxy resins. However, they do not appear to accumulate in concentrations sufficient to affect the sensory characteristics of wine (Larroque *et al.*, 1989).

Fiberglass tanks occasionally retain a small amount of free styrene. If the styrene migrates into the wine at levels above 0.1 to 0.2 mg/liter, it produces a plastic odor and taste (Brun, 1984).

Residues from degreasing solvents and improperly formulated paints can be a source of various off-odors. When winery equipment is disinfected with hypochlorite, the chlorine may react with degreasing solvents or phenols found in paint (Strauss *et al.*, 1985), and one or more chlorophenols may be generated. Wine coming in contact with the affected equipment may show vague medicinal, chemical, or disinfectant odors. Chlorophenols also may be used as a preservative on cooperage oak. These compounds may be microbially modified to produce 2,4,6-trichloroanisole, which is associated with a "corky" (musty) off-odor.

Finally, prolonged storage of water in tanks can lead to the development of a strong musty, earthy taint. This is associated with the growth of actinomycetes in the storage tank. If the water is subsequently used to dilute distilled spirits in fortifying dessert wines, the wine will show an undesirable earthy aspect (see Gerber, 1979).

Suggested Readings

General Texts

Amerine, M. A., Berg, H. W., Kunkee, R. E., Ough, C. S., Singleton, V. L., and Webb, A. D. (1980). "The Technology of Wine Making," 4th Ed. AVI Publ., Westport, Connecticut.

Farkaš, J. (1988). "Technology and Biochemistry of Wine," Vols. 1 and 2. Gordon & Breach, New York.

Margalit, Y. (1992). "Winery Technology and Operations: A Handbook for Small Wineries." The Wine Appreciation Guild, San Francisco, California.

Ough, C. S. (1992). "Winemaking Basics." Food Products Press, Binghamton, New York.

Ribéreau-Gayon, J., Peynaud, E., Ribéreau-Gayon, P., and Sudraud, P. (eds.) (1972–1977). "Traité d'Oenologie: Sciences et Techniques du Vin," Vols. 1–4. Dunod, Paris.

Troost, R. (1988). "Technologie des Weines: Handbuch der Lebensmitteltechnologie," 2nd Ed. Ulmer, Stuttgart.

Vine, R. P. (1981). "Commercial Winemaking." AVI Publ., Westport, Connecticut.

Adjustments

Beelman, R. B., and Gallander, J. F. (1979). Wine deacidification. *Adv. Food Res.* **25**, 1–53.

Boulton, R. (1985). Acidity modification and stabilization. *Proc. Int. Symp. Cool Climate Vitic. Enol. June 25–28, 1984, Corvallis, OR* (D. A. Heatherbell, P. B. Lombard, F. W. Bodyfelt, and S. F. Price, eds.), OSU Agric. Exp. Stat. Tech. Publ. No. 7628, pp. 482–495. Oregon State Univ., Corvallis, Oregon.

Champagnol, F. (1986). L'acidité des moûts et des vins. 1ère Partie. Facteurs physico-chimiques et technologiques de variation. *Rev. Fr. Oenol.* **104**, 26–30 and 51–57.

Singleton, V. L., and Ough, C. S. (1962). Complexity of flavor and blending of wines. *J. Food Sci.* **12**, 189–196.

Stabilization and Clarification

Beech, F. W. (ed.) (1984). "Tartrates and Concentrates: Eighth Wine Subject Day." Long Ashton Research Station, Univ. of Bristol, Bristol.

Blade, W. H., and Boulton, R. (1988). Adsorption of protein by bentonite in a model wine solution. *Am. J. Enol. Vitic.* **39**, 193–199.

García-Ruiz, J. M., Alcántara, R., and Martín, J. (1991). Evaluation of wine stability to potassium hydrogen tartrate precipitation. *Am. J. Enol. Vitic.* **42**, 336–340.

Goswell, R. W. (1983). Tartrate stabilization trials. *In* "Quality Control: Proceedings of the Sixth Wine Subject Day" (F. W. Beech and W. J. Redmond, eds.), pp. 62–65. Long Ashton Research Station, Univ. of Bristol, Bristol.

Lüdemann, A. (1987). Wine clarification with a crossflow microfiltration system. *Am. J. Enol. Vitic.* **38**, 228–235.

Lüthi, H., and Vetsch, U. (1981). "Practical Microscopic Evaluation of Wines and Fruit Juices." Heller Chemie- und Verwaltsingsgesellschaft mbH, Schwäbisch Hall, Germany.

Matteson, M. J., and Orr, C. (eds.) (1986). "Filtration Principles and Practices," 2nd Ed., Chemical Industries Ser. No. 27. Dekker, New York.

Rankine, B. C. (1983). How to gauge filtration choices. *Wines Vines* **64**(6), 82–84.

Aging

Marais, J. (1986). Effect of storage time and temperature of the volatile composition and quality of South African *Vitis vinifera* L. cv. Colombar wines. *In* "The Shelf Life of Foods and Beverages" (G. Charalambous, ed.), pp. 169–185. Elsevier, Amsterdam.

Marais, J., van Wyk, C. J., and Rapp, A. (1992). Effect of storage time, temperature and region on the levels of 1,1,6-trimethyl-1,2-dihydronaphthalene and other volatiles, and on quality of Weisser Riesling wines. *S. Afr. J. Enol. Vitic.* **13,** 33–44.

Rapp, A., and Güntert, M. (1986). Changes in aroma substances during the storage of white wines in bottles. *In* "The Shelf Life of Foods and Beverages" (G. Charalambous, ed.), pp. 141–167. Elsevier, Amsterdam.

Simpson, R. F., and Miller, G. C. (1983). Aroma composition of aged Riesling wines. *Vitis* **22,** 51–63.

Oak

Chatonnet, P. (1989). Origines et traitements des bois destinés à l'élevage des vins de qualité. *Rev. Oenologues* **15,** 21–25.

Chatonnet, P. (1991). Incidences du bois de chêne sur la composition chimique et les qualitiés organoleptiques des vins. Applications technologiques. Thèse. Université de Bordeaux II, Bordeaux. Published by P. Chatonnet, Bordeaux, France.

Guimberteau, G. (ed.) (1992). Le bois et la qualité des vins et eaux-de-vie. *Connaiss. Vigne Vin,* numéro spécial.

Hoey, A. W., and Codrington, J. D. (1987). Oak barrel maturation of table wine. *Proc. 6th Aust. Wine Ind. Tech. Conf.* (T. Lee, ed.), pp. 261–266. Australian Industrial Publ., Adelaide, Australia.

Keller, R. (1987). Différentes variétés de chênes et leur répartition dans le monde. *Connaiss. Vigne Vin* **21,** 191–229.

Maga, J. A. (1985). Flavor contribution of wood in alcoholic beverages. *In* "Progress in Flavor Research—Proceedings of the Fourth Weurman Flavour Research Symposium" (J. Adda, ed.), pp. 409–416. Elsevier, Amsterdam.

Naudin, R. (1991). "L'Élevage des Vins de Bourgogne en Fûts Neuf de Chêne." Institut Technique de la Vigne et du Vin, Beaune, France.

Singleton, V. L. (1974). Some aspects of the wooden container as a factor in wine maturation. *In* "Chemistry of Winemaking" (A. D. Webb, ed.), Advances in Chemistry Ser. No. 137, pp. 254–278. American Chemical Society, Washington, D.C.

Swan, J. S. (1993). What's best for barrels: Air or kiln-drying? *Wine Vines* **74**(7), 43–49.

Cork, Closures, and Bottles

Anderson, T. C. (1987). Understanding and improving bag-in-box technology. *Proc. 6th Aust. Wine Ind. Tech. Conf.* (T. Lee, ed.), pp. 272–274. Australian Industrial Publ., Adelaide, Australia.

Anonymous (1987). Comptes-rendus de la table ronde international "Le Chêne-Liège et son Utilisation: Problèmes Économiques et Qualitatifs de l'Emploi du Liège," April 6, 1987, Sassari, Italy. *Boll. CIDESO* **4,** 1–140.

Cooke, G. B. (1961). "Cork and the Cork Tree." Pergamon, New York.

Dessain, G., and Tondelier, M. (1991). "Liège de Méditerranée." Éditions Edisud/Narration, La Calade, Aix-en-Provence, France.

Guimberteau, G., Lefebvre, A., and Serrano, M. (1977). Le conditionnement en bouteilles. *In* "Traité d'Oenologie: Sciences et Techniques du Vin" (J. Ribéreau-Gayon, E. Peynaud, P. Ribéreau-Gayon, and P. Sudraud, eds.), Vol. 4, pp. 579–643. Dunod, Paris.

Riboulet, J. M., and Alegoët, C. (1986). "Aspects Pratiques du Bouchage des Vins." Collection Avenir Oenologie, Bourgogne Publ., Chaintre, France.

Romat, H. (1988). Le bouchage des vins tranquilles. *Connaiss. Vigne Vin,* Numéro hors Série.

Vieira Natividade, J. (1956). "Subériculture." Éditions de l'École Nationale du Génie Rural des Eaux et Forêts, Nancy, France.

Cork-Induced Spoilage

Amon, J. M., and Simpson, R. F. (1986). Wine corks: A review of the incidence of cork related problems and the means for their avoidance. *Aust. Grapegrower Winemaker* **268,** 66–68, 70–72, 75–77, and 79–80.

Amon, J. M., Simpson, R. F., and Vandepeer, J. M. (1987). A taint in wood-matured wine attributable to microbiological contamination of the oak barrel. *Wine Ind. J.* **2**(Aug.), 35–37.

Boidron, J. N., Lefebvre, A., Riboulet, J. M., and Ribéreau-Gayon, P. (1984). Les substances volatiles susceptibles d'être cédées au vin par le bouchon de liège. *Sci. Aliments* **4,** 609–616.

Lee, T. H., and Simpson, R. F. (1991). Cork taints. *Pract. Winery Vineyard* July/August, 9–16.

Riboulet, J. M. (1991). Cork tastes. *Pract. Winery Vineyard,* July/August, 9 and 16–19.

Riboulet, J. M., and Alegoët, C. (1986). "Aspects Pratiques du Bouchage des Vins." Collection Avenir Oenologie, Bourgogne Publ., Chaintre, France.

Simpson, R. F. (1990). Cork taint in wine: A review of the causes. *Aust. N.Z. Wine Ind. J* **5**(Nov.), 286–287, 289, and 293–296.

Other Forms of Wine Spoilage

Bidan, P., and Collon, Y. (1985). Métabolisme du soufre chez la levure. *Bull. O.I.V.* **58,** 545–561.

Brun, S. (1984). Toxicology of materials during the transportation of wine. *Int. Symp. Wine Transportation Proc., April 1984, Montréal, Canada,* pp. 72–87.

Drysdale, G. S., and Fleet, G. H. (1988). Acetic acid bacteria in winemaking: A review. *Am. J. Enol. Vitic.* **39,** 143–154.

Fleet, G. (1992). Spoilage yeasts. *Crit. Rev. Biotechnol.* **12,** 1–44.

Lafon-Lafourcade, S., and Ribéreau-Gayon, P. (1984). Les altérations des vins par les bactéries acetiques et les bactéries lactiques. *Connaiss. Vigne Vin* **18,** 67–82.

Radler, F. (1986). Microbial biochemistry. *Experientia* **42,** 884–893.

Rapp, A., Pretorius, P., and Kugler, D. (1992). Foreign and undesirable flavours in wine. *In* "Off-Flavors in Foods and Beverages" (G. Charalambous, ed.), pp. 485–522. Elsevier, London.

Strauss, C. R., Wilson, B., and Williams, P. J. (1985). Taints and off-flavours resulting from contamination of wines: A review of some investigations. *Aust. Grapegrower Winemaker* **256,** 20, 22, and 24.

References

Agulló, C., and Seoane, E. (1981). Free hydroxyl groups in the cork suberin. *Chem. Ind. (London)* **5**(Sept.), 608–609.

Aiken, J. W., and Noble, A. C. (1984). Comparison of the aromas of oak- and glass-aged wines. *Am. J. Enol. Vitic.* **35,** 196–199.

Aldercreutz, P. (1986). Oxygen supply to immobilized cells. 5. Theoretical calculations and experimental data for the oxidation of glycerol by immobilized *Gluconobacter oxydans* cells with oxygen or *p*-benzoquinone as electron acceptor. *Biotechnol. Bioeng.* **28**, 223–232.

Amerine, M. A., and Joslyn, M. A. (1970). "Tables Wines: The Technology of Their Production," 2nd Ed. Univ. of California Press, Berkeley.

Amerine, M. A., and Roessler, E. B. (1983). "Wines—Their Sensory Evaluation." Freeman, New York.

Amerine, M. A., Berg, H. W., Kunkee, R. E., Ough, C. S., Singleton, V. L., and Webb, A. D. (1980). "The Technology of Wine Making." AVI Publ., Westport, Connecticut.

Amon, J. M., Simpson, R. F., and Vandepeer, J. M. (1987). A taint in wood-matured wine attributable to microbiological contamination of the oak barrel. *Aust. N.Z. Wine Indust. J.* **2**, 35–37.

Anderson, T. C. (1987). Understanding and improving bag-in-box technology. *Proc. 6th Aust. Wine Ind. Tech. Conf.* (T. Lee, ed.), pp. 272–274. Australian Industrial Publ., Adelaide, Australia.

Anonymous (1983). Bouchon Tage is a synthetic wine stopper. *Wines Vines* **64**(9), 47.

Anonymous (1984). *Proc. Int. Symp. Wine Transportation.* April 17–19, 1984.

Anonymous (1991). Wieviel Styrol geben Kunststofftanks ab? *Weinwirtsch. Tech.* **127**(6), 23–25.

Anonymous (1992). A winery evaluates synthetic cork. *Wines Vines* **73**(7), 44–46.

Armstrong, D. N. (1987). Leakers, equipment failure and quality assurance procedures for defect supervision. *Proc. 6th Aust. Wine Ind. Tech. Conf.* (T. Lee, ed.), pp. 275–276. Australian Industrial Publ., Adelaide, Australia.

Asensio, A., and Seoane, E. (1987). Polysaccharides from the cork of *Quercus suber*, I. Holocellulose and cellulose. *J. Nat. Prod.* **50**, 811–814.

Barillère, J. M., Bidan, T., and Dubois, C. (1983). Thermorésistance de levures et des bactéries lactique isolées du vin. *Bull. O.I.V.* **56**, 327–351.

Bach, H. P. (1982). Der Einfluss verschiedener Verschlüsse auf den Wein während der Lagerung in Abhängigkeit vom Füllverfahren und der Lagermethóde. 1. Mitt.: Der Einfluss auf die Analytik. *Wein-Wiss.* **37**, 400–429.

Badler, V. R., McGovern, P. E., and Michel, R. H. (1990). Drink and be merry! Infrared spectroscopy and ancient Near Eastern wine. *MASCA Res. Pap. Sci. Archaeol.* **7**, 25–36.

Bass, G. F. (1971). A Byzantine trading venture. *Sci. Am.* **225**(2), 22–33.

Beckmann, J. (translated by W. Johnson) (1846). "History of Inventions, Discoveries and Origins," Vol. 2, 4th Ed. Henry G. Bohn, London.

Beelman, R. B., and Gallander, J. F. (1979). Wine deacidification. *Adv. Food Res.* **25**, 1–53.

Bertuccioli, M., and Viani, R. (1976). Red wine aroma: Identification of headspace constituents. *J. Sci. Food Agric.* **27**, 1035–1038.

Bobet, R. A. (1987). Interconversion reactions of sulfides in model solutions. M. S. Thesis, Univ. of California, Davis.

Bobet, R. A., Noble, A. C., and Boulton, R. B. (1990). Kinetics of the ethanethiol and diethyl disulfide interconversion in wine-like solutions. *J. Agric. Food Chem.* **38**, 449–452.

Boidron, J. N., Lefebvre, A., Riboulet, J. M., and Ribéreau-Gayon, P. (1984). Les substances volatiles susceptibles d'être cédées au vin par le bouchon de liège *Sci. Aliments* **4**, 609–616.

Bonorden, W. R., Nagel, C. W., and Powers, J. R. (1986). The adjustment of high pH/high titratable acidity wines by ion exchange. *Am. J. Enol. Vitic.* **37**, 143–148.

Borges, M. (1985). New trends in cork treatment and technology. *Beverage Rev.* **5**, 15, 16, 19, and 21.

Brezovesik, L. (1986). Mould fungi as possible cause for corkiness in wine. *Borgazdasag* **34**, 25–31 (in Hungarian).

Brock, T. D. (1983). "Membrane Filtration." Science Tech., Madison, Wisconsin.

Brun, S. (1980). Pollution du vin par le bouchon et le dispositif de surbouchage. *Rev. Fr. Oenol.* **77**, 53–58.

Brun, S. (1984). Toxicology of materials during the transportation of wine. *Int. Symp. Wine Transportation Proc., April 17–19, 1984, Montréal, Canada,* pp. 72–87.

Bui, K., Dick, R., Moulin, G., and Galzy, P. (1986). Reverse osmosis for the production of low ethanol content wine. *Am. J. Enol. Vitic.* **37**, 297–300.

Bureau, G. M., Charpentier-Massonal, M., and Parsee, M. (1974). Étude des goûts anormaux apportés par le bouchon sur le vin de Champagne. *Rev. Fr. Oenol.* **56**, 22–24.

Calisto, M. C. (1990). DMDC's role in bottle stability. *Wines Vines* **71**(10), 18–21.

Cantarelli, C. (1964). Il difetto da idrogeno solforato dei vini, la natura, le cause, i trattamenti preventivi e di risanamento. *Atti Accad. Ital. Vite Vino Siena* **16**, 163–175.

Carey, R. (1988). Natural cork: The "new" closure for wine bottles. *Vineyard Winery Management* **14**(5), 38.

Carpentier, N., and Maujean, A. (1981). Light flavours in champagne wines. *Flavour '89: 3rd Weurman Symp. Proc. Int. Conf., Munich April 28–30, 1981* (P. Schreier, ed.), pp. 609–615. de Gruyter, Berlin.

Casey, J. A. (1990). A simple test for cork taint. *Aust. Grapegrower Winemaker* **324** (Dec.), 40.

Casey, J. A. (1993). Cork as a closure material for wine. *Aust. Grapegrower Winemaker,* in press.

Cattaruzza, A., Peri, C., and Rossi, M. (1987). Ultrafiltration and deep-bed filtration of a red wine: Comparative experiments. *Am. J. Enol. Vitic.* **38**, 139–142.

Champagnol, F. (1986). L'acidité des moûts et des vins. 1ere Partie. Facteurs physico-chimiques et technologiques de variation. *Rev. Fr. Oenologie* **104**, 26–30 and 51–57.

Chatonnet, P. (1989). Origines et traitements des bois destinés à l'élevage des vins de qualité. *Rev. Oenologues* **15**, 21–25.

Chatonnet, P. (1991). Incidences du bois de chêne sur la composition chimique et les qualitiés organoleptiques des vins. Applications technologiques. Thèse. Univ. de Bordeaux II, P. Chatonnet, Bordeaux, France.

Chaves das Neves, H. J., Vasconcelos, A. M. P., and Costa, M. L. (1990). Racemization of wine free amino acids as function of bottling age. *In* "Chirality and Biological Activity," pp. 137–143. Alan R. Liss, New York.

Chen, C.-L. (1970). Constituents of *Quercus rubra. Phytochemistry* **9**, 1149.

Chen, C.-L., and Chang, H. M. (1985). Chemistry of lignin biodegradation. *In* "Biosynthesis and Biodegradation of Wood Components" (T. Higuchi, ed.), pp. 535–556. Academic Press, New York.

Cillers, J. J. L., and Singleton, V. L. (1990). Nonenzymatic autooxidative reactions of caffeic acid in wine. *Am. J. Enol. Vitic.* **41**, 84–86.

Cole, J., and Boulton, R. (1989). A study of calcium salt precipitation in solutions of malic acid and tartaric acid. *Vitis* **28**, 177–190.

Crowell, E. A., and Guymon, J. F. (1975). Wine constituents arising from sorbic acid addition, and identification of 2-ethoxyhexa-3,5-diene as source of geranium-like off-odor. *Am. J. Enol. Vitic.* **26**, 97–102.

Cuénat, Ph., and Kobel, D. (1987). La diafiltration en oenologie: Un example d'application à des vins vieux. *Rev. Suisse Vitic. Arboric. Hortic.* **2**, 97–103.

Cuénat, Ph., Zufferey, E., and Kobel, D. (1990). La prévention des odeurs sulfhydriques des vins par la centrifugation après la fermentation alcoolique. *Rev. Suisse Vitic. Arboric. Hortic.* **22**, 299–303.

Datta, S., and Nakai, S. (1992). Computer-aided optimization of wine blending. *J. Food Sci.* **57**, 178–182.

Davenport, R. R. (1982). Sample size, product composition and microbial spoilage. *In* "Shelf Life—Seventh Wine Subject Day" (F. W. Beech, ed.), Long Ashton Research Station, Univ. of Bristol, Bristol.

de la Garza, F., and Boulton, R. (1984). The modelling of wine filtrations. *Am. J. Enol. Vitic.* **35**, 189–195.

Delfini, C. (1987). Observations expérimentales sur l'origine et la disparition de l'alcool benzylique et de l'aldéhyde benzoïque dans les moûts et les vins. *Bull. O.I.V.* **60**, 463–473.

Delfini, C., Gaia, P., Bardi, L., Mariscalco, G., Contiero, M., and Pagiara, A. (1991). Production of benzaldehyde, benzyl alcohol and benzoic acid by yeasts and *Botrytis cinerea* isolated from grape musts and wines. *Vitis* **30**, 253–263.

de Mora, S. J., Eschenbruch, R., Knowles, S. J., and Spedding, D. J. (1986). The formation of dimethyl sulfide during fermentation using a wine yeast. *Food Microbiol.* **3**, 27–32.

Di Stefano, R., and Ciolfi, G. (1985). L'influenza della luce sull'Asti Spumante. *Vini Ital.* **27**, 23–32.

Dubois, P. (1983). Volatile phenols in wines. *In* "Flavour of Distilled Beverages" (J. R. Piggott, ed.), pp. 110–119. Ellis Horwood, Chichester.

Dubourdieu, D., Serro, M., Vannier, A. C., and Ribéreau-Gayon, P. (1988). Étude comparée des tests de stabilité protéique. *Connaiss. Vigne Vin* **22**, 261–273.

Duerr, P., and Cuénat, P. (1988). Production of dealcoholized wine. *Proc. 2nd Int. Symp. Cool Climate Vitic. Oenol., Jan. 11–15, 1988, Auckland, N.Z.* (R. E. Smart, S. B. Thornton, S. B. Rodriguez, and J. E. Young, eds.), pp. 363–364. New Zealand Soc. Vitic. Oenol., Auckland, New Zealand.

Edelényi, M. (1966). Study on the stabilization of sparkling wines. *Borgazdasag* **12**, 30–32 (in Hungarian). (reported in Amerine *et al.*, 1980).

Edwards, T., Singleton, V. L., and Boulton, R. (1985). Formation of ethyl esters of tartaric acid during wine aging: Chemical and sensory effects. *Am. J. Enol. Vitic.* **36**, 118–124.

Eschenbruch, B., and Dittrich, H. H. (1986). Stoffbildungen von Essigbakterien in bezug auf ihre Bedeutung für die Weinqualität. *Zentralbl. Mikrobiol.* **141**, 279–289.

Evans, C. S. (1987). Lignin degradation. *Process Biochem.* **22**, 102–105.

Fabre, S. (1989). Bouchons traités au(x) peroxyde(s): Détection du pouvoir oxydant et risques d'utilisations pour le vin. *Rev. Oenologues* **15**, 11–15.

Farris, G. A., Fatichenti, F., and Deiana, P. (1989). Incidence de la température et du pH sur la production d'acide malique par *Saccharomyces cerevisiae*. *J. Int. Sci. Vigne Vin* **23**, 89–93.

Fletcher, J. (1978). Dating the geographical migration of *Quercus petraea* and *Q. robur* in Holocene times. *Tree-Ring Bull.* **38**, 45–47.

Flores, J. H., Heatherbell, D. A., Henderson, L. A., and McDaniel, M. R. (1991). Ultrafiltration of wine: Effect of ultrafiltration on the aroma and flavor characteristics of White Riesling and Gewürztraminer wines. *Am. J. Enol. Vitic.* **42**, 91–96.

Frommberger, R. (1991). Cork products—A potential source of polychlorinated dibenzo-*p*-dioxins and polychlorinated dibenzofurans. *Chemosphere* **23**, 133–139.

Gerber, N. N. (1979). Volatile substance from actinomycetes: Their role in the odor pollution of water. *CRC Crit. Rev. Microbiol.* **7**, 191–214.

Gibson, L. J., Easterling, K. E., and Ashby, M. F. (1981). The structure and mechanics of cork. *Proc. R. Soc. London A* **377**, 99–117.

Gini, B., and Vaughn, R. H. (1962). Characteristics of some bacteria associated with the spoilage of California dessert wines. *Am. J. Enol. Vitic.* **13**, 20–31.

Goswell, R. W. (1981). Tartrate stabilization trials. *In* "Quality Control: Proceedings of the Sixth Wine Subject Day, October 26–27, 1981" (F. W. Beech and W. J. Redmond, eds.), pp. 62–65. Long Ashton Research Station, Univ. of Bristol, Bristol.

Guimberteau, G., Lefebvre, A., and Serrano, M. (1977). Le conditionnement en bouteilles. *In* "Traité d'Oenologie: Sciences et Techniques du Vin" (J. Ribéreau-Gayon, E. Peynaud, P. Ribéreau-Gayon, and P. Sudraud, eds.), Vol. 4, pp. 579–643.

Gulson, B. L., Lee, T. H., Mizon, K. J., Korsch, M. J., and Eschnauer, H. R. (1992). The application of lead isotope ratios to determine the contribution of the tin-lead to the lead content of wine. *Am. J. Enol. Vitic.* **43**, 180–190.

Gunata, Y. Z., Bayonove, C. L., Baumes, R. L., and Cordonnier, R. E. (1986). Stability of free and bound fractions of some aroma components of grapes cv. Muscat during the wine processing: Preliminary results. *Am. J. Enol. Vitic.* **37**, 112–114.

Guymon, J. F., and Crowell, E. A. (1977). The nature and cause of cap–liquid temperature differences during wine fermentation. *Am. J. Enol. Vitic.* **28**, 74–78.

Hammond, S. M. (1976). Microbial spoilage of wines. *In* "Wine Quality—Current Problems and Future Trends" (F. W. Beech, A. G. H. Lea, and C. F. Timberlake, eds.), pp. 38–44. Long Ashton Research Station, Univ. of Bristol, Bristol.

Harding, F. L. (1972). The development of colors in glass. *In* "Introduction to Glass Science" (L. D. Pye, H. J. Stevens, and W. C. LaCourse, eds.), pp. 391–431. Plenum, New York.

Hasuo, T., and Yoshizawa, K. (1986). Substance change and substance evaporation through the barrel during whisky ageing. *Proc. 2nd Aviemore Conf. Malting Brewing Distillation, May 19–23, 1986* (I. Campbell and F. G Priest, eds.), pp. 404–408. Institute of Brewing, London.

Hasuo, T., Saito, K., Terauchi, T., Tadenuma, M., and Sato, S. (1983). Influence of environmental humidity on aging of whiskey (Studies on aging of whiskey, Part III). *J. Brew. Soc. Jpn.* **78**, 966–969 (in Japanese).

Heimann, W., Rapp, A., Volter, I., and Knipser, W. (1983). Beitrag zur Entstehung des Korktons in Wien. *Dtsch. Lebensm-Rundsch.* **79**, 103–107.

Helmcke, J. G. (1954). Neue Erkenntnisse uber den Aufbau von Membranfiltern. *Kolloid Zeitschrift* **135**, 29–43.

Heresztyn, T. (1986a). Formation of substituted tetrahydopyridines by species of *Brettanomyces* and *Lactobacillus* isolated from mousy wines. *Am. J. Enol. Vitic.* **37**, 127–131.

Heresztyn, T. (1986b). Metabolism of volatile phenolic compounds from hydroxycinnamic acids by *Brettanomyces* yeast. *Arch. Microbial.* **146**, 96–98.

Hoadley, R. B. (1980). "Understanding wood." Taunton Press, Newtown, Connecticut.

Hoey, A. W., and Codrington, J. D. (1987). Oak barrel maturation of table wine. *Proc. 6th Aust. Wine Ind. Tech. Conf.* (T. Lee, ed.), pp. 261–266. Australian Industrial Publishers, Adelaide, Australia.

Hsu, J., and Heatherbell, D. A. (1987). Heat-unstable proteins in wine. I. Characterization and removal by bentonite fining and heat treatment. *Am. J. Enol. Vitic.* **38**, 11–16.

Hsu, J., Heatherbell, D. A., Flores, J. H., and Watson, B. T. (1987). Heat-unstable proteins in wine. II. Characterization and removal by ultrafiltration. *Am. J. Enol. Vitic.* **38**, 17–22.

Ireton, D. C. (1990). Spinning cone column: What's it all about? *Wines Vines* **71**(1), 20–21.

Joncheray, J.-P. (1976). L'épave grecque, or étrusque, de Bon-Ponté. *Cah. Archaeol. Subaquatique* **5**, 5–36.

Joyeux, A., Lafon-Lafourcade, S., and Ribéreau-Gayon, P. (1984). Evolution of acetic acid bacteria during fermentation and storage of wine. *Appl. Environ. Microbiol.* **48**, 153–156.

Kilby, K. (1971). "The Cooper and His Trade." John Baker, London. (Republished by Linden Publ. Co., Fresno, CA).

Kösebalaban, F., and Özilgen, M. (1992). Kinetics of wine spoilage by acetic acid bacteria. *J. Chem. Tech. Biotechnol.* **55,** 59–63.

Langhans, E., and Schlotter, H. A. (1985). Ursachen der Kupfer-Trüng. *Dtsch. Weinbau* **40,** 530–536.

Larroque, M., Brun, S., and Blaise, A. (1989). Migration des monomères constitutifs des résines époxydiques utilisées pour revêtir les cuves à vin. *Sci. Aliments* **9,** 517–531.

Larue, F., Rozes, N., Froudiere, I., Couty, C., and Perreira, G. P. (1991). Influence du développement de *Dekkera/Brettanomyces* dans les moûts et les vins. *J. Int. Sci. Vigne Vin* **25,** 149–165.

Lavigne, V., Boidron, J. N., and Dubourdieu, D. (1992). Formation des composés lourds au cours de la vinification des vins blancs secs. *J. Int. Sci. Vigne Vin* **26,** 75–85.

Lefebvre, A. (1981). Le bouchage liège des vins. *In* "Actualités Oenologiques et Viticoles" (P. Ribéreau-Gayon and P. Sudraud, eds.), pp. 335–349. Dunod, Paris.

Lefebvre, A. (1988). Le bouchage liege des vins tranquilles. *Connaiss. Vigne Vin Numéro hors Série,* 11–35.

Lefebvre, A., Riboulet, J. M., Boidron, J. N., and Ribéreau-Gayon, P. (1983). Incidence des micro-organismes du liège sur les altérations olfactives du vin. *Sci. Aliments* **3,** 265–278.

Leistner, L., and Eckardt, C. (1979). Vorkommen toxinogener Penicillien bei Fleischerzeugnissen. *Fleischwirtschaft* **59,** 1892–1896.

Lesko, L. H. (1977). "King Tut's Wine Cellar." Albany Press, Albany.

Levreau, R., Lefebvre, A., Serrano, M., and Ribéreau-Gayon, P. (1977). Étudies du bouchage liège. I. Rôle des surpressions dans l'apparition des "bouteilles couleuses," bouchage sous gas carbonique. *Connaiss. Vigne Vin* **11,** 351–377.

Little, A. C., and Liaw, M. W. (1974). Blending wines to color. *Am. J. Enol. Vitic.* **25,** 79–83.

Lubbers, S., Leger, B., Charpentier, C., and Feuillat, M. (1993). Effet colloide-protecteur d'extraits de parois de levures sur la stabilité tartrique d'une solution hydro-alcoolique modele. *J. Int. Sci. Vigne Vin* **27,** 13–22.

Maarse, H., Nijssen, L. M., and Angelino, S. A. G. F. (1988). Halogenated phenols and chloranisoles: Occurrence, formation and prevention. "Characterization, Production and Application of Food Flavours: Proceedings of the Second International Wartburg Aroma Symposium 1987," pp. 43–61. Akademie Verlag, Berlin.

McKearin, H. (1973). On stopping, bottling and binning. *Int. Bottler Packer* (April), 47–54.

Macpherson, C. C. H. (1982). Life on the shelf. *In* "Shelf Life— Seventh Wine Subject Day" (F. W. Beech, ed.), Long Ashton Research Station, Univ. of Bristol, Bristol.

Marais, J. (1986). Effect of storage time and temperature of the volatile composition and quality of South African *Vitis vinifera* L. cv. Colombar wines. *In* "The Shelf Life of Foods and Beverages" (G. Charalambous, ed.), pp. 169–185. Elsevier, Amsterdam.

Marsal, F., and Sarre, Ch. (1987). Étude par chromatographie en phase gazeuse de substances volatiles issues du bois de chêne. *Connaiss. Vigne Vin* **21,** 71–80.

Maujean, A., and Seguin, N. (1983). Contribution à l'étude des goûts de lumière dans les vins de Champagne. 3. Les réactions photochimiques responsables des goûts de lumière dans le vins de Champagne. *Sci. Aliments* **3,** 589–601.

Maujean, A., Haye, B., and Bureau, G. (1978). Étude sur un phénomène de masque observé en Champagne. *Vigneron Champenois* **99,** 308–313.

Maujean, A., Millery, P., and Lemaresquier, H. (1985). Explications biochimiques et métaboliques de la confusion entre goût de bouchon et goût de moisi. *Rev. Fr. Oenol.* **99,** 55–67.

Médina, B. (1981). Metaux polluants dans les vins. *In* "Actualitiés Oenologiques et Viticoles" (P. Ribéreau-Gayon and P. Sudraud, eds.), pp. 361–372. Dunod, Paris.

Moreau, M. (1978). La mycoflore des bouchons de liège. Son évolution au contact du vin: Conséquences possibles du métabolisme des moisissures. *Rev. Mycol.* **42,** 155–189.

Moutounet, M., Rabier, P., Puech, J.-L., Verette, E., and Barillère, J. M. (1989). Analysis by HPLC of extractable substances in oak wood: Application to a Chardonnay wine. *Sci. Aliments* **9,** 35–51.

Muraoka, H., Watanabe, Y., Ogasawara, N., and Takahashi, H. (1983). Trigger damage by oxygen deficiency to the acid production system during submerged acetic fermentation with *Acetobacter aceti. J. Ferment. Technol.* **61,** 89–93.

Murrell, W. G., and Rankine, B. C. (1979). Isolation and identification of a sporing *Bacillus* from bottled brandy. *Am. J. Enol. Vitic.* **30,** 247–249.

Nagel, C. W., Weller, K., and Filiatreau, D. (1988). Adjustment of high pH–high TA musts and wines. *Proc. 2nd Int. Symp. Cool Climate Vitic. Oenol., Jan. 11–15, 1988, Auckland, N.Z.* (R. E. Smart, S. B. Thornton, S. B. Rodriguez, and J. E. Young, eds.), pp. 222–224. New Zealand Soc. Vitic. Oenol., Auckland, New Zealand.

Negueruela, A. I., Echavarri, J. F., Los Arcos, M. L., and Lopez de Castro, M. P. (1990). Study of color of quaternary mixtures of wines by means of the Scheffé design. *Am. J. Enol. Vitic.* **41,** 232–240.

Neradt, F. (1984). Tartrate-stabilization methods. *In* "Tartrates and Concentrates: Eighth Wine Subject Day" (F. W. Beech, ed.), pp. 13–25. Long Ashton Research Station, Univ. of Bristol, Bristol.

Nishimura, K., Ohnishi, M., Masuda, M., Koga, K., and Matsuyama, R. (1983). Reactions of wood components during maturation. *In* "Flavour of Distilled Beverages: Origin and Development" (J. R. Piggott, ed., pp. 241–255. Ellis Horwood, Chichester.

Nykänen, L. (1986). Formation and occurrence of flavor compounds in wine and distilled alcoholic beverages. *Am. J. Enol. Vitic.* **37,** 84–96.

Nykänen, L., Nykänen, I., and Moring, M. (1985). Aroma compounds dissolved from oak chips by alcohol. *In* "Progress in Flavor Research—Proceedings of the Fourth Weurman Flavour Research Symposium" (J. Adda, ed.), pp. 339–346. Elsevier Science Publ., Amsterdam.

Ough, C. S. (1985). Some effects of temperature and SO$_2$ on wine during stimulated transport or storage. *Am. J. Enol. Vitic.* **36,** 18–22.

Peck, E. C. (1957). How wood shrinks and swells. *For. Prod. J.* **7,** 234–244.

Peleg, Y., Brown, R. C., Starcevich, P. W., and Asher, A. (1979). Methods for evaluating the filterability of wine and similar fluids. *Am. J. Enol. Vitic.* **30,** 174–178.

Pereira, H., Rosa, M. E., and Fortes, M. A. (1987). The cellular structure of cork from *Quercus suber* L. *IAWA Bull.* **8,** 213–218.

Perin, J. (1977). Compte rendu de quelques essais de réfrigération des vins. *Vigneron Champenois* **98,** 97–101.

Pes, A., and Vodret, A. (1971). Il gusto di tappo nei vini in bottiglia. Stazione Sperimentale del Sughero, Tempo-Pausania, Sardinia.

Peterson, R. G. (1976). Formation of reduced pressure in barrels during wine aging. *Am. J. Enol. Vitic.* **27,** 80–81.

Petrich, H. (1982). Untersuchungen über den Einfluss des Fungizids N-Trichloromethylthio–Phthalimide auf Geruchs- und Geschmacksstoffe von Wiener. Dissertation, in *Vitis: Enol. Vitic. Abstr.* **23,** 1 M 13). F. R. Inst. für Lebensmittelchemie, Univ. of Stuttgart, Stuttgart.

Pocock, K. F., Strauss, C. R., and Somers, T. C. (1984). Ellagic acid deposition in white wines after bottling: A wood-derived instability. *Aust. Grapegrower Winemaker* **244,** 87.

Pontallier, P., Salagoïty-Auguste, M., and Ribéreau-Gayon, P. (1982).

Intervention du bois de chêne dans l'évolution des vins rouges élevés en barriques. *Connaiss. Vigne. Vin* **16**, 45–61.

Postel, W. (1983). Solubilité et inhibition du crystallisation de tartrate de calcium dans le vin. *Bull. O.I.V.* **56**, 554–568.

Poulard, A., Leclanche, A., and Kollonkai, A. (1983). Dégradation de l'acide tartrique de moût par une nouvelle espèce: *Exophiala jeanselmei* var. *heteromorpha*. *Vignes Vins* **323**, 33–35.

Puech, J.-L. (1984). Characteristics of oak wood and biochemical aspects of Armagnac aging. *Am. J. Enol. Vitic.* **35**, 77–81.

Puech, J.-L. (1987). Extraction of phenolic compounds from oak wood in model solutions and evolution of aromatic aldehydes in wines aged in oak barrels. *Am. J. Enol. Vitic.* **38**, 236–238.

Puerto, F. (1992). Traitment des bouchons: Lavage au péroxide contrôlé. *Rev. Oenologenes* **64**, 21–26.

Quinn, M. K., and Singleton, V. L. (1985). Isolation and identification of ellagitannins from white oak wood and an estimation of their roles in wine. *Am. J. Enol. Vitic.* **35**, 148–155.

Rankine, B. C. (1966). Decomposition of L-malic acid by wine yeasts. *J. Sci. Food Agric.* **17**, 312–316.

Rankine, B. C., and Pilone, D. A. (1973). *Saccharomyces bailii*, a resistant yeast causing serious spoilage of bottled table wine. *Am. J. Enol. Vitic.* **24**, 55–58.

Rapp, A., and Güntert, M. (1986). Changes in aroma substances during the storage of white wines in bottles. *In* "The Shelf Life of Foods and Beverages" (G. Charalambous, ed.), pp. 141–167. Elsevier, Amsterdam.

Ribéreau-Gayon, P. (1986). Shelf-life of wine. *In* "Handbook of Food and Beverage Stability: Chemical, Biochemical, Microbiological and Nutritional Aspects" (G. Charalambous, ed.), pp. 745–772. Academic Press, Orlando, Florida.

Ribéreau-Gayon, P., and Glories, Y. (1987). Phenolics in grapes and wines. *Proc. 6th Aust. Wine Ind. Tech. Conf.* (T. Lee, ed.), pp. 247–256. Australian Industrial Publ., Adelaide, Australia.

Ribéreau-Gayon, P., and Ribéreau-Gayon, J. (1980). La vinification en rouge par "macération finals à chaud." *C.R. Acad. Agric. Fr.* **66**, 207–215.

Ribéreau-Gayon, J., Peynaud, E., Ribéreau-Gayon, P., and Sudraud, P. (eds.) (1976). "Traité d'Oenologie: Sciences et Techniques du Vin," Vol. 3. Dunod, Paris.

Ribéreau-Gayon, J., Peynaud, E., Ribéreau-Gayon, P., and Sudraud, P. (eds.) (1977). "Traite' d'Oenologie: Sciences et Techniques du Vin," Vol. 4. Dunod, Paris.

Riboulet, J. M., and Alegoët, C. (1986). "Aspects Pratiques du Bouchage des Vins." Collection Avenir Oenologie, Bourgogne Publ., Chaintre, France.

Rodriguez-Clemente, R., and Correa-Gorospe, I. (1988). Structural, morphological and kinetic aspects of potassium hydrogen tartrate precipitation from wines and ethanolic solutions. *Am. J. Enol. Vitic.* **39**, 169–179.

Rous, C., and Alderson, B. (1983). Phenolic extraction curves for white wine aged in French and American oak barrels. *Am. J. Enol. Vitic.* **34**, 211–215.

Schanderl, H. (1971). Korkgeschmack von Weinen. *Dtsch. Wein-Ztg.* **107**, 333–336.

Schneyder, J. (1965). Die Behebung der durch Schwefelwasserstoff und Merkaptane versursachten Geruchsfehler der Weine mit Silberchlorid. *Mitt. Rebe Wein, Ser. A (Klosterneuburg)* **15**, 63–65.

Schreier, P., Drawert, F., Kerènyl, Z., and Junker, A. (1976). Gaschromatographisch–massenspektrometrische Untersuchung flüchtiger Inhaltsstoffe des Weines. VI. Aromastoffe in Tokajer Trockenbeerenauslese (Aszu)-Weinen a) Neutralstoffe. *Z. Lebensm. Unters. Forsch.* **161**, 249–258.

Schütz, M., and Kunkee, R. E. (1977). Formation of hydrogen sulfide from elemental sulfur during fermentation by wine yeast. *Am. J. Enol. Vitic.* **28**, 137–144.

Sefton, M. A., Francis, I. L., and Williams, P. J. (1990). Volatile norisoprenoid compounds as constituents of oak woods used in wine and spirit maturation. *J. Agric. Food Chem.* **38**, 2045–2049.

Siegrist, J., Léglise, M., and Lelioux, J. (1983). Caractères analytiques sécondaires de quelques vins atteints de la maladie de l'amertume. *Rev. Fr. Oenol.* **23**, 47–48.

Simpson, R. F. (1990). Cork taint in wine: A review of the causes. *Aust. N.Z. Wine Ind. J.* Nov., 286–287, 289, and 293–296.

Simpson, R. F., and Miller, G. C. (1983). Aroma composition of aged Riesling wines. *Vitis* **22**, 51–63.

Singleton, V. L. (1962). Aging of wines and other spirituous products, acceleration by physical treatments. *Hilgardia* **32**, 319–373.

Singleton, V. L. (1974). Some aspects of the wooden container as a factor in wine maturation. *In* "Chemistry of Winemaking" (A. D. Webb, ed.), Advances in Chemistry Ser. No. 137, pp. 254–278. American Chemical Society, Washington, D.C.

Singleton, V. L. (1976). Wine aging and its future. First Walter and Care Regnal Memorial Lecture, July 28, 1976, Tanunda Institute, Roseworthy Agricultural College, South Australia.

Singleton, V. L. (1987). Oxygen with phenols and related reactions in musts, wines, and model systems: Observation and practical implications. *Am. J. Enol. Vitic.* **38**, 69–77.

Singleton, V. L., and Ough, C. S. (1962). Complexity of flavor and blending of wines. *J. Food Sci.* **12**, 189–196.

Singleton, V. L., Sullivan, A. R., and Kramer, C. (1971). An analysis of wine to indicate aging in wood or treatment with wood chips or tannic acid. *Am. J. Enol. Vitic.* **22**, 161–166.

Sitte, P. (1961). Zum Feinbau der Suberinschichten in Flaschenkork. *Protoplasma* **54**, 555–559.

Sneyd, T. N. (1988). Tin lead capsules. *Aust. Wine Res. Inst. Tech. Rev.* **56**, 1.

Somers, T. C., and Evans, M. E. (1986). Evolution of red wines. I. Ambient influences on colour composition during early maturation. *Vitis* **25**, 31–39.

Somers, T. C., and Pocock, K. F. (1990). Evolution of red wines. III. Promotion of the maturation phase. *Vitis* **29**, 109–121.

Somers, T. C., and Ziemelis, G. (1985). Flavonol haze in white wines. *Vitis* **24**, 43–50.

Spedding, D. J., and Raut, P. (1982). The influence of dimethyl sulphide and carbon disulphide in the bouquet of wines. *Vitis* **21**, 240–246.

Strauss, C. R., Wilson, B., and Williams, P. J. (1985). Taints and off-flavours resulting from contamination of wines: A review of some investigations. *Aust. Grapegrower Winemaker* **256**, 20, 22, and 24.

Swan, J. S. (1986). Maturation of potable spirits. *In* "Handbook of Food and Beverage Stability: Chemical, Biochemical, Microbiological and Nutritional Aspects" (G. Charalambous, ed.), pp. 801–833. Academic Press, Orlando, Florida.

Tchernia, A. (1986). "Le Vin de l'Italie Romaine: Essai d'Histoire Economique d'après les Amphores." École Française de Rome, Rome.

Tyson, P. J., Luis, E. S., and Lee, T. H. (1982). Soluble protein levels in grapes and wine. *Grape Wine Centennial Symp. Proc.*, pp. 287–290. Univ. of California, Davis.

Vaughn, R. H. (1955). Bacterial spoilage of wines with special reference to California conditions. *Adv. Food Res.* **7**, 67–109.

Vieira Natividade, J. V. (1956). "Subericulture." Éditions de l'École nationale du Génie rural des Eaux et Forêts, Nancy, France.

Villettaz, J. C., Steiner, D., and Trogus, H. (1984). The use of a beta glucanase as an enzyme in wine clarification and filtration. *Am. J. Enol. Vitic.* **35**, 253–256.

Voilley, A., Lamer, C., Dubois, P., and Feuillat, M. (1990). Influence of macromolecules and treatments on the behavior of aroma compounds in a model wine. *J. Agric. Food Chem.* **38**, 248–251.

Wainwright, T. (1970). Hydrogen sulphide production by yeast under conditions of methionine, pantothenate or vitamine B_6 deficiency. *J. Gen. Microbiol.* **61**, 107–119.

Webb, M. (1987). Cardboard box quality, finish and glues. *Proc. 6th Aust. Wine Ind. Tech. Conf.* (T. Lee, ed.), pp. 277–280. Australian Industrial Publishers, Adelaide, Australia.

Weetal, H. H., Zelko, J. T., and Bailey, L. F. (1984). A new method for the stabilization of white wine. *Am. J. Enol. Vitic.* **35**, 212–215.

Winterhalter, P., Sefton, M. A., and Williams, P. J. (1990). Volatile C_{13}-norisoprenoid compounds in Riesling wine are generated from multiple precursors. *Am. J. Enol. Vitic.* **41**, 277–283.

Wucherpfennig, K., Otto, K., and Wittenschläger, L. (1984). Zur Möglichkeit des Einsatzes von Pektin- und Alginsäure zur Entfernung von Erdalkali- und Schwermetallionen aus Weinen. *Wein-Wiss.* **39**, 132–139.

Wucherpfennig, K., Millies, K. D., and Christmann, M. (1986). Herstellung entalkoholisierter Weine unter besonderer Berüchsichtigung des Dialyseverfahrens. *Weinwirtsch. Tech.* **122**, 346–354.

Young, R. (1982). It's worth complaining! *Decanter* 7(7), 21.

Zambonelli, C., Soli, M. G., and Guerra, D. (1984). A study of H_2S non-producing strains of wine yeasts. *Ann. Microbiol.* **34**, 7–15.

Zoecklein, B., Fugelsang, K. C., Gump, B. H., and Nury, F. S. (1990). "Production Wine Analysis." Van Nostrand-Rheinhold, New York.

9

Specific and Distinctive Wine Styles

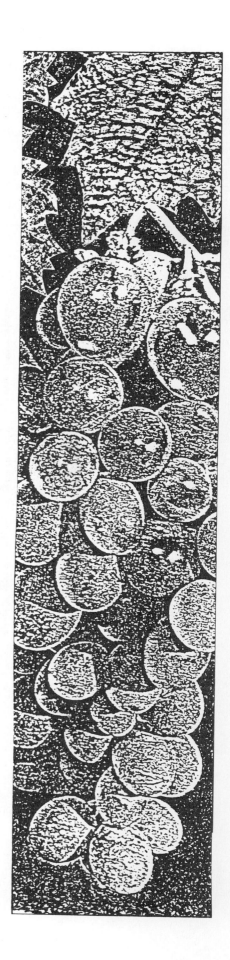

Introduction

Many of the wine styles currently available have been produced for millennia. However, several current styles apparently have no ancient equivalent. Wine styles often reflect the unique climatic and politicoeconomic environment in which they have arisen. For example, botrytized wines emerged in regions favoring the selective development of "noble rot"; sparkling wine evolved in a region (Champagne) usually unsuitable for the production of red wines; and port arose out of expanded trade between England and Portugal, associated with a marked increase in duties paid on French wines imported into England.

Some of the new wine styles have spread throughout the world, while others have remained local specialities. This chapter covers some of the more important and unique wine styles.

Sweet Table Wines

Sweet table wines encompass a wide diversity of styles containing little in common other than sweetness. They may be white, rosé, or red and may range from aromatically simple to complexly fragrant. The most famous are those made from noble-rotted, *Botrytis*-infected grapes.

Botrytized Wines

Wines unintentionally made from grapes infected by *Botrytis cinerea* undoubtedly have been made since time immemorial. The fungus is omnipresent and produces a destructive bunch rot wherever climatic conditions permit. Typically the resultant wine is unpalatable, but under unique climatic conditions the infected grapes develop what is called a noble rot. These grapes produce one of the most exquisite white wines.

When noble-rotted grapes were first intentionally used for wine production is unknown. Historical evidence favors the Tokaji region of Hungary about the mid-seventeenth century. The production of botrytized wine in Germany is reported to have begun at Schloss Johannisberg in 1750. When the deliberate production of botrytized wines began in France is uncertain, but production appears to have been well established in the Sauternes region by about 1830 to 1850. Botrytized wines currently are produced throughout much of Europe, wherever conditions are favorable to noble-rot development. The idea of selectively using infected grapes for wine has been slow to catch on in the New World, but botrytized wines are now produced to a limited extent in Australia, Canada, New Zealand, South Africa, and the United States.

INFECTION

Most infections early in the season develop from spores produced on overwintered fungal tissue in the vineyard. The initial inoculum probably develops from mycelia overwintered on infected tissues or from resting structures called *sclerotia*. Infections begin on aborted and senescing flower parts, notably the stamens and petals. Infected flower parts located within developing clusters may initiate fruit infection later on in the season, although reactivation of latent fruit infections can be important under dry conditions. Latent infections develop when hyphae invade nascent green fruit and then cease to progress. Under moist conditions, new infections induced from external conidial sources are probably more important than latent infections.

Disease susceptibility and its direction (bunch versus noble rot) depend on several factors. Skin toughness and open fruit clusters reduce disease incidence, while heavy rains, protracted damp periods, and shallow rooting increase susceptibility. Conditions such as shallow rooting and high humidity induce berry splitting and favor bunch-rot development. In contrast, noble rot develops under conditions of fluctuating humidity late in the season where humid nights are followed by dry days. Such conditions permit infection but limit fungal growth and modify fungal metabolism.

Depending on the temperature and humidity, new spores (**conidia**) are produced within a few days to several weeks. The spores are borne on elongated, branched filaments called **conidiophores**. The shape of the spore clusters (Fig. 9.1) so resembles grape clusters that the scientific name of the fungus comes from the Greek term βοτσυζ, meaning grape cluster. Because the microclimate of the fruit cluster markedly affects fungal development, various stages of healthy, noble- and bunch-rotted grapes may occur in the same cluster (Fig. 9.2).

On infection, several hydrolytic enzymes are released by *Botrytis*. Particularly destructive are the pectolytic enzymes that degrade the pectinaceous component of the berry cell wall. The enzymes also cause the collapse and death of the affected tissues. With loss of physiological control, the fruit dehydrates under dry conditions. Because the vascular connections with the vine become disrupted as the fruit reaches maturity, moisture lost is not replaced. Additional water is lost via evaporation from the conidiophores.

The loss of moisture appears to be a crucial factor in determining the direction of disease development. Drying retard berry invasion and appears to modify the metabolism of *Botrytis cinerea*. Drying also concentrates the juice, a feature crucial in the development of the sensory properties of the wine. Finally, drying limits secondary invasion by bacteria and fungi. Invasion by fungi including *Penicillium*, *Aspergillus*, and *Mucor* may generate off-tastes and odors in *Botrytis*-infected grapes. For example, *Penicillium frequentans* produces plastic and moldy odors via the synthesis of styrene (Jouret *et al.*, 1972) and 1-octen-3-ol, respectively (Kaminiski *et*

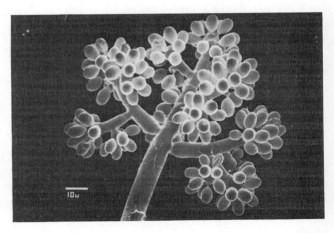

Figure 9.1 Grapelike cluster of *Botrytis cinerea* spores (conidia) produced on the spore-bearing structure, the conidiophore. (Photograph courtesy of Dr. D. H. Lorenz, Neustadt, Germany.)

Figure 9.2 Cluster of botrytized grapes showing berries in different stages of noble rot. (Photograph courtesy of Dr. D. H. Lorenz, Neustadt, Germany.)

CHEMICAL CHANGES INDUCED BY NOBLE ROTTING

As noted above, some changes result from the action of secondary invaders, such as the synthesis of gluconic acid by acetic acid bacteria. Although an indicator of *Botrytis* infection, gluconic acid has no known sensory significance in wine. The sweet taste of the cyclic ester (lactone) of gluconic acid is apparently too slight to be perceptible. However, several by-products of *Botrytis* metabolism and the juice concentration associated with fruit drying are very significant.

Some changes appear to be due solely to the concentrating effects of drying, such as the increase noted in citric acid content (Table 9.1). With other compounds, fungal metabolism is sufficiently active to result in a decrease in concentration, despite the concentrating effect of evaporative water loss. This is particularly noticeable with tartaric acid and ammonia (Table 9.1). The selective metabolism of tartaric acid, versus malic acid, is important in avoiding a marked decrease in pH, or an excessive increase in perceived sweetness.

One of the most notable changes during noble rotting is an increase in sugar concentration. This occurs in spite of a reduction in total sugar content of about 35 to 45%. The decrease in osmotic potential, induced by concentration of the sugar content, may explain the inability of *B. cinerea* to metabolize most of the sugars during infection (Sudraud, 1981). Occasionally, glucose is selectively metabolized, which is reflected in the high fructose/glucose ratio of botrytized grapes and wines.

The production and accumulation of glycerol during noble rotting potentially augment the smooth mouth-feel of botrytized wines. This potential may be enhanced by

al., 1974). Saprophytic fungi are commonly present during the development of Botrytis bunch rot. The unpleasant phenolic flavor commonly associated with wines produced from bunch-rot infected grapes (Boidron, 1978) is rarely observed in noble-rot wines. This may result from the low phenol content of the skins and separation of the juice from the skins before fermentation.

One of the typical chemical indicators of *Botrytis* infection is the presence of gluconic acid. Although *B. cinerea* produces gluconic acid, the acetic acid bacterium *Gluconobacter oxydans* is particularly active in its synthesis. Because acetic acid bacteria frequently invade grapes infected by *Botrytis*, they are probably responsible for most of the gluconic acid found in diseased grapes (Sponholz and Dittrich, 1985). Acetic acid bacteria also produce acetic acid and ethyl acetate, and they are the most likely source of the elevated contents of these compounds in botrytized wines.

Table 9.1 Comparison of Juice from Healthy and *Botrytis*-Infected Grapes[a]

Component	'Sauvignon' berries		'Sémillon' berries	
	Healthy	Infected	Healthy	Infected
Fresh weight/100 berries (g)	225	112	202	98
Sugar content (g/liter)	281	326	247	317
Acidity (g/liter)	5.4	5.5	6.0	5.5
Tartaric acid (g/liter)	5.2	1.9	5.3	2.5
Malid acid (g/liter)	4.9	7.4	5.4	7.8
Citric acid (g/liter)	0.3	0.5	0.26	0.34
Gluconic acid (g/liter)	0	1.2	0	2.1
Ammonia (mg/liter)	49	7	165	25
pH	3.4	3.5	3.3	3.6

[a] After Charpentié (1954), reproduced by permission.

the simultaneous synthesis of other polyols, such as arabitol, erythritol, mannitol, *myo*-inositol, sorbitol, and xylitol.

A distinctive feature of noble rotting is the loss in varietal aroma. This is particularly noticeable with 'Muscat' varieties. The aroma loss is explained largely by destruction of the terpenes that give the varieties their distinctive fragrance. Several aromatic terpenes, such as linalool, geraniol, and nerol, are metabolized by *B. cinerea* to less volatile compounds such as β-pinene, α-terpineol, and various furan and pyran oxides (Bock *et al.*, 1986, 1988). The latter may produce the phenolic and iodinelike odors reported in some botrytized wines (Boidron, 1978). *Botrytis* also produces esterases that can degrade the esters that give many white wines their fruity character (Dubourdieu *et al.*, 1983). The significance of the losses depends on the relative importance of the degraded compounds to the varietal aroma of the wine. 'Muscat' varieties often lose more fragrance than they gain, whereas 'Riesling' and 'Sémillon' generally gain more in aromatic complexity than they lose in varietal distinctiveness.

In addition to destroying certain aromatic compounds, fungal metabolism synthesizes others. The most distinctive of *Botrytis* flavorants appears to be sotolon. Sotolon has a sweet fragrance and may contribute to the characteristic aroma of botrytized wines. In combination with other aromatic compounds found or produced in botrytized wines, sotolon helps to give the wine a distinctive honeylike fragrance (Masuda *et al.*, 1984). Infected grapes also contain the "mushroom" alcohol, 1-octen-3-ol. Among additional compounds are derivatives generated from grape terpenes. Over 20 terpene derivatives have been isolated from infected grapes (Bock *et al.*, 1985).

Infection by *B. cinerea* not only affects the taste and fragrance of the resultant wine, but it also influences the ease of grape picking, yeast and bacterial activity in the juice, and the filterability and aging properties of the wine. Hydrolytic enzymes from the fungus disrupt the integrity of the grape skin, causing it to split more easily during rainy spells or rupture during harvesting. Surprisingly, infection retards separation of the berry from the pedicel during ripening (Fregoni *et al.*, 1986).

Infection by *Botrytis* may affect the activity of other microorganisms on and in grapes, during fermentation, and in the wine. In the vineyard, infection facilitates secondary invasion by acetic acid bacteria and several saprophytic fungi. *Botrytis* infection also affects the surface yeast flora, both increasing cell numbers and modifying species compositions. For example, *Candida stellata*, *Torulaspora delbrueckii*, and *Saccharomyces cerevisiae* f. *bayanus* generally dominate the yeast flora of infected grapes. The dependence of *T. delbrueckii* on

oxygen availability may partially explain its early demise during fermentation (Mauricio *et al.*, 1991). *Saccharomyces cerevisiae* f. *bayanus* appears to be particularly able to withstand the high sugar concentration and low nitrogen, thiamine, and sterol conditions found in botrytized juice.

Because thiamine deficiency increases the synthesis of sulfur-binding compounds, notably acetaldehyde, during fermentation, the free SO_2 content of the wine may be seriously reduced. In addition, acetic acid bacteria augment the content of sulfur-binding compounds, such as 2- and 5-oxoglutaric acids, in *Botrytis*-infected grapes. Consequently, more sulfur dioxide may be needed for sufficient antimicrobial and antioxidant protection. To avoid the need for additional sulfur dioxide, small amounts of thiamine (0.5 mg/liter) commonly are added to the juice before fermentation (Dittrich *et al.*, 1975).

There is considerable variation in the influence of different *Botrytis* strains on yeast growth. Suppression may result from the release of toxic fatty acids by fungal esterases. In addition, some *Botrytis* strains produce compounds that can either stimulate (Minárik *et al.*, 1986) or suppress (Dubourdieu, 1981) yeast metabolism.

Little is known about the specific effects noble rotting has on malolactic fermentation. The presence of high residual sugar and glycerol contents would presumably favor development. Therefore, one of the advantages of sulfur dioxide addition is to limit off-flavor production and the reduction in acidity caused by malolactic fermentation.

Laccases are one of the most important groups of enzymes produced by *Botrytis cinerea*. Their function during infection is unknown, but laccases are suspected to inactivate antifungal phenols such as pterostilbene and resveratrol in grapes (Pezet *et al.*, 1991). Different laccases are induced by grape juice, gallic acid, and *p*-coumaric acid. In addition, pectin may augment laccase synthesis in the presence of phenolic compounds (Marbach *et al.*, 1985). In wine, laccases can oxidize a wide range of important grape phenols, for example, *p*-, *o*-, and some *m*-diphenols, diquinones, anthocyanins, and tannins, and a few nonphenolics, such as ascorbic acid. The oxidation of 2(*S*)-glutathionylcaftaric acid may generate much of the golden coloration of botrytized wine (see Macheix *et al.*, 1991). Unlike grape polyphenol oxidase, fungal laccases are still quite active at wine pH values and in the presence of sulfur dioxide at levels typically found in must and wine. Concentrations of about 50 mg/liter SO_2 at pH 3.4 are required to inhibit the action of laccase in wine (about 125 mg/liter in must) (Kovač, 1979). Such levels can be used in white juice but would excessively bleach red musts. Alternately, hydrogen sulfide as low as 1 to 2.5 mg/liter (added as a sulfide salt) completely inhibit laccase activity (see Macheix *et al.*, 1991).

The activity and stability of laccase in must and wine have serious consequences for red wines. Because the enzyme rapidly and irreversibly oxidizes anthocyanins, even small amounts of infection can generate considerable color loss and browning. Unlike grape polyphenol oxidase, fungal laccases are little inhibited by oxidized phenolic compounds (Dubernet, 1974). The gold color produced from the few phenolic compounds present in white grapes, in contrast, is considered a positive quality feature. The degree of browning depends largely on the grape variety and the level of infection.

During infection, *Botrytis* synthesizes a series of high molecular mass polysaccharides. These fall into two distinct classes. One group consists primarily of polymers of mannose and galactose, with small amounts of glucose and rhamnose. They vary between 20,000 and 50,000 daltons and induce increased production of acetic acid and glycerol during fermentation (Dubourdieu, 1981). The other groups consist of β-glucans, polymers of glucose ranging between 100,000 and 1,000,000 daltons. They have little, if any, effect on yeast metabolism but form strandlike lineocolloids in the presence of alcohol that can plug filters during clarification. As little as 2 to 3 mg/liter can seriously retard filtration (Wucherpfennig, 1985). The availability of commercial sources of β-glucanases should permit the degradation of the compounds and ease filtration. Botrytized grapes usually are harvested and pressed gently to minimize the release of β-glucans into the juice. The β-glucans are located predominantly just under the skin, in association with fungal cells that produce them.

Botrytis also may induce a form of calcium salt instability. *Botrytis cinerea* produces an enzyme that oxidizes galacturonic acid (a breakdown product of pectin) to mucic (galactaric) acid. Mucic acid slowly binds with calcium, forming an insoluble salt. This may produce the sediment that occasionally forms in bottles of botrytized wines.

Because of berry volume loss during dehydration, the risks of leaving the grapes on the vine to overmature, and the difficulties associated with fermentation and clarification, botrytized wines cannot be produced inexpensively.

TYPES OF BOTRYTIZED WINES

Tokaji Aszú As noted previously, botrytized wines appear to have evolved independently in several European regions. Tokaji may have been the first deliberately produced botrytized wine. Its most famous version, Aszú Eszencia, is derived from juice that spontaneously seeps out of highly botrytized (*aszú*) berries placed in small tubs. About 1 to 1.5 liters of *eszencia* may be obtained from 30 liters of *aszú* grapes. After several weeks, the collected *eszencia* is placed in small wooden barrels for fermentation and maturation in cellars at about 9°C. Because of the very high sugar content of the juice, occasionally over 50%, fermentation occurs slowly and often produces little more than 5 to 7% alcohol before termination. After fermentation ceases, the bungs are usually left slightly ajar. The degree of oxidation this permits may be limited by the reported growth of the common cellar mold *Rhinocladiella* (synonyms *Cladosporium*, *Racodium*) *cellaris* on the surface of the wine (Sullivan, 1981). Nevertheless, the concentration of acetaldehyde, acetals, and acetoin suggests a considerable degree of oxidation (see Schreier, 1979). After a variable period of maturation in this manner, the barrels are bunged tight. Up to an additional 20 years of in-barrel maturation may ensue before bottling.

In contrast, most Aszú wines are made from mixing young white wine, or juice derived from healthy grapes, with "paste" made from pulverized *aszú* berries. The technique is very old, having been demanded in a law passed in 1655 (Asvany, 1987). The *aszú* paste may be made from the grapes before or after the *eszencia* has drained away. The mixture is placed in open vats for several days for an initial period of fermentation. After filtration, the fermenting mixture is placed in barrels (*Gönci*) with a capacity of about 136 liters. The amount of the *aszú* paste added to the mixture is indicated in terms of an old unit of measure, the *puttony*, which equals about 28 to 30 liters. Fermentation continues slowly and may take several weeks or months to finish. The alcohol content may reach 14% (Table 9.2). Occasionally, a distillate made from Tokaji wine may be added to terminate fermentation prematurely to achieve a sweeter finish. Maturation in barrels of 136 to 240 liters capacity occurs in a manner similar to that described for Tokaji Eszencia, except for the shorter dura-

Table 9.2 Chemical Composition of Tokaji Wines[a]

Quality grade	Total extract (g/liter)	Extract residue (g/liter)	Sugar content (g/liter)	Ethanol content (%, v/v)
Two puttonyos	55	25	30	14
Three puttonyos	90	30	60	14
Four puttonyos	125	35	90	13
Five puttonyos	160	40	120	12
Six puttonyos	195	45	150	12
Eszencia	300	50	250	10

[a] After Farkaš (1988), reproduced by permission.

tion. Volume lost by evaporation during aging may be replaced with high-strength wine spirits. Aszú wines are commonly pasteurized before bottling.

German Botrytized Wines   German botrytized wines come in a variety of categories. The basic characteristics are often indicated by the Prädikat designation given on the label. *Auslesen* wines are derived from specially selected clusters of late-harvested fruit. *Beerenauslesen* (BA) and *trockenbeerenauslesen* (TBA) wines are similarly derived but come from individually selected berries or dried berries, respectively. Although the fruit is typically botrytized, this is not obligatory (Anonymous, 1979). Thus, wines within the categories may or may not come from botrytized grapes. Each category also is characterized by an increasingly high minimum sugar content in the grapes used (see Table 10.1).

BA and TBA juices typically contain more sugar than is converted to alcohol during fermentation. The wines are correspondingly sweet and low in alcohol strength—commonly 6 to 8% (Table 9.3). Auslesen wines may be fermented dry or may retain residual sweetness, depending on the desires of the winemaker.

The other main Prädikat wine categories, namely, *Kabinett* and *Spätlese*, may be derived from botrytized juice but seldom are. Their sweetness is usually derived from the addition of **süssreserve**, unfermented or partially fermented juice kept aside for sweetening after the crush. *Süssreserve* is added just before bottling to the dry wine produced from the major portion of the juice. Various techniques have been used to restrict microbial growth in the *süssreserve* before use. These include high doses of sulfur dioxide, storage at temperatures near or below freezing, and the use of high carbon dioxide pressures.

Alternately, sugars may be retained in the juice by prematurely terminating fermentation. This may be achieved by filtering out the yeasts or by allowing the buildup of CO_2 pressure in sealed, reinforced fermentors.

Because the residual sugar content makes the wine microbially unstable, stringent measures must be taken to avoid microbial spoilage. Sterile filtration of the wine into sterile bottles sealed with sterile corks is common. Sterile bottling has supplanted the previous use of high sulfiting at bottling.

French Botrytized Wines   In France, the most well-known botrytized wines are produced in the Sauternes region of Bordeaux. Here, over a period of several weeks, noble-rotted grapes may be selectively removed from clusters on the vine. Because of the cost of multiple selective harvesting, most producers harvest but once and separate the botrytized berries from the bunches. Uninfected grapes are used in the production of a dry wine.

Typically, only one sweet style is produced in Sauternes, in contrast to the many botrytized styles produced in Germany. A major stylistic difference between French and German botrytized wines is the alcohol level achieved. French styles commonly exceed 11% alcohol, while German versions seldom do. Sweet botrytized wines are produced, more or less similarly, in other regions of France, notably Alsace and the Loire.

VARIETIES USED

Many grape varieties are used in the production of botrytized wines. Their appropriateness depends on several factors. Essentially all are white varieties, thus avoiding the brown coloration produced by anthocyanin oxidation. Most varieties mature late, thus timing ripening to the development of conditions favorable to noble rotting. They are also relatively thick skinned. Because of the softening induced by *B. cinerea*, harvesting soft-skinned cultivars is very difficult. Thick-skinned varieties are also less susceptible to splitting and bunch rot.

'Riesling' and 'Sémillon' are the primary cultivars used

Table 9.3   Composition of Some 1971 *Beerenauslesen* and *Trockenbeerenauslesen* Wines[a]

Wine type[b]	Total extract (g/liter)	Sugar content (g/liter)	Alcohol content (%, w/v)	Total acidity (g/liter)	Glycerol content (g/liter)	Acetaldehyde content (mg/liter)	Tannin content (mg/liter)	pH
BA	163	74	7.9	8.7	12.0	73	250	3.2
BA	152	103	6.3	9.4	13.6	62	390	3.2
BA	119	78	7.9	7.9	10.9	139	390	3.0
TBA	299	224	5.3	11.4	13.0	56	291	3.6
TBA	303	194	6.4	10.5	40.0	163	446	3.5

[a] Data from Watanabe and Shimazu (1976).
[b] BA, *beerenauslese*; TBA, *trockenbeerenauslese*.

in the production of botrytized wines, for which they seem ideally suited. A local Hungarian variety, 'Furmint,' is the predominant variety used in the production of Tokaji Aszú. Other varieties, such as 'Picolit,' 'Gewürztraminer,' 'Chenin blanc,' and 'Pinot blanc,' are used depending on tradition and adaptation to local conditions.

INDUCED BOTRYTIZATION

The production of botrytized wines is both a risky and expensive procedure. Leaving mature grapes on the vine increases the likelihood of bird damage, bunch rot, and other fruit losses. These dangers may be partially diminished by successive harvesting, but the process is too expensive to be used commonly. In Germany, much of the crop is harvested to produce nonbotrytized wine and only a small portion left on the vine for the development of noble-rot or *eiswein* production.

Where climatic conditions are unfavorable for noble-rot development, harvested fruit may be artificially exposed to conditions that favor development (Nelson and Amerine, 1957). The fruit is first sprayed with a solution of *Botrytis* spores. The fruit is then placed in trays and held at about 90 to 100% humidity for 24 to 36 hr at 20° to 25°C. These conditions permit spore germination and fruit penetration. Subsequently, cool dry air is passed over the fruit to induce partial dehydration and restrict invasion. After 10 to 14 days, the infection has developed sufficiently that the fruit can be pressed and the juice fermented. Induced botrytization has been used successfully, but on a limited scale, in California and Australia.

Inoculation of juice with spores or mycelia of *B. cinerea*, followed by aeration, apparently induces many of the desirable sensory changes that occur in vineyard infections (Watanabe and Shimazu, 1976). Whether the artificial nature of production might limit consumer acceptance is unknown.

Nonbotrytized Sweet Wines

Sweet nonbotrytized wines are produced in most, if not all, wine-producing regions of the world. Most have evolved slowly into their modern forms. On the other hand, several modern styles have developed quickly in response to perceived consumer demand.

DRYING

Drying is probably the oldest procedure employed to produce sweet wines. It involves placing fruit to dehydrate on mats or trays in the shade. After several weeks or months, the grapes are crushed and the concentrated juice fermented. Variations on the procedure are prevalent throughout southern Europe.

Variation in style often depends on the grape variety or varieties used and the treatments applied before, during, or after fermentation. For example, 'Moscato' grapes in Sicily may be cured in a solution of salt water and volcanic ash, before being crushed and fermented. For *vin santo*, the grapes are dried for between 50 and 100 days, and the juice is typically placed in small barrels located in attics (see Stella, 1981). Here, the wines ferment and age under fluctuating extremes of heat and cold for 2 to 6 years.

HEATING

Another ancient technique entails concentrating the juice, or semisweet wine, by gentle heating or boiling. The treatment induces a loss of varietal character and the acquisition of a caramelized or baked odor. The use of heat in the production of Madeira is described later in this chapter.

FREEZING, CRYOEXTRACTION, AND REVERSE OSMOSIS

For the production of *eiswein*, juice concentration results from freezing. Grapes are left on the vine until winter temperatures fall to about −6° to −7°C. As freezing occurs, the juice becomes concentrated as water is extracted from the cells to form ice crystals. By harvesting and pressing the grapes while frozen, most of the water remains in the press as ice, and the concentrated juice is isolated largely undiluted. The sugar concentration is typically so high that fermentation ends prematurely, leaving the wine with a high residual sugar content. The termination of fermentation is probably due to the combined stresses of the high sugar content and the accumulation of ethanol and toxic C_8 and C_{10} carboxylic acids. Although prolonged overripening may result in a net loss of some aromatic compounds, concentration appears to more than make up for the loss. The golden color of *eisweins* probably results from the combined effects of juice concentration, caftaric acid oxidation, and the release of catechins on freezing. Although *eiswein* is a relatively new style, the process has already spread beyond the confines of Germany. Conditions in the vineyards of southern Ontario, Canada seem particularly favorable to the production of fine eisweins.

Cryoextraction is the technological equivalent of eiswein production. It has the advantage of permitting the degree of juice concentration to be selected in advance (Chauvet *et al.*, 1986). It also avoids the risks of leaving the fruit on the vine for months, and the difficulties of harvesting and crushing grapes during frigid winter conditions. Reverse osmosis has also been used to produce concentrated juices for making *eiswein*-like wines. These techniques, however, may lack desirable flavor changes that develop during the long vineyard overripening.

JUICE CONCENTRATE (SWEET RESERVE)

The addition of unfermented grape juice (sweet reserve or *süssreserve*) to dry wine is the most widespread technique for producing sweet wines. It was first perfected in Germany and has the advantages of neither losing varietal distinctiveness nor producing additional flavors. If the juice is derived from the same grapes used in making the wine, varietal, vintage, and geographical origins are not compromised. It also may add to the varietal fruitiness of the wine, which is occasionally lost during fermentation. Furthermore, the procedure is technically simpler and more easily controlled than most other sweetening producers.

Red Wine Styles

Recioto Style Wines

Unlike white wines, few red wines are produced with a sweet finish. Even fewer are made from grapes infected with *Botrytis cinerea*. Nevertheless, the most famous red wine from Veneto, Italy, is made from grapes partially infected with *Botrytis*. These are the *recioto* wines from Valpolicella. Recioto Valpolicella is made from a blending of musts from 'Corvina,' 'Molinara,' and 'Rondinella' grapes. Similar wines are made in Lombardy from 'Nebbiolo' or 'Groppello.'

Recioto wines develop much of their distinctive fragrance and flavor from the processing of the fruit prior to fermentation. The fragrance contains elements that resemble the sharp smell of tulip blossoms. The wines typically have a high alcohol content and a smoother, more harmonious taste than the majority of full-bodied red wines. The wine may be made in a sweet (*amabile*), a dry (*amarone*), or a sparkling (*spumante*) style. Because of the unique fragrance of *recioto* wines, the process supplies winemakers with an additional means of producing wines with distinctive character.

PRODUCTION OF RECIOTO WINES

Healthy, fully mature clusters, or the most mature portions, are placed in single layers on trays designed to ease air flow around the fruit. The trays are stacked in rows several meters high in well-ventilated warehouses (Fig. 9.3). Natural ventilation may be augmented with fans to keep the relative humidity below 90%. The grapes are left to dry slowly for several months under cool ambient temperatures. The fruit is usually turned every few weeks to favor more uniform drying. Cool temperatures (3° to 12°C), and humidity levels below 90%, are crucial to restricting microbial spoilage.

During the 3 to 4 months of storage, the physical and chemical characteristics of the grapes undergo major changes. The most obvious is the 25 to 40% drop in moisture content. Other significant changes appear to accrue from the action of *B. cinerea*. Fungal growth probably develops from latent fruit infections acquired in the spring during flowering. Under the dry, cool conditions of storage, the fungus develops slowly. Sporulation

Figure 9.3 Warehouse for the slow drying of grapes in the production of *recioto* wines. (From Usseglio-Tomasset *et al.*, 1980, reproduced by permission.)

seldom occurs on the fruit unless conditions of high humidity develop.

The percentage of fruit showing infection usually increases relative to the duration of storage. The actual percentage depends on the variety. For example, 'Corvina' is quite susceptible, with up to 45% of the fruit showing signs of infection after 4 months. In contrast, 'Rondinella' may show less than 5% infection after the same period.

As infection progresses, the red/blue coloration of the fruit turns a pale purplish brown. The grapes also become flaccid, and the skin loses its strength. The visual and mechanical manifestations of infection are reflected in equally marked chemical alterations (Table 9.4). These changes show the distinctive effects of noble rotting by *B. cinerea*. Although fungal metabolism reduces the sugar content, the proportionally greater loss of water by evaporation results in a marked increase in sugar concentration. The Brix can thus rise from 25° to over 40° in heavily noble-rotted grapes. Owing to selective metabolism of glucose by the fungus, the fructose concentration rises more than that of glucose. The grapes also show a marked (~10- to 20-fold) increase in gluconic acid and glycerol concentration. Total acidity declines marginally if at all. Despite dehydration, the tartaric acid concentration remains relatively constant, while that of malic acid declines. These data may indicate that selective acid metabolism by *B. cinerea* during storage is somewhat different from that occurring in the vineyard. Alternately, the atypically low malic acid content may reflect the metabolic action of the grapes during prolonged storage.

Browning and loss of color are noticeable but less marked than might be expected considering the 50% infection rate by *B. cinerea*. This may result from the suppression of laccase activity by the high sugar content of the grapes (Doneche, 1991), poorly understood factors limiting laccase activity (Guerzoni *et al.*, 1979), or the resistance of some varieties to infection. However, the most resistant variety, 'Rondinella,' usually constitutes only about 25 to 30% of the blend in Valpolicella.

Botrytis cinerea probably generates the distinctive fragrance of recioto-style wines. The fungus has been correlated with the production of a marked phenolic flavor in red grapes (Boubals, 1982).

Following storage, the grapes are stemmed, crushed, and allowed to ferment under the action of indigenous yeasts. When a noticeable residual sweetness is desired, the must is kept cool ($\leq 12°C$) throughout alcoholic fermentation. For the dry *amarone* style, the must is warmed to, or allowed to rise to, about 20°C during fermentation. The fermentation temperature influences the relative dominance of different indigenous yeast strains derived from the fruit. *Saccharomyces cerevisiae* f. *uvarum* predominates at cool temperatures, leaving the wine with a perceptibly sweet finish. In contrast, *Sacch. cerevisiae* f. *bayanus* metabolizes the fermentable sugars at warmer temperatures and promotes the production of *amarones*. Fermentation typically lasts about 40 days.

Alcoholic fermentation may recommence when ambient temperatures rise in the spring. This may be either prevented by filtration for the *amabile* style or encouraged for production of *spumante*-style wines.

Figure 9.4 shows that the concentrations of glycerol, 2,3-butanediol, and gluconic acid are roughly compa-

Table 9.4 Changes in Must Composition during the Drying of Grapes for *Recioto* Wines[a]

	Duration of drying (days)				
	0	19	40	73	101
Sugar (g/liter)	181	188	193	204	230
Glucose (g/liter)	92	90	91	95	103
Fructose (g/liter)	89	98	102	109	128
Total extract (g/liter)	205	216	228	232	261
Total acidity (g/liter)	6.9	7.5	6.4	6.5	6.6
pH	3.1	3.0	3.1	3.1	3.2
Volatile acidity (mg/liter)	50	90	90	100	120
Tartaric acid (g/liter)	7.3	7.4	7.7	5.9	7.1
Malic acid (g/liter)	5.7	5.2	3.9	4.4	3.8
Glycerol (g/liter)	0.2	0.7	0.8	2.9	5.1
Gluconic acid (g/liter)	0.2	0.7	1.1	1.0	1.7

[a] Data from Usseglio-Tomasset *et al.* (1980).

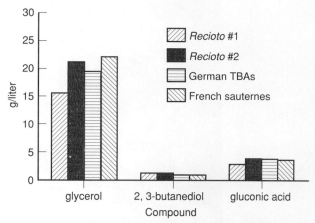

Figure 9.4 Comparison between two *recioto* wines and the averages of three botrytized wines each from Germany and France for glycerol, 2,3-butanediol, and gluconic acid. (Data for *recioto* wines from Usseglio-Tomasset *et al.*, 1980; those for TBAs and sauternes from Yunome *et al.*, 1981.)

rable to those expected for highly botrytized wines. The high levels of both glycerol and alcohol contribute to the smooth texture of the wine. The smooth sensation is undoubtedly aided by the limited extraction of tannins during cool fermentation temperatures. Cool temperatures also favor the production/retention of fragrant esters, augmenting the distinctive bouquet of recioto wines.

Whether malolactic fermentation plays a significant role in the development of recioto wines is unclear. Published data on the chemical composition of *recioto* wines do not suggest its occurrence. However, malolactic fermentation may take place in the sporadic refermentation of sweet *recioto* wines in the spring and summer months. Malolactic fermentation may be the source of the carbon dioxide that occasionally gives the *amabile* style a slight effervescence.

Amarone-style wines commonly are aged in oak for several years prior to bottling to improve the fragrance and harmony of the wine. Old cooperage is preferred to avoid giving the wine a marked oaky character.

Carbonic Maceration Wines

In its simplest from, carbonic maceration may be as old as winemaking itself. Its use was probably widespread until the introduction of mechanical crushers in the nineteenth century. Mechanical crushers permitted complete rather than partial crushing of the fruit and improved pigment and tannin extraction. Subsequently, stemming and crushing before fermentation became dominant and considered "traditional." Decline in the full or partial use of carbonic maceration also correlates roughly with the increased use of bottles and prolonged aging. In only a few regions, such as Beaujolais, did carbonic maceration remain the dominant technique in producing red wines.

Grapes have long been known to metabolize malic acid, especially during ripening under warm conditions. Although carbonic maceration favors malic acid decarboxylation, the process is not specifically used as a deacidification technique. The primary advantage has been in producing early maturing wines. The production of a unique and distinctly fruity aroma has led the technique to develop a new and growing following. Possibly because such wines have an image as light "quaffable" wines, connoisseurs often malign the wines the technique generates.

Because of limited regional use, carbonic maceration has received little attention outside southern France and Italy. With the current popularity of *nouveau*-type wines, interest in the technique is redeveloping. The Institut National de la Recherche Agronomique in Montfavet has carried out extensive studies on carbonic mac-

eration since the 1940s (Flanzy *et al.*, 1987). The Institute coined the term "carbonic maceration" for the technique, and most of this section is based on their work.

Figure 9.5 compares carbonic maceration with "traditional" vinification. Fundamentally, carbonic maceration differs from traditional vinification in that grape berry fermentation precedes yeast and malolactic fermentation. For this to occur, it is essential that the fruit be harvested in a manner that minimizes breakage.

Carbonic maceration was initially proposed to refer to the anaerobic maceration of whole berries placed in an atmosphere of carbon dioxide. This has subsequently been extended to include grape cell alcoholic fermentation—whether or not air is initially removed from the fermentor by flushing with carbon dioxide.

Typically, carbonic maceration takes place in the presence of a small amount of must, released when a portion of the fruit ruptures during loading of the fer mentor. Thus, berry fermentation typically occurs in association with limited yeast fermentation.

If the fruit is not dumped into a fermentor, anaerobic maceration can occur in the absence of free juice. For example, containers of grapes may be placed in specially designed sealed chambers after harvesting (Càstino and Ubigli, 1984) or wrapped in plastic for the maceration period. Because pigment extraction is poor, the juice is left to ferment with the seeds and skins after crushing until the desired color has been achieved. Under normal conditions, however, juice fermentation with the seeds and skins after carbonic maceration is unnecessary. Sufficient berries break open during the maceration phase to allow adequate pigment extraction.

Typically, the whole harvest undergoes carbonic maceration, though in certain regions the technique may be used with only part of the crop. Partial carbonic maceration may be employed when the varietal aroma is masked if the whole crop is treated. In Bordeaux, for example, if more than 85% of the fruit undergoes carbonic maceration, the 'Cabernet' character is suppressed (Martinère, 1981). Modifying the percentage of the crop undergoing carbonic maceration can be used to regulate the relative contribution of carbonic maceration to the character of the wine.

The intensity of the carbonic maceration aroma also may be modified by adjusting the duration and temperature of the process. Alternately, the wine may be blended with wine derived by traditional procedures.

At the end of carbonic maceration, the grapes are pressed and the juice allowed to ferment to dryness by yeast action. Malolactic fermentation typically occurs shortly after the termination of alcoholic fermentation. After completion of malolactic fermentation, the wine typically receives a light dosing with sulfur dioxide (20 to

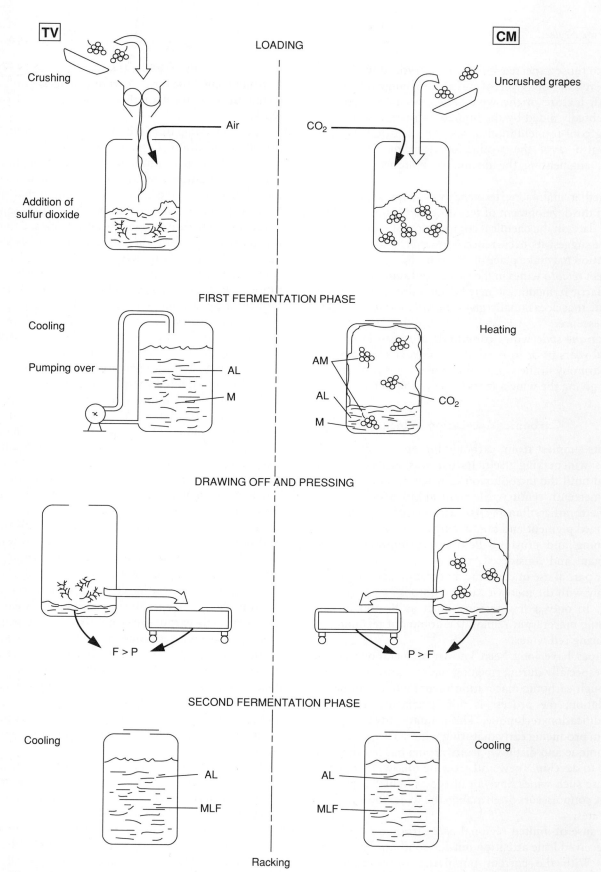

Figure 9.5 Schematic comparison between carbonic maceration (CM) and traditional vinification (TV) with crushing before fermentation. AL, Alcoholic fermentation; MLF, malolactic fermentation; F, free-run; M, maceration, AM, anaerobic metabolism; P, press-run. (After Flanzy and André, 1973, reproduced by permission.)

50 mg/liter). This helps prevent further microbial action. Racking commonly occurs at the same time. Racking may be delayed for several weeks, however, as some winemakers believe that substances released by yeast autolysis enhance wine flavor.

Although used most extensively in the production of light, fruity, red wines, carbonic maceration can yield wines capable of long aging. The procedure has also been used to a limited extent in producing rosé and white wines.

ADVANTAGES AND DISADVANTAGES

One of the advantages of carbonic maceration is the generation of a unique and pronounced fruity aroma. The fragrance has been described variously as showing kirsch, cherry, or raspberry aspects. Additional descriptors are given in Table 9.5.

For relatively neutral varieties, such as 'Aramon,' 'Carignan,' and 'Gamay,' carbonic maceration gives an appealing fruitiness that traditional vinification does not provide. With varietally distinctive cultivars such as 'Cabernet Sauvignon,' 'Merlot,' and 'Concord,' carbonic maceration either masks or destroys the varietal character. This may be desirable or not, depending on the attractiveness of the varietal aroma. With other cultivars, such as 'Syrah' and 'Maréchal Foch,' carbonic maceration adds complexity to the varietal fragrance. Carbonic maceration is even reported to enhance the varietal aroma of some white grapes (Bénard *et al.*, 1971).

Because traditional vinification extracts more tannins than carbonic maceration, the process has advantages when used with highly tannic grapes. Deacidification might also justify its use with acidic grapes.

Table 9.5 Sample Descriptors Used to Describe Certain Carbonic Maceration Wines[a]

	Full carbonic maceration	Semicarbonic maceration (Beaujolais)
Visual appearance	Ruby red	Ruby red
Fragrance	Kirsch	Hyacinth
	Coffee	Coffee
	English candy	English candy
	Vanilla	Vanilla
	Grilled almonds	Cherry
	Russian leather	Banana
	Resin	Raspberry
Quality	Fine (predominantly vegetal and lactic)	Rich (predominantly winy and phenolic)
Taste	Subtle	Rough
	Buttery	Tannic

[a] After Flanzy *et al.* (1987), reproduced by permission.

Carbonic maceration wines seldom demonstrate a yeasty bouquet after the completion of vinification. The wine, which has a smoother taste, can thus be enjoyed sooner. This has financial benefits, as the wines can be bottled and sold within weeks of vinification. Thus, capital is not tied up for years in cellar stock.

Customarily, the carbonic maceration aroma does not improve on aging and fades relatively quickly. Unless a varietal aroma or aged bouquet replaces the fading carbonic maceration fragrance, the wine commonly has a shelf life of only about 6 months to 1 year. Carbonic maceration does not itself limit shelf life. The aging potential depends primarily on the quality and properties of grapes fermented. Consequently, some carbonic maceration wines show long aging potential, notably those from northern Beaujolais, the Rhône Valley, and Rioja. Prolonged maceration presumably favors the extraction of sufficient aromatics, anthocyanin, and tannins to give the wine aging potential.

In the past, the comparative simplicity of the carbonic maceration process added to the benefits of early drinkability. It required neither destemming nor treading of the grapes, and whole grape clusters could be loaded directly into wide, shallow vats. The process of fermentation developed slowly, resulting in a lower maximum temperature during fermentation. Thus, there was less likelihood of fermentation sticking in warm regions. Crushing and pressing were easier, since the skins became weak and the grapes flaccid during carbonic maceration.

Some of these advantages still apply, for example, early drinkability and the easier pressing of pulpy grapes. However, other aspects of carbonic maceration are incompatible with current harvesting and winemaking practices. In many areas, mechanical harvesting is both cost effective and compatible with fruit quality. However, mechanical harvesters may rupture many berries and reduce the proportion of fruit in which carbonic maceration is possible. In addition, most modern fermentation tanks are tall. This would result in excessive rupture of the grapes during loading and by the cumulative mass of the fruit in the tank.

Care must be taken at all stages, from harvesting to loading, to minimize fruit rupture. Broad, shallow (~2.5 m) vats are preferred, as they reduce berry rupture and ease loading. Vats permit the ready displacement of air with carbon dioxide at the beginning of maceration. Subsequently, the vat opening is covered to restrict air access, while permitting the escape of excess carbon dioxide.

In some regions, notably Beaujolais, the must in the vat is periodically pumped over the fruit. Although frequently practiced, pumping over is no longer recommended. The procedure increases oxidation and may cause undesirably high concentrations (≥ 150 mg/liter) of ethyl acetate (Descout, 1986). By temporarily removing the bouyant action of must, pumping over induces fruit rupture. Ethanol produced in the must also favors fruit rupture. Alcohol kills and weakens fruit cells. Rupture curtails both the duration of and proportion of the fruit undergoing carbonic maceration.

One of the more serious drawbacks of carbonic maceration is the high demand it places on fermentor capacity at harvest time. Because the fruit is neither stemmed nor crushed, it takes up considerably more volume than the must derived from the same amount of fruit. Furthermore, the initial grape fermentation signficantly prolongs the total fermentation period (Fig. 9.6). While malolactic fermentation typically follows shortly after yeast fermentation, this does not offset the need for increased fermentor capacity at a critical time of the year.

The problem of fermentor capacity can be sidestepped with the use of specially designed storage containers (Càstino and Ubigli, 1984) or by adequately wrapping the fruit in plastic (Rankine *et al.*, 1985). This is easiest when the grapes are left in the containers in which they were harvested. Either procedure avoids the berry rupture inevitable during vat loading.

FERMENTATION—PHASE I

Whole-Grape Fermentation In the absence of oxygen, grape cells change from respiratory to fermentative metabolism. This shift is more rapid if air is flushed out with carbon dioxide. Carbon dioxide, being more dense than air, forces air to float out the vat as the CO_2 is added. Carbon dioxide is customarily preferred to nitrogen as it has uniquely desirable properties. It directly induces leakage of ions from cells (Yurgalevitch and Janes, 1988) and shifts the equilibria of decarboxylation reactions (Isenberg, 1978). Carbon dioxide also may accelerate the breakdown of pectins, by inducing the synthesis of grape pectinases.

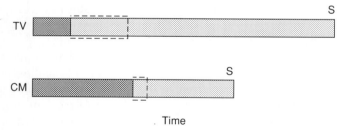

Figure 9.6 Representative duration of vinification by carbonic maceration (CM) and traditional procedures (TV). Darker bars represent the first fermentation phase: yeast-induced alcoholic fermentation in TV and anaerobic berry metabolism plus alcoholic fermentation of the free-run juice in CM. Lighter bars represent the second fermentation phase: the end of alcoholic fermentation and malolactic fermentation. Dashed boxes represents the period during which malolactic fermentation occurs. S, Biological stability. (After Flanzy *et al.*, 1987, reproduced by permission.)

Alcoholic fermentation in grape cells appears to be similar to that found in yeasts. The primary end product is ethanol, with smaller accumulations of glycerol, acetaldehyde, acetic acid, and succinic acid.

Although alcohol dehydrogenase induces the reduction of acetaldehyde to ethanol, the sensitivity of the enzyme to ethanol inactivation is insufficient to explain the limited alcoholic fermentation of grape cells (Molina *et al.*, 1986). Instead, enzyme activity may cease because of ethanol-induced membrane disruption and cell death (Romieu *et al.*, 1989). The latter would result in the release of organic acids stored in cell vacuoles, lowering the cytoplasmic pH and inhibiting alcohol dehydrogenase activity. Grape ethanol synthesis usually is less than 2%.

During grape-cell fermentation, malic acid is metabolized to other acids (primarily oxaloacetic, pyruvic, and succinic acids), as well as to ethanol. Depending on the grape variety and fermentation temperature (Flanzy *et al.*, 1987), upward of 15 to 60% of the malic acid may be metabolized during carbonic maceration. Significant decarboxylation to lactic acid does not occur. The other major grape acids (tartaric and citric) are occasionally metabolized. Their metabolism appears to depend predominantly on the grape variety.

Associated with grape fermentation is a modification in the operation of the shikimic acid pathway. Shikimic acid accumulates along with volatile products of its metabolism, such as ethyl cinnamate, benzaldehyde, vinylbenzene, and salicylic acid. The latter is not itself volatile, but it reacts with ethanol to form an aromatic ethyl ester. Higher concentrations of ethyl decanoate, eugenol, methyl and ethyl vanillates, ethyl and vinyl guaiacols, and ethyl and vinyl phenols develop during carbonic maceration than in traditional vinification (Ducruet, 1984). The high ethyl cinnamate and ethyl decanoate levels may be sufficiently distinctive to serve as indicators of carbonic maceration.

The precise chemical nature of the characteristic aroma of carbonic maceration wines remains unclear. However, some elements of the fragrance have been tentatively ascribed to ethyl cinnamate and benzaldehyde. Respectively, they may generate some of the characteristic strawberry/raspberry (Versini and Tomasi, 1983) and cherry/kirsch (Ducruet, 1984) fragrances found in carbonic maceration wines.

Low hexyl acetate and hexanol contents are generally associated with carbonic maceration. Without prefermentation crushing, there is less chance for fatty acid oxidation.

One of the distinctive consequences of carbonic maceration is a reduction in the amount of free ammonia and a rise in the concentration of amino acids (Flanzy *et al.*, 1987). Some of the amino acids undoubtedly arise from the enzymatic breakdown of proteins. Others may form from the synthesis of amino acids via the TCA (tricarboxylic acid) cycle and glycolysis intermediates and ammonia. Although the total concentration of amino acids rises, the contents of specific amino acids vary independently. For example, the concentrations of aspartic and glutamic acids fall, owing to their metabolism during carbonic maceration (Nicol *et al.*, 1988).

The release of organic nitrogen during maceration helps explain the rapid onset and completion of both alcoholic and malolactic fermentation after pressing. Whether the high amino acid content plays a role in the development of the characteristic carbonic maceration aroma is unknown.

During carbonic maceration, pectins break down in the fruit. Consequently, the attachment of cells to one another weakens, and the pulp loses its solid texture. If the carbon dioxide produced inside intact fruit escapes, the berries become flaccid. Otherwise, the CO_2 pressure maintains the shape, but not the strength, of the fruit.

At the beginning of carbonic maceration, the fruit absorbs carbon dioxide from the surrounding environment. The amount dissolved depends on the temperature, varying from about 50% of berry volume at 18°C to 10% at 35°C. As berries become saturated, carbon dioxide liberated during fermentation begins to be released. Production of carbon dioxide by grapes ceases when the cells die due to alcohol toxicity, or when fermentation is no longer sufficient to sustain cellular integrity.

As grape cells die, regulation of the movement of substances across the membrane ceases. This enables the release of various substances from the cells, notably phenolic compounds. The extraction of phenols is complex and often highly specific. Major influencing factors are the temperature and duration of carbonic maceration, as well as the presence of fermenting juice around the fruit.

Anthocyanins are more rapidly and extensively dissolved than tannins. As high temperatures speed color stability by favoring anthocyanin–tannin polymerization, winemakers prefer a short maceration at temperatures above 30°C. Extraction of phenolic compounds appears to be primarily from the skins, with little coming from the seeds.

Nonflavonoid phenolics are both extracted and modified structurally during maceration. Chlorogenic acid dissolves, and the tartrate esters of *p*-coumaric and caffeic acids hydrolyze rapidly. As a result, small quantities of free *p*-coumaric and caffeic acid accumulate.

Submersion of the fruit in fermenting must, common at the base of the vat, markedly increases anthocyanin and tannin extraction. This undoubtedly is due to the elevated alcohol content that develops in the freed, fermenting must. As alcohol diffuses into intact grapes,

ethanol dissolves phenols in the fruit. Thus, pigment and tannin extraction is not limited just to those berries that break open at the bottom of the vat.

Of the factors influencing grape fermentation, temperature is probably the most significant and easily controlled. The first phase of carbonic maceration is commonly conducted at between 30° and 32°C to shorten the duration of carbonic maceration, promote grape fermentation, and favor pigment and tannin extraction. To favor the rapid onset of carbonic maceration, fruit is often picked late in the afternoon on warm sunny days. Alternately, the fruit may be heated to the desired temperature. The preferred initial temperature in Beaujolais, however, is reported to be between 18° and 22°C (Descout, 1983).

Fermentation of Released Juice Typically, the fruit is fermented in shallow vats. The breakage and amount of juice released depend on the maturity and health of the fruit, grape variety, tank depth, and mechanism of fruit unloading. To minimize crushing, tanks in Beaujolais are commonly no more than 2.5 m high. Nevertheless, about 10 to 20% of the grapes break during loading. This level increases during maceration as the berries weaken and the weight of the fruit above ruptures those below (Fig. 9.7). Pumping over augments fruit rupture and the amount of free-run liberated. The proportion of juice released by fruit rupture varies widely, but by the end of maceration about 35 to 55% of the juice has been freed.

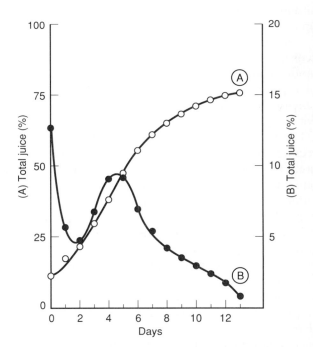

Figure 9.7 Release of free-run must and wine during carbonic maceration: A, percentage free-run released; B, percentage free-run of total released per day. (After André *et al.*, 1967, reproduced by permission.)

If the must is low in acidity, tartaric acid may be added to the free-run. Sulfur dioxide addition at this point usually brings the level of SO_2 in the free-run up to 20 to 50 mg/liter. Sulfiting is kept to a minimum to avoid the production of hydrogen sulfide and any delay in the onset of malolactic fermentation. Early completion of malolactic fermentation is essential to the production of *primeur* or *nouveau* wines, which need microbial stability within a few weeks of vinification.

Occasionally, chaptalization (the addition of sugar) is conducted before pressing. If desired, however, sugar addition normally occurs after pressing, when alcoholic fermentation is at its apex (Descout, 1983).

Because juice inoculation occurs spontaneously following rupture, some yeast fermentation occurs concurrently with carbonic maceration. This has a marked effect on the course and duration of grape cell fermentation. Yeast fermentation is most marked on fruit submerged in the fermenting juice. Even in the absence of released juice, however, the population of yeasts and bacteria on the grapes increases. The grape flora can grow on nutrients already present on the skin. The absence of oxygen is not a limiting factor. The microbial flora is in direct contact with long-chain unsaturated fatty acids needed for growth and membrane production under anaerobic conditions. These compounds are part of the waxy cuticular layer of grape skins.

Few studies on the composition of the yeast flora during carbonic maceration have been conducted. *Saccharomyces cerevisiae* appears to be the dominant species, although *Schizosaccharomyces pombe* may constitute up to 25% of the yeast population (Barre, 1969).

The yeast population reaches about 8 to 12×10^7 cells by the time of pressing. Fungicides on the fruit can modify this value significantly though, especially because of the small juice volume initially released. To offset potential yeast suppression, the juice may be inoculated with an active yeast culture.

Use of yeast strains adapted to high temperatures is required because of the high fermentation temperatures preferred. Nevertheless, it is important to prevent heat buildup above 35°C. Temperatures at or about 35°C can induce yeast death and leave the must open to the growth of spoilage yeast and bacteria.

Yeast inoculation tends to reduce the accumulation of ethyl acetate (Descout, 1986). The origin of this compound is not precisely known. As its increase is not directly correlated with the simultaneous buildup of acetic acid, it presumably is synthesized directly by the grapes or indigenous yeasts. Pumping over and periodic chaptalization (frequent but small additions of sugar) encourage ethyl acetate accumulation.

Yeast activity has a considerable influence on the course of carbonic maceration. Where carbon dioxide is not used to displace the air in the fermentor, yeast action generates most of the carbon dioxide that eventually blankets the fruit. If the fruit is cool, and not heated artificially, yeast metabolism also liberates most of the heat that warms the fruit during carbonic maceration. Yeasts quickly convert released sugars to ethanol, in contrast to the limited conversion in whole grapes.

Alcohol vapors generated during carbonic maceration are partially absorbed by the fruit. Not surprisingly, more ethanol diffuses into fruit immersed in fermenting juice (Fig. 9.8). By acting as a sink for the alcohol produced in the released juice, intact berries aid yeast fermentation by slowing the buildup of ethanol in the juice. Malic acid also tends to diffuse inward, permitting its continued metabolism by grape cells. If lactic acid bacteria are active in the juice, however, the flow of malic acid may reverse. Sugar slowly diffuses out of intact berries, adding to the sugars released by progressive rupture of the fruit. The sugars provide a continuing nutrient source for yeast metabolism. By the end of carbonic maceration, the sugar content of intact fruit usually has fallen to about 50 to 70 g/liter (Descout, 1986). Throughout carbonic maceration, juice release associated with fruit rupture helps minimize the accumulation

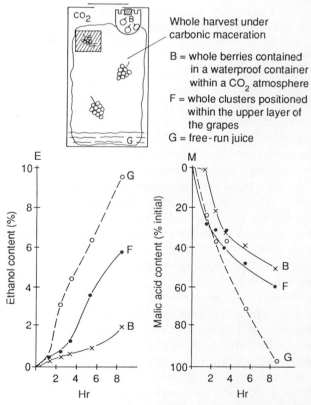

Figure 9.8 Changes in ethanol and malic acid contents in grapes during carbonic maceration in a carbon dioxide atmosphere. (After Flanzy *et al.*, 1987, reproduced by permission.)

of octanoic and decanoic acids and supplies amino nitrogen.

Although alcohol accumulation inhibits grape fermentation, it does not disrupt development of the carbonic maceration aroma (Tesnière *et al.*, 1991). Conversely, ethanol appears to be required only for inducing the production of the aroma precursors of carbonic maceration.

The end of carbon dioxide release is one of the primary indicators winemakers use to determine when carbonic maceration has ceased. Another indicator is the drop in juice specific gravity to 1.02 or lower. Alternately carbonic maceration may be halted when the juice color or flavor development has reached the desirable intensity.

Maceration typically lasts 6 to 8 days but may last up to 2 weeks. Long maceration is more common when there is no simultaneous yeast fermentation. Extended contact between the juice and fruit often leads to development of a bitter, herbaceous character. This presumably results from the extraction of phenolic compounds from the stems.

FERMENTATION—PHASE II

Once the decision has been taken to terminate carbonic maceration, the free-run is allowed to escape and the intact grapes and pomace pressed to extract the remaining juice. If the free-run shows no signs of active malolactic fermentation, it is typical to combine all juice fractions for yeast alcoholic fermentation. However, if the free-run juice is undergoing malolactic fermentation, the press-run juice is usually fermented separately. There is the concern that fermentable sugars in the press-run may be metabolized by the lactic acid bacteria, producing high levels of volatile acidity.

The free-run and press-run fractions also may be fermented separately to permit blending based on their respective qualities. Free-run juice, unlike the case for traditionally vinified wines, are of lower quality than the press-run juice. Free-run juice produces wine less alcoholic (by about 1 to 2% ethanol) than the press run juice. Free-run wine is also lighter in color, while being more herbaceous, bitter tasting, and higher in acetaldehyde and 2,3-butanediol contents than its press-run counterpart. The press-run fraction generates wine that is more aromatic, alcoholic, and colored. It also contains most of the esters, fusel alcohols, and aromatic compounds that give carbonic maceration wines their distinctive fragrance. Although the total phenolic contents in both fractions are nearly identical, the specific composition is different. Tannins in the press-run are softer tasting and less bitter than are those in the free-run.

For the lighter *primeur*-style typical of Beaujolais nouveau, more free-run wine is used. When a wine with longer aging potential is desired, the blend contains a higher proportion of the press-run fraction.

For the second phase of vinification, a temperature of 18° to 20°C is preferred. This is believed to help retain the distinctive fragrance donated by carbonic maceration. If, as usual, the primary phase has taken place or reached temperatures considerably above 18° to 20°C, cooling is required. Some cooling occurs spontaneously when carbonic maceration comes to completion and during pressing. Nevertheless, additional cooling is often required. Even at 18° to 20°C, fermentation is tumultuous and customarily complete within 48 hr.

In addition to consuming fermentable sugars, yeasts modify the concentration of volatile phenols. Malolactic fermentation further alters the phenol concentration. The effects are more pronounced in carbonic maceration wines than in traditionally produced wines. Both 4-vinyl guaiacol and 4-vinyl phenol contents increase, while 4-ethyl phenol decreases during alcoholic fermentation (Etiévant *et al.*, 1989). All volatile phenols increase during malolactic fermentation.

Malolactic fermentation typically begins immediately following the termination of alcoholic fermentation, if not before. This is favored by the limited use of sulfur dioxide, storage at warm temperatures, reduced acidity, and the availability of nitrogenous and other nutrients. If malolactic fermentation is slow to commence, the wine is commonly inoculated with lactic acid bacteria, usually derived from wines already having undergone malolactic fermentation. Natural inoculation appears to induce quicker fermentation than inoculation with commercially available cultures.

AGING

Most carbonic maceration wines are produced for rapid consumption, with only a small proportion vinified for long aging. A few of the changes that can occur during aging are shown in Fig. 9.9; they are similar to those occurring in traditionally vinified wines, but are more pronounced. Subjectively, it is known that the fruity aroma induced by carbonic maceration soon fades. If a desirable varietal aroma or aging bouquet replaces the grape fermentation aroma, the wine is more likely to age well.

Maturation in oak usually has been considered inappropriate for carbonic maceration wines. However, short exposure to oak can add complexity to the wine. Winemakers differ considerably in opinion concerning whether oak benefits or harms the fruity character of the wine.

USE WITH ROSÉ AND WHITE WINES

Although carbonic maceration is predominantly used for the production of red wines, rosé and white wines are

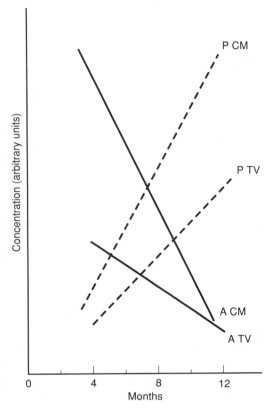

Figure 9.9 Development of aromatic compounds during maturation. 0, End of malolactic fermentation; A, isoamyl acetate and benzaldehyde; P, volatile phenols; CM, carbonic maceration; TV, traditional vinification. (After Flanzy *et al.*, 1987 from data published in Etiévant *et al.*, 1989, reproduced by permission.)

occasionally vinified using the technique. In the production of rosé wines, grapes are kept from being submerged in free-run juice. This limits pigment and tannin extraction, while maximizing development of the fruity aroma. If pigment extraction is insufficient, the grapes are crushed and the second phase of fermentation conducted briefly in contact with the seeds and skins. Once sufficient color has been obtained, the fermenting must is pressed to separate the juice from the pomace.

Similarly, white grapes treated to carbonic maceration are kept isolated from fermenting juice. The duration of the process for white wines is commonly shorter than usual for either red or rosé wines, often being little more than 48 hr. The precise duration and temperature chosen depend on winemaker preferences and how maceration affects the varietal aroma. Carbonic maceration can either suppress or enhance varietal character (Bénard *et al.*, 1971).

An alternative to the typically short carbonic maceration at warm temperature is maceration at 5°C for about 3 days (Montedoro *et al.*, 1974). The procedure favors ester synthesis/retention and reduces phenolic extrac-

tion. Centrifugation also may be used to reduce the phenol content and diminish color intensity.

For the second phase of vinification, juice fermentation is conducted without skin contact. If low in pH, the wine may be acidified on pressing. Alternately, the wine may be cooled and sulfited to prevent deacidification by malolactic fermentation.

Occasionally, the harvest is divided into lots, one treated according to traditional procedures and the other treated to carbonic maceration. The fractions may be blended together to provide a wine with enriched fragrance and improved acid balance.

Sparkling Wines

Sparkling wine owes much of its development to technical advancements unrelated to the production of the wine itself. These involved major improvements in the manufacture of glass in the late 1700s and the reintroduction of cork bottle-closures to France about the same time. The availability of strong glass was as important to withstand the high pressures that develop in sparkling wine, as was cork to retain the high carbon dioxide content found. These developments, along with an atypically long spell of cold weather in Europe (Ladurie, 1971), combined to favor the evolution of sparkling wine. Instead of producing inferior quality red wines (*vin gris*), Don Perignon (1638–1715) introduced techniques that culminated in the production of champagne. However, the procedure took an additional 150 years to develop into its near modern form. Subsequently, the method spread throughout most of the winemaking world. The twentieth century has seen major improvements in the method, as well as the development of major and minor adaptations designed to minimize production costs.

A classification of sparkling wines is given in Table 1.2. The three major processes used are diagrammatically compared in Fig. 9.10.

Most sparkling wines obtain the carbon dioxide supersaturation from a second alcoholic fermentation, which is typically induced by the addition of yeast and sugar to dry white wine. Infrequently, the sparkle may come from the continuation of the primary alcoholic fermentation in the wine after bottling. Excess CO_2 also may be generated by delayed malolactic fermentation. The latter two processes, jointly or combined, were undoubtedly involved in the production of the first sparkling wines. The **traditional** method began to approach its modern form when it was discovered that the addition of sugar favored the occurrence of the second, in-bottle fermentation. Subsequently, knowledge of the impor-

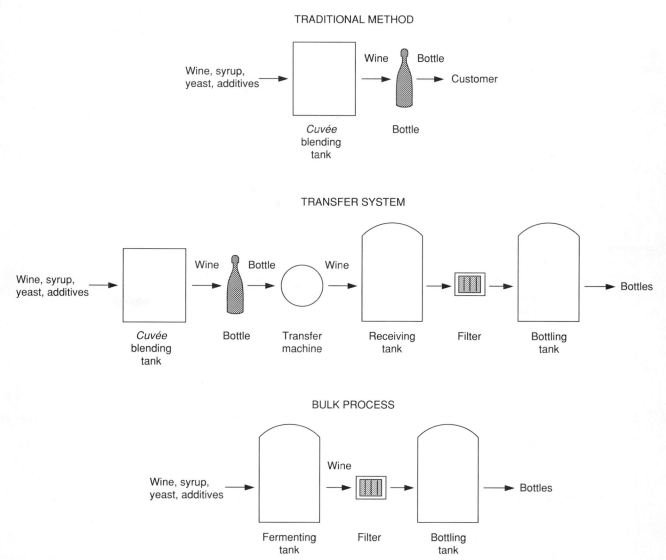

Figure 9.10 Flow diagram for three methods of sparkling wine production. (Reprinted with permission from Berti, 1981. Copyright 1981 American Chemical Society.)

tance of yeasts in fermentation led to their concurrent addition.

Although sparkling wines are usually classified by the means of production (Table 1.2), this is of little practical value to consumers. Wines produced by different techniques are often distinguishable only by careful sensory evaluation. More obvious sensory differences develop from the color and aroma of the base wines, the degree of carbon dioxide supersaturation, the duration of lees (yeast sediment) contact, and the sweetness given the finished wine.

For the purpose of this text, sparkling wines are discussed in the typical manner, that is, by the method of production. Figure 9.11 outlines the traditional method described below.

Traditional Method

GRAPE CULTIVARS EMPLOYED

Although both white and red grapes may be used, most sparkling wines are white. Thus, particular attention must be taken during the harvest and pressing to avoid pigment extraction from red grapes.

In the Champagne region of France, three grape varieties are used, one white ('Chardonnay') and two red ('Pinot noir' and 'Meunier'). Although varieties are occasionally used separately, most champagnes are produced from a blend of all three cultivars. Each variety is deemed to contribute a special character to the blend (**cuvée**): 'Chardonnay' providing finesse and elegance, 'Pinot noir' donating body, and 'Meunier' giving fruit-

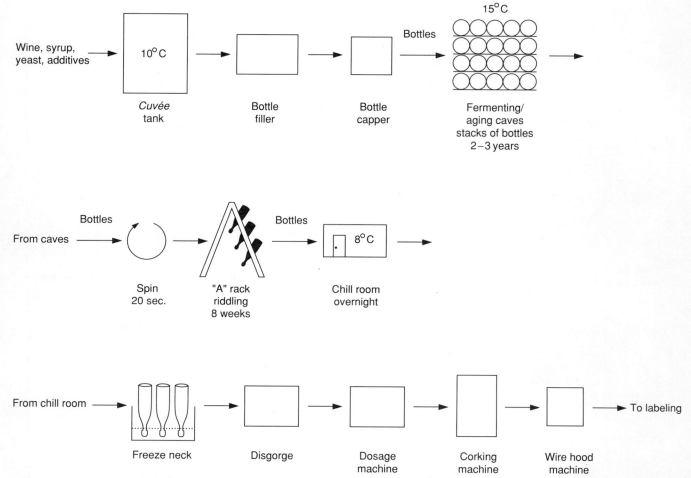

Figure 9.11 Traditional method of sparkling wine production. (Reprinted with permission from Berti, 1981. Copyright 1981 American Chemical Society.)

iness and roundness. Each also is considered to mature at different rates, with 'Chardonnay' being the slowest and 'Meunier' the most rapid. Thus, 'Meunier' features prominently in nonvintage blends aged in-bottle for about 1 year before disgorging (removal of the yeasts). Conversely, 'Chardonnay' is an important component in vintage blends, aged for at least 3 years in-bottle before disgorging.

In other areas of France, regional cultivars, such as 'Chenin blanc' in the Loire Valley, are often the dominant or only varieties employed. Outside France, either indigenous varieties or locally grown French cultivars are used in making sparkling wines. In Spain, the native varieties 'Parellada,' 'Xarel-lo,' and 'Viura' are employed. Each variety is considered to contribute a different important characteristic to the blend, with 'Parellada' providing fragrance and softness, 'Viura' donating finesse and elegance, and 'Xarel-lo' imparting strength and color.

HARVESTING

Where financial return permits, harvesting of both white and red grapes occurs manually. Manual harvesting permits both pre- and postfruit selection to exclude moldy grapes. This is especially critical when red grapes are used as pigment can diffuse into the juice. Laccases also can cause serious oxidative browning. Manual picking also minimizes fruit rupture, the release of juice, subsequent oxidation of the juice, and the extraction of pigments and tannins. However, because of the slowness of manual harvesting, the fruit may not be picked at its optimal quality if rainy conditions develop. Inability to harvest quickly can lead to considerable loss in quality under poor harvest conditions.

Harvesting of the fruit commonly occurs earlier than is usual for table wine production. Early picking yields grapes higher in total acidity and lower in pH. Acidity and pH are important in providing the freshness desired in most traditionally produced sparkling wines. Early picking also assures lower °Brix levels, yielding wines of lower alcohol content. An alcohol content of between 9 and 10.5% is desired for base wines. In addition, slightly immature fruit have less varietal character, favored in the production of most sparkling wines. Moreover, the grapes are more likely to be healthy. It is not suprising that Champagne, the most northerly wine-producing region in France, is associated with the evolution of sparkling wine. Grapes from the region routinely yielded poorly colored acidic juice, low in sugar content and varietal aroma—all features currently considered desirable in the production of most sparkling wines.

PRESSING

In traditionally produced sparkling wines, the grapes are pressed whole, without prior stemming or crushing. This minimizes the extraction of pigments from red grapes, as well as the release of solids, grape polyphenol oxidases, and potassium from the fruit. Because the latter remain with the pomace, the juice is less oxidized, and the need for sulfur dioxide addition is reduced. Pressing intact grapes is considered to promote early malolactic fermentation and favors the onset of the second, in-bottle, yeast fermentation. Furthermore, pressing limits the extraction of varietal aroma compounds that could mask the subtle aged fermentation bouquet so desired in champagne. Finally, the technique yields juice with the highest acidity and lowest potassium levels.

Pressing whole grapes takes considerably longer than conventional pressing. It often takes 2 hr for the first pressing in the large-diameter vertical presses historically preferred in Champagne. This inadvertently exposes the juice to oxidation. While brown pigments form, they generally precipitate during fermentation. The limited oxygen exposure gives the wine a degree of insensitivity to oxidative browning by promoting the early removal of the most easily oxidized phenols from the juice.

The grapes are generally unstemmed and pressed as whole clusters. The stems provide channels for juice escape, thus minimizing the pressure required. The large shallow presses historically used provide a large surface area, further minimizing the pressure needed for juice release. Nevertheless, the traditional wooden presses are slowly being replaced by horizontal presses. Several horizontal presses work well with unstemmed, uncrushed grapes and have the advantages of taking less space, being easier to load and unload, and permitting more efficient pomace crumbling between successive pressings.

About 66 liters of juice are extracted per 100 kg of grapes in Champagne. The largest portion (~75%) consists of free-run juice, the remainder coming from the first two press fractions. Similar yields are permissible in most French regions producing sparkling wine. Quality and pricing constraints, not legislation, are the primary factors influencing juice yield in much of the New World. Increasing the volume pressed from the grapes typically compromises quality by extracting more flavor and phenolic compounds, which detract from the subtle bouquet so highly prized in dry sparkling wine. Phenols also increase the tendency of the wine to gush on opening.

Sulfur dioxide is commonly added to the juice as it comes from the press. Levels of 40 to 60 mg/liter are typical in Champagne (Moulin, 1987). Juice not already cool is routinely cooled to about 10°C and left to clarify for 12 to 15 hr before fermentation.

As noted above, grapes pressed whole liberate fewer solids than those pressed after crushing (~0.5% versus 2–4%) (see Randall, 1987). Where the grapes are crushed before pressing, the extra solids are normally removed prior to fermentation using bentonite-facilitated settling, centrifugation, or filtration. The use of peristaltic pumps in transporting the juice to temporary storage tanks minimizes particulate extraction following crushing and pressing.

Even under optimal conditions, the juice obtained from red grapes often contains a slight pinkish tinge. The anthocyanins involved usually coprecipitate with yeasts during fermentation, or later during fining. Thus, anthocyanase addition, or other forms of decolorization, are customarily unnecessary.

PRIMARY FERMENTATION

Juice fermentation is typical of most modern white wine vinification. It usually occurs about 15°C; lower temperatures are considered to give a grassy odor, and higher temperatures yield wines lacking in finesse (Moulin, 1987). Bentonite and/or casein may be added to aid fermentation. Inoculation with selected yeast strains is almost universal. It helps avoid the production of perceptible amounts of sulfur dioxide, acetaldehyde, acetic acid, or other undesired volatiles by indigenous yeasts.

If the juice is too low in pH (≤3.0), malolactic deacidification is commonly encouraged. Some producers also believe that malolactic fermentation donates a subtle bouquet that they desire. It also reduces excessive acidity, permitting a greater proportion of the wine to be left dry (*brut*). As the bacterial sediment produced is difficult to remove by riddling, it is important that malolactic fermentation be complete before the second, in-bottle fermentation. Malolactic fermentation is favored by the use of minimal sulfur dioxide addition and maturation at or above 18°C.

Before preparing the blend (*cuvée*), the individual base wines are separately clarified and stabilized by cultivar, site, and vintage. Maturation may last from a few months to several years. Aging typically occurs in stainless steel but occasionally takes place in large or small oak cooperage. Certain producers in Champagne mature some of the base wines on lees under light CO_2 pressure (100 to 150 kPa) in 1.5 liter bottles (Randall, 1987).

PREPARATION OF THE *CUVÉE*

The blending of wines produced from different sites, varieties, and vintages is one of the hallmarks of the traditional method. Because single wines seldom possess all the features producers desire, samples of different base wines are blended to obtain a small number of basic blends. Blending is based solely on sensory evalution of the wines concerned. Based on the evaluations, the formula for the *cuvée* is developed. Besides improving the quality of the sparkling wine, blending helps minimize yearly variations in supply and quality. This is essential in producing the consistency required for brand name wines.

Because blending disrupts tartrate stability, the *cuvée* is typically cold stabilized to reestablish tartrate equilibrium. The *cuvée* is loose-filtered cold to remove the tartrate crystals that may form. Tight filtration is less desirable as it may remove molecules important to the formation of a stable and fine effervescence.

TIRAGE

Tirage involves adding a concentrated (50 to 65%) sucrose solution, containing other nutrients, to the *cuvée*. The solution is added just prior to yeast inoculation. The tirage may be made up in water or the *cuvée* itself. When wine is not the solvent, citric acid may be added at about 1 to 1.5% to activate sucrose hydrolysis into glucose and fructose.

Sufficient tirage solution is added to supply about 24 g sucrose per liter. On fermentation, this produces a pressure considered appropriate for most sparkling wines, namely, 600 kPa (~6 atm). Because the pressure exerted by carbon dioxide varies with temperature and other factors, the concentration of CO_2 is occasionally expressed in terms of grams of gas contained. Most sparkling wines contain about 15 g CO_2.

If the *cuvée* contains residual fermentable sugars, the amount must be subtracted from the quantity of sucrose added with the tirage. About 4.2 g of sugar is required to generate 2 g of carbon dioxide. During fermentation, the alcohol content generally rises by 1%.

Thiamine and diammonium hydrogen phosphate [$(NH_4)_2HPO_4$] often are added with the tirage to supply 0.5 and 100 mg/liter, respectively. Thiamine appears to counteract the alcohol-induced inhibition of sugar uptake by yeast cells (Bidan *et al.*, 1986). Nitrogen addition is unnecessary to yeast activity if the concentration of

assimilable nitrogen in the *cuvée* is above 15 mg/liter. Nevertheless, it may help suppress the production of hydrogen sulfide. Occasionally, trace amounts of copper salts (≤ 0.5 mg/liter) are added to further reduce hydrogen sulfide production (Berti, 1981). Some producers incorporate bentonite, gelatin, or isinglass to aid yeast flocculation at the end of fermentation. However, evidence suggesting the negative effect of several fining agents on effervescence indicates that their use may be ill-advised (Maujean *et al.*, 1990).

If the base wines have not undergone malolactic fermentation, the cuvée may be sterile filtered. Providing a sulfur dioxide content of greater than 10 mg/liter free SO_2 is effective, but less preferable.

YEASTS AND CULTURE ACCLIMATION

The second fermentation requires inoculation of the *cuvée* wine with a special yeast strain. The strains were previously classified under the designation of *Saccharomyces bayanus* but are currently considered physiological variants of *Sacch. cerevisiae*.

Because of the special and exacting conditions that prevail during the second fermentation, yeasts must be capable of commencing fermentation at alcohol contents between 8 and 12%, at temperatures of about 10°C, at pH values as low as 2.8, and with free sulfur dioxide contents up to 25 mg/liter. The suppressive influence of sulfur dioxide on fermentation rate is illustrated in Fig. 9.12. The cells also need to flocculate readily to produce a coarse sediment for efficient removal during riddling (see below). Developments in the use of encapsulated yeast may avoid both the need for flocculation and the expense of the riddling/disgorging process. The yeast also must have low tendencies to produce hydrogen sulfide, sulfur dioxide, acetaldehyde, acetic acid, and ethyl acetate. The presence of an active proteolytic ability after fermentation aids amino acid and oligopeptide release during yeast autolysis.

Because of the unfavorable conditions in the *cuvée*, the inoculum is acclimated before addition. Otherwise, most of the yeast cells die and a prolonged latency results before fermentation commences. Acclimation usually starts with inoculation of a glucose solution at about 20° to 25°C. The culture is aerated to assure adequate production of unsaturated fatty acids and sterols required for cell division and proper membrane function. Once growing actively, the culture may be added to *cuvée* wine to produce a 60:40 mixture. Over the next few days, *cuvée* wine is added to reach a 80 to 90% *cuvée* mixture. Simultaneously, the culture is cooled to the desired fermentation temperature (Markides, 1987).

The *cuvée* is inoculated with the acclimated culture to reach a concentration of about 3 to 4×10^6 cells/ml (~2–5% the *cuvée* volume). Higher inoculation levels

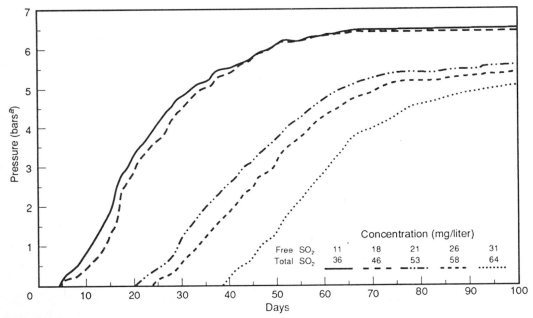

Figure 9.12 Influence of sulfur dioxide concentration on carbon dioxide production during the second fermentation in the traditional method of sparkling wine production. In-bottle fermentation at 11°C started with an initial yeast population of 1.5×10^6 cells/ml. [a]1 bar = 1 atm = 101 kPa. (After Bidan *et al.*, 1986, reproduced by permission.)

are thought to increase the likelihood of hydrogen sulfide production, while lower levels increase the risk of failed or incomplete fermentation.

SECOND FERMENTATION

Once the *cuvée* has been mixed with the tirage and yeast inoculum, the wine is bottled. Formerly, the bottle was sealed with a cork stopper, held by a reusable metal clamp called the *agrafe*. Currently, crown caps are used as they are as effective but less expensive and more easily removed than cork stoppers.

Occasionally 375 ml, 1500 ml, and larger volume bottles are used, but the 750 ml bottle is standard. Except where a brand-distinctive shape or color is used, the bottle typically has pronounced sloping shoulders and a green tint. The glass is thicker than usual to withstand the high pressures that develop during the second fermentation. Special care is taken during annealing of the glass to minimize the possibility of bottle explosion.

Filled bottles may be stacked on their sides in large, free-standing piles, in cases, or in specially designed containers ready for mechanical riddling. The wine is kept at a stable temperature, preferably between 10° and 15°C, for the second fermentation. Cooler temperatures may result in premature termination of fermentation, while warmer temperatures may result in both a rapid rise in alcohol content and a drop in redox potential. The former may prematurely terminate fermentation and the latter increase hydrogen sulfide production (Markides, 1987). A stable temperature also helps to maintain yeast viability under difficult fermentation conditions.

At 11°C, a common fermentation temperature in Champagne, the second fermentation may last about 50 days (Fig. 9.13). During the early stages of fermentation, the yeast population goes through about three to four cell divisions, reaching a final concentration of about 1 to 1.5×10^7 cells/ml. The rate of fermentation is largely dependent on the temperature, pH, and sulfur dioxide content of the *cuvée* wine.

After fermentation, the bottles may be transferred to a new site for maturation. Storage typically occurs at about 10°C. Yeast contact during maturation commonly lasts about 9 months but may continue for 3 or more years, depending on the characteristics desired.

During in-bottle maturation, the number of viable cells drops rapidly. After about 80 days, the viable yeast population drops to below 10^6 cells/ml. By disgorgement, normally 9 months to 1 year after tirage, few if any viable cells remain. Even within 6 weeks, the cells show atypical, large, expanded vesicles. By 3 months, the cells become plasmolyzed and most typical membrane-bound organelles have disappeared (Piton *et al.*, 1988). This is associated with equally marked changes in membrane lipid content. Changes in cell wall structure also occur, notably the disappearance of the innermost layer. The rapid decline in viability contrasts greatly with the slow decline and relative stabilization of the yeast population following the primary fermentation (Fig. 7.10).

The structural changes noted above are associated with major metabolic perturbations. As the wine becomes depleted in fermentable sugars, the cells begin to metabolize internal energy reserves such as glycogen.

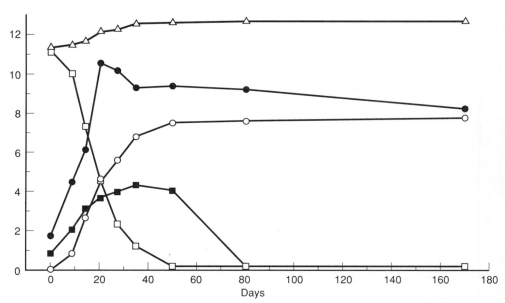

Figure 9.13   Changes during sparkling wine production. ●, Total yeast cells ($\times 10^6$); ■, viable yeast cells ($\times 10^6$); □, sugars/2 (g/liter); ○, pressure (bars); △, ethanol (%). (After Bidan *et al.*, 1986, by permission.)

When these are no longer sufficient to sustain cells at a "resting" level, the cells begin to degenerate. One of the first indicators of degeneration is disruption of membrane function and the leakage of nutrients from the cell. As the cells die (complete about 6 months after tirage), autolysis continues the release and activates cellular hydrolytic enzymes that digest structural components of the cell (Leroy *et al.*, 1990).

Yeast strain, grape variety, storage conditions, and duration of lees contact all influence the release of nitrogenous compounds from yeast cells. Yeast strains differ not only in the amount, but also in the specific amino acids released. It is generally considered that the optimal temperature for lees contact is about 10°C. At higher temperatures, the rate of nitrogen release increases, and the nature of the nitrogenous compounds liberated changes. Temperature also influences the rate and types of changes in aromatic compounds produced (see Chapter 8).

The release of amino acids and various oligopeptides during yeast autolysis has frequently been associated with the development of a "toasty" bouquet. Certain amino acids may be precursors for various aromatic compounds. For example, sotolon may be derived from threonine, ethoxy 5-butyrolactone from glutamic acid, benzaldehyde from phenylalanine, and vitispirane from methionine (see Bidan *et al.*, 1986). Nevertheless, changes in the concentrations of the compounds are not readily correlated with changes in the level of the pertinent amino acids.

Changes in the concentrations and types of fatty acids and lipids have been noted during lees contact. The level of fatty acids may increase initially but subsequently declines. Polar lipids decrease in concentration, while neutral lipids increase. These changes continue for at

least 11 years (Troton *et al.*, 1989). The triacylglycerol accumulated may act as an important precursor for aromatic compounds during aging.

Modification in the concentration of esters has been reported during maturation. Similar to the aging of still table wine, most acetate and ethyl esters of fatty acids decline, while those formed from the major organic acids increase (Silva *et al.*, 1987). Some changes in the volatile composition of champagne are shown in Fig. 9.14.

Correlations between bubble size and foam stability

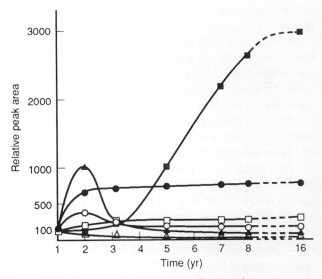

Figure 9.14   Concentration (relative peak area of traces on a gas chromatogram) of aromatic compounds during the aging of champagne. ■, Benzaldehyde; ●, unknown; □, vitispirane; ▲, nerolidol; △, hexyl acetate and isoamyl butyrate; ○, total volatile acidity. (From Loyaux *et al.*, 1981, reproduced by permission.)

and the presence of certain polysaccharides and hydrophobic proteins have been demonstrated (Maujean *et al.,* 1990). Colloidal proteins can increase two to three times in concentration within 1 year. Nevertheless, the physicochemistry of effervescence is still insufficiently understood to assess the precise role of the compounds in effervescence development.

RIDDLING

One of the most involved and expensive procedures in sparkling wine production involves the removal of yeast sediment (lees). The first step entails loosening and suspending the cells in the wine. Subsequent positioning of the bottle moves the lees toward the neck. The agitation produced is essential for optimal yeast flocculation (Stratford, 1989).

Historically, riddling was done by hand and typically took 3 to 8 weeks. It involved positioning of the bottles, neck down, in A racks (*pupitres*). Initially, the bottles were positioned at about 30° from vertical. Subsequently, the sides of the *pupitres* were moved so that by the end of riddling the bottles were about 10° to 15° from vertical. By vigorously twisting the bottle back and forth, about one-eighth of a turn, the sediment was dislodged. The bottle was then dropped into the rack, a quarter turn from its original position. This action was generally repeated two to three times, at 2-day intervals. Subsequently, rotation occurred alternately, one-eighth of a turn to the right or to the left.

Manual riddling is rapidly disappearing because of its cost, duration, and space demands. Automated mechanical riddling is less expensive, takes only about 1 week to 10 days, and requires much less space. When fermentation and storage occur in the same container as riddling, less bottle handling is required. Various systems for automated riddling are commercially available.

DISGORGING, DOSAGE, AND CORKING

After riddling, the bottles may be left neck down for several weeks in preparation for sediment removal. For disgorging, the bottles are cooled to about 7°C and the necks immersed in an ice/$CaCl_2$ or ice/glycol solution (about −20°C) to freeze the sediment. Cooling increases the solubility of carbon dioxide and reduces the likelihood of gushing on opening. Freezing the yeast plug at the neck facilitates removal of the sediment. Freezing commonly occurs while the bottles are being transported to the disgorging machine.

The disgorging machine rapidly removes the cap and permits ejection of the frozen yeast plug. The mouth of the bottle is rapidly covered in sequence by several devices. These adjust the fluid to the desired volume by either adding or removing wine. Adjustment is often

necessary, as the amount of wine lost during disgorging can vary considerably. Furthermore, most sparkling wines have a *dosage liqueur* added before corking.

The *dosage* typically consists of a concentrated sucrose solution (~60%) dissolved in high quality aged white wine. Occasionally, brandy is incorporated. A small quantity of sulfur dioxide or ascorbic acid may be added to prevent subsequent in-bottle fermentation and limit oxidation. The volume of *dosage* added depends on the sweetness desired and the sucrose concentration of the *dosage*.

A few sparkling wines receive no *dosage*. These are generally referred to as *nature*. They are rare as the *cuvée* seldom has sufficient balance to be harmonious when bone dry. *Brut* wines are adjusted with dosage to possess a final sugar content of up to 1.5%; *extra-sec* wines generally contain between 1.2 and 2% sugar; *sec* wines commonly possess between 2 and 4% sugar; *demi-sec* wines obtain between 3 and 5% sugar; and *doux* styles possess more than 5% residual sugar. The range of sugar found in each category may vary beyond that indicated above.

Once volume adjustment and dosage are complete, the bottles are sealed with special corks 31 mm in diameter and 48 mm long. They are commonly composed of agglomerate cork, to which two disks of natural cork have been glued. Once the cork is inserted, and just before addition of the wire hood, the upper 10 mm of the cork is compressed into the standard rounded shape. Once the wire hood has been fastened, the bottle is agitated to disperse the dosage throughout the wine. After cleaning, the bottle is decorated with its capsule and various labels. Special glues are commonly used to retard loosening of the label in water. The bottles are subsequently stored for about 3 months to allow the corks to "set" in the neck. Before setting, cork extraction is particularly difficult.

YEAST ENCLOSURE

The incorporation of yeasts and other microbes into a matrix of a stable gel is increasingly being used in industrial fermentations. The application of this technology to enology is comparatively recent (Fumi *et al.,* 1988). By injecting a yeast/gel mixture through fine needles into a fixing agent, small beads of encapsulated yeasts are generated (Fig. 9.15). Each bead contains several hundred yeast cells. Because of the mass of the beads, inversion of the bottles results in rapid settling of the beads to the neck, thus eliminating the need for riddling.

Wines produced and aged with encapsulated yeasts show only subtle chemical differences from traditionally produced counterparts (Hilge-Rotmann and Rehm, 1990). These differences appear not to influence the sensory properties of the wine. Because of the substantial cost saving, yeast encapsulation may become the method

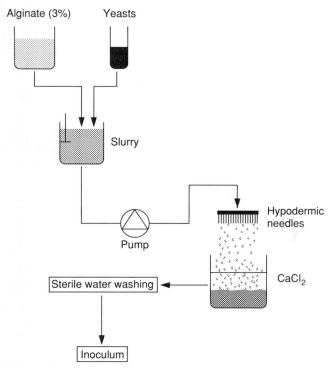

Figure 9.15 Encapsulation of yeasts in alginate. (From Fumi *et al.*, 1988, reproduced by permission.)

of choice for makers of traditionally produced sparkling wines.

An alternative technique involves the use of a Millispark cartridge. The tirage of yeasts, sugar, and nutrients are contained within a special cylinder composed of water- and gas-permeable microporous membranes. The former permits the exchange of nutrients and ethanol between the cartridge and *cuvée*, while the latter allows carbon dioxide to escape into the surrounding wine. The Millispark technique eliminates the expense involved with either manual or automatic riddling. Once the wine is to be disgorged, uncapping can occur at cellar temperature. Current tests indicate that the cartridge does not alter the traditional physical, chemical, or sensory characteristics of the wine (Lemonnier and Duteurtre, 1989).

Transfer Method

The transfer method was developed in the 1940s as a means of avoiding both the expense of manual riddling and the low quality of the wines initially produced by the bulk method. With advancements in automated riddling, and the developments in yeast encapsulation, most of the advantages of the transfer method have been negated. Furthermore, advances in the bulk method have eliminated the sources of the poor quality that initially plagued the technique. Because the transfer system is

capital intensive, but does not have the prestige and pricing advantage of the traditional method, its continued existence is in doubt. The remaining advantage of the transfer method may consist only of blending out bottle-to-bottle differences that arise in the traditional method.

Preparation of the wine up to riddling is essentially identical to that described for the traditional method. Because the wines are not riddled, fining agents need not be added to aid yeast sedimentation. Typically, the bottles are stored neck down in cartons for aging. After aging, the wine is chilled to below 0°C, before discharge into a transfer machine and passage to pressurized receiving tanks. The wine is usually sweetened and sulfited at this stage. Subsequently, the wine is clarified by filtration and decolored if necessary. The wine is typically sterile filtered just before bottling.

Bulk Method

Current forms of the bulk method are variations on the technique initially developed by Charmat about 1907 (Charmat, 1925). The procedure works well with sweet sparkling wines stressing varietal character. The most well-known examples are those produced from 'Muscat' grape varieties. The marked varietal character of 'Muscat' grapes would mask the subtle bouquet generated at considerable cost by the traditional method.

Occasionally, the bulk method is used where the traditional aged bouquet is desired. In this situation, the wine is stored on the yeasts for up to 9 months. However, as expensive pressure tanks are tied up for months, many of the economic advantages of the system are lost.

One of the features generally thought to characterize bulk-processed sparkling wine is its poorer effervescence. However, as an accurate means of assessing this property has only recently become available (Maujean *et al.*, 1988), objective proof of this assertion is presently lacking.

Fermentation of the juice for base wine production may go to dryness or be terminated prematurely. Frequently, the primary fermentation is terminated at about 6% alcohol to retain sugars for the second fermentation. Termination is either by exposure to cold, followed by yeast removal, or by yeast removal directly. Yeast removal is achieved by a combination of centrifugation and filtration, or by a series of filtrations. Once the *cuvée* has been formulated, the wines are combined with yeast additives (ammonia and vitamins) and sugar if necessary. The second fermentation takes place in reinforced stainless steel tanks, similar to those employed in the transfer process.

Typically, removal of the lees occurs at the end of fermentation. Where desired for bouquet development, the lees are intermittently stirred during the contact period. Left undisturbed, a thick layer of yeast cells would form, producing a low redox potential that could generate reduced-sulfur taints. Stirring also helps to release amino acids from yeast cells that may be involved in the evolution of a toasty aspect in the bouquet. However, stirring also releases fat particles from yeast cells that are not easily removed by filtration and may interfere with effervescence production (Schanderl, 1965).

At the end of fermentation, or lees contact, the wine is cold stabilized to precipitate tartrates. Yeast removal may be achieved by a combination of centrifugation and filtration, or by a series of filtrations. It is imperative that the operations, as in the transfer method, be conducted at an isobarometric pressure. Otherwise, carbon dioxide may be lost, or gained if the pressurizing gas is carbon dioxide. Sugar and sulfur dioxide contents are adjusted just before sterile filtration and bottling.

Occasionally, still wine may be added to the sparkling wine before final filtration and bottling. This technique may be used to produce wines of reduced carbon dioxide pressure, such as cold duck.

Other Methods

Small amounts of sparkling wine are produced by the **rural** or **natural** method. The primary fermentation is terminated prematurely by repeatedly removing the yeasts by filtration. This also removes essential nutrients from the juice, notably nitrogen. Formerly, fermentation was stopped by repeatedly skimming off the cap from the fermenting juice. Once fermentation ceased, the wine was bottled, and a second in-bottle refermentation slowly converted the residual sugars to carbon dioxide. Yeast removal, then as now, typically employs manual riddling and disgorging.

Other wines have derived their sparkle from malolactic fermentation. The primary example is, or was, *vinho verde* from northern Portugal. The grapes commonly are harvested low in sugar but high in acidity, and they produce wines correspondingly low in alcohol and high in acidity. As little sulfiting traditionally was used during winemaking, and racking occurred late, conditions favored the development of malolactic fermentation. As the wines were kept tightly bunged after fermentation, the small volume of carbon dioxide produced during the winter and spring months was trapped. The *pétillant* wine was consumed directly from the barrel. When maturation shifted to large tanks, much of the carbon dioxide liberated by malolactic fermentation escaped from the wine. This was especially marked when the wine was filtered to produce a stable, crystal-clear wine for bottling. Currently, the wine may be carbonated to reintroduce the characteristic *pétillance*. Occasionally,

still *vinho verde* wines are produced without carbonation or malolactic fermentation when the wines are low in malic acid content.

In Italy, some red wines become *pétillant* following in-bottle malolactic fermentation. Often the same wine is produced in both still and *spumante* versions.

In the former Soviet Union, sparkling wines are commonly produced in a continuous fermentation process. Though extensively used in Russia, it has been used only comparatively recently outside the former Soviet Union, for example, Portugal. Multistage bioreactor continuous fermentors also are being investigated in Japan (Ogbonna *et al.*, 1989).

Carbonation

The injection of carbon dioxide under pressure is undoubtedly the least expensive means of producing a sparkling wine. It is also the least prestigious. Correspondingly, carbonation is used only for the least expensive effervescent wines. The base wine needs to be of good quality as carbonation can accentuate faults the wine may possess.

Although carbonated wines are generally discounted as unworthy of the serious attention of connoisseurs, carbonation has the advantage of leaving the aromatic and taste characteristics of the wine unmodified. No secondary microbial activity affects the sensory properties of the wine.

Production of Rosé and Red Sparkling Wine

Although red grapes are often used in producing sparkling wines, they are customarily processed to produce white wines. Occasionally, the grapes are fermented with the skins to produce a rosé or light red wine. However, the tannins extracted along with the pigments complicates the second fermentation and accentuates gushing. Because of the tendency of such wines to gush, the bulk method of production is preferred. The base wines are almost universally encouraged to go through malolactic fermentation to give the wine a smoother taste.

Rosé sparkling wines may be produced from rosé base wines. However, rosé champagnes are commonly produced by blending small amounts of red wine into the white cuvée.

Rosé and red sparkling wines are commonly finished sweet, with low carbon dioxide pressures. They are typically either pétillant (≥ 7 g CO_2/liter) or crackling (≥ 9 g CO_2/liter). Most sparkling wines contain at least 12 g CO_2/liter. The specific carbon dioxide levels applying to each of these terms vary from jurisdiction to jurisdiction.

Effervescence and Foam Characteristics

Bubble size, foam (*mousse*) characteristics, and the degree, duration, and stability of effervescence, are important aspects in the perception of sparkling wine. Not surprisingly, the origin and factors affecting the development of effervescence and foam have come under considerable scrutiny (Jordan and Napper, 1987).

Carbon dioxide may exist in five states in water, namely, microbubbles of gas, dissolved gas, carbonic acid, carbonate ions, and bicarbonate ions. Within the normal pH range of wine, carbon dioxide exists predominantly in the form of dissolved gas.

Many factors affect the solubility of carbon dioxide in wine and, thereby, the pressure exerted by the gas contained (Lonvaud-Funeland and Matsumoto, 1979). The most significant factors are the sugar and ethanol contents of the wine and its temperature (Fig. 9.16). Increasing these factors decreases gas solubility and increases the pressure exerted. Once the bottle is opened, the ambient atmospheric pressure becomes a critical factor. The low external pressure of the surrounding air decreases solubility and promotes bubble nucleation.

On opening, a typical sparkling wine experiences a pressure drop from about 600 to 100 kPa (ambient atmospheric pressure). This decreases carbon dioxide solubility from about 14 to 2 g/liter and results in the eventual liberation of about 5 liters of carbon dioxide (from a 750 ml bottle) (Jordan and Napper, 1987). The gas usu-ally does not escape immediately, as there is insufficient free energy for bubble formation. Most of the carbon dioxide enters a *metastable* state, from which it is slowly released.

Carbon dioxide escapes from the wine by a number of mechanisms. The slowest, and least significant, is diffusion. Most of the loss results from the formation of bubbles. Bubble nucleation occurs both spontaneously and through the action of various physical forces. Spontaneous effervescence is the primary source of the continuous stream of bubbles so desired in sparkling wines. Provoked effervescence is undesirable as it enhances gushing and wine loss.

Spontaneous effervescence results from what is called **heterogeneous** nucleation. The nucleation step in bubble formation takes considerable free energy. Rough surfaces on the glass, or suspended particles in the wine, typically catalyze bubble formation. This results indirectly from entrapped microbubbles in the wine, produced when the wine is poured into a glass. Nucleation sites remain active as long as gas remains entrapped in microbubbles. Heterogeneous nucleation accounts for the slow release of about 60% of the CO_2 over a period of about 1 hr (Fig. 9.17).

Spontaneous effervescence is well known to be inhibited by the presence of even traces of detergent. Detergent appears to restrict the induction or enlargement of bubbles on nucleation sites on the glass surface.

Gushing, when a bottle is opened or the wine poured,

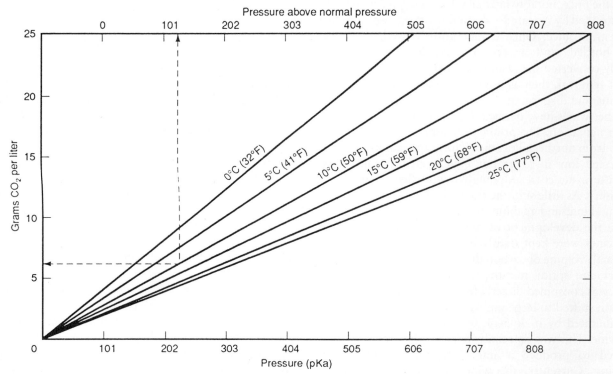

Figure 9.16 Effect of pressure and temperature on the carbon dioxide content of wine. For atm values divide by 101. (After Vogt, 1977, reproduced by permission.)

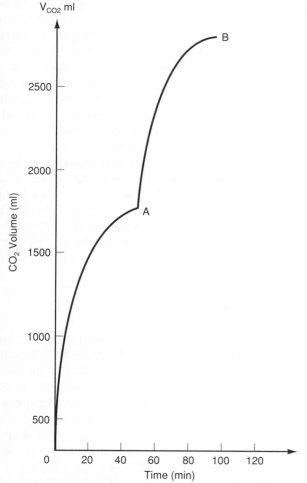

V_{CO_2} ml

Figure 9.17 Example of the slow effervescence of a sparkling wine on opening of the bottle (0–A) and after swirling agitation (A–B). (After Maujean *et al.*, 1988, reproduced by permission.)

bubbles rise to the surface. Gushing from this source takes a few seconds to develop.

Gushing also may result from **semistabilized microbubbles** formed shortly after rough handling of the wine. Shaking incorporates bubbles that act as sites for bubble growth. Gushing from semistabilized microbubbles and homogeneous nucleation is well known from its association with sport celebrations.

The formation of durable, continuous chains of small bubbles is an important feature for sparkling wines. The factors which regulate this property are still poorly understood. Cool temperatures during fermentation and aging, and long contact with the lees, are thought to favor this property. Colloidal glycoproteins released during yeast autolysis are probably particularly important in the formation of sustained chains of fine bubbles (Feuillat *et al.*, 1988; Maujean *et al.*, 1990).

Another property in the perceived quality of sparkling wines is the persistence of a small ring of bubbles (*cordon de mousse*) around the edge of the glass. In contrast to beer, the foam rapidly collapses and must be continuously replenished. The durability of the *mousse* is largely dependent on the nature of the surfactants and on the type and number of metallic ions in the wine. Gravity tends to remove fluid from between the bubbles, causing them to fuse with one another. Thinning of the fluid layer between the bubbles forces the bubbles to take on polyhedral shapes. As a result, uniformity of the pressure against the sides of the bubble is lost. This forces fluid into the angled corners of the bubbles and induces further compaction. Carbon dioxide in small bubbles increasingly comes under more pressure than in larger bubbles, promoting the diffusion of gas from smaller to larger bubbles. As the size of the remaining bubbles enlarges, they become increasingly susceptible to rupture.

The presence of large proteinaceous or polysaccharide surfactants can restrict compression of the bubbles. Interaction between surfactants may give a degree of rigidity and elasticity to the *mousse*. The latter can absorb the energy of mechanical shocks, limiting fusion and bubble rupture.

results from a number of separate nucleation processes. The mechanical shock of opening or pouring provides sufficient free energy to weaken the bonds between water and carbon dioxide. Disruption of the van der Waals forces permits carbon dioxide molecules to form nascent bubbles. The process is called **homogeneous** nucleation. If the bubbles reach a critical size, they incorporate more CO_2 than they lose. They continue to grow and begin the ascent to the surface. Because the source of free energy for homogeneous nucleation is transient, it does not induce continued effervescence.

Another potential source of gushing comes from **stabilized microbubbles.** These develop from bubbles incorporated into the wine from agitation during handling. Most of the bubbles so formed float to the surface and break. Tiny bubbles may lose carbon dioxide to the wine and dissolve. However, surfactants in the wine may coat the face of the bubble, producing a gas-impermeable membrane that stabilizes the bubble. On opening, the

Fortified Wines

Fortified wines are classified together because of their elevated alcohol content, such wines usually having had wine spirits added at some stage in production. The marked flavor of fortified wines gives the grouping an additional unifying property. Because of flavor intensity, they are seldom consumed with meals, being served instead as aperitifs or dessert wines. Regrettably, several governments also combine them for the purposes of higher taxation.

Most of the well-known fortified wines evolved comparatively recently, during the last two to three centuries. Southern Europe developed distinctive styles in most regions, namely, sherry in Spain, port in Portugal, marsala in Sicily, madeira on the islands of Madeira, and vermouth in northern Italy. The production of some of these styles is discussed below.

Sherry and Sherrylike Wines

Sherry evolved into its present form in southern Spain, possibly as late as the early 1800s. The details of its development from a young table wine transported to England in the 1600s, are unclear (Gonzalez Gordon, 1972). The solera system is thought to have evolved in the nineteenth century (Jeffs, 1982). In the present form of the system, only white wines are used.

In Spain, the designation *sherry* is used as a geographical appellation. It is restricted to wines produced in and around Jerez, Andalucia. Similar wines produced elsewhere in Europe are not permitted to use the sherry appellation. Nevertheless, similar wines may use the stylistic terms *fino, amontillado,* and *oloroso.*

Outside Europe, the designation "sherry" is used generically for wines that, to varying degrees, may resemble Spanish sherries. The name of the country or region of origin is typically appended to the term sherry. Such sherries may be produced by techniques similar to those employed in Jerez. More commonly, though, they are produced by different methods.

Because three distinctly different techniques are used worldwide, each is described separately. These include the traditional Spanish **solera** technique, the **submerged fino** procedure, and the **baked** method.

SOLERA SYSTEM

The solera system developed in Spain is a form of fractional blending. That is, young wine is sequentially added to older wine in proportion to the amount of older wine removed (Fig. 9.18). Fractional blending is conducted periodically in all stages (*criaderas*) in the system. The technique is ideally suited to the production of wine that is both brand distinctive and consistent from year to year.

The frequency and proportion of wine transferred must be adjusted to the style desired. The number of *criaderas* is equally important as each influences the development of the wine. For example, *fino* sherries require frequent transfers and many *criaderas* stages, while *oloroso* sherries develop best with infrequent transfers and few *criadera* stages.

These factors also influence the average age of the sherry produced. When a *solera* is begun, the average age

of the wine increases rapidly (Fig. 9.19). It subsequently slows and finally reaches what approximates a constant age. The constant age is reached more quickly when either, or both, the frequency or proportion of wine transferred is increased. The number of *criaderas* in a solera system also influences the rate and maximal age achieved. The greater the number of *criaderas*, the higher the average age reached, but the slower it is attained. Formulas for calculating the effects of these factors on average age are discussed in Baker *et al.* (1952).

Spanish sherry is subdivided into three major categories: **finos, amontillados,** and **olorosos.** They also may be subdivided, based on where the wines are produced and matured (e.g., Sanlúcar de Barrameda versus Jerez de la Frontera), by the sensory characteristics (e.g., *palo cortados* versus *raya olorosos*), or on how they are blended (e.g., cream-type sherries).

BASE WINE PRODUCTION

Whereas the development of sherry into a *fino* or *oloroso* once seemed arbitrary, it currently can be largely predicted and directed. Experience has shown that juice derived from grapes grown in cooler vineyards, or in cooler years, are more predisposed to become *finos*. Vineyards containing higher proportions of chalk in the soil also tend to favor *fino* development. Gentle grape pressing, and the inclusion of little press-run juice, directs development toward a *fino*. Conversely, juices derived from grapes grown under hot conditions and on soils containing less chalk, pressed in hydraulic presses, and incorporating press-run fractions generally develop into *olorosos*. Slightly higher initial phenolic contents are desired in wines designed for *oloroso* production as they favor oxidation of the wine. These tendencies can be further directed by the level of fortification employed. Levels of 15 and 18% alcohol favor *fino* and *oloroso* development, respectively. The level of cask filling and maturation temperature also influence development of the wine.

Production of the base wine generally follows standard procedures, but fermentation takes place at higher temperatures (20° to 27°C) than currently preferred elsewhere for white wines. Pressing almost immediately follows crushing to minimize tannin extraction. Tannins give a roughness inconsistent with accepted sherry norms. Because the juice often has an undesirably high pH, tartaric acid may be added to correct the deficiency. The older plastering procedure involved adding *yeso*, a crude form of gypsum (calcium sulfate). Plastering both lowered the pH and provided a source of sulfate that, on conversion to sulfite, could inhibit the growth of spoilage bacteria, notably *Lactobacillus trichodes*. Inoculation with specific yeast strains is still uncommon, with fer-

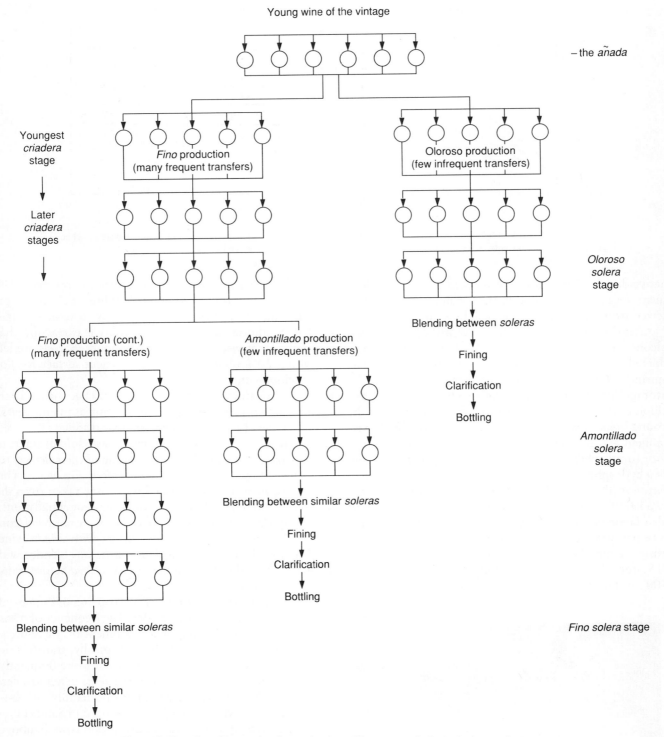

Figure 9.18 Flow diagram for the production of *fino*, *amontillado*, and *oloroso* sherries.

mentation developing spontaneously from the indigenous grape and bodega flora.

To avoid interference with the sherry flavor, the base wine should be of neutral aroma. In Spain, the neutral flavored 'Palomino' and 'Pedro Ximénez' varieties are preferred.

STYLISTIC FORMS OF JEREZ SHERRY

Finos *Finos* are the lightest, most subtly flavored sherries, and they are characterized by a *flor* bouquet. The latter develops from the action of a film of yeast cells that grow on the surface of the wine (Fig. 9.20). The film-forming yeast (*flor*) typically is the same as that

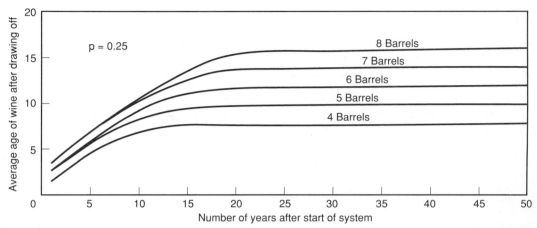

Figure 9.19 Average wine age in the oldest container of four-, five-, six-, seven-, and eight-solera systems when 25% of the wine is transferred biennially. (From Baker *et al.*, 1952, reproduced by permission.)

active during fermentation of the wine. If *flor* development does not occur rapidly, an inoculum may be transferred from casks containing an active film.

After the first racking, the wine is fortified to bring the alcohol content up to 15 to 15.5%. Fortification is conducted with a 50:50 blend of relatively rectified wine spirits (~95% ethanol) and aged sherry, called *miteado*. Storage for about 3 days permits settling of the heavy cloud that forms and avoids production of a haze in the young (*añada*) wine. At 15%, alcohol favors *flor* development and restricts the growth of acetic acid bacteria. *Flor* forms when the alcohol content favors production of a hydrophobic cell wall in *Saccharomyces cerevisiae* f. *bayanus* (Iimura *et al.*, 1980). Low pH, biotin (Iimura *et al.*, 1980), and phenolic compounds (Cantarelli, 1989) also favor film development. The hydrophobic cell surface permits the yeast cells to float on the wine, generating the filmlike growth.

Sulfur dioxide is commonly adjusted to about 100 mg/liter to limit growth of lactic acid bacteria. The

wine is then stored in American oak cooperage. The casks, called *butts*, hold about 500 liters of wine. Typically, they have been used previously to ferment wine. The conditioning minimizes oak flavor extraction that might otherwise mask the *fino* bouquet. Barrels are left with a 20% ullage to provide sufficient surface for *flor* development.

During storage, the *añada* wine is checked to determine its progress. If it is not developing as intended, it is either used in another sherry style or distilled.

When wine is removed from the last (*solera*) stage, in preparation for bottling, the volume in the *solera* is replenished from the next oldest *criadera* (Fig. 9.18). This volume is in turn replenished with wine taken from the second oldest *criadera*. This continues sequentially, until the youngest *criadera* is reached. The wine removed from the youngest *criadera* is replenished with *añada* wine. The wine drawn from each *butt* is generally blended with wine from other *butts* in the same *criadera* before transfer to the next stage.

About one-quarter of the wine (100 liters) is removed, and replenished, at each transfer. The frequency of transfer depends on the rate of development of the wine, as determined by sensory analysis. Typically, transfers occur about twice a year, but may occur more frequently. There generally are four or five *criaderas* in a fino solera system. There may, however, be considerably more, especially with Manzanilla *finos* produced in Sanlúcar.

The *butts* of a *criadera* are arrayed in rows in aboveground buildings called *bodegas*. Generally, the *criaderas* are stacked no more than three to four layers high to avoid structural damage to the *butts*. Additional *criaderas* in a solera system are housed in other regions of the *bodega*. Large firms generally have several solera systems operational at any one time for each style of sherry produced.

Currently, wine transfer from *criadera* to *criadera* is

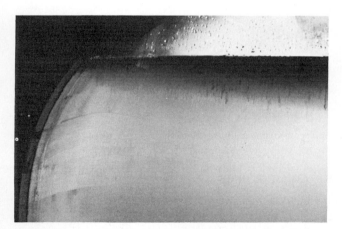

Figure 9.20 Growth of *flor* yeasts on the surface of wine in a sherry butt. (Photograph courtesy of Bodegas Pedro Domecq, S.A.)

being automated. Formerly, transfer was labor intensive, involving manual siphoning, blending, and subsequent pouring. Several ingenious devices involving perforated tubes (*rociadors*) and wedge-shaped funnels with an angled spout (*canoas*) were developed to minimize disturbance to the yeast film during filling.

Frequent wine transfer is critical to the development and maintenance of an active *flor* growth, presumably providing nutrients, such as proline for yeast growth and biotin for generation of a hydrophobic cell wall. Providing a favorable surface area to volume ratio (SA/V) is also important to *flor* activity. Leaving the butts about 20% empty creates a SA/V ratio of about 15 cm^2/liter that appears optimal (Fornachon, 1953). The practice provides both sufficient contact for yeasts with the primary energy sources (ethanol, glycerol, and acetic acid) and an adequate supply of oxygen for respiration. The bung hole is left slightly ajar to allow air exchange. In the presence of oxygen, yeast mitochondrial aldehyde dehydrogenase is produced and oxidizes ethanol to acetaldehyde (Millán and Ortega, 1988). This generates much of the substrate for function of the TCA cycle.

The taxonomic nature of the *flor* population is still unclear. This probably reflects the wide range of yeasts that can participate and inhabit the film. The dominant *flor* yeasts are strains of either *Saccharomyces cerevisiae* or *Torulaspora delbrueckii*. Currently *Sacch. bayanus*, *Sacch. beticus*, *Sacch. chevalieri*, *Sacch. italicus*, and *Sacch. prostoserdovii* are considered physiological races of *Sacch. cerevisiae*, and *Sacch. fermentati* is classified as *Torulaspora delbrueckii*. Different strains vary in higher alcohol, ester, and terpene production, but whether these differences are of sensory significance is unclear (Cabrera *et al.*, 1988). However, mixed cultures appear to form more uniform film growths than pure cultures (Criddle *et al.*, 1981).

Flor yeasts play many important roles in the development of *fino* sherries. As the film grows over the wine, yeast respiration limits diffusion of oxygen into the wine. Thus, the redox potential of the wine decreases, although the wine is seemingly exposed to air. Metabolism of ethanol, glycerol, acetic acid, and possibly other nutrients results in the liberation of acetaldehyde and other aromatics. Fermentation in the lower, submerged portion of the film probably metabolizes residual sugars in the wine.

The acetaldehyde produced by yeast respiration gives sherry its oxidized bouquet. Subsequent reaction of acetaldehyde with ethanol, glycerol, and other polyols generates acetals. Of these, only 1,1-diethoxyethane is likely to accumulate sufficiently to add a "green" note to the fragrance of the wine (see Etiévant, 1991). Small amounts of terpenes, such as linalool, *cis*- and *trans*-nerolidol, and *trans,trans*-farnesol, are synthesized by *flor* yeasts (Fagan *et al.*, 1981). Several lactones, notably substituted γ-butyrolactones, have been isolated from *fino* sherries. They are generally regarded as important in the development of a *fino* character (Kung *et al.*, 1980). The lactone sotolon may be particularly important as it possess a walnutlike fragrance. Sotolon has been isolated from *vin jaune*, a sherrylike wine produced in the south of France (Dubois *et al.*, 1976). Nevertheless, several researchers believe that the *fino* character is generated by the combined effects of many aromatics, including lactones, acetals, terpenes, and aldehydes.

In addition to the oxidative metabolism of compounds such as ethanol and acetic acid by film yeasts, volatile compounds are lost through the sides and bung hole of the cooperage as well. This has been estimated to result in the loss of up to 5% of the sherry per year (Gonzalez Gordon, 1972). The loss of water from the *butts* is partially reduced by sprinkling water on the floor of the sherry *bodega*. Evaporation of compounds from the *butts* not only results in lost volume; it also produces a concentration of other compounds in the sherry (Martínez de la Ossa *et al.*, 1987).

Although *flor* coverage is commonly complete, activity is not constant. Growth is usually most active in the spring and fall months, when the ambient temperatures in the *bodega* are between 15° and 20°C. During the winter and summer months, unfavorable temperature conditions slow growth and *flor* coverage may become patchy.

As the film growth thickens, lower sections break off and fall to the bottom. Because of yeast autolysis, sediment accumulates only slowly, and the casks seldom need cleaning. Nutrients released by autolysis are probably important to continued *flor* growth. Substances released also may be important in the development of the typical *fino* bouquet, similar to the situation with traditionally produced sparkling wines.

Wine removed from a *solera* may be blended with wines from other *soleras* in the *bodega*. The alcohol content is commonly adjusted up to 16.5% alcohol, or a level desired for the market concerned. Increasing the alcohol content stops further *flor* activity. During maturation, the malic acid level falls, which may leave the wine with insufficient acidity. If so, addition of tartaric acid is common. A polishing clarification and cold stabilization prepare the wine for bottling. *Fino* sherries are seldom blended with sweetening or *color* wines, and they are sold as dry, pale-colored, aperitif wines.

Amontillados *Amontillado* sherries begin development similar to that of fino sherry. Subsequently, the frequency of transfer is slowed, decreasing the rate of nutrient replenishment and thus slowly terminating *flor* growth. The slower rate of transfer also accentuates the loss of water through the cooperage, increasing the alcohol content and inhibiting *flor* metabolism. As an exposed wine surface is no longer required, the butts are usually filled when the decision is made to shift the development of the wine toward *amontillado* development. Without *flor* protection, the wine becomes darker colored and develops a richer oxidized flavor. There are few *criaderas* stages in *amontillado* soleras, the number depending on the flavor desired. Most *amontillado* soleras are initiated intentionally, rather than by accident as in past.

When drawn from the solera, *amontillado* sherries may be sweetened and fortified to meet particular market demands. In Spain, the wine is usually left unmodified. After cold stabilization, and a polishing clarification, the wine is ready for bottling. *Amontillado* sherries also may be used in preparing cream-type sherry blends.

Olorosos The first step in production of an *oloroso* sherry involves fortification of the *añada* wine to about 18% alcohol. This inhibits yeast and bacterial growth and makes *oloroso* maturation less sensitive to temperature fluctuation compared to other sherry types. Correspondingly, the *butts* are placed in areas of the *bodega* showing the least temperature stability. The *butts* are commonly filled to about 95% capacity, and irregular topping limits the rate and degree of oxidation. This may partially explain the minimal increase in acetaldehyde content observed during the aging of *oloroso* sherries. However, an additional factor is probably the conversion of acetaldehyde to acetic acid and subsequent esterification with ethanol to ethyl acetate. This is indicated by the progressive increase in the concentration of acetic acid and ethyl acetate as *oloroso* sherries are aged (Martínez de la Ossa *et al.*, 1987).

Because of the longer maturation in oak, the concentration of phenolic compounds is generally higher in *amontillado* and *oloroso* sherries (Estrella *et al.*, 1986). Sugar and alcohol contents have also been noted to rise during the aging of these two types of sherries.

There are typically few *criadera* stages in an *oloroso* solera. Transfer rates are slow, often amounting to only 15% per year. Because fractional blending is limited, the wine shows considerable barrel-to-barrel variation. The cellarmaster maintains brand consistency through subsequent blending.

Unblended dry *oloroso* sherries are seldom found on the international market. They are usually brought up to about 21% alcohol and blended with sweetening and *color* wines. After clarification and stabilization they are ready for bottling.

Palo cortado and *raya* sherries are special *oloroso* sherries. They are more subtle and rougher versions, respectively.

SWEETENING AND *COLOR* WINES

Sweetening is typically achieved by adding one of two special sweetening wines, *PX* or *mistela*. PX is juice extracted from sun-dried berries of the 'Pedro Ximénez' variety. The juice is fortified to about 9% alcohol, allowed to settle, and placed in special *soleras* for aging. The "wine" so produced possesses about 40% sugar. *Mistela* is produced from 'Palomino,' the variety used in sherry production. The free-run juice and first pressing are fortified to about 15% alcohol, allowed to settle, and aged in casks or tanks. *Mistela* is not fractionally blended through a solera system. It generally contains about 16% sugar.

Color wine is usually obtained from the second pressings of 'Palomino' grapes. Boiling brings the volume down to about one-fifth of the original volume. The froth that forms during boiling is periodically removed. The product, called *arrobe,* is a thick, dark, highly caramelized 70% sugar solution. Addition of *arrobe* to fermenting 'Palomino' juice successively slows the rate of fermentation after each addition, until fermentation ceases. The end product (*vino de color*) is about one part *arrobe* to two parts partially fermented juice. It possesses an alcohol strength of about 8% and contains about 22% sugar. The wine is raised to about 15% alcohol and solera aged.

EUROPEAN SHERRYLIKE WINES

The major source of sherrylike wines, other than Jerez, is Montilla-Moriles, which lies about 160 km northeast of Jerez. Its wines were once transported to Jerez for maturation and used in the production of Jerez sherry. In Montilla-Moriles, 'Pedro Ximénez' is the predominant variety grown. Grapes of this variety can, without special drying in the sun, yield wines of up to 15.5 to 16% alcohol. Thus, *flor* develops spontaneously without fortification.

Finos are produced from free-run and first press-run fractions. *Olorosos* are produced primarily from free-run juice plus additional press fractions. Fermentation traditionally occurs in large earthenware vessels (*tinajas*) possessing capacities between 6000 and 9000 liters. They resemble the storage vessels, called *pithoi*, used in ancient Greek and Roman times. The wines are solera aged in a manner analogous to that used in Jerez.

Small amounts of solera-aged sweet wine also are produced in Malaga, about 180 km east of Jerez. Most Malaga wine is produced without solera aging, while winemakers that use fractional blending employ fermentation procedures distinct from those practiced in Jerez and Montilla. Grapes of the 'Pedro Ximénez' and 'Moscatel' cultivars are placed on mats to dehydrate and overripen in the sun. Juice fortified to 7% alcohol may be added to freshly pressed juice before fermentation. The subsequent fermentation is slow and often incomplete. The resultant wine has an alcohol level of 15 to 16% and a residual sugar content of 160 to 200 g/liter. The wine may be further sweetened with *PX* and *mistela*. The wine may be colored with *sancocho*, a product similar to the *color* wine of Jerez. As *sancocho* is concentrated to only one-third of the original volume, it is lighter in color and has a lower degree of caramelization than *color* wine. Solera aging, when employed, occurs without the interaction of *flor*, in a manner similar to that of *oloroso*.

European sherrylike wines produced outside Spain include Vernaccia di Oristano and Malvasia di Bosa from Sardinia as well as *vin jaune* from France. Both Sardinian wines are *flor*-matured wines made from fully mature grapes of the 'Vernaccia' and 'Malvasia' varieties, respectively. Natural ripening on the vine commonly yields grapes with sufficient sugar concentration to produce wines having over 15% alcohol. The wines often need no fortification to favor *flor* development.

Vin jaune is usually produced from the 'Savagnin' cultivar, a mild-flavored strain of 'Traminer.' The grapes are harvested late and allowed to dry for several months to develop a high sugar concentration. The wine produced from the grapes is aged in-barrel for at least 6 years. *Flor* development occurs spontaneously, without inoculation or fortification.

NON-EUROPEAN SHERRYLIKE WINES

Solera-Aged Sherries Production of solera-aged wine, similar to that practiced in Spain, is uncommon in the New World. The expense of fractional blending and the prolonged maturation undoubtedly explain this situation. Up to ten times the volume of wine may be maturing as is sold each year.

South African sherries are produced with solera blending, but the details are quite different from those in Spain. 'Palomino' and 'Chenin blanc' ('Steen') are the varieties normally used. The juice is inoculated with selected yeast strains, chosen for their excellent fermentation and film-forming habits.

Wines designed to become *flor*-matured sherries are fortified to 15 to 15.5% alcohol. They are placed, without clarification, in 450 liter *butts* for 2 to 4 years. A 10% ullage provides surface for *flor* development. After the initial maturation, storage of both lees and wine occurs in casks containing about 1500 liters. There are generally two *criaderas* and one *solera* stage. Each stage has but one or two casks. Correspondingly, little fractional blending occurs between transfers. Wine is generally drawn off in 450 liter lots, equivalent to the contents of the *añada* barrels. Owing to proportionally higher evaporation of water through the wood, the alcohol content reaches a level that inhibits *flor* activity. The wine generated is apparently intermediate in character between a *fino* and an *amontillado*.

Wines intended for *oloroso* production are fortified to about 17% alcohol after fermentation. Subsequent storage occurs for about 10 years in butts without fractional blending. Sweetening *mistela*, derived from 'Palomino' or 'Chenin blanc' juices, also is fortified to 17% alcohol and matured for upward of 10 years in oak casks. *Color* wines are produced from *arrobe*, blended into young sherry, and stored in *butts* for prolonged periods.

In Australia, *flor* sherries are seldom fractionally blended. Typically, the wine is inoculated with a film-forming yeast. After fortification, the wine is matured for upward of 2 years in barrels (~275 liters) or cement tanks (~1000 liters). When the desired *flor* character has been obtained, the wine is fortified to 18 to 19% alcohol. Further maturation occurs in oak for 1 to 3 years.

Submerged-Culture Sherries A *flor* procedure markedly different from that used in Spain has been pioneered in Australia, California, and Canada. It involves a submerged-culture technique where the respiratory growth of the *flor* yeasts is maintained with agitation and aeration throughout the whole volume of the wine processed.

The base wine is fortified to about 15% alcohol and inoculated with an acclimated culture of *flor* yeast. Optimal conditions for growth are a pH of about 3.2, a temperature of 15°C, and an SO_2 content close to 100 mg/liter. Oxygen is provided by bubbling filtered air or oxygen through the wine. Use of porcelain sparging bulbs finely disperses the gas, improving oxygen adsorption and minimizing loss of aldehydes and other aromatics. The yeasts are kept suspended and highly dispersed by mechanical agitation.

The process has the advantage of rapidly producing high levels of acetaldehyde. By adjusting the duration of yeast action, slightly aldehydic wines (~200 mg/liter acetaldehyde) to heavily aldehydic wines (>1000 mg/liter) can be obtained.

After *flor* treatment, fortification with relatively neutral spirits raises the alcohol content to 17 to 19%. Fortification appears to intensify the *flor* character. Because the wine generally lacks the complexity and finesse of solera-aged wines, it is customarily used to enhance the complexity of baked sherries (see below), rather than used alone. The lack of finesse may be due to the absence of the reductive phase that occurs under the *flor* growth and the release of aromatic by-products during yeast autolysis.

Baked Sherries Baking has been the most popular technique for producing sherries in Canada and the United States. It involves a process that resembles more the production of madeira than Jerez sherry. Not surprising, the wines resemble madeira more than Spanish sherry.

Varieties that oxidize fairly readily are preferred in the production of baked sherries. In eastern North America the variety 'Niagara' is routinely used, while in California varieties such as 'Thompson seedless,' 'Palomino,' 'Tokay,' and 'Sultana' are commonly employed. Both white and red grape varieties may be used, as baking destroys the original color of the wine. Posson (1981) suggests that juice possessing a pH no higher than 3.4 is preferable for submerged-culture sherries, with pH values between 3.4 and 3.6 optimal for baked sherries.

Slow "baking" occurs when storage occurs in barrels exposed to the sun. More rapid baking is achieved in artificially heated rooms. Heating coils also may be inserted directly into wine storage tanks; heating is variously provided by passing steam or hot water through the coils. California winemakers appear to prefer baking at 49°C for 4 weeks, rather than the former 10 weeks of exposure at 60°C (Posson, 1981).

Heating induces the formation of a wide variety of oxidative and Maillard compounds, include furfurals, caramelization compounds, and browning by-products. Baking also promotes ethanol oxidation to acetaldehyde (Kundu *et al.*, 1979). Air or oxygen gas may be bubbled through the heated wine to accelerate oxidation.

The desired level of baking may be measured chemically, by the production of 5-(hydroxymethyl)-2-furaldehyde, or colorimetrically by the development of brown pigments. Nevertheless, the generally preferred method is by sensory perception.

After baking, especially by rapid heating, the wine requires maturation to lose some of the resulting strong flavor and rough mouth-feel. Although oak maturation is preferred, used barrels are employed to avoid giving the wine an oaky flavor. Aging may last from 6 months to more than 3 years.

Baked wines are always finished sweet. The sweetness may come from fortified grape juice added to a base wine. Alternately, premature termination of fermentation by fortification can retain residual sweetness in the base wine.

Port and Portlike Wines

The beginnings of port development, or *porto* as it is called in Portugal, are unclear. Fortification may have been used as early as 1670. Nevertheless, the practice seems not to have become standard until the mid-eighteenth century. The premature termination of fermentation by the addition of brandy is essential in modern port production. The retention of a high sugar content and the higher alcohols added during fortification give port two of its most distinguishing features. Subsequent aging and blending differentiate the various port styles.

PORTO

Porto is produced primarily from red grapes grown and fermented in the upper Douro Valley in northern Portugal. Although originating in the upper Douro, the wine is transported downriver to Oporto for maturation and aging. These processes occur in buildings called *lodges* in Vila Nova de Gala, located at the mouth of the Douro River, opposite the city of Oporto. Small amounts of white port also are produced.

Most *porto* is not vintage dated. Producers blend samples from several vintages and localities to produce the formula for brand name wines of consistent character. After 2 to 3 years maturation, most *porto* is bottled and sold as **ruby** port. Blending small quantities of white port into a ruby port may be used to produce inexpensive brands of **tawny** ports. However, only long aging in oak produces high quality tawny port. During aging, the bright red color fades to a tawny hue, and a mild, complex, oxidized character develops. Wines of superior quality from a single vintage, bottled between the second and third year of maturation, become **vintage** port. After long bottle aging, vintage port develops a distinctive and highly complex fragrance. Wine from a single vintage, aged in cooperage for about 5 years before bottling, may be designated **late-bottled vintage** (LBV) port. LBV port matures more rapidly than vintage port, is correspondingly less expensive, and generates no sediment. A few single estate (*quinta*) ports are produced, usually from a single vintage. **Vintage-character** ports often are produced from finer quality ruby ports coming from the Cima Corgo region of the Douro.

BASE WINE PRODUCTION

Port wine potentially may be produced from a wide range of grape varieties. There are 28 red and 19 white cultivars commonly grown in the Douro. Although white and red cultivars may be planted separately, varieties of each group are generally intermixed. Grapes from the cultivars grown on single site usually are harvested, crushed, and vinified together.

The major red varieties are 'Touriga Nacional,' 'Mourisco,' 'Mourisco de Semente,' 'Tinta Roriza,' 'Tinta Cão,' and 'Tinta Francisco.' They have the stable coloration, fruity aromas, and high sugar content required to produce good port. 'Codega,' 'Malvasia,' and 'Rabigato' are the preferred white varieties.

Formerly, the grapes were vinified on the premises of the vineyard in shallow stone vats called *lagars*. Currently, most of the wine is vinified by regional cooperatives using modern crushing, pressing, and fermenting equipment. Little wine is produced by the old foot-treading procedure. *Lagars* are still used, but autofermentors are now standard equipment. Inoculation of the must with a specific strain of yeast is becoming more common, except in the traditional treading process where fermentation by the indigenous grape and lagar flora is preferred.

A major problem in port production is extracting sufficient color before the must is fortified. Pigment extraction is largely dependent on the heat and ethanol generated during the short fermentation period. Because the wine is separated from the pomace when the sugar level falls to about 14.5° Brix, opportunity for pigment extraction is limited. Extensive mixing of the juice and pomace during fermentation aids pigment extraction. Autofermentors achieve this mixing automatically, while the long treading traditionally employed in lagar fermentation achieved the same end. Use of deeply pigmented varieties, such as 'Sousão' and 'Tinta Cão,' and the addition of sulfur dioxide (100 mg/liter) further help release sufficient pigmentation.

Stopping fermentation midstream (after about 24 to 48 hr) retains the wine's high acetaldehyde content present at this stage of fermentation (Fig. 7.13). This likely aids color stability by favoring the production of anthocyanin–tannin polymers. The high sugar content retained tends to mask the bitterness of tannins, but not their astringency.

When the fermenting must is run off, it is fortified with wine spirits. By the time fermentation stops, the must has dropped about 2° Brix, reflecting the combined effects of the alcohol (18%) and sugar contents (9 to 10%) of the wine. Alcohol decreases the specific gravity, while the sugar content increases it. Figure 9.21 illustrates the relationship between the initial and final °Brix of a wine fortified to 20.5% alcohol at different °Brix values during fermentation.

The actual amount of wine spirits required depends on the volume of fermenting must, its alcohol content at fortification, the alcoholic strength of the wine spirits, and the desired degree of fortification. The proportion of spirit to must can be determined from the respective alcoholic strengths (Joslyn and Amerine, 1964). The proportion of spirit required is calculated by subtracting the alcoholic strength of the must being fortified (i.e., 8%) from the desired alcoholic strength (i.e., 18%). The corresponding must proportion is calculated by subtracting the desired alcoholic strength (i.e., 18%) from that of the fortifying spirit (i.e., 78%). In this example, 10 parts of the spirit (18 − 8 = 10) would be required per 60 parts (78 − 18 = 60) of the fermenting must to achieve the desired 18% alcohol.

The first press fraction from red port is fortified to the same level as the free-run. The press fractions may be kept separate for independent aging or blended immediately with the free-run. The press fractions are an important source of anthocyanins and phenolic flavors.

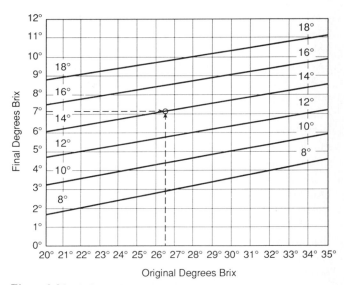

Figure 9.21 Relation between initial and final degrees Brix of musts fortified to 20.5% ethanol. For example, if the initial reading were 26.5° Brix and the desired final residual value 7.2° Brix, then fortification should occur at 14° Brix. (From Joslyn and Amerine, 1964, reproduced by permission.)

Previously, white wines were fermented on the skins in a manner similar to that for red port. However, the current trend is for a short maceration period. As with red ports, most white ports are fortified when half the original sugar content has been fermented. Semidry and dry white ports are fortified later or when fermentation is complete.

Fortification uses spirits distilled from wine produced in Portugal. Unlike most fortifying spirits, it is neither highly rectified nor concentrated. It is only about 77% alcohol and contains many flavorants, notably higher alcohols. These give the spirit, and port, a distinctive character. This feature is seldom present in non-Portuguese ports. The latter are customarily fortified with highly rectified neutral spirits at about 95% alcohol.

To assure the complete termination of fermentation, the wine is thoroughly mixed with the fortifying spirit. Storage occurs in wood or cement cooperage. The first racking usually occurs between November and March. Additional fortification at this time brings the ethanol concentration up to 19 to 20.5%.

Transportation to the lodges in Vila Nova de Gala occurs in the spring. Here, the wine receives most of its aging and blending. In contrast, most white port is aged in the upper Douro.

MATURATION AND BLENDING

Maturation occurs in large wooden or cement tanks, or in oak casks of about 525 liter capacity (*pipes*). The type and duration of aging depend largely on the style of wine desired. Racking may vary from quarterly to yearly. Slight fortification after each racking may bring the alcoholic strength up to 21%, and compensate for alcohol lost via evaporation from the cooperage.

Aging of ruby and tawny ports commonly occurs in pipes left partially empty. This enables the development of a slightly oxidized character. In contrast, vintage port is protected from oxygen exposure. Vintage port derives much of its distinctive bouquet from the long reductive in-bottle aging.

Because of the large number of small producers in the Douro, blending of individual wines usually begins shortly after transfer to Oporto. As the character of each blend becomes more evident, further blending reduces the number of blends to a more manageable figure. Blending in the first 2 years is usually confined to wine produced from a single vintage. Later, wines not used in one of the vintage-style ports may be blended fractionally with older wine. For example, in the development of ruby port, wine from *reserve* blends is commonly added to 2- or 3-year-old ruby ports. Much of the blend becomes the *reserve* for the following year. The remainder is used in the preparation of the final blend, which may contain several reserve blends, plus optional amounts of sweeter and drier wines. The final blend is left to mature in oak cooperage for several months prior to fining, stabilization, and bottling. Inexpensive tawny ports are blended similarly, but with a portion of aged white port incorporated. Quality long-aged tawny ports are produced in a manner similar to ruby port, but with extended maturation in pipes. White port is not added in the development of aged tawny ports. Tawny ports may indicate the minimum average age of the wine contained: 10, 20, 30, or 40 plus years.

Most ruby and tawny ports are tartrate stabilized by rapidly cooling and holding the wine at −10°C for about 2 weeks. The addition of Kieselsol to the cold wine before filtration helps yield a stably clear wine.

Vintage ports are not filtered before bottling. The thick sediment that forms is considered important in the development and aging potential of the wine.

SWEETENING AND BLENDING WINES

During racking, blending, and aging, the sugar level of the port may decline. To bring the sugar content back to the desired level, special sweetening wines may be employed. The main sweetening wine is called *jeropiga*. It is port wine fortified to 20% alcohol when a cap begins to form on the fermenting must. Both white (*branca*) and reddish (*loira*) jeropigas are produced. Fully red (*tinta*) jeropigas produced with the addition of elderberry juice are no longer authorized. Juice concentrated under vacuum occasionally may be used for sweetening.

In addition, special wines may be used for coloration.

The process for producing these wines is called *repisa*. After half the must is run off in the usual manner, the remaining must is treaded or extensively pumped over to extract additional color.

PORTLIKE WINES

Many countries produce wines by techniques more or less similar to those used in Portugal. In only a few instances, however, are the wines serious competitors to Portuguese porto internationally. Australian and South African ports are the primary *porto* alternatives.

In regions where intensely colored varieties are not grown, extracting sufficient pigmentation can be a serious problem. One solution is thermovinification. Various procedures have been used, including exposing the fruit to steam, plunging fruit into boiling water, or heating the juice and pomace. Exposure to steam or boiling water is common in Australia and eastern North America.

Occasionally, the must may be fermented dry before fortification. This improves pigment extraction but can lead to excessive tannin extraction. Sweetening comes from must fortified shortly after fermentation has begun, similar to *jeropiga* production. Alternately, a must concentrate may be used. Where it is important to resemble *porto*, avoiding prolonged heating during must concentration is essential. This is unimportant when "baked" port is produced.

Baking may take various forms, from storing wine on the tops of wineries to direct heating with oxygen sparging. The duration of baking is generally shorter than that used in producing baked sherries. Baking gives the wine a distinctive oxidized/caramelized bouquet.

Many cultivars are used in producing New World ports. 'Shiraz,' 'Grenache,' and 'Carignan' have often been used in Australia. 'Hermitage' ('Cinsaut') and Portuguese varieties are commonly employed in South Africa. 'Carignan,' 'Petite Sirah,' and 'Zinfandel' are typically used in the cooler regions of California, while 'Sousão,' 'Rubired,' and 'Royalty' are generally employed in hotter regions. 'Concord' is customarily used in the eastern parts of Canada and the United States.

AROMATIC CHARACTER OF PORTS

The chemical nature of the fragrance of port has received little attention. The common view is that the portlike bouquet comes from the combined effects of many compounds, not a single or a few unique substances (see Williams *et al.*, 1983). As previously noted, higher alcohols derived from fortifying spirits are important in the distinctiveness of Portuguese ports. Ports given extensive wood aging show high concentrations of diethyl and other succinate esters. These may contribute to the basic port fragrance. "Oak" lactone (β-methyl-γ-octalactone isomers) and other oxygen heterocyclic compounds have also been isolated. Some of the latter are furan derivatives, such as dihydro-2-(3H)-furanone, and may donate a sugary oxidized fragrance. Esters of 2-phenylethanol also may generate part of the fruity sweet fragrance of ports. Many acetals have been isolated from tawny ports, but their contribution to the oxidized character of the wine is unclear.

Madeira

Madeira wine evolved on the island of the same name, some 640 km off the coast of Morocco. Madeira is primarily characterized by its distinct "baked" bouquet. This is obtained from intentional heating of the wine. Subsequent maturation occurs in wooden cooperage for several years.

Heat processing of wine has not been widely adopted in other parts of the world. Outside Madeira, it is most commonly used in North America for the production of baked sherries and some ports. Not surprisingly, such wines resemble madeira more than sherries or *porto*.

Madeira wines are produced in an incredible range of styles. Some are very sweet, others almost dry. They range from versions produced from a single grape cultivar, and vintage-dated, to those that are highly blended and carry only a brand name. Some are fractionally blended, using a soleralike system, others not. Although the variations produce subtle differences in style and character, the predominate factor that distinguishes madeira from most other fortified wines is the exposure to heating, called *esteufagem*.

BASE WINE PRODUCTION

Better madeiras are produced almost exclusively from white grapes. The preferred varieties are 'Malvasia,' 'Sercial,' 'Verdelho,' and 'Bual de Madeira.' 'Listrão' and two red varieties, 'Tinta Negra Mole' and 'Negra,' are commonly used for inexpensive versions. Grapes from better sites and preferred cultivars are crushed, fermented, and stored separately to retain their distinctive characters, at least until blending.

Fermentation typically occurs in large cement fermentors, containing about 200 to 300 hl. Fermentation develops spontaneously from indigenous yeasts. The duration of fermentation depends on the style desired. Very sweet madeiras, commonly called malmsey, are fortified early to retain a high sugar content. Buals are fortified when about half the sugars have been fermented. Verdelho and especially sercial styles are fermented to or near dryness. Fermentation to dryness may take upward of 4 weeks under the cool conditions prevailing in the wineries. Regrettably, the style names, which are similar if not identical to grape varietal names, do not necessar-

ily refer to the grape variety or varieties used in the production of the wine.

Fortification presently involves the addition of neutral wine spirits (~95% alcohol). Sufficient spirit is added to raise the alcohol content to 14 to 18% alcohol. After fortification, clarification occurs with *Spanish earth*, a form of bentonite. At this point, the wine, called *vinho claro*, is ready for heat processing.

HEAT PROCESSING

Where quantities permit, wine from different varieties are separately sealed in large capacity cement tanks. Smaller lots are placed in elongated wooden casks (*charuto*) or shorter casks (*ponche*), for heating. The size and type of cooperage appear to have little influence on the quality of wine produced.

The temperature in the heating room is slowly raised over a period of about 2 weeks to a maximum of around 41°C. The wine is customarily processed at that temperature for at least 3 months. After baking, the wine is cooled slowly to ambient temperatures. Occasionally, cooling is speeded by passing cold water through the heating coils. Additional heating at a cooler temperature may take place in wooden casks positioned directly above the heating rooms.

Alternately, small lots of wine may be heated in butts stored in non-air-conditioned warehouses. Depending on positioning, the wine is variously heated or cooled for upward of 8 or more years. This old technique is called the *canterio* system.

FURTHER MATURATION

Fining removes most of the heavy brown sediment produced during heating. Use of charcoal achieves any additional decolorization deemed necessary.

Further aging occurs in wood cooperage of differing capacities. Oak is frequently used, but other woods have also been used, such as chestnut, satinwood, and mahogany. Addition of wine spirits supplies the alcohol lost during heating and raises the alcohol content to 18 to 20%.

Small lots of wine from exceptionally fine vintages may be aged in wood for at least 20 years. After a further 2 years in-bottle, the wine may be called **vintage madeira** and mention the vintage date on the label. Such wines are commonly designated as *garrafeira* (or *frasqueira*) wines. Lower quality madeiras often are aged for only 13 months before being released. Much of this apparently goes into producing madeira sauce. Better quality madeiras are matured for at least 5 years after baking.

SWEETENING AND BLENDING WINES

Juice from grapes produced on the adjacent island of Porto Santo are commonly used to produce a special sweetening wine called *surdo*. The hotter climate of Porto Santo yields grapes of higher sugar content than typical for the main island of Madeira. The juice is fortified shortly after it begins to ferment. The *surdo* is customarily heated similar to madeira wine. Occasionally, though, some may be left unheated. This leaves the *surdo* with a fresh fruity flavor useful in producing certain proprietary blends. Fortified juice, without fermentation, is called *abafado*.

Coloring wine is produced from must heat-concentrated to about one-third the original volume. It is dark colored and has a distinct caramelized fragrance.

BLENDING

Wines from different vintages and varieties usually are kept separate for at least the first 2 years of wood maturation. Subsequently, producers begin the process of blending. Further maturation and blending eventually produce the final blend. *Surdo* and coloring wines are added to madeira as required.

Suggested Readings

Sweet Table Wines

Nelson, K. E., and Amerine, M. A. (1957). The use of *Botrytis cinerea* Pers. in the production of sweet table wines. *Hilgardia* 26, 521–563.

Ribéreau-Gayon, P. (1988). *Botrytis:* Advantages and disadvantages for producing quality wines. *Proc. 2nd Int. Symp. Cool Climate Vitic. Oenol., Jan. 11–15, 1988, Auckland, N.Z.* (R. E. Smart, S. B. Thornton, S. B. Rodriguez, and J. E. Young, eds.), pp. 319–323. Soc. Vitic. Oenol., Auckland, New Zealand.

Ribéreau-Gayon, J., Ribéreau-Gayon, P., and Seguin, G. (1980). *Botrytis cinerea* in enology. *In* "The Biology of *Botrytis*" (J. R. Coley-Smith, K. Verhoeff, and W. R. Jarvis, eds.), pp. 251–274. Academic Press, London.

Carbonic Maceration

Descout, J.-J. (1986). Specificites de la vinification beaujolaise et possibilités d'évolution des productions. *Rev. Fr. Oenol.* 101, 19–26.

Flanzy, C., Flanzy, M., and Bernard, P. (1987). "La Vinification par Macération Carbonique." Institute National de la Recherche Agronomique, Paris.

Rankine, B. C., Ewart, A. J. W., and Anderson, J. K. (1985). Evaluation of a new technique for winemaking by carbonic maceration. *Aust. Grapegrower Winemaker* 256, 80–83.

Ribéreau-Gayon, J., Peynaud, E., Ribéreau-Gayon, P., and Sudraud, P. (1976). Vinification avec macération carbonique. *In* "Traité d'Oenologie: Sciences et Techniques du Vin" (J. Ribéreau-Gayon, E. Peynaud, P. Ribéreau-Gayon, and P. Sudraud, eds.), Vol. 3, pp. 289–314. Dunod, Paris.

Recioto Process

Usseglio-Tomasset, L., Bosia, P. D., Delfini, C., and Ciolfi, G. (1980). I vini Recioto e Amarone della Valpolicella. *Vini d'Italia* 22, 85–97.

Sparkling Wines

Berti, L. A. (1981). Sparkling wine production in California. *In* "Wine Production Technology in the United States" (M. A. Amerine, ed.), ACS Symp. Ser. No. 145, pp. 85–121. American Chemical Society, Washington, D.C.

Bidan, P., Feuillat, M., and Moulin, J. Ph. (1986). Vins mousseux et pétillants. Rapport de la France. *Bull. O.I.V.* 59, 565–623.

Markides, A. J. (1987). The microbiology of methode champenoise. *Proc. 6th Aust. Wine Ind. Tech. Conf.* (T. Lee, ed.), pp. 232–236. Australian Industrial Publ., Adelaide, Australia.

Moulin, J. P. (1987). Champagne: The method of production and the origin of the quality of this French wine. *Proc. 6th Aust. Wine Ind. Tech. Conf.* (T. Lee, ed.), pp. 218–223. Australian Industrial Publ., Adelaide, Australia.

Randall, W. D. (1987). Options for base wine production. *Proc. 6th Aust. Wine Ind. Tech. Conf.* (T. Lee, ed.), pp. 224—231. Australian Industrial Publ., Adelaide, Australia.

Fortified Wines

Criddle, W. J., Goswell, R. W., and Williams, M. A. (1981). The chemistry of sherry maturation. I. The establishment and operation of a laboratory-scale sherry solera. *Am. J. Enol. Vitic.* 32, 262–267.

Fornachon, J. C. M. (1953). "Studies on the Sherry Flor." Australian Wine Board, Adelaide, Australia.

Gonzalez Gordon, M. M. (1972). "Sherry. The Noble Wine." Cassell, London.

Goswell, R. W. (1986). Microbiology of fortified wines. *Dev. Food Microbiol.* 2, 1–20.

Goswell, R. W., and Kunkee, R. E. (1977). Fortified wines. *In* "Economic Microbiology. Volume I. Alcoholic Beverages" (A. H. Rose, ed.), pp. 477–535. Academic Press, New York.

Joslyn, M. A., and Amerine, M. A. (1964). "Dessert, Appetizer and Related Flavored Wines." Univ. of California Press, Berkeley.

Posson, P. (1981). Production of baked and submerged culture Sherry-type wines in California 1960–1980. *In* "Wine Production Technology in the United States" (M. A. Amerine, ed.), ACS Symp. Ser. No. 145, pp. 143–153. American Chemical Society, Washington, D.C.

References

André, P., Bénard, P., Chambroy, Y., Flanzy, C., and Jouret, C. (1967). Méthode de vinification par macération carbonique. I. Production de jus de goutte en vinification par macération carbonique. *Ann. Technol. Agric.* 16, 109–116.

Anonymous. (1979). "The Wine Industry in the Federal Republic of Germany." Evaluation and Information Service for Food, Agriculture and Forestry, Bonn, Germany.

Asvany, A. (1987). Désignation des vins à appellation d'origine de la région de Tokay-Hegyalja. *Symp. "Les Appellations d'Origine Historiques," Jerez de la Frontera, Spain, March 16–18,* pp. 187–196. Office International de la Vigne et du Vin, Paris.

Baker, G. A., Amerine, M. A., and Roessler, E. B. (1952). Theory and application of fractional-blending systems. *Hilgardia* 21, 383–409.

Barre, P. (1969). Rendement en levure des jus provenant de baies de raisins placée en atmosphère carbonique. *C. R. Acad. Agric. Fr.* 55, 1274–1277.

Bénard, P., Bourzeix, M., Buret, M., Flanzy, C., and Mourgues, J. (1971). Méthode de vinification par macération carbonique. *Ann. Technol. Agric.* 20, 199–215.

Berti, L. A. (1981). Sparkling wine production in California. *In* "Wine Production Technology in the United States" (M. A. Amerine, ed.), ACS Symp. Ser. No. 145, pp. 85–121. American Chemical Society, Washington, D.C.

Bidan, P., Feuillat, M., and Moulin, J. Ph. (1986). Vins mousseux et pétillants. Rapport de la France. *Bull. O.I.V.* 59, 565–623.

Bock, G., Benda, I., and Schreier, P. (1985). Biotransformation of linalool by *Botrytis cinerea*. *J. Food Sci.* 51, 659–662.

Bock, G., Benda, I., and Schreier, P. (1986). Metabolism of linalool by *Botrytis cinerea*. *In* "Biogeneration of Aromas" (T. H. Parliment and R. Crouteau, eds.), ACS Symp. Ser. No. 317, pp. 243–253. American Chemical Society, Washington, D.C.

Bock, G., Benda, I., and Schreier, P. (1988). Microbial transformation of geraniol and nerol by *Botrytis cinerea*. *Appl. Microbiol. Biotechnol.* 27, 351–357.

Boidron, J. N. (1978). Relation entre les substances terpéniques et la qualité du raisin (rôle du *Botrytis cinerea*). *Ann. Technol. Agric.* 27, 141–145.

Boubals, D. (1982). Progress and problems in the control of fungus diseases of grapevines in Europe. *Grape Wine Centennial Symp. Proc. 1980*, pp. 39–45. Univ. of California, Davis.

Cabrera, M. J., Moreno, J., Ortega, J. M., and Medina, M. (1988). Formation of ethanol, higher alcohols, esters, and terpenes by five yeast strains in musts from Pedro Ximénez grapes in various degrees of ripeness. *Am. J. Enol. Vitic.* 39, 283–287.

Cantarelli, C. (1989). Phenolics and yeast: Remarks concerning fermented beverages. *Yeast* 5, S53–61.

Càstino, M., and Ubigli, M. (1984). Prove di macerazione carbonica con uve Barbera. *Vini Ital.* 26, 7–23.

Charmat, P. (1925). Ten tank continuous Charmat production of champagne. *Wines Vines* 6(5), 40–41.

Charpentié, Y. (1954). Contribution à l'étude biochimique des facteurs de l'acidité des vins. Thèse Ingénieur-Docteur, Univ. of Bordeaux, Bordeaux, France.

Chauvet, S., Sudraud, P., and Jouan, T. (1986). La cryoextraction sélective des moûts. *Rev. Oenologues* 39, 17–22.

Criddle, W. J., Goswell, R. W., and Williams, M. A. (1981). The chemistry of sherry maturation. II. The establishment and operation of a laboratory-scale sherry solera. *Am. J. Enol. Vitic.* 32, 262–267.

Descout, J.-J. (1983). Particularités de la vinification en raisins entiers. Problèmes posés par la chaptilization. *Rev. Oenologues* 29, 16–19.

Descout, J.-J. (1986). Specificites de la vinification beaujolaise et possibilites d'évolution des productions. *Rev. Fr. Oenol.* 101, 19–26.

Dittrich, H. H., Sponholz, W. R., and Göbel, H. G. (1975). Vergleichende Untersuchungen von Mosten und Weinen aus gesunden und aus *Botrytis*-infizierten Traubenbeeren. *Vitis* 13, 336–347.

Doneche, B. (1991). Influence des sucres sur la laccase de *Botrytis cinerea* dans le cas de la pourriture noble du raisin. *J. Int. Sci. Vigne Vin* 25, 111–115.

Dubernet, M. (1974). Recherches sur la tyrosinase de *Vitis vinifera* et la laccase de *Botrytis cinerea:* Applications technologiques. Thèse Doctorat de 3ᵉᵐᵉ Cycle, Univ. of Bordeaux II, Bordeaux, France.

Dubois, P., Rigaud, J., and Dekimpe, J. (1976). Identification de la diméthyl-4,5-tétrahydrofuranedione-2,3 dans le vin jaune du Jura. *Lebensm. Wiss. Technol.* 9, 366–368.

Dubourdieu, D. (1981). Les polysaccharides secrétés par *Botrytis cinerea* dans la baie de raisin, leur incidence sur le métabolisme de la levure. *In* "Actualités Oenologiques et Viticoles" (P. Ribéreau-Gayon and P. Sudraud, eds.), pp. 224–230. Dunod, Paris.

Dubourdieu, D., Koh, K. H., Bertrand, A., and Ribéreau-Gayon, P. (1983). Mise en évidence d'une activité estérase chez *Botrytis cinerea*. Incidence technologique. *C.R. Acad. Sci. Paris, Ser. C* 296, 1025–1028.

Ducruet, V. (1984). Comparison of the headspace volatiles of carbonic maceration and traditional wine. *Lebensm. Wiss. Technol.* 17, 217–221.

Estrella, M. I., Hernández, T., and Olano, A. (1986). Changes in polyalcohol and phenol compound contents in the aging of Sherry wines. *Food Chem.* 20, 137–152.

Etiévant, P. X. (1991). Wine. *In* "Volatile Compounds in Foods and Beverages" (H. Maarse, ed.), pp. 483–546. Dekker, New York.

Etiévant, P. X., Issanchou, S. N., Marie, S., Ducruet, V., and Flanzy, C. (1989). Sensory impact of volatile phenols on red wine aroma: Influence of carbonic maceration and time of storage. *Sci. Aliments* 9, 19–33.

Fagan, G. L., Kepner, R. E., and Webb, A. D. (1981). Production of linalool, *cis-* and *trans-*nerolidol, and *trans,trans-*farnesol by *Saccharomyces fermentati* growing as a film on simulated wine. *Vitis* 20, 36–42.

Farkas, J. (1988). "Technology and Biochemistry of Wine," Vol. 1. Gordon & Breach, New York.

Feuillat, M., Charpentier, C., Picca, G., and Bernard, P. (1988). Production de colloïdes par les levures dans les vins mousseux élaborés selon la méthode champenoise. *Rev. Fr. Oenol.* 111, 36–45.

Flanzy, M., and André, P. (eds.) (1973). "La Vinification par Macération Carbonique." Étude No. 56. Editions S.E.I., Centre Nationale de la Recherche Agronomique, Versailles.

Flanzy, C., Flanzy, M., and Bernard, P. (1987). "La Vinification par Macération Carbonique." Institute National de la Recherche Agronomique, Paris.

Fornachon. J. C. M. (1953). "Studies on the Sherry Flor." Australian Wine Board, Adelaide, Australia.

Fregoni, M., Iacono, F., and Zamboni, M. (1986). Influence du *Botrytis cinerea* sur les caractéristiques physico-chimiques du raisin. *Bull. O.I.V.* 59, 995–1013.

Fumi, M. D., Trioli, G., Colombi, M. G., and Colagrande, O. (1988). Immobilization of *Saccharomyces cerevisiae* in calcium alginate gel and its application to bottle-fermented sparkling wine production. *Am. J. Enol. Vitic.* 39, 267–272.

Gonzalez Gordon, M. M. (1972). "Sherry: The Noble Wine." Cassell, London.

Guerzoni, M. E., Zironi, R., Flori, P., and Bisiach, M. (1979). Influence du degré d'infection par *Botrytis cinerea* sur les caractéristiques du moût et du vin. *Vitivinicoltura* 11, 9–14.

Hilge-Rotmann, B., and Rehm, H.-J. (1990). Comparison of fermentation properties and specific enzyme activities of free and calcium-alginate entrapped *Saccharomyces cerevisiae*. *Appl. Microbiol. Biotechnol.* 33, 54–58.

Iimura, J., Hara, S., and Otsuka, K. (1980). Cell surface hydrophobicity as a pellicle formation factor in film strain of *Saccharomyces*. *Agric. Biol. Chem.* 44, 1215–1222.

Isenberg, F. M. R. (1978). Controlled atmosphere storage of vegetables. *Hortic. Rev.* 1, 337–394.

Jeffs, J. (1982). "Sherry." Faber and Faber, London.

Jordan, A. D., and Napper, D. H. (1987). Some aspects of the physical chemistry of bubble and foam phenomena in sparkling wine. *Proc. 6th Aust. Wine Ind. Tech. Conf.* (T. Lee, ed.), pp. 237–246. Australian Industrial Publ., Adelaide, Australia.

Joslyn, M. A., and Amerine, M. A. (1964). "Dessert, Appetizer and Related Flavored Wines." Univ. of California Press, Berkeley.

Jouret, C., Moutounet, M., and Dubois, P. (1972). Formation du vinylbenzène lors de la fermentation alcoolique du raisin. *Ann. Technol. Agric.* 21, 69–72.

Kaminiski, E., Stawicki, S., and Wasowicz, E. (1974). Volatile flavour compounds produced by moulds of *Aspergillus, Penicillium* and Fungi Imperfecti. *Appl. Microbiol.* 27, 1001–1004.

Kovać, V. (1979). Étude de l'inactivation des oxydases du raisin par des moyens chimiques. *Bull. O.I.V.* 52, 809–826.

Kundu, B. S., Bardiya, M. C., and Tauro, P. (1979). Sun-baked sherry. *Process Biochem.* **14**(4), 14–16.

Kung, M. S., Russell, G. F., Stackler, B., and Webb, A. D. (1980). Concentration changes in some volatiles through six stages of a Spanish-style solera. *Am. J. Enol. Vitic.* **31**, 187–191.

Ladurie, E. L. (1971). "Times of Feast, Times of Famine: A History of Climate since the Year 1000." Doubleday, Garden City, New York.

Lemonnier, J., and Duteurtre, B. (1989). Un progrés important pour le champagne et les vins de "méthode traditionelle." *Rev. Fr. Oenol.* **121**, 15–26.

Leroy, M., Charpentier, M., Duteurtre, B., Feuillat, M., and Charpentier, C. (1990). Yeast autolysis during champagne aging. *Am. J. Enol. Vitic.* **41**, 21–28.

Lonvaud-Funel, A., and Matsumoto, N. (1979). Coefficient de solubilité du gas carbonique dans les vins. *Vitis* **18**, 137–147.

Loyaux, D., Roger, S., and Adda, J. (1981). The evolution of champagne volatiles during aging. *J. Sci. Food Agric.* **32**, 1254–1258.

Macheix, J. J., Sapis, J. C., and Fleuriet, A. (1991). Phenolic compounds and polyphenoloxidases in relation to browning grapes and wines. *Crit. Rev. Food Sci. Nutr.* **30**, 441–486.

Marbach, I., Harrel, E., and Mayer, A. M. (1985). Pectin, a second inducer for laccase production by *Botrytis cinerea*. *Phytochemistry* **24**, 2559–2561.

Markides, A. J. (1987). The microbiology of methode champenoise. *Proc. 6th Aust. Wine Ind. Tech. Conf.* (T. Lee, ed.), pp. 232–236. Australian Industrial Publ., Adelaide, Australia.

Martínez de la Ossa, E., Pérez, L., and Caro, I. (1987). Variations of the major volatiles through aging of sherry. *Am. J. Enol. Vitic.* **38**, 293–297.

Martinière, P. (1981). Thermovinification et vinification par macération carbonique en bordelais. *In* "Actualités Oenologiques et Viticoles" (P. Ribéreau-Gayon and P. Sudraud, eds.), pp. 303–310. Dunod, Paris.

Masuda, M., Okawa, E., Nishimura, K., and Yunome, H. (1984). Identification of 4,5-dimethyl-3-hydroxy-2(5*H*)-furanone (sotolon) and ethyl 9-hydroxynonanoate in botrytised wine and evaluation of the roles of compounds characteristic of it. *Agric. Biol. Chem.* **48**, 2707–2710.

Maujean, A., Gomerieux, T., and Garnier, J. M. (1988). Étude de la tenue et de la qualité de mousse des vins effervescents. *Bull. O.I.V.* **61**, 24–35.

Maujean, A., Poinsaut, P., Dantan, H., Brissonnet, F., and Cossiez, E. (1990). Étude de la tenue et de la qualité de mousse des vins effervescents. II. Mise au point d'une technique de mesure de la moussabilité, de la tenue et de la stabilité de la mousse des vins effervescents. *Bull. O.I.V.* **63**, 405–427.

Mauricio, J. C., Guijo, S., and Ortega, J. M. (1991). Relationship between phospholipid and sterol contents in *Saccharomyces cerevisiae* and *Torulaspora delbrueckii* and their fermentation activity in grape musts. *Am. J. Enol. Vitic.* **42**, 301–308.

Millán, C., and Ortega, J. M. (1988). Production of ethanol, acetaldehyde and acetic acid in wine by various yeast races: Role of alcohol and aldehyde dehydrogenase. *Am. J. Enol. Vitic.* **39**, 107–112.

Minárik, E., Kubalová, V., and Silhárová, Z. (1986). Further knowledge on the influences of yeast starter amount and the activator *Botrytis cinerea* on the course of fermentation of musts under unfavourable conditions. *Kvasny Prum.* **28**, 58–61 (in Polish).

Molina, I., Nicolas, M., and Crouzet, J. (1986). Grape alcohol dehydrogenase. I. Isolation and characterization. *Am. J. Enol. Vitic.* **37**, 169–173.

Montedoro, G., Fantozzi, P., and Bertuccioli, M. (1974). Essais de vinification de raisins blancs avec macération à basse température et macération carbonique. *Ann. Technol. Agric.* **23**, 75–95.

Moulin, J. P. (1987). Champagne: The method of production and the origin of the quality of this French wine. *Proc. 6th Aust. Wine Ind Tech. Conf.* (T. Lee, ed.), pp. 218–223. Australian Industrial Publ., Adelaide, Australia.

Nelson, K. E., and Amerine, M. A. (1957). The use of *Botrytis cinerea* Pers. in the production of sweet table wines. *Hilgardia* **26**, 521–563.

Nicol, M.-Z., Romieu, C., and Flanzy, C. (1988). Catabolisme de l'aspartate et du glutamate dans les baies de raisin en anaérobiose. *Sci. Aliments* **8**, 51–65.

Ogbonna, J. C., Amano, Y., Nakamura, K., Yokotsuka, K., Shimazu, Y., Watanabe, M., and Hara, S. (1989). A multistage bioreactor with replaceable bioplates for continuous wine fermentation. *Am. J. Enol. Vitic.* **40**, 292–298.

Pezet, R., Pont, V., and Hoang-Van, K. (1991). Evidence for oxidative detoxification of pterostilbene and resveratrol by laccase-like stilbene oxidase produced by *Botrytis cinerea*. *Physiol. Mol. Plant Pathol.* **39**, 441–450.

Piton, F., Charpentier, M., and Troton, D. (1988). Cell wall and lipid changes in *Saccharomyces cerevisiae* during aging of champagne wine. *Am. J. Enol. Vitic.* **39**, 221–226.

Posson, P. (1981). Production of baked and submerged culture Sherry-type wines in California 1960–1980. *In* "Wine Production Technology in the United States" (M. A. Amerine, ed.), ACS Symp. Ser. No. 145, pp. 143–153. American Chemical Society, Washington, D.C.

Randall, W. D. (1987). Options for base wine production. *Proc. 6th Aust. Wine Ind. Tech. Conf.* (T. Lee, ed.), pp. 224—231. Australian Industrial Publ., Adelaide, Australia.

Rankine, B. C., Ewart, A. J. W., and Anderson, J. K. (1985). Evaluation of a new technique for winemaking by carbonic maceration. *Aust. Grapegrower Winemaker* **256**(Apr.), 80–83.

Romieu, C., Sauvage, F. X., Robin, J. P., and Flanzy, C. (1989). Évolution de diverses activités enzymatiques au cours du métabolisme anaérobie de la baie de raisin. *Connaiss. Vigne Vin* **23**, 165–173.

Schanderl, H. (1965). Über die Entstehung voh Hefefett bei der Schaumweingärung. *Mitteil Klostern.* **15**, 1–13.

Schreier, P. (1979). Flavor composition of wines: A review. *Crit. Rev. Food Sci. Nutr.* **12**, 59–111.

Silva, A., Fumi, M. D., Montesissa, G., Colombi, M., and Colagrande, O. (1987). Incidence de la conservation en présence de levures sur la composition des vins mousseux. *Connaiss. Vigne Vin* **21**, 141–162.

Sponholz, W. R., and Dittrich, H. H. (1985). Über die Herkunft von Gluconsäure, 2- und 5-Oxo-Gluconsäure sowie Glucuron- und Galacturonsäure in Mosten und Weinen. *Vitis* **24**, 51–58.

Stella, C. (1981). Tecnologia di produzione dei vini bianchi passiti. *Quad. Vitic. Enol. Univ. Torino* **5**, 47–58.

Stratford, M. (1989). Yeast flocculation—The influence of agitation. *Yeast* **5**, S97–103.

Sullivan, C. L. (1981). Tokaj. *Vintage* **10**(3), 29–35.

Tesnière, C., Nicol, M.-Z., Romieu, C., and Flanzy, C. (1991). Effect of increasing exogenous ethanol on the anaerobic metabolism of grape berries. *Sci. Aliment* **11**, 111–124.

Troton, D., Piton, F., Charpentier, M., and Duteurtre, B. (1989). Changes of the lipid composition of *Saccharomyces cerevisiae* during aging of Champagne yeast. *Yeast* **5**, 141–143.

Usseglio-Tomasset, L., Bosia, P. D., Delfini, C., and Ciolfi, G. (1980). I vini Recioto e Amarone della Valpolicella. *Vini Ital.* **22**, 85–97.

Versini, G., and Tomasi, T. (1983). Confronto tra i componenti volatili dei vini rossi ottenuti con macerazione tradizionale e macerazione carbonica. Importanza differenziante del cinnamato di etile. *Enotecnico* **19**, 595–600.

Vogt, E. (1977). "Der Wein, seine Bereitung, Behandlung und Untersuchung," 7th Ed. Ulmer, Stuttgart.

Watanabe, M., and Shimazu, Y. (1976). Application of *Botrytis cinerea* for wine making. *J. Ferment. Technol.* **54**, 471–478.

Williams, A. A., Lewis, M. J., and May, H. V. (1983). The volatile flavor components of commercial port wines. *J. Sci. Food Agric.* **34**, 311–319.

Wucherpfennig, K. (1985). The influence of *Botrytis cinerea* on the filterability of wine. *Wynboer Tegnies* **13**, 18–19.

Yunome, H., Zenibayashi, Y., and Date, M. (1981). Characteristic components of botrytised wines—sugars, alcohols, organic acids, and other factors. *J. Ferment. Technol.* **59**, 169–175.

Yurgalevitch, C. M., and Janes, H. W. (1988). Carbon dioxide enrichment of the root zone of tomato seedlings. *J. Hortic. Sci.* **63**, 265–270.

10

Wine Laws, Authentication, and Geography

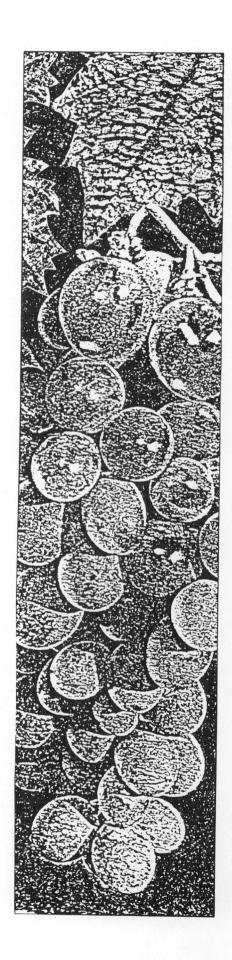

Appellation Control Laws

Basic Concepts and Significance

In modern society, governments often pass laws designed to regulate the health and safety of commercial products. Legislators also pass laws regulating the quality standard of merchandise. Wines are no exception. Possibly because of their diversity, wines are subject to more regulations than most commodities. Statutes may cover aspects ranging from how grapes are grown to when and where wine may be consumed. Because of this complexity, only aspects dealing with geographical origin and style are discussed below. These aspects are of more general interest as they affect labeling and reflect cultural differences in what are considered to be the origins of wine quality.

Appellation Control (AC) laws regulate the use of geographic names relating to wine origin. The limits of designated regions often correspond to the boundaries of specific geological and geographic features of the region. This reflects the common conviction that soil type and climate produce conditions that give wines from particular regions unique sensory characteristics. This opinion is so widely held that most countries have, or are developing, Appellation Control laws. Although the laws can be justifiably used to prevent fraudulent use of the name and reputation of a region, the marketing advantage obtained by wines produced in renowned regions appears to have been the prime factor promoting the establishment of Appellation Control laws. A corollary of the concept that geographic features establish potential wine quality is that the smaller and more unique the region, the more likely the wines are to be distinctive in sensory properties. However, this also tends to increase the scarcity and prestige of the wine from those regions, potent attributes in increasing both the price and profitability of the wines.

Although soil type is often used to define the boundaries of a wine region, there is little evidence to show that it significantly influences grape or wine characteristics (see Chapter 5). Far more significant, but less easily delimited, is the regional mesoclimate, of which soil structure (not type) is an important component.

An additional and important component of many AC laws is the regulation of grape varietal use. Such legislation is appropriate where the intention is to maintain the traditional regional style that may be crucial to commercial success. Nevertheless, strict adherence to such regulation slows and can inhibit stylistic evolution and improvement.

Most Appellation Control (AC) laws regulate viticultural and enological practice. Some, such as the total ban against irrigation, are unjustified. The standard explanation for the irrigation ban is that irrigation increases yield and thereby reduces grape quality. While the objection is valid if irrigation is used improperly, regulations on permissible grape yield in some AC regions seem so flexible as to make them appear almost meaningless. Modern developments in vine training are permitting simultaneous improvements in both yield and grape quality (see Chapter 4). The greatest damage caused by the overzealous imposition of AC laws is the potential stifling of grape grower and winemaker experimentation. In countries such as the United States, legislation is limited to aspects of geographic and varietal authenticity. As fully objective assessment of wine excellence is impossible, governments are best advised to avoid quality endorsements.

Although not the first to institute Appellation Control laws, France was the first to establish a country-wide AC system in the 1930s. Currently, all wine producing members of the European Economic Community (EEC) possess their own AC laws, within the framework of general EEC regulations. Outside Europe, South Africa has established the most comprehensive set of Appellation Control laws.

In the simplest form, Appellation Control laws apply only to geographic origin. More complex forms often dictate grape varietal use, prohibit certain vineyard practices, limit maximal yield per hectare, designate required production procedures, and specify maturation conditions. Inherent in the laws is the intent to assure stylistic authenticity and quality. This is often misinterpreted as guaranteeing wine quality; wine excellence can never be guaranteed. Quality can be easily compromised by conditions beyond the control of the producer or shipper. The best that can be done is to have representative samples assessed yearly for their current character. How specific bottles of wine may improve or deteriorate cannot be predicted with precision.

Although sensory evaluations are subjective, and therefore difficult to quantify, they have greater significance than strictly objective chemical analyses. Currently, only Germany has a system assessing all quality-designated wines before permission is granted to use a particular QbA or QmP designation. In a few countries, use of certain quality designations requires the wine of each vintage pass a sensory evaluation. Examples are the DOCG and VQA designations of Italy and Canada, respectively. French AOC wines require sensory evaluation, but not all wines are sampled yearly.

Although Appellation Control laws probably have helped to maintain or raise average quality standards in the areas concerned, the importance given ranking by wine critics suggests a lack in consumer confidence in AC "guarantees" of quality. Similarly, the popularity of vintage guides also attests to a failure of Appellation Control designations as adequate indicators of excellence.

As noted above, an aspect of AC legislation is the implication that the smaller the region, the greater the potential sensory individuality of the wine. This has led to the view that blending wines from different sites or regions is inimical to quality. The fallacy of this view is obvious from the quality of wines produced in the Douro and Champagne regions of Portugal and France, respectively. In these regions, "house" style usually replaces precise vineyard location as the major indicator of distinctiveness and quality. Skillful blending of wines from different sites often enhances the desirable features of the component wines. The value of blending is also evident in the quality of the wines from Australia, where blending wines from different vineyards and regions is common.

While Appellation Control laws promote authentic labeling, they can disadvantage wines of equal or better quality but produced in regions not possessing an established prestige. By promoting the sale of wine from well-known regions, AC laws tend to increase profitability and, thereby, the use of costly practices that may enhance quality. On the negative side, they also may promote the continued use of inefficient and outmoded technologies for the sake of tradition. As with other products, success often tends to breed success. The importance of a desirable AC designation is reflected in the land values of Burgundy vineyards (Fig. 10.1).

Appellation Control laws have tended to fix the locations in which particular grape varieties are grown. This is reflected in the highly localized distribution of many grape varieties in Europe. Regrettably, the same ordinances can retard the adoption of practices that might enhance wine excellence. Outside Europe, legislation regulating viti- and vinicultural varieties and practices is less restrictive and allows greater flexibility for experimentation and change. This is particularly useful in the New World where consumer tastes have changed dramatically since the 1960s. In addition, winemaking styles are not so inexorably linked with particular regions in the New World that the style must be maintained to retain market acceptance.

While views on what constitutes quality may fluctuate fairly quickly, legislative change seems inexorably slow. As a consequence, traditional varieties accepted at the inception of regulations may not possess the properties currently recognized as optimally desirable (Pouget, 1988), but they remain permitted or required. This has led several producers to dissociate themselves from the legal constraints and accept the penalty of using the "lowest" official wine designation. Where the producers have sufficient repute and skill, their wines often gain a prestige above those following officially prescribed procedures. This occurrence has been particularly noticeable with some of the finest winemakers in Italy. Innovation, though not always successful, is surely the route to further improvement.

Appellation and related control laws also regulate the use of terms placed on wine labels. Besides geographical origin, labels often show the vintage, cultivar(s) used, and whether the wine was estate bottled. Regrettably, consumers may assume that greater precision is indicated than warranted. For example, geographic designation often does not require that 100% of the wine come

from the named appellation. Equally, the regulations defining vintage designation, varietal composition, and estate bottling often vary from jurisdiction to jurisdiction. For example, varietal designation in the United States generally requires that more than 75% of the wine come from the named cultivar. In Oregon, however, state regulations require 90% of the wine to be from the variety designated on the label. In the eastern states, wine made from *Vitis labrusca* cultivars can be varietally labeled when possessing no more than 51% of the stated variety. Although this may be considered deception by some, the information provided can be useful to the consumer without being absolute precise.

In most jurisdictions, wines varietally designated must contain wine more than 75% derived from the named cultivar. This is usually more than sufficient to assure that the designation reflects the essential character of the named cultivar. Flexibility is useful in making adjustments to maintain or improve wine character. This is well known in Bordeaux, where wines from related varieties are often blended with 'Cabernet Sauvignon' to improve quality. Excellence should be the ultimate goal of winemaker and consumer alike. Label data are there to guide consumer purchase, not provide technical production detail.

Geographical Expression

FRANCE

As France was the first country to establish a country-wide set of Appellation Control laws, its approach has been the model for several other control legislations. Development of the laws in France took 30 years (1905 to 1935) to approach their near-modern expression. Fine-tuning and subsequent integration into EEC regulations have not modified the basic intent and viewpoint of the laws. The laws were intended to assure geographic authenticity and maintain the traditional character and repute of the wines and names of the designated region.

Part of the goal has involved intense legal efforts to inhibit the use of French geographic names on wines produced outside the officially designated regional boundaries. In this regard, France has been largely successful in all but a few countries, notably Argentina, Australia, and the United States. Restriction of European geographic names to wines produced within the officially designated region has also become part of the mandate of EEC Appellation Control regulations.

The French AC legislation established in 1935 recognized only a single category, the *Appelation d'Origine Contrôlée* designation (AOC). Subsequently, the categories *Vins Délimités de Qualité Supérieure* (VDQS), *Vins de Pays* (country wines), and *Vins de Table* (table wines) were established.

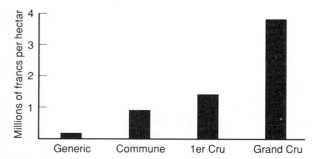

Figure 10.1 Average land values for vineyards carrying the Côte d'Or appellations for white wine. (From Moran, 1988, reproduced by permission.)

The least regulated and uniform category is the *Vins de Table* category. A similar category is found in all EEC countries. The EEC "table wine" category should not be confused with the common English distinction between table and fortified wines. The EEC designation refers to wines coming from one or several of the EEC member countries. Where the wine comes from a single country, it may originate from one or more regions not possessing, or using, an official AC designation. Only one country, or a general EEC origin, may be specified. Beyond limits on permitted grape varieties, alcoholic strength, and acidity level, there are few viti- or vinicultural restrictions on the production or characteristics of the wine.

The *Vins de Pays* category was first created in 1968. It is similar in concept to the *Vins de Table*, except for the permission to designate the regional origin of the wine. The label also may state the vintage date and varietal origin of the wine.

The VDQS category, established in 1949, is gradually being phased out in conformance with the two basic categories of wines recognized by EEC regulations—namely, Table and VQPRD (*Vins de Qualité produits dans les Régions Déterminées*). Most VDQS regions are being upgraded to the AOC category (the VQPRD category in France), with slight reduction in delimited area. The VDQS category covers "quality" wines produced in regions without a "long tradition of renown." Otherwise, production regulations are similar to those of AOC regions.

AOC designated regions are those considered to possess a long tradition of producing fine wines. To maintain the traditional character of the wines of a region, laws regulate most aspects of varietal use, grape growth, wine production, and maturation. As most wine styles evolved in specific regions, protection of the character and associated geographic name of the wine has been the cornerstone of European Appellation Control legislation.

An important notion in the development of AOC legislation has been the concept of *terroir* (local soil and climatic conditions) (Laville, 1990). Inherent in the *terroir* concept is the view that small regions are more likely to be homogeneous and produce grapes of more consistent and potentially high quality. In Bordeaux, for example, this has produced a hierarchy of appellations, progressing from a general regional classification, to subregional designations, such as Pomerol, to individual commune appellations, such as Pauillac. In Burgundy, classification may pass from general regional appellations to individual vineyard designations. Controls on grape growing, and especially winemaking, generally become more specific and limiting as the AOC designation becomes smaller.

Wines produced within AOC regions are required to undergo both chemical and sensory testing. However, the tests are designed only to remove faulty or atypical wines, not to assign a quality ranking (Marquet, 1987). Because tastings usually occur a few months after fermentation, they give little indication of the potential of the wine to develop. In addition, not all wines receive governmental sensory assessment every year.

A humorous, but illustrative, example of a hypothetical AOC designation is given in Fig. 10.2. Wine from within the basic enclosed region (solid line) would have the right to the *Celliers* (fermentation room) appellation. Wine from Château Cider, located outside the region, could at best be designated as a *Vin de Pays*. The Celliers AOC is subdivided (dashed line) into three AOCs: *Carafe* (decanter), *Charnu* (full-bodied), and *Bouchon* (cork stopper). The appellation regions are traversed by the *Rivière Levure* (yeast river). The Charnu appellation itself possesses several smaller AOC designations (dotted lines). *Appellation Cave Contrôlée*, the most central appellation, is located close to town of Cave (cellar). The highest AOC ranking is accorded the vineyard called *Chalon-Collage* (Dragnet-glue).

With few exceptions, individual vineyards are not given AOC recognition. However, vineyard ranking is very popular among wine connoisseurs. Although vineyard ranking is not a component of the Appellation Control system, France sanctions it. The best-known of these are the *cru classé* rankings of Bordeaux and Burgundy.

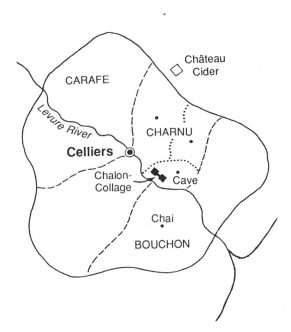

Figure 10.2 Hypothetical AOC region according to *Appellation d'Origine Contrôlée* laws in France. (From de Blij, 1983, reproduced by permission.)

GERMANY

The German Appellation Control system is conceptually different from the French counterpart. German legislation separates its geographic and quality aspects. This probably reflects the traditional interest in distinguishing *naturweins,* wines made from fully mature grapes without chaptalization.

Unlike the French system, German law designates all its major wine-producing regions. The regions are subdivided by size, but without any official implication that wines produced in smaller regions are necessarily of better or more distinctive character. The largest appellations are called *bestimmte Anbaugebiete* (designated growing regions). These are divided into one or more areas designated as *Bereich.* Further divisions are based either on group vineyard sites (*Grosslage*) or individual vineyard sites (*Einzellage*). The latter divisions are usually associated with the name of the closest village or suburb (*Orsteil*). This greatly facilitates identification of the exact location of the origin of a wine. There is no simple way of distinguishing between *grosslage* and *einzellage* sites, however. This may be intentional to avoid giving a marketing advantage to the level of official designation.

While differences in maximum allowable per-hectare yield occur between regions, the differences are particularly marked between the various *Qualitätswein* (VQPRD) levels. It is primarily this feature that differentiates the German VQPRD system from its French equivalent. Also, samples of every *Qualitätswein* must pass several tests before use of the designation is permitted. If the wine fails to reach a minimum grade, it may be declassified to a *Landwein.* This is the German equivalent to the *Vins de Pays* designation in France. Geographic origin may be no more precise than one of several specified regional names. The basic EEC table wine designations, equivalent to *Vins de Table* in France, is called *Tafelwein* in Germany.

Within the *Qualitätswein* designation, Germany possesses seven categories. The least demanding in terms of grape maturity is the QbA (*Qualitätswein eines bestimmeten Anbaugebietes*) category. It has more stringent yield/hectare limits than *Landweins* but is less restrictive than the higher categories, grouped under the QmP (*Qualitätswein mit Prädikat*) designation. The maximum yield per hectare varies depending on the region and cultivar used. In addition, QbA juice may be chaptalized, while QmP juice may not.

QmP (*Prädkat*) wines are subdivided into six categories, based largely on the degree of grape ripeness at harvest. Each designated wine region (*Anbaugebiete*) establishes for the cultivars growing in the region the official harvest commencement date. The dates vary yearly, depending on the prevailing climatic conditions. Once harvesting of the *kabinett* quality is complete, and after a specified period, permission must be requested to gather late-harvest grapes. Typically, the grapes must be inspected before crushing. Each late-harvest designation must be made from grapes possessing specific minimum levels of soluble solids (Oechsle) (Table 10.1). Thus, depending on the maturity, late-harvested grapes may produce *spätlese, auslese, beerenauslese* (BA), or *trockenbeerenauslese* (TBA) wines. *Spätlese* is the basic late-harvest category. *Auslese* wines come from individually selected grape clusters, possessing overripe fruit. *Beerenauslese* wines come from individually selected grapes that are overripe and preferably botrytized.

Table 10.1 Characteristics of Juice in the Rheingau and Mosel-Saar-Ruwer Regions of Germany for *Tafelwein, Qualitätswein,* and *Qualitätswein mit Prädikat* Wines[a]

	Starting soluble solids (Oechsle)/alcohol potential (%)						
Cultivar	Table wine	QbA wine	Kabinett wine	Spätlese wine	Auslese wine	Beerenauslese wine	Trockenbeerenauslese wine
Rheingau							
'Riesling'	44/5	60/7.5	73/9.5	85/11.4	95/13.0	125/17.7	150/21.5
Other white grapes	44/5	60/7.5	73/9.5	85/11.4	100/13.8	125/17.7	150/21.5
Weissherbst varieties	44/5	60/7.5	78/10.3	85/11.4	105/14.5	125/17.7	150/21.5
'Pinot noir'	44/5	68/8.8	80/10.6	90/12.2	105/14.5	125/17.7	150/21.5
Red cultivars	44/5	60/7.5	80/10.6	90/12.2	105/14.5	125/17.7	150/21.5
Mosel-Saar-Ruwer							
'Riesling'	44/5	57/7	70/9.1	76/10.0	83/11.1	110/15.3	150/21.5
'Elbling'	44/5	57/7	70/9.1	80/10.6	88/11.9	110/15.3	150/21.5
Other varieties	45/5	60/7.5	73/9.5	80/10.6	88/11.9	110/15.3	150/21.5

[a] From Anonymous (1979), reproduced by permission.

Trockenbeerenauslese wines come from individually selected grapes that have shriveled on the vine and are preferably botrytized. A special late-harvested category is the *eiswein*. For *eiswein* production, grapes must be picked and crushed frozen (below −6° and −7°C) and must possess Oechsle values equivalent to that of a *beerenauslese*.

Before the wine can use the designations, samples must pass both chemical and sensory evaluation tests administered by regional authorities. The sensory evaluation assesses whether the wine is typical of the cultivar indicated (if specified) and the desired QmP category. Failure to reach established norms for each category may prevent use of the cultivar name and result in demotion to a lower category, or declassification to a *Landwein*. During the 1970s, when the system was first introduced, about 1.5 and 3% of tested wines were demoted and declassified, respectively. Wines passing both the chemical and sensory analyses receive an official control number (*Amtliche Prüfungsnummer* or *A.P.Nr.*) for a specified amount of wine.

Because quality designations are associated with grape maturity, which can vary from year to year, Germany possesses no official hierarchical ranking of vineyard sites or regions. Individual producers may band together, imposing more restrictive regulations to obtain a higher market profile. The most famous of these is the VDP (*Verein Deutscher Prädikatsweingüter*), using an eagle symbol. Increased profile also may be achieved by winning quality seals at regional and national competitions.

ITALY

The Italian Appellation Control system bears greater resemblance to its French than its German equivalents. The table wine category possesses two levels, namely, the *Vino da Tavola* and newly established *Indicazione Geografica Tipica* (IGT) categories. *Vino da Tavola* refers to wines possessing only a general regional designation, wines from areas not currently possessing an official appellation, or wines made in contravention of appellation regulations. IGT wines possess the right to use specific geographic designations, mention cultivar names, and state the year of production. Because some of the most prestigious wine producers in Italy are in disagreement with the restrictions imposed by existing local AC regulations, some of the finest Italian wines carry the lowest official AC designation.

Italy possesses two categories of VQPRD wines. The first, introduced in 1966, was the DOC (*Denominazione di Origine Controllata*). It currently covers about 10 to 12% of all Italian wine production. While resembling the French AOC, the DOC differs in several respects. In the Italian system there is no specific division into smaller, more restrictive designations. However, multiple appellations possessing the same geographic name may occur, differing only in style or the variety stipulated on the label. In other designations, varietal name(s) may not be used in the appellation name. This is especially common with well-known designations, such as Soave, Barolo, and Chianti. A third appellation grouping employs traditional, but nongeographic, names as part or the whole of the designation, for example, Lacryma Christi del Vesuvio and Est! Est!! Est!!!

Since 1982, selected DOC designations have been raised to the DOCG (*Denominazione di Origine Controllata e Garantita*) designation. Wines in this category must submit to several sensory analyses at various points during production and maturation, besides the standard chemical analyses.

One of the unique features of the Italian AC system is the potential pyramidal interconnection of the categories. Wines with the DOCG category are recommended not to exceed 55 hl/ha for red and rosé wines (60 hl/ha white wines). If these values are exceeded by 20%, the wine is demoted to the next lower (DOC) category. DOC regions are recommended to not exceed 70 hl/ha for red and rosé wines (85 hl/ha for whites). Exceeding the latter limit by 20% would demote the wine to the IGT classification. Official recommendations for IGT regions are no more than 100 hl/ha for red and rosé wines (115 hl/ha for whites). Overproduction by a further 20% would declassify the wine to the *Vino de Tavola* category. The law provides only recommended maximum production levels, as these are set by committees associated with each designated AC region. Additional information on the new Italian AC system may be found in a general discussion by Fregoni (1992).

SOUTH AFRICA

The South African Appellation Control system reflects many aspects found in European legislation but has incorporated distinctive aspects. The system recognizes five levels of geographic designation. These progress from Regional Viticultural Designations to Areas, Districts, Wards (vineyard groupings), and finally Estates (individual vineyards). Most vineyards have not been granted officially designated status.

Wines produced within the appellations may be submitted for official evaluation and certification. Passing the certification tests permits the producer to apply an official neck seal from the Wine and Spirit Board. The neck seal possesses a certification number and bands that note the vintage, the varietal origin, and whether the wine is estate bottled. Furthermore, if the wine achieves an especially high ranking in the sensory evaluation test, the rarely used designation "Superior" appears on the seal.

UNITED STATES

Since 1978, the BATF (Federal Bureau of Alcohol, Tobacco, and Firearms) has been empowered to designate *viticultural area* appellations in the United States. There had been previous regulations affecting the use of geographic names, but they had involved only country, state, or county designations. Viticultural areas need not reflect existing state or county boundaries.

For designation of a viticultural area, a proposal must be submitted to the BATF. Various types of supporting documentation must be supplied with the proposal. These include data establishing the local or national recognition of the named area, historical or current evidence for the proposed boundaries, climatological, geologic, and topographic features distinguishing the region from surrounding regions, and precise boundaries noted on topographic maps from the U.S. Geological Survey. Geographic size is not a critical factor. Existing viticultural areas may have smaller appellations designated within their borders, and larger appellations may be formed out of part or all of several existing appellations. There is no intent that size should imply anything about potential wine quality.

Designation of a viticultural area does not impose any special regulations on cultivar use, viticultural practices, or winemaking procedures. Thus, the American expression is intended only to regulate the authenticity and distinctiveness of the origin of the wine.

While foreign geographic names such as Chablis and Burgundy still appear on some American wines, producers of premium wines scrupulously avoid their use. Varietal name and vineyard location have proved a more effective means of obtaining recognition and market position. The primary exceptions are the appellations sherry, port, and champagne. For sherry and port, the terms have been used as both generic and regional designations. Although English possesses a legitimate generic term for champagnelike wines—sparkling—the latter has not achieved the prestige or market recognition it deserves, nor will it if producers of sparkling wines continue to promote the champagne appellation on their labels. Inability to use an established geographic wine designation need not limit marketability as some fear. Absence of the *sherry* designation on similar wines, such as Montilla or Château-Chalon, seems not to have limited their marketability.

CANADA

Currently, only one province in Canada possesses a set of Appellation Control laws, namely, British Columbia. While the other major wine-producing province, Ontario, possesses a similar set of regulations, they are voluntary and administered by an independent body, the Vint-ners Quality Alliance (VQA). In its administration of the regulations, the Alliance is aided by provincial government agencies.

The systems in both provinces have regulations affecting the cultivars permitted, and they require wines to pass minimum sensory analysis standards to earn the right to use the VQA designation. While both provinces have designated several viticultural areas, there is no stated or implied superiority to smaller appellations. The same is true for single vineyard and estate bottling. Nevertheless, more stringent regulations affecting cultivar use apply to the regional than province-wide appellations in Ontario.

AUSTRALIA

Although Australia does not possess, and appears generally opposed to, an Appellation Control concept, the Hunter Valley region of New South Wales has a voluntary accreditation system. It is unique in its simplicity and degree of governmental support. Wines may be awarded one of two seals. These distinguish between wines of superior quality on the basis of current maturity (Classic) and potential to improve (Benchmark). Currently, each year's winners are presented in a public ceremony and tasting at the Parliament House in Sydney. Open endorsement by the state government is credited with improving the profile of quality wines in the region.

Detection of Wine Misrepresentation and Adulteration

In the previous section, laws designed to assure authenticity of wine origin were discussed. The effectiveness of the regulations depends largely on the willingness and ability of enforcing agencies to assess compliance. Enforcement of the laws is a political/economic decision beyond the scope of this text. However, the technical ability to assess compliance is within the realm of science and is an appropriate topic for discussion.

Establishing conformity with many of the viticultural constraints of AC legislation is usually relatively simple. Features such as use of irrigation, cultivars grown, training systems employed, or percentage of land planted can be directly assessed by on-site inspection. Yield per hectare, total soluble solids, and harvest date often can be checked against cellar records, simple chemical analyses, or corroborative testimony. Until recently, however, validation of the geographic and varietal nature of wine located in a winery, or the degree of chaptalization and fortification, was often impossible. Sensory analysis is insufficiently precise and cannot be automated.

Detection of contraventions of several AC regulations has depended on the development of modern analytical instrumentation, including scintillation counters, mass spectrometry, nuclear magnetic resonance (NMR), and the latest forms of liquid and gas chromatography. Detection of more serious, and toxic, adulteration of wine requires even more extensive use of analytical instrumentation. Because of the incredible range of adulteration possible, it is usually necessary to suspect the type of duplicity involved.

Although most modern wine falsification is nontoxic, it can do irreparable harm to the wine sales of a producer or region. The image of wine as a natural and wholesome beverage is important in marketing and pricing. Geographic and varietal identity is also important, and is often considered synonymous with quality by many consumers. For premium wines, the prestige associated with the appellation of the wine can be as or more important than the sensory quality of the wine contained. Thus, the more famous the region, the more critical it becomes to assure consumers of the authenticity of the origin of the wine.

Besides marketing concerns related to fraudulent misrepresentation, circumvention of AC regulations places conscientious producers at an economic disadvantage. Governments are also quite concerned as they have a financial interest in avoiding the loss of tax revenue associated with wine scandals.

The methods used to detect violation of Appellation Control statutes depend on the doctoring suspected. Because the instrumentation is expensive and requires considerable skill in use and interpretation, the methods cannot be used routinely. Current use is justified only when falsification is suspected. Nevertheless, the ability of current instrumentation in detecting fraudulent activity is impressive.

Validation of Geographic Origin

The major discriminating procedures are based on differences in the isotopes of hydrogen (H), carbon (C), and oxygen (O), the major elements in wine. In addition, differences in the local distributions of microelements (Frank and Kowalski, 1984) and distinctive phenol and aromatic characteristics of grape varieties (Etiévant *et al.*, 1988; Rapp *et al.*, 1993) have potential in detecting noncompliance with Appellation Control regulations.

Isotopes of an element possess similar chemical properties but differ in mass. The mass differences come from the number of neutrons present in the nucleus of the atom. For example, the most common isotope of hydrogen possesses a single neutron (^{1}H), while deuterium (^{2}H) contains two and tritium (^{3}H) three neutrons. Although seven carbon isotopes are known, only three are sufficiently common to be of practical use. The most common carbon isotope, representing about 99% of the

total, possesses 12 neutrons (^{12}C). Carbon-13 (^{13}C) contains 13 neutrons, and carbon-14 (^{14}C) possesses 14 neutrons. The two important oxygen isotopes possess 16 and 18 neutrons, respectively (^{16}O and ^{18}O). While most of the isotopes are stable, some such as carbon-14 are unstable and release radioactive particles on decay.

The value of isotopes in wine authentication comes from the influence of environmental and biotic factors on the distribution and relative proportions of the isotopes. Enzymes and biological processes such as transpiration often discriminate between isotopes, differentially using or favoring the lightest isotope. In addition, enzymes in different organisms, or diverse enzymes in the same organism, may show characteristic and divergent levels of preferential isotope use. Thus, the proportions of isotopes can differ from species to species, variety to variety, region to region, year to year, and with irrigation. Differences in specific isotopic proportion are measured relative to the deviation (δD $^0/_{00}$) from an international standard. (Sample ratio minus the standard ratio, divided by the standard, multiplied by 1000).

Because climatic and enzymatic factors differentiate between isotopes, distinctive and stable variations between wines produced in different locations may be precisely and reproducibly detected. For example, regional precipitation and temperature differences affect the rate of evapotranspiration, and thus the ^{2}H/^{1}H ratio of the water and ethanol of wines produced at particular sites (Martin, 1988). Local environmental and biotic factors also can generate distinctive deviations in the ^{18}O/^{16}O ratio of water and the ^{13}C/^{12}C ratio of the organic constituents of grapes (Fig. 10.3). These differences may be

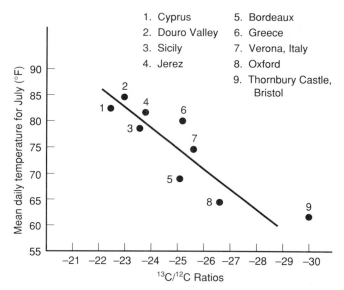

Figure 10.3 Relationship between ^{13}C/^{12}C ratios in grapes and temperature for selected growing areas in 1980. (From Stacey, 1984, reproduced by permission.)

sufficient to allow tentative identification of the country (Fig. 10.4A), region (Fig. 10.4B), and vintage (Fig. 10.4C) of the wines. They also may permit the identification and determine the proportion of unauthorized juice or wine added to a particular wine.

Not only is the isotopic signature of a wine influenced by the regional climate, it also is affected by the grape cultivar and the yeasts and bacteria involved in fermentation. These factors can either enhance or diminish the distinctive isotopic ratio produced by local weather conditions. Even juice concentration is likely to affect the isotopic ratio. Thus, precise information on standard wine production procedures in the region is required when assessing fraudulent adulteration. Correspondingly, authentic examples of the wine in question are

needed to recognize the effects of viti- and vinicultural practices on isotopic ratios.

Under most circumstances, the ^{14}C content of a wine provides little useful information concerning its origin. The rate of carbon-14 decay only permits precise measurement of units of time greater than 80 years. However, the atmospheric testing of nuclear weapons during the late 1950s and early 1960s raised the normally constant level of carbon-14 in the atmosphere. This marked, but temporary, atmospheric perturbation is reflected in the ^{14}C level of wines produced during that period and subsequently (Fig. 10.5). Thus, from 1956 to the present, the ^{14}C level of a wine permits one or more potential vintage dates to be proposed. Precise vintage pinpointing requires additional information as any measurement may indicate two or occasionally three possible vintage dates. For example, a measurement of 17.5 cpm (counts per minute) could be generated from wine produced in 1959, 1963, or 1979 (Fig. 10.5).

Validation of Conformity to Wine Production Regulations

Although radiocarbon (^{14}C) dating has limited value in authenticating the origin of wine, it can be very useful in detecting the presence of "synthetic" additives. Synthetic organic compounds often are derived from fossil fuels and may either be "nature identical" or "artificial." The former are chemically and isotopically identical to naturally occurring compounds. Artificial compounds may be either isomers of natural compounds or unrelated compounds possessing sensory properties similar to natural compounds, for example, the sugar substitute

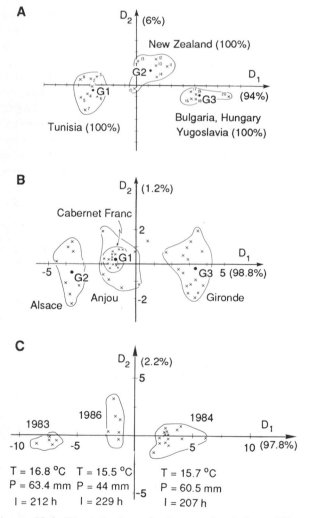

Figure 10.4 Discrimination of wines produced from different (A) countries, (B) geoclimatic regions, and (C) 'Cabernet Sauvignon' vintages from Saumur-Champigny. T, Temperature; P, precipitation; I, incident radiation for C. (After Martin, 1988, reproduced by permission.)

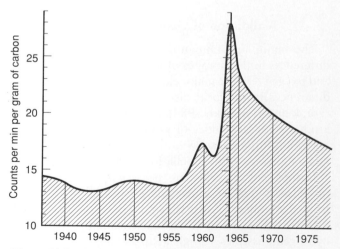

Figure 10.5 Evolution of the ^{14}C radioactivity of alcohols in Bordeaux wine from 1935 to 1979. (After Martinière, 1981, reproduced by permission.)

saccharine. While the additives may not be dangerous to human health, their addition to wine is typically illegal. Synthetic compounds usually can be recognized as such as they are generally derived from fossil fuels. Because of their "age," they possess no detectable ^{14}C. Organic compounds lose detectable carbon radioactivity after about 50,000 years. Thus, a port fortified with synthetic alcohol, rather than the required grape spirits, could be recognized by appearing to be several thousand years old according to ^{14}C measurements. The synthetic alcohol depresses the average ^{14}C age of the ethanol content of the wine.

Until recently, ^{14}C dating could detect only the addition of major organic wine constituents. However, the combination of liquid/liquid extraction and gas chromatography, with improvements in ^{14}C assessment, can permit differentiation between even milligram samples of natural versus synthetic compounds (McCallum *et al.*, 1986). Use of carbonation versus secondary yeast fermentation can be detected with ^{14}C analysis if the carbon dioxide is derived from a fossil fuel source.

In most countries, the addition of water to wine is illegal. However, assessment of compliance has been fraught with difficulties (Ribéreau-Gayon *et al.*, 1982). Of older techniques, depression of the relatively stable magnesium content of wine has been the most accepted indication of watering (Robertson and Rush, 1979). However, this difference could be corrected by adding the requisite amount of magnesium with the dilution water.

Currently, modifications in the $^2H/^1H$ and $^{18}O/^{16}O$ ratios of water are being investigated as more accurate indicators of dilution. Even the groundwater from the vineyard location usually has a detectably different isotopic balance compared to the water content of the wine. Water absorbed from the soil is exposed to selective transpirational loss, use in photosynthesis, and involvement in cellular hydrolytic reactions.

In many countries, addition of sugar to must or wine is forbidden. In others (e.g., Germany), its use in chaptalization is regulated but permissible only in the lower wine categories. Sugar is added either as sucrose or invert sugar (enzymatically or acid hydrolyzed sucrose). In either case, the sugar is quickly assimilated and converted to ethanol. Thus, detection of chaptalization requires a means of distinguishing between grape- and nongrape-derived ethanol, which is most easily accomplished by measuring the $^{13}C/^{12}C$ ratio of the alcohol in the wine. Ethanol derived from different sugar sources can be differentiated. Addition of any readily available sugar would shift the $^{13}C/^{12}C$ ratio of the ethanol produced during fermentation. The level of discrimination would depend on the amount of sugar added and the source of

the nongrape sugar. Detection of levels of chaptalization down to 0.3% is possible (Martin, 1988).

An additional indicator of chaptalization with invert sugar is 2-hydroacetylfuran (Rapp *et al.*, 1983). The compound appears not to occur in either must or wine without the addition of invert sugar (acid hydrolyzed cane sugar).

The $^{13}C/^{12}C$ ratio also can be used to detect the addition of other plant- and microbial-derived constituents, such as glycerol and acids. Each compound possesses a $^{13}C/^{12}C$ ratio characteristic of its biological origin. While the differences are most marked between plants possessing distinctive photosynthetic processes (C_3, C_4, and CAM pathways), differences also occur between different plant species and the same species in different climatic regions.

Adulteration with most nongrape pigments, such as cochineal, poke, or orseille, can be readily detected by the distinctive chemical properties of the pigments. Even the addition of oenocyamine, a color extract from grapes, can be detected. High-performance liquid chromatography permits the separation and quantitative measurement of the many anthocyanins present in red wine. A pattern atypical for the cultivar and year would indicate the likelihood of color addition or the incorporation of must/wine from unauthorized grape varieties. Chromatography has been used for years in Europe to verify the absence of must or wine from non-*V. vinifera* cultivars in Appellation Control wines. Anthocyanins from other *Vitis* species or from interspecific hybrids with *V. vinifera* typically possess diglucoside anthocyanins not found in *V. vinifera* grapes. However, some complex *V. vinifera* interspecific hybrids do not synthesize diglucoside anthocyanins (van Buren *et al.*, 1970; Guzun *et al.*, 1990). In addition, the absence of particular anthocyanins may be used to partially verify the varietal origin of a wine. For example, 'Pinot noir' grapes do not possess acylated anthocyanins (Rankine *et al.*, 1958).

Other harmless but unauthorized adulterants relatively easily detected in wine are sorbitol and apple juice (Burda and Collins, 1991). Sorbitol may be added with the intention of giving the wine a smoother mouth-feel. Apple juice may be added as an inexpensive grape juice extender before fermentation. Sorbitol rarely occurs naturally in wine at concentrations greater than 1 g/liter. Apple juice addition is likely if the wine possesses higher than usual amounts of sorbitol and chlorogenic acid. Wine usually possesses less than 2 mg/liter chlorogenic acid.

Although modern instrumental analysis will not stop wine adulteration, it can make detection more certain and definitive. By making avoidance of detection so complex and costly, such analysis may make fraudulent adulteration financially unremunerative.

World Wine Regions

In the remainder of the chapter, the distinctive climatic, soil, varietal, viticultural, and enological features of the various wine regions of the world are highlighted. Western Europe constitutes the most significant region, both in terms of vineyard area and quantity of production. Correspondingly, more detail is available about European wine regions than other viticultural areas. Nevertheless, where possible, other regions are covered similarly.

A map indicating some of the major European wine regions is given in Fig. 10.6. Grape hectarage and wine production data for several European and New World viticultural regions are listed in Table 10.2.

Western Europe

The importance of Western Europe in current wine production is a reflection of both the physical and social climate. Richness in natural resources combined with historical and geographic factors in Western Europe to favor industrialization and the accumulation of free capital, conditions conducive to developing an expanding and increasingly discriminating class of wine connoisseurs. The latter became more important as transport improved and vineyard proximity to the consuming public became less significant. Increased profit both permitted and spurred the use of more costly and complex practices that could enhance grape and wine quality. The improvements complemented the inherent climatic advantages of the respective regions. Technical developments also permitted the potential preservation and development of the finer characteristics of the wine. Consequently, distinctive regional features become more pronounced, as a social class willing and desirous to appreciate these aspects grew. In contrast, repeated invasions, limited natural resources and the imposition of Islamic injunctions against wine consumption largely prevented an equivalent development from occurring in Eastern Europe for the last 500 years.

Although Western Europe is not inherently more suited to grape growing and winemaking than some other European regions, or parts of the world, it has evolved wines well suited to highlight the qualities of indigenous cultivars and take positive advantage of local

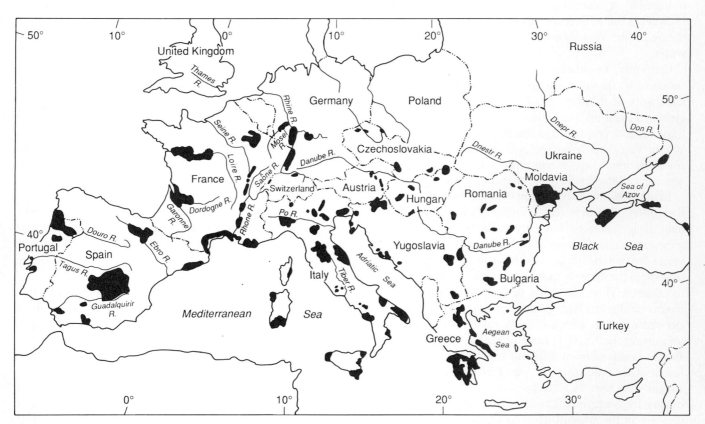

Figure 10.6 Map showing the approximate location of some of the wine regions of European countries.

Table 10.2 Comparison of Viticultural Surface Areas and Wine Production[a]

Country	Wine production (10³ hl)	Vineyard area (10³ ha)	Total land area (10⁶ ha)	Wine yield per total land area (hl/10³ ha)	Vineyard area per land area (ha/10³ ha)	Wine yield per vineyard hectarage (hl/ha)
France	65,529	940	55	1197	17.2	69.7
Italy	59,000	1050	30	1958	34.8	56.3
Spain	40,377	1473	51	800	29.2	27.4
Argentina	20,250	267	278	73	1.0	75.8
Soviet Union	17,984	1077	2228	8	0.5	16.8
United States	15,998	322	953	17	0.3	49.7
Portugal	11,372	379	9	1236	41.2	30.0
South Africa	9988	106	112	80	0.9	84.8
Germany	8514	105	35	238	2.9	81.1
Romania	5900	245	24	248	10.3	24.1
Hungary	5472	138	9	588	14.8	40.0
Yugoslavia	5170	225	26	202	23.0	23.0
Australia	4446	59	768	6	0.08	75.4
Chile	3978	120	76	53	1.6	32.2
Greece	3525	150	13	267	11.4	23.4
Austria	3166	58	8	377	6.9	54.6
Brazil	3108	57	821	4	0.07	54.5
Bulgaria	2925	140	11	263	12.6	20.9
Czechoslovakia	1550	47	13	121	3.7	33.0
Switzerland	1313	15	4	318	3.6	87.6
New Zealand	490	5	27	18	0.19	98.0
Canada	455	6	997	0.5	0.01	75.8
United Kingdom	14	1	24	0.5	0.04	12.3

[a] Data for wine yield and vineyard area for 1990 are from OIV (1991). Note that the production figures per hectare have not been adjust to reflect the variable proportion of the total grape harvest used as table grapes or in raisin production, nor do they reflect wine produced from imported grapes or juice. For many countries, these data are unavailable.

climatic conditions. Cultural tastes throughout the world have developed largely in relation to these styles. In addition, climatic vicissitudes have been used to create the mystique of vintage and regional variability, often important in marketing fine (or at least expensive) wines.

CLIMATE

The climate of Western Europe is significantly modified by the influence of adjacent bodies of water, namely, the Atlantic Ocean and the Mediterranean and Baltic seas. Slow heating and cooling of the large water bodies, combined with the predominant westerly winds, retard rapid or marked seasonal changes in temperature over much of Europe. The general east/west orientation of the mountain ranges does not impede the wind flow off large bodies of water onto the landmass. These influences are heightened by the warming action of the Gulf Stream during the winter (Fig. 10.7). Consequently, Europe's

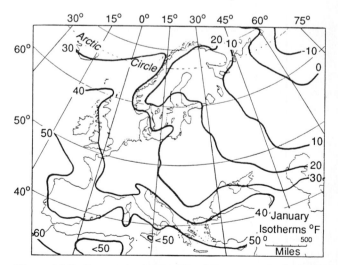

Figure 10.7 January isotherms (°F) of Europe. Note the marked warming influence of the Gulf Stream on the western portion of Europe. (From Shackleton, 1958, reproduced by permission.)

climate is milder than latitude alone would suggest, and this permits commercial grape culture at higher latitudes than on any other continent.

In Western Europe, the limit of commercial wine production angles northward from the Loire Valley (47°30′ N) in France up into the Rhine Valley (51° N) in Germany (Fig. 10.8). The increasingly northern limit for grape culture as one passes eastward through France results from the progressive warming of the cool winds off the Atlantic by passage over the continental landmass. Nevertheless, the northernmost viticultural regions of Germany hug river valleys, and the most favored sites occur on south-facing slopes. East of the Rhine and its tributaries, the northern limit of grape cultivation turns slowly southward, as the moderating effect of the Gulf Stream and surrounding seas diminishes. Here, frigid winter cold becomes more limiting than reduced light intensity or accumulated heat units of the northern latitude. Thus, the limit of viticulture moves south, through Czechoslovakia and Romania, to just north of the Black Sea.

The southern extent of commercial viticulture is limited by warm winters that ineffectively break bud dormancy and the increasing arid character of the climate.

Although Western Europe shares many common cultural, historical, and climatic features related to proximity to several seas and the Atlantic Ocean, there are equally important differences. Concerning grape growing and winemaking practices, the climatic differences are the most significant. Climatically, Europe may be divided into several broad zones. The area encompassing most of the western portion of southern France possesses both mild winters and summers. More precipitation falls during the autumn and winter months than the rest of the year. Nevertheless, adequate precipitation for vine development usually occurs throughout the growing season. The climates of north central and western France and southwestern Germany possess several common features. While the growing season is shorter, abundant spring and summer rain and longer day lengths spur growth. Dry sunny autumn days favor fruit maturation. Austria and Hungary possess similar climatic conditions but tend to have higher humidities and colder winters. Southeastern France and much of Italy, Spain, Portugal, and Greece possess a Mediterranean climate characterized by dry hot summers, with most prescipitation occurring during the mild winter months.

Geographic and topographical features can significantly modify local climatic conditions. For example, Table 10.3 shows the effect of latitude on the climate of regions spanning the major north/south limits of 'Pinot noir' cultivation in Western Europe. The influence of lattitude is reflected in an increase in permitted chaptalization as one progresses northward. Vineyard altitude

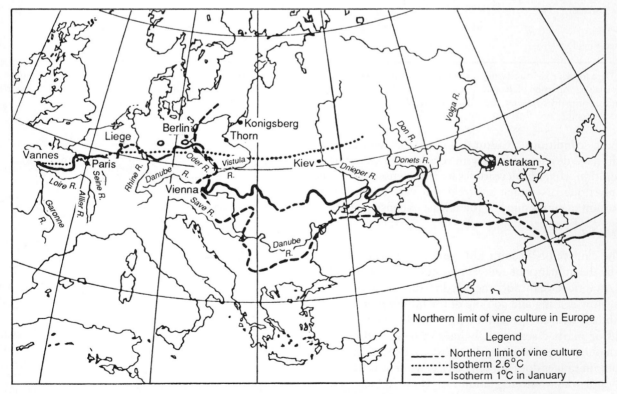

Figure 10.8 Northern limit of most grapevine culture in continental Europe (Afer Branas *et al.*, 1946, reproduced by permission.)

Table 10.3 Climatic Change Spanning the Major Regions Supporting Commercial Planting of 'Pinot Noir'[a]

Region	Latitude	Frost-free period (Celcius degree-days)	Average temperature of warmest month (°C)	Must minimal potential alcohol required (%)	Maximum permissible chaptalization (% potential alcohol increase)
Ahr	50°50′ N	887	17.8	7.5	3.5
Champagne	49°20′ N	988	18.3	8.5	2.5
Kaiserstuhl	48°10′ N	1045	18.8	8.9	2.5
Mâcon	46°20′ N	1223	20.0	10.0	2.0

[a] After Becker (1977), reproduced by permission.

also is important in both northern and southern regions. Lower altitudes usually are preferable at high latitudes, and higher altitudes at lower latitudes. In general, the average annual temperature (isotherm) decreases by about 0.5°C per 100 m elevation in altitude, or degree latitude increase (Hopkin's Bioclimatic Law). In practice, though, these preferences may be significantly modified by regional factors, such as site openness or valley width (Seeman *et al.*, 1979).

Besides affecting the distribution of grape culture, climate also has a marked influence on the potential of regions to produce fine wines. For example, most of the well-known wine regions of Europe are in mild to cool climatic zones. The absence of hot weather during ripening favors the retention of grape acidity. This gives the resulting wine a fresh taste and helps restrict microbial spoilage. Cool harvest conditions also favor the development or retention of varietal flavors and minimize overheating during fermentation. In addition, storage in cool cellars represses microbial growth that could induce spoilage. Finally, cool conditions have required that most vineyards be situated on south- or west-facing slopes to obtain sufficient heat and light exposure. This incidentally has positioned vineyards on less fertile but better drained sites. These features have restrained excessive vine vigor, promoted fruit ripening and provided a degree of frost protection.

In contrast, the hot conditions typical of southern regions favor acid metabolism and a rise in juice pH. Besides producing a 'flat' taste, the low acidity makes the wines more susceptible to oxidation and microbial spoilage. Although only approximately 0.004% of grape phenols are in a readily oxidized state at pH 3.5 (Cillers and Singleton, 1990), they are so unstable that oxidative reactions occur readily. Even minor increases in pH can significantly increase the tendency of a wine to oxidize. Thus, protection from oxidation is more critical in warm areas than in cooler regions. In addition, grapes tend to accumulate higher sugar contents under warm condi-

tions. These, along with harvesting and fermentation under warm conditions, increase the likelihood of premature termination of fermentation. By retaining fermentable sugars, the wine is much more susceptible to undesirable forms of malolactic fermentation and microbial spoilage. Warmer cellar conditions further enhance the risks of spoilage.

While modern advancements in viticulture and enology have increased the potential to produce a wider range of wine styles, prevailing conditions have had a profound effect on the evolution of wine styles in European regions. Cool climates favored the production of white wines, low in alcohol content but fruity and tart. Such wines normally have been consumed alone as sipping wines, before or after meals. Where white wines have been more alcoholic they have functioned primarily as a food beverage. In warmer regions, red wines have tended to predominate. Here, higher sugar contents in the grapes permitted the production of full-bodied wines, well suited to consumption with meals. In hot Mediterranean regions, high sugar and low acid grapes favored the production of alcoholic wines that tended to oxidize readily. These features have tended to promote the production of oxidized, sweet or artificially flavored high-alcohol wines, appropriate for use as aperitifs or dessert wines.

CULTIVARS

In Europe, notably in France, cultivation of grape varieties has tended to be highly localized. This has given rise to the view that cultivar distribution reflects a conscious selection of cultivars particularly suited to local climatic conditions. At sites where religious orders have produced wine for centuries, empirical trials may have found indigenous or imported grape cultivars especially suited to the climate. In most localities, however, wine was consumed within the year of its production, a situation incompatible with assessing aging ability (the *sine qua non* of wine quality). Also, varieties were commonly planted more or less randomly within vineyards, as well as harvested and vinified together. Thus, assessment of the quality of one cultivar relative to another would have been essentially impossible.

Finally, there is little solid evidence documenting the continued cultivation of specific varieties in particular regions. Important exceptions are 'Riesling' and 'Pinot noir,' for which information may go back 500 years. Even in Bordeaux, 'Cabernet Sauvignon' is reported to have been rare, or nonexistent, until the early 1800s (Penning-Rowsell, 1985). More likely, unintentional selection resulted among vines of local origin, derived from indigenous strains of *Vitis vinifera*, or accidental crossing with imported cultivars. Most selection would have been for obvious traits, such as compatibility with local climatic and soil conditions, higher sugar content, adequate acidity, and aroma. Subtleties such as aging potential, development of delicate bouquets, and complexity would have been selected more by accident than design. Only within the last two centuries have conditions become more conducive to the intentional selection of premium-quality cultivars.

VITICULTURE

In Europe, traditional cultivation ranged from dense plantings, with about 5000 to 10,000 vines/hectare, to interplanting with field crops with trees as supports. In densely planted vineyards, each vine occupied about 1 m^2, resulting in considerable intervine competition and restrained vigor. This had the advantage of reducing the level of pruning and manual labor required in the vineyard. It also was compatible with the use of single horse-drawn equipment in vineyards. Restrained vigor also resulted from the confinement of most vineyards to poor soils where cereal and vegetable crops would not grow well. This applied equally to sloped sites, which generally were, and still are, ill-suited to annual crop production.

The previous relegation of most grape culture to poor and sloped agricultural sites, and the convenience of dense plantings, led to the erroneous view that these conditions were necessary for grape quality. That these conditions reduce individual vine size and productivity, and favor fruit maturation, is not in question. However, as noted in Chapter 4, new training systems involving canopy management permit cultivation of widely spaced vines on rich soils without sacrificing fruit quality. In addition, fruit yield is increased and mechanization facilitated. Some of the techniques currently are being integrated into European vineyard practice.

ENOLOGY

In Europe, more variation is found in winemaking than in viticultural practice. The differences reflect the wine styles that have evolved in response to climatic or marketing factors. In addition, major changes have occurred since the mid-nineteenth century. Previously, long-aging wines were matured in large volume oak, chestnut, or other tight-wood cooperage, including both white and red wines. Furthermore, red and occasionally white wines were produced by fermenting the juice with the seeds and skins for up to several weeks. The resulting wines were often partially oxidized and possessed higher volatile acidity levels than now acceptable (Sudraud, 1978). The current tendency is to reduce the maceration period for both white wines (before fermentation) and red wines (during fermentation). In addition, shorter maturation in wood also is currently favored. Premium white wines may receive up to 6 months of maturation in oak barrels, while premium reds often receive up to 2 years of in-barrel aging. Limited oak exposure is used to preserve more of the fresh aroma of the wine. Greater emphasis is currently placed on the development of a reductive, in-bottle aging bouquet than in the past.

FRANCE

France has a diverse topography with few homogeneous regions. Most agricultural regions of similar geographic character are comparatively small and specialized in crop production. This is reflected in the localized cultivar plantings and regional wine styles throughout France. Although no one soil type or geologic origin distinguishes French vineyards, several regions are either on calcareous soils or soils covering chalk substrata.

While the effect of Appellation Control laws has tended to stabilize cultivar distribution in AOC- and VDQS-designated regions, marked changes in the varietal composition of French vineyards are still occurring, notably in nondesignated regions. For example, the proportion of French–American hybrid varieties dropped from about 30% in 1958 to less than 5% in 1988. Remarkable in terms of worldwide trends is the increase in red cultivar plantings. In France, the hectarage of red *Vitis vinifera* cultivars has grown by 9% since 1958, while the planting of white cultivars has declined by 7%. Currently, white cultivars cover only about 30% of French vineyard area. Of these, about 40% of the yield is used in the production of brandies, such as Cognac and Armagnac.

Over the period 1958–1988 there has been an increase in the cultivation of premium red cultivars (Fig. 10.9). Even several well-known white cultivars have lost ground (Fig. 10.10), and others are close to extinction. For example, 'Sémillon' and 'Chenin blanc' plantings have declined by about 50%, and 'Viognier' is cultivated on only 82 ha (Boursiquot, 1990). 'Chardonnay' and 'Sauvignon blanc' are two of the few famous white cultivars to see cultivation increase. The most marked in-

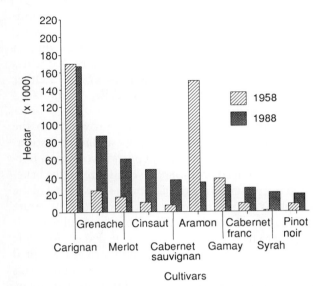

Figure 10.9 Planting of the 10 most common red *Vitis vinifera* cultivars in France in 1958 and 1988. (Data from Cadastre Viticole, Institute des Vins de Consummation Courante (IVCC), Recensement Agricole, Service Central des Enquiêtes et Études Statistiques (SCEES)–Institute National de la Statistique et des Études Economiques. (INSEE) reported in Boursiquot, 1990.)

crease in cultivar coverage has occurred with 'Syrah,' where hectarage expanded about 16-fold. Other red varieties with significantly expanded plantings are 'Cabernet Sauvignon,' 'Cabernet franc,' 'Merlot,' and 'Pinot noir.'

Although France is famous for one of its sparkling wines (champagne), the vast majority of French wines are still. As in other major wine producing and consum-

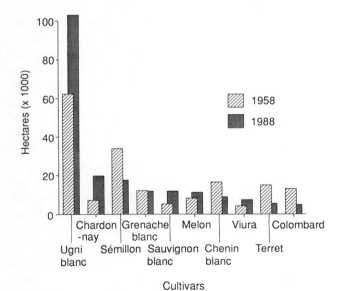

Figure 10.10 Planting of the 10 most common white *Vitis vinifera* cultivars in France in 1958 and 1988. (Data from Cadastre Viticole, IVCC, Recensement Agricole, SCEES–INSEE reported in Boursiquot, 1990.)

ing countries, most French wines are red. Only a few sweet or fortified wines are produced.

While wine is produced commercially in most regions of France, only a few regions are widely represented in world trade, notably Alsace, Bordeaux, Burgundy, Champagne, Loire, and the Rhône. These regions are briefly discussed below, as well as the less prestigious but most important wine-producing region in terms of volume, the Midi (Languedoc-Roussillon).

Alsace   Alsace is one of the most culturally distinctive regions of France. This reflects its German/French heritage and alternate French and German rule. Not surprisingly, wines from Alsace bear a varietal resemblance to German wines but a stylistic similarity to French wines. In addition, Alsace is the one French region where varietal origin is typically and prominently displayed on the wine label.

Alsatian vineyards run from north to south along the eastern side of the Vosges Mountains in northeastern France (47°50′ to 49° N). The vineyard region possesses three structurally distinctive zones. The zone running along the edge of the mountains has excellent drainage and benefits from solar warming of the shallow, rocky, siliceous soil. The foothill region is predominantly calcareous and generally possesses the best microclimate for grape cultivation. The soils of the plains are of more recent alluvial origin and possess excellent water-retention properties. Most vineyards occur at an altitude ranging from 170 and 360 m.

Alsace yields predominantly dry white wines, although some sweet and sparkling wines are produced. Yearly production averages about 0.9 million hl. Predominant cultivars are 'Gewürztraminer' (20%), 'Pinot blanc' (18%), 'Riesling' (20%), and 'Silvaner' (21%), with small plantings of 'Chasselas,' 'Muscat,' 'Pinot gris,' and 'Pinot noir.'

Because of the cool climate, grapes are often harvested high in acidity and the wine treated to favor malolactic fermentation. The local strains of lactic acid bacteria involved in the latter may generate some of the distinctive regional flavor found in most wines from Alsace.

Bordeaux   Bordeaux is the largest of the famous French viticultural regions (44°20′ to 45°30′ N). It runs southeast for about 150 km along the banks of the Gironde, Dordogne, and Garonne Rivers. Vineyards cover about 112,000 ha, and annual production averages over 6 million hl.

The Bordeaux region, located near the mouth of the Aquitaine Basin, is trisected by the junction of the Dordogne and Garonne Rivers to form the Gironde. These zones are divided into some 30 variously sized AOC areas. The best-known are those on the western banks of the Gironde (Haut Médoc) and Garonne (Graves and Sauternes) and the eastern bank of the Dordogne (Pomerol and Saint-Émilion). Although Bordeaux is best known for its red wines, about 40% of production is white. White wines are produced primarily in Graves and Entre-Deux-Mers. The best vineyard sites are generally on shallow slopes or alluvial terraces adjacent to the Gironde, or low lying areas along the Garonne and Dordogne Rivers at altitudes between 15 and 120 m.

Geologically, Bordeaux shows relatively little diversity. The bedrock is predominantly composed of Tertiary marls or sandstone intermixed with limestone inclusions. The substrata are usually covered by alluvial deposits of gravel and sand of Quarternary origin, topped with silt. The soils are generally poor in humus and exchangeable cations. This is partially offset at better sites by the soil depth, often 3 m or more. Deep soils also provide vines with access to water during periods of drought and good drainage in heavy rains.

The presence of extensive forests to the east and south protects Bordeaux vineyards from direct exposure to winds off the Atlantic. Proximity to the ocean and rivers also provides some protection from rapid temperature changes, but it limits summer warmth. Wet autumns occasionally cause difficulty during harvest by favoring fruit splitting, bunch rot, and dilution of the sugar and flavor content of the grapes. Soil depth and drainage, along with local microclimate, appear to be more significant to quality than soil type or geologic origin (Sequin, 1986).

Unlike the wines of some French viticultural regions, Bordeaux wines typically are blends of wines produced from two or more cultivars. Depending on the AOC, the predominant cultivar may vary. In the Haut Médoc the predominant variety in red wines is 'Cabernet Sauvignon' while in Pomerol it is 'Merlot.' This partially results from 'Cabernet Sauvignon' grapes being "herbaceous" on the clay soils of Pomerol, but not on the sandy/gravelly soils of the Haut Médoc. 'Merlot' is more forgiving of the slower warming of the clay soils in Pomerol.

The percentage of each cultivar in a vineyard can vary considerably between estates. In addition, the wine blend produced often changes from year to year. This permits the winemaker to compensate for deficiencies in the character of the different varietal base wines. Wine not incorporated into the final blend may be bottled under an alternate label or sold for use in a general Bordeaux blend. The standard red cultivars are 'Cabernet Sauvignon,' 'Cabernet franc,' 'Merlot,' 'Petit Verdot,' and 'Malbec.' The first three constitute about 90% of red cultivar plantings in Bordeaux.

White Bordeaux wines also tend to be blends of wines produced from two or more cultivars. In most areas, the predominant cultivar is 'Sauvignon blanc,' with 'Sémil-lon' coming second. The 2 : 1 proportion of these cultivars is reversed in Sauternes and Barsac, where sweet, occasionally botrytized, wines are produced. In contrast to German botrytized wines, Bordeaux botrytized wines are usually high in alcohol content (14 to 15%). Other permitted white cultivars are 'Muscadelle,' 'Ugni blanc,' and 'Colombard.'

Because of good harbor facilities, proximity to the climate-moderating ocean, and long-established association with discriminating wine-importing countries, Bordeaux has been well positioned to capitalize on the benefits of many winemaking developments. It was one of the first regions to initiate the modern practice of estate bottling and in-bottle aging. It also influenced the shift from tank to barrel maturation of wine. Except for some white wines, Bordeaux wines are tank or vat fermented rather than fermented in-barrel as in Burgundy. This possibly reflects the considerably larger size of many Bordeaux vineyards (*châteaux*). Most cover about 5 to 20 ha, with some encompassing 40 to 80 ha.

Burgundy　Burgundy is often considered to include several regions beyond the confines of Burgundy proper (the Côte d'Or). The ancillary regions include Chablis to the northwest and the more southern areas of Challonais, Mâconnais, and Beaujolais. Total wine production in all Burgundian regions averages about 1.5 million hl per year.

For all its fame, the Côte d'Or consists of only a narrow strip of vineyards, seldom more than 2 km wide. The strip runs about 50 km in a northeasterly direction from Chagny to Dijon (46°50′ to 47°20′ N) along the western edge of the broad Saône Valley. Although the vineyards are in a river valley, the Saône River is too distant ($\geq$20 km) to have a significant effect in directly modifying vineyard microclimate. A major physical feature favoring viticulture in the region is the southeast inclination of the valley wall. The porous soil structure and 5 to 20% slope promote good drainage and favor early spring warming of the soil. Sites located partially up the slope, at an altitude between 250 and 300 m, are generally preferred. Less distinguished wines are normally produced from vines grown on higher ground or on the alluvial soils of the valley floor.

The Côte d'Or is divided into two subregions, the northern Côte de Nuits and the southern Côte de Beaune. Although there are exceptions, the Côte de Nuits is more famous for its red wines, and the Côte de Beaune for its whites. This difference is commonly ascribed to the steeper slopes and limestone-based soils of the Côte de Nuits and the shallower slopes and marly clays of the Côte de Beaune.

The predominant cultivars planted in the Côte d'Or are both early maturing, 'Pinot noir' and 'Chardonnay.'

These cultivars produce some of the most sought-after wines in the world.

'Pinot noir' can produce delicately fragrant, subtle, smooth wines of great quality under ideal conditions. Regrettably, these conditions are rare even in Burgundy. Several factors, besides climatic variation and the notorious clonal diversity of 'Pinot noir,' are probably responsible. Most vineyards are under multiple ownership, with individual owners cultivating and harvesting only several rows of vines at each site. Vignerons generally possess small portions of many vineyards scattered throughout the region. Thus, the wine tends to be fermented in small lots (to maintain the site identity) by producers whose technical skill and equipment are highly variable. Although the wines may be fermented and matured in oak cooperage, new oak barrels for maturation are not considered the quality feature here that it is in Bordeaux.

Although 'Pinot noir' matures early, it is not intensely pigmented. Thus, to improve color extraction, part of the crop may be subjected to thermovinification (heating to 90°C for several hours) before fermentation. Frequent punching down of the cap during fermentation also is necessary. Because of the onerous nature of the latter, considerable interest has been shown in the use of small capacity (~50 hl) tanks that frequently and automatically mix the pomace and fermenting juice.

White wine is produced primarily from 'Chardonnay' grapes, though some comes from 'Aligoté.' The latter must be so designated on the label. Most white wines are fermented in barrels or small tanks.

Because of the cool climate, chaptalization is commonly required to reach the alcohol content (12 to 13%) considered typical for Burgundian wines. Malolactic fermentation is promoted for its beneficial effect in deacidification. As a consequence, the wines usually are racked infrequently, and the associated long contact with the lees tends to influence the character of Burgundian wines. It is thought by some that the accumulation of yeasts and tartrate on the insides of the barrel limits excessive uptake of an oak flavor from the cooperage.

Chablis is a delimited region some 120 km northwest of the Côte d'Or (47°48' to 47°55' N) just east of Auxerre. The region is characterized by a marly subsoil topped by a limestone/flint-based clay. Sites located on well-exposed slopes (15 to 20%) are preferred, to achieve better sun exposure and drainage. This is especially important as the region frequently suffers killing late-spring frosts. To further enhance protection from frost damage, the vines are trained low to the ground. *Cordon de Royat* and short-trunk double *Guyot* training systems are common. Shoot growth seldom reaches more than 1.5 m above ground level. Thus, the vines remain close to heat radiated from the soil.

'Chardonnay' is the only authorized cultivar in Chablis. Yield varies from about 50,000 to 100,000 hl from 1500 ha of vines.

Beaujolais is the most southerly region in Burgundy (45°50' to 46°10' N). It runs about 70 km as a broad strip of hilly land from just north of Lyon to just south of Mâcon. Most vineyards are located on slopes that are part of the eastern edge of the Massif Central. Here, the subsoil is deep and derived from granite and schists. The soil has considerable clay content and may be admixed with calcareous and black shale deposits.

The most distinctive feature of Beaujolais has been its retention of the old production technique called carbonic maceration (see Chapter 9). The procedure can generate wines that are pleasantly drinkable within a few months of production. It also results in the synthesis of a distinctively fresh, fruity fragrance. The light style generated by carbonic maceration has become very popular since the late 1960s as Beaujolais Nouveau. The red cultivar grown in the region, 'Gamay noir,' responds well to carbonic maceration. Nevertheless, the technique can yield wine that ages well, and such wines are produced predominantly in several villages in the northern portion of Beaujolais. Possibly to distance themselves stylistically from the *nouveau* style, most producers from these villages (*crus*) do not mention the name Beaujolais on the wine labels. Beaujolais produces about 1 million hl of wine per year, with more than 60% going into Beaujolais Nouveau.

Champagne Champagne is probably the best known French wine. So many producers in other parts of the world have used the term champagne in a semigeneric manner that champagne has become identified, if not synonymous, with sparkling wines in the minds of most consumers.

The designated region of Champagne is quite large, comprising 30,000 ha of vines (3% of French vineyards). The annual production of about 2 million hl is largely, but not exclusively, used in the production of sparkling wine. Most of the region lies east-northeast of Paris, spanning out equally on both sides of the Marne River for about 120 km. The other main section lies to the southeast, in the Aube *département*. Nevertheless, the greatest concentration of vineyards (~50%) and the best sites occur within the vicinity of Épernay (49°02' N). Here lie two prominences (*falaises*) that rise above the valley floor. The Montagne de Reims creates steep south- and east-facing slopes along the Marne River, and more gentle slopes northward toward Reims, some 6 km away. The Côte des Blancs, just south of Épernay, provides steeply sloped vineyard sites facing eastward. Soil cover is shallow (15 to 90 cm) and overlies a hard bedrock of chalk. Because of the slope, the topsoil needs to be periodically restored.

All three authorized grape cultivars are planted in the Épernay region, but the pattern of distribution varies between regions. The north and northeastern slopes of the Montagne de Reims are planted almost exclusively to 'Pinot noir,' while along the eastern and southern inclines 'Pinot noir' and 'Chardonnay' are cultivated. On the eastern ascent of the Côte des Blancs, essentially only 'Chardonnay' is grown. The best vineyards tend to lie between 140 and 170 m altitude (about halfway up the slopes) and possess eastern to southern orientations. 'Meunier' may be grown on the *falaises* as well, but it is primarily cultivated along the Marne Valley and other delimited regions. In the valley, soils are more fertile and less calcareous than those of the *falaises,* but the area is more susceptible to frost damage. Although the cultivar is less preferred, about 48% of Champagne plantings are 'Meunier,' with the rest divided about equally between 'Pinot noir' and 'Chardonnay.'

In Champagne, the northernmost of French vineyard regions, the vines are trained low to the ground. As noted above, the best sites are on slopes which direct the flow of cold air away from the vines and out into the valley. The inclination of the sites also can provide conditions that enhance spring and fall warming. Surprisingly, some excellent 'Pinot noir' vineyards face north. Although the slopes tend to be shallow, solar warming will still be less than on level sites. In Champagne, however, good coloration and phenol synthesis are not essential to wine quality. In contrast, low color content simplifies the extraction of uncolored juice from the grapes. In addition, delayed fruit ripening probably aids in the harvesting of healthy grapes, which is a necessity in producing white wines from red grapes. 'Pinot noir' becomes very susceptible to *Botrytis* bunch rot at maturity. With delayed maturation under the cool conditions, resistance to *Botrytis* infection remains relatively high. This is especially important as precipitation occurs predominantly in the late summer and fall.

That the grapes often are harvested somewhat immature is not the problem it would be with other wine styles. It is actually desirable that the grapes possess only sufficient sugar to permit the production of a wine with about 9% alcohol. This facilitates initiation of the second in-bottle fermentation so integral to champagne production. If the acidity is excessive, deacidification can be achieved with malolactic fermentation or blending. Although training vines close to the ground makes mechanical harvesting impossible, this is acceptable as the red varieties must be handpicked to avoid berry rupture. Mechanical harvesting unavoidably would result is some berry rupture and the associated diffusion of pigment into the juice. Manual harvesting also permits the removal of diseased fruit by hand.

Although regions are ranked relative to the potential quality of the fruit produced, champagnes are rarely vineyard designated. Champagnes usually are blended from wines from different sites and vintages to generate consistent house styles. Each champagne firm ("house") creates its own styles. The procedure also helps to cushion variations in annual yield and quality, and it stabilizes prices. In exceptional years, vintage champagnes may be produced. In vintage champagnes, at least 80% of the wine must come from the indicated vintage.

Loire   The Loire marks the northern boundary of commercial viticulture in western France (~47° N), a full degree latitude south of Champagne. This apparent anomaly results from the proximity and access of the Loire Valley to cooling winds off the Atlantic Ocean. The region consists of several distinct subregions, stretching from the mouth of the Loire River near Nantes to Pouilly-sur-Loire, some 450 km upstream. Most regions specialize in varietal wines produced from one or a few grape varieties. Loire vineyards cover some 61,000 ha and annually provide about 3.5 million hl of wine.

Nearest the Atlantic Ocean is the Pays Nantais which produces white wines from the 'Muscadet' ('Melon') or 'Gros Plant' varieties. About 100 km upstream is the region of Anjou-Saumur. Here 'Chenin blanc' is the dominant cultivar grown. Although 'Chenin blanc' usually produces dry white wines, noble-rotted fruit produce sweet wines high in alcohol content (~14%). Rosés also are a regional speciality coming from the 'Cabernet franc' and 'Groslot' varieties. In the central district of Touraine, light red wines are derived from 'Cabernet franc' and 'Cabernet Sauvignon' (Chinon and Bourgueil appellations) and white wines from 'Chenin blanc' (appellation Vouvray). In Vouvray, dry, sweet (botrytized), and sparkling wines are produced. The best-known upper Loire appellations are Sancerre and Pouilly-sur-Loire; the wines come primarily from 'Sauvignon blanc,' though some wine also is made from 'Chasselas.'

In the Loire Valley, vineyard slope becomes significant only in the upper reaches of the river around Sancerre and Pouilly-sur-Loire. Here, chalk cliffs rise to an altitude of 350 m. In most regions, moisture retention, soil depth, and drainage are the important factors influencing the microclimate (Jourjon *et al.,* 1991).

Southern France   Progressing south from the union of the Saône and Rhône Rivers, just below Beaujolais, the climate progressively takes on a Mediterranean character. Total precipitation declines, and peak rainfall shifts from the summer to winter months. The average temperature also rises considerably. Here, red grapes consistently develop full pigmentation, and red wines are the dominant type produced. Because of the longer growing season, cultivars adapted to those conditions predominate.

'Syrah' is generally acknowledged as the finest of red cultivars from southern France. 'Syrah' has shown a marked increase in cultivation after the decline associated with and following the phylloxera devastation of the 1870s. A similar fate befell many other highly respected cultivars, such as 'Roussanne,' 'Marsanne,' 'Viognier,' and 'Mataro' ('Mourvèdre'). The fruitfulness of French–American hybrids, developed initially to avoid the expense of grafting in phylloxera control, induced further displacement of indigenous varieties. The shift of vine culture from cool highland slopes to the hotter rich plains resulted in overcropping and a reduction in wine quality. Thus, wines from southern regions ranked as highly as the best Bordeaux wines in the mid 1800s are little known today.

Generally, the best-known regions currently are those in the upper Rhône Valley. In regions such as the Côte Rôtie (45°30' N) and Hermitage (45° N), the best sites are on steep slopes. In some areas, the slopes are terraced, for example, Condrieu. One exception is Châteauneuf-du-Pape (44°05' N), which is in the center of the lower Rhône Valley and situated on shallow slopes. The Rhône Valley possesses some 38,000 ha of vines and produces about 2 million hl of wine per year.

In the upper Rhône, most of the wines are produced from a single cultivar. Progressing southward, the tendency shifts to blending with several to many cultivars. Also, the predominant cultivar changes from 'Syrah' in the upper Rhône, to 'Garnacha' in the lower Rhône, to 'Carignan' in the Midi (primarily Languedoc and Roussillon).

With the tendency to long hot, dry summers in the south, vegetative growth ceases early, producing short, sturdy shoots. This, and the value of fruit shading, have promoted the continued use of the *Goblet* training system. The bushy form developed has been thought to help minimize water loss by ground shading. However, data from van Zyl and van Huyssteen (1980) do not support this view. The system also obviates the need for a trellis and yearly shoot positioning. In addition, the short vine stature and sturdy shoots are less vulnerable to the strong southerly *mistral* winds common in the lower Rhône and Rhône Delta. Windbreaks are a common feature in the area, but surprisingly are absent in most of its viticultural regions.

Although the upper Rhône Valley, and to a lesser extent the lower, produces several wines of international repute, this is rare in the Midi (along the Mediterranean coast). Only some sweet wines, such as Banyuls, appear to have gained an international clientele. Nevertheless, these regions generate more than 40% of the 65 million hl yearly wine production in France. Grapes are the single most important crop in the region.

Improvement in wine quality in the Midi will depend on planting better cultivars, eliminating overcropping, and adopting the more widespread use of modern winemaking equipment and techniques. However, this is difficult to achieve in an economically depressed area accustomed to subsidized prices for wine destined largely for distillation into industrial alcohol.

GERMANY

Germany's reputation for quality wines far exceeds its significance in terms of quantity; Germany produces only about 3% of the world's supply. However, much of the international repute comes from a small quantity of botrytized and drier 'Riesling' *Prädikat* wines. Nevertheless, much fine wine comes from the less well known cultivar 'Müller-Thurgau.' The fine reputation of German wines, despite the high latitude, partially reflects the high technical skill of its grape growers and winemakers, and the assistance provided by the many excellent research facilities throughout the country.

The high latitude of vineyard regions in Germany (47°40' to 50°40' N), and resulting cool climate, favor the retention of fruit flavors and a refreshing acidity. The "liability" of low °Brix levels in the grapes has been turned into an asset in the production of naturally light, low alcohol wines (7.5 to 9.5%). With the advent of sterile filtration in the late 1930s, crisp semisweet wines with fresh fruity/floral fragrances could be produced in quantity. These wines ideally fit the role they have typically played in Germany, namely, as light sipping wines consumed before or after meals. The botrytized speciality wines have for several centuries been favorite dessert replacements. As befits their use, over 85% of all German wines are white. Even the red wines are made in a light style, often resembling a rosé more than standard red wines. While most German wines generally are not considered ideal meal accompaniments, several producers are developing dry wines to meet the growing demand for this style in Germany.

Although German viticultural regions reflect the typical European regional specilization with particular cultivars, many varieties grow in all areas. In only a few regions does one cultivar predominate. Unlike the case in most European countries, modern cultivars are grown extensively. For example, 'Müller-Thurgau' is the most extensively grown German cultivar (24%), ahead of 'Riesling' (21%). Nearly half the vineyards are planted with varieties developed in ongoing German grape-breeding programs. Their earlier maturity, higher yield, and floral fragrance have made them valuable in producing wines at the northern limit of commercial viticulture. Both new cultivars and clonal selection of established varieties have played a significant role in raising vineyard productivity without resulting in a loss in wine quality. Of red cultivars, the most frequently planted are 'Spätburgunder' ('Pinot noir'), 'Portugieser,' and 'Ruländer' ('Pinot gris').

One of the most distinctive features of German viticulture is the high proportion of vineyards on slopes (Fig. 10.11). This has meant that viticulture usually has not competed with other crops for land. Most famous sites are on valley walls unsuitable for other crop cultivation. While steep inclinations may produce favorable microclimates for grape growth, they are often incompatible with mechanized vineyard activities. Thus, to facilitate viticulture, vineyard consolidation has been encouraged in several regions, as well as structural modification to produce terraces suitable for mechanization (Luft and Morgenschweis, 1984).

Formerly, vines were trained on short trunks (10 to 30 cm), as is still common in northern France. Current thought now recommends trunks be about 50 to 80 cm high. Not only is cultivation and harvesting easier with taller trunks, but the vines are less susceptible to disease, owing to better air circulation and drying. Trellising also helps position shoot growth and leaf production for optimal light exposure. This is important, as the long summer days ($\geq$16.5 hr) and abundant precipitation promote the development of a large assimilation area for photosynthesis. In addition, by locating shoot-bearing wood low on the vine, and directing shoot growth up-

ward, grapes are kept as close as possible to heat radiated from the soil. Only on particularly steep locations, notably along the Mosel-Saar-Ruwer, are vines still trained to individual stakes with two arched canes.

On all but steep slopes, narrow-gauge tractors permit most vineyard activities to be mechanized. The small tractors are needed because the vines are planted more densely than is common in the New World.

While German vineyard regions have a cool temperate climate, the southernmost viticultural portions of Baden possess warmer but wetter conditions. They occur on the windward side of the Black Forest across from Alsace. The Baden region consists of a series of noncontiguous areas spanning 400 km, from the northern shores of Lake Constance (Bodensee) (47°40' N) to the Tauber River, just south of Würzburg (49°44' N). Most of the vineyards occur in the 130 km stretch between Freiburg and Baden-Baden. The best sites apparently are located on the volcanic slopes of the Kaiserstuhl and Tuniberg.

'Spätburgunder' ('Pinot noir') is often grown on well-drained south-facing slopes of Baden and produces about 90% of Germany's 'Pinot noir' wines. On heavier loamy soils, 'Müller-Thurgau,' 'Ruländer,' and 'Gutendal' ('Chasselas') tend to predominate. 'Gutendal' is especially well adapted to the more humid portions of the region. Wine production is largely under the control of several large, skilled cooperatives. The region has been particularly active in vineyard consolidation and terrace construction.

Directly north of Alsace, on the west side of the Rhine Valley, is the Rheinpfalz. It continues as the Rheinhessen to where the Rhine River turns westward at Mainz. Combined, these regions possess about half of all German vineyards (48,000 ha) and produce most of the wine

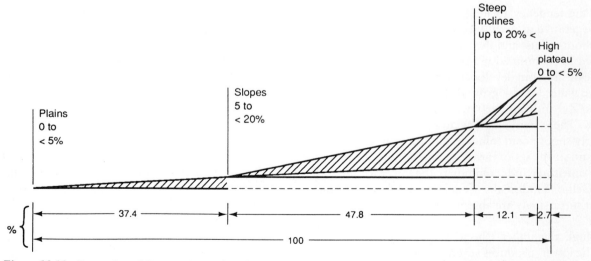

Figure 10.11 Proportion of German vineyards on land within different slope ranges in 1977. Note: Angles of 10° and 20° correspond to inclines of 18 and 37%, respectively. (From Anonymous, 1979, reproduced by permission.)

exported from Germany. The region consists largely of rolling fertile land at the northern end of the Vosges (Haardt) Mountains. *Liebfraumilch* is the best-known regional wine. The regions span latitudes from 49° to 50° N.

As with other German wine regions, the best vineyard sites occupy south- and east-facing slopes and are planted primarily with 'Riesling' vines. Occasional black basalt outcrops, as found around Deidesheim, are thought to improve the local microclimate and provide extra potassium. The longer growing cycle of 'Riesling' demands the warmest, sunniest microclimates to mature fully. In the Rheinpfalz, 'Riesling' constitutes the second most commonly cultivated variety (17%), whereas in the Rheinhessen it covers only 7% of vineyard hectarage. In both regions, 'Müller-Thurgau' is the dominant variety, covering about 22% of vineyard sites. Many new cultivars are grown and add needed fragrance to the majority of wines produced. Soil type varies widely over both regions. Combined, the regions are up to 30 km wide and together stretch 150 km in length.

At Mainz, the Rhine River turns and flows southwest for about 25 km until it turns northwest again past Rüdesheim (~50° N). Along the northern slope of the valley are some of the most famous vineyard sites in Germany. The region, called the Rheingau, possesses about 3000 ha of vines. The soil type varies along the length of the region and up the slope. Along the river-banks are alluvial sediments; further up, the soil becomes more clayey, finally changing to a brown loess. The soil is generally deep, well-drained, and calcareous. While soil structure is important, factors such as slope aspect and inclination, wind direction and frequency, and sun exposure are of equal or greater importance. As the river widens to about 800 m in the Rheingau region, it significantly moderates the climate. Mists that arise from the river on cool autumn evenings, combined with dry sunny days, favor the development of noble-rotted grapes (*edelfäule*). This development is crucial to the production of the most prestigious wines of the region.

'Riesling' is the predominant cultivar grown in the Rheingau (82%), followed by 'Spätburgunder' (6.7%). The latter is cultivated almost exclusively on the steep slopes between Rüdesheim and Assmannshausen, where the course of the river turns northward again. On less favored sites, new cultivars such as 'Müller-Thurgau,' 'Ehrenfelser,' and 'Kerner' may be grown.

Vineyards continue to be planted along the slopes of the Rhine and its tributaries up to Bonn, as the river continues its northwesterly flow to the Atlantic. Red wines from 'Spätburgunder' and 'Portugieser' are produced along one of the tributaries, the Ahr. Heat retained by the volcanic slate and tufa of the steep slopes may be critical in permitting red wines to be produced in this northerly location (50°34′ N). However, most of the region, called the Mittelrhein, cultivates 'Riesling' along the steep banks. As in the more southerly regions of the Nahe and Neckar valleys, most of the wine is sold locally and seldom enters export channels.

A major artery joining the Rhine at Koblenz is the Mosel River. Along the banks of the Mosel are vineyards with a reputation as high as those of the Rheingau. The Mosel-Saar-Ruwer region stretches from the French border to Koblenz (49°30′ to 50°18′ N) and has 13 million ha of vines. Although the overall direction is northwest, the Mosel snakes extensively for about 65 km in its middle section. This section possesses many steep slopes with southern aspects between the Eifel and Hunsruck mountains. This stretch of the river, and its tributaries, the Saar and Ruwer, are planted almost exclusively with 'Riesling' vines. The central section, called the Mittel-mosel, also is blessed with a slate-based soil that favors heat retention. Thus, not only does the soil possess excellent drainage, but it also provides frost protection and promotes grape maturation. The Mittelmosel contains most of the region's most famous vineyards.

The final German river valley particularly associated with viticulture is that of the Main River. The Main flows westward and joins the Rhine at Mainz, the beginning of the Rheingau. Vineyards in the Main Valley are upstream and centred around Würzburg (~49° N). The region, called Franken, is noted for 'Silvaner' wines and the squat green bottle, the *bocksbeutel*. Franken is the only German region not using slim bottles for wine. Although best known for 'Silvaner' wines, Franken has a continental climate more suited to the cultivation of the newer, early maturing cultivars, such as 'Müller-Thurgau.' The latter covers almost half of the 5500 vineyard hectarage in Franken, while 'Silvaner' covers only about 20%.

SWITZERLAND

Although a small country, Switzerland has vineyards covering approximately 15,000 ha, with an annual wine production of about 1.3 million hl. The ethnic diversity and distribution of the inhabitants are reflected in the wines produced. Along the northern border, Switzerland produces light wines resembling those in neighboring Baden. In the southwest, wines are made more alcoholic, similar to those across the frontier in France. In the southeast, the small Italian-speaking region specializes in the production of full-bodied red wines.

While the southern vineyard regions of Switzerland are parallel to the Côte de Nuits (~46°30′ N), the altitude, between 400 and 800 m, produces a cooler climate than its latitude, shared with Burgundy, would suggest. Consequently, most vineyards are on south-facing slopes adjacent to lakes or river valleys. The southwestern vineyards of the Vaud and Valais hug the south-facing slopes around Lake Geneva and the head of the Rhône River; those of the west occur along the eastern slopes of Lake Neuchâtel and associated tributaries; those in northern Switzerland congregate around the Rhein, Rheintal, and Thurtal river valleys and the south-facing slopes of Lake Zürich; and finally those in the southeast (Ticino) embrace the slopes around lakes Lugano, Maggiore, and their tributaries. All the regions are associated with the headwaters of three great viticultural river valleys, the Rhine, the Rhône, and the Po. As in Germany, use of steeper slopes has avoided competition with most other agricultural land use.

Because of the frequent strong winds in most of the vineyard regions in Switzerland, the shoots are tied early and topped to promote stout cane growth. Abundant rainfall has fostered the use of open canopy configurations to favor foliage drying and minimize fungal infection.

The moist, cool climate usually necessitates chaptalization if "normal" alcohol levels are to be achieved. Malolactic fermentation is used widely to reduce excessive wine acidity. The short growing season requires cultivation of early maturing varieties, such as 'Müller-Thurgau' in the north and 'Chasselas doré' in the southwest. Although most cultivars are white, there is a growing interest in red wine production. The demand has resulted in a marked increase in the planting of 'Pinot noir' and 'Gamay' in western regions. In the southeast (Ticino), 'Merlot' is the dominant cultivar.

CZECHOSLOVAKIA

The former Czechoslovakia covers a primarily mountainous region, divided along most of its length by the Carpathian Mountains in the east and the Moravian Heights and Sumava in the west. The latter separate the small northern vineyard area of Bohemia (50°33′ N) from the warmer, more protected, southern regions of Moravia and Slovakia (~48° N).

In Bohemia, the vineyards are located on slopes along the Elbe River Valley, about 40 km northwest of Prague. Low altitude and the use of a variety of rootstocks differentially affecting bud break help to cushion the influence of the cool and variable climate of the region (Hubáčková & Hubáček, 1984). The predominant cultivars are 'Riesling,' 'Traminer,' 'Müller-Thurgau,' and 'Chardonnay.' The small size of the viticultural region, and its proximity to Germany, probably explains the resemblance of the cultivars and wine styles to those of its neighbor to the west.

In south central Czechoslovakia, most vineyard sites are positioned on the slopes of the Moravia River valley, north of Bratislava. The Moravian vineyards (12,000 ha) cultivate primarily white varieties, such as 'Riesling,'

'Traminer,' 'Grüner Veltliner,' and 'Müller-Thurgau,' similar to their Austrian neighbors. Small amounts of red wine are produced from 'Portugieser,' 'Limberger,' and 'Vavřinecké.'

Most of Slovakia's vineyards (35,000 ha) adjoin the Danube basin along the southern border of the country. In common with other viticultural regions, most of Slovakia's wines are white. While many Western European cultivars are grown, the junction of the region with Eastern Europe is reflected in the cultivation of eastern cultivars. This is particularly marked in the most easterly viticultural region. The area is an extension of the Tokaj region of Hungary. Here, the varieties 'Furmint' and 'Hárslevelű' are the dominant cultivars. The wines are identical in style to those of the adjoining Hungarian region.

AUSTRIA

The mountains of Austria restrict the 58,000 ha of Austrian vineyards to the eastern flanks of the country. Much of the vine-growing area is in low-lying regions along the Danube River and Danube Basin to the north (~48° N) and the Hungarian Basin to the east. A small section, Steiermark, is located across from Slovenia in former Yugoslavia (~46°50′ N).

The best-known viticultural regions are the Burgenland (Rust-Neusiedler-See), Gumpoldskirchen, and Wachau. The Burgenland vineyards adjacent to the Neusiedler See (47°50′ N) occur on rolling slopes at an altitude between 150 and 250 m. The shallow, 30 km long lake creates autumn evening mists that, combined with sunny days, promote the development of highly botrytized grapes and lusciously sweet wines. Here, 'Riesling,' 'Müller-Thurgau,' 'Muscat Ottonel,' 'Weissburgunder' ('Pinot blanc'), 'Ruländer,' and 'Traminer' predominate. Vineyards located north and south of Vienna produce light dry wines. These are derived primarily from the popular Austrian variety 'Grüner Veltliner.' Other varieties grown are 'Riesling,' 'Traminer,' and the local specialities 'Zierfändler' and 'Rotgipfler' in Gumpoldskirchen, south of Vienna. Most of the vineyards in the latter region are positioned on southeastern slopes at an altitude between 200 and 400 m. About 65 km west of Vienna is the Wachau region. Vineyards in Wachau lie on steep south-facing slopes of the Danube, just west of Krems. They occur at altitudes of about 200 to 300 m. The predominant cultivar is 'Grüner Veltliner.'

While most Austrian wine is white, some red wine is produced. The most important red cultivar is the new variety 'Zweigelt' (Mayer, 1990). Although Austria annually produces about 3 million hl of wine, only a small proportion is exported. Most wine production is designed for early and local consumption. As in Germany, the wines usually have the varietal origin and grape maturity noted on the label. The designations from *quali-tätswein* to *trockenbeerenauslese* are the same, except that the requirements for total soluble solids (Oechsle) are higher than in Germany.

UNITED KINGDOM

Although the United Kingdom is not a viticulturally important region on the world scene, U.K. vineyards indicate that commercial wine production is possible up to 54°45′ N. The extension of wine production further north than German vineyards results from both the moderating influence of the Gulf Stream and the cultivation of early maturing varieties, such as 'Müller-Thurgau' and 'Seyval blanc.' Other potentially valuable cultivars are 'Auxerrois,' 'Chardonnay,' 'Madeleine Angevine,' and several of the new German cultivars. Improved viticultural and enological practices also have played important roles in reestablishing winemaking in England. Cooling of the European climate in the thirteenth century, combined with England's possession of Bordeaux, probably explains the demise of viticulture in England about that period.

Southern Europe

Although immigrants from France and Germany were seminal in the development of much of the wine industry in Australia, the New World, and South Africa, their basic practices in turn were introduced from Southern Europe. During the ancient classical period, Greece and Italy were considered, albeit by themselves, as the most famous of wine countries. With the fall of these ancient civilizations, wine production declined both in quantity and quality.

At the end of the Medieval period, social conditions north of the Alps, combined with the cooler climate, favored the production of fine wine. In contrast, grapes continued to be grown with other crops in much of Southern Europe, often trained up trees in the field (Fregoni, 1991). Without monoculture, insufficient time could be devoted to controlling yield and optimizing fruit excellence. The subsistence economy in Southern Europe was incompatible with the exacting demands of quality viticulture. In addition, the absence of a large, discriminating middle class led to the persistence of low quality wines. Only in the northern parts of Italy, associated with the Renaissance, did fine wine production again develop. Regrettably, political division, associated with the rise of aggressive nation-states in central Europe, resulted again in a stagnation of quality wine production. In Greece, repeated invasion by oppressive foreign powers severely restricted the redevelopment of the wine industry, as well as that in much of Eastern Europe. Although Spain and Portugal did not experience similar events, their colonial activities did not translate into lasting economic benefits at home. The absence of a growing, prosperous middle class may explain why wine skills remained comparatively simple in most of Southern Europe until recently.

Mediterranean wines tended to be alcoholic, oxidized, and low in total but high in volatile acidity. Inadequate temperature control often resulted in wines high in residual sugar content and susceptible to spoilage. In contrast, wines produced north of the Alps tended to be less alcoholic, fresh, dry, flavorful, and microbially stable.

With the growth and spread of technical skill, the standards of winemaking have improved dramatically throughout Southern Europe. Regrettably, implementation of some procedures is limited as the economic return is insufficient to pay for the improvements. Better known regions, producing wines able to command higher prices, have been those best able to benefit from technical advances.

ITALY

Italy and France often exchange top ranking as the world's major wine-producing country. Each annually produces about 60 to 70 million hl, approximately one-fifth of the world's wine production. Italy's vineyard area is more extensive than that of France (1,050,000 versus 940,000 ha), but more table grapes are grown than in France (7% versus 2%).

Vineyards in Italy predominantly are exposed to a Mediterranean climate, receiving most of the rain during the winter months. Nevertheless, nearly one-third of Italy (the Po River valley and the Italian Alps) possesses a more continental climate, without a distinct dry summer season. Also, the Apennines running down the middle of the Italian Peninsula provide a cooler, more moist climate along leeward slopes of the mountains. The mountains in Italy produce considerable variation in yearly precipitation, from over 170 cm on some slopes to under 40 cm in southern areas. The predominantly north/south axis of Italy results in coverage of over 10° latitude, from 46°40′ N, parallel with Burgundy in France, to 36°30 N, parallel with the northern tip of Africa. Nonetheless, a common feature of the country, apart from mountainous regions, is a mean July temperature of about 21° to 24°C. Combined with a wide range of soil types and exposures, Italy possesses an incredible variety of viticultural microclimates.

Italy possesses one of the largest and oldest collections of grape varieties known. Thus, it is not surprising that Italy produces an incredible range of wine style within its borders. Many regions produce several white and red wines in dry, sweet, and slightly sparkling (*frizzante*) versions. Many regions also produce limited quantities of wine using distinctive, often ancient, techniques found nowhere else.

The wide diversity of wine styles found within many Italian regions may have hindered the acceptance of Italian wines internationally. Also confusing to many consumers is the lack of consistency in wine designation. In some regions, the wines are varietally designated, while in others geographic origin, producer, or mythological name is used. Often the province of origin is not stated on the label. This situation may be explained by the long division of Italy into many separate city states, duchies, kingdoms, and papal dominions. The difficulty of land travel until recent time also probably limited the development of a common wine-designation system.

For many centuries, poverty in most Italian regions resulted in a system of subsistence farming based on polyculture, as noted above. Only in a few regions, such as northwestern Italy and Castelli Romani in central Italy, was pure viticulture practiced. Today, the old polyculture is essentially gone, and Italian viticulture is similar to that in other parts of the world. There is, however, a remarkable range of training systems in practice, many used only locally. In several areas, high pergola or tendone training is used, either to favor light and air exposure for disease control or to limit sun- and windburning of the fruit. Low training is more common in the hotter, drier south. This may minimize water stress and promote sugar accumulation desirable in the production of sweet fortified wines. Irrigation may be practiced in southern regions, where protracted periods of drought occur during the hot summer months.

In most regions, enological practice is modern and has improved wine quality. Modernization also is effecting a shift from traditional long maturation in large wood cooperage (≥ 5 years) to shorter aging in barrels (≤ 2 years). There is considerable debate concerning the relative merits of oak flavor in Italian wines. Nevertheless, several unique and distinctive winemaking styles are still being practiced in Italy. Regrettably, most have been little investigated, but it is hoped that they will be studied adequately before they fall victim to standardized winemaking procedures.

Northern Italy   Much of the Italian wine sold internationally, excepting that sold in bulk for blending with other wines, comes from northern Italy. The regions involved are situated on the slopes of the Italian Alps and the North Italian Plain. These regions possess a mild continental climate without a distinct dry period. The arch of Alps that forms Italy's northern frontier usually protects the region from cold weather systems coming from the north, east and west.

The most northerly region is that of Trentino-Alto Adige. Although the vineyard area covers only 13,000 ha, and annually produces about 1.5 million hl of wine, production in Trentino-Alto Adige constitutes about 35% of Italy's total bottled-wine exports. The region also leads in the proportion of DOC appellations. Unlike most Italian wines, the label usually states the name of the grape variety followed by the regional DOC appellation.

The vineyards in the northern half of Trentino-Alto Adige often line the narrow portion of the Adige River valley on steep slopes at altitudes between 450 and 600 m. The region called Alto Adige (South Tyrol) stretches about 30 km, both north and south from Bolzano (46°31′ N). The considerable German-speaking population of the region is reflected in the style and care with which the wines are produced. The region produces many white and red wines. Whites may be produced from cultivars such as 'Riesling,' 'Traminer,' 'Silvaner,' 'Pinot blanc,' 'Chardonnay,' and 'Sauvignon blanc.' Red wines may be produced from distinctively local cultivars, such as 'Schiava' and 'Lagrein,' or French cultivars, such as 'Pinot noir' and 'Merlot.'

In contrast to Alto Adige, Trentino covers a 60 km strip of gravelly alluvial soil on the valley floor that widens about 20 km north of Trento (46°04′ N). Here the vineyards are at no higher than 200 m altitude. The most famous wine of the region comes from the local red cultivar 'Teroldego.' The vines are trained on supports resembling an inverted L (*pergola trentina*) to increase canopy exposure to light and air. Many different white and red wines also are produced, largely from the same varieties cultivated in Alto Adige. Besides standard table wines, the region produces a *vin santo* from the local white cultivar 'Nosiola' and considerable quantities of dry sparkling wine (*spumante*) from 'Pinot noir' and 'Chardonnay.'

Veneto is a major wine-producing region adjacent to Trentino, where the Alps taper off into foothills and the broad Po Valley. The vineyard area in Veneto covers 85,000 ha and annually produces about 8 million hl of wine. Internationally, the best known wines come from hilly country above Verona (45°28′ N), where the Adige River turns eastward. The dry white wine Soave comes primarily from the 'Garganega' grape cultivated on the slopes east of the city. The vineyards producing the grapes used in Valpolicella are situated north and east of Verona, while those involved in making Bardolino are further west, along the eastern shore of Lake Garda. Both red wines are produced primarily from 'Corvina' grapes, with slightly different proportions of 'Molinara' and 'Rondinella' involved. 'Negrara' also may be part of the Bardolino blend. Grapes for Valpolicella are generally grown on higher ground (200 to 500 m) than those for Bardolino (50 to 200 m) and produce a darker red wine. The most distinctive Valpolicellas come from specially selected and partially dried grape clusters using the *recioto* process described in Chapter 9. During vinification, the fermentation may be stopped prematurely to retain a detectable sweetness or continued to dryness to produce an *amarone.*

Two other regions in Veneto are also fairly well known outside Italy. These are Breganze, about 50 km northeast of Verona, and Conegliano, 50 km north of Venice. Although Breganze produces both red and white wines, it is most famous for a sweet white made from the local cultivar 'Vespaiolo.' The grapes are processed similarly to those used to produce red *recioto* wines. In Conegliano, another local white cultivar, 'Prosecco,' is used to produce still, *frizzante,* and *spumante* wines. The latter two may contain some 'Pinot bianco' and 'Pinot grigio.' Besides regional cultivars, Conegliano also cultivates several widely grown Italian and French varieties.

Although fine wines are produced in Friuli-Venezia Giulia and Lombardy, provinces respectively to the east and west of Veneto, Piedmont in the northwest is the most recognized internationally. The largest and best known of Piedmontese wine areas is centered around Asti (44°54′ N), in the Monferrato Hills. This subalpine region receives abundant rainfall, but well distributed throughout the year. Precipitation peaks in the spring and fall. The latter is often associated with foggy autumn days at lower altitudes. The area is famous for full-bodied reds, sweet spumante, and bittersweet vermouths. As a whole, Piedmont possesses a vineyard area of 66,000 ha and yields about 3 million hl of wine yearly.

The most famous red wines of the Piedmont region come from the 'Nebbiolo' grape, grown north and south of Alba (44°41′ N) and associated with the villages of Barbaresco and Barolo. In both appellation regions, the best sites are located on higher portions of either east- or south-facing slopes. Those associated with Barolo generally are at a higher altitude, and less likely to be fog covered in the fall than the slopes around Barbaresco. 'Nebbiolo' is grown in other areas of Piedmont, notably the northwest corner, and in northern Lombardy. In both areas, local regional names such as 'Spanna' and 'Chiavennasca' are used instead of 'Nebbiolo.' Many 'Nebbiolo' wines carry the name of the town from which they come but not the cultivar name.

Lombardy's 'Nebbiolo' region (Valtellina) occurs on steep, south-facing slopes lining the Adda River, where it leaves the Alps and flows westward into Lake Como. The vineyards along the 50 km strip often occur on 1 to 2 m wide, man-made terraces. Owing to strong west winds, canes of the vines are often intertwined. In addition, the vines are trained low to gain extra warmth from the soil and garner protection from winter winds. Recioto-like wines are occasionally produced in Valtellina and are called *Sfursat* or *Sforzato.*

Most Piedmontese wines are produced from indigenous grape varieties. Most are red, such as 'Barbera,' 'Dolcetto,' 'Bonarda,' 'Grignolino,' and 'Freisa,' though some whites are grown, such as 'Moscato bianco' and 'Cortese.' 'Moscato bianco' is used primarily in the production of the sweet, aromatic, sparkling wine Asti Spumante. 'Cortese' is commonly used to produce a dry still white wine in southern Piedmont, but it also is employed in the production of some *frizzante* and *spumante* wines.

Pure varietal wines, rather than blends, are normally produced in Piedmont. Modern techniques of fermentation are common, though some wines, notably those from 'Nebbiolo,' may be aged for several years in large oak casks. Extensive experimentation with barrel aging is in progress.

Carbonic maceration is used to a limited extent to produce *vino novello* wines, similar to those of Beaujolais Nouveau.

Piedmont is the center of vermouth production in Italy. The fortified herb-flavored aperitif was initially produced from locally grown 'Moscato bianco' grapes. Currently, most of these are used in the production of Asti Spumante. Although most of the base wines for producing vermouth come from further south, processing still occurs in Piedmont.

Central Italy Although wine is produced in all Italian provinces, Chianti is the Italian wine most associated with Italy in the mind of many wine drinkers. Chianti is Tuscany's, and Italy's, largest appellation. It includes seven separate subregions that constitute about 70% of the 83,000 ha of vines in Tuscany. These cover a 180 km wide area in central Tuscany. The most famous section, Chianti Classico, incorporates the central hilly region between Florence and Siena (~43°30′ N). Vineyards grow mostly 'Sangiovese,' with smaller quantities of two other red cultivars, 'Canaiolo' and 'Colorino,' and the white cultivars 'Trebbiano' and 'Malvasia.' These are added within specified limits to make Chianti. In addition, up to 10% of the must may come from other cultivars.

The total region encompassed by Chianti includes a wide range of soils, from clay to gravel, slopes with considerable variation in aspect and inclination, and altitudes extending from about 200 to 500 m. These geographic differences, combined with the flexibility in varietal content, confers on Chianti the potential for as much variation as in Bordeaux.

While most versions of Chianti are made using standard vinification techniques, several winemakers still use the ancient *governo* process. In the *governo* process, from 3 to 10% of the grape harvest is kept aside. During a 2-month storage period, the grapes undergo slow partial drying. In addition, there is a change in the yeast flora of the grapes. While the population of *Kloeckera apiculata* declines markedly, the proportion of *Saccharomyces* cerevisiae strains increases (Messini et al., 1990). After storage, the grapes are crushed and allowed to commence fermentation. At this point, the fermenting must is added to wine previously made from the main portion of the crop. The cellar may be heated to facilitate the slow refermentation of the wine/must mixture. The second yeast fermentation, induced primarily by Sacch. cerevisiae, donates a light frizzante that enhances the early drinkability of the wine. The process also delays the onset, if not the eventual occurrence, of malolactic fermentation.

Besides Chianti, several other Tuscan red wines are made from the 'Sangiovese' grape. These wines may involve the use of non-Italian grape varieties, such as the inclusion of 'Cabernet Sauvignon' in Carmignano, or the production of a pure varietal wine, such as Brunello di Montalcino. Although 'Sangiovese' is the most common name for the variety, local vernacular names such as 'Brunello' and 'Prugnolo' are commonly used, and superior clones, such as 'Sangioveto,' have been individually named.

Southern Italy   Southern Italy produces most of the country's wine, much of it going into inexpensive Euroblends or being distilled into industrial alcohol. The application of modern viticultural and enological practices is gradually increasing the production of better wines, bottled and sold under their own name. Examples of local cultivars producing fine red wines are 'Aglianico' in Basilicata and Campania, 'Gaglioppo' in Calabria, 'Negro Amaro' and 'Primitivo' and Apulia, 'Cannonau' in Sardinia, and 'Nerello Mascalese' in Sicily. Several indigenous cultivars also can generate interesting dry white wines, such as 'Greco,' 'Fiano,' and 'Torbato.'

In reflection of the hot dry summers in Southern Italy, sweet and fortified wines have long been the best wines of the region. These can vary from dessert wines made from 'Malvasia,' 'Moscato,' and 'Aleatico' grapes, to sherrylike wines from 'Vernaccia di Oristano,' to Marsala produced in Sicily.

SPAIN

Spain, along with Portugal, forms the most westerly of the three major Mediterranean peninsulas. The Iberian Peninsula is the largest and consists primarily of a mountain-ribbed plateau averaging 670 m in altitude. The altitude and mountain ranges along the northern and northwestern edges produce a long rain shadow over most of the plateau (Meseta). With its Mediterranean climate, most of Spain experiences hot, arid to semiarid conditions throughout the summer.

Climatic conditions have markedly influenced Spanish viticultural practice. Because of a shortage of irrigation water, most vines are trained with several low trunks or arms (≤50 cm), each pruned to two short spurs. The short, unstaked bushy vines limit water demand and shade the soil where most surface roots are located. Vine density is kept as low as 1200 vines/ha in the driest areas of the Meseta (La Mancha) to assure sufficient moisture during the long hot summers. Daytime maxima of 40° to 44°C are common in the region. In areas with more adequate and uniform precipitation, such as Rioja, planting densities average about 3000 to 3600 vines/ha. In most of Europe, densities of 5000 to 10,000 vines/ha are common. Thus, although Spain possesses more vineyard area than any other country (~1.5 million ha), it ranks third in wine production (~40 million hl/year). Its yield per hectare is about one-third that of France.

In the northern two-thirds of Spain, table wines are the primary vinous product, while in the southern third (below La Mancha), fortified sweet or sherrylike wines are produced. Until fairly recently, wine production in most regions was unpretentious. Although the wines were acceptable as an inexpensive local beverage, they had little appeal outside Spain, except for blending because of the high alcohol content and deep color. Reminders of ancient practices are evident in the continuing but diminishing use of tinajas. These large volume amphoralike clay fermentors are still employed in southern Spain (Valdepeñas and Montilla-Moriles). Currently, the Penedés region south of Barcelona is the major center for enological and viticultural innovation. Nevertheless, traditional centers of wine excellence, such as Rioja in the north and the Sherry region in the south, also are involved in considerable experimentation and modernization. Modern trends also are affecting grape and wine production in the major producing areas of central Spain and are beginning to permit the wines to compete internationally under their own appellations.

While the reputation of fine Spanish wines is based largely on indigenous cultivars such as 'Tempranillo,' 'Viura,' and 'Palomino,' most wines are produced from the varieties 'Airen' and 'Garnacha.' So extensively are these grown that they constitute the two most important cultivars (by coverage) in the world (Robinson, 1986).

Rioja   Rioja has a long tradition of producing the finest table wines in Spain. The region spans a broad 120 km section of the Ebro River valley and its tributaries from northwest of Haro (42°35′ N) to east of Alfaro (42°08′ N). The upper (western) Alta and Alavesa regions are predominantly hilly and possess vineyards mostly on slopes rising about 300 m above the valley floor. The altitude of the valley floor drops steadily eastward, from 480 m in the west to about 300 m at Alfaro. The lower (eastern) Baja region generally possesses a rolling landscape with many vineyards on level expanses of valley floor.

The south-facing slopes of the northern Sierra Cantábrica possess a primarily stony calcareous clay soil, while the south side of the valley and the north-facing slopes of the southern Sierra de la Demanda possess a calcareous subsoil overlain by stony ferruginous clay or alluvial silt. The parallel sets of mountain ridges that run east/west along the valley help to shelter the region from cold north winds and hot blasts off the Meseta to the south.

The western portions of Rioja generally yield more delicately flavored wines than the eastern region. This probably results from the cooler climate generated by the higher altitude. The altitude also produces a higher and more uniform annual precipitation. In contrast, the eastern Baja region possesses a more distinctly Mediterranean climate, with precipitation averaging about 60% that of Haro in Rioja Alta. Differences in varietal composition in the various regions also affect the characteristics of the wines. In the upper Rioja 'Tempranillo' is the predominant red cultivar, while in the Baja 'Garnacha' ('Grenache') is dominant. Soil and microclimate differences between the southern and northern slopes in the upper Rioja may explain the presence of more 'Graciano,' 'Mazuelo' ('Carignan'), and 'Garnacha' in the Alta and more 'Viura' in the Alavesa.

The wines of Rioja are generally blends of wines from several cultivars and vineyards. 'Tempranillo' is the predominant cultivar in red wines, with various amounts derived from the other red cultivars and occasionally the white 'Viura.' Most traditional white wines are blends of 'Viura' and 'Malvasia,' with small quantities of 'Garnacha blanco.' Modern white wines are nearly pure 'Viura.'

In the past, vineyards were planted in the proportions desired in the finished wine. The grapes also were picked and fermented together. Current trends are to separate the different cultivars in the vineyard to permit each variety to be picked at optimal maturity. Separate fermentation of each variety also permits blending based on the properties of each wine.

In the blend, each cultivar is considered to add a component deficient in the others. For red wines, 'Tempranillo' provides the acid balance, aging potential, and distinctive fragrance desired, 'Graciano' donates additional subtle flavors, 'Mazuelo' adds color and tannins, and 'Garnacha' confers alcoholic strength. With white Riojas, 'Viura' provides a light fruitiness and resistance to oxidation, while 'Malvasia' donates fragrance, color, and body.

Vinification procedures often vary considerably between bodegas (wineries). For red wines, traditional procedures involve a form of carbonic maceration (see Chapter 9) where whole grape clusters are placed in *lagars* (open concrete or stone vats) for fermentation.

Large bodegas may employ a semi-carbonic maceration process, where the grapes are stemmed before being added uncrushed to fermentation vats. The modern tendency involves the standard processes of stemming and crushing before fermentation in either wooden vats or stainless steel tanks. The latter more easily permit cooling to keep the fermenting must below 25°C.

Traditionally produced white wines involve maceration on the skins for several hours before pressing. Usually fermentation occurs at warm temperatures (25° to 28°C). Modern white wines are separated from the skins shortly after crushing and fermented at cool temperatures (14° to 18°C).

The preferred method of maturation also varies considerably. The traditional procedure for white and red wines is prolonged lees contact, associated with infrequent racking, and several years of aging in oak barrels (225 liters). American oak barrels are preferred and used repeatedly to avoid a new-oak flavor. During aging, the wines mellow, acquire a distinctive vanilla flavor, and develop a complex, slightly oxidized bouquet. The modern trend is to use shorter aging, a higher proportion of new oak barrels, and longer in-bottle aging.

Penedés Catalonia in northeastern Spain possesses several viticultural regions along its coast. In the past, Catalonia was known largely for its alcoholic, dark red wines added to weaker French wine. It also produced a special, sweet, oxidized red wine that exhibits a *rancio* odor, namely, *priorato* from Tarragona. However, Penedés, located just southwest of Barcelona, is now the most productive and important wine-producing region of Catalonia.

Despite a relatively small size, Penedés (~41°20′ N) produces about 1.5 million hl of wine per year from 25,000 ha of vines. In the late 1800s, it began the transformation that now makes Penedés a major producer of sparkling (*cava*) wine. In addition, innovations begun by Bodega Torres in the 1960s have resulted in a break with tradition. Considerable commercial success followed the introduction of cold fermentation for white wines. Red wines with limited oak maturation also are being produced in increasing quantities. Penedés now produces a greater range of wines from both indigenous and foreign cultivars than any other Spanish wine region.

Although Penedés consists of a narrow strip of land along the Mediterranean coast, it shows considerable climatic variation across its 30 km width. Penedés changes westward from a hot coastal zone to cool slopes in the coastal Catalonian Mountains. Most of the traditional red cultivars, such as 'Monastrell,' 'Ull de Llebre' ('Tempranillo'), 'Cariñena,' and 'Garnacha,' are grown on the coastal plain. However, most newer vineyards are located along the more temperate central zone at an altitude of about 200 m. The majority of grapes are traditional cultivars, such as 'Xarel-lo' and 'Macabeo' ('Viura'). Nevertheless, several foreign red cultivars are grown in the region, notably 'Cabernet Sauvignon.' The *cava*-producing facilities utilize most of the local white grape harvest. The mountainous foothills further west have an even cooler, more moist environment. This favors the growth of local cultivars, such as 'Parellada,' and foreign cultivars, such as 'Chardonnay,' 'Sauvignon blanc,' 'Riesling,' and 'Pinot noir.' This region possesses slopes ranging in altitude from 500 to 800 m above sea level.

Although training still favors *Goblet* systems commonly used throughout Spain, the trunks are often higher, with the first branches allowed to develop 50 cm above ground level. Nevertheless, most foreign cultivars are trained using trellising systems common throughout most of central Europe. Because of the above average precipitation in much of Penedés, vine density is higher (4000 to 5000 vines/ha) than usual in Spain. Yield per hectare correspondingly is considerably above the national average.

Southwestern Spain   Sherry is produced in southwestern Spain, where the lowlands of the Guadalquivir River valley meet the Atlantic Ocean. The region encompasses an area about 60 km in diameter, centered around Jerez de la Frontera (36°42' N). The best sites are mostly aggregated in the north and northwest, from around Jerez to the coastal town of Sanlúcar de Barrameda. The area contains about 19,000 ha of vines and annually produces about 1.2 million hl of wine.

The most significant factors influencing the quality and stylistic features of sherry are the procedures that follow vinification (see Chapter 9). Nonetheless, microclimatic features influence the quality of the base wine used in sherry production.

One of the more important features involves modification of the vine microclimate by the soil. *Albariza* sites, possessing a high chalk content (30–60%), exhibit several desirable properties. The soils are very porous and permit the rapid uptake of the primarily winter rains. Under the hot summer sun, the soil forms a hard, noncracking crust. This increases the water available by restricting evaporation from the soil surface. Since the water-holding capacity of the soil is only about 35% by weight, soil depth is important to sustained water supply through the typically long summer drought. Level land and low hills tend to limit excessive water loss by drainage. Because of the low latitude (36°42' N), vineyard slope and aspect are relatively unimportant. Although the low altitude of the region provides no cooling, proximity to the Atlantic Ocean has a moderating influence on the climate. As a consequence, both winters and sum-

mers are comparatively mild, with maximum temperatures rarely rising above 37°C (versus 45°C and above on the Meseta). Ocean breezes also increase the relative humidity of the region.

While *albariza* soils generally possess desirable properties, they also have drawbacks. The low fertility and susceptibility to nematode infestation limit fruit yield.

The predominant cultivar grown is the *fino* clone of 'Palomino.' It has the advantage of higher yield than standard clones. Other cultivars grown in limited amounts, especially on poorer sites, are 'Pedro Ximénez' and 'Moscatel.' The latter is used only for a varietally designated sweet wine. The main advantages of the 'Palomino' variety are its tough skin, disease resistance, and low varietal aroma. The disadvantages of its late maturity are partially offset by the heat and light reflected from the white *albariza* soil.

PORTUGAL

Portugal produces an amazing range of wines for a comparatively small country. In addition, wine production is largely limited to the northern half of the country. Nonetheless, it annually produces about 11 million hl of wine from 380,000 ha of vines. Thus, Portugal ranks among the top 10 wine-producing countries of the world. Among the wines produced are possibly the world's most popular wine (Mateus rosé), some of the most aromatically intricate wines (*porto*), a delicately effervescent white wines (*vinho verde*), dark tannic long-aging red wines (Bairrada and Dão), and a complex baked wine (Madeira, from the island of the same name). All of these are produced almost exclusively from indigenous Portuguese cultivars, rarely planted elsewhere, even in neighboring Spain. Although producing some of the most skillfully blended wines for mass distribution, Portugal has retained much of its vinous heritage unaffected by outside influences.

The Upper Douro   The eastern portion of the Douro River Valley is the origin of Portugal's most presitigious wine, port. Port is produced in what was one of the most inaccessible and inhospitable regions of Portugal. The delimited port region stretches for about 120 km along the banks of the Douro and its tributaries from Barqueiros in the west to Barca d'Alva near the Spanish border. The region roughly parallels 41° N latitude. Although the present area has expanded extensively eastward from the delimitation in 1761 (Fig. 10.12), most of the 34,000 ha of vines are still centered around Pêso da Régua. The region currently produces about 2.2 million hl of wine annually, but only about 40% is used in port production.

Because of the steep mountainous terrain through

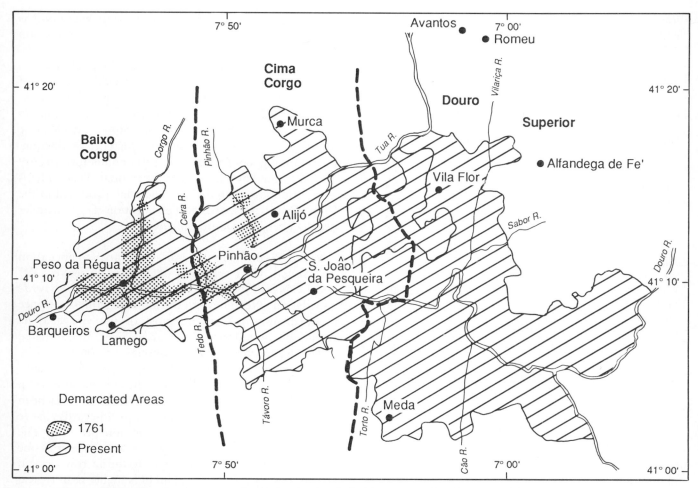

Figure 10.12 Map showing current (hatched regions) and 1761 boundaries (dotted regions) of the port wine area (Upper Douro) and the three major subdivisions of the Upper Douro. (Adapted from LANDSCAPES OF BACCHUS: THE VINE IN PORTUGAL, by Dan Stanislawski, Copyright © 1970, p. 113. Reproduced by permission of the publisher.)

which the Douro River passes, the slopes along the valley wall may possess inclinations of 60°. Because of the steep slopes and prevailing dry conditions, little soil has accumulated. Thus, most vineyards are a series of man-made sloped terraces held by stone retaining walls. The soil itself consists primarily of schist. While high in potassium (~12%), the rocky soil is low in phosphorus and organic material. These deficiencies have usually been offset by the addition of manure, often from the animals used to work the vineyard. Because most old terraces are too steep to permit mechanization, many new and some old terraces have been graded to permit partial mechanization.

As a wine style, port was developed by British shippers for an increasingly discriminating market at home. Thus, quality very early became an important aspect in the evolution of port. This culminated in a highly detailed and critical analysis of the vineyards. Each vineyard site is assigned to one of six quality categories, based on features affecting microclimate, soil conditions, and vine characteristics. Of a maximum of 1680 points, about two-thirds reflect environmental features, while the remaining third concerns only viticultural influences. More points may be subtracted for negative influences than are granted for desirable features.

Many of the factors indicated in Table 10.4 reflect the importance assigned to regional temperature and moisture conditions in the development of grape quality. For example, the Upper Douro is divided into five geographic subregions, of initially increasing rank, moving eastward. This mirrors the higher temperatures and lower precipitation in the upper regions of the valley. The most easterly subregion, beyond the watershed dividing the Tua and Vilariça rivers, often receives insufficient rainfall to offset the baking summer drought. Because the grapes may raisin before harvest, the wines rarely possess the characteristics deemed appropriate for port production. The bias for low altitude in site ranking also reflects the general desirability of the warmer, drier conditions near the valley floor versus the cooler, wetter conditions

Table 10.4 Apportionment of Points to Features Considered Important to the Quality of Grapes for Port Elaboration[a]

Trait	Penalty points	Award points	Spread	Percentage of spread
Primary importance				
Low productivity	−900	150	1050	20.6
Altitude	−900	240	1140	22.4
Physical nature of soil	−400	100	500	9.8
Locality	−50	600	650	12.8
Type of training	−500	100	600	11.8
Subtotal	−2750	1190	3940	77.4
Secondary importance				
Cultivars used	−400	200	600	11.8
Degree of slope	−1	101	102	2.0
Subtotal	−401	301	702	13.8
Tertiary importance				
Site aspect	−30	100	130	2.6
Vine spacing	−50	50	100	2.0
Soil texture	0	80	80	1.6
Vine age	0	60	60	1.2
Shelter	0	80	80	1.6
Subtotal	−80	370	450	8.8
Total	−3231	1861	5092	100.0

[a] Data from Instituto do Vinho do Porto, Ministério da Agricultura, Porto (1992).

of the upper slopes. Schistose soils are preferred over granitic soils, possible because of the fractured structure of the former. Schists more readily permit rain and root penetration, features vital to a steady and sufficient supply of water through the summer drought. The narrow sections of fertile alluvial deposits on the valley floor are given 600 demerit points, almost assuring that grapes grown there will be excluded from use in port production. Aspect and degree of slope are calculated into the ranking, but they are considered of minor importance.

Of viticultural features, the training system is considered of the greatest importance, earning up to 12.1% of possible points. The highest number of points is assigned for single-cane *Guyot* training. It has the advantage of restricting vine vigor on nutrient-poor soils. Arbors are discouraged by being penalized 500 demerit points. In contrast, cultivar composition is awarded only a maximum of 150 points. Although some cultivars, such as 'Bastardo,' 'Mourisco,' 'Tinta-Francisca,' 'Tinta-Cão,' and 'Touriga-Nacional,' are considered finer than others, the granting of only 6.1% of points to cultivar composition is revealing. It acknowledges the need for the grower to be able to adjust the cultivar composition of the vineyard to the demanding and variable conditions of the region and site. It also recognizes the contribution of the varietal mix to the final quality of port. While almost as many white cultivars are permitted as red,

most ports are red. Thus, although some white port is produced, most white grapes are used in the production of table wines.

Depending on the ranking of the vineyard, and the current market for port, port officials set the quantity of wine permitted from each vineyard category. For example, vineyards in category A normally can sell up to about 700 liters/1000 vines to port shippers, while those in category F seldom can sell any for port production. Each ranking, commencing at 1200 points for category A, is separated by a 200-point differential. This explains why little more than 40% of the wine produced in the Upper Douro is used for port elaboration. The remainder usually goes into Douro table wine.

Vinho Verde   *Vinho verde* is the largest appellation in Portugal. The five regions of the zone lie between 41° and 42° N latitude. While the borders of the *vinho verde* appellation extend beyond the limits of Minho province in northwestern Portugal, most of the vineyards lie within its boundaries. Vineyards cover an estimated 71,000 ha and yearly produce about 2 million hl of wine.

In contrast to the Upper Douro, where schistose soils are predominant, granite forms the soil substratum in Minho. While rich in potassium, the soil is deficient in phosphorus. Depending on the region, vineyards receive about 120 to 160 cm of precipitation per year. The weather becomes progressively drier and warmer as one moves from the eastern highlands to the narrow western coastal belt. Although the rainfall is higher than typical for the Iberian Peninsula, the distribution is distinctly Mediterranean, peaking in the winter and being minimal during the summer.

Legal impositions designed to prevent competition with Upper Douro producers and increases in population density led to the elimination of most vineyards on fertile soils during the mid 1700s. Thus, vines were largely relegated to polyculture with other crops on terraced sites or were allowed to grow up and between trees along roadsides and fields. The latter method, though significantly complicating viticultural practice and harvesting, did free land for the cultivation of food crops. Unintentionally, the change in viticultural practice may have favored the production of the wine style for which the region is now known. Because the growth habit limited sugar accumulation and malic acid respiration, grapes were often harvested high in acidity and low in sugar content. In contrast, grapevines trained on the horizontal T trellises being used in new vineyards are more likely to reach standard sugar and acidity levels by harvest.

Due to cool winter conditions, malolactic fermentation generally occurs in late winter or spring. Because the wine historically was stored in sealed barrels or casks, the carbon dioxide released by malolactic fermentation was trapped in the wine. The slight fizz, combined with the low alcohol content of the wine (~8 to 9.5%), produced a light, refreshingly tart wine. Aromatic substances released during malolactic fermentation also may contribute to the distinctive fragrance of the wine. Under the demands of commercial success, large producers rarely depend on malolactic fermentation for the *pétillance* and commonly use carbonation.

As in port elaboration, most *vinho verdes* are blends of wines produced from several grape cultivars. Only rarely are varietal *vinho verdes* produced. An exception is the *vinho verde* produced from 'Alvarinho' grapes in Monção, across from Spanish Galicia. The wine is usually less effervescent than most *vinho verdes* and has traditionally been grown in low arbors.

Although most *vinho verdes* are red (~70%), the white version is almost exclusively the style exported. Thus, most exported *vinho verde* comes from the Northern Monção and southern Penafiel regions, areas specializing in cultivating white varieties. 'Alvarinho' cultivation is mainly centered in Monção, while other white varieties, such as 'Loureiro,' 'Trajudura,' 'Azal,' 'Avesso,' 'Bataco,' and 'Pedernã,' occur throughout Minho.

Madeira   Grape cultivation occurs around much of the island of Madeira but is concentrated in the south. Madeira is the largest and most significant of the three volcanic islands in the Madeira Archipelago, which is situated about 640 km off the western coast of Morocco. The latitude of 32°40′ gives Madeira a subtropical climate that is moderated by the surrounding Atlantic Ocean. Its ancient volcanic peak, rising to a height of 1800 m, induces sufficient precipitation to favor luxurious plant growth, but also produces a humid climate. This resulted in a decimation of the vine population when powdery mildew and phylloxera reached the island in the late 1800s. This partially explains the displacement of many traditional grape varieties with French–American hybrids, such as 'Jacquez.' With modern chemical control and phylloxera-resistant rootstock, traditional cultivars are regaining vineyard space lost over a century ago. The smaller island of Porto Santo, without significant highlands, is considerably hotter and drier. No grapes appear to be grown on the smallest island of Desertas.

The steep slopes of Madeira have required the construction of a tiered series of narrow terraces up the slopes of the volcano. The volcanic ash which forms most of the soil is claylike and rich in potassium, phosphorus, and nitrogen. Although much of the island is cultivated, vineyards are largely restricted to lower and mid altitudes. Of traditional varieties, 'Malmsey' ('Mal-

vasia') is concentrated closest to the shore, generally at altitudes up to about 300 m; 'Verdelho' is usually cultivated between 300 and 600 m; 'Bual' is planted between 400 and 1000 m, but down to the coast on the north side; 'Tinta Negra' is generally grown between 300 to 1000 m; and 'Sercial' cultivation is limited mostly to 800 to 1100 m.

GREECE

Greece occupies the most easterly of the three major Mediterranean peninsulas. Although Greece produces about 4.5 million hl of wine from 78,000 ha of wine grapes, its major importance comes from its role in the ancient dispersal of winemaking knowledge throughout the western Mediterranean.

In ancient times, highly regarded wines were produced in northern Greece, notably Thrace and the island of Thásos (40°41′ N), and especially the Aegean islands of Khios (38°20′ N) and Lemnos (39°58′ N) off the coast of Turkey. Today, only Samos, a 'Muscat'-based dessert wine from the island of Samothráki (40°23′ N), reflects the former vinous glory of the Aegean Islands.

Many current Greek wines still use an ancient winemaking technique, namely, the addition of resin. Its addition during fermentation gives *retsina* a terpenelike character. The preferred resin source is the Aleppo pine (*Pinus halepensis*) that grows south of Athens in central Greece. Most of the wine (~85%) is produced from the 'Savatiano' and 'Rhoditis' cultivars, which produce white and rosé *retsinas*, respectively. Although *retsinas* have maintained broad popularity in Greece, the appreciation of this wine style has not spread significantly. The other distinctive Greek wine commonly seen internationally is a red fortified dessert wine made from 'Mavrodaphne.' The appellation has the same name, Mavrodaphne. It comes from Patras (38°15′ N) on the northern coast of Peloponnisos.

Like Portugal, Greece still produces the vast majority of its wines from indigenous, possibly ancient, varieties. Thus, Greece retains a wealth of grape varieties whose merits in most cases are ill-explored. For example, the 'Rhoditis' grape can produce a delicate and uniquely flavored dry white wine quite appreciated internationally. The loss of political independence and economic wealth for almost two millennia, combined with Ottoman oppression, has deprived Greece of the potential vinous excellence inherent in its climate and early winemaking expertise.

Eastern Europe

The viticulturally important countries of Eastern Europe include Hungary, Romania, Bulgaria, the states that formed Yugoslavia, and the western regions of the former Soviet Union. The region includes two major arched mountain chains, the Carpathian and Dinaric. Together they form the Hungarian and Romano-Bulgarian Basins. Unlike the Alps, the mountains seldom rise above 1500 m and have rounded, tree-covered tops. The region also includes the lowlands north and between the Black and Caspian Seas.

With the exception of the Dalmation (Yugoslavian) coast, the entire region has cold winters. Hot summers occur everywhere except in the high mountains. Along the Dalmatian coast, precipitation averages over 100 cm per year, and summer drought is experienced. Moving eastward, rainfall declines to less than 50 cm north of the Black Sea, occurring primarily in the spring and summer months.

Because the region has served as a gateway between Western Europe and Asia, it has borne the brunt of many incursions from both the east and west. It was also decimated by Ottoman attack and rule. This periodically forced much of the population out of the plains to seek refuge in the mountains. The vinous effect was severe retardation in the development of viticulture and wine production for centuries. The least affected was Hungary, the region most distant from Turkey. It was the last country invaded and the first free from foreign rule.

The lack of significant resources of coal, iron, and water power for rapid industrial growth also favored subsistence farming, which has persisted throughout much of the region into modern times. Wines were produced and consumed locally as unsophisticated but safe food beverages.

Until recently, comparative isolation from Western Europe resulted in wine production being based on indigenous cultivars. Thus, Eastern Europe contains a wonderfully complex collection of grape varieties. How many diamonds-in-the-rough occur among the varieties is little known, at least outside the respective regions. Among the vast collection of cultivars, it is reasonable to expect some to be of equal quality to those in Western Europe. It is hoped that their cultivation will not be supplanted by western cultivars in the rush to appease the international bias for recognized cultivar names. The world already seems amply supplied with 'Cabernet Sauvignon,' 'Chardonnay,' and 'Riesling' wines.

The area also possesses the largest assortment of freely growing wild *V. vinifera*. These may constitute a valuable genetic resource for future cultivar improvement.

Because of the major vineyard-expansion program following the Second World War, Eastern Europe had become, by the early 1990s, a major wine producer. The former Soviet Union was fifth in world ranking, with 18 million hl, and Romania, Hungary, and Yugoslavia in tenth, eleventh, and twelfth place, respectively.

HUNGARY

Hungarian vineyards cover only 2% of the cultivated land area and are dispersed throughout the country. The total vineyard area is 138,000 ha and annually produces about 5 million hl of wine.

The northern latitude (45°50' to 48°40' N) and cold winters of Hungary are reflected in the cultivation of white cultivars. However, the warm summers often permit the grapes to develop high °Brix values, permitting the frequent production of slightly sweet white wines high in alcohol (13 to 14%). These apparently suit the Hungarian preference for spicy food. While most wines are intended to be consumed with meals, some dessert wines are produced. The most famous of these are the Tokaji Aszú wines.

The majority of wine (60 to 70%) comes from the sandy Hungarian Basin in the south central region of the country. Most of the production is consumed locally, with higher quality wine coming from the slopes bordering the basin usually exported. Examples of some of the better sites are Pécs and Vilány from the southwest, Lake Balaton to the west, Sopron and Mór from the northwest, and Eger and Tokaj in the northeast.

Tokaji comes from an ancient volcanic region in the northeast corner of the country across from eastern Slovakia. Vineyards occur on the sandy loam of the southeast-facing slopes bounded by the Szerencs and Bodrog Rivers (48°06' N) at altitudes between about 100 and 300 m. The wine usually comes from a blending of several cultivars. The dominant varieties are 'Furmint' (60 to 70%), 'Hárslevelű (15 to 20%), and small plantings of 'Sárga Muskotály' ('Yellow Muscat'). Additional cultivars grown are 'Leányka,' 'Traminer,' and 'Wälschriesling.' While the sweet, botrytized *aszú* styles are the most renowned (Chapter 9), dry versions are more common.

One of the red wines frequently seen internationally is Egri Bikavér. The wine comes from Eger (47°53' N), about 100 km southwest of Tokaj and northeast of Budapest. It is produced from a blend of several varieties, including 'Kadarka,' the most significant Hungarian red cultivar. Also potentially involved are 'Merlot,' 'Pinot noir,' and 'Oporto.'

Other wines commonly seen internationally come from the northwestern slopes of Lake Balaton (46°46' to 47° N). In the central zone, vineyards are scattered over the steeper hills that rise about 150 to 200 m above the shoreline. The lake moderates rapid climatic change and reflects light up onto the vines. Some protection from north winds is derived from the Bakony Forest, which reaches a maximum altitude of 700 m. The sandy slopes drain well and warm early in the season. The wines are almost exclusively white, alcoholic (13 to 15%), and frequently semisweet. Several indigenous cultivars are grown, such as 'Kéknyelü' and 'Zöldszilváni,' as well as foreign cultivars, such as 'Pinot gris' ('Szürkebarát'), 'Muscat Ottonel,' and 'Wälschriesling' ('Olaszrizling').

Wines coming from the plains are produced from a wide variety of cultivars, the most common being 'Olaszrizling' and 'Ezerjo.'

YUGOSLAVIA

Combined, the states that formerly composed Yugoslavia show a superficial resemblance to Italy, their neighbor across the Adriatic. Both are elongated along a northwest/southeast axis, have mountainous regions along the northern frontiers, and are divided along the length by mountain ranges. However, the extended eastern connection (41° to 46°50′ N) of Yugoslavia with the landmass of Eurasia permits the ready access of continental climatic influences. Thus, most of the country experiences cold, snowy winters and hot, moist summers. The Mediterranean influence is limited to the western (Dalmatian) coastline, between the Adriatic Sea and the Dinaric Mountains. Even here, rainfall is higher than is typical for most Mediterranean regions.

Except where the Hungarian Basin extends into northeastern Croatia, there is very little lowland. Most of the country is mountainous and generally above 400 m in altitude, with extensive areas above 1000 m. Because of this feature, most vineyards (225,000 ha) are arranged around the edges of the former country, namely, the Dalmatian coast and associated islands, the Hungarian Basin in the northeast, and the Morava–Vardar Corridor in the southeast. Combined the Yugoslav states annually produce about 5 million hl of wine. Because of the highly diverse climatic, topographic, and ethnic divisions, it is not surprising that the wine styles Yugoslavia produce are equally diverse.

Partially owing to the proximity of Slovenia to Austria and Italy, and its once being a part of Austria, its wines and grape varieties are similar to those of its neighbors. 'Graševina' ('Wälschriesling'), 'Silvaner,' 'Sauvignon blanc,' 'Traminer,' Šipon' ('Furmint'), and the indigenous 'Plavać' are the most common varieties. The Adriatic moderates the alpine climate, and the region benefits from mild winters and temperate summers. The best known wine region is situated around Ljutomer in the northeast (46°25′ N).

The wines of Croatia fall into two groups, those from the Hungarian Basin and the mountains north of Zagreb and those from the Dalmatian coast. The highlands in the northeast, adjacent to Slovenia, continue to show an Austrian influence. The dominance of white cultivars continues onto the Hungarian Basin lowlands. However, on the plains the prevalence of 'Wälschriesling' reflects the similar dominance of the cultivar in the neighboring region of Hungary. In contrast, the Dalmatian coastal region produces predominantly red wines. It also cultivates primarily indigenous Yugoslavian cultivars. The most distinctive of these is the dark red 'Plavać mali.' The most extensively grown red cultivar in much of Yugoslavia is 'Prokupac.' Indigenous white cultivars favored are 'Plavać,' 'Maraština,' 'Grk,' 'Vugava,' and 'Zilavka.' The Istrian Peninsula in the northeast reflects in its varietal plantings a strong historical association with Italy.

Serbia is the main region for bulk-wine production. Most of the production is red and comes from 'Prokupac.' 'Smederevka' is the main white variety. However, Western European cultivars are tending to replace native cultivars, presumably to gain acceptance in foreign markets.

Although Macedonian vineyards were decimated during the Ottoman domination and subsequently destroyed by phylloxera, they were reestablished after the Second World War. Most of the cultivars chosen were indigenous cultivars, including such red varieties as 'Prokupac,' 'Kadarka,' and 'Stanusina.' There also are smaller plantings of the white cultivars 'Žilavartea' and 'Smederevka.'

ROMANIA

Romania is another Eastern European country having most of the vineyards distributed around the periphery of the country. The eastern curved arch of the Carpathian Mountains divides the country in two. Mountains reach up to 2400 m and enclose the central Transylvanian Plateau at an elevation of about 600 m.

The 245,000 vineyard ha in Romania are divided about equally between red and white cultivars. From these the country annually produces about 6 million hl of wine. Extensive plantings of both native and Western European cultivars occurred following the Second World War. The better, local, white cultivars appear to be 'Fetească alba,' 'Grasă de Cotnari,' 'Tămîioasa romînească,' and 'Frîncuşa.' The most well-known red varieties are 'Fetească neagră' and 'Babeasca neagră.' As with other indigenous Eastern European cultivars, their evaluation elsewhere could expand the variety and interest of wines produced worldwide.

Few Romanian wines are well known internationally. This partially relates to most trade having been conducted with its Eastern European neighbors. This may change with the dramatic political transformation of former Eastern Bloc countries in the 1990s. The most highly regarded of Romanian wines, at least historically, are those coming from around Cotnari and Grasă (47°27′ N), along the lower slopes of the Carpathian Mountains in the northeast.

BULGARIA

Like so many other Eastern European countries, in Bulgaria, Ottoman domination brought wine production to a virtual halt for several centuries. Viticulture was reestablished following the First World War, but the vineyards were extensively destroyed during the Second World War. The major replanting that followed, combined with an emphasis on export to western countries, helps explain the predominance of western cultivars in Bulgaria. Bulgaria currently exports over 60% of its wine production of nearly 3 million hl from 140,000 ha of vines.

Because winery facilities also had to be reconstructed, the wine industry in Bulgaria is modern. The emphasis on quantity, however, has been at the expense of fine wine production. The vineyards lie largely between latitudes 41°45' and 43°40' N, in a land ideally situated to viticulture.

Of native cultivars, approximately 8000 ha are devoted to 'Dimiat,' a grape commonly used for the production of sweet wines. The Georgian cultivar 'Rkatzietéli' is the dominant white cultivar. It is used extensively in wines sold in Eastern Europe. Indigenous red cultivars are 'Pamid,' 'Shiroka Melnishka Losa,' and 'Mavrud.'

SOVIET UNION (COMMONWEALTH OF INDEPENDENT STATES)

The countries that formed the western portion of the former Soviet Union lie in an immense plain. The East European (Russian) Plain stretches from the Caucasus 2400 km north to the Arctic Ocean and an equal distance eastward to the Ural Mountains. Without moderation by large bodies of water or mountains to deflect air flow, continental weather systems move largely unimpeded over the region. Thus, summers may be very hot and winters bitterly cold, with marked and rapid changes in temperature throughout much of the year. Precipitation falls off moving eastward, especially north of the Black Sea. These climatic influences limit commercial viticulture primarily to the more moderate climates north of the Black Sea, Moldavia, Georgia, and the southern region of the Ukraine and Russia. Other Russian vineyards, and those of Azerbaijan, are found along the eastern edges of the Caspian Sea.

Owing to a decision taken in the early 1950s, the Soviet Union embarked on a massive vineyard expansion program. Vineyard area increased from about 400,000 to over 1.1 million ha by the 1970s. This propelled the Soviet Union from tenth position, in terms of vineyard hectarage, to second in importance worldwide. The annual yield of about 18 million hl places the region fifth globally in terms of production. This partially reflects the considerable proportion of its vineyard area devoted to producing table and raisin grapes.

About 50% of the vineyard area is located in the European portion near the Black and Caspian seas; 30% is positioned in the Transcaucasian zone between the Black and Caspian seas, with the remainder occurring in south central Asia. In the latter region, production of raisins and table grapes is the main viticultural activity.

Although 70 foreign cultivars are grown throughout the former Soviet Union, they constitute only about 30% of plantings. Some 100 indigenous varieties occupy most of the remaining hectarage. Over 200 additional varieties are cultivated, but only in limited quantities. The most extensively grown white cultivar is 'Rkatzietéli,' covering 250,000 ha. This makes 'Rkatzietéli' second only to 'Airen' as the most extensively grown (by hectarage) of white grape cultivars (Robinson, 1986). Other commonly grown local varieties are 'Mtsvane,' originally coming from Georgia, and 'Fetească.' Popular red cultivars are 'Saperavi,' 'Khindogny,' and 'Tsimyansky.' Many western cultivars, such as 'Traminer,' 'Riesling,' 'Aligoté,' and 'Cabernet Sauvignon,' are grown in Moldavia and the southern Ukraine.

Although the former Soviet Union (C.I.S.) is a major wine producing region, only about 2.5% of the annual production is exported. Of the now independent states, Moldavia has the largest production, consistent with possessing the most extensive vineyard area, about 200,000 ha. It is followed by Azerbaijan, with 180,000 ha. In addition to producing brandies and fortified wines, Azerbaijan grows a large proportion of its grapes to be used as a fresh fruit crop. The Ukraine, including the Crimea, possesses 174,000 ha of vines, while the vineyard area in Russia covers 147,000 ha. Although possessing smaller total vineyard areas, Georgia and Armenia are important in the production of fine wines. In these regions, vineyards may date back to the first attempts by humans to establish viticulture.

Unlike the wines of most wine-producing regions, about three-quarters of all wine produced in the former Soviet Union possesses a distinctly sweet taste (over 15% sugar). Sparkling wines also are especially appreciated, and constitute about 10% of the total wine production (about 256 million bottles in 1989). To economize production costs, a continuous fermentation system was developed. Much of the production of sparkling wine is centered close to the Black Sea, around Krasnodar in the Kubar Valley (45°03' N) and near Rostov-na-Donu along the Don River (47°16 N) and in southwestern Russia.

One of the major factors limiting viticulture in the former Soviet Union is the continental climate. The bitterly cold winters require that the vines of about 50% of the vineyards be laid down and covered with soil each winter. The annual practice is not only expensive, but exposes the vines to mechanical damage and additional disease problems. Therefore, an extensive breeding program directed at increasing cultivar hardiness has been in progress for several decades. Central to the success has been the incorporation of frost resistance from *Vitis amurensis*. Improved cultivars, such as 'Burmunk,' 'Mertsavani,' 'Karmreni,' and 'Nerkarat,' are able to survive adequately without burial during the winter. Development of high trunks also is beneficial in raising buds above the coldest zone near ground level. Restricting irrigation late in the season and applying cryoprotectants such as mirval, migugen, and krezatsin can improve bud survival (Kirillov *et al.*, 1987).

North Africa and the Near East

In both North Africa and the Near East, wine production has been declining for decades. This has been particularly marked in several Moslem countries, where religious interdiction against wine consumption often has been reimposed on independence from European control. For example, wine production in Algeria has dropped to 3% of the preindependence 1962 value. However, even in Israel, where wine has religious significance, current production is about 40% of the 1971–1975 level.

The region is potentially capable of producing fine wines, for example, the wines from Château Musar in the Bekaa Valley of Lebanon. However, most wines have been either excessively sweet or flat and unbalanced, at least to those accustomed to standard wines.

Because wine consumption can be a serious felony in Islamic states, most grapevines grown are table or raisin varieties. The rapid loss of acidity in such grapes during maturation is appropriate for a fresh fruit crop or in raisining, but it is undesirable where wine is the intended product. The colossal cluster of grapes reportedly carried back to Moses from Canaan (Numbers 13:23) was indicative of the agricultural fertility of the region, not the quality of its wine grapes.

Far East

Both China and Japan have been introduced to winemaking on several occasions. Although *Vitis vinifera* cultivars have continued to be cultivated in these countries, wine has not become part of the cultural fabric as it did in the West. Various cultural and genetic hypotheses have been presented to explain this lack, but none seem adequate. In Japan, the generally inappropriate climate is likely the dominant reason. In China,

however, several regions are suitable for the cultivation of *V. vinifera*. Regardless of the reason(s), viticulture and especially winemaking have attracted limited interest in the Far East.

CHINA

Surprisingly, the majority of Chinese vineyards occur in one of the most rigorous climatic regions, the far northwestern province of Xinjiang Uygur (~40°15′ N). Its location in central Asia, north of Tibet, exposes the vines to extremes of drought, summer heat, and frigid cold. The grapes grown on 27,000 ha are used almost exclusively for raisin production. Limited cultivation of table and wine grapes occurs in the east, north of the Yangtze River (~30°30′ N). Monsoon rains and summer heat south of the Yangtze are unfavorable to most grape cultivars, while north of 35° latitude cold winters often require *V. vinifera* vines to be covered with soil during the winter. However, the alkaline nature of the soil makes it suitable for growing *V. vinifera* on its own root system. This is possible as phylloxera is of limited occurrence in China.

Cultivation of *V. labrusca* cultivars and hybrids occurs in the north central portions of Manchuria (~44° N). Here, abundant rainfall produces more acidic soils, suitable to *V. labrusca* cultivars. However, use of cold-hardy *Vitis rupestris* or *V. amurensis* rootstock is generally necessary to limit frost-induced root damage during the frigid winters. Temperatures can frequently dip to −30° to −40°C. The indigenous *V. amurensis* is the most cold tolerant of *Vitis* species, surviving temperatures down to −50°C without significant harm. Several pure *V. amurensis* cultivars are grown, such as 'Tonghua' and 'Changbeisan,' but their unisexual habit makes yield erratic. The bisexual cultivar 'Shuanqing' is a major improvement. However, *V. vinifera* hybrids, such as 'Beichum' and 'Gongniang,' possessing cold hardiness derived from *V. amurensis*, are more popular. Older *V. vinifera* varieties still widely cultivated are 'Longyan,' 'Niunai,' and 'Wuhebai' ('Sultana').

Owing to the cold sensitivity of most cultivars, the vines are trained to ease removal from the trellis for winter burial. In hilly terrain, trellising has usually been on sloping elongated pergolas. On level ground, fan training has been common. Both systems use multiple cordons or bearing shoots to maintain sufficient wood subtlety to permit the annual lying down and raising of the vines. A modern adaptation of existing systems, incorporating concepts of canopy management, is the single Dragon training system using a vertical T-trellis (Fig. 10.13).

The most important winemaking area of China is the Liaoning Peninsula in Shandong province (~37° N).

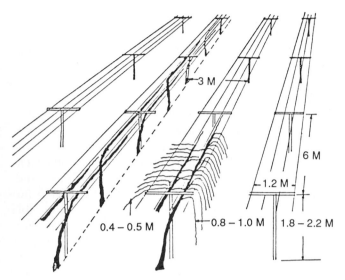

Figure 10.13 Diagrammatic representation of the single Dragon training of vines in China. (By Xiu Deren from Luo, 1986, reproduced by permission.)

Shandong possesses about 14,000 ha of vines. The area experiences both maritime and continental climatic influences. About 12,700 ha of vines are cultivated in the coastal region of the Bo Hai Bay, and west and north of Beijing (~39°55′ N). The other main regions include 11,500 ha in Henan province (33° to 34° N), and the adjoining northern portions of Jiangsu and Anhai provinces, as well as 3600 ha in Manchuria (42°30′ to 46° N).

Traditionally, Chinese wines have been fortified and sweet. Often only 30% of the contents have been grape derived (Hua, 1990). The government is beginning to improve quality standards, but the required purchase of local grapes, regardless of maturity, keeps wine quality minimal. Joint ventures with European firms have more freedom in choosing grapes, and wines up to international standards are being produced largely for export. Total wine production in China is approximately 900,000 hl per year.

JAPAN

While winemaking experience in Japan goes back at least to the eighth century A.D., acceptance of wine has been slow. Viticulture currently occurs throughout the main island of Honshu and the northern island of Hokkaido. Nevertheless, activity is still mostly concentrated around Kofu (35°41′ N), about 100 km southwest of Tokyo. The drier foothills climate of Mount Fuji provides the most suitable conditions in Japan for the growth of European *V. vinifera* cultivars.

The major problems facing Japanese viticulture are the monsoon rains that often occur during flowering and harvesting and the cold winters. To counteract these undesirable conditions, the vines typically are trained high onto vertically branched pergolas. Sloped sites also are preferred. These provide both better drainage and sun exposure, and avoid competition for the limited supply of level arable land as well.

Of *Vitis vinifera* cultivars, the indigenous 'Koshu' variety is the most well adapted to Japanese conditions. The other dominant cultivars are *V. labrusca* hybrids such as 'Delaware,' 'Campbell's Early,' 'Neo-Muscat,' and 'Muscat Bailey A.' Both disease resistance and acid soil tolerance make *V. labrusca* hybrids more suitable for cultivation than most *V. vinifera* varieties. Currently, Japan produces about 550,000 hl of wine from 27,000 ha of grapevines. This is about 1/100 the volume of sake produced from fermented rice.

Australia and New Zealand

AUSTRALIA

Although Australia is not among the leading top 10 wine-producing countries, the quality of Australian wines makes it significant on the world scene. The range of climatic conditions, from the cool, moist highlands of Tasmania to the hot, arid conditions of the Murray Valley, creates many opportunities for producing distinctive wines. The judicious selection of premium European cultivars has given even the simplest Australian wines a quality seldom found elsewhere.

For over 100 years, local preference for sweet fortified wines dictated wine production in Australia. However, consumer preference has shifted since the 1960s to dry table wines. This led to major changes in both viticultural and enological practice, as well as the expansion of grape growing back into cooler regions. Currently, viticulture is concentrated in the southeastern portion of the continent. The region forms a triangle from the Clare Valley north of Adelaide (South Australia), to Muswellbrook in the Hunter Valley (New South Wales), and south to Geelong below Melbourne (Victoria). The region incorporates most of the eastern Australian continent between the 10° and 20°C annual isotherms (Fig. 1.1). However, the latitude range (32° to 38° S) is equivalent to that between southern Spain and Madeira. While much of Australia is hot and arid, southern coastal regions are significantly influenced by cold Antarctic currents flowing eastward below the continent. The high pressure systems that prevail over the South Indian Ocean generate westerly winds that cool southern and western coastal regions. In addition, the Southeast Trades and low pressure weather systems coming down the eastern Australian coast provide associated regions with precious moisture. Thus, Australian viticultural regions are influenced by a range of climatic conditions from Mediterranean to temperate maritime.

The mid latitudes and generally dry climate in Australia provide longer growing seasons, with higher light intensities than typical in Europe. This means that planting on sloped sites is less important than in Europe. Also, the vineyards are rarely subject to frost. Rainfall in coastal regions is usually adequate for viticulture and well distributed throughout the year. Going inland, precipitation rapidly declines toward the arid interior. Most Australian vineyards lie in the transition zone between the coast and interior where irrigation is typically necessary.

With the exception of some regions in Victoria and adjacent New South Wales, grafting for phylloxera control is unnecessary (Ruhl, 1990). Although phylloxera was accidentally introduced into Australia in the nineteenth century, it has not spread widely. Nevertheless, grafting to resistant rootstock is required in most areas to control the potentially serious damage caused by nematodes.

Because of the former importance of sweet fortified wines and brandy in Australia, grape growing became, and still remains, centered along the Murray and Murrumbidgee river valleys. This region contains contiguous areas of South Australia, New South Wales, and Victoria. With the shift to dry wines, replanting focused on aromatically distinctive cultivars such as 'Chardonnay,' 'Riesling,' and 'Traminer,' and away from neutral flavored varieties such as 'Sultana' and 'Trebbiano.' Even 'Cabernet Sauvignon' has joined 'Shiraz' ('Syrah') as the most important red varieties grown in these hot regions. The proportion of these premium cultivars in the cooler viticultural regions to the south is even more marked.

While varieties such as 'Cabernet Sauvignon,' 'Pinot noir,' and 'Chardonnay' are being more extensively grown, 'Shiraz' remains the main red variety cultivated throughout Australia. 'Sémillon' ('Hunter Riesling') has long been the major white variety cultivated in the Hunter Valley of New South Wales. 'Muscat of Alexandria' is another important cultivar long grown in New South Wales and northern Victoria.

Amazing from the European perspective is the diversity of climatic regimes in which European cultivars excel. In the moderate climate of northeastern Victoria (Rutherglen), flinty chablislike wines have been made from 'Pedro Ximénez.' Fine 'Cabernet Sauvignon' and 'Chardonnay' wines are regularly produced in the hot, arid climate of the Murray River Valley. Excellent 'Marsanne' wines are produced from the cool Yarra Valley in Victoria to the subtropical region of the Hunter Valley in New South Wales. Even award-winning 'Pinot noir' wines have come from the Hunter Valley. Thus, Australia illustrates how viticultural and enological practice can offset the commonly assumed limitations of macroclimate.

Australia also illustrates that irrigation does not necessarily lead to reduced grape and wine quality. On the contrary, irrigation provides the grower with a potent means of controlling vine vigor and ripening. Furthermore, Australia has amply demonstrated that low yield is not an obligatory prior condition for high grape and wine quality.

One of the more distinctive characteristics of winemaking in Australia is the dominance of several major producers. They not only possess major holdings in widely dispersed wine regions, but they also produce a full range of wines, from bag-in-box to prestigious estate-bottled wines.

Australia is particularly unique in its general disinterest in Appellation Control laws. This permits producers to blend wines from different regions without penalty. With the variety of wines available in Australia, winemakers can blend to bring out the best features of the wines at their disposal. Even the wine often accorded the premier rank among Australian red wines, Penfold's Grange Hermitage, comes from a blending of several wines from different vineyards in South Australia. For decades, blends designated by Bin Number have become the quality hallmark of several producers, rather than vineyard names. In Australia, blending has been raised to the art associated with champagne and port production in Europe. Smaller producers of premium wines, as in other parts of the world, usually accentuate regional distinctiveness rather than ideal harmony. It is to the credit of the Australian people that both views are accepted and appreciated equally.

Other extensions of Australian inventiveness are the novel training systems and attention to harvest criteria. Australia, and its distant neighbor New Zealand, have both generated new training systems. These have been designed to improve early productivity (Tatura), increase yield (Lincoln), improve canopy exposure (RT2T), or achieve better pruning economy (minimal pruning). In determining the harvest date, greater concern is shown about fruit flavor than in most countries. Combined with advanced enological practices, both light aromatic 'Traminer' and smooth, dark, full-flavored 'Shiraz' wines can be produced from grapes grown in hot, arid climates.

The 25,000 ha of vines in South Australia yield about 2.5 million hl of wine annually. This constitutes about 60% of all Australian wine. The majority of vineyards occur on irrigated lands associated with the Murray Valley (~34°10′ S), some 250 km northeast of Adelaide. The best known wine region in South Australia is situated 55 km northeast of Adelaide, in the 30 km long Barossa Valley (~34°35′ S). While most of the vines occur on the valley floor, vineyards are increasingly being planted up the western slopes of the valley. Here, the altitude of the

Mount Lofty Range retards ripening and is thought to enhance flavor development. Vineyards of the Eden Valley occur on the eastern slopes of the mountain. Smaller, but increasingly important, regions occur in the Clare Valley northeast of Adelaide, the Southern Vales regions south of the city, and the milder regions of Padthaway (36°30′ S) and Coonawarra (37°15′ S) in the southeastern corner of the state.

New South Wales is the second most important Australian state in terms of wine production. It produces about 1.2 million hl of wine from 12,000 ha of vines. As with South Australia, the largest production comes from the irrigated regions along the Murray and Murrumbidgee valleys. Nevertheless, the fame of the state rests with the Hunter Valley vineyards. The original area is centered just northeast of Cessnock (32°58′ S). While initially famous for 'Shiraz' and 'Sémillon,' it has developed a reputation for excellent 'Chardonnay' and 'Cabernet Sauvignon' wines. Breezes off the South Pacific bring precipitation and cooling to the subtropical latitude. Cloud cover during the hottest part of the day also tends to moderate the heat and intensity of the sun. Frequent drying winds from the interior help to offset the disease-favoring humidity of the sea breezes. While rains brought by Pacific breezes obviate the need for irrigation, they occasionally produce problems during harvest, making the quality of Hunter Valley wine one of the most variable in Australia. Further inland to the northwest is the Muswellbrook region (32°15′ S). Its drier climate usually makes irrigation necessary. Further inland again, in the high valleys of the Great Dividing Range, are the vineyards of the Mudgee region.

Victoria, the third major wine-producing Australian state, is beginning to regain some of its former viticultural importance. As with the other southeastern states, much wine comes from irrigated vineyards along the Murray Valley. Nevertheless, particular interest is being given to plantings in the cooler maritime south. Examples are the Yarra valley (37°50′ S) just east of Melbourne, Geelong (38°06′ S) about 70 km southwest of the city, and Drumborg in the southwestern corner, near Portland (38°20′ S).

Some 2000 km to the west are the limited vineyard regions of Western Australia. Most vineyard area is situated northeast of Perth (31°45′ S) in the arid Swan Valley. However, cooler regions in the southwestern corner of the state are favored for premium table-wine production. In the Margaret River district, westerlies off the Indian Ocean provide both cooling and summer rains. However, owing to the closeness to the ocean, spring gale-force winds can inflict serious damage to varieties that undergo break bud early. There also can be considerable salt transport and accumulation by the winds. The other prime region occurs around Mount Barker. Vine-

yard altitude typically provides climatic moderation and year-round precipitation.

The coolest of Australian vineyard areas is the island of Tasmania. The vineyard area, while small, is expanding. Currently, most of the vines are planted around the capital city of Hobart (37°50′ S) and in the northern portion of the island near Launceston (37°50′ S).

NEW ZEALAND

New Zealand was the last region in the Southern Hemisphere to see a major expansion in its wine industry. Although New Zealand has produced wine for more than 150 years, the industry underwent most of its development and considerable expansion in the 1970s. In 1965, vineyards covered only about 300 ha. By the mid 1980s, coverage had expanded to about its modern size, 5000 ha. While the North Island possesses most of the vineyard hectarage, plantings on the South Island are increasing. From these vineyards, New Zealand produces about 500,000 hl of wine per year. The latitude and position of the islands, and their distance from any large landmass, provide a moderate to cool environment conducive to the production of fine quality wines.

Most of the vineyards in the North Island occur on the east side of the island. Located on the leeward side of the mountains dividing the island, they possess a drier, sunnier climate than western vineyards. The western portion of the island may receive up to 200 cm of rain per year. Combined with fertile volcanic soils, vines are likely to grow very vigorously, produce dense canopies, and be particularly susceptible to fungal diseases. These conditions promoted the development of new training systems, such as the TK2T and RT2T. With enhanced canopy openness, disease control is facilitated and fruit quality increased, without negating the natural benefits of the region's fertile soils. Ground covers further help to restrain excessive vigor and enhance evapotranspiration, minimizing some of the potential dangers of the region's high rainfall.

The largest vineyard area of the North Island is centered in the Poverty Bay region (38°40′ S), in the northeastern portion of the island. Second in importance is the Hawke's Bay region situated around Napier (39°30′ S). The west coast vineyard regions are located north and south of Auckland (~37° S).

The newest vineyard regions are located on the South Island. The largest region is in the northeastern district of Marlborough, around Blenheim (41°30′ S). Sheltering provided by the enclosing branches of the Spencer Mountains limits annual precipitation to about 50 cm per year. Additional vineyards are being planted on the Canterbury Plain, west of Christchurch (43°30′ S).

Because of the disease severity associated with the original vineyard regions in New Zealand, French–

American hybrids covered most of the vineyards. With improved chemical control and the shift toward table wines, *Vitis vinifera* cultivars have replaced most French–American hybrids. The most widely planted cultivar is 'Müller Thurgau,' followed by 'Palomino.' The latter reflects the former importance of sweet fortified wines in New Zealand. Smaller but increasingly significant plantings of premium European cultivars are transforming the varietal composition of New Zealand vineyards. These include 'Cabernet Sauvignon,' 'Pinot noir,' 'Chardonnay,' 'Riesling,' 'Traminer,' and 'Sauvignon blanc.'

South Africa

South Africa has a long history of grape culture and winemaking, dating back to 1655. It initially became famous for Constantia, a sweet fortified wine, once one of the most sought-after wines in Europe. The specialization in fortified wine production is still evident today. About half of South African wines become sherry and port, or are distilled into brandy. The recent shift to table wine production, so pronounced in most New World countries, has been less marked in South Africa.

South Africa ranks second only to Argentina as the largest wine producer in the Southern Hemisphere. The 106,000 ha of vines annually yield about 10 million hl of wine. Viticultural activity is largely concentrated in the southwestern coastal region of Cape Province (Fig. 10.14). This area, spanning from 31° to 34°15′ S, is equivalent to the latitude spread from southern California (Los Angeles) to the Gulf of California. The other viticultural region is centered around Upington (28°25′ S), along the Orange River Valley in central South Africa.

Given the subtropical latitude, the early excellence of South African fortified wines is not surprising. Most of the coastal Cape region is influenced by a Mediterranean climate, with limited rainfall occurring primarily in the winter months. The climate is relatively stable and endures few erratic variations. Consequently, South African vintages are almost consistently good to excellent.

While not possessing the largest vineyard area (14,000 ha), Stellenbosch has the highest concentration of grape cultivation of any district in South Africa. The

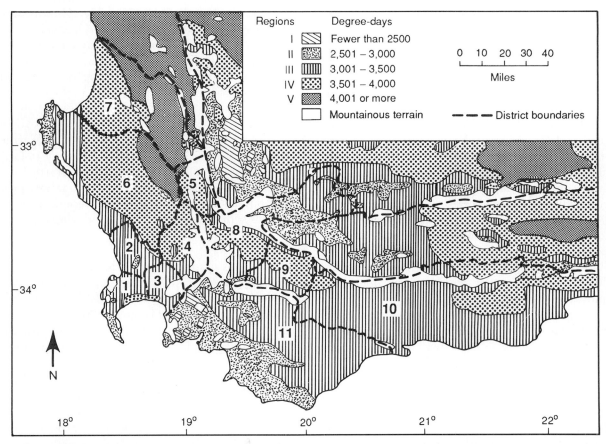

Figure 10.14 Mountainous and Fahrenheit degree-day climatic regions superimposed on viticultural districts of Cape Province in South Africa. (1) Constantia, (2) Durbanville, (3) Stellenbosch, (4) Paarl, (5) Tulbagh, (6) Swartland, (7) Piketberg, (8) Worcester, (9) Robertson, (10) Swellendam, (11) Overberg. (After Le Roux, 1974, reproduced by permission.)

pivotal location of Stellenbosch in the L-shaped viticultural region of the Cape has made it central to viticultural and enological research in South Africa.

Stellenbosch is only 40 km east of Cape Town (33°48′ S). The position of Stellenbosch on the north shore of False Bay shields it from direct exposure to Atlantic influences. Nonetheless, the cooling effect of the Atlantic Ocean reaches the vineyards and slows vine growth. This postpones ripening into the milder autumn and reduces excessive vineyard evapotranspiration. Mountain chains to the north and east of Stellenbosch favor cloud and rain formation. These conditions result in 20 to 25% of the 50 cm annual precipitation falling during the growing season. Nevertheless, some of the finest sites still require periodic irrigation. The famed red wines of the region generally come from vines grown on the moister west-facing slopes of Mt. Simonsberg and the Stellenbosch Mountains. White cultivars tend to do best on the sandy soils of the western lowlands. 'Cabernet Sauvignon' appears to do better in the warm climate of South Africa than in equivalent climates in Europe.

Viticulture becomes progressively less important both north and east of Stellenbosch. Grape culture also declines through a complex patchwork of mountains and valleys into the *highveld* interior. This reflects the decline in precipitation and increase in temperature in the associated regions. One exception is the most southerly extension of the Cape, the Overberg district. Here, conditions are cooler than in Stellenbosch, but high winds and sand dunes make viticulture excessively difficult throughout much of the district.

The L-shaped stretch of coastline viticulture, extending north and east of Stellenbosch, also shows a shift from table to fortified and distilling wine production. This change is already apparent on entering the Paarl district just north of Mt. Simonsberg. While the Paarl district is larger and possesses more vineyard area (18,000 ha), the vineyards are less concentrated than those of Stellenbosch. Even larger in size is the adjacent Worcester district to the east. It has about 17,000 vineyard ha, located largely in the western region. Winds funneling through the valley provide some cooling and extra moisture, especially to vineyards on higher slopes. The western region annually collects about 75 cm of rain, while the eastern portion receives about 25 cm. Shadows cast by mountain peaks over 850 m high limit the duration of exposure to the hot subtropical sun. Vines producing table wines usually benefit from east-facing slopes, while those producing fortified wines are favored by west-facing slopes. Similar, but drier, is the adjacent Tulbagh district to the north of Paarl. Further west again, and along the Atlantic coast, is Swartland. Cool sea breezes and heavy dew help partially compensate for the low annual rainfall (25 cm).

Additional districts extend viticulture northward and eastward to form a region up to 130 km deep and respectively 250 and 400 km to the north and east of Cape Town. The Cape Peninsula contains the remaining Constantia vineyards. The Constantia district is permitted only a single agricultural activity, viticulture.

One of the more important viticultural problems in the regions is the high proportion of acidic soils. It is estimated that about 70% of the vines of the Cape grow in soils below pH 5. Toxic levels of available aluminum probably explain much of the observed poor root growth in acidic soils. This influence may be counteracted by the incorporation of sufficient lime to raise the soil pH. Liming also probably helps root growth by improving the soil structure. Use of acid-tolerant rootstocks can further counter the detrimental effects of acidic soils.

Although the production of different wine styles has not changed as dramatically as in other non-European countries, the varietal composition of South African vineyards has changed considerably in the past 30 years. The most marked transformation has been the striking increase in 'Steen' ('Chenin blanc'), which covers about 30% of the vineyard hectarage in South Africa. A corresponding decline has occurred in the cultivation of 'Hermitage' ('Cinsaut'). Other cultivars registering decreased use are 'Palomino,' 'Green Grape' ('Sémillon'), 'Hanepoot' ('Muscat of Alexandria'), and the local cultivar 'Pinotage.' In contrast, cultivation of varieties such as 'Cabernet Sauvignon' and 'Riesling' has increased.

South America

South America has been associated with viticulture and winemaking almost since its discovery and colonization by the Spanish 500 years ago. Nevertheless, emergence of South America as an important wine producing region has been comparatively recent. Figure 10.15 shows some of the main wine-producing areas.

CHILE

Chile is unique among world nations in being almost exclusively coastal. While spanning almost 40° of latitude, from the Peruvian border (17°24′ S) to the Cape of Good Hope (56° S), it is only on average 180 km wide. Along the 4500 km coastline of Chile, only the zone between 32° and 38° S is amenable to viticulture and premium wine production. Chilean vineyard area (120,000 ha) and yearly wine production (~4 million hl) are considerably less than that of its eastern neighbor, Argentina. Within this region, though, conditions are more favorable for premium wine production than in Argentina.

The best viticultural regions in Chile lie at latitudes roughly equivalent to those of southern California in the

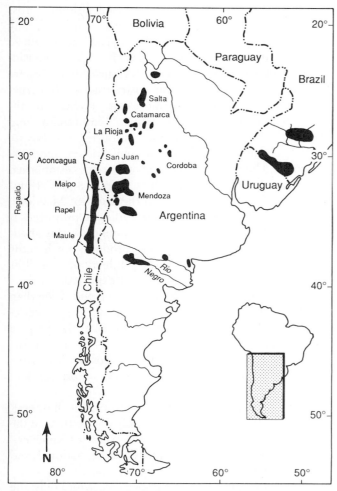

Figure 10.15 Main wine-producing regions of Chile, Argentina, and Brazil. (After de Blij, 1985, reproduced by permission.)

Northern Hemisphere. The cool Humboldt Current and the altitude of the vineyards in the central valley (*Nucleo Central*) provide a temperate Mediterranean climate. Precipitation increases rapidly along the 600 km stretch from north of Santiago to the Bio Bio River. Further south, the climate becomes maritime, without a summer drought period. At the northern end of the *Nucleo Central* annual precipitation averages about 25 cm, while at the southern end near Concepcíon it averages more than 75 cm annually.

Of the several viticultural zones in Chile, the most highly regarded is the *Regadio*. It encompasses central Chile, from north of the Aconcagua River to south of the Maule River. Except for the Aconcagua region, the best sites occur within the broad *Nucleo Central,* formed by the coastal *cordillera* and the Andes to the east. Most of the vineyards are associated with river valleys that cross the *cordillera*. From north to south they include the Maipo, Cachopoal, Tinguiririca, Lontué, and Maule riv-

ers. The Regadio zone is subdivided from north to south into the Aconcagua, Maipo, Rapel, and Maule regions.

Within the Regadio zone, sites north and south of Santiago (33°26′ S) in the Maule River valley are considered superior. The regional average rainfall of about 40 cm provides moisture to the deep loamy/gravelly soils. This, combined with irrigation water from the Maipo River, provides adequate moisture for vine growth and grape ripening. The gentle rolling landscape facilitates excellent drainage. Prevalent dry sunny conditions and moderate temperatures help limit disease development. The calcareous layer in the soil favors *Vitis vinifera* root growth. Chile is the major wine-producing country still unafflicted with phylloxera, and most cultivars can be grown on their own root system. The winter season is sufficiently cool to satisfy bud dormancy requirements and permit bud break in the spring. Maipo's moderate, stable climate usually provides conditions optimal for the production of fine quality fruit. The adjacent Rapel region, south of Maipo, is slightly cooler and yearly receives about 50 to 60 cm of rain. It also produces excellent quality fruit and wine. Annually, both regions produce about 200,000 and 500,000 hl of wine, respectively.

While most grape varieties in Chile are red, the cultivars grown in the central Regadio regions are different from the 'País' variety that dominates most of the country's vineyards. For example, 'Cabernet Sauvignon' comprises close to half of all red cultivars in the Maipo and Rapel regions but constitutes less than 17% of all red cultivars in Chile. In the Maipo region 'Cot' ('Malbec') is the second most widely grown red variety, while in the Rapel region 'Petit Verdot' is the second most cultivated red cultivar. Of white cultivars, 'Sémillon' is by far the dominant cultivar, with 'Sauvignon blanc' a distant second. The Spanish variety 'Torontel' comes in third. The predominance of French cultivars in a region colonized by Spanish immigrants reflects the effect of Silvestre Ochagavia, an influential viticulturist and politician in the 1850s.

Of the other two regions in the Regadio zone, Maule, south of Rapel, is the largest. It possesses 29,000 ha of vines, in contrast to the 15,000 ha each in Maipo and Rapel. Some sites are as fine as those further north, but others possess a more unstable, moist environment. This is reflected in the increased planting of the 'País' variety. Nevertheless, the conditions are still favorable for premium white cultivars. 'Sémillon' constitutes 80% of all white varieties grown in Maule.

In the northern Regadio region of Aconcagua, 'País' dominates red plantings as 'Sémillon' does the whites. However, the warmer climate is reflected in the increased cultivation of varieties used in the production of Pisco brandy.

The shift of cultivars to those used in Pisco production is particularly noticeable in the adjacent northern Pisquera viticultural zone. Here, 'Muscat' and 'País' constitute about 80% of the plantings of white and red cultivars, respectively. Table grapes also increase in importance in the Pisquera zone. The primary deficit of the zone is neither its hot climate nor dry environment, but the erratic and violent storms that can ravage the area.

South of the Regadio zone, one enters a transition zone where irrigation becomes progressively unnecessary. In the Secano zone, south of the Bio Bio River, the increasing rainfall and cool climate make viticulture increasingly problematic. Nevertheless, the zone still has an extensive vineyard area, covering 27,000 ha, and produces about 1.7 million hl of wine per year, slightly more than the Maule region.

ARGENTINA

Grape growing and winemaking have been conducted in Argentina since the mid-1500s, as in Chile. However, the growth of the Argentinean wine industry into the fifth largest in the world occurred only within the twentieth century. The current vineyard area of 267,000 ha and annual wine production around 20 million hl make it the largest wine producer in both the Southern and Western Hemispheres. Viticulture constitutes the third largest industry in Argentina. As in Russia, most of the production is consumed locally. Argentina exports only about 1.5 to 3% of the annual production. However, export of concentrated grape juice for fermentation in other countries adds significantly to its actual contribution to the world's wine supply.

Vineyard regions in Argentina occur almost exclusively in the rain shadow of the Andes, along the western border of the country. Wines are grown from south of the province of Jujuy (24° S) to along the Negro River (40° S). However, the major concentration of vineyards occurs in the province of Mendoza (~32° to 36° S).

The location of Mendoza in the lee of the Andes gives it an arid climate, annually receiving about 20 cm of rain. Irrigation of the extensive vineyards in the region is possible only because of the relative flatness of the land and the ready availability of river or artesian water. The deep, loosely compacted soils permit good drainage, water retention, and root penetration. Irrigation costs are partially offset by the savings derived from the disease-limiting dry air. The clear skies and altitude (~500 to 750 m) of the region generate day/night temperature fluctuations up to 25°C. Thus, the heat summation of some 1900 to 2100 Celsius degree-days gives the impression that the region is cooler than it is. To avoid excessive acid loss during grape ripening, it is usual to harvest early and prevent malolactic fermentation. While the summers are hot, the winter period is adequate to permit bud break in the spring.

Regional differences in latitude and altitude across Mendoza produce significant differences in fruit ripening and quality. The basins of the Tunuyána and Mendoza rivers, southwest of Mendoza (32°54' S), are considered the best viticultural areas. Regional differences are partially reflected in cultivar distributions. 'Cabernet Sauvignon' and 'Malbec' are more common in the northern portions of Mendoza, while 'Tempranillo' and 'Sémillon' are more frequent in central regions. Nearly 50% of all cultivars are premium red varieties, and 20% are white. The remainder consists of varieties established in Argentina for centuries, such as 'Criolla' and 'Cereza.'

The province of San Juan, to the north of Mendoza, ranks second in terms of quantity of wine produced. It has about 58,000 ha of vines and annually yields about 5 million hl of wine. San Juan also produces much of the wine used in sherry and brandy production, as well as most of the concentrated grape juice exported.

The San Juan area is slightly hotter and drier than Mendoza but more frequently subjected to strong desiccating winds from the Andes. Consequently, pergola training systems are commonly used to protect the grapes from intense sun and wind exposure. The same training system has been used in Mendoza for hail protection.

In San Juan, premium red cultivars, such as 'Barbera,' 'Nebbiolo,' and 'Malbec,' cover only about 10% of the vineyard area. White varieties, such as 'Pedro Ximénez,' 'Muscat,' and 'Torrontés,' and the reds 'Criolla' and 'Cereza' are the dominant cultivars planted.

Of the other provinces that produce wine, only the south central Rio Negro is of considerable importance. The more southerly latitude (~38° S) provides Rio Negro with a cooler climate. Thus, grapes mature more slowly and generally develop a better acid/sugar balance during ripening.

In Rio Negro, the important viticultural area occurs on the broad floodplain of the Negro River, east of the junction of the Limay and Neuquén rivers. With an annual precipitation around 20 cm, irrigation is essential. Textural soil differences may generate the features that distinguish wines produced on opposite sides of the river. The north side possesses more sand and gravel, while the southern portion has finer, more fertile soils.

Argentinean vineyards have a much wider diversity of cultivars than neighboring Chile. The original European settlers brought Spanish varieties and techniques that still dominate the wine industry. French cultivars came later, primarily via Chile, and finally Italian varieties with the influx of Italian immigrants beginning in the middle of the nineteenth century.

As noted above, winemaking procedures are predominantly Spanish. Thus, the finer wines are given several years of aging in casks before bottling. With red wines, the results meet with widespread approval, both in Argentina and abroad. The slightly oxidized character given white wine by the procedure is appreciated in Argentina but has not developed a major following in foreign markets.

BRAZIL

Although not widely recognized as a wine-producing nation, Brazil possesses approximately 57,000 ha of vines and annually produces about 3 million hl of wine. Of the several states involved, only the southernmost Rio Grande do Sul is of considerable significance. It contains about 70% of Brazil's vineyard area.

Most of the vineyards are congregated north of the Jacui River (~29°30′ S), about 120 km northwest of Porto Alegre. The other main region in Rio Grande do Sul lies along the border with Uruguay (~31° S). Both regions show moist, warm summers and mild winters. The moist climate generally is unfavorable to the cultivation of *Vitis vinifera*. Nevertheless, modern chemical disease control has permitted the expansion of *V. vinifera* cultivation in Brazil and neighboring Uruguay. The predominant red cultivars are either Italian, such as 'Barbera,' 'Bonarda,' or 'Nebbiolo,' or French cultivars like 'Cabernet franc' or 'Merlot.' 'Muscat' varieties, along with some 'Trebbiano' and 'Sémillon,' are the main white cultivars. Nevertheless, about 80% of all cultivars are *Vitis labrusca* varieties or French–American hybrids. Important cultivars are 'Isabella,' 'Dutchess,' 'Niagara,' 'Delaware,' 'Concord,' and Seibel hybrids. Their main advantage in the humid climate of Brazil is their superior disease resistance.

North America

The North American market has experienced the same dramatic shift in consumer preference as noted above in Australia and New Zealand. In addition to a move away from fortified wines, white wines have become the preference with the majority of consumers. These changes have induced considerable adjustment in viticultural practice, cultivar planting, and winemaking and have spawned the opening of wineries specializing in premium wines. It also has forged greater communication and cooperation between grape growers and winemakers. High yield, with minimal concern for sugar/flavor/acid balance, is unacceptable where fine table wines are produced.

These changes also have spurred legal changes that have induced many grape growers to start their own wineries. Consequently, the North American industry has begun to resemble that typical of Europe. The effect has been an improvement in the level of grape and wine quality. Nevertheless, North American winery conglomerates still retain a dominate position in wine production. While continuing to produce fortified wines, they have adopted and championed technological advancements that have made North America one of the best quality/price wine regions of the world.

UNITED STATES

The experimental and technological innovation generated by the rapid adjustment of winemaking to new preferences has thrust the United States into the forefront of wine research. This has occurred despite its producing only about 25 to 40% of the wine generated by any of the three major wine-producing countries, Italy, France, and Spain. Currently, U.S. production stands at about 16 million hl. Vineyards covers about 322,000 ha.

California Although wine is produced in nearly every state in the Union, California is the major producer. It currently produces over 90% of all American wine, about 15 million hl annually. Vineyard area covers about 258,000 ha.

Since the shift toward table wines, the focus of interest has moved from the Central Valley to the numerous valleys that directly open to the Pacific Ocean. Nevertheless, about 60% of Californian vines grow in the southern portion of the Central Valley, the San Joaquin. In addition, nearly 80% of Californian wines come from this strip of land 650 km long and up to 150 km wide. The Central Valley lies approximately between latitudes 35° and 38° N. The reliably warm Mediterranean climate, rich soils, and flat landscape of the region are ideal for most forms of agriculture. The high Coastal Range effectively separates the valley from moisture-carrying sea breezes.

Although hot and arid, the San Joaquin valley provides the table wines consumed by most Americans. This has been possible with practices such as early harvesting, in-field crushing, cool fermentation, prevention of malolactic fermentation, and protection from oxidation. These procedures are comparable to those used in similar regions in Australia, Chile, and South Africa. New cultivars specifically bred for Central Valley conditions helped its transformation into a producer of consistently good, inexpensive, table wines. Particularly valuable were 'Rubired,' 'Ruby Cabernet,' and 'Emerald Riesling.' However, around Lodi (38°07′ N), cultivars like 'Chenin blanc,' 'Colombard,' 'Chardonnay,' 'Zinfandel,' and 'Cabernet Sauvignon' are grown extensively. Here, the opening of the Central Valley to the Pacific produces cooler nights, which favor acid retention and flavor development in premium cultivars. The southern portion of the San Joaquin Valley remains the center for sherry and port production. It also is the main location of the prominent raisin and table grape industries. Average rainfall declines in the valley from about 45 cm per year around Sacramento (38°35′ N) to about 13 cm at the southern tip. Because most of the precipitation comes during the winter, irrigation is essential in most years.

The best-known coastal valley is the Napa Valley, located northeast of San Francisco (37°45′ N). It is typical of several valleys affected by an influx of cool, moist air from the Pacific Ocean. This results as rising air currents, generated by heating of interior parts of the valley, draw in cooler air from the ocean. This often is associated with fog development. As the air is heated, however, the humidity drops and a drying influence is produced inland. The effect is less marked along the valley walls.

In addition to declining precipitation as one progresses up the valleys, there is a similar reduction moving down the coast. For example, average precipitation decreases from about 100 cm per year in Mendocino (39°18′ N), to roughly 45 cm around Santa Barbara (34°26′ N), and just over 20 cm in San Diego (32°43′ N). Since most of the rain comes during the winter, coastal valley viticulture typically requires irrigation. While irrigation increases the cost of production, it provides an opportunity to regulate vine growth selectively and limit several diseases and pests. For example, California is generally unaffected by downy mildew, black rot, and several grape and berry moths. The absence of the winged stage of phylloxera may partially explain the freedom of most central coastal vineyards from phylloxera, years after its accidental introduction into California.

Although viticulture is practiced over nearly 6° of latitude, temperature changes are often more marked along the length of the coastal valleys than down much of the coastline. For example, several coastal valleys pass from heat summation Region I (cool) to Region III (mild) along their length (see Fig. 5.6). This change can occur over a distance of less than 40 km. Temperature regimes near the opening of coastal valleys can be similar to those found 13° latitude further north in Europe. Altitude influences up valley slopes further enhance short-distance climatic diversity. However, valleys in California possess longer growing seasons, have milder winters, and experience higher light intensities than European counterparts. Greater proximity of valley vineyards to the coast partially explains the apparent anomaly that some of the coolest viticultural regions of California are in the south (San Luis Obispo and Santa Barbara, ~35° N).

The combination of diverse temperature regimes with local differences in moisture and soil conditions has provided California with a remarkably varied range of growing conditions juxtaposed next to one another. Consequently, cultivars typically separated by hundreds of kilometers in Europe may grow within sight of one another in California. This has influenced the production of stylistically different wines within kilometers of one another. It has also fostered an acceptance of a much wider range of varietal expressions than in Europe. Ab-

sence of a traditional style for particular regions in California has left winemakers free to experiment and create their own distinctive wines. Rejection of the view that grape varieties succeed only under a limited range of climatic conditions advances the development of new and better regional and varietal wines.

The shift in consumer preference to dry white wines has had marked effects on Californian viticulture. One solution has been the production of blush and "white" wines from excess red grape capacity. Another solution has been the grafting over to, or replanting with, white cultivars. This has been particularly marked with 'Chardonnay,' which currently covers 21,000 ha, slightly less than the coverage of 'Colombard.' Most of the latter is grown in the San Joaquin Valley. Since 'Colombard' is seldom produced as a varietal wine, 'Chardonnay' is the most widely produced varietal wine in California. Other major white cultivars grown extensively in California are 'Chenin blanc,' 'Sauvignon blanc,' and 'Riesling.'

Another indicator of current trends is the importance of 'Cabernet Sauvignon' in California. It currently covers 135,000 ha, about the same as the unique Californian red variety 'Zinfandel.' Even the enigmatic 'Pinot noir' is cultivated on almost 4000 ha. It is sixth in coverage, following 'Zinfandel,' 'Cabernet Sauvignon,' 'Grenache,' 'Barbera,' and 'Carignan.'

The proportion of premium cultivars grown in coastal valley vineyards is significantly higher than for the state as a whole. This reflects the cooler environment of the coastal valleys and greater similarity of the climate to the European homeland of the cultivars.

As in Australia, extensive research has been directed toward improving grape quality. One aspect of this research has been the isolation of cultivar clones with distinctive flavor characteristics. Some producers are combining the must or wine from several clones to enhance wine complexity (Long, 1987). Such activity may produce, under precise control, some of the clonal diversity that tends to exist in many European vineyards.

One of the few disappointing aspects of wine production in California, and in most other New World countries, is the inordinate space given a few grape varieties. Although several Spanish and Italian cultivars are grown widely, they are seldom used in varietally designated wines. They are essential to the quality of inexpensive wines, but their anonymity prevents their receiving the recognition they deserve. This is probably one of the unfortunate legacies of the prevalent English view that French wines, or at least their cultivars, are superior to all others. Thus, as usual, it will depend on a few, dedicated, skilled visionaries to convince conservative consumers and wine critics of the merit of other grape cultivars.

As noted above, the best known of the coastal valleys in California is the Napa Valley. The adjacent Sonoma

Valley is only slightly less well known. Both are about equal in vineyard area and wine production. The vineyards in the northern region of Mendocino county are more dispersed and scattered through the many valleys. The Monterey region, south of San Francisco, only recently has become an important viticultural area. Other regions, while important for quality, are minor in terms of production. Several regions possess viticultural areas possessing a reputation and consumer following exceeding that of the county name. Examples are the Alexander Valley in Sonoma and the Santa Ynez Valley in Santa Barbara.

Pacific Northwest Of the two states separating California and Canada, Washington is the more significant in wine production. It produces about 140,000 hl from 12,000 vineyard ha. This qualifies it as the third most important wine-producing state in the United States. By comparison, Oregon yields only about 30,000 hl of wine from 1700 vineyard ha.

In Washington, the primary vineyard area is situated in the south central region, approximately between 45° and 48° N latitude. The region encompasses the connecting valleys of the Columbia River and its tributaries, the Yakima and Snake. It is bounded on all sides by mountains: the Rocky Mountains to the east, the Cascades to the west, the Okanagan highlands to the north, and the Blue to the south. The mountains provide protection both from cold north and east winds and from moisture-laden winds from the west. The area possesses a dry, sunny climate, with much of the limited precipitation coming in winter in the form of snow. The summers are warm with cool nights. The rapid decline in temperature in mid-September helps retain fruit acidity. Heat summation varies from about 1220 to 1500 Celsius degree–days (Regions I to II). The soils are primarily sandy loams of various depths and are commonly underlain by a calcareous hardpan, typical of most dryland regions. Irrigation is typically necessary. Cold winter temperatures occasionally reach −25°C. Sites midway up slopes provide the optimal frost protection by draining away cool air while avoiding the cold at the slope apex. South-facing orientations are preferred because of the extra spring and fall light received.

About two-thirds of the grapes grown in central Washington are used in juice production, notably from 'Concord.' Since the late 1960s, plantings of *Vitis vinifera* cultivars have increased considerably. In areas experiencing the coldest winters, varieties possessing cold tolerance such as 'Riesling,' 'Gewürztraminer,' 'Pinot noir,' and 'Chardonnay' are cultivated. 'Cabernet Sauvignon' and other longer season cultivars grow better on warmer sites. Currently the dominant cultivar is 'Riesling.' As phylloxera is absent, a major cost saving is achieved by avoiding the necessity of grafting.

A second viticultural region occurs in the area surrounding Seattle and extending to the Canadian border (~47° to 49° N). The temperate maritime climate of the Puget Sound area is markedly different from the arid conditions of the central and eastern parts of the state. Although freezes are seldom experienced, the cool moist climate (heat summation of 850 to 1050 Celsius degree–days) retards growth and grape ripening. Thus, short season varieties such as 'Müller-Thurgau,' 'Madeleine Angevine,' and 'Okanagan Riesling' are preferred. The vineyard area is highly dispersed and small.

In Oregon, commercial viticulture exists in four regions: the southern portion of the Columbia River valley across from the state of Washington and a string of three valleys formed between the Coast Range and Cascade Mountains. Of the latter, the largest and most significant is the Willamette Valley. It extends southward from the Washington border for about 220 km and can be 100 km wide. It spans a latitude from about 44° to 46° N. The Umpqua Valley lies south of the Willamette Valley, from which it is separated by a semimountainous divide. An even smaller region situated within the Klamath Mountains is the Rogue Valley. It lies close to the Californian border.

Although the Willamette Valley receives up to 150 cm of rain per year, only about one-third falls during the growing season. Long sunny days during the spring and summer tend to compensate for the moderate to cool temperatures (heat summation of 1050 to 1250 Celsius degree–days). Nevertheless, south-facing slopes are preferred for the extra light and heat received. As with other cool climatic regions, the midlevel of the slope provides the optimal frost protection consistent with heat gain. Deep soils are preferred to provide protection against occasional summer droughts. However, excellent drainage is necessary to avoid excessive early shoot vigor in such sites.

Preferred cultivars are 'Pinot noir,' 'Chardonnay,' and 'Riesling.' Of these, Oregon has had its greatest success with 'Pinot noir.' The wine is produced in a manner similar to that in France. This may involve early harvesting to retain sufficient acidity, even though it requires chaptalization. During fermentation, some whole clusters may be incorporated with the must and the temperatures permitted to rise to 29°C. Fermentation usually is preferred in small cooperage, with gentle punching down rather than pumping over. Malolactic fermentation is favored by infrequent racking and minimal addition of sulfur dioxide. There is considerable interest in the use of special strains of lactic acid bacteria to enhance flavor complexity during malolactic fermentation. Furthermore, some vineyards are planted at high densities (10,000 vines/ha). The narrow rows require the use of special machinery designed to work under such conditions (Adelsheim, 1988).

Oregon has the advantages of a climate that accentuates the varietal character of 'Pinot noir' and a fairly consistent concept among producers on how the wine should be made. The combination has helped the state achieve rapid international recognition for its 'Pinot noir' wines. It is a good example of how a small region can quickly establish an identity in the crowded and competitive world wine market.

East of the Rockies The most significant wine-producing region after California is New York State. New York annually produces about 1 million hl of wine from 14,000 vineyard ha. Most eastern states as well as those in the south and southwest produce wine. While the wines are of considerable local interest and pride, production capacity is small, and the wines seldom are found outside their home state. Production in Georgia, Texas, Ohio, Michigan, Florida, Virginia, and Pennsylvania is about 57,000, 22,000, 19,000, 15,000, 14,000, 12,000, and 11,000 hl, respectively.

The northeastern states receive precipitation throughout the year. During the winter, snow cover often provides needed frost protection. However, high humidity during the growing season favors several severe fungal pathogens, notably downy mildew and both black and bunch rots. In addition, the region is endemically infested with phylloxera and insect pests absent west of the Rockies.

Southern and coastal states seldom suffer from vine-damaging frost conditions, but they are more humid. Humid conditions demand greater fungicide use or require the cultivation of varieties derived from indigenous *Vitis* species or varieties containing resistance genes. The bacterium *Xylella fastidiosa*, the causal agent of Pierce's disease, limits cultivation of *V. vinifera* cultivars in the southern and coastal states.

The central portion of the United States has little viticultural activity. That which does occur is located either in the southwestern states of Texas, New Mexico, and Arizona or in the east central states of Michigan and Ohio. Viticulture in Wisconsin usually requires that the vines be laid down and covered with soil each winter. Throughout the northern states, vines are commonly trained with two or more trunks. This minimizes the damage caused by death of one trunk from crown gall, Eutypa dieback, or other related problems.

One of the most distinctive features of the eastern wine industry is the tremendous diversity of cultivars. Because of repeated failures of *Vitis vinifera* to survive west of the Rockies, the industry developed using *Vitis labrusca* cultivars and hybrids in the northeast, *V. aestivalis* cultivars in the midwest, and *V. rotundifolia* cultivars in the southeast. *Vitis labrusca* cultivars, such as 'Concord,' 'Niagara,' 'Catawba,' 'Isabella,' and 'Ives,' are well adapted to the climate and prevalent pathogens and pests of the northeast. They produce good fortified and sparkling wines which were the staple of the wine industry for more than a century. In the midwest, cultivars such as 'Norton' and 'Cynthiana' were popular. In the south, the predominant cultivars are muscadine varieties such as 'Scuppernong.'

With an increase in the popularity of dry wines, the northeastern states and adjacent Canada began to explore the use of French–American hybrids. These possessed many of the winemaking properties of *V. vinifera*, combined with some of the disease resistance of one or more indigenous American *Vitis* species. The success of the trials added cultivars like 'de Chaunac,' 'Baco noir,' 'Vidal blanc,' 'Seyval blanc,' and 'Aurora' to the list of commonly grown cultivars. Breeding programs in both New York and neighboring Ontario produced new *V. vinifera* hybrids such as 'Cayuga white' and 'Ventura.' In the southeast, renewed breeding work has generated new muscadine cultivars, such as 'Carlos,' 'Noble,' and 'Magnolia.' Further south, the cultivars 'Stover' and 'Suwannee' have been bred to the climate and disease situation in Florida.

To the new varieties are now being added European *V. vinifera* cultivars. With appropriate rootstocks, pesticide application, site selection, training, and winter protection, European cultivars can now survive and prosper after more than two centuries of failure. While the success is gratifying, the long-term benefit of *V. vinifera* cultivars to the wine industry in the eastern and central states is unclear. One short-term effect is to direct interest away from hybrids inherently more suited to the local climate. Locally developed cultivars could give regional wines a distinctiveness that European cultivars cannot. Nevertheless, the acceptance and prestige derived from successfully cultivating familiar European varieties may encourage the production of fine examples of cultivars possessing distinctive flavors.

Wine production in New York State is largely centered in the Finger Lakes region (~42°25' to 42°50' N), about 100 km south of Lake Ontario. The vineyards are primarily situated on slopes adjacent to a series of narrow elongated lakes oriented north/south. Thus, the vines receive either an eastern or western exposure. While such sites receive less light than south-facing slopes, this can be beneficial in delaying the loss of insulating snow cover and retarding premature bud burst in the spring. The lakes moderate temperature fluctuations, a feature especially important in the late winter and spring when the vines are losing their cold acclimation. The lakes also help to prolong autumn warmth and favor fruit ripening.

During the summer, cloud cover and precipitation can reduce temperature maxima. These factors also limit solar intensity and increase the humidity. The Geneva

Double Curtain (GDC) training system developed at Geneva, New York, was designed primarily to counteract these effects (see Chapter 4). By opening the canopy and allowing the shoots to grow pendulously, the system greatly improved the degree and uniformity of fruit exposure to air and sun. In addition to increased fruit quality and health, vineyard yield was boosted owing to improved leaf photosynthetic efficiency.

Other increasingly significant wine regions in the Empire State occur about 150 km north of New York City, along the Hudson River, and on the northeastern branch of Long Island, east of the city. The maritime climate of the latter lengthens the growing season and raises the average winter temperature. However, the additional cloud cover delays ripening.

In Ohio, the vineyards are concentrated along the southern side and islands of Lake Erie. These are desirably close to the large population center of Cleveland. Similarly, the majority of Michigan vineyards are located near Chicago, along the southeastern portion of Lake Michigan.

In Virginia, most vineyards are situated in close proximity to Washington, D.C. This permits consumers to quickly reach the farm wineries by car. Although avoiding the colder climate of more northern states, Virginia suffers erratic winter temperatures that occasionally cause severe bud damage to *V. vinifera* and French–American hybrid cultivars. Otherwise, Virginia possesses a desirably mild climate and occurs north of the natural distribution of Pierce's disease. Considerable interest has been shown recently in the viticultural potential of Virginia by Californian, New York, Canadian, and European wine enterprises.

Georgia is the largest wine producer of southern states. However, this may be due to state laws permitting up to 60% imported content in its wines. Because of its size and diverse climatic regions, Texas may soon become the major wine-producing area of the Southeastern and Gulf States.

CANADA

Although the wine industry of Canada is small on a world scale (Table 10.2), it has considerable regional economic significance. Most production in Canada is located in southern Ontario and south central British Columbia. However, the wine industry has been forced into a major restructuring due to a General Agreement on Tariffs and Trade (GATT) ruling in 1987 and implementation of the Free-Trade Agreement with the United States in 1989. Lack of protective tariffs have made most wines based on *Vitis labrusca* and French–American hybrids unprofitable. For example, vineyard hectarage in British Columbia shrank from a high of 1375 ha in 1988 to 460 ha in 1989. The vineyards that remain are planted primarily with *V. vinifera* cultivars, or are being replanted with them. French–American hybrids now cover only about a quarter of the vineyard hectarage. In Ontario, the larger fresh fruit and juice market partially cushioned the effects of these changes. Nevertheless, the shift to *V. vinifera* from French–American hybrids and *V. labrusca* is also a major trend in Ontario vineyards. In both provinces, there is a marked preference for white cultivars, notably 'Riesling,' 'Chardonnay,' and 'Gewürztraminer.' For red wines, 'Cabernet Sauvignon' and 'Pinot noir' are favored.

In Ontario, the largest vineyard area is located along the southwestern edge of Lake Ontario, between Hamilton and Niagara Falls (~43 N). The region is bounded on the south by the Niagara Escarpment, a prominent geologic feature that markedly affects the climate of the region (Fig. 5.11). Sites on 4 to 10% slopes, about 4 to 6 km from the lake, are the most favored. Cold air drainage draws warmer air down from the temperature inversion layer that often develops during calm, cold nights. The vines also are sufficiently far from the lake to avoid a marked chilling by the summer flow of cool air off the water. Although the north-facing slopes of the Escarpment limit sun exposure, it also delays bud burst, further minimizing the likelihood of frost damage in the spring, but shortens the growing season. The location of the vineyards between Lakes Ontario and Erie greatly cushions the effect of the continental climate. The long, mild autumn produced usually supplies ample time for ripening most short-season *V. vinifera* cultivars. Nevertheless, the region often experiences a hard freeze in late November or December. While normally undesirable, it has favored the production of *eisweins*, a style that is becoming an Ontario speciality. The other main region for wine production in Ontario is along the northern portion of Lake Erie.

On Canada's west coast, the vineyards are considerably further north than those in Ontario. The vines in British Columbia grow primarily between 49° and 50° versus 43° N in Ontario. Nevertheless, dry, sunny conditions, cold protection of the surrounding mountains, and the influence of the Pacific Coast provides the Okanagan Valley with a moderate climate. Semiarid conditions make irrigation necessary in most locations. The vineyards are located primarily on slopes and plateaus lining a series of narrow elongated lakes in the Okanagan River Valley. Vines also grow along the Similkameen River, a tributary joining the Okanagan about 25 km from the U.S. border. While the southern portion of the Okanagan Valley is warmer, its drier environment provides less snow cover. Thus, the vines are about as vulnerable to cold damage as in the northern portion of the valley. Nonetheless, about 60% of the grape production occurs in the south, between Penticton (49°30′ N) and the border with Washington State (49° N).

Nova Scotia, on the eastern coast of Canada, possesses several vineyards and local wineries (~45° N). As the province is almost entirely surrounded by water, the vines are exposed to a maritime climate. The Bay of Fundy and the Atlantic Ocean moderate continental influences from the west but retard early bud break. However, the typically long autumn helps ripen short-season cultivars grown in Nova Scotia. In addition to familiar French–American hybrids, the region also grows several new German *V. vinifera* cultivars. Particularly interesting is the cultivation of *V. vinifera* × *V. amurensis* cultivars from Russia, notably 'Michurinetz' and 'Severnyi.'

Suggested Readings

Appellation Control Laws

Anonymous. (1979). "The Wine Industry in the Federal Republic of Germany." Evaluation and Information Services for Food, Agriculture and Forestry, Bonn, Germany.

Becker, W. (1987). L'Historique des indications de provenance et appellations d'origine allemandes pour les vins et les vins mousseux. *Symp. "Les Appellations d'Origine Historiques," Jerez de la Frontera, Spain, March 16–18, 1987,* pp. 151–173. Office International de la Vigne et du Vin, Paris.

Benecke, F. S. (1990). A "quality" accreditation. *Wines Vines* 71(5), 19–22.

de Blij, H. J. (1983). "Wine: A Geographic Appreciation." Rowman & Allanheld, Totowa, New Jersey.

Denis, D. (1989). "La Vigne et Le Vin—Régime Juridique." Sirey, Paris.

Marquet, P. (1987a). L'importance des facteurs naturels et humains dans le développement des appellation d'origine françaises. *Symp. "Les Appellations d'Origine Historiques," Jerez de la Frontera, Spain, March 16–18, 1987,* pp. 33–41. Office International de la Vigne et du Vin, Paris.

Marquet, P. (1987b). L'évolution de la notion d'appellation d'origine en France. *Symp. "Les Appellations d'Origine Historiques," Jerez de la Frontera, Spain, March 16–18, 1987,* pp. 175–185. Office International de la Vigne et du Vin, Paris.

Mendelson, R. (1987). Appellations of Origin for the wines in USA. *Symp. "Les Appellations d'Origine Historiques," Jerez da la Frontera, Spain, March 16–18, 1987,* pp. 237–248. Office International de la Vigne et du Vin, Paris.

Moran, W. (1988). The wine appellation: Environmental description or economic device? *Proc. 2nd Int. Symp. Cool Climate Vitic. Oenol., Jan. 11–15, 1988, Auckland, N.Z.* (R. E. Smart, S. B. Thornton, S. B. Rodriquez, and J. E. Young, eds.), pp. 356–360. New Zealand Soc. Vitic. Oenol., Auckland, New Zealand.

Sills, D. (1987). The evolution of protection in English law. *Symp. "Les Appellations d'Origine Historiques," Jerez de la Frontera, Spain, March 16–18, 1987,* pp. 275–285. Office International de la Vigne et du Vin, Paris.

Tinlot, M. R. (1987). La définition de l'Appellation d'Origine. *Symp.*

"*Les Appellations d'Origine Historiques," Jerez de la Frontera, Spain, March 16–18, 1987,* pp. 129–138. Office International de la Vigne et du Vin, Paris.

Detection of Wine Misrepresentation and Adulteration

McCallum, N. K., Rothbaum, H. P., and Otlet, R. L. (1986). Detection of adulteration of wine with volatile compounds. *Food Technol. Aust.* **38**, 318–319.

Martin, G. J. (1988). Les applications de la mesure par résonance magnétique nucléaire du fractionnement isotopique naturel spécifique (RMN–Fins) en viticulture et en oenologie. *Rev. Fr. Oenol.* **114**, 23–24 and 53–60.

Martin, G. J., and Martin, M. L. (1988). The site-specific natural isotope fractionation–NMR method applied to the study of wines. *In* "Wine Analysis" (H. F. Linskens and J. F. Jackson, eds.), pp. 258–275. Springer-Verlag, Berlin.

Ribéreau-Gayon, J., Peynaud, E., Sudraud, P., and Ribéreau-Gayon, P. (1982). Recherche des fraudes. *In* "Traité d'Oenologie: Sciences et Techniques du Vin" (J. Ribéreau-Gayon, E. Peynaud, P. Sudraud, and P. Ribéreau-Gayon, eds.), Vol. 1, 2nd Ed., pp. 575–617. Dunod, Paris.

Stacey, R. J. (1984). Isotopic analysis: Its application in the wine trade. *In* "Tartrates and Concentrates—Proceedings of the Eighth Wine Subject Day" (F. W. Beech, ed.), pp. 45–51. Long Ashton Research Station, Univ. of Bristol, Long Ashton, England.

Geographic Regions — General

de Blij, H. J. (1983). "Wine, A Geographic Appreciation." Rowman & Allanheld, Totowa, New Jersey.

de Blij, H. J. (1985). "Wine Regions of the Southern Hemisphere." Rowman & Allanheld, Totowa, New Jersey.

de Blij, H. J. (ed). (1992). "Viticulture in Geographic Perspective." *Proc. 1991 Miami AAG Symposium.* Miami Geographical Society, Univ. of Miami, Coral Gables, Florida.

Dickenson, J. (1990). Viticultural geography: An introduction to the literature in English. *J. Wine Res.* **1**, 5–24.

Gladstone, J. (1992). "Viticulture and Environment". Winetitles, Adelaide, Australia.

Johnson, H. (1985). "The World Atlas of Wine." Simon & Schuster, New York.

Lichine, A. (1987). "New Encyclopedia of Wines and Spirits." Alfred A. Knopf, New York.

Prescott, J. A. (1965). The climatology of the vine (*Vitis vinifera* L.): The cool limits of cultivation. *Trans. R. Soc. South Aust.* **89**, 5–22.

Argentina

Catania, C., and de del Monte, S. A. (1986). Détermination des aptitudes oenologiques des différents cépages dans la république Argentine. *XIX Int. Vitic. Oenol. Congr., Santiago, Chile, November 1986,* Vol. 2, pp. 17–33. Office International de la Vigne et du Vin, Paris.

Australia

Coombe, B. G., and Dry, P. T. (eds). (1988). "Viticulture. Volume I. Resources in Australia." Australian Industrial Publishers, Adelaide, South Australia.

Helm, F. K., and Cambourne, B. (1988). The influence of climatic variability on the production of quality wines in the Canberra district of South Eastern Australia. *Proc. 2nd Int. Symp. Cool Climate Vitic. Oenol., Jan. 11–15, 1988, Auckland, N.Z.* (R. E. Smart, S. B. Thornton, S. B. Rodriguez, and J. E. Young, eds.), pp. 17–20. New Zealand Soc. Vitic. Oenol., Auckland, New Zealand.

Jordan, A. D., Croser, B. J., and Lee, T. H. (1985). Australia. *Int. Symp. Cool Climate Vitic. Enol.* (B. A. Heatherbell, P. B. Lombard, F. W. Bodyfelt, and S. F. Price, eds.), OSU Agric. Exp. Stn. Tech. Publ. No. 7628, Oregon State Univ., Corvallis, Oregon.

Brazil

Mathes, K. P., and Almada, F. (1989). The emerging patterns of viticulture in Brazil. *Wines Vines* 70(6), 52–56 and 58.

Canada

Anonymous. (1984). "Atlas of Suitable Grape Growing Locations in the Okanagan and Similkameen Valleys of British Columbia." Agric. Canada & Assoc. B.C. Grape Growers, Kelowna, Canada.

Reynolds, A. G., and Strachan, G. E. (1985). Canada. *Int. Symp. Cool Climate Vitic. Enol.* (B. A. Heatherbell, P. B. Lombard, and S. F. Price, eds.), OSU Agric. Exp. Stn. Tech. Publ. No. 7628, pp. 358–380. Oregon State Univ., Corvallis, Orgeon.

Sayed, H. (1992). "Vineyard Site Suitability in Ontario." Ministry Agric. Food Ontario & Agric. Canada, Queen's Printer for Ontairo, Toronto, Canada.

Chile

Santibañez, M. M. F., Diaz, F., Gaete, C., and Daneri, D. (1986). Bases climatiques pour le zonage de la région viti-vinicole chilienne. *XIX Int. Vitic. Oenol. Cong., Santiago, Chile, November 1986*, Vol. 2, pp. 93–124. O.I.V., Paris.

China

Hua, L. (1990). Les indications géographiques des vins et leur protection en Chine. *Bull. O.I.V.* 63, 282–287.

Huang, H. (1981). Viticulture in China. A 21-century legacy of cultivation. *Wines Vines* 62(10), 28–39.

Huang, S. B. (1990). Agroclimatology of the major fruit production in China: A review of current practice. *Agric. For. Methods* 53, 125–142.

Czechoslovakia

Davenport, R. R. (1978). Wines and wine research in Czechoslovakia. *In* "Third Wine Subject Day," pp. 7–14. Long Ashton Research Station, Univ. of Bristol, Long Ashton, England.

England

Ordish, G. (1977). "Vineyards in England and Wales." Faber and Faber, London.

Pearkes, G. (1985). England. *Int. Symp. Cool Climate Vitic. Enol.* (B. A. Heatherbell, P. B. Lombard, F. W. Bodyfelt, and S. F. Price, eds.), OSU Agric. Exp. Stn. Tech. Publ. No. 7628, pp. 347–357. Oregon State Univ., Corvallis, Oregon.

France

Boursiquot, J. M. (1990). Évolution de l'encépagement du vignoble français au cours des trente dernières années. *Prog. Agric. Vitic.* 107, 15–20.

duPuy, P. (1985). Endeavors to produce quality in the wines of Burgundy: Control of soil, clonal selection, maturity, processing techniques and commercial practices. *Int. Symp. Cool Climate Vitic. Enol.* (B. A. Heatherbell, P. B. Lombard, F. W. Bodyfelt, and S. F. Price, eds.), OSU Agric. Exp. Stn. Tech. Publ. No. 7628, pp. 292–314. Oregon State Univ., Corvallis, Oregon.

Forbes, P. (1979). "Champagne, The Wine, The Land, and the People." Victor Gollancz, London.

Pomeral, C. (ed.) (1984). "Terroirs et Vins de France." Total-Edition-Presse, Paris.

Schaeffer, A. (1985). Wine quality as influenced by grape maturity, clonal selection and processing techniques. Experience with Alsace grapes. *Int. Symp. Cool Climate Vitic. Enol.* (B. A. Heatherbell, P. B. Lombard, F. W. Bodyfelt, and S. F. Price, eds.), OSU Agric. Exp. Stn. Tech. Publ. No. 7628, pp. 274–296. Oregon State Univ., Corvallis, Oregon.

Germany

Anonymous. (1979). "The Wine Industry in the Federal Republic of Germany." Evaluation and Information Services for Food, Agriculture and Forestry, Bonn, Germany.

Hallgarten, S. F. (1981). "German Wines." Publivin, London.

Wahl, K. (1988). Climate and soil effects on grapevine and wine: The situation on the northern borders of viticulture—the example Franconia. *Proc. 2nd Int. Symp. Cool Climate Vitic. Oenol., Jan. 11–15, 1988, Auckland, N.Z.* (R. E. Smart, S. B. Thornton, S. B. Rodriguez, and J. E. Young, eds.), pp. 1–5. New Zealand Soc. Vitic. Oenol., Auckland, New Zealand.

Hungary

Halász, Z. (1981). "The Book of Hungarian Wines." Corvina, Budapest.

Italy

Anderson, B. (1980). "Vino—The Wines and Wine Makers of Italy." Little & Brown, Boston, Massachusetts.

Hazan, V. (1982). "Italian Wine." Alfred A. Knopf, New York.

New Zealand

Eschenbruch, R. (1985). New Zealand. *Int. Symp. Cool Climate Vitic. Enol.* (B. A. Heatherbell, P. B. Lombard, F. W. Bodyfelt, and S. F. Price, eds.), OSU Agric. Exp. Stn. Tech. Publ. No. 7628, pp. 345–346. Oregon State Univ., Corvallis, Oregon.

Portugal

Gonçalves, F. E. (1984). "Portugal—A Wine Country." Editora Portuguesa de Livros Técnicos e Cientificos, Lisbon.

Stanislawski, D. (1970). "Landscapes of Bacchus: The Vine in Portugal." Univ. of Texas Press, Austin.

Soviet Union

Titov, A. P., and Djeneev, S. U. (1991). Viticulture et oenologie de l'U.R.S.S. *Bull. O.I.V.* **64**, 575–583.

Tuchnott, O. G. (1978). Wine research in Russia. *In* "Third Wine Subject Day" pp. 15–38. Long Ashton Research Station, Univ. of Bristol, Long Ashton, England.

Spain

Read, J. (1984). "Wines of the Rioja." Sotheby Publ., London.

Read, J. (1986). "The Wines of Spain." Faber and Faber, London.

United States

Adelsheim, D. B. (1985). Recent developments in the wine industry of the Pacific Northwest west of the Cascade Mountains. *Int. Symp. Cool Climate Vitic. Enol.* (B. A. Heatherbell, P. B. Lombard, F. W. Bodyfelt, and S. F. Price, eds.), OSU Agric. Exp. Stn. Tech. Publ. No. 7628, pp. 100–118. Oregon State Univ., Corvallis, Oregon.

Baxevanis, J. J. (1992). "The Wine Regions of America: Geographical Reflections and Appraisals." Vinifera Wine Growers Journal, Stroudsburg, Pennsylvania.

Klein, J. K. (1981). Wine production in Washington State. *In* "Wine Production Tehcnology in the United States" (M. A. Amerine, ed.), ACS Symp. Ser. No. 145, pp. 193–224. American Chemical Society, Washington, D.C.

Long, Z. (1985). Recent developments in California North Coast viticulture and enology. *Symp. Cool Climate Vitic. Enol.* (B. A. Heatherbell, P. B. Lombard, F. W. Bodyfelt, and S. F. Price, eds.), OSU Agric. Exp. Stn. Tech. Publ. No. 7628, pp. 119–127. Oregon State Univ., Corvallis, Oregon.

Morton, L. T. (1985). "Winegrowing in Eastern America." Cornell Univ. Press, Ithaca, New York.

Olien, W. C. (1990). The Muscadine grape: Botany, viticulture, history and current industry. *HortScience* **25**, 732–739.

Wagner, P. (1981). Grapes and wine production in the East. *In* "Wine Production Technology in the United States" (M. A. Amerine, ed.), ACS Symp. Ser. No. 145, pp. 193–224. American Chemical Society, Washington, D.C.

Wolfe, W. H., and Hirschfelt, D. J. (1985). Recent develoments of the wine industry: East of the Cascade Mountains, Pacific Northwest, U.S.A. *Int. Symp. Cool Climate Vitic. Enol.* (B. A. Heatherbell, P. B. Lombard, F. W. Bodyfelt, and S. F. Price, eds.), OSU Agric. Exp. Stn. Tech. Publ. No. 7628, pp. 81–99. Oregon State Univ., Corvallis, Oregon.

References

Adelsheim, D. (1988). Oregon experiences with Pinot noir and Chardonnay. Proc. 2nd *Int. Symp. Cool Climate Vitic. Oenol.*, Jan. 11–15, 1988, Auckland, N.Z. (R. E. Smart, S. B. Thornton, S. B. Rodriguez, and J. E. Young, eds.), pp. 264–269. New Zealand Soc. Vitic. Oenol., Auckland, New Zealand.

Anonymous. (1979). "The Wine Industry in the Federal Republic of Germany." Evaluation and Information Service for Food, Agriculture and Forestry, Bonn, Germany.

Becker, N. (1977). The influence of geographical and topographical factors on the quality grape crop. *O.I.V. Int. Symp. Quality Vintage*, pp. 169–180. Oenol. Vitic. Res. Inst., Stellenbosch, South Africa.

Boursiquot, J. M. (1990). Évolution de l'encépagement du vignoble français au cours des trente dernières années. *Prog. Agric. Vitic.* **107**, 15–20.

Branas, J., Bernon, G., and Levadoux, L. (1946). "Eléments de Viticulture Générale." Published by the authors. L'École nationale de Agriculture, Montpellier, France.

Burda, K., and Collins, M. (1991). Adulteration of wine with sorbitol and apple juice. *J. Food Protect.* **54**, 381–382.

Cillers, J. J. L., and Singleton, V. L. (1990). Nonenzymatic autooxidative reactions of caffeic acid in wine. *Am. J. Enol. Vitic.* **41**, 84–86.

de Blij, H. J. (1983). "Wine: A Geographic Appreciation." Rowman & Allanheld, Totowa, New Jersey.

de Blij, H. J. (1985). "Wine Regions of the Southern Hemisphere." Rowman & Allanheld, Totowa, New Jersey.

Etiévant, P., Schlich, P., Bertrand, A., Symonds, P., and Bouvier, J.-C. (1988). Varietal and geographic classification of French red wines in terms of pigments and flavonoid compounds. *J. Sci. Food Agric.* **42**, 39–54.

Frank, I. E., and Kowalski, B. R. (1984). Prediction of wine quality and geographical origin from chemical measurements by partial least-squares regression modelling. *Anal. Chim. Acta* **162**, 241–251.

Fregoni, M. (1991). "Origines de la Vigne et de la Viticulture." Musumeci Editeur, Quart (Vale d'Aosta), Italy.

Fregoni, M. (1992). The new millennium opens for the wines of Italy. *Ital. Wines Spirits* **16**(1), 9–18.

Guzun, N. I., Nedov, P. N., Usatov, V. T., Kostrikin, I. A., and Meleshko, L. F. (1990). Grape selection for resistance to biotic and abiotic environmental factors. *Proc. 5th Int. Symp. Grape Breeding*, Sept. 12–16, 1989, pp. 219–222. St. Martin, Pfalz, Germany (Special Issue of *Vitis*).

Hua, L. (1990). Les indications géographiques des vins et leur protection en Chine. *Bull. O.I.V.* **63**, 282–287.

Hubáčkova, M., and Hubáček, V. (1984). Frost resistance of grapevine buds on different rootstocks. *Vinohrad* **22**, 55–56 (in Czech).

Jourjon, F., Morlat, R., and Seguin, G. (1991). Caractérisation des terroirs viticoles de la moyenne vallée de la Loire. Parcelles expérimentales, climat, sols et alimentation en eau de la vigne. *J. Int. Sci. Vigne Vin* **25**, 179–202.

Kirillov, A. F., Levit, T. Kh., Skurtul, A. M., Koz'mik, R. A., Grozova, V. M., Syli, V. N., Khanin, Y. D., Baryshok, V. P., Semenova, N. V., and Voronkov, M. G. (1987). Enhancement of the cold hardiness in grapevine as affected by cryoprotectors. *Sadovod. Vinograd. Mold.* **42**, 36–39 (in Russian).

Laville, P. (1990). Le terroir, un concept indispensable à l'élaboration et à la protection des appellations d'origine comme à la gestion des vignobles: Le cas de la France. *Bull. O.I.V.* **63**, 217–241.

Le Roux, E. (1974). Kilmaatsreek Indeling van die Suidwes Kaaplandse Wynbougebiede. Master's Thesis, Univ. of Stellenbosch, Stellenbosch, South Africa.

Long, Z. R. (1987). Manipulation of grape flavour in the vineyard: California, North Coast region. *Proc. 6th Aust. Wine Ind. Tech. Conf.* (T. Lee, ed.), pp.. 82–88. Australian Industrial Publ., Adelaide, Australia.

Luft, G., and Morgenschweis, G. (1984). Zur Problematik grossterrassierter Flurbereinigung im Weinbaugebiet des Kaiserstuhls. *Z. Kulturtech. Flurbereinig.* **25**, 138–148.

Luo, G. (1986). Dragon system of training and pruning in China's viticulture. *Am. J. Enol. Vitic.* **37**, 152–157.

McCallum, N. K., Rothbaum, H. P., and Otlet, R. L. (1986). Detection of adulteration of wine with volatile compounds. *Food Technol. Aust.* **38**, 318–319.

Marquet, P. (1987). L'importance des facteurs naturels et humains dans le développement des appellations d'origine françaises. *Symp. "Les Appellations d'Origine Historiques,"* Jerez de la Frontera, Spain, March 16–18, 1987, pp. 33–41. Office International de la Vigne et du Vin, Paris.

Martin, G. J. (1988). Les applications de la mesure par résonance magnétique nucléaire du fractionnement isotopique natural spécifique (RMN–Fins) en viticulture et en oenologie. *Rev. Fr. Oenol.* **114**, 23–24 and 53–60.

Martinière, P. (1981). Évolution de la radioactivité par le carbone 14 des vins de Gironde—Détermination du millésime. *In* "Actualités Oenologiques et Viticole" (P. Ribéreau-Gayon and P. Sudraud, eds.), pp. 384–390. Dunod, Paris.

Mayer, G. (1990). Results of cross-breeding. *Proc. 5th Int. Symp. Grape Breeding, Sept. 12–16, 1989*, pp. 148. St. Martin, Pfalz, Germany (Special Issue of *Vitis*).

Messini, A., Vincenzini, M., and Materassi, R. (1990). Evolution of yeasts and lactic acid bacteria in "Chianti" wines as affected by the refermentation according to Tuscan usage. *Ann. Microbiol.* **40**, 111–119.

Moran, W. (1988). The wine appellation: Environmental description or economic device? *Proc. 2nd Int. Symp. Cool Climate Vitic. Oenol., Jan. 11–15, 1988, Auckland, N.Z.* (R. E. Smart, S. B. Thornton, S. B. Rodriguez, and J. E. Young, eds.), pp. 356–360. New Zealand Soc. Vitic. Oenol., Auckland, New Zealand.

OIV. (1991). The state of vitiviniculture in the world and the statistical information in 1990. *Bull. O.I.V.* **64**, 893–945.

Penning-Rowsell, E. (1985). "The Wines of Bordeaux," 5th Ed. Wine Appreciation Guild, San Francisco, California.

Pouget, R. (1988). L'encépagement des vignobles français d'appellation d'origine contrôlée: Historique et possibilités d'évolution. *Bull. O.I.V.* **61**, 185–195.

Rankine, B. C., Kepner, R. E., and Webb, A. D. (1958). Comparison of anthocyanin pigments of vinifera grapes. *Am. J. Enol. Vitic.* **9**, 105–110.

Rapp, A., Mandery, H., and Heimann, W. (1983). Flüchtige Inhaltsstoffe aus "Flüssigzucker." *Vitis* **22**, 387–394.

Rapp, A., Volkmann, C., and Niebergall, H. (1993). Untersuchung flüchtiger Inhaltsstoffe des Weinaromas: Beitrag zur Sortencha-rakterisierung von Riesling und Neuzüchtungen mit Riesling—Abstammung. *Vitis* **22**, 171–178.

Ribéreau-Gayon, J., Peynaud, E., Sudraud, P., and Ribéreau-Gayon, P. (1982). Recherche des fraudes. *In* "Traité d'Oenologie: Sciences et Techniques du Vin" (J. Ribéreau-Gayon, E. Peynaud, P. Sudraud, and P. Ribéreau-Gayon, eds.), Vol. 1, 2nd Ed., pp. 575–617. Dunod, Paris.

Robertson, J. M., and Rush, G. M. (1979). Chemical criteria for the detection of winemaking faults in red wine. *Food Technol. N.Z.* **14**, 3–11.

Robinson, J. (1986). "Vines, Grapes and Wines." Knopf, New York.

Ruhl, E. H. (1990). Better rootstocks for wine grape production. *Aust. Grapegrower Winemaker* **304**, 113–115.

Seeman, J., Chirkov, Y. I., Lomas, J., and Primault, B. (1979). "Agrometerology." Springer-Verlag, Berlin.

Seguin, G. (1986). "Terroirs" and pedology of wine growing. *Experientia* **42**, 861–873.

Shackleton, M. R. (1958). "Europe: A Regional Geography," 6th Ed. Longmans, London.

Stacey, R. J. (1984). Isotopic analysis: Its application in the wine trade. *In* "Tartrates and Concentrates—Proceedings of the Eighth Wine Subject Day" (F. W. Beech, ed.), pp. 45–51. Long Ashton Research Station, Univ. of Bristol, Long Ashton, England.

Stanislawski, D. (1970). "Landscapes of Bacchus: The Vine in Portugal." Univ. of Texas Press, Austin.

Sudraud, P. (1978). Evolution des taux d'acidité volatile depuis le début du siècle. *Ann. Technol. Agric.* **27**, 349–350.

van Buren, J. P., Bertino, J. J., Einset, J., Remaily, G. W., and Robinson, W. B. (1970). A comparative study of the anthocyanin pigment composition in wines derived from hybrid grapes. *Am. J. Enol. Vitic.* **21**, 117–130.

van Zyl, J. L., and van Huyssteen, L. (1980). Comparative studies on wine grapes on different trellising systems. I. Consumptive water use. *S. Afr. J. Enol. Vitic.* **1**, 7–14.

11

Sensory Perception and Wine Assessment

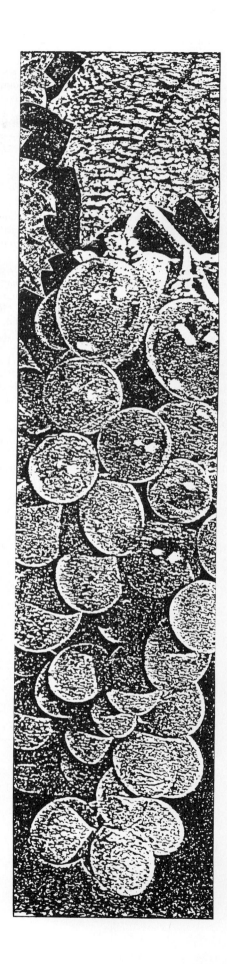

Visual Sensations

Color

The visual characteristics of a wine depend on how its chemical and particulate nature transmit, absorb, and reflect visible radiation. Although such characteristics can be accurately measured with a spectrophotometer (Fig. 8.11), the significance to human color perception is far from simple. Spectrophotometric measurements assess the intensity of individual wavelengths, while the eye responds to light by combining the responses from different types of receptor cells (cones and rods). The cones differentially respond to light within the blue, green, or red ranges of visible spectrum, while the rods respond to low light intensity more uniformly across the visible spectrum. Consequently, there is no simple relationship between spectrophotometric measurements and human color perception.

Currently, the best objective approximation of human color perception is tristimulus colorimetry (Court and Ford, 1981), but several researchers have proposed changes in measuring wine color (Negueruela *et al.*,

1990; Heredia and Guzmán Chozas, 1992). Tristimulus colorimetry involves measuring the light transmission at 445, 495, 550, and 625 nm, followed by calculation of the tristimulus values *X, Y,* and *Z.* Tristimulus colorimeters, in contrast to spectrophotometers, directly correlate these values in terms of color hue and depth. Such information could be used to improve blending wines to a predetermined color (Negueruela *et al.,* 1990).

Other than the pleasure it may give, observation of wine color yields little information beyond a rough indication of grape pigmentation, duration of skin contact, wine age, and some wine faults. Even here, caution is required to avoid being influenced unjustly, especially if wines of different ages or winemaking processes are assessed together. Color also can influence the perception of wine quality (Tromp and van Wyk, 1977; Williams *et al.,* 1984). To offset this influence, wines may be tasted in dark glasses or under red light.

Often people learn to associate particular colors with certain wines. Young dry white wines generally range from nearly colorless to pale straw colored. A more obvious yellow tint is often suspect, unless associated with maturation in oak cooperage. Sweet white wines may vary from a pale straw color to yellow-gold. Sherries vary from pale straw to golden-brown in color, depending on the style. Rosé wines are expected to be pale pink, without shades of blue. Hints of brown, purple, or orange usually indicate oxidation. Red wines vary from deep purple to pale tawny red. Initially, most red wines have a purplish-red hue. Varieties such as 'Gamay' and 'Pinot noir' seldom yield wines with deep colors and rapidly develop a ruby color. More intensely pigmented varieties, such as 'Nebbiolo' and 'Cabernet Sauvignon,' may remain deep red for decades. Red ports, depending on style, may be deep red, ruby, or tawny colored.

Because wines eventually take on brown hues, brown is often used as an indicator of wine age (Somers and Evans, 1977). However, a brownish cast may indicate oxidation or heating. Therefore, wine age, type, and style must be known before interpreting the meaning and significance of a brownish hue. Brownish shades are acceptable only if associated with the development of a desirable processing or aging bouquet. The heating of madeira, which gives the wine its brown coloration and baked bouquet, is an example of process-produced browning. Because most wines fail to develop a desirable aging bouquet, brown casts are typically an indicator that the wine is past its prime.

Clarity

In contrast to the complexity of interpreting the significance of wine color, haziness is always considered a fault. With modern quality control, consumers expect perfectly clear beverages and long shelf life. Thus, considerable effort is expended in producing wines stable in terms of clarity (see Chapter 8).

Most wines are initially supersaturated with tartrate salts. During maturation, physicochemical isomerization reduces tartrate solubility, and cool conditions enhance crystallization. Crusty, flakelike crystals are usually potassium bitartrate, while fine crystals are typically calcium tartrate (Lüthi and Vetsch, 1981). Additional crystalline deposits may consist of calcium malate, calcium oxalate, calcium sulfate, and calcium mucate. Consumers occasionally mistake crystalline deposits in wine for glass fragments.

Another potential source of haziness is the resuspension of sediment. **Sediment** occurs most frequently in older red wines and may consist of polymerized and precipitated anthocyanins, tannins, proteins, tartrate crystals, fining agents, and cell fragments. The presence of sediment often is considered a sign of quality by many wine connoisseurs. Depending on the chemical composition, sediment may or may not have a bitter taste.

Casse is an infrequent cause of haziness resulting from a reaction between metallic ions and soluble proteins. As the components of casse coalesce and reach colloidal size, a milky haze develops. Although unacceptable, casse does not affect the taste or aromatic character of the wine.

Microbial spoilage may be an additional source of haziness. Although either bacteria or yeasts may be involved, bacteria are the most frequent causal agents. For example, some lactic acid bacteria held together by extracellular polysaccharides form long macroscopic filaments, producing a condition called *ropiness.* Disruption of the filaments generates turbidity and an oily texture. The condition usually is associated with the production of off-odors.

When oxygen has access to wine, microaerobic yeasts and acetic acid bacteria may grow in or on the wine. Occasionally, they produce a thick film on the wine's surface that liberates variously sized particles on disruption. Such growths taint the wine with off-odors and off-tastes.

Viscosity

Although viscosity is often mentioned in the popular wine literature, perceptible increases usually occur only when sugar and/or alcohol contents are high (Burns and Noble, 1985), or in cases of wine showing *ropiness.* The glycerol content apparently needs to be relatively high (≥ 25 g/liter) to have a detectable sensory influence on viscosity (Noble and Bursick, 1984).

Sparkle (Effervescence)

In sparkling wines, numerous chains of minute, densely arranged bubbles are an important quality feature. Bubble formation is usually associated with long yeast autolysis following the second yeast fermentation (see Chapter 9).

Still wines may occasionally contain sufficient carbon dioxide to produce bubbles along the sides and bottom of a wine glass. Usually this is caused by wine's being bottled before the excess carbon dioxide dissolved in young wine has had sufficient time to escape. Occasionally, bubbles may result from microbial metabolism in the wine following bottling.

Tears

Tears ("legs") is another phenomenon often given considerable attention in the popular wine press. Tears form after the wine has been swirled and a film of wine coats the inner surface of the glass. Because ethanol evaporates from the film more rapidly than the main volume of wine in the glass, the surface tension of the film increases rapidly relative to that in the rest of the glass. As the water molecules in the film pull closer together, they produce droplets at the rim. Because the mass of the drops increase relative to their size, drops start to sag downward, producing "arches." Finally, the drops slide down, forming the "tears." When the drops reach the surface of the wine in the bowl, fluid is lost and the drop pulls back.

Once formed, tears continue to develop as long as alcohol evaporation draws sufficient wine up the film to counteract the action of gravity in pulling the film down. Cooling generated by alcohol evaporation further helps generate convection currents that draw wine up the glass (Neogi, 1985). Thus, factors affecting the rate of evaporation, such as temperature, alcohol content, and the liquid/air interface, influence tears formation. Contrary to popular belief, glycerol neither significantly affects nor is required for formation of tears. Movement of wine up the sides of a wine glass can be demonstrated by adding a drop of food coloring, or nonwettable powder, to wine after tears have formed.

Taste and Mouth-Feel

Taste and mouth-feel are perceptions derived from two different sets of chemoreceptors in the mouth. Taste is initiated by specialized **receptor** neurons located in **taste buds.** They generate at least four basic tastes, namely, sweet, sour, salt, and bitter. Mouth-feel is activated by **free nerve endings** and gives rise to the sensa-

tions of astringency, heat, body, prickling, and pain. The cells also produce the "burning" impression generated by hot spices and the "coolness" of menthol.

Taste

Taste buds are located primarily on the tongue, but they may also occur on the soft palate, pharynx, epiglottis, larynx, and upper portion of the esophagus. On the tongue, taste buds form depressions on raised growths called **papillae.** Taste buds resemble pear-shaped structures possessing up to 50 neuroepithelial cells (Fig. 11.1). Individual receptor cells remain active for only about 10 days before being replaced by differentiating adjacent epithelial cells. Each receptor (**gustatory**) cell terminates in a receptive **dendrite** or several **microvilli** that project into the oral cavity. Impulses initiated from the receptive endings pass down to the cell body to connect with one of several cranial nerves innervating the oral cavity. Nerve stimulation not only generates impulses sent to the brain, but also maintains the integrity of the taste buds. The distribution pattern of cranial nerves in the tongue partially reflects the differential sensitivity of the different areas of the tongue to taste substances (Fig. 11.2).

Four taste perceptions are usually recognized: sweet, sour, salty, and bitter. Some researchers have proposed an expansion of the list to recognize two types each of

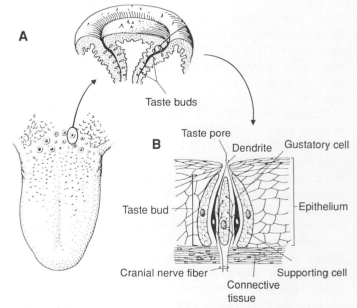

Figure 11.1 Taste receptors. (A) Location of taste buds in a papilla and (B) structure of a taste bud. (From PRINCIPLES OF ANATOMY AND PHYSIOLOGY 1/ed. by G. Tortora and N. Anagnostakos. Copyright © 1975 by Biological Sciences Textbooks, Inc., and Nicholas Peter Anagnostakos. Reprinted by permission of HarperCollins Publishers.)

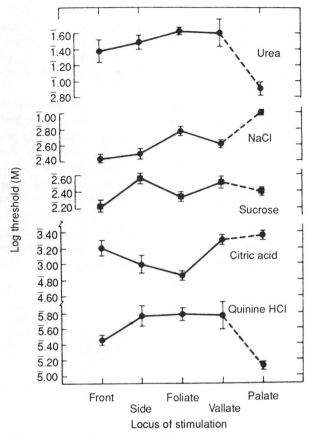

Figure 11.2 Threshold for five compounds as a function of locus on the tongue and soft palate. (From Collins, 1974, by permission.)

sweetness and bitterness, and additional sensations such as pleasant, umami ("good taste" in Japanese), and metallic (see Boudreau *et al.*, 1979). Nevertheless, others suggest that distinct tastes are artificial, and only figments created out of a complex continuum of perceptions (Erickson, 1985). Still others have presented data indicating that some nontraditional tastes, for example metallic, may be unrecognized aspects of olfaction (Hettinger *et al.*, 1990). Currently, psychophysical, neurophysiological, cytological, biochemical, and genetic techniques are being used to study the precise nature of taste perception.

Genetic analysis of specific taste deficiencies (**ageusia**) has shown that some forms are associated with recessive genes, for example, the bitter perception of phenylthiocarbamide (Kalmus, 1971) and saccharin (Bartoshuk, 1979). This is considered evidence for the association of specific receptor molecules with particular taste (**sapid**) compounds. Recently, a protein has been isolated from circumvallate papillae that bonds selectively and reversibly with the sweet-tasting protein thaumatin (Sato, 1987). Glands closely associated with the taste buds produce a protein (von Ebner's gland protein) that may

play a role in promoting taste reception (Schmale *et al.*, 1990).

Some studies have noted that receptors of similar sensitivity tend to be grouped together (Scott and Giza, 1987). Nevertheless, individual receptor neurons appear to react, though differentially, to more than one sapid substance (Scott and Chang, 1984). In addition, taste buds generally respond to more than one type of taste compound (Beidler and Tonosaki, 1985). These factors may partially explain poor taste localization with complex solutions such as wine.

SWEET AND BITTER TASTES

Although sweetness and bitterness may appear to be opposed sensations, their modes of activation appear to be related. Both sensations appear to be partially dependent on the formation of van der Waals attractions and hydrogen bonds with sites on receptor neurons (Beidler and Tonosaki, 1985). It is not known if this similarity is involved in the partial and mutual suppression of the sensations.

Glucose and fructose are the primary sources of a sweet taste in wine, with fructose being sweeter. The perception of sweetness may be enhanced by the glycerol and ethanol content of the wine.

Flavonoid phenolics are the primary bitter compounds in wines, with tannin monomers (catechins) being more bitter than the polymeric forms (condensed tannins) (Robichaud and Noble, 1990). In tannic red wines, the bitterness of the tannins is often confused with (Lee and Lawless, 1991) or masked by (Arnold and Noble, 1978) the astringent sensation of tannins. During aging, wine often develops a smoother taste as tannin polymers precipitate, resulting in a decline in bitterness and astringency. However, if smaller phenolics remain in solution, or larger tannins hydrolyze, the perceived bitterness of the wine may increase.

Other bitter compounds occasionally found in wine are glycosides, terpenes, and alkaloids. Naringin is one of the few bitter glycosides occurring in wine. Bitter terpenes rarely occur in wine except when added with pine resin (e.g., *retsina*). Similarly, bitter alkaloids rarely occur in wine except where they come from herbs and barks used in flavoring wines such as vermouth.

SOUR AND SALTY TASTES

Sourness and saltiness are commonly called the *electrolytic* tastes since small, soluble, inorganic cations (positively charged ions) induce both sensations. Sourness is induced primarily by H^+ ions, saltiness by metal and metalloid cations.

Because the tendency of acids to dissociate into ions is markedly influenced by pH and anionic and cationic components of the acids, both factors significantly affect the sourness of wine. While undissociated acid molecules are relatively inactive in stimulating receptor neurons, they do have an effect on perceived acidity (Ganzevles and Kroeze, 1987). The major acids affecting wine acidity are tartaric, malic, and lactic acids. Additional acids occur in wine, but, with the exception of acetic acid, few occur in sufficient concentration to contribute significantly to the perception of sourness.

Salts also dissociate into positively and negatively charged ions. Salt cations are typically a metal ion, for example, K^+ and Ca^{2+}, while anions may be either inorganic or organic, such as Cl^- and bitartrate, respectively. As with sourness, salt perception is not solely influenced by the activating salt cation. The tendency of a salt to ionize greatly affects perceived saltiness, as does the size of the associated anion. For example, large organic anions suppress the sensation of saltiness. Because the major salts in wine possess large organic anions (i.e., tartrates and bitartrates) that dissociate poorly at wine pH, their common cations (K^+ and Ca^{2+}) do not actively stimulate salt receptors. In addition, the comparative scarcity of Na^+ in wine, the primary cation inducing saltiness is a contributing factor in explaining the general absence of salty sensations in wine.

Factors Influencing Taste Perception

Many factors affect the ability of a person to detect and identify taste sensations. These are conveniently divided into four categories: physical, chemical, biological, and psychological.

Interest in the effect of temperature on taste has existed for over a century. Regardless, the role of temperature in perception is still unclear. The general view is that perception is optimal at normal mouth temperature. However, a recent study suggests that cooling has little effect on the perception of some acids or salts, but reduces sensitivity to sugars and bitter alkaloids (Green and Frankmann, 1987). Because the perception of bitterness (and astringency) of red wines increases at cool temperatures, the disagreement with the previous study may relate to the different receptors involved in alkaloid- versus phenolic-induced bitterness (Boudreau *et al.,* 1979). Another important physicochemical factor affecting taste perception is pH. Through its effect on organic and amino acid ionization, and their salts, pH influences the solubility, shape, and biological activity of the compounds. Modification of the shape of receptor molecules on gustatory neurons also could affect taste perception.

Sapid substances not only directly stimulate receptor neurons, but also influence the perception of other sapid molecules. For example, mixtures of different sugars suppress the perception of sweetness, especially at high concentrations (McBride and Finlay, 1990). In addition,

members of one group of sapid substances can affect the perception of another, as in the suppression of bitterness and sourness by sugars and phenolics. Although suppression of perception is common, ethanol enhances the sweetness of sugars and the bitterness of alkaloids, while suppressing the sourness of some acids.

Sapid substances often have more than one sensory quality. For example, tannins may be both bitter and astringent; glucose can be sweet and mildly tart; potassium salts are salty and bitter; and alcohol both possesses a sweet taste and produces sensations of "burning" and "weight." In heterogeneous mixtures, these "side tastes" can significantly affect overall taste perception (Kroeze, 1982). The intensity of a mixture apparently reflects the intensity of the dominant component, not an integration of the separate intensities of the individual components (McBride and Finlay, 1990). The

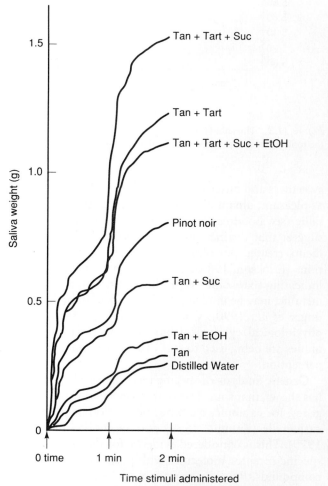

Figure 11.3 Amount of parotid saliva secreted in response to tasting 'Pinot noir' wine and selected constituents of the wine, singly and in combination. Tan, tannin; Tart, tartaric acid; Suc, sucrose; EtOH, ethanol. (From Hyde and Pangborn, 1978, reproduced by permission.)

origin of these interactions may be various and complex (Avenet and Lindemann, 1989).

The interaction of sapid compounds on perception is further complicated by the changing chemical nature of wine in the mouth. Wine stimulates salivary flow (Fig. 11.3), which both dilutes the wine and modifies its chemistry. Because saliva chemistry can change temporally, and often differs between individuals, the effect of saliva can be quite variable.

Several studies have noted a loss in sensory acuity with age (Bartoshuk *et al.,* 1986; Stevens and Cain, 1993). There is a general reduction in the number of both taste buds and sensory receptors per taste bud. Nevertheless, age-related sensory loss is not known to seriously limit wine tasting ability. Certain medications also reduce taste sensitivity (see Schiffman, 1983). In addition, taste perception can be disrupted by agents such as sodium lauryl sulfate (sodium dodecyl sulfate) used in toothpaste (DeSimone *et al.,* 1980).

Acuity loss generally increases the detection threshold, the lowest concentration at which a substance can be sensed. Chronic oral and dental ailments may create lingering mouth tastes, complicating discrimination at low concentrations (Bartoshuk *et al.,* 1986). This could explain why detection thresholds usually are higher in elderly people with natural dentation than those with dentures. Acuity loss also appears to depress the ability to identify sapid substances in mixtures (Stevens and Cain, 1993).

Although recessive genetic traits can produce specific taste deficiencies, subtle individual variations in taste acuity are more common (Fig. 11.4). Individual acuity also can vary over time. Sensitivity to phenylthiocarbamide (PTC) has been found to vary by a factor of 100 over several days (Blakeslee and Salmon, 1935).

Taste **adaptation** is the transient acuity loss associated with extended exposure to a sapid substance. At moderate levels, adaptation can become complete. Correspondingly, it is usually recommended that wine tasters cleanse the palate between samples with water or bread. **Cross-adaptation** is the effect of adaptation to one compound on reduced sensitivity to another.

Color appears to influence the quality perception of wine (Tromp and van Wyk, 1977), as well as taste perception (Maga, 1974; Clydesdale *et al.,* 1992). Most of the data suggest that these influences are learned associa-

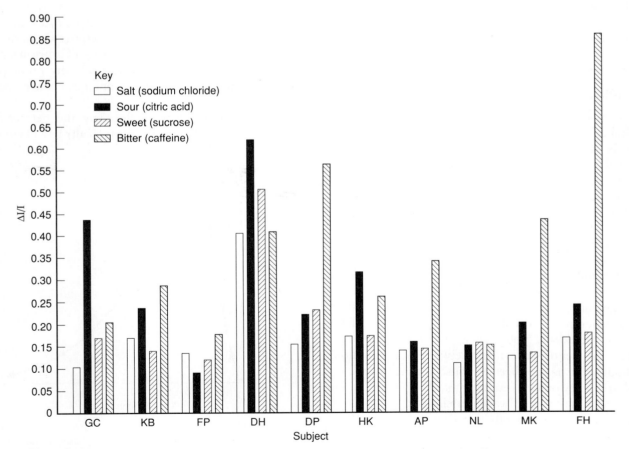

Figure 11.4 Different sensitivity for the four basic taste qualities for each of ten subjects. (From Schutz and Pilgrim, 1957, reproduced by permission.)

tions (see Clydesdale *et al.*, 1992), although results from Pangborn *et al.* (1963) suggest otherwise.

A "taste" aspect of some volatile compounds often has been noted by wine tasters as well as researchers (Murphy *et al.*, 1977; Enns and Hornung, 1985). For example, fruity fragrances are commonly described as possessing a "sweet" aspect. Whether this results from the stimulation of taste buds on the back of the epiglottis or from an enhancement of the perception of sweetness is unknown (Frank and Byram, 1988). Not all odorants enhance the perception of sweetness. In addition, where there is apparent conflict between expected gustatory and olfactory sensations, the olfactory stimulus appears to have priority and represses "aberrant" taste perceptions (Murphy *et al.*, 1977).

Cultural and historical background also can affect sensory perception, or at least the development of preferences (Barker, 1982). Historically developed biases probably explain the oft-mentioned matching of locally produced foods and wines.

Mouth-Feel

Mouth-feel refers to sensations activated by **free nerve endings** of the trigeminal nerve. The distribution of free nerve endings throughout the oral cavity generates diffuse, poorly localized sensations. In wine, mouth-feel includes the perceptions of astringency, temperature, burning, prickling, "body" (weight), and metallic.

ASTRINGENCY

Astringency is a dry, puckery, dust-in-the-mouth sensation typically experienced with red wines. The sensation is induced primarily by phenolic compounds extracted from grape seeds and skins and wood cooperage. White wines seldom show astringency as they generally possess low concentrations of phenolic compounds.

Although astringency may be confused with bitterness (Lee and Lawless, 1991), and both are induced by related groups of compounds, they are distinct sensations. Adding to the potential confusion are their similar response curves: both perceptions develop comparatively slowly and possess a lingering aftertaste (Figs. 11.5 and 11.6). At high concentrations of condensed tannins, the strong astringency may partially mask the perception of bitterness (Arnold and Noble, 1978). Trained tasters may be able to differentiate accurately between these sensations, although this has not been established conclusively.

Astringency in wine results primarily from the binding and precipitation of salivary proteins and glycoproteins with phenolic compounds (Haslam and Lilley, 1988). The two major phenolic groups involved are flavonoid and nonflavonoid tannins (Chapter 6). The main reaction between tannins and proteins involves bonds be-

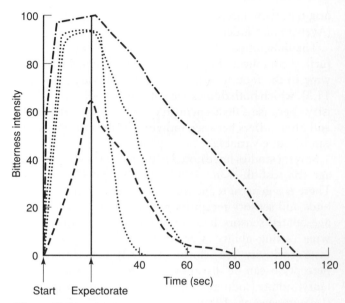

Figure 11.5 Individual time–intensity curves for four judges in response to bitterness of 15 ppm quinine in distilled water. [From E. J. Leach and A. C. Noble (1986). Comparison of bitterness of caffeine and quinine by a time–intensity procedure. *Chem. Senses* 11, 339–345. Reproduced by permission of Oxford University Press.]

tween —NH$_2$ and —SH groups of proteins and *o*-quinone groups of tannins (Haslam *et al.*, 1992). Other tannin–protein reactions are known (see Guinard *et al.*, 1986b) but apparently are of little significance in wine. In the binding of tannins with one or more proteins, the mass, shape, and electrical properties change and can lead to precipitation.

An important factor influencing the perceived astringency of white wines is the typically low pH of these

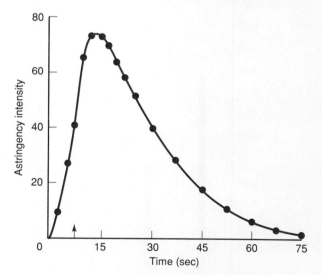

Figure 11.6 Average time–intensity curve for astringency of 500 mg/liter tannic acid in white wine. The sample was held in the mouth for 5 sec (↑) and then expectorated. (From Guinard *et al.*, 1986a, reproduced by permission.)

wines (Guinard *et al.*, 1986b). Hydrogen ions in the wine can affect protein hydration as well as phenol and protein ionization, and thereby influence phenol–protein bonding and precipitation. Low pH also is likely to induce some salivary glycoprotein precipitation and directly induce astringency (Dawes, 1964).

It is believed that the dry mouth-feel of astringency comes partially from precipitated salivary proteins and glycoproteins coating the teeth and oral cavity. On the teeth, the coating generates a rough texture. On the mucous epithelium, precipitated proteins and tannins could force water away from the cell surface, simulating dryness. Reactions with cell-membrane glycoproteins and phospholipids of the mucous epithelium may be even more important than those with salivary proteins. This is likely as perceived astringency increases with the frequency of repeated exposure to tannins (Guinard *et al.*, 1986a). The astringent sensation may result from reversible malfunctioning of the cell membrane, such as disruption of catechol amine methylation. In addition, the relatedness of certain tannin constituents to adrenaline and noradrenaline could stimulate localized blood vessel constriction, further enhancing a dry, puckery mouth-feeling.

As noted above, astringency is one of the slowest in-mouth sensations to develop. For tannic acid, maximal perception develops within about 15 sec (Fig. 11.6). The subsequent decline occurs even more slowly. Different tasters have similar but nonidentical response curves to various tannins.

As noted above, the intensity and duration of an astringent response often increases during a tasting. This phenomenon is less likely to occur when wine is consumed with food, owing to the reaction between food proteins and tannins. However, if wines are tasted in quick succession, and without adequate palate cleansing, the increase in apparent astringency could produce tasting sequence errors. Sequence errors are differences in perception due to the order in which wines are tasted. Although tannins stimulate secretion of saliva, production may be insufficient to restrict an increase in perceived astringency.

One of the most important factors influencing astringency is the molecular size of the tannin. Initially, tannins become increasingly astringent as molecular size (polymerization) increases. Eventually the polymers become so large that they precipitate and lose their astringent properties. Polymerization with acetaldehyde, anthocyanins, and polysaccharides also enhances tannin precipitation and a loss in astringency.

BURNING

Wines high in ethanol content produce a "burning" sensation in the mouth, which is especially noticeable at the back of the throat. It is not known whether the receptors involved are the same as those generating the "hot" sensation associated with pungent condiments. Wines particularly high in sugar content (e.g., eisweins and Tokaji Eszencia) possess a sensation described as a sugar "burn." Whether the sensation is fundamentally different from the burning and hot sensations of alcohol and pungent condiments is unknown.

TEMPERATURE

The cool mouth-feel produced by chilled sparkling or dry white wine seems to add an element of interest and pleasure to wines of subtle flavor. Cool temperatures also help retain the effervescence of sparkling wines. In contrast, red wines typically are served at room temperature. This preference may be based on a reduction in the perception of bitterness and astringency, and an increased volatility of aromatic compounds in the wine. Nevertheless, the preferred serving temperature of wine may reflect custom as much as any other factor. This is suggested by the nineteenth century preference in England for consuming red Bordeaux wines cold (Saintsbury, 1920).

PRICKLING

Bubbles bursting in the mouth produce a prickling, tingling, occasionally burning sensation. The feeling is elicited by wines containing more than 3 to 5‰ carbon dioxide. The sensation appears partially related to bubble size and temperature, and it is augmented at cold temperatures. Carbon dioxide stimulates acid, salt, and temperature receptors on the tongue (Kawamura and Adachi, 1967) and enhances significantly the perception of cold in the mouth (Green, 1992). Carbon dioxide also may suppress odor perception (Cain and Murphy, 1980).

BODY (WEIGHT)

Although "body" is a desirable aspect in most wines, the precise origin of the sensation remains unclear. It is roughly correlated with the alcohol and sugar contents of the wine. Glycerol appears to increase the perception of body, while acidity reduces the sensation. Body also appears to involve aspects of wine fragrance and fragrance intensity.

METALLIC

A metallic mouth-feel is occasionally associated with some dry wines. It is especially noticeable as an aftertaste. The origin of the sensation is unknown, but it may be induced by iron and copper ions. Although the concentrations required (>20 and 2 mg/liter, respectively) are normally well above those found in wine, the detection threshold may be increased by sugar, citric acid, and ethanol but lowered by tannins (Moncrieff, 1964). Some acetamides also are reported to have a metallic aspect (Rapp, 1987).

Taste and Mouth-Feel Sensations in Wine Tasting

To distinguish between the various taste and mouth-feel sensations, tasters may concentrate sequentially on the sensations of sweetness, sourness, bitterness, astringency, and balance. The temporal response to these substances can be useful in confirming the identification of taste sensations (Kuznicki and Turner, 1986). Localization of the sensations in the mouth and on the tongue also can be useful in affirming taste characterization. Balance is a summary perception, derived from the interaction of sapid and mouth-feel sensations.

Sweetness is probably the most rapidly detected taste perception. Sensitivity to sweetness appears to be optimal at the tip of the tongue (Fig. 11.2). Possibly because of the low intensity of the sensation in dry wines, sweetness tends to be the first taste to show adaptation.

Sourness is also detected rapidly. The rate of adaptation to sourness may be slower, and it may generate a lingering aftertaste. Acid detection is commonly strongest along the sides of the tongue but varies considerably between individuals, with some people detecting sourness more distinctly on the back of the lips or the insides of the cheeks. Strongly acidic wines can induce astringency, giving the teeth a rough feel.

Detection of bitterness usually follows, as it typically takes several seconds to develop and the peak intensity may not be reached for 10 to 15 sec (Fig. 11.5). On expectoration, the sensation declines gradually and may linger for several minutes. Most bitter-tasting compounds found in wine, primarily phenolics, are perceived at the back, central portion of the tongue. In contrast, sensitivity to the bitterness of alkaloids, as found in vermouths, is highest on the soft palate and the front of the tongue (Boudreau *et al.*, 1979). Because bitter sensations at the back of the tongue develop more rapidly, and most bitter sensations in wine are induced by phenolic compounds, bitter sensations in wine are detected most commonly at the back of the tongue (McBurney, 1978). The bitterness of wine is often more difficult to assess when the wine is very astringent. High levels of astringency may partially mask the perception of bitterness.

Astringency is often the last sensation detected as it can take 15 sec or more for perception to develop fully (Fig. 11.6). On expectoration, the sensation slowly declines over a period of several minutes. Astringency is poorly localized as it is sensed by free nerve endings distributed throughout the mouth. Because perceived intensity and duration increase with repeated sampling,

some judges recommend that astringency be based on the first taste. This would give a perception more closely approximating the astringency detected on consuming the wine with food. Others consider that judging astringency should occur only after several samplings, when the modifying affects of saliva are less marked.

The increase in perceived astringency that can occur when tasting a series of wines (Guinard *et al.*, 1986a) could seriously affect the validity of wine assessment. This is especially true with red wines, where the first wine in a series often appears the smoothest. A similar situation could occur in a series of dry wines, as well as making a sweeter wine appear overly sweet. These influences are sufficiently well known that tastings are organized to avoid the comparison of wines of markedly different character. However, design errors still can have significant effects on well-conceived comparative tastings. The effect of sequence error may be partially offset by arranging that each taster sample the wines in random order. Lingering taste effects can be minimized by assuring adequate palate cleansing between each wine tasted.

Odor

Olfactory System

NASAL PASSAGES

The olfactory tissue is localized in two small areas in the upper regions of the nasal passages. Volatile compounds may reach the olfactory epithelium either directly via the nostrils, or indirectly from the back of the throat. The latter route is especially important in the generation of flavor, the combination of both taste and olfactory sensations.

The nasal passage is bilaterally divided into right and left halves by a central septum. The receptors in each cavity send signals to the corresponding halves of the olfactory bulb, located directly above at the base of the skull. Some experiments suggest that the left hemisphere of the brain may possess greater odor discrimination than the right hemisphere. Because impulses from the olfactory bulbs partially crossover in the brain, this may explain the greater discriminatory skill of the right nostril (Zatorre and Jones-Gotman, 1990).

Each nasal cavity is further incompletely subdivided transversely by three outgrowths, the **turbinate bones** (Fig. 11.7). These increase the surface area contact between air and the epithelial linings of the nasal passages. While inducing turbulence, warming, and cleansing of the air, the folds restrict air flow past the recessed olfactory regions. It is estimated that only 5 to 10% of the inhaled air moves past the olfactory mucosa. Even at high rates of inspiration, the value may increase by 20%

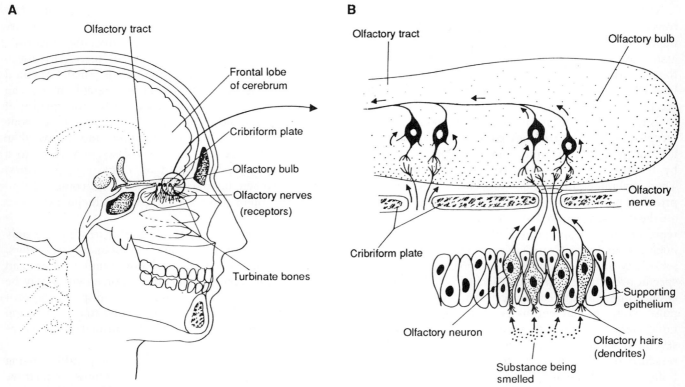

A

Olfactory tract

Frontal lobe
of cerebrum

Cribriform plate

Olfactory bulb

Olfactory nerves
(receptors)

Turbinate bones

B

Olfactory tract

Olfactory bulb

Olfactory
nerve

Cribriform plate

Supporting
epithelium

Olfactory neuron

Olfactory hairs
(dendrites)

Substance being
smelled

Figure 11.7 Receptors for olfaction. (A) Location of the receptors of olfaction in the nasal cavity. (B) Enlarged aspect of olfactory receptors. (After PRINCIPLES OF ANATOMY AND PHYSIOLOGY 1/ed. by G. Tortora and N. Anagnostakos. Copyright © 1975 by Biological Sciences Textbooks, Inc., and Nicholas Peter Anagnostakos. Reprinted by permission of HarperCollins Publishers.)

(De Vries and Stuiver, 1961). Although higher flow rates may slightly enhance odor perception, the vigor of breathing apparently does not affect perceived odor intensity (Laing, 1983). The usual recommendation to take short whiffs of the fragrance of a wine during tasting probably has more to do with avoiding odor adaptation than enhancing odor perception.

Studies suggest that lateral nasal glands in the nose may discharge an odor-binding protein (**olfactomedin**) into the air as it passes into the nose (Snyder *et al.*, 1988). The protein binds nonspecifically to volatile compounds and is carried with the air stream into the nasal passages. The binding protein is lipophilic and may aid the deposition and absorption of aromatic compounds into the olfactory epithelium. The protein also is synthesized by glands associated with the olfactory epithelium. This may partially explain the concentration of odorant molecules up to 1000 to 10,000 times in the nasal mucus (Senf *et al.*, 1980).

Only a fraction of the aromatic molecules that reach the olfactory regions is absorbed into the mucus that coats the olfactory epithelium. Of these, only a proportion is likely to diffuse through the mucus and reach reactive sites on the receptor neurons. In some animals, high concentrations of cytochrome-dependent oxygen-

ases accumulate in the olfactory mucus (Dahl, 1988). The enzymes catalyze a wide range of reactions that may increase the hydrophilic properties of odorants and facilitate their release from the mucus coating of the olfactory epithelium.

OLFACTORY, EPITHELIUM, RECEPTOR NEURONS, AND CONNECTION WITH THE BRAIN

The olfactory epithelium consists of a thin layer of tissue covering an area of about 2.5 cm^2 on each side of the nasal septum. Each region contains about 10 million **receptor** neurons and associated **supporting** and **basal cells.** Receptor neurons are specialized nerve cells that respond to aromatic compounds; supporting cells (and glands underlying the epithelium) produce a special mucus and an odor-binding protein (olfactomedin) that coats the olfactory epithelium (Snyder *et al.*, 1991), while basal cells differentiate into receptor neurons as the latter degenerate. Receptor neurons remain active for only a few weeks before they degenerate and are replaced. Differentiating basal cells produce extensions that grow through openings in the skull (**cribriform plate**) to connect with the olfactory bulb at the base of the brain. In humans, olfactory and gustatory neurons are the only nerve cells known to regenerate regularly.

As with gustatory cells in the mouth, olfactory neurons show a common cellular structure. Odor quality, the distinctive aromatic character of any odorant, is not associated with obvious morphological differentiation of the receptor neurons. The presence of several, long, hairlike cellular projections called **cilia** is thought to facilitate receptor contact with aromatic compounds. Odor quality is thought to arise from the differential sensitivity of receptor neurons to aromatic compounds. This may reflect the presence of a unique family of proteins produced by the olfactory epithelium (Buck and Axel, 1991). These guanine nucleotide regulatory proteins (**G-proteins**) bear several variable regions that could generate the diversity required of receptor specificity. Selective reproduction of a specialized subclass of basal cells producing a particular G-protein may explain the increased sensitivity of some individuals on repeated exposure to specific aromatic compounds (Wysocki *et al.*, 1989).

On stimulation, a receptor cell sends an electrical impulse along its filamentous extension to the olfactory bulb. Here, receptor cells synapse with one or more of several types of nerve cells in the olfactory bulb. The olfactory bulb is a small, bilaterally lobed portion of the brain that collects and edits the information received from the olfactory receptors in the nose. From here, impulses are sent via the lateral olfactory tract to the hypothalamus and several higher centers in the brain. Feedback impulses may also pass downward to the olfactory bulb and regulate its response to incoming signals from the olfactory epithelium. Interaction between the olfactory bulb and other centers of the brain is undoubtedly of great importance to odor perception, but knowledge of the interactions still is limited (Freeman, 1991).

Odorants and Olfactory Stimulation

There is no precise definition of what constitutes an olfactory compound. Based on human perception, there are thousands of olfactory compounds, including a wide range of chemical groups. For air-breathing animals, an odorant must be volatile (pass into a gaseous phase at normal temperatures). Although this places limitations on the molecular size of an odorant (≤ 300 daltons), low molecular mass implies neither volatility nor aromaticity. Most aromatic compounds have strong hydrophobic (fat-soluble) and weak polar (water-soluble) sites. They also tend to bond weakly with cellular constituents and dissociate readily. Volatility may be influenced by the presence of other constituents in wine such as sugars (Sorrentino *et al.*, 1986), ethanol (Williams and Rosser, 1981), and macromolecules (Voilley *et al.*, 1991).

Although the chemical nature of odor quality has been studied for decades, no general theory has found widespread acceptance. Current thought favors the view that several molecular properties may be involved in olfactory stimulation and the perception of odor quality (Ohloff, 1986). These include electrostatic attraction, hydrophobic bonds, van der Waals forces, hydrogen bonds, and dipole–dipole interactions. Small structural modifications, such as found in stereoisomers, can markedly affect the perceived intensity and quality of related compounds. For example, the D- and L-carvone stereoisomers possess spearmintlike and carawaylike qualities, respectively. In a few cases, sensitivity to a series of chemically related odorants (e.g., pyrazines) appears to be related to the ability of the compounds to bind to proteins from the olfactory epithelium (Pevsner *et al.*, 1985).

Some aromatic compounds appear to stimulate only one type of receptor and are classed as **primary odors.** However, most odorants probably stimulate more than one type of receptor, and their odor qualities may be derived from the duration, number, and specific types of receptors stimulated. This is analogous to the derivation of the color spectrum from the three primary color receptors of the eye.

Compounds possessing similar odors, and belonging to the same chemical group, appear to show competitive inhibition. This phenomenon, called **cross-adaption,** suppresses perception of an aromatic compound by prior exposure to a related odorant. While mixtures of different odorant groups usually retain their distinct and separate odor qualities, they may suppresses the intensity of the perception of one another. However, at subthreshold concentrations they may act synergistically and induce their mutual perception (Selfridge and Amerine, 1978). Occasionally, a mixture of aromatic compounds generates a fragrance whose quality seems unrelated to the component molecules (Laing and Panhuber, 1978).

These diverse reactions, combined with the variability of individual sensitivity to aromatic compounds, or the ability to use descriptive terms precisely, help to explain the common divergence of opinion people express about particular wines.

Sensations from Trigeminal Nerve

The free nerve endings of the trigeminal nerve are derived from the same cranial nerve that innervates the oral cavity and generates the sensations of mouth-feel. In the nose, the free nerve endings are scattered throughout the nasal epithelium, excepting the olfactory epithelia. The nerve endings respond to a wide range of pungent and irritant chemicals, often at very low concentrations. At higher concentrations, the receptors appear to respond to all odorant molecules.

Most strong pungent chemicals react nonspecifically with protein sulfhydryl (—SH) groups or break protein disulfide (—S—S—) bridges (see Cain, 1985). The resultant reversible structural changes in membrane proteins may stimulate firing of free nerve endings. Most pungent compounds apparently have a net positive charge, while putrid compounds commonly possess a net negative charge (Amoore *et al.*, 1964).

At high concentrations, most if not all aromatic compounds stimulate trigeminal nerve fibers. Those that are strongly hydrophobic may dissolve into the lipid component of the cell membrane, disrupting cell permeability and inducing nerve firing (Cain, 1985). Unlike the reduced excitability of olfactory neurons on prolonged odorant exposure, free nerve endings become more sensitive on repeated or prolonged exposure (Cain, 1976). Stimulation of free nerve endings by an odorant may be associated with a change in odor quality. For example, hydrogen sulfide produces a yeasty bouquet and adds an aspect of fruitiness to wine at low concentrations (~1 μg/liter) but generates a rotten-egg odor at slightly higher concentrations (MacRostie, 1974). In addition, ethyl acetate, which often is thought to add to the aromatic complexity of a wine at low concentrations (<50 mg/liter), produces an off-odor at higher concentrations (>150 mg/liter) (Amerine and Roessler, 1983). Most fragrances eventually lose their pleasantness at high concentrations.

Odor Perception

Individual variation in odorant perception has long been known. What is new is the knowledge of the extent of the variation (Pangborn, 1981). Variation can affect the ability to detect, identify, and sense the intensity of odors, as well as the emotional or hedonic response to odors.

The **detection threshold** is the concentration at which the presence of a substance becomes noticeable. Human sensitivity to odorants varies over 10 orders of magnitude, from ethane at 2×10^{-2} *M* to 10^{-10}–10^{-12} *M* for mercaptans. Even sensitivity to chemically related compounds may show tremendous variation. For example, pyrazines show detection thresholds spanning 9 orders of magnitude (Seifert *et al.*, 1970). Pyrazines contribute to the bell pepper and moldy aspects of some wines.

When the detection threshold of an individual is markedly below normal, the condition is called **anosmia** (odor blindness). Anosmia can be general or may affect only a small range of related compounds (Amoore, 1977). The occurrence of specific anosmias varies widely in the population. For example, it is estimated that about 3% of the human population is anosmic to isovalerate (sweaty), while 47% is anosmic to 5α-androst-16-en-3-one (urinous). The neurological nature of most specific anosmias is unknown.

Hyperosmia, the detection of odors at abnormally low concentrations, is little understood. One of the most intriguing accounts of hyperosmia relates to a student's 3-week experience of suddenly being able to recognize people and objects by their odors (Sachs, 1985). Also unestablished is the origin of the normally limited olfactory skills of humans, compared with most animals. However, the limited skills may arise from the comparatively small size of the human olfactory epithelium and olfactory bulbs. For example, the olfactory epithelium in dogs can be as large as 150 cm^2, as compared with about 5 cm^2 in humans. Comparative measurements of the odor thresholds in dogs and humans indicate that those of dogs are about 100 times lower (Moulton *et al.*, 1960). Nevertheless, Stoddart (1986) proposes that repression of odorant sensitivity was of selective advantage in the evolution of the nuclear family and gregarious life-style of humans. This could occur by feedback suppression of impulses from the olfactory tissues, or diminished production of olfactomedin, G-proteins, or other molecular causes.

The detection threshold may be temporarily influenced by the presence of other volatile substances. As mentioned above, at subthreshold concentrations, two or more compounds may act synergistically in increasing perceived intensities. Another influence of mixture interaction is the effect of solutes on the volatility of aromatics and, therefore, their apparent threshold in beverages such as wine (Fig. 11.8). Because the ethanol effect on odorant volatility is greatest at about 1% alcohol, its importance may be limited to the "finish" of a wine. At this point, the alcohol content may have fallen sufficiently to increase the volatility of some fragrant molecules. The loss of alcohol from wine coating the sides of the glass, associated with swirling, also may enhance the liberation of some aromatics.

The **recognition threshold** refers to the minimum concentration at which an aromatic compound can be correctly identified. The recognition threshold is typically higher than the detection threshold.

It is generally recognized that people have considerable difficulty in correctly identifying odors in the absence of visual clues (Engen, 1987). Nevertheless, it is often thought that expert tasters and perfumers have superior odor acuity. While this may be true, winemakers often fail to recognize their own wine in blind tasting, and experienced wine tasters frequently misidentify the varietal and geographical origin of wines (Noble *et al.*, 1984; Winton *et al.*, 1975). Thus, the superiority of some individuals in detecting and naming fragrances and wines is unsubstantiated, or at least unquantified.

While odors commonly are organized into groups based on origin, such as "floral," "smoky," and "resinous," odor memories appear to be arranged relative to

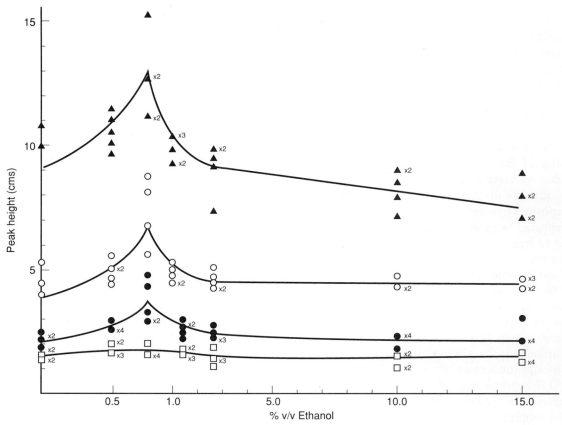

Figure 11.8 Effect of ethanol on the headspace composition of a synthetic mixture. ●, Ethyl acetate; ○, ethyl butrate; ▲, 3-methylbutyl acetate; □, 2-methylbutanol. [From A. A. Williams and P. R. Rosser (1981). Aroma enhancing effect of ethanol. *Chem. Senses* **6**, 149–153. Reproduced by permission of Oxford University Press.]

the events with which they have been associated. The more significant the event, the more intense and stable the association. Engen (1987) views this memory pattern as equivalent to the use of words by young children. Children tend to categorize objects and events functionally, rather than as abstract concepts; for example, a chair is something on which one sits versus a piece of furniture. This view may help to explain why it is so difficult to use unfamiliar terms for familiar odors. Currently, the language of wine fragrance is relatively impoverished and relies heavily on terms describing other aromatic objects. Often the words people use to describe wine indicates little more than their emotional response to the wine (Lehrer, 1975; Dürr, 1985). At best, this expresses feelings in a poetic manner; at worst, it may be used to impress or intimidate. The difficulty of correctly naming familiar odors has been called the "tip-of-the-nose" phenomenon (Lawless and Engen, 1977). The phenomenon is often experienced when tasting wines blind. Suggestions can improve identification but also can unduly influence opinions during a tasting.

The **perceived intensity** of aromatic compounds, as compared to their detection or recognition, often varies considerably. For example, compounds such as hydrogen sulfide or mercaptans are perceived as being intense even at the recognition thresholds. In addition, the rate of change in perceived intensity varies widely between compounds. For example, a 3-fold increase in perceived intensity is correlated with a 25-fold increase in the concentration of propanol but a 100-fold increase in the concentration of amyl butyrate (Cain, 1978). A rapid increase in perceived intensity is characteristic of most off-odors.

Sources of Variation in Olfactory Perception

There are small sex-related differences in olfactory acuity. Women are generally more sensitive to and more skilled at identifying odors than are men (Fig. 11.9). There are also sex-related (or experience-related) differences in the types of odors identified. In addition, women between puberty and menopause experience changes in olfactory discrimination that are correlated with cyclical changes in hormone levels (see Doty, 1986).

Age also affects olfactory acuity by increasing both the detection and recognition thresholds of perception

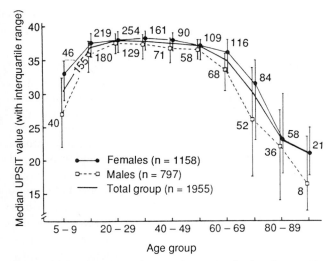

Figure 11.9 University of Pennsylvania Smell Identification Test (UPSIT) scores as a function of age and gender. Numbers by data points indicate sample sizes. (From Doty *et al.*, 1984, reproduced by permission. Copyright 1984 by the AAAS.)

(Stevens *et al.*, 1984; Cowart, 1989). While these effects may begin in the 20s, they tend to increase markedly only during the 70s and 80s (Fig. 11.9). Considerable diversity in individual sensitivity occurs in all age groups (Stevens *et al.*, 1984).

Reduction in the turnover rate of olfactory receptor neurons may partially explain age-related olfactory loss (Doty and Snow, 1988). However, degeneration of the olfactory bulb and nerve connections in the rhinencephalon (olfactory cortex) may be more important. Olfactory regions frequently degenerate earlier than other parts of the brain (Schiffman *et al.*, 1979). This probably explains why smell is often the first of the chemical senses to show an age-related acuity loss. Although wine judging ability may decline with age, experience and mental concentration may compensate for sensory loss.

Nasal and sinus infections may accelerate certain degenerative changes, and upper respiratory tract infections can have persistent effects long after the infection has passed. Short-term effects involve a massive increase in mucus secretion that both blocks air access to the olfactory regions and retards diffusion to the receptor neurons. Loss of olfactory ability also is associated with several major diseases, such as polio, meningitis, and osteomyelitis. Both may destroy the olfactory nerve and cause generalized anosmia. In addition, some genetic diseases are associated with generalized anosmia, such as Kallmann's syndrome. Certain medications and illicit drugs such as cocaine disrupt the olfactory epithelium and diminish the sense of smell (see Schiffman, 1983).

It is commonly believed that hunger increases olfac-

tory sensitivity and, conversely, that satiation lowers it. This view is supported by the report that hunger and thirst increase the general reactiveness of the olfactory bulb and cerebral cortex (Freeman, 1991). Nevertheless, one study has shown an increase in olfactory acuity following food intake (Berg *et al.*, 1963).

Smoking produces long-term, but reversible, impairment of olfactory discrimination (Frye *et al.*, 1990). Thus, the response of smokers and nonsmokers to wine is likely to differ. However, smoking has not prevented some individuals from becoming highly skilled winemakers and cellarmasters. Nevertheless, the short-term impairment of olfactory skill has made smoking in tasting rooms unacceptable.

Adaptation is an additional source of altered olfactory perception. Adaptation may result from either a loss in the excitability of olfactory receptors or a decline in sensitivity at some higher center in the brain. Thus, on continued exposure the apparent intensity of an odorant may fall rapidly (Cain, 1974). It appears that the more intense the odor, the longer it takes for adaptation or habituation to develop.

The effect of adaptation is very important in wine tasting. Because the phenomenon occurs rapidly with subtle odors, the character of the fragrance of a wine may change quickly. Thus, wine tasters usually are counseled to take only short whiffs of wine. An interval of 30 sec to 1 min usually allows reestablishment of normal acuity. However, with aromatically complex wines such as vintage ports, it can be beneficial to smell the wine for a prolonged period. The progressive adaptation to various aspects of the bouquet reveals different components of the complex fragrance of the wine and may present new and pleasurable experiences.

Emotional response is particularly important to the overall perception of wine quality. Regrettably, little is known about how and why people develop hedonic reactions to odors. However, it appears that threshold intensity is not directly related to preference (Trant and Pangborn, 1983). The development of odor preferences appears to be particularly marked between the ages of 6 to 12 (Garb and Stunkard, 1974). There also appear to be significant, but later, age-related changes in wine partiality (Williams *et al.*, 1982). However, these changes may relate more to experience rather than to age itself.

Psychological factors play an important role in response to odors. Knowledge of the prestige and origin of a wine often affects a person's opinion of the wine. Comments from tasters, especially those with distinguished reputations, can have a profound influence on the views of others. If these psychological influences enhance the appreciation of the wine, they may be beneficial. However, psychological influences must be reduced to a minimum during critical wine assessment.

Significance of Odor Assessment in Wine Tasting

Although modern analytical instrumentation is superior to human sensory acuity in chemical measurement and identification, the days of the human wine taster are far from over. Tasters are needed in developing and maintaining standards that go beyond simple assessment of the typical character of a regional wine. Trained tasters are required to detect faults still difficult to assess instrumentally. While instrumental analysis provides the precision required for toxic and regulated substances, for most off-odors it is the perceived, rather than absolute, concentration that is important. Trained tasters are also valuable in detecting the compounds that give varietal and regional wines their distinctive character.

In wine tasting, one is usually more interested in the positive, pleasure-giving aspects of the fragrance of a wine. Regrettably, little progress has been made in describing varietal, regional, or stylistic features in words. Although proposals for a more meaningful range of wine descriptors have been presented, several studies suggest that only a small set of descriptive terms usually are necessary (Lawless, 1984).

Improvement in wine terminology can be beneficial even at the consumer level. It can help focus attention on the aromatic characteristics of wine that may enhance appreciation. Vital in this goal is the availability of representative odor samples. Published lists are available (see Appendix 11.1), but samples are often difficult to prepare, maintain, or standardize. Even "pure" chemicals may contain trace amounts of contaminants that can alter both the perceived intensity and quality of the odor. "Scratch-and-sniff" samples of representative wine flavors and off-odors would be useful but currently are unavailable.

Currently, more is known about the chemical nature of wine faults than about the positive fragrance characteristics of wine. The following section briefly summarizes the characteristics of several important off-odors in wine. Discussion of the volatile compounds important in wine fragrance, and the origin of off-odors, is given in Chapters 6 and 8. Formulas for making faulty wine samples are given in Appendix 11.2.

Off-Odors

Quick and accurate identification of off-odors is advantageous to both winemakers and wine merchants. For the winemaker, early remedial action often can correct the situation before the fault becomes serious or irreversible. For the wine merchant, avoiding losses associated with faulty wines can improve the profit margin. Consumers should also know more about wine faults so that rejection, if justified, be based on genuine faults, not

unfamiliarity, the presence of bitartrate crystal, or the wine being "too" dry (often incorrectly called "vinegary").

There is no precise definition of what makes a wine faulty—human perception is too variable. Furthermore, compounds that produce off-odors at certain concentrations are often deemed desirable at low concentrations, where they may produce interesting and subtle nuances. In addition, faults in one wine may not be undesirable aspects in another, for example, the complex oxidized bouquets of sherries and the fusel odors of *porto*. Some faults, such as barnyardy, may be considered pleasingly "rustic" in certain wines. For some individuals, noticeable oakiness is a fault, while others consider it a highly prized aspect. It is intriguing that faults in inexpensive wines may be more acceptable in expensive wines, for example, the ethyl acetate odor often noticeable in prestigious Sauternes. Nevertheless, there is some general agreement on what constitutes aromatic faults in wine, in contrast to the comparative absence of such agreement for taste and mouth-feel faults.

Once one of the most frequently occurring faults in wine, **oxidation** is now comparatively rare in commercial table wines. Oxidation produces a "flat," acetaldehyde off-odor. In bottled wine, it can develop from the use of faulty corks, or improper cork insertion. Rapid temperature changes also can permit oxygen to gain access to the wine by putting sufficient pressure on the cork to loosen the seal with the neck. However, the major cause of premature oxidation of bottled wine is positioning of the bottle upright. In this position, drying and shrinkage of the cork can allow air to seep into the wine. Even under ideal storage conditions, sufficient oxygen may pass through the cork, or between it and the sides of the bottle, to induce most white wines to oxidize within 4 to 5 years. Wine placed in bag-in-box packages may oxidize because of the slow oxygen penetration through the sides of the bag or through faulty spigots. Use of modern winemaking procedures has largely eliminated oxidative spoilage during vinification.

Factors influencing the tendency of bottled wine to oxidize are the phenol content, notably o-diphenols, level of free sulfur dioxide, pH, and temperature and light conditions during storage.

A mild form of oxidation develops within several hours of opening a bottle of wine. In recently bottled wine, this condition is called "bottle sickness." Such mild forms of oxidization dissipate if the wine is isolated from oxygen for several weeks or months. During this period, the acetaldehyde generated on oxidation slowly reacts with other compounds in the wine, such as tannins and sulfur dioxide. When the level of acetaldehyde falls below the recognition threshold, the wine loses the oxidized odor.

Like all odors, the oxidized character is difficult to express in words. It has a rough, chemical, aldehydic, sherrylike character. In mild forms, oxidation gives wine a "flat" aspect. The wine also changes color, taking on more golden shades in a white wine and brown shades in a red wine. Red wines are far less likely to suffer from oxidation because of their tannin content (see Chapter 6).

The off-odor generated by **ethyl acetate** is currently less common than in the past (Sudraud, 1978). At concentrations below 50 mg/liter, ethyl acetate can add a subtle fragrance; however, above about 150 mg/liter, it tends to generate an unpleasant "sour" vinegary odor. Threshold values vary with the type and intensity of the wine fragrance.

Although ethyl acetate is produced early in fermentation, the concentration usually falls below the recognition threshold by the end of fermentation. Concentrations of ethyl acetate sufficiently high to generate an off-odor usually result from the metabolism of acetic acid bacteria. An infrequent source of undesirable levels of ethyl acetate comes from the metabolism of *Hansenula* during fermentation.

The accumulation of **acetic acid** to detectable levels also is largely the result of the metabolism of acetic acid bacteria. The thresholds of detection and recognition are about 100 times higher than those of ethyl acetate. Vinegary wines typically are sharply acidic, with an irritating odor derived from the combined effects of acetic acid and ethyl acetate.

Although the beneficial actions of **sulfur dioxide** in winemaking are multiple, excessive addition can produce an irritating burnt-match odor. This fault usually dissipates rapidly if the wine is swirled in the glass for a short period. However, because sulfur dioxide can initiate asthmatic attacks in sensitive individuals (Taylor *et al.*, 1986), there is a concerted effort worldwide to minimize its use in wine.

A **geranium-like** odor can develop from the use of another wine preservative, sorbate. People hypersensitive to sorbate detect a butterlike odor when it is used. However, the most important off-odor occasionally associated with sorbate use is a geranium-like off-odor. The sharp penetrating odor is produced by 2-ethoxyhexa-3,5-diene, production of which involves the action of certain lactic acid bacteria (see Chapter 8).

During fermentation, yeasts produce limited amounts of **higher (fusel) alcohols,** and these may add to the complexity of the fragrance of a wine. If the alcohols accumulate to levels greater than about 300 mg/liter, they become a negative quality factor in most wines. However, detectable levels are an expected and characteristic feature of Portuguese ports (*porto*), where the elevated fusel aspect comes from the addition of largely unrectified wine spirits.

Diacetyl usually is found at low levels as a result of yeast and bacterial metabolism. It also may accumulate during malolactic fermentation or from oak barrels as an oxidation by-product. While typically considered butterlike, diacetyl is a negative quality factor above the recognition threshold.

Wines may develop **corky** or **moldy** odors from a variety of compounds (see Chapter 8). One of the most common, 2,4,6-trichloroanisole (2,4,6-TCA), usually develops as a consequence of fungal growth on or in the cork following the application of PCP (a pentachlorophenol fungicide) to cork trees or the bleaching of cork stoppers with chlorine. 2,4,6-TCA produces a distinctive chlorophenol odor at a few parts per trillion. Another off-odor in cork can develop as a consequence of the growth of filamentous bacteria such as *Streptomyces*. Additional moldy odors in cork can result from the production of guaiacol by fungi such as *Penicillium* and *Aspergillus*.

While most moldy (corky) taints come from cork, oak cooperage also is a potential source of corky off-odors. It is critical that cooperage be properly treated to restrict the growth of microbes on the inner surfaces during storage.

Some fortified wines are purposely heated to over 45°C for several months. Under such conditions, the wine develops a distinctive, **baked,** caramellike odor. While characteristic and expected in wines such as madeira, a baked odor is a negative quality feature in table wines. In table wines, a baked odor is indicative of excessive exposure to heat during transit or storage.

Hydrogen sulfide and **mercaptans** may be produced in wine during fermentation and aging. Their presence may be undetected as they can occur at levels below the recognition threshold. Whereas hydrogen sulfide usually can be eliminated by aeration, sulfide by-products such as mercaptans are more intractable. Removal of mercaptans usually requires the addition of trace amounts of silver chloride or copper sulfate. Mercaptans impart off-odors reminiscent of farmyard manure or rotten onions. Disulfides are formed under similar reductive conditions as mercaptans and generate cooked-cabbage and shrimplike odors. Related compounds such as 2-mercaptoethanol and 4-(methylthio)butanol produce intense barnyardy and chive–garlic odors.

Light-struck (*goût de la lumière*) refers to a reduced-sulfur odor that can develop in wine during exposure to light (see Chapter 8). This fault is but one of several undesirable consequences of the exposure of wine to light. It results from the combined odors of dimethyl sulfide, methyl mercaptan, and dimethyl disulfide.

Several **herbaceous** off-odors have been detected in wines. The best known are associated with the presence of "leaf" (C_6) aldehydes and alcohols. The compounds are derived from the oxidation of grape lipids. Fruit shading, particular climates, fruit maceration, and some strains of lactic acid bacteria can influence the development of vegetable odors of unknown chemical nature. Light can also induce an off-odor in Asti Spumante, referred to as a "strange, vegetable-like" odor (Di Stefano and Ciolfi, 1985). The formation of the latter is inhibited by the presence of sorbate. Depending on one's reaction to the bell pepper aroma of 2-methoxy-3-isobutylpyrazine, a characteristic fragrance of many 'Cabernet Sauvignon' and 'Sauvignon blanc' wines, it is an off-odor or a desirable aroma compound.

The presence of a **mousy** taint is associated with the metabolism of spoilage microbes such as several *Lactobacillus* and *Brettanomyces* species. The odor is caused by one of several tetrahydropyridines.

Bitter-almond odors in wine may have several origins. One involves residual ferrocyanides following blue fining. Decomposition of ferrocyanides can release small quantities of hydrogen cyanide, which possesses a bitter-almond odor. A more common source of a bitter-almond odor in wine is the microbial conversion of benzyl alcohol to benzaldehyde. The precursor, benzyl alcohol, may come from cooperage covered with or containing epoxy resins, from gelatin used as a fining agent, or from grapes.

Additional off-odors of known origin include **raisined** (use of sun-dried grapes), **cooked** (wines fermented at high temperatures), **stemmy** (presence of green grape stems during fermentation), and **rancio** (old oxidized red wines). Other "off-odors" may be aroma compounds of particular grape cultivars, for example the foxy and strawberry-like aroma of some *Vitis labrusca* hybrids. Other off-odors of unknown chemical nature are **rubbery**, **weedy**, and **earthy** (*goût de terroir*).

Wine Assessment and Sensory Analysis

Wine assessment and sensory analysis cover various aspects of wine evaluation. They may include preference determination, assessment of individual properties in comparative tastings, or the development of flavor profiles for particular varietal and/or regional wines.

The intent of a tasting profoundly affects both its design and the type and degree of analysis appropriate. Wine society tastings often involve wines whose origin and price are known in advance. Analysis, if any, involves little more than ranking the wines in order of preference. In regional and international tastings, wines usually are grouped by region, variety, and style. Simple numerical averaging of the scores is used to develop a ranking, usually without statistical analysis of significance. Tastings intended to assess vini- and vicultural practices require more exacting conditions, including appropriate design considerations essential for the valid application of statistical tests. The latter can estimate the degree and significance of unavoidable taster variation and unsuspected interactions that could obscure valid conclusions. Computers have made sophisticated statistical procedures available to all requiring their use.

Conditions for Sensory Analysis

TASTING ROOM

The ideal lighting for visual assessment of wine usually is considered to be natural north-lighting. However, in most tasting situations this is impossible. In addition, the light source may be less important than previously thought (Brou *et al.*, 1986). Any bright, white light source probably is acceptable. In situations where wines of different hues must be tasted together, it can be advantageous to have the option of using red light to disguise the color of the wines. Use of red or black wine glasses is an alternate technique. Normally, however, white tabletops or countertops as well as white or neutral colored walls, are used to facilitate color differentiation.

Tasting rooms need to be adequately air-conditioned, both for taster comfort and to limit the development of a background odor. Covers on the glasses also help to limit the escape and accumulation of wine odors in the tasting room. Watch glasses are commonly used for this purpose, but 60 mm plastic petri dish bottoms are often superior. Because the bottom may fit snugly over the mouth of the wine glass, the need to hold the cover on with the forefinger during swirling is avoided. Use of dentist-type sinks or cuspidors with tops for expectoration at each tasting station further minimize odor buildup.

Each tasting station should be physically isolated (cubicles) to limit taster interaction (Fig. 11.10). Silence also limits between—taster influence and facilitates concentration. Where tasters cannot be physically separated, the order of wine presentation may be varied among the tasters to negate the influence of taster interaction.

NUMBER OF WINES

The number of wines adequately evaluated per session depends on the level of assessment required for each sample. If rejection of faulty samples is the only intent, 20 to 50 wines can easily be assessed at one sitting. However, if the wines are similar, and must be compared critically, 5 to 6 wines may be the normal limit. The wines should be tasted at a relaxed pace to avoid odor adaptation or an increase in perceived astringency. Frequent breaks are desirable when wines are assessed criti-

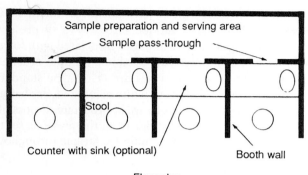

Floor plan

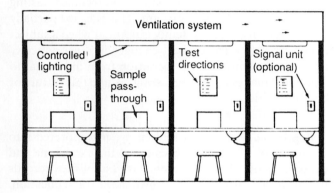

Front elevation

Figure 11.10 Schematic diagram of the booth area in a sensory evaluation facility (not drawn to scale). (From Stone and Sidel, 1985, reproduced by permission.)

cally. Detailed, written sensory analysis may require 10 min or more per wine. The evaluation of each wine may be spread over 30 min to observe fragrance development and duration.

PRESENTATION OF SAMPLES

Glasses There is an amazing variety in wine glasses, most of which appear to be designed for artistic merit, not suitability for wine evaluation. The basic requirements for a wine tasting glass are simple. The bowl should be broader at the base than the top, and the material should be clear, uncolored glass. The sloping sides concentrate the fragrance and permit vigorous swirling of the wine to release the aromatic compounds. For swirling, the glass should not be filled more than one-third full. Thin crystal is more aesthetic but unnecessary. The International Standards Organization (ISO) tasting glass illustrated in Fig. 11.11 is particularly adequate, and far superior to most. For sparkling wines, flute-shaped glasses are preferred as the long tapering sides allow the effervescence characteristics to be adequately viewed.

It is essential that wine glasses be properly washed and rinsed between tastings. Residual odors can easily alter the fragrance of a wine, and traces of oil residue or

detergent can limit the effervescence of sparkling wines. Foreign odors may originate from cabinets in which wine glasses are stoed. Cabinets made from aromatic wood should be avoided.

Temperature There is general agreement that most red wines taste best between 18° and 20°C. Young carbonic maceration wines, such as most *nouveau* wines, are preferred at about 14° to 16°C. With white wines, there is less agreement, some preferring 11° to 13°C, while others suggest 16°C. Generally, the sweeter the wine, the cooler the "optimal" temperature. There also is divergence in opinion concerning the ideal temperature for sparkling wine, varying between 8° to 13°C. Sweet fortified wines are commonly served at about 18°C, while dry fortified wines, notably *fino* sherries, are taken cool (14°C) to cold (8°C). In large public tastings, it may be impossible to serve wines at the optimal temperature. If they are served at room temperature (~20°C), off-odors may be accentuated. Where wine temperature can

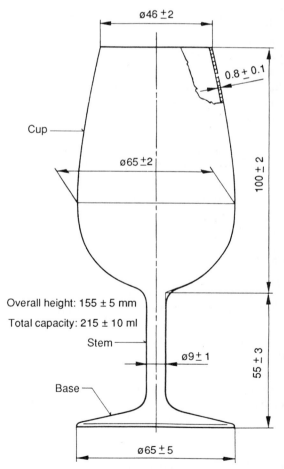

Figure 11.11 ISO wine tasting glass. Dimensions are in millimeters. (Courtesy of International Standards Organization, Geneva, Switzerland.)

be controlled, it might be preferable to obtain a consensus from the judges concerning the serving temperature. If the wines are to be assessed over a protracted period, it is preferable to present the wines cooler than "ideal" so that the wine passes through the optimal temperature range during the tasting.

Concealing Wine Identity   Except at informal tastings, wines are served "blind" to avoid bias based on wine origin. Although the preferred technique is to pour the wine into decanters before the tasting, this is often unnecessary. Placing the bottles in paper bags, with capsules removed, usually is adequate if bottles are of the same size, shape, and color. Necks may need to be cleaned with a damp cloth to remove fungal growth, dried wine, or lead salts, not only for health reasons, but also to remove clues as to the age and character of the wines.

The information provided to the tasters about the wines depends on the purpose of the tasting. If wines are to be judged relative to a particular style, variety, or region, this should be known in advance. If wines from various regions, varieties, or styles are being assessed together, however, it may be inappropriate for the tasters to know the details beforehand. If simple hedonic preferences are being assessed, no information concerning origin or price should be given. Double-blind tests may be useful in especially critical tastings, where neither the tasters nor the administers of the samples know the origin or nature of the wines.

Decanting and Breathing   If wines contain sediment, they should be decanted in advance. Otherwise, repeated tipping of the bottle during pouring will suspend the sediment, resulting in subsequent samples becoming cloudy.

Opening bottles in advance to "breathe" is unnecessary. The limited wine/air interface in the neck is insufficient to result in any prompt change in the sensory character of the wine. Even decanting, which more effectively exposes the wine to oxygen, seldom produces marked changes in sensory characteristics within an hour. The main exception involves old wines, where the minimal fragrance can dissipate rapidly. However, prolonged exposure to air results in table wines developing a stale aspect within several hours, presumably owing to the formation of acetaldehyde. The view that breathing has a beneficial effect may reflect the nineteenth century custom of decanting the wine to separate the wine from the sediment. The procedure also allowed time for the so-called bottle stink, with which the wines of the period were often afflicted, to dissipate.

Presentation Sequence   To avoid unintended or perceived ranking by the sequence of presentation, a two-digit code for each wine is recommended. Amerine and Roessler (1983) suggest numbers from 14 to 99, presumably to avoid psychological biases associated with lower numbers. Code numbers are used in a random order.

If different groups of wines are tasted, the standard serving recommendations of white before red, dry before sweet, and young before old are appropriate. If possible each set of wines should differ from the previous set to help maintain interest and minimize fatigue throughout the tasting session.

Time of Day   It is common to hold technical tasting in the late morning or afternoon. This is based partially on the view that sensory acuity is optimal when a subject is hungry. Although cyclical changes in sensory acuity occur throughout the day, variation from individual to individual makes designing tasting around this feature of dubious value.

Replicates   Replicates seldom are incorporated into the protocol of a tasting because of the extra time and expense involved. If the tasters have established records of consistency, there may be little need for replication. Nevertheless, duplicate bottles should be available as substitutes for faulty wines.

Wine Score Cards

While many score cards are available, few have been studied sufficiently to establish how quickly they come to be used consistently. This obviously is a critical aspect if important decisions are based on the results. Generally, the more detailed the score card, the longer it takes for it to be used consistently (Ough and Winton, 1976). Thus, score cards should be as simple as possible, consistent with the intent of the tasting. Incorporation of unnecessary detail reduces the consistency of use and, therefore, the potential value of the data obtained. However, insufficent choice may result in "dumping," where tasters express an important opinion about the wine in an unrelated category on the card (Lawless and Clark, 1992). Possibly the most widely used scoring system is the modified Davis score card (Table 11.1).

Where ranking and detailed sensory analyses are desired, two separate score cards may be valuable (Tables 11.2 and 11.3). Not only does this simplify assessment, but it may avoid the "halo" effect of one assessment prejudicing another (Lawless and Clark, 1992). For example, astringent bitter wines might be scored poorly on overall quality and drinkability (Table 11.2) but be rated more leniently when marked separately on the various sensory aspects of the wine (Table 11.3). Separating the two assessments temporally decreases the likelihood of one assessment influencing the other. This view is consis-

Table 11.1 Modified Davis Score Card for Wine Grading[a]

Characteristic	Maximum points
Appearance Cloudy 0, clear 1, brilliant 2	2
Color Distinctly off 0, slightly off 1, correct 2	2
Aroma and bouquet Vinous 1, distinct but not varietal 2, varietal 3 Subtract 1 or 2 for off-odors, add 1 for bottle bouquet	4
Vinegary Obvious 0, slight 1, none 2	2
Total acidity Distinctly high or low 0, slightly high or low 1, normal 2	2
Sweetness Too high or too low 0, normal 1	1
Body Too high or low 0, normal 1	1
Flavor Distinctly abnormal 0, slightly abnormal 1, normal 2	2
Bitterness Distinctly high 0, slightly high 1, normal 2	2
General quality Lacking 0, slight 1, impressive 2	2

[a] From Amerine and Singleton (1977), reproduced by permission.

tent with recent studies which indicate that the perception of total taste intensity reflects the intensity of the strongest component, while the perception of the individual components may be differentially influenced by interaction between the components (McBride and Finlay, 1990).

No one score card is universally applicable to all

Table 11.2 General Wine Scoring Chart[a]

General quality ranking		
Unacceptable	0	
Average quality with some defects	−	
	0	
Average quality	−	
	0	
	+	
Above average quality with some superior qualities	0	
	+	
Superior	0	

[a] From Tromp and Conradie (1979), reproduced by permission.

wines. For example, effervescence is very important in assessing sparkling wines but irrelevant for still wines. In addition, authors in different countries seem to rate features differently. For example, in England, taste and smell are respectively given 9 and 4 marks out of 20 by Johnson (1985) while taste, smell, and flavor are assessed 6, 6, and 2 marks, respectively, by Amerine and Singleton (1977) in the United States. For the descriptive analysis of the wines (see below), score cards specific of each wine type need to be prepared.

Number of Tasters

Where a series of wines are to be sampled repeatedly, the same tasters should be present at all tastings. However, in most commercial tastings it may be impossible or unnecessary for all tasters to be present at all tastings. Usually, a nucleus of 12 to 15 tasters is assembled so that at least 10 tasters can be present at every tasting. If continual monitoring of the tasters is performed, this number should assure that an adequate number of tasters will be "in form" to permit statistically valid data analysis. Results from tasters who were temporarily "out of form" (see below) should be excluded prior to data analysis.

Because tasters genuinely perceive tastes and odors differently, they also may diverge considerably in their concepts of wine quality. Therefore, the number of tasters should be sufficiently large to buffer individual idiosyncrasies, or the tasters should be selected and trained for the particular skills necessary. Which approach is preferred will depend on the purpose and nature of the tasting. Detailed descriptive analysis requires extensive and specific training, whereas consumer acceptance testing should involve no special training.

In large public tastings, the number of tasters usually is correspondingly large. Individual tasters can taste only a limited number of wines accurately and must be asked to assess only wines within their range of experience. Tasters having little experience with dry sherries cannot be expected to evaluate these wines adequately.

When the purpose of the tasting is to establish consumer preference, it is important to have a representative cross section of the target group, as considerable divergence exists within and between consumer subgroups. For example, sweetness and freedom from bitterness and astringency are particularly important to members of less affluent social groups and to infrequent wine drinkers (Williams, 1982). Woodiness and spiciness are disliked by infrequent wine drinkers but are appreciated by most wine experts. The importance of color and aroma to general acceptability increases with consumer age and the frequency of wine consumption. Not surprisingly, differences in experience also influence the terms used by people to describe wine (Dürr, 1984; Solomon, 1990).

Table 11.3 Ten-Category Score Card for Wine Grading[a]

		Detailed Assessment					
		Superior to outstanding	Very good to above average	Good to average	Acceptable	Unacceptable	Motivation
Eye	Clarity						
	Color						
Nose	Trueness to type						
	Maturation bouquet						
	Purity						
Palate	Acidity						too much / too little
	Fullness						
	Flavor						
	Astringency/ bitterness						too much / too little
Overall impression	Harmony						

[a] From Tromp and Conradie (1979), reproduced by permission.

Tasters

Training

Until fairly recently, wine assessment was done primarily by professional winemakers or those in wine marketing. As all individuals are likely to possess biases and "blind spots," most wine assessment is now done by teams of tasters. This has required the training of more tasters than produced by the former, relatively informal, in-house training system. There also is the desire to have some standardization in the training and assessment of tasting skills.

Sensory training often consists of an extensive series of tastings of a wide variety of wines. This rapidly gives the trainee a base of experience on which to build. Because the purpose of training and the availability of wines vary so widely, no specific suggestions for appropriate wines can be given.

For economy and convenience, aspects of varietal grape aromas may be simulated by producing "standard" odor samples. These have the advantage of being continuously available as a reference. Standards may be prepared and stored under wax in sample bottles (Williams, 1975, 1978) or in wine (Noble *et al.*, 1987). Appendix 11.1 provides recipes for the preparation of standard odor samples. Because "pure" compounds may contain trace impurities that can modify the odor quality of the test compound, it may be necessary to conduct initial purification before use (Meilgaard *et al.*, 1982).

While training improves the consistency with which descriptive terms are used, it may not eliminate idiosyncrasies in the use of terms (Lawless, 1984). Thus, adequate identification of odors in wine often requires the use of several tasters (Clapperton, 1978).

Besides correct identification of varietal odors, identification of odor faults is a vital component of taster training. In the past, faulty samples usually were obtained from wineries, but samples prepared in the laboratory can be more standardized and readily available. In addition, they can be presented in any wine and at any concentration desired. Appendix 11.2 provides recipes for the preparation of off-odor samples.

Training usually includes taste samples prepared in either aqueous solutions or wine. As with odor training, sample trials allow the student to discover personal idiosyncrasies concerning the detection of sapid substances. Appendix 11.3 provides recipes for preparing taste and mouth-feel samples.

When selecting tasters for training, it is often more important to select for interest and ability to learn rather than initial "skill." Motivation is critical to both learning and consistent attendance at tastings. Because initial skill in recognizing odors usually reflects previous exposure (Cain, 1979), not innate ability, measures of learning ability are important in screening potential tasters (Stahl and Einstein, 1973). In addition, some people initially anosmic to a compound develop the ability to smell the compound on repeated exposure (Wysocki *et al.*, 1989).

Basic screening tests usually are designed to eliminate tasters with insufficient sensory acuity. Subsequently, ability to learn identification and differentiation skills, and consistent use of important sensory terms, may be assessed. Appendix 11.4 describes procedures for giving wine discrimination tests. Basic screening tests are also discussed in Amerine and Roessler (1983) and Basker (1988).

Measuring Tasting Acuity and Consistency

Measuring tasting acuity has two primary aspects. It assesses the skills learned during training and monitors consistency during and between actual tasting sessions.

Depending on the purpose of the training, various skills are required. In descriptive wine analysis odor and taste acuity are essential, whereas in quality evaluation discrimination between subtle differences is more important. However, regardless of the task, correct and consistent use of terms is required. Because tests of consistency require repeated sampling, the tests may involve only olfactory sensations. Taste and mouth-feel acuity are the only tests requiring actual *tasting*. Tests should be conducted on different days as acuity often varies from day to day.

Assessment of performance with a series of simple tests before or during each tasting session is valuable for several reasons. As individual acuity varies, data from tasters having "off" days can be removed before analysis of the tasting results is conducted. Furthermore, the tests show whether individual tasters are using terms consistently and if they require a refresher course.

Brien *et al.* (1987) distinguish five aspects of taster consistency. **Discrimination** is defined as a measure of the ability to distinguish between wines of distinct character; **stability** refers to scoring similarity between tastings; **reliability** assesses the reproducibility of score differences between replicate sets of wines; **variability** gauges the range of scores between replicate wine samples; and **agreement** evaluates scoring differences between tasters. Of these measures of consistency, two require that identical wines be sampled repeatedly, either on one occasion (reliability) or on separate occasions (variability).

Measures of discrimination, stability, and variability are derived from analyses of variance between scores in successive tastings. Measures of agreement and reliability are derived from correlation matrices obtained from the scoring results. For details of these and other statistical tests, the reader is directed to Brien *et al.* (1987), Amerine and Roessler (1983), and Stone and Sidel (1985).

While analyses of consistency are useful, caution must be used in their interpretation. While a high degree of agreement may appear desirable, it may indicate uniform lack of skill or an inadequate reflection of normal variability in perceptive ability. Also, measures of consistency which require replicate tastings may be invalid if the samples are not actually identical. In addition, reliability may be affected by the number of replicates, improving as the tasters learn from repeated sampling. Finally, variability may be higher with experienced tasters than with inexperienced tasters, because of the greater confidence of the former in using the full range of marks permitted.

Wine Tasting Technique

There is no universally accepted procedure for wine assessment. Often tasters evolve a unique sequence of steps that they follow almost unconsciously. What follows is a synthesis of techniques gleamed from Johnson (1985), Broadbent (1979), and Peynaud (1987) as well as discussions with skilled tasters and experience gained in training tasters. The recommendations given below are intended as a guide for experimentation and assessment.

Some authorities such as Peynaud (1987) recommend that the mouth be rinsed with wine before beginning a tasting session. This is intended to familiarize the senses with the basic properties of the wines to be assessed, so that the taster can concentrate on the unique properties that distinguish the wines to be judged. Peynaud also counsels against rinsing the palate between samples. He feels that rinsing the palate with water alters the sensitivity and increases the difficulty of comparing wines. Only when the palate seems "tired" does he suggest rinsing the palate with water, or consuming small pieces of bread. In this recommendation, Peynaud is at variance with most other authorities.

Appearance

In visual assessment, the glass of wine is angled against a light background. Except for samples taken directly from barrels, the wine should be brilliantly clear. Haziness in barrel samples is of less concern as it is eliminated before bottling.

Color correctness can only be evaluated if the tasters have had sufficient experience with the wines being assessed, and if the type and age of the wine are known. Abnormal color can be a sign of several wine faults. While considerable attention has been paid to wine "tears," presence of tears is but a rough indicator of the alcohol content of a wine. Effervescence is important in assessing the quality of sparkling wine, but it is rarely of concern in still wines. Bubble formation at the bottom of a glass of still wine usually suggests early bottling, before the wine has lost its carbon dioxide supersaturation; occasionally it may indicate spoilage arising from in-bottle fermentation.

Odor In-Glass

Occasionally tasters are advised to smell the wine before swirling. Where a series of wines are to be compared, it may be easier to position the head over the glass and smell each wine in sequence than to raise each glass to the nose. The procedure focuses the attention of the taster on the most freely volatile aspects of the wine. Some authorities find this exercise of little practical value (Peynaud, 1987).

Subsequently, the wine is swirled to aid the liberation of aromatic compounds from the wine. Tulip-shaped glasses permit vigorous swirling and help to concentrate the aromatics in the glass. Swirling the wine facilitates the release of aromatics by increasing air contact with the wine. Successive whiffs of the fragrance may be taken at the mouth of the glass and in the bowl. This assesses the fragrance at different degrees of dilution and may produce different sensations. Considerable attention is usually required in recognizing the varietal fragrance of wines, which may involve repeated assessment to confirm the initial impression. Fragrance deserves the attention it demands as it generates most of the unique character and complexity of a wine.

Although swirling wine may be simple, it usually requires some practice. It is recommended to commence swirling by rotating the glass gently in a circle on the table or countertop, until the technique is acquired. Watch glasses or small petri dish bottoms may be placed over the mouth of the wine glass to permit vigorous swirling of the wine.

The technique people normally use in smelling objects appears to be fully adequate in recognizing odors (Laing, 1986) and probably need not be modified for wine assessment. Inhalation periods longer than about half a second do not improve odor identification of most compounds (Laing, 1982). However, Peynaud (1987) suggests that odor adaptation may occasionally be of value

is revealing compounds initially masked by other odors. This appears to be true for complex wines such as fine ports. Regardless of the technique found most effective, it is important to record the impressions.

Where possible, fragrance assessment should be conducted over about 30 min, as this permits evaluation of important aspects such as the duration and development of the fragrance. Both aspects are highly regarded and considered important to quality, especially in premium wines. However, as with most aspects of tasting, there are those who feel that repeated sampling is unnecessary. Peynaud (1987) considers that repeated sampling only "results in a total loss of sensitivity".

In technical assessment, tasters are usually trained using samples specifically designed for the purpose, and odor samples are made available during each session.

Flavor and aroma wheels (Meilgaard et al., 1979; Noble et al., 1987) can be useful aids in teaching introductory sensory evaluation. These tools help focus student attention on the distinctive flavor elements of different wines. Once tasters have developed sufficient experience with wine, description of wine in terms of fruits, flowers, vegetables, etc., usually becomes unnecessary and counterproductive. It is generally more meaningful to characterize wines by their production style, varietal origin, and aging process.

In-Mouth Sensations

After the initial assessment of fragrance, focus usually turns to taste and mouth-feel. Samples of about 5 to 10 ml are sufficient to coat all sensory receptors in the mouth. The wine usually is rotated in the mouth to achieve contact with all oral sensory receptors.

Sweetness and sourness generally are detected first, although the perception of acidity tends to linger longer than that of sweetness. Because bitterness and astringency take longer to develop, they may be assessed following sweetness and sourness. Subsequently, the taster may concentrate on other mouth-feel sensations, such as prickling and the "burning" sensation of alcohol. Finally, the integrated perceptions of balance, "body," and flavor should be noted.

Differences in the time sequence of perception can serve to confirm individual tastes in a complex mixture (Kuznicki and Turner, 1986). However, the duration of the sensations apparently cannot be relied on in identifying sapid substances. Persistence of a sensation depends on the concentration of the compounds involved and the maximum intensity achieved (Robichaud and Noble, 1990). However, for the majority of wine connoisseurs and wine tastings, the value of conscientious assessing of sapid sensations is not to identify the sensations, but to focus on how they interact with other aspects of the wine.

There are differing opinions on whether taste and mouth-feel should be assessed on the first taste or during subsequent samplings. Reaction between tannins and saliva proteins, which diminishes the sensations of bitterness and astringency in the first sample, is limited in subsequent samples. Thus, the first sample more closely approximates the sensation produced when wine is consumed with a meal. This will be important if the tasting involves assessment of the suitability of the wine as a food accompaniment. Otherwise, the taster must make a mental adjustment to the taste of the wine to predict its food compatability.

To enhance the detection of fragrance in the mouth, tasters may draw air through the wine with the mouth closed, but the lips slightly ajar. The procedure has the effect of liberating aromatic compounds, analogous to that of swirling wine in the glass.

Finish

The lingering sensations of taste, mouth-feel, and fragrance following expectoration or swallowing of the wine are termed the "finish" of the wine. Most lingering flavor sensations are considered desirable, and the longer the finish, the more highly rated the wine. Exceptions include lingering acidic, metallic, bitter, and astringent sensations. Fortified wines generally exhibit a longer finish than table wines because of their more intense flavors.

Assessment of Overall Quality

After the individual sensory aspects of a wine have been studied, attention shifts to integrating the various sensations. This aspect places the greatest demands on the previous wine-tasting experience of the taster. It may involve aspects of the conformity and distinctiveness of the wine within regional standards; the development, duration, and complexity of the fragrance; the duration and character of the finish; and the memorableness of the tasting experience.

Most European authorities feel that quality should be assessed only within regional appellations, counseling against comparative tastings between regions or grape varieties. Although these restrictions make tastings simpler, they negate much of their value in promoting quality improvement. When tasting concentrates on artistic quality, rather than stylistic purity, comparative tasting can be especially revealing and informative. Comparative tastings are more popular in the New World, where artistic merit tends to be considered more highly than purity of regional style.

Statistical and Descriptive Analysis of Tasting Results

Simple Tests

For most tastings, simple statistical tests are usually adequate in assessing whether tasters can distinguish between any, or all, of the sampled wines. One measure of significance is based on the range of scores for each wine and the cumulative score range. An example of this use is given below (Table 11.4).

For the wines to be considered distinguishable from one another, the range in scores must be greater than the statistic given in Appendix 11.5, multiplied by the sum of the score ranges for individual wines. In this example, the pertinent statistic for five tasters and five wines is 0.81, for significance at 5%. For the tasters to be considered capable of distinguishing between the wines, the range of total scores must be greater than the product (10.5) of the statistic (0.81) multiplied by the sum of score ranges (13). Because the range of total scores (11) is greater than the calculated product noted above (10.5), the tasters are considered able to distinguish differences among the wines.

To determine which wines were distinguished, the second (lower) statistic in Appendix 11.5 (0.56 in this instance) is multiplied by the sum of score ranges (13) to produce a product (7.3). When the difference between the total scores of any pair of wines is greater than the calculated product (7.3), the wines may be considered significantly different. Table 11.5 shows that Wine 1 is distinguishable from wines 3 and 5, but not from Wines 2 and 4, while Wines 2, 3, 4, and 5 are indistinguishable from one another.

Caution must be taken in interpreting the results of these tests. For example, Wine 1 might have been faulty,

Table 11.4 Hypothetical Scores of Five Tasters for Five Wines

	Judging results				
	Wine 1	Wine 2	Wine 3	Wine 4	Wine 5
Taster 1	5	6	7	5	8
Taster 2	6	6	8	6	5
Taster 3	4	7	7	4	6
Taster 4	5	4	6	6	6
Taster 5	4	6	7	7	7
Total scores	24	29	35	28	32
Range (max–min)	2	3	2	3	3

Sum of score ranges = $(2 + 3 + 2 + 3 + 3) = 13$
Range of total scores = $(35 - 24) = 11$

Table 11.5 Difference between the Sum of Scores of Pairs of Wines in Table 11.4

	Wine 1	Wine 2	Wine 3	**Wine 4**	Wine 5
Wine 1	—				
Wine 2	5	—			
Wine 3	11[a]	6	—		
Wine 4	4	1	7	—	
Wine 5	8[a]	3	3	4	—

a Values are significant at a 5% probability.

and had an appropriate sample been substituted none of the wines might have been considered significantly different. In addition, there is no means of determining if the tasters were marking consistently in a single test. Thus, had more competent tasters been involved, significant differences might have been detected. Even if one or two tasters were "out of form," incorrect conclusions might be drawn. Conclusions can be no more valid than the quality of the data on which they are based.

Numerous examples of the use of this, and other statistical techniques, are given by Amerine and Roessler (1983). Their text is an excellent reference for those wishing details on the use of statistics in wine analysis.

Analysis of Variance

For a more detailed evaluation of tasting results, an analysis of variance (ANOVA) may be used. While ANOVA techniques are more complicated, computers have made them available to essentially all those requiring their use. Direct electronic incorporation of data further eases analyzing large amounts of data. Amerine and Roessler (1983) give several examples of the use of analysis of variance in wine tasting.

Analysis of variance can assess not only whether any two or more wines are detectably different, but also whether any two or more tasters are scoring significantly differently. In addition, the analysis permits evaluation of the significance of the interaction between various factors in the tasting. Furthermore, it can provide measures of taster discrimination, stability, and variability.

Multivariate Analysis and Descriptive Analysis of Wine

Another powerful statistical tool is the application of multivariate analysis to the descriptive (sensory) analysis of wine. The intent of the analysis is to determine the features that distinguish wines made from specific variet-

ies (Guinard and Cliff, 1987), wines from within particular regions (Williams *et al.*, 1982), or wines made by particular processes.

For use of the technique, prospective tasters undergo screening, followed by a sampling of a series of wines representing the variety, region, or style to be investigated. During the tastings, the tasters work toward consensus on descriptors that adequately represent the distinctive features of the wines. Subsequently, wine samples are judged using these descriptors to assess their adequacy, and whether the number could be reduced. Analysis of consistent and correct term usage is typically performed, and inconsistent tasters are removed from the panel before the formal study begins.

In most studies, there is no attempt to directly associate quality of the wines with chemical or sensory characteristics. However, a study by Herraiz and Cabezudo (1980/1981) has attempted to make such an association. The results from descriptive sensory analysis were used to develop a Quality Index, based on variation in a set of characters considered important to wine quality. Quality Index values subsequently were correlated with gas chromatograms of each wine. The Quality Index was closely correlated with the concentrations of eight volatile wine constituents.

In descriptive sensory analysis, extensive training and discussion are required before conducting the test. This has led some researchers to wonder about the polarization of views involved in developing agreement on term usage (Myers and Lamm, 1975), the tendency of people to be highly idiosyncratic in the use of odor terms (Lawless, 1984), and even whether "correct" sets of descriptors are possible (Solomon, 1991). To avoid such concerns, Williams and Langron (1983) propose that tasters be allowed to use their own vocabulary to describe wine appearance, aroma, taste, and flavor. A scale is used to measure the intensity of each attribute, and these are subjected to a multidimensional mathematical model (Procrustes analysis) that permits direct comparison of the data (Oreskovich *et al.*, 1991). Although the latter technique avoids some of the problems of descriptive sensory analyses, it presumably still requires that the tasters have prior experience with the wines.

Of particular interest is the combination of chemical analysis, descriptive sensory analysis, and consumer subgroup preferences. Such a combination (Williams *et al.*, 1982; Williams, 1984) may permit the correlation of preference data with particular aromatic and sapid substances. If the chemical nature of consumer preferences could be defined, it would allow a more precise selection and blending of wines for particular consumer groups. Such designing of wines may not be consistent with the romantic image cultivated by wine merchants, but providing the right wine at the right price can be decidedly profitable!

While designing wines for a particular market can be profitable, it is happily not the only means of providing consumers with what they want. As Vernon Singleton (1976) has said,

Wine is, and must remain I feel, one of the few products with almost unlimited diversity. . . . keeping the consumer forever intrigued, amused, pleased, and never bored.

Appendix 11.1

Aroma and Bouquet Samples[a]

Sample	Amount per 300 ml of base wine	Sample	Amount per 300 ml of base wine
Temperate tree fruit		Green bean	100 ml Canned green bean juice
Apple	15 mg Hexyl acetate	Herbaceous	3 mg 1-Hex-3-enol
Cherry	3 ml Cherry brandy essence (Noirot)	**Spice**	
		Anise/licorice	1.5 mg Anise oil
Peach	100 ml Juice from canned peaches	Peppermint	1 ml Peppermint extract (Empress)
Apricot	2 Drops of undecanoic acid γ-lactone plus 100 ml juice from canned apricots	Black pepper	2 g Whole black peppercorns
		Cinnamon	15 mg *trans*-Cinnamaldehyde
		Nuts	
Tropical tree fruit		Almond	5 Drops of bitter almond oil
Litchi	100 ml Litchi fruit drink (Leo's)	Hazelnut	3 ml Hazelnut essence (Noirot)
Banana	10 mg Isoamyl acetate	Coconut	1.0 ml Coconut essence (Club House)
Guava	100 ml Guava fruit drink (Leo's)		
Lemon	0.2 ml Lemon extract (Empress)	**Woody**	
Vine fruit		Oak	3 g Oak chips (aged $\geq$1 month)
Blackberry	5 ml Blackberry essence (Noirot)	Vanilla	24 mg Vanillin
Raspberry	60 ml Raspberry liqueur	Pine	7.5 mg Pine needle oil (1 drop) 、
Black currant	80 ml Black currant nectar (Ribena)	Eucalyptus	9 mg Eucalyptus oil
		Pyrogenous	
Passion fruit	10 ml Ethanolic extract of one passion fruit	Incense	1/2 Stick of Chinese incense
Melon	100 ml Melon liqueur	Smoke	0.5 ml Hickory liquid smoke (Colgin)
Floral		**Mushroom**	
Rose	6 mg Citronellol	Agaricus	Juice from 200 g microwaved mushrooms
Violet	1.5 mg β-Ionone		
Orange blossom	20 mg Methyl anthranilate	Truffle	30 ml Soy sauce
Iris	0.2 mg Irone	**Miscellaneous**	
Lily	7 mg Hydroxycitrolellal	Chocolate	3 ml Chocolate liqueur
Vegetal		Butterscotch	1 ml Butterscotch flavor (Wagner)
Beet	25 ml Canned beet juice		
Bell pepper	5 ml 10% Ethanolic extract from dried bell peppers (2 g)		

[a] The recipes are given only as a guide as adjustments will be required based on both individual needs and material availability. (Additional recipes may be found in Meilgaard, 1988; Noble *et al.*, 1987; Peynaud, 1980; Williams, 1978.) Pure chemicals have the advantage of providing highly reproducible samples, while "natural" sources are more complex but more difficult to standardize. Readers requiring basic information for preparing samples may find Stahl (1973), Furia and Bellanca (1975), and Heath (1981) especially useful. Most specific chemicals can be obtained from major chemical suppliers, while sources of fruit, flower, and other essences include wine supply, perfumery, and flavor supply companies.

Note I: With whole fruit, the fruit is ground in a blender with 95% alcohol. The solution is left for about a day in the absence of air, filtered through several layers of cheesecloth, and added to the base wine. Several days later, the sample may need to be decanted to remove excess precipitates.

Note II: To limit oxidation, about 20 mg of potassium metabisulfite may be added per sample.

Note III: Since only 30 ml samples are required at any one time, it may be convenient to disperse the original sample into 30 ml screw-capped test tubes for storage. Parafilm can be stretched over the cap to further prevent oxygen penetration. Samples stored in a refrigerator usually remain good for several months.

Appendix 11.2

Basic Off-Odor Samples[a]

Sample	Amount per 300 ml of base wine
Cork	
2,4,6-TCA	3 μg 2,4,6-Trichloroanisole
Guaiacol	3 mg Guaiacol
Actinomycete	Extract from *Streptomyces griseus*[b]
Penicillium	Extract from *Penicillium* sp. on cork[c]
Chemical	
Fusel	120 mg Isoamyl and 300 mg isobutyl alcohol
Geranium-like	40 mg 2,4-Hexadienol
Buttery	12 mg Diacetyl[d]
Plastic	1.5 mg Styrene
Sulfur	
Sulfur dioxide	200 mg Potassium metabisulfite
Goût de la lumière	4 mg Dimethyl sulfide[d] and 0.4 mg ethanethiol
Mercaptan	4 mg Ethanethiol
Hydrogen sulfide	2 ml Solution with 1.5 mg NaS
Miscellaneous	
Oxidized	120 mg Acetaldehyde
Baked	1.2 g Fructose added and sample baked 4 weeks at 55°C
Vinegary	3.5 g Acetic acid
Ethyl acetate	60 mg Ethyl acetate

[a] Other off-odor sample preparations are noted in Meilgaard *et al.* (1982).

Note I: To limit oxidation, about 20 mg of potassium metabisulfite may be added per 300 ml of base wine.

Note II: Since only 30 ml samples are required at any one time, it may be convenient to disperse the original sample into 30 ml screw-capped test tubes for storage. Parafilm can be stretched over the cap to further prevent oxygen penetration. Samples stored in a refrigerator usually remain good for several months.

[b] *Streptomyces griseus* is grown on nutrient agar in 100 cm diameter petri dishes for 1 week or more. The colonies are scraped off and added to the base wine. Filtering after a few days should provide a clear sample.

[c] *Penicillium* sp. isolated from wine corks is inoculated on small chunks (1–5 mm) of cork soaked in wine. The inoculated cork is placed in a petri dish and sealed with Parafilm to prevent the cork from drying out. After about 1 month obvious growth of the fungus should be noticeable. Chunks of the overgrown cork are added to the base wine. Within a few days, the sample can be filtered to remove the cork. The final sample should be clear.

[d] Because of the likelihood of serious modification of the odor quality of these chemicals by contaminants, Meilgaard *et al.,* (1982) recommend that they be purified prior to use: for diacetyl, use fractional distillation and absorption (in silica gel, aluminum oxide and activated carbon); for dimethyl sulfide, use absorption.

Appendix 11.3

Taste and Mouth-Feel Sensations[a]

Sample[b]	Amount per 750 ml water	Sensations
Sugar solution	15 g Sucrose	Sweet
Acid solution	2 g Tartaric acid	Sour
Bitter solution 1	10 mg Quinine sulfate	Bitter
Bitter solution 2	0.5 g Caffeine	Bitter
Astringent solution 1	1 g Tannic acid	Astringent, woody
Astringent solution 2	3 g Aluminum sulfate	Astringent, sweet
Alcohol solution	48 g Ethanol	Sweet, hot, body, alcoholic odor

[a] Samples are commonly made up in water, but an artificial wine base may be used. If the samples are made up in wine, one must calculate the level of the compounds in the wine initially, if the final concentration needs to be known. These solutions are similar to those proposed by Marcus (1974) and Peynaud (1980).

[b] All solutions, with the exception of the tannin samples, can be stored in the refrigerator for weeks without degeneration.

Appendix 11.4

Training and Testing of Wine Tasters

The following tests were designed for the Manitoba Liquor Commission in 1986 for the selection and training of an in-house tasting panel. Members are chosen from those having taken the course. Current members are periodically retested.

I. ODOR RECOGNITION TESTS

1. Aroma and Bouquet Odors

The Aroma and Bouquet Test is designed to assess the ability of trainees to correctly identify several aroma and bouquet odors. Because learning is a component of the test, trainees are given a training session with the standards several hours before being tested. In these sessions, all standards (Appendix 11.1) are presented in 200 ml ISO or Libby 8470 wine glasses (Fig. 11.11) with tops covered with 60 mm plastic petri dish bottoms. Sheets containing the names of each sample are provided, with space for the trainees to write notes on features that may help them to recognize each standard.

In each test, the samples are colored randomly or presented in dark-colored glasses. This is required as preparation of some samples unavoidably affects the appearance of the base wine, especially white wines. Most standards are presented in each test to allow a change in the ability of the trainee to correctly identify the samples to be determined.

The answer sheet contains the names of all the standards plus space for four or more control samples. The names are present on the answer sheet as term use, not memory recall, is being assessed. Control samples (unadjusted base wine) are used to determine if the trainees can distinguish the unadjusted wines from the standards. For simplicity, trainees mark the number of the sample glass across from the appropriate odor term or control indication on the answer sheet.

After each test, the trainees are told the identity of the samples and allowed time to go over the samples again. The test is performed several times on different days.

2. Off-Odors

Basic test. The basic Off-Odor Test is similar in design to the Aroma and Bouquet Test except that it assesses the ability of the trainee to learn and correctly identify several odor faults. Training sessions are held on the same day as the test. During the training sessions, trainees are encouraged to make notes on the origin and characteristics of each off-odor. During the tests, trainees are permitted to use the notes in identifying the faults. The samples (see Appendix 11.2) used in the training and testing sessions are identical.

The answer sheet contains the names of all the faults present plus places for four control samples. Trainees mark the number of the sample glass across from the appropriate off-odor term or control indication on the answer sheet.

Three training/testing sessions are usually adequate to assess the ability of a trainee to learn and correctly identify wine odor faults, although additional training is usually required to maintain the ability.

Off-odors in different wines. One of the problems with the previous test is that as trainees come to recognize a increasing number of off-odors, the remaining faults become easier to identify by a simple process of elimination. Another weakness is that the off-odors are presented at only one concentration, and against a common wine background. To provide a more accurate measure of the ability of a trainee to identify faults under more "natural" conditions, the following test may be used.

The test consists of preparing faulty wines in a series of red and white wines. The wines provide a diverse aromatic background for the off-odors. Two or more levels of the faults are prepared in one or more wines. In the example given below, only a selection of the more important and easily prepared faults are used.

In the test, faulty samples are placed in random order with control (unmodified) samples of the wines used. The trainees are provided with an answer sheet containing a list of all the off-odors studied, plus space for an undefined number of controls. The trainees smell each

sample and determine if it is faulty or not. If not, the number of the sample is placed in the control row. If faulty, the number of the sample is placed in the row corresponding to the perceived off-odor.

The following is an example of a typical setup.

Off-Odors in Four Types of Wines at Two Concentrations

Wine	Off-odor	Chemical added	Amount per 300 ml
Gewürztraminer	Oxidized	Acetaldehyde	20, 60 mg
	Sulfur dioxide	Potassium metabisulfite	67, 200 mg
	2,4,6-TCA	2,4,6-Trichloroanisole	1.5, 4.5 µg
	Plastic	Styrene	1.5, 4.5 mg
Sauvignon blanc	Vinegary	Acetic acid	0.5, 2 g
	Buttery	Diacetyl	2, 6 mg
	Ethyl acetate	Ethyl acetate	20, 60 mg
	Geranium-like	2,4-Hexadienol	10, 40 mg
Beaujolais	Geranium-like	2,4-Hexadienol	10, 40 mg
	Buttery	Diacetyl	5, 24 mg
	Ethyl acetate	Ethyl acetate	20, 60 mg
	Oxidized	Acetaldehyde	20, 60 mg
Pinot noir	Guaiacol	Guaiacol	0.2, 0.6 mg
	Mercaptan	Ethanethiol	5, 24 µg
	2,4,6-TCA	2,4,6-Trichloroanisole	1.5, 4.5 µg
	Plastic	Styrene	1.5, 4.5 mg

II. TESTS FOR DISCRIMINATING ABILITY

1. Varietal Dilution Test

The intention of the Varietal Dilution Test is to obtain a measure of the ability of a trainee to distinguish slight differences between wine samples. For this purpose, one or more wines possessing distinctive varietal aromas are chosen. The wines are diluted with an aromatically neutral base wine of similar color. If neutral base wines of similar color are unavailable, the wines may be presented in dark-colored glasses or under red light. The dilution series can be at any level desired, but dilutions of 4, 8, 16, and 32% provide a reasonable range for discrimination.

Five sets of three glasses for each dilution is the minimum requirement. Diluted (or undiluted) wine is poured into two of the three glasses. The remaining glass has the opposite control (or diluted) sample poured into it. Thus, each set of three glasses has at least one "different" sample, but not consistently the diluted or control sample. This testing procedure is called the **triangle test**. The sets of glasses are arranged at random for the test.

The trainees move past each set of glasses, remove their covers, and smell each sample. They are to record for each set which glass contains the "different" sample.

Using Appendix 11.6, one can determine the level at which trainees begin to distinguish that the wine has been diluted. Under the conditions described above (five replicates of each dilution), a trainee must identify correctly the different sample 4 out of 5 times to be considered able to distinguish between the diluted and undiluted samples.

2. Wine Differentiation Test

The purpose of the Wine Differentiation Test is to obtain a measure of the ability of trainees to distinguish between similar wines. As in the previous test, the triangulation procedure is used. Two distinctive examples of several types of wine are chosen. For each pair of wines, ten sets of glasses are prepared with two glasses containing one of the wines and a third glass containing the other wine. If the two wines are noticeably different in color, adjusting the color with food coloring may eliminate this identification feature. Alternately, dark-colored glasses or red illumination may be used.

Use of three distinctively different types of wines has the advantage that triangulation sets of each can more easily be arranged to present different types of wine in sequence. This limits adaptation to a particular wine aroma. With ten replicates of each pair of wines, the trainee needs to obtain seven correct responses to indicate identification to a $p = 0.05$ level (Appendix 11.6).

Before the test, the trainees may study each wine to be differentiated. Otherwise, it is likely that the test will be measuring both the ability to differentiate between the wines and the ability to learn their differences during the test. In addition, one may request that the trainees identify the grape variety used or the regional origin of each triangulation set. For this, the wines should be identified in advance so that identification, not prior experience, is being assessed.

3. Wine Recognition Test

The final test is to designed to determine the ability of the trainees to recognize wines previously assessed. In the first part of the Wine Recognition Test, each trainee is given a set of five glasses each containing a different wine. Each of the five wines should be sufficiently different to be easily distinguishable. Dark-colored glasses, or dim red lighting, can be used to eliminate color as a criterion for distinguishing between the wines. Each wine is identified by number or letter code only.

The trainees assess each wine for odor, taste, and flavor, using a score card for assessing wine quality (e.g., Tables 11.1, 11.2, and 11.3). They retain the score cards for reference in part two of the test.

In the second part of the test, a set of glasses containing seven wines is presented to each trainee. The trainees are told that within the seven are the five wines they had judged just previously. They are told that the other two glasses may contain repeats of the wines tasted previously or may contain wines different from those just assessed. The trainees sample each wine and identify it by code as one of the previously sampled wines or as a new sample.

Appendix 11.5

Multipliers for Estimating Significance of Difference by Range: One-Way Classification, 5% Level[a,b]

Number of judges	Number of wines								
	2	3	4	5	6	7	8	9	10
2	3.43	2.35	1.74	1.39	1.15	0.99	0.87	0.77	0.70
	3.43	1.76	1.18	0.88	0.70	0.58	0.50	0.44	0.39
3	1.90	1.44	1.14	0.94	0.80	0.70	0.62	0.56	0.51
	1.90	1.14	0.81	0.63	0.52	0.44	0.38	0.33	0.30
4	1.62	1.25	1.01	0.84	0.72	0.63	0.57	0.51	0.47
	1.62	1.02	0.74	0.58	0.48	0.40	0.35	0.31	0.28
5	1.53	1.19	0.96	0.81	0.70	0.61	0.55	0.50	0.45
	1.52	0.98	0.72	0.56	0.47	0.40	0.34	0.30	0.27
6	1.50	1.17	0.95	0.80	0.69	0.61	0.55	0.49	0.45
	1.50	0.96	0.71	0.56	0.46	0.40	0.34	0.30	0.27
7	1.49	1.17	0.95	0.80	0.69	0.61	0.55	0.50	0.45
	1.49	0.96	0.71	0.56	0.47	0.40	0.35	0.31	0.28
8	1.49	1.18	0.96	0.81	0.70	0.62	0.55	0.50	0.46
	1.49	0.97	0.72	0.57	0.47	0.41	0.35	0.31	0.28
9	1.50	1.19	0.97	0.82	0.71	0.62	0.56	0.51	0.47
	1.50	0.98	0.73	0.58	0.48	0.41	0.36	0.31	0.28

(continues)

Appendix 11.5 (*Continued*)

Number of	Number of wines								
judges	2	3	4	5	6	7	8	9	10
10	1.52	1.20	0.98	0.83	0.72	0.63	0.57	0.52	0.47
	1.52	0.99	0.74	0.59	0.49	0.42	0.37	0.32	0.29
11	1.54	1.22	0.99	0.84	0.73	0.64	0.58	0.52	0.48
	1.54	0.99	0.75	0.60	0.49	0.42	0.37	0.32	0.29
12	1.56	1.23	1.01	0.85	0.74	0.65	0.58	0.53	0.49
	1.56	1.00	0.75	0.60	0.50	0.43	0.38	0.32	0.30

[a] Entries in the table are to be multiplied by the sum of ranges within wines. The upper value must be exceeded by the range in wine totals to indicate significance. If significance is indicated, the lower value must be exceeded by pairs of wine totals to indicate a significance between individual wines.

[b] After T. E. Kurtz, R. F. Link, J. W. Tukey, and D. L. Wallace. Short-cut multiple comparisons for balanced single and double classifications: Part I, Results. *Technometrics* 7, 95–165 (1965). Reprinted with permission from *Technometrics*. Copyright (1965) by the American Statistical Association and the American Society for Quality Control. All rights reserved. From Amerine and Roessler (1983).

Appendix 11.6

Minimum Numbers of Correct Judgments to Establish Significance at Various Probability Levels for the Triangle Test (One-Tailed, $p = 1/3$)[a]

Number of	Probability level							Number of	Probability level						
trials (*n*)	0.05	0.04	0.03	0.02	0.01	0.005	0.001	trials (*n*)	0.05	0.04	0.03	0.02	0.01	0.005	0.001
5	4	5	5	5	5	5	—								
6	5	5	5	5	6	6	—	31	16	16	16	17	18	18	20
7	6	6	6	6	7	7	7	32	16	16	17	17	18	19	20
8	6	7	7	7	7	7	8	33	17	17	17	18	18	19	20
9	7	7	7	7	8	8	8	34	17	17	18	18	19	20	21
10	7	7	7	7	8	8	9	35	17	18	18	19	19	20	22
11	7	7	8	8	8	9	10	36	18	18	18	19	20	20	22
12	8	8	8	8	9	9	10	37	18	18	19	19	20	21	22
13	8	8	9	9	9	10	11	38	19	19	19	20	21	21	23
14	9	9	9	9	10	10	11	39	19	19	20	20	21	22	23
15	9	9	10	10	10	11	12	40	19	20	20	21	21	22	24
16	9	10	10	10	11	11	12	41	20	20	20	21	22	23	24
17	10	10	10	11	11	12	13	42	20	20	21	21	22	23	25
18	10	11	11	11	12	12	13	43	20	21	21	22	23	24	25
19	11	11	11	12	12	13	14	44	21	21	22	22	23	24	26
20	11	11	12	12	13	13	14	45	21	22	22	23	24	24	26
21	12	12	12	13	13	14	15	46	22	22	22	23	24	25	27
22	12	12	13	13	14	15	16	47	22	22	23	23	24	25	27
23	12	13	13	13	14	15	16	48	22	23	23	24	25	26	27
24	13	13	13	14	15	15	16	49	23	23	24	24	25	26	28
25	13	14	14	14	15	16	17	50	23	24	24	25	26	26	28
26	14	14	14	15	15	16	17	60	27	27	28	29	30	31	33
27	14	14	15	15	16	17	18	70	31	31	32	33	34	35	37
28	15	15	15	16	16	17	18	80	35	35	36	36	38	39	41
29	15	15	16	16	17	17	19	90	38	39	40	40	42	43	45
30	15	16	16	16	17	18	19	100	42	43	43	44	45	47	49

—, data not available.

[a] Abridged from tables compiled by Roessler *et al.* (1978). From Amerine and Roessler (1983), reproduced by permission.

Suggested Readings

Visual Sensations

Clydesdale, F. M. (1993). Color as a factor in food choice. *Crit. Rev. Food Sci. Nutr.* **33**, 83–101.

Walker, J. (1983). What causes the "tears" that form on the inside of a glass of wine? *Sci. Am.* **248**, 162–169.

Taste and In-Mouth Sensations

Akabas, M. H. (1990). Mechanisms of chemosensory transduction in taste cells. *Int. Rev. Neurobiol.* **32**, 241–279.

Bartoshuk, L. M. (1985). Chemical sensation: Taste. *In* "Nutrition in Oral Health and Disease" (R. L. Pollack and E. Kravitz, eds.), pp. 53–67. Lea & Febiger, Philadelphia, Pennsylvania.

Beidler, L. M., and Tonosaki, K. (1985). Multiple sweet receptor sites and taste theory. *In* "Taste, Olfaction and the Central Nervous System" (D. W. Pfaff, ed.), pp. 1–15. Rockefeller Univ. Press, New York.

Birch, G. G. (1987). Chemical aspects of sweetness. *In* "Sweetness" (J. Dobbing, ed.), pp. 282–291. Springer-Verlag, London.

Haslam, E., and Lilley, T. H. (1988). Natural astringency in foodstuffs —A molecular interpretation. *Crit. Rev. Food Sci. Nutr.* **27**, 1–40.

Langstaff, S. A., and Lewis, M. J. (1993). The mouthfeel of beer—A review. *J. Inst. Brew.* **99**, 31–38.

Noble, A. C. (1988). Analysis of wine sensory properties. *In* "Wine Analysis" (H. F. Linskens and J. F. Jackson, eds.), pp. 9–28. Springer-Verlag, Berlin.

Pangborn, R. M. (1987). Selected factors influencing sensory perception of sweetness. *In* "Sweetness" (J. Dobbing, ed.), pp. 49–66. Springer-Verlag, London.

Schiffman, S. S. (1983). Taste and smell in disease. *N. Engl. J. Med.* **308**, 1275–1279.

Shamil, S., and Birch, G. G. (1990). A conceptual model of taste receptors. *Endeavour, New Ser.* **14**, 191–193.

Stevens, J. C., and Cain, W. S. (1993). Changes in taste and flavor in aging. *Crit. Rev. Food Sci. Nutr.* **33**, 27–37.

Trant, A. S., and Pangborn, R. M. (1983). Discrimination, intensity, and hedonic responses to color, aroma, viscosity, and sweetness of beverages. *Lebensm. Wiss. Technol.* **16**, 147–152.

Olfactory Sensations

Cain, W. S. (1985). Chemical sensation: Olfaction. *In* "Nutrition in Oral Health and Disease" (R. L. Pollack and E. Kravitz, eds.), pp. 68–83. Lee & Febiger, Philadelphia, Pennsylvania.

Doty, R. L., and Snow, J. B., Jr. (1988). Age-related alterations in olfactory structure and function. *In* "Molecular Neurobiology of the Olfactory System" (F. L. Margolis and T. V. Getchell, eds.), pp. 355–374. Plenum, New York.

Engen, T. (1988). The acquisition of odour hedonics. *In* "Perfumery: The Psychology and Biology of Fragrance" (S. van Toller and G. H. Dodd, eds.), pp. 79–90. Chapman & Hall, London.

Gilbert, A. N., and Kare, M. R. (1991). A consideration of some psychological and physiological mechanisms of odour perception. *In* "Perfumes: Art, Science and Technology" (P. M. Müller and D. Lamparsky, eds.), pp. 127–149. Elsevier, Applied Science, London.

Halpen, B. P. (1985). Environmental factors affecting chemoreceptors: An overview. *In* "Toxicology of the Eye, Ear, and Other Special Senses" (A. W. Hayes, ed.), pp. 195–211. Raven, New York.

Laing, D. G. (1986). Optimum perception of odours by humans. *Proc. 7th World Clean Air Congr.*, Vol. 4, pp. 110–117. Clear Air Society of Australia and New Zealand. Sydney, Australia.

Margolis, F. L., and Getchell, T. (1991). Receptors—Current status and future directions. *In* "Perfumes: Art, Science and Technology" (P. M. Müller and D. Lamparsky, eds.), pp. 481–498. Elsevier Applied Science, London.

Ohloff, G. (1986). Chemistry of odor stimuli. *Experientia* **42**, 271–279.

Ohloff, G., Winter, B., and Fehr, C. (1991). Chemical classification and structure—odour relationships. *In* "Perfumes: Art, Science and Technology" (P. M. Müller and D. Lamparsky, eds.), pp. 287–330. Elsevier Applied Science, London.

Scott, J. W. (1986). The olfactory bulb and central pathways. *Experientia* **42**, 223–229.

Shirley, S. G. (1992). Olfaction. *Int. Rev. Neurobiol.* **33**, 1–53.

Sensory Analysis

Amerine, M. A., and Roessler, E. B. (1983). "Wines: Their Sensory Evaluation," 2nd Ed. Freeman, San Francisco, California.

Laing, D. G. (1986). Optimum perception of odours by humans. *Proc. 7th World Clean Air Congr.*, Vol. 4, pp. 110–117. Clear Air Society of Australia and New Zealand. Sydney, Australia.

Moskowitz, H. (ed.) (1988). "Applied Sensory Analysis of Foods." CRC Press, Boca Raton, Florida.

Noble, A. C. (1988). Analysis of wine sensory properties. *In* "Wine Analysis" (H. F. Linskens and J. F. Jackson, eds.), pp. 9–28. Springer-Verlag, Berlin.

Stone, H., and Sidel, J. (1985). "Sensory Evaluation Practices." Academic Press, Orlando, Florida.

Stone, H., Sidel, J., Oliver, S., Woolsey, A., and Singleton, R. C. (1974). Sensory evaluation by quantitative descriptive analysis. *Food Technol. Chicago* **28**(Nov.), 24–34.

Williams, A. A., Rogers, C., and Noble, A. C. (1984). Characterization of flavour in alcoholic beverages. *Found. Biotech. Ind. Ferm. Res.* **3**, 235–253.

Zervos, C., and Albert, R. H. (1992). Chemometrics: The use of multivariate methods for the determination and characterization of off-flavors. Developments in Food Science 28. *In* "Off-Flavors in Foods and Beverages." (G. Charalambous, ed.), pp. 669–742. Elsevier Sci. Publ., Amsterdam.

References

Amerine, M. A., and Roessler, E. B. (1983). "Wines: Their Sensory Evaluation," 2nd Ed. Freeman, San Francisco, California.

Amerine, M. A., and Singleton, V. L. (1977). "Wine—An Introduction." Univ. of California Press, Berkeley.

Amoore, J. E. (1977). Specific anosmia and the concept of primary odors. *Chem. Senses Flavours* **2**, 267–281.

Amoore, J. E., Johnson, J. W., Jr., and Rubin, M. (1964). The stereochemical theory of odor. *Sci. Am.* **210**(2), 42–49.

Arnold, R. A., and Noble, A. C. (1978). Bitterness and astringency of grape seed phenolics in a model wine solution. *Am. J. Enol. Vitic.* **29**, 150–152.

Avenet, P., and Lindemann, B. (1989). Perspective of taste reception. *J. Membr. Biol.* **112**, 1–8.

Barker, L. M. (ed). (1982). "The Psychobiology of Human Food Selection." AVI Publ., Westport, Connecticut.

Bartoshuk, L. M. (1979). Bitter taste of saccharin related to the genetic ability to taste the better substance 6-*n*-propylthiouracil. *Science* **205**, 934–935.

Bartoshuk, L. M., Rifkin, B., Marks, L. E., and Bars, P. (1986). Taste and aging. *J. Gerontol.* **41**, 51–57.

Basker, D. (1988). Assessor selection: Procedures and results. *In* "Applied Sensory Analysis of Foods" (H. Moskowitz, ed.), Vol. 1, pp. 125–143. CRC Press, Boca Raton, Florida.

Beidler, L. M., and Tonosaki, K. (1985). Multiple sweet receptor sites and taste theory. *In* "Taste, Olfaction and the Central Nervous System" (D. W. Pfaff, ed.), pp. 47–64. Rockefeller Univ. Press, New York.

Berg, H. W., Pangborn, R. M., Roessler, E. B., and Webb, A. D. (1963). Influence of hunger on olfactory acuity. *Nature (London)* **197,** 108.

Blakeslee, A. F., and Salmon, T. N. (1935). Genetics of sensory thresholds: Individual taste reactions for different substances. *Proc. Natl. Acad. Sci. U.S.A.* **21,** 84–90.

Boudreau, J. C., Oravec, J., Hoang, N. K., and White, T. D. (1979). Taste and the taste of foods. *In* "Food Taste Chemistry" (J. C. Boudreau, ed.), pp. 1–30. American Chemical Society, Washington, D.C.

Brien, C. J., May, P., and Mayo, O. (1987). Analysis of judge performance in wine-quality evaluations. *J. Food Sci.* **52,** 1273–1279.

Broadbent, M. (1979). "Wine Tasting." Christie's Wine Publ., London.

Brou, P., Sciascia, T. R., Linden, L., and Lettvin, J. Y. (1986). The colors of things. *Sci. Am.* **255**(3), 84–91.

Buck, L., and Axel, R. (1991). A novel multigene family may encode odorant receptors: A molecular basis for odor recognition. *Cell (Cambridge, Mass.)* **65,** 175–187.

Burns, D. J. W., and Noble, A. C. (1985). Evaluation of the separate contributions of viscosity and sweetness to perceived viscosity, sweetness and bitterness of Vermouth. *J. Texture Stud.* **16,** 365–381.

Cain, W. S. (1974). Perception of odour intensity and the time-course of olfactory adaption. *Trans. Am. Soc. Heating Refrigeration Air-Conditioning Engineers* **80,** 53–75.

Cain, W. S. (1976). Olfaction and the common chemical sense: Some psychophysical contrasts. *Sens. Processes* **1,** 57.

Cain, W. S. (1978). The odoriferous environment and the application of olfactory research. *In* "Handbook of Perception" (E. C. Carterette and P. M. Friedman, eds.), Vol. 6A, pp. 197–229. Academic Press, New York.

Cain, W. S. (1979). To know with the nose: Keys to odor identification. *Science* **203,** 467–469.

Cain, W. S. (1985). Chemical sensation: Olfaction. *In* "Nutrition in Oral Health and Disease" (R. L. Pollack and E. Kravitz, eds.), pp. 68–83. Lee & Febiger, Philadelphia, Pennsylvania.

Cain, W. S., and Murphy, C. L. (1980). Interaction between chemoreceptive modalities of odour and irritation. *Nature (London)* **284,** 255–257.

Clapperton, J. F. (1978). Sensory characterisation of the flavour of beer. *In* "Progress in Flavour Research" (D. G. Land and H. E. Nursten, eds.), pp. 1–20. Applied Science Publ., London.

Clydesdale, F. M., Gover, R., Philipsen, D. H., and Fugardi, C. (1992). The effect of color on thirst quenching, sweetness, acceptability and flavor intensity in fruit punch flavored beverages. *J. Food Qual.* **15,** 19–38.

Collins, V. B. (1974). Human taste response as a function of locus on the tongue and soft palate. *Percept. Psychophys.* **16,** 169–174.

Court, E. A., and Ford, M. A. (1981). Colour measurement and specification. *In* "Quality Control—Proceedings of the Sixth Wine Subject Day," (F. W. Beech and W. J. Redmond, eds.), pp. 78–88. Long Ashton Research Station, Univ. of Bristol, Bristol.

Cowart, B. J. (1989). Relationship between taste and smell across the adult life span. *Ann. N.Y. Acad. Sci.* **561,** 39–55.

Dahl, A. R. (1988). The effect of cytochrome *P*-450 dependent metabolism and other enzyme activities on olfaction. *In* "Molecular Neurobiology of the Olfactory System" (F. L. Margolis and T. V. Getchell, eds.), pp. 51–70. Plenum, New York.

Dawes, C. (1964). Is acid-precipitation of salivary proteins a factor in plaque formation? *Arch. Oral Biol.* **9,** 375–376.

DeSimone, J. A., Heck, G. L., and Bartoshuk, L. M. (1980). Surface active taste modifiers: A comparison of the physical and psychophysical properties of gymnemic acid and sodium lauryl sulfate. *Chem. Senses* **5,** 317–330.

De Vries, H., and Stuiver, M. (1961). The absolute sensitivity of the human sense of smell. *In* "Sensory Communications" (W. A. Rosenblith, ed.), pp. 157–167. Wiley, New York.

Di Stefano, R., and Ciolfi, G. (1985). L'influenza dell luce sull'Asti Spumante. *Vini Ital.* **27,** 23–32.

Doty, R. L. (1986). Reproductive endocrine influences upon olfactory perception: A current perspective. *J. Chem. Ecol.* **12,** 497–511.

Doty, R. L., Shaman, P., Applebaum, S. L., Giberson, R., Siksorski, L., and Rosenberg, L. (1984). Smell identification ability changes with age. *Science* **226,** 1441–1443.

Doty, R. L., and Snow, J. B., Jr. (1988). Age-related alterations in olfactory structure and function. *In* "Molecular Neurobiology of the Olfactory System." (F. L. Margolis and T. V. Getchell, eds.). pp. 355–374. Plenum, New York.

Dürr, P. (1984). Sensory analysis as a research tool. *Found. Biotech. Ind. Ferment. Res.* **3,** 313–322.

Dürr, P. (1985). Gedanken zur Weinsprache. *Alimentia* **6,** 155–157.

Engen, T. (1987). Remembering odors and their names. *Am. Sci.* **75,** 497–503.

Enns, M. P., and Hornung, D. E. (1985). Contributions of smell and taste to overall intensity. *Chem. Senses* **10,** 357–366.

Erickson, R. P. (1985). Definitions: A matter of taste. *In* "Taste, Olfaction and the Central Nervous System" (D. W. Pfaff, ed.), pp. 129–150. Rockefeller Univ. Press, New York.

Frank, R. A., and Byram, J. (1988). Taste–smell interactions are tastant and odorant dependent. *Chem. Senses* **13,** 445–455.

Freeman, W. J. (1991). The physiology of perception. *Sci. Am.* **264**(2), 78–85.

Frye, R. E., Schwartz, B. S., and Doty, R. L. (1990). Dose-related effects of cigarette smoking on olfactory function. *J. Am. Med. Assoc.* **263,** 1233–1236.

Furia, T. E., and Bellanca, N. (eds.) (1975). "Fenaroli's Handbook of Flavor Ingredients," 2nd Ed., Vols. 1 and 2. CRC Press, Cleveland, Ohio.

Ganzevles, P. G. J., and Kroeze, J. H. A. (1987). The sour taste of acids. The hydrogen ion and the undissociated acid as sour agents. *Chem. Senses* **12,** 563–576.

Garb, J. L., and Stunkard, A. J. (1974). Taste aversions in man. *Am. J. Psychiatry* **131,** 1204–1207.

Green, B. G. (1992). The effects of temperature and concentration on the perceived intensity and quality of carbonation. *Chem. Senses* **17,** 435–450.

Green, B. G., and Frankmann, S. P. (1987). The effect of cooling the tongue on the perceived intensity of taste. *Chem. Senses* **12,** 609–619.

Guinard, J., and Cliff, M. (1987). Descriptive analysis of Pinot noir wines from Carneros, Napa, and Sonoma. *Am. J. Enol. Vitic.* **38,** 211–215.

Guinard, J., Pangborn, R. M., and Lewis, M. J. (1986a). The timecourse of astringency in wine upon repeated ingestion. *Am. J. Enol. Vitic.* **37,** 184–189.

Guinard, J., Pangborn, R. M., and Lewis, M. J. (1986b). Preliminary studies on acidity–astringency interactions in model solutions and wines. *J. Sci. Food Agric.* **37,** 811–817.

Haslam, E., and Lilley, T. H. (1988). Natural astringency in foods: A molecular interpretation. *Crit. Rev. Food Sci. Nutr.* **27,** 1–40.

Haslam, E., Lilley, T. H., Warminski, E., Liao, H., Cai, Y., Martin, R., Gaffney, S. H., Goulding, P. N., and Luck, G. (1992). Polyphenol complexation. A study in molecular recognition. *In* "Phenolic Compounds in Food and Their Effects on Health I: Analysis, Occurrence, and Chemistry" (C.-T. Ho, C. Y. Lee, and M.-T. Huang, eds.), ACS Symp. Ser. No. 506, pp. 8–50. American Chemical Society, Washington, D.C.

Heath, H. B. (1981). "Source Book of Flavor." AVI Publ., Westport, Connecticut.

Heredia, F. J., and Guzmán Chozas, M. (1992). Proposal of a novel formula to calculate dominant wavelength of colour of red wines. *Food Chem.* **43**, 125–128.

Herraiz, J., and Cabezudo, M. D. (1980/1981). Sensory profile of wines, quality index. *Process Biochem.* **16**, 16–19 and 43.

Hettinger, T. P., Myers, W. E., and Frank, M. E. (1990). Role of olfaction in perception of non-traditional 'taste' stimuli. *Chem. Senses* **15**, 755–760.

Hyde, R. J., and Pangborn, R. M. (1978). Parotid salivation in response to tasting wine. *Am. J. Enol. Vitic.* **29**, 87–91.

Johnson, H. (1985). "The World Atlas of Wine." Simon & Schuster, New York.

Kalmus, H. (1971). Genetics of taste. *In* "Handbook of Sensory Physiology" (L. M. Beidler, ed.), pp. 165–179. Springer-Verlag, Berlin.

Kawamura, Y., and Adachi, A. (1967). Electrophysiological analysis of taste effectiveness of soda water and CO_2 gas. *In* "Olfaction and Taste II" (T. Hayashi, ed.), pp. 431–437. Pergamon, Oxford.

Kroeze, J. H. A. (1982). The relationship between the side taste of masking stimuli and masking in binary mixtures. *Chem. Senses* **7**, 23–37.

Kurtz, T. E., Link, T. E., Tukey, R. F., and Wallace, D. L. (1965). Short-cut multiple comparisons for balanced single and double classifications. *Technometrics* **7**, 95–165.

Kuznicki, J. T., and Turner, L. S. (1986). Reaction time in the perceptual processing of taste quality. *Chem. Senses* **11**, 183–201.

Laing, D. G. (1982). Characterization of human behaviour during odour perception. *Perception* **11**, 221–230.

Laing, D. G. (1983). Natural sniffing gives optimum odour perception for humans. *Perception* **12**, 99–117.

Laing, D. G. (1986). Optimum perception of odours by humans. *Proc. 7th World Clean Air Congr.*, Vol. 4, pp. 110–117. Clear Air Society of Australia and New Zealand. Sydney, Australia.

Laing, D. G., and Panhuber, H. (1978). Application of anatomical and psychophysical methods to studies of odour interactions. *Progr. Flavour Res. Proc. 2nd Weurman Flavour Res. Symp. Norwich, Apr. 2–6, 1978* (D. G. Land and H. E. Nursten, eds.), pp. 27–47. Applied Science Publ., London.

Lawless, H. T. (1984). Flavor description of white wine by "expert" and nonexpert wine consumers. *J. Food Sci.* **49**, 120–123.

Lawless, H. T. and Clark, C. (1992). Psychological biases in time-intensity scaling. *Food Technol Chicago.* **46**(11), 81, 84–86, 90.

Lawless, H., and Engen, T. (1977). Associations of odors: Interference, mnemonics and verbal labeling. *J. Exp. Psychol. Hum. Learn. Memory* **3**, 52–59.

Leach, E. J., and Noble, A. C. (1986). Comparison of bitterness of caffeine and quinine by a time–intensity procedure. *Chem. Senses* **11**, 339–345.

Lee, C. B., and Lawless, H. T. (1991). Time-course of astringent sensations. *Chem. Senses* **16**, 225–238.

Lehrer, A. (1975). Talking about wine. *Language* **51**, 901–923.

Lüthi, H., and Vetsch, U. (1981). "Practical Microscopic Evaluation of Wines and Fruit Juices." Heller Chemie-und Verwaltsingsgesellschaft mbH, Schwäbisch Hall, Germany.

McBride, R. L., and Finlay, D. C. (1990). Perceptual integration of tertiary taste mixtures. *Percept. Psychophys.* **48**, 326–336.

McBurney, D. H. (1978). Psychological dimensions and perceptual analyses of taste. *In* "Handbook of Perception" (E. C. Carterette and M. P. Friedman, eds.), Vol. 6A, pp. 125–155. Academic Press, New York.

MacRostie, S. W. (1974). Electrode measurement of hydrogen sulfide in wine. M.S. Thesis, Univ. California, Davis.

Maga, J. A. (1974). Influence of color on taste thresholds. *Chem. Senses Flavour* **1**, 115–119.

Marcus, I. H. (1974). "How to Test and Improve Your Wine Judging Ability." Wine Publ., Berkeley, California.

Meilgaard, M. C. (1988). Beer flavor terminology—A case study. *In* "Applied Sensory Analysis of Foods," Vol. 1. (H. Moskowitz, ed.). pp. 73–87. CRC Press, Boca Raton, Florida.

Meilgaard, M. C., Dalgliesch, C. E., and Clapperton, J. F. (1979). Beer flavor terminology. *J. Am. Soc. Brew. Chem.* **37**, 47–52.

Meilgaard, M. C., Reid, D. S., and Wyborski, K. A. (1982). Reference standards for beer flavor terminology system. *J. Am. Soc. Brew. Chem.* **40**, 119–128.

Moncrieff, R. W. (1964). The metallic taste. *Perfum. Essent. Oil Rec.* **55**, 205–207.

Moulton, D. E., Ashton, E. H., and Eayrs, J. T. (1960). Studies in olfactory acuity. 4. Relative detectability of *n*-aliphatic acids by the dog. *Anim. Behav.* **8**, 117–128.

Murphy, C., Cain, W. S., and Bartoshuk, L. M. (1977). Mutual action of taste and olfaction. *Sens. Processes* **1**, 204–211.

Myers, D. G., and Lamm, H. (1975). The polarizing effect of group discussion. *Am. Sci.* **63**, 297–303.

Negueruela, A. I., Echavarri, J. F., Los Arcos, M. L., and Lopez de Castro, M. P. (1990). Study of color of quaternary mixtures of wines by means of the Scheffé design. *Am. J. Enol. Vitic.* **41**, 232–240.

Neogi, P. (1985). Tears-of-wine and related phenomena. *J. Colloid Interface Sci.* **105**, 94–101.

Noble, A. C., and Bursick, G. F. (1984). The contribution of glycerol to perceived viscosity and sweetness in white wine. *Am. J. Enol. Vitic.* **35**, 110–112.

Noble, A. C., Williams, A. A., and Langron, S. P. (1984). Descriptive analysis and quality ratings of 1976 wines from four Bordeaux communes. *J. Sci. Food Agric.* **35**, 88–98.

Noble, A. C., Arnold, R. A., Buechsenstein, J., Leach, E. J., Schmidt, J. O., and Stern, P. M. (1987). Modification of a standardized system of wine aroma terminology. *Am. J. Enol. Vitic.* **36**, 143–146.

Ohloff, G. (1986). Chemistry of odor stimuli. *Experientia* **42**, 271–279.

Oreskovich, D. C., Klein, B. P., and Sutherland, J. W. (1991). Procrustes analysis and its applications to free-choice and other sensory profiling. *In* "Sensory Science Theory and Application in Foods" (H. T. Lawless and B. P. Klein, eds.), pp. 353–393. Dekker, New York.

Ough, C. S., and Winton, W. A. (1976). An evaluation of the Davis Wine Score Card and individual expert panel members. *Am. J. Enol. Vitic.* **27**, 136–144.

Pangborn, R. M. (1981). Individuality in responses to sensory stimuli. *In* "Criteria of Food Acceptance" (J. Solms and R. L. Hall, eds.), pp. 177–219. Forster Publ., Zurich.

Pangborn, R. M., Berg, H. W., and Hansen, B. (1963). The influence of colour on discrimination of sweetness in dry table wines. *Am. J. Psychiatry* **76**, 492–495.

Pevsner, J., Trifiletti, R. R., Strittmatter, S. M., and Snyder, S. H. (1985). Isolation and characterization of an olfactory receptor protein for odorant pyrazines. *Proc. Natl. Acad. Sci. U.S.A.* **82**, 3050–3054.

Peynaud, E. (1980). "Le Goût du Vin." Bordas, Paris.

Peynaud, E. (1987). "The Taste of Wine." MacDonald, London.

Rapp, A. (1987). Veränderung der Aromastoffe während der Flaschenlagerung von Weissweinen. *Primo Simposio Internazionale: Le Sostanze Aromatiche dell'Uva e del Vino, S. Michele all'Adige, Trentino, Istituto Agrario Provinciale, June 25–27, 1987*, pp. 286–296.

Robichaud, J. L., and Noble, A. C. (1990). Astringency and bitterness of selected phenolics in wine. *J. Sci. Food Agric.* **53**, 343–353.

Roessler, E. B., Pangborn, R. M., Sidel, J. L., and Stone, H. (1978). Expanded statistical tables for estimating significance in prepared-preference, paired-difference, duo–trio and triangle tests. *J. Food Sci.* **43**, 940–943 (1978).

Sachs, O. (1985). The dog beneath the skin. *In* "The Man Who Mistook His Wife for a Hat," pp. 149–153. Duckworth, London.

Saintsbury, G. (1920). "Notes on a Cellar-Book." Macmillan, London.

Sato, M. (1987). Taste receptor proteins. *Chem. Senses* **12**, 277–283.

Schiffman, S. S. (1983). Taste and smell in disease. *N. Engl. J. Med.* **308**, 1275–1279.

Schiffman, S., Orlandi, M., and Erickson, R. P. (1979). Changes in taste and smell with age: Biological aspects. *In* "Sensory Systems and Communication in the Elderly" (J. M. Ordy and K. Brizzee, eds.), pp. 247–268. Raven, New York.

Schmale, H., Holtgreve-Grez, H., and Christiansen, H. (1990). Possible role for salivary gland protein in taste reception indicated by homology to lipophilic ligand carrier proteins. *Nature* (*London*) **343**, 366–369.

Schutz, H. G., and Pilgrim, F. J. (1957). Differential sensitivity in gustation. *J. Exp. Psychol.* **54**, 41–48.

Scott, T. R., and Chang, G. T. (1984). The state of gustatory neural coding. *Chem. Senses* **8**, 297–313.

Scott, T. R., and Giza, B. K. (1987). Neurophysiological aspects of sweetness. *In* "Sweetness" (J. Dobbing, ed.), pp. 15–32. Springer-Verlag, London.

Seifert, R. M., Buttery, R. G., Guadagni, D. G., Black, D. R., and Harris, J. G. (1970). Synthesis of some 2-methoxy-3-alkylpyrazines with strong bell pepper-like odors. *J. Agric. Food Chem.* **18**, 246–249.

Selfridge, T. B., and Amerine, M. A. (1978). Odor thresholds and interactions of ethyl acetate and diacetyl in an artificial wine medium. *Am. J. Enol. Vitic.* **29**, 1–6.

Senf, W., Menco, B. P. M., Punter, P. H., and Duyvesteyn, P. (1980). Determination of odour affinities bases on the dose–response relationships of the frog's electro-olfactogram. *Experientia* **36**, 213–215.

Singleton, V. L. (1976). Wine aging and its future. "The First Walter and Carew Reynell Memorial Lecture, July 28, 1976," pp. 1–39. Tanunda Institute, Roseworthy Agricultural College, South Australia.

Snyder, S. H., Sklar, P. B., and Pevsner, J. (1988). Molecular mechanism of olfaction. *J. Biol. Chem.* **263**, 13971–13974.

Snyder, S. H., Rivers, A. M., Yokoe, H., Menco, B. Ph. M., and Anholt, R. R. H. (1991). Olfactomedin: Purification, characterization and localization of a novel olfactory glycoprotein. *Biochemistry* **30**, 9143–9153.

Solomon, G. E. A. (1990). Psychology of novice and expert wine talk. *Am. J. Psychol.* **103**, 495–517.

Solomon, G. E. A. (1991). Language and categorization in wine expertise. *In* "Sensory Science Theory and Applications in Foods" (H. T. Lawless and B. P. Klein, eds.), pp. 269–294. Dekker, New York.

Somers, T. C., and Evans, M. E. (1977). Spectral evaluation of young red wines: Anthocyanin equilibria, total phenolics, free and molecular SO_2 "chemical age" *J. Sci. Food Agric.* **28**, 279–287.

Sorrentino, F., Voilley, A., and Richon, D. (1986). Activity coefficients of aroma compounds in model food systems. *AIChE J.* **32**, 1988–1993.

Stahl, W. H. (1973). "Compilation of Odor and Taste Threshold Value Data." Am. Soc. Testing Materials, Data Series 48, Philadelphia, Pennsylvania.

Stahl, W. H., and Einstein, M. A. (1973). Sensory testing methods. *In* "Encyclopedia of Industrial Chemical Analysis" (F. D. Snell and L. S. Ettre, eds.), Vol. 17, pp. 608–644. Wiley, New York.

Stevens, J. C., and Cain, W. C. (1993). Changes in taste and flavor in aging. *Crit. Rev. Food Sci. Nutr.* **33**, 27–37.

Stevens, J. C., Bartoshuk, L. M., and Cain, W. S. (1984). Chemical senses and aging: Taste versus smell. *Chem. Senses* **9**, 167–179.

Stoddart, D. M. (1986). The role of olfaction in the evolution of human sexual biology: An hypothesis. *Man* **21**, 514–520.

Stone, H., and Sidel, J. (1985). "Sensory Evaluation Practices." Academic Press, Orlando, Florida.

Sudraud, P. (1978). Évolution des taux d'acidité volatile depuis le début du siècle. *Ann. Technol. Agric.* **27**, 349–350.

Taylor, S. L., Higley, N. A., and Bush, R. K. (1986). Sulfites in foods: Uses, analytical methods, residues, fate, exposure assessment, metabolism, toxicity, hypersensitivity. *Adv. Food Res.* **30**, 1–76.

Tortora, G. J., and Anagnostakos, N. P. (1975). "Principles of Anatomy and Physiology." Canfield Press, San Francisco, California.

Trant, A. S., and Pangborn, R. M. (1983). Discrimination, intensity, and hedonic responses to color, aroma, viscosity, and sweetness of beverages. *Lebensm. Wiss. Technol.* **16**, 147–152.

Tromp, A., and Conradie, W. J. (1979). An effective scoring system for sensory evaluation of experimental wines. *Am. J. Enol. Vitic.* **30**, 278–283.

Tromp, A., and van Wyk, C. J. (1977). The influence of colour on the assessment of red wine quality. *In: Proc. S. Afr. Soc. Enol. Vitic.*, Nov. 21–22, 1977. 107–117.

Voilley, A., Beghin, V., Charpentier, C., and Peyron, D. (1991). Interactions between aroma substances and macromolecules in a model wine. *Lebensm. Wiss. Technol.* **24**, 469–472.

Williams, A. A. (1975). The development of a vocabulary and profile assessment method for evaluating the flavor contribution of cider and perry aroma constituents. *J. Sci. Food Agric.* **26**, 567–582.

Williams, A. A. (1978). The flavour profile assessment procedure. *In* "Sensory Evaluation—Proceedings of the Fourth Wine Subject Day," Sept. 6–7, 1978, pp. 41–56. Long Ashton Research Station, Univ. of Bristol, Bristol.

Williams, A. A. (1982). Recent developments in the field of wine flavour research. *J. Inst. Brew.* **88**, 43–53.

Williams, A. A. (1984). Measuring the competitiveness of wines. *In* "Tartrates and Concentrates—Proceedings of the Eighth Wine Subject Day Symposium" (F. W. Beech, ed.), pp. 3–12. Long Ashton Research Station, Univ. of Bristol, Bristol.

Williams, A. A., and Langron, S. P. (1983). A new approach to sensory profile analysis. *In* "Flavor of Distilled Beverages: Origin and Development" (J. R. Piggott, ed.), pp. 219–224. Ellis Horwood, Chichester.

Williams, A. A., and Rosser, P. R. (1981). Aroma enhancing effects of ethanol. *Chem. Senses* **6**, 149–153.

Williams, A. A., Bains, C. R., and Arnold, G. M. (1982). Towards the objective assessment of sensory quality in less expensive red wines. *Grape Wine Centennial Symp. Proc. 1980*, pp. 322–329. Univ. of California, Davis.

Williams, A. A., Langron, S. P., and Noble, A. C. (1984). Influence of appearance of the assessment of aroma in Bordeaux wines by trained assessors. *J. Inst. Brew.* **90**, 250–253.

Winton, W., Ough, C. S., and Singleton, V. L. (1975). Relative distinctiveness of varietal wines estimated by the ability of trained panellists to name the grape variety correctly. *Am. J. Enol. Vitic.* **26**, 5–11.

Wysocki, C. J., Dorries, K. M., and Beauchamp, G. K. (1989). Ability to perceive androsterone can be acquired by ostensibly anosmic people. *Proc. Natl. Acad. Sci. U.S.A.* **86**, 7976–7978.

Zatorre, R. J., and Jones-Gotman, M. (1990). Right-nostril advantage for discrimination of odors. *Percept. Psychophys.* **47**, 526–531.

Index

Acetaldehyde, 197, 199, 208, 237, 446
Acetals, 199–200
Acetic acid, 187–188, 237, 340
Acetic acid bacteria, 340
    induced wine spoilage, 326–328
Acidification, 280
Aging, 294–299, 353
    appearance effects, 295
    carbonic maceration wines, 353
    factors affecting
        light, 298–299
        oxygen, 298
        temperature, 298
        vibration, 299
    fragrance effects
        bottle-aged bouquet, origin, 297
        loss of aroma and fermentation
            bouquet, 296–297
    overview, 294–295
    rejuvenation of old wine, 299
    taste effects, 296
Air pollution, 138

Alcohols, *see also* Ethanol; Methanol
    diols, 186
    higher, 185–186, 447
    polyols, 186
    sugar, 186
Aldehydes, 199
Ammonia, 61, 64, 113–114
Anthocyanin, 59, 192–195, 231,
    295
Appellation Control laws
    Australia, 386
    basic concepts and significance,
        380–382
    Canada, 386
    France, 382–383
    Germany, 384–385
    Italy, 385
    South Africa, 385
    United States, 386
Australia, 417–418
    Appellation Control laws,
        386

Bacteria
    acetic acid, 340
        induced wine spoilage, 326–328
    lactic acid
        induced wine spoilage, 325–326
        malolactic fermentation role, 188,
            259–264
    pathogens, 128–130
Barrels, *see* Cooperage
Basal leaf removal, benefits, 73, 75,
    80–81, 121
Bench grafting, 100
Berry, *see also* Fruit
    maturation
        chemical changes
            acids, 56–58
            aromatic compounds, 61–62
            lipids, 60
            nitrogen-containing compounds,
                60–61
            pectins, 60
            phenols, 59–60

Berry (*continued*)
  potassium, 58–59
  sugars, 55–56
  climatic influences and cultural
      influences
    inorganic nutrients, 66
    sunlight, 63–65
    temperature, 65–66
    water, 66
    yield, 63
  seed morphology, 55
  structure, 50–54
Biological control
  pests, 120–121
  pathogens, 120–121
  weeds, 140
Blending, 232, 282, 358, 366, 374, 375
Boron, 117
Botrytis cinerea, see Bunch rot; Noble
    rot
Botrytized wines, *see* Table wines,
    botrytized
Bottle closures, 318–319; *see also*
    Cork
Bottles, *see also* Containers
  disadvantages, 321
  glass
    filling
      production, 321–322
      shape and color, 322–323
Bunch rot, *Botrytis cinerea*, 123–126
  fruit quality effects, 123–124
Bunch-stem necrosis, 137

Cabernet Sauvignon, 26
Calcium, 55, 115, 254
California, 423–425
  climatic regions, 162–163
Canada, 427–428
  Appellation Control laws, 386
Canes, 82
  cuttings, 99
Canopy
  characteristics, 92
  division, 88
  height, 88–89
Canopy management
  definition, 75
  fruit yield and quality effects, 76
  principles, 90–91
Carbohydrates, 206, 264–265
Carbon dioxide, 207–208, 251–253
Carbonic maceration, *see* Maceration
Chardonnay, 27
Chenin blanc, 27
Chip budding, 100–102
Chlorine, 117, 254

Chlorosis
  lime-induced, 95, 116
Clarification
  centrifugation, 291
  filtration, *see* Filtration
  must, 228–229
  racking, 290–291
Climate, *see also* Temperature
  solar radiation, 159, 169–172
  temperature, 163–164
    berry maturation effects, 65–66
    chilling and frost injury, 161, 166–169
  Western Europe, 391–393
  wind, 160, 172–174
  water, 174
Clonal selection, 23–24
Color adjustment, 281–282
Compost, 118–119
Containers, *see also* Bottles
  bag-in-box, 323
Cooperage
  aeration, 309
  barrel production
    assembly, 302–303
    conditioning and care, 305
    life span, 305–306
    size, 304–305
    staves, 301–302
  in-barrel fermentation, 309–310
  materials
    cement, 310
    fiberglass, 310–311
    oak
      chemical composition
        cell lumen constituents, 307
        cell wall constituents,
            306–307
        extracted compounds,
            308–309
      disadvantages, 310
      primacy, 300–301
      species and properties, 299–300
    stainless steel, 310–311
Copper, 117, 211, 287
Cork, *see also* Bottle closures
  agglomerate cork, 316
  alternatives, 318–319
  cellular structure, 313
  cork oak, 311
  culture and harvest, 311–313
  faults
    deposits, 317
    leakage, 316–317, 320
    taints, 317–318, 323–325
  history, 311
  insertion, 319–320
  physiochemical properties, 313–314
  stopper production, 314–316
  structure, 311

Cover crops, 139–140
Crown gall, 128–129
Cryoextraction, 230
Crushing, 223
Cultivars
  identification, 25, 192
  improvement, 23–25
  interspecies hybrids
    American, 28
    French-American, 28–29
  origins, 18–20
  Vitis vinifera, 26–29
    red, 26–27
    white, 27–28

Dagger nematode, 132
Daktulosphaira vitifoliae, see
    Phylloxera
Deacidification
  biological, 280
  column ion-exchange, 280
  precipitation, 279–280
Dealcoholization, 281
Dejuicing, 226
Disease, *see* Pathogens
Downy mildew, 127
Drainage, 157, 161–162
Drip irrigation, 106, 108

Epidermis
  berry, 50, 52–54
  leaf, 41
  root, 33–34
  shoot, 40
  ultraviolet radiation, 44
Esters
  chemical nature, 200
  origin, 200–201
Ethanol, 184, 214, 265–266
Ethyl acetate, 200, 201, 327, 352
Ethyl carbamate, 204, 251
Eutypa dieback, 127–128
Euvitis, 12–14; *see also* Vitis

Fanleaf degeneration, 131
Fermentation
  alcoholic
    biochemistry, 236–240
    fermentors, *see* Fermentors
    overview, 234–236
    stuck, 258–259
    yeast, factors influencing
      alcohols, 247–248
      carbon dioxide and pressure,
          251–253

carbon and energy sources, 246–247
  inorganic elements, 253–254
  lipids, 249
  nitrogen-containing compounds, 248–249
  oxygen and aeration, 251
  pesticide residues, 257–258
  phenols, 249–250
  sulfur dioxide, 250–251
  temperature, 254–257
  vitamins, 253
  yeast succession during, 243–244
carbonic maceration, 350–354
in-barrel, 309–310
madeira, 375
malolactic
  biological factors
    bacterial interactions, 267
    viral interactions, 267
    yeast interactions, 266–267
  chemical factors
    carbohydrates, 264–265
    ethanol, 265–266
    gases, 266
    nitrogen-containing compounds, 265
    organic acids, 265
    organic compounds, 266
  control
    inhibition, 268–269
    inoculation, 268
  lactic acid bacteria, 188, 259–264
  physiochemical factors
    cellular practices, 264
    pH, 264
    temperature, 264
  wine quality effects
    acidity, 261–262
    amine production, 263
    flavor modification, 262–263
    microbial stability, 262
porto, 374–375
recioto wines, 345–346
sparkling wines, 357–358, 359–361, 363
Fermentors
  batch-type, 232–233
  continuous, 233
  size, 233–234
Fertilization
  macronutrients, 108–109
    calcium, 115
    magnesium, 115–116
    nitrogen, 58, 113–114
    phosphorus, 115
    potassium, 58–59, 115
    sulfur, 116
  micronutrients, 108–109
    boron, 117

chlorine, 117
  copper, 117
  iron, 116
  manganese, 116
  molybdenum, 117
  zinc, 116
nutrients
  need, 111
  requirements, 111–113
  supply and acquisition, 109–111
organic, 117–118
  animal manure, 118
  compost, 118–119
  green manure, 118–119
  ground cover, 118–119
Field grafting, 100
Fining
  activated charcoal, 288–289
  albumin, 289
  bentonite, 289
  casein, 289–290
  gelatin, 290
  isinglass, 290
  kieselsol, 290
  polyvinylpyrrolidone, 290
  tannin, 290
Flavescence dorée, 130
Flavonoids, 188–192, 196
Flowers, *see also* Inflorescence
  cluster thinning, 73, 76–77, 81–82
  development, 46–48
  types, 49–50
Fortified wines
  classification, 7–8, 365
  madeira
    base wine production, 375
    heat processing, 375
    maturation, 375
    sweetening and blending, 375–376
  port
    aromatic character, 374–375
    base wine production, 372–373
    maturation and blending, 373–374
    portlike wines, 374
    porto, 372, 409–410
    sweetening and blending, 374
  sherry
    amontillados, 369–370
    base wine production, 366–367
    European sherrylike wines, 370–371
    finos, 367–369
    non-European sherrylike wines
      baked sherries, 371–372
      solera-aged sherries, 371
      submerged-culture sherries, 371
    olorosos, 370
    solera system, 366

France, 394–395
  Alsace, 395
  Appellation Control laws, 382–383
  Bordeaux, 395–396
  Burgundy, 396–397
  Champagne, 397–398
  Loire, 398–399
  southern, 398–399
Fruit, *see also* Berry; Grapes
  cluster thinning, 73, 76, 81–82
  pruning and, 76
  quality, pathogenesis and, 123–124
  trimming and, 80
  yield, 76
    plant spacing and, 89
Fungi, 123, 124–128
Fungicides, 120
Furrow irrigation, 106–107
Fusel alcohols, *see* Alcohols, higher

Geneva Double Curtain, 88, 91
Germany, 399–401
  Appellation Control laws, 384–385
Glycerol, 186, 239–240, 341
Grafting
  bench, 100
  chip budding, 100–102
  field, 100
  green, 100
  older techniques, 100, 102
  rootstock, 95–98
  T-budding, 100–101
  whip, 100
Grapes, *see also* Berry; Fruit
  acetals, 199–200
  acids, 186–188, 215
  aldehydes, 199
  aromas, chemical nature, 211–212
  carbohydrates, 206
  chemical functional groups, 179–182
  commercial importance, 3–5
  crushing, 223
  esters
    chemical nature, 200
    origin, 200–201
  harvesting, *see* Harvesting
  hydrocarbons, 205–206
  ketones, 199
  lactones, 201–202
  lipids, 206
  minerals, 211
  nitrogen-containing compounds, 203–204
  nucleic acids, 207
  organosulfur compounds, 204–205
  pectins and gums, 183
  phenols, *see* Phenols
  proteins, 207

Grapes (*continued*)
  stemming, 222–223
  sugars, 182–183
  terpenes, 202–203
  water, 182
  vitamins, 207
  yield/quality ratio, 63
Grapevine, *see also* Vineyard
  air pollution and, 138
  berry, *see* Berry
  breeding
    disease and pest resistance, 122
    nonstandard, 23–25
    rootstock, 20–21
    scion, 21–23
  cold tolerance, 165–166
  cultivation
    plant density, 89–90
    row orientation, 90
  diseases, *see* Pathogens
  fertilization, *see* Fertilization
  flowering, 47–49
  grafting, *see* Grafting
  growth
    cycle, 72–74, 76
    sunlight effects, 170–172
    vine vigor, 94–95
  harvesting, *see* Harvesting
  inflorescence, 45–50
    cluster morphology, 46
    flower development, 46–47, 73
    flower type, 49–50
    pollination and fertilization,
      47–49
  origin, 1–3
  permanent wilting percentage, 103
  pests, *see* Insects; Spider mites
  physiological disorders, 137–138
  propagation, 98–99
  pruning, *see* Pruning
  rootstock, *see* Rootstock
  root system
    development, 35–37, 73
    secondary tissue development,
      35
    young root, 33–35
  shoot system
    buds, 37–38
    growth, 38–41, 73–74
    leaves, *see* Leaves
    tendrils, 40, 73
    tissue development, 39–40
    species and varieties, 11–29
  training, *see* Training
  varieties, 25–29
  vine cycle, 72–75
Green grafting, 100
Gums, 183

Harvesters
  efficiency, factors affecting, 143–144
  horizontal impactor, 143
  lateral striker, 143
  striker/shaker combination, 143
  trunk-shaker, 143
Harvesting, 140
  harvesters, *see* Harvesters
  manual, 142
  mechanical, 143–145
  pruning and, 77
  sampling, 142
  timing, 141–142
Health, wine consumption effects, 4, 9
Heavy metals, 211, 287–288
Hedging, 80
  vigor control, 94
Herbicides, 139
Higher alcohols, 185–186
Hybrids
  American, 28, 426–427
  French-American, 28–29, 426–427
Hydrocarbons, 205–206
Hydrogen fluoride, 138
Hydrogen sulfide, 204–205

Inflorescence, 45–46
  cluster morphology, 46
  flower development, 46–47, 73
  flower type, 49–50
  pollination and fertilization, 47–49
Inflorescence necrosis, 137
Insects, 132–133
  biological control, 120–121
  leafhoppers, 134–136
  phylloxera, 133–134
  tortricid moths, 135–137
Integrated pest management, 119
Iron, 116, 211, 254, 287
Irrigation
  berry quality effects, 66, 73
  drip, 106, 108
  furrow, 106–107
  soil structure and, 103–104,
    157–158
  sprinkler, 107–108
  timing and level, 105–106
  water quality, 106
  water stress, 104–105
Italy, 403–404
  Appellation Control laws, 385
  central, 405–406
  northern, 404–405
  southern, 406

Juice, *see* Must

Ketones, 199

Laccases, 207, 341–342, 346
Lactic acid, 188, 259–262
Lactic acid bacteria
  control, 267–269
  induced wine spoilage, 325–326
  malolactic fermentation, 259–264
  origin and growth, 263–264
Lead, 211
Leafhoppers, 134–136
Leafroll, 131–132
Leaves, 73
  leaf scorch, 115
  leaf wilt, 104–105
  photosynthesis and other
    light-activated processes, 41–44
  removal, *see* Basal leaf removal
  structure, 40–41
  transpiration and stomatal
    function, 44
Lipids, 60, 206, 249
Lyre training system, 88, 91, 93

Madeira, *see* Fortified wines, madeira
Magnesium, 115–116
Malic acid, 44, 56–58, 64, 188, 260
Maceration
  carbonic
    advantages and disadvantages,
      349–350
    fermentation, 350–353
    red wines, 347–349
      aging, 353
    rosé and white wines, 353–354
  red wines, 225–226
  white wines, 223–225
Malolactic fermentation, *see*
  Fermentation, malolactic
Manganese, 109, 116, 254
Manure, 117–118
Mercaptans, *see* Organosulfur
  compounds
Methanol, 184–185
Mildew, *see* Downy mildew; Powdery
  mildew
Mites, spider, *see* Spider mites
Molybdenum, 117, 254
Mosel arch, 85
Mouth-feel
  assessment in wine tasting, 440
  astringency, 438–439
  body, 439
  burning, 439
  metallic, 211, 439–440

phenol effects, 196, 438–439
prickling, 439
temperature, 439
Mulches, 139
Müller-Thurgau, 27
Muscadinia, 12–14; *see also Vitis rotundifolia*
Muscat blanc, 27–28
Must
  adjustments
    acidity and pH, 229
    sugar content, 229–231
  clarification, 228–229
Mycorrhizal associations, 34–35
  root hair formation, 33
  spontaneous infection, 35

Nebbiolo, 26
Necrosis, *see* Bunch-stem necrosis; Inflorescence necrosis
Nematodes, 132
Nitrogen
  compounds containing, 60–61, 203–204, 248–249, 265
  fertilization, 58, 108, 113–114
Noble rot, *Botrytis cinerea*
  chemical changes induced by, 340–342
  infection, 339–340
  wines produced from affected grapes, 339, 342–344
Nonflavonoids, 188–192
Nucleic acids, 207

Oak
  chemical composition
    cell lumen constituents, 307
    cell wall constituents, 306–307
    extracted compounds, 308–309
  cooperage
    disadvantages, 310
    primacy, 300–301
  cork oak, 311
  species and properties, 299–300
Odor, 196–197
  aroma and bouquet samples, 457
  assessment in wine tasting, 446, 454
  odorants, olfactory stimulation and, 442
  olfactory perception, 443–446
  olfactory system, 440–443
  trigeminal nerve sensations, 442–443
Off-odors, 317–318, 328–329, 446–448, 458, 459
Oidium, *see* Powdery mildew

Organosulfur compounds, 204–205, 328–329
Overcropping, 63, 73, 77
  prevention, 81
Oxidation, *see* Phenols; Spoilage
Oxygen, 208, 251, 298
Ozone, 138

Pathogens
  bacterial
    crown gall, 128–129
    flavescence dorée, 130
    Pierce's disease, 129–130
  control
    biological, 120–121
    chemical, 120
    environmental modification, 121–122
    eradication and sanitation, 122–123
    genetic, 122
    quarantine, 123
  fruit quality effects, 123–124
  fungal
    bunch rot, *Botrytis cinerea*, 123, 124–126
    downy mildew, 127
    Eutypa dieback, 127–128
    powdery mildew, 126–127
  nematodes
    dagger, 132
    root-knot, 132
  viruses and viroids
    fanleaf degeneration, 131
    leafroll, 131–132
    yellow speckle, 132
Pectins, 60, 183–184, 285
Peronospora, *see* Downy mildew
Pesticide residues, fermentation effects, 257–258
Pests, *see also* Insects; Spider mites
  biological control, 120–121
Phenols
  antimicrobial action, 198
  berry maturation, effects, 59–60
  clarification role, 198
  color effects
    red wines, 192–195
    white wines, 195–196
  flavonoids and nonflavonoids, 188–192
  odor effects, 196–197
  oxidant and antioxidant action, 197–198
  oxidizing enzymes, 207, 228, 341–342
  taste and mouth-feel effects, 196, 435, 438–439

yeast fermentation effects, 249–250
Phloem, 33, 35, 40, 55, 73, 79
Phosphorus, 113, 115, 254
Photosynthesis, 41–44
  leaf maturation and, 42–43
  phytochrome-induced phenomena and, 42, 44
Phylloxera, 20, 95–96, 133–134
Phytochrome, induced phenomena, 42, 44
Pierce's disease, 129–130
Pinot blanc, 28
Pinot gris, 28
Pinot noir, 26–27
  plant spacing, 90
Pollen, 16, 47
Port, *see* Fortified wines, port
Portugal, 408
  Madeira, 411
  Upper Douro, 408–410
  Vinho Verde, 410–411
Potassium, 58–59, 108, 115
Powdery mildew, 126–127
Pressing, 226–229, 357
Proteins, 207
Pruning
  cane, 85–86
  options, 77–78
  physiological effects, 75–77
  principles, 77
  procedures and timing, 78–83
  spur, 86
  training and, 84–85

Red wines, 221
  effects of phenols on color, 195–196
  maceration
    carbonic, see Maceration, carbonic
    overview, 223–225
  recioto style, production, 345–347
  sparkling, 363
Reverse osmosis, 230, 231
Riesling, 28
Root-knot nematode, 132–133
Rootstock, 20
  breeding programs, 21
  characteristics, 95–97
  cost, 98
  fruit quality and, 97–98
  genetic strategies, 22
  grafting, 95–102
  varieties, 21
Root system
  development, 35–37, 73
  secondary tissue development, 35
  young root, 33–35

Rosé wines
  carbonic maceration, 353–354
  overview, 221
  sparkling, 363
Ruakura Twin Two Tier, 88, 91,
  93–94

Saccharomyces cerevisiae, see Yeast
Sangiovese, 27
Sauvignon blanc, 28
Seed morphology, 55
Sémillon, 28
Sherry, *see* Fortified wines, sherry
Shoot system
  buds, 37–38
  growth, 38–41, 73–74
  leaves, *see* Leaves
  tendrils, 40, 73
  tissue development, 39–40
Skin contact, *see* Maceration
Soil
  calcareous, 116
  color, 158–159
  depth, 158
  geologic origin, 155
  drainage, 157
  moisture, 104–105, 157
  nitrate, 114
  nutrient content, 158
  organic content, 159
  pH, 158
  preparation, 102
  structure, 156–157
  texture, 155–156
  water availability, 104–105,
    157–158
South Africa, 419–420
  Appellation Control laws, 385
Spain
  Penedés, 407–408
  Rioja, 406–407
  southwestern, 408
Sparkling wines, 354–355
  bulk method, 362–363
  carbonation, 363
  classification, 7
  effervescence and foam, 364–365
  malolactic fermentation, 363
  red, 363
  rosé, 363
  rural method, 363
  traditional method
    cultivars, 355–356
    cuvée preparation, 358
    corking, 361
    disgorging, 361
    dosage, 361
    fermentation

primary, 357–358
secondary, 359–361
harvesting, 356–357
pressing, 357
riddling, 361
tirage, 358
yeast
  culture acclimation and, 358–359
  enclosure, 361–362
transfer method, 362–363
Spider mites, 137
Spoilage
  accidental contaminants, 330–331
  bacteria-induced
    acetic acid bacteria, 326–328
    lactic acid bacteria, 325–326
    other bacteria, 328
  cork-related, 317–318, 323–325
  heat effects, 330
  light effects, 329
  oxidation effects, 329–330
  sulfur off-odors, 328–329
  temperature effects, 330
  yeast-induced, 325
Sprinkler irrigation, 107–108
Stabilization
  calcium salt instability, 284
  calcium tartrate instability, 284
  masque, 287
  metal casse, 286–287
    copper, 287
    iron, 286–287
  microbial, 287–288
  oxidative casse, 286
  polysaccaride removal, 285–286
  potassium bitartrate instability,
    283–284
  protein, 284–285
  tannin removal, 286
Stemming, 222–223
Still wines, *see also* Red wines; Rosé;
  Table wines; White wines
  classification, 5–7
Stomatal apparatus, transpiration
  regulation, 44
Succinic acid, 188
Sugar, 182–183, 229–231, 246–247
Sugar alcohols, 106
Sulfur, 108, 116
Sulfur dioxide, 208–210, 250–251
Sultana, 77
Sunlight, 159–160, 169–170
  aging effects, 298–299
  berry maturation effects, 298–299
  physiological effects, 170–172
  spoilage effects, 329
Supraextraction, 223
Sweetening, 280–281, 374
Syrah, 27

Table wines, *see also* Red wines; Rosé;
  White wines
  classification, 5–7
  botrytized sweet wines, 339–344
    French, 343
    German, 343
    grape variety, 343
    induced botrytization, 344
    infection, 339–340
    noble rotting, chemical changes
      induced by, 340–342
    Tokaji Aszú, 342
  nonbotrytized sweet wines,
    production, 344–345
Tartaric acid, 56–58, 188
Taste, 196, 434–435
  assessment in wine tasting, 440
  perception factors, 436–438
  phenol effects, 196
  sour/salty, 435–436
  sweet/bitter, 435
Tatura Trellis, 88, 94
T-budding, 100–102
Temperature, *see also* Climate
  aging effects, 298
  malolactic fermentation effects, 264
  as mouth-feel sensation, 439
  significance in wine tasting, 449–450
  wine spoilage effects, 330
  yeast fermentation effects, 254–257
Tempranillo, 27
Terpenes, 202–203, 296, 341
Thermotherapy, 122
Thermovinification, 231
Tillage, 139
Topography
  drainage, 161–162
  frost and winter protection, 161
  solar exposure, 159-160
  wind direction, 160
Tortricid moths, 135–137
Training
  bearing wood length, 85–86
  bearing wood origin, 83–85
    cordon training, 83, 85
    head training, 83–84
  canopy
    characteristics, 92
    division, 88
    height, 88–89
    management principles, 90–91
    planting density and, 89–90
  Geneva Double Curtain, 88, 91
  Lyre, 91
  pruning and, 84–85
  row orientation and, 91
  Ruakura Twin Two Tier, 88, 91,
    93–94
  shoot positioning, 86–88

Tatura Trellis, 88, 94
trunk number, 89
Traminer, 27
Transpiration, regulation by stomatal apparatus, 44
2,4,6-Trichloroanisole, 317–318, 326

United States, 423–427
Appellation Control laws, 386

Vigor, control, 94–95
Vineyard, *see also* Grapevine
air pollution effects, 138
disease control, *see* Pathogens
drainage, 157, 161–162
fertilization, *see* Fertilization
irrigation, *see* Irrigation
pest control
chemical methods, 120
biological control, 120–121
environmental modification, 121–122
integrated pest management, 119–120
leafhoppers, 134–135
phylloxera, 133–135
quarantine, 123
spider mites, 137
tortricid moths, 135–136
planting and establishment, 102
rainfall, 155
soil
color, 158–159
depth, 158
geologic origin, 155
nutrient content, 158
organic content, 159
pH, 158
preparation, 102
structure, 156–157
texture, 155–156
sunlight, 159–160, 169–172
physiological effects, 170–172
temperature effects, 163–164
frost injury, 161, 164–169
vine cycle, 72–75
vine growth management, 75–98
vine propagation, 98–102
water availability, 157–158, 174
weed control, *see* Weed control
wind, 160, 172–174
Vinification, *see* Winemaking
Viruses, 131–132
Vitamins, 207, 253
Vitis, 11–14
cultivars
American hybrids, 28

French–American hybrids, 28–29
origins, 18–20
distribution, 14–18
geographic origin, 14–18
Vitis rotundifolia
rootstock breeding, 23
varieties, 20, 28–29
Vitis vinifera, 12, 20, 23
bisexuality, 12, 16, 49
cultivars, 26–29
American hybrids, 28, 426–427
French-American hybrids, 28–29, 426–427
origins, 18–20
red, 26–27, 29
white, 27–28, 29
geographic distribution, 15
geographic origin, 14–15
rootstock, 95

Water
availability and stress, 66, 103–106, 157–158
as chemical constituent of grapes and wines, 180, 182
as climatic influence in site selection, 174
Weed control, 138–140
biological control, 140
cover crops, 139–140
herbicides, 139
mulches, 139
tillage, 139
Whip grafting, 100
White wines
effects of phenol on color, 195–196
maceration, 225–226, 353–354
overview, 221
Wine
adulteration and misrepresentation, detection, 386–387
validation of geographic origin, 387–388
validation of conformity to wine production regulations, 388–390
aging, *see* Aging
aroma and bouquet samples, 457
blending, 232, 282, 358, 366, 374, 375
chemical constituents, 178–215
acetals, 199–200
acids, 186–188, 215
alcohols, 184–186
aldehydes, 199
carbohydrates, 206
carbon dioxide, 207–208

esters
chemical nature, 200
origin, 200–201
hydrocarbons, 205–206
hydrogen sulfide, 204–205
ketones, 199
lactones, 201–202
lipids, 206
minerals, 211
nitrogen-containing compounds, 203–204
nucleic acids, 207
organosulfur compounds, 204–205
oxygen, 208
pectins and gums, 183
phenols, *see* Phenols
proteins, 207
sugars, 182–183, 213
sulfur dioxide, 208–210
terpenes, 202–203
vitamins, 207
water, 182
chemical functional groups, 179–182
clarity, 433
classification, 5–8
color, 432–433
red, 192–195
white, 195–196
commercial importance, 3–5
health effects, 4, 9, 210
mouth-feel, *see* Mouth-feel
odor, 196–197, 440–448
olfactory perception, 443–446
olfactory stimulation, 442
olfactory system, 440–443
origin, 1–3
off-odors, 317–318, 328–329, 446–448, 458, 459
quality, 8
sparkle, 364–365, 434
special styles, 338–376
spoilage, *see* Spoilage
taste, 196, 434–438
perception factors, 436–438
sour/salty, 435–436
sweet/bitter, 435
tasters, *see* Wine tasters
tasting, *see* Wine tasting
tears, 434
viscosity, 433
Winemaking, *see also specific procedures*
acidity and pH adjustments, 229, 278
basic procedures, 221–222
origin, 1–3
postfermentation treatments, 277–331
prefermentation practices, 222–232

Wine presses
  horizontal, 226
  pneumatic, 226–227
  screw, 227–228
Wine tasters
  acuity, 453
  consistency, 453
  testing
    odor recognition
      aroma and bouquet odors,
        458–459
      off-odors, 459
    discriminating ability
      varietal dilution test, 459–460
      wine differentiation test, 460
      wine recognition test, 460
Wine tasting
  conditions for
    number of wines, 448–449
    number of tasters, 451
    presentation of wines
      concealment of identity, 450
      decanting and breathing, 450
      glasses, 449
      replicates, 450
      sequence, 450
      temperature, 449–450
      time of day, 450
    score card, 450–451
    tasting room, 448
  sensations
    mouth-feel, 440
    odor, 446
    taste, 440
    visual
      clarity, 443
      color, 432–433
      sparkle, 434

  tears, 434
  viscosity, 433
  statistical analysis
    simple tests, 455–456
    analysis of variance, 456
    multivariate analysis, 456–457
  tasters, *see* Wine tasters
  technique
    appearance, 453
    descriptive analysis, 456–457
    finish, 455
    in-mouth sensations, 454–455
    odor in-glass, 454
    overall quality assessment, 455
World wine regions
  Australia, 417–418
  Eastern Europe, 411–412
    Bulgaria, 414
    Hungary, 412–413
    Romania, 413
    Soviet Union, 414–415
    Yugoslavia, 413
  Far East
    China, 415–416
    Japan, 416
  Near East, 415
  New Zealand, 418–419
  North Africa, 415
  North America
    Canada, 427–428
    United States, 423–427
  South Africa, 419–420
  South America
    Argentina, 422
    Brazil, 423
    Chile, 420–421
  Southern Europe, 402–403
    Greece, 411

  Italy, 403–406
  Portugal, 408–411
  Spain, 406–408
  Western Europe, 390–391
    Austria, 402–403
    climate, 391–393
    cultivars, 393–394
    Czechoslovakia, 402
    enology, 394
    France, 394–399
    Germany, 399–401
    Switzerland, 401–402
    United Kingdom, 401–402
    viticulture, 394

Xylem, 33, 35, 55, 73

Yeast, *see also* Fermentation
  characteristics, 244
  classification, 241
  ecology, 241–242
  genetic modification, 244–246
  induced wine spoilage, 325
  life cycle, 241
  physiological races, 270
  succession during fermentation,
    243–244
  synonymy, 269–270
Yellow speckle, 132

Zinc, 116, 254
Zinfandel, 27

FOOD SCIENCE AND TECHNOLOGY

International Series

Maynard A. Amerine, Rose Marie Pangborn, and Edward B. Roessler, *Principles of Sensory Evaluation of Food*. 1965.

Martin Glicksman, *Gum Technology in the Food Industry*. 1970.

Maynard A. Joslyn, *Methods in Food Analysis*, second edition. 1970.

C. R. Stumbo, *Thermobacteriology in Food Processing*, second edition. 1973.

Aaron M. Altschul (ed.), *New Protein Foods: Volume 1, Technology, Part A*–1974. *Volume 2, Technology, Part B*–1976. *Volume 3, Animal Protein Supplies, Part A*–1978. *Volume 4, Animal Protein Supplies, Part B*–1981. *Volume 5, Seed Storage Proteins*–1985.

S. A. Goldblith, L. Rey, and W. W. Rothmayr, *Freeze Drying and Advanced Food Technology*. 1975.

R. B. Duckworth (ed.), *Water Relations of Food*. 1975.

John A. Troller and J. H. B. Christian, *Water Activity and Food*. 1978.

A. E. Bender, *Food Processing and Nutrition*. 1978.

D. R. Osborne and P. Voogt, *The Analysis of Nutrients in Foods*. 1978.

Marcel Loncin and R. L. Merson, *Food Engineering: Principles and Selected Applications*. 1979.

J. G. Vaughan (ed.), *Food Microscopy*. 1979.

J. R. A. Pollock (ed.), *Brewing Science, Volume 1*–1979. *Volume 2*–1980. *Volume 3*–1987.

J. Christopher Bauernfeind (ed.), *Carotenoids as Colorants and Vitamin A Precursors: Technological and Nutritional Applications*. 1981.

Pericles Markakis (ed.), *Anthocyanins as Food Colors*. 1982.

George F. Stewart and Maynard A. Amerine, *Introduction to Food Science and Technology*, second edition. 1982.

Malcolm C. Bourne, *Food Texture and Viscosity: Concept and Measurement*. 1982.

Hector A. Iglesias and Jorge Chirife, *Handbook of Food Isotherms: Water Sorption Parameters for Food and Food Components*. 1982.

Colin Dennis (ed.), *Post-Harvest Pathology of Fruits and Vegetables*. 1983.

P.J. Barnes (ed.), *Lipids in Cereal Technology*. 1983.

David Pimentel and Carl W. Hall, *Food and Energy Resources*. 1984.

Joe M. Regenstein and Carrie E. Regenstein, *Food Protein Chemistry: An Introduction for Food Scientists*. 1984.

Maximo C. Gacula, Jr., and Jagbir Singh, *Statistical Methods in Food and Consumer Research*. 1984.

Fergus M. Clydesdale and Kathryn L. Wiemer (eds.), *Iron Fortification of Foods*. 1985

Robert V. Decareau, *Microwaves in the Food Processing Industry*. 1985.

S. M. Herschdoerfer (ed.), *Quality Control in the Food Industry*, second edition. *Volume 1*–1985. *Volume 2*–1985. *Volume 3*–1986. *Volume 4*–1987.

F. E. Cunningham and N. A. Cox (eds.), *Microbiology of Poultry Meat Products*. 1987.

Walter M. Urbain, *Food Irradiation*. 1986.

Peter J. Bechtel, *Muscle as Food*. 1986.

H. W.-S. Chan, *Autoxidation of Unsaturated Lipids*. 1986.

Chester O. McCorkle, Jr., *Economics of Food Processing in the United States*. 1987.

Jethro Jagtiani, Harvey T. Chan, Jr., and William S. Sakai, *Tropical Fruit Processing*. 1987.

J. Solms, D. A. Booth, R. M. Dangborn, and O. Raunhardt, *Food Acceptance and Nutrition*. 1987.

R. Macrae, *HPLC in Food Analysis*, second edition. 1988.

A. M. Pearson and R. B. Young, *Muscle and Meat Biochemistry*. 1989.

Dean O. Cliver (ed.), *Foodborne Diseases*. 1990.

Marjorie P. Penfield and Ada Marie Campbell, *Experimental Food Science*, third edition. 1990.

Leroy C. Blankenship, *Colonization Control of Human Bacterial Enteropathogens in Poultry*. 1991.

Yeshajahu Pomeranz, *Functional Properties of Food Components*, second edition. 1991.

Reginald H. Walter (ed.), *The Chemistry and Technology of Pectin*. 1991.

Herbert Stone and Joel L. Sidel, *Sensory Evaluation Practices*, second edition. 1993.

Robert L. Shewfelt and Stanley E. Prussia, *Postharvest Handling: A Systems Approach*. 1993.

R. Paul Singh and Dennis R. Heldman, *Introduction to Food Engineering*, second edition. 1993.

Tilak Nagodawithana and Gerald Reed, *Enzymes in Food Processing*, third edition. 1993.

Dallas G. Hoover and Larry R. Steenson, *Bacteriocins*. 1993

Takayaki Shibamoto and Leonard Bjeldanes, *Introduction to Food Toxicology*. 1993.

John A. Troller, *Sanitation in Food Processing*, second edition. 1993.

Ronald S. Jackson, *Wine Science: Principles and Applications*. 1994.

Robert G. Jensen and Marvin P. Thompson, *Handbook of Milk Composition*. In Preparation.